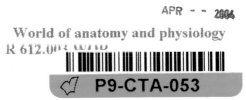

WORLD *of* ANATOMY AND PHYSIOLOGY

WORLD *of*

ANATOMY AND PHYSIOLOGY

K. Lee Lerner and Brenda Wilmoth Lerner, *Editors*

Volume 1
A-J

GALE®

THOMSON

™

GALE

Detroit • New York • San Diego • San Francisco • Cleveland • New Haven, Conn. • Waterville, Maine • London • Munich

THOMSON

GALE

World of Anatomy and Physiology

K. Lee Lerner and Brenda Wilmoth Lerner

Project Editor
Kimberley A. McGrath

Editorial
Mark Springer

Permissions
Kim Davis

Imaging and Multimedia
Randy Bassett, Mary K. Grimes, Lezlie Light, Dan Newell, David G. Oblender, Robyn V. Young

Product Design
Cindy Baldwin, Michelle DiMercurio, Michael Logusz

Manufacturing
Wendy Blurton, Evi Seoud

LIBRARY OF CONGRESS CATALOG-IN-PUBLICATION DATA

World of anatomy and physiology / K. Lee Lerner and Brenda Wilmoth Lerner, editors.
  p. cm.
Includes bibliographical references and index.
  ISBN 0-7876-5684-4 (set : hardcover)
  1. Physiology—Encyclopedias. 2. Anatomy—Encyclopedias. I. Lerner, K. Lee. II. Lerner, Brenda Wilmoth.
QP11.W67 2002
612'.003—dc21
                                                2002005517

ISBN 0-7876-5684-4 (set), ISBN 0-7876-5685-2 (v.1), ISBN 0-7876-5686-0 (v.2)

Printed in the United States of America
10 9 8 7 6 5 4 3 2 1

CONTENTS

VOLUME 2: K-Z

V

W

Y

Z

INTRODUCTION

World of Anatomy and Physiology is devoted to the study of the intricate relationships of form and function within the human body. Although anatomy is the most ancient of all medical studies, the early years of the twenty-first century are an exciting time to undertake the study of the structure and function of the human body. Around the world, thousands of dedicated research and clinical science specialists provide a constant stream of insights into the most intimate aspects of our anatomy and physiology.

Recent rapid progress in cell biology allows, for the first time, insight into the fundamental mechanisms of development. Never before in human history has information moved so rapidly from the laboratory to the clinical setting. The development of reproductive medicine and gene therapy promises to propel us into a new and revolutionary era of biotechnology and biomedical science.

World of Anatomy and Physiology is a collection of 650 entries on topics covering a range of interests—from biographies of the pioneers of anatomy and physiology to explanations of the latest developments and advances in embryology and developmental biology. Despite the complexities of terminology and advanced knowledge of biochemistry needed to fully explore some of the topics in physiology, every effort has been made to set forth entries in everyday language and to provide accurate and generous explanations of the most important terms. The editors intend *World of Anatomy and Physiology* for a wide range of readers. Accordingly, *World of Anatomy and Physiology* articles are designed to instruct, challenge, and excite less-experienced students, while providing a solid foundation and reference for more advanced students.

The very essence of the attributes that define life is the province of two complementary branches of science, anatomy and physiology. All of the sub-disciplines of medical science are built on the fundamental foundations of anatomy and physiology.

Anatomy studies the structure of body parts and their relationships to one another, while physiology concerns the function of the body's structural machinery and how all the body parts work to carry out life-sustaining activities. Questions such as how the living body is able to see, hear, keep warm, or digest food are at the core of anatomical and physiological research.

As our most accomplished athletes and artists prove, the human body has the ability to move gracefully, to lift an arm accurately, and perform many various tasks. The anatomist and physiologist observe and question how body's nervous system, muscles, and joints perform these tasks. To date, the investigation of such questions has depended largely on anatomical studies, which involved examination of anatomical structure by dissection. Technological advances in microcopy and imaging now allow anatomists to view living body and cellular structure on the molecular scale.

Advances in molecular biology allow researchers to investigate a broad range of phenomena, including the physiochemical processes taking place in the cells and tissues of the body, the electrical events underlying the actions of the nervous system, and the feedback mechanisms that allow fine control of complex physiological processes.

With the rapid expansion of scientific knowledge, various disciplines have split off from the parent disciplines of anatomy and physiology. Biochemistry, the study of chemical processes in cells and tissues was probably the first discipline to diverge, then also biophysics, which deals with physical processes within cells. More recently, neurophysiology became established as a separate and specialized area of study. Quite often, anatomical or physiological research is carried out in departments of medicine, where basic scientific knowledge is required for the understanding of disease. The reverse situation is also common: basic research in anatomy or physiology reveals how a disease occurs. For example, basic research into the pancreas and the hormone insulin led to the discovery of successful treatment for the disease diabetes mellitus. Finally, anatomists and physiologists have long been interested in the science underlying such things as athletic performance, the function of the respiratory and nervous systems under anesthesia, the effects of low or high barometric pressures, lack of oxygen, and so on. This all comes under the heading of applied anatomy and physiology.

In all these aspects of the study of anatomy and physiology, it is important to appreciate the role of mechanisms that control bodily functions. For example, we cannot divorce the study of muscular contraction from the feed-back systems that control the structural elements of muscle. All muscles contain structures that signal back to the central nervous system the muscle length and the degree of contractile force. At the spinal level, reflex actions occur to control the performance as precisely as possible in terms of length and force. Without these sophisticated control systems, it becomes very difficult to use muscles properly, precisely, or accurately.

Similar mechanisms exist to control arterial blood pressure. Obviously, if blood pressure is too high, there will be an increased likelihood of rupture of blood vessels and consequent hemorrhage as occurs in strokes. If blood pressure falls too low, the blood supply to the brain will be impaired and consciousness may be lost. When one considers the fact that the human body contains trillions of cells in nearly constant activity, and that remarkably little goes wrong with it, one appreciates what a marvelous system it is. Walter Cannon, an American physiologist of the early twentieth century, spoke of the "wisdom of the body" and he coined the term "homeostasis" to describe its ability to maintain relatively stable internal conditions even though the outside world changes continuously. Although the literal translation of homeostasis is "unchanging," the term does not actually mean a static, or unchanging, state in the body. Rather, it indicates a dynamic state of equilibrium, or a balance, in which the internal conditions vary, but always within relatively narrow limits. In general, the body is in homeostasis when its needs are adequately met and it is functioning smoothly.

Maintaining homeostasis is much more complicated than it appears at first glance. Virtually every organ plays a role in maintaining the constancy of the internal environment. Adequate blood levels of vital nutrients must be continuously present, and heart activity and blood pressure must be constantly monitored and adjusted so that the blood is propelled to all body tissues. Metabolic wastes must not be allowed to accumulate and the body temperature must be precisely controlled. A wide variety of chemical, thermal, and neural factors act and interact in complex ways—sometimes helping and sometimes hindering the body as it works to maintain its "steady rudder" to steer carefully between the dangers of anatomical and physiological extremes.

A system of communication is essential for homeostasis to function. Communication is accomplished chiefly by the nervous and endocrine systems, which use electrical and chemical mechanisms to transmit impulses, information, and instructions. All homeostatic control mechanisms have at least three interdependent control mechanisms: receptors, control centers and effectors. Most homeostatic control systems function as negative feedback mechanisms, where the output of the system shuts off the original stimulus or reduces its intensity. In positive feedback systems, the response enhances the original stimulus so that the activity, or output, is accelerated.

Blood clotting is an example of a process controlled by positive feedback mechanisms.

Although *World of Anatomy and Physiology* concentrates on topics in classical anatomy and physiology, the editors have tried to provide insight into important areas of developmental and reproductive biology. Almost daily, new discoveries extend our understanding of reproductive biology and embryology. At an equally rapid pace, biotechnologies emerge to expand applications of those discoveries. The pace of change and innovation is challenging to all publications. For example, during the writing of *World of Anatomy and Physiology*, researchers announced the creation of cloned human embryos that grew to the six-cell stage. Days before going to press, researchers at the Whitehead Institute for Biomedical Research announced results in the British journal *Nature* that indicated that fully differentiated adult cells can be used to form clones.

Accordingly, *World of Anatomy and Physiology* has attempted to incorporate references and basic explanations of the latest findings and applications. Although certainly not a substitute for in-depth study of important topics such as stem cell research or cloning, the editors hope to provide students and readers with the basic information and insights that will enable a greater understanding of the news and stimulate critical thinking regarding current events in biomedicine.

In the Classical world, the Greek philosopher Aristotle (384–322 B.C.) raised fundamental questions about form, function, reproduction, and development. The quest for insight and answers into the embryological development of humans continues to fascinate and challenge modern scientists. We hope that *World of Anatomy and Physiology* inspires a new generation of scientists who will join in the exciting quest to unlock the remaining secrets of life. It is our modest wish that this book provide valuable information to students and readers regarding topics that play an increasingly prominent role in our civic debates, and a fundamentally important and intimate role in our everyday lives.

K. Lee Lerner and Brenda Wilmoth Lerner, editors
New Orleans
February, 2002

How to Use the Book

Students who are new to the study of anatomy and physiology should start their studies with a careful reading of the article titled, "Anatomical nomenclature," and with an inspection of the article's accompanying diagrams. Over the centuries, anatomists developed a standard nomenclature, or method of naming anatomical structures. In order to standardize nomenclature, anatomical terms relate to the *standard anatomical position*. When the human body is in the standard anatomical position it is upright, erect on two legs, facing frontward, with the arms at the sides each rotated so that the palms of the hands turn forward. Based upon this position, the terms superior, inferior, anterior, posterior and a number of other terms gain clarity and precision. The editors recommend that the reader bookmark the "Anatomical nomenclature" article so quick reference may be made during the reading of other articles. With a little practice and patience, seemingly daunting anatomical nomenclature becomes understandable and readers will achieve comfort with its use.

Although we have attempted to include a number of stimulating photos and informative diagrams *World of Anatomy and Physiology* is not an atlas of anatomy. The deepest understanding of anatomical form requires an atlas that can visually reinforce descriptive writings. There are a number of excellent anatomical atlases available to students and readers that will greatly enhance the reading of *World of Anatomy and Physiology* articles. In particular, the editors recommend the McMinn and Hutchings classic work titled *"Color Atlas of Human Anatomy* and Marjorie England's *Color Atlas of Life Before Birth*, both published by Year Book Medical Publishers, Chicago.

The articles in the book are meant to be understandable to anyone with a curiosity about topics in anatomy or physiology. Cross-references to related articles, definitions, and biographies in this collection are indicated by **bold-faced type**, and these cross-references will help explain and expand the individual entries. Although far from containing a comprehensive collection of topics related to genetics, *World of Anatomy and Physiology* carries specifically selected topical entries that directly impact topics in classical anatomy and physiology. For those readers interested in genetics, the editors recommend Gale's *World of Genetics* as an accompanying reference. For those readers interested in specific genetic disorders, the editors highly recommend the *Gale Encyclopedia of Genetic Disorders*.

This first edition of *World of Anatomy and Physiology* has been designed with ready reference in mind:

- **Entries are arranged alphabetically**, rather than by chronology or scientific field. In addition to classical topics, *World of Anatomy and Physiology* contains many articles addressing the impact of advances in anatomy and physiology on history, ethics, and society.
- **Bold-faced terms** direct reader to related entries.
- **"See also" references** at the end of entries alert the reader to related entries not specifically mentioned in the body of the text.
- A **Sources Consulted** section lists the most worthwhile print material and web sites we encountered in the compilation of this volume. It is there for the inspired reader who wants more information on the people and discoveries covered in this volume.
- The **Historical Chronology** includes many of the significant events in the advancement of anatomy and physiology. The most current entries date from just days before *World of Anatomy and Physiology* went to press.
- A **comprehensive General Index** guides the reader to topics and persons mentioned in the book. Bolded page references refer the reader to the term's full entry.

Although there is an important and fundamental link between the composition and shape of biological molecules and their functions in biological systems, a detailed understanding of biochemistry is neither assumed or required for *World of Anatomy and Physiology*. Accordingly, students and other readers should not be intimidated or deterred by the complex names of biochemical molecules (especially the names for particular proteins, enzymes, etc.). Where necessary, sufficient information regarding chemical structure is provided. If desired, more information can easily be obtained from any basic chemistry or biochemistry reference.

Advisory Board

In compiling this edition, we have been fortunate to rely upon the expertise and contributions of the following scholars who served as academic and contributing advisors for *World of Anatomy and Physiology*, and to them we would like to express our sincere appreciation for their efforts to ensure that *World of Anatomy and Physiology* contains the most accurate and timely information possible:

Robert G. Best, Ph.D.
Director, Division of Genetics, Department of Obstetrics and Gynecology
University of South Carolina School of Medicine
Columbia, South Carolina

Antonio Farina, M.D., Ph.D.
Visiting Professor, Department of Pathology and
Laboratory Medicine
Brown University School of Medicine
Providence, Rhode Island
and
Department of Embryology, Obstetrics, and Gynecology
University of Bologna
Bologna, Italy

Brian D. Hoyle, Ph.D.
Microbiologist
Nova Scotia, Canada

Eric v.d. Luft, Ph.D., M.L.S.
Curator of Historical Collections
SUNY Upstate Medical University
Syracuse, New York

Danila Morano, M.D.
Department of Embryology, Obstetrics, and Gynecology
University of Bologna
Bologna, Italy,

Judyth Sassoon, Ph.D., ARCS
Department of Biology & Biochemistry
University of Bath
Bath, England

Constance K. Stein, Ph.D.
Director of Cytogenetics, Assistant Director of Molecular
Diagnostics
SUNY Upstate Medical University
Syracuse, New York

Acknowledgments

It has been our privilege and honor to work with the following contributing writers and scientists who represent scholarship in anatomy and physiology spanning five continents: Mary Brown; Sherri Chasin Calvo; Bryan Cobb, M.S., *Ph.D. (Department of Genetics, University of Alabama, Birmingham); Sandra Galeotti, M.S.; Larry Gillman, Ph.D.; Brook Hall, Ph.D.; Nicole LeBrasseur, Ph.D.; Adrienne Wilmoth Lerner, M.A.* (Graduate Student, Department of History, Vanderbilt University); Agnieszka Lichanska, Ph.D.; Jill Liske, M.Ed.; Kelli Miller; Lissa Rotundo; Tabitha Sparks, Ph.D.; Susan Thorpe-Vargas, Ph.D.; and David Tulloch, Ph.D.

(*) Anticipated by date of publication

Many of the academic advisors for *World of Anatomy and Physiology*, along with others, authored specially commissioned articles within their field of expertise. The editors would like to specifically acknowledge the following special contributions:

Antonio Farina, M.D., Ph.D.
Ethical issues in embryological research

Danila Morano, M.D.
Pregnancy, maternal physiological and anatomical changes

Judyth Sassoon, Ph.D.
Biochemistry

Constance K. Stein, Ph.D.
Stem cells

The editors would like to extend special thanks Dr. Judyth Sassoon for her contributions to the introduction to *World of Anatomy and Physiology*. The editors also wish to acknowledge Dr. Eric v.d. Luft for his diligent and extensive research related to his compilation of the "History of anatomy and physiology" series of articles.

The editors gratefully acknowledge the assistance of Ms. Robyn Young and the Gale Imaging and Multimedia team for their guidance through the complexities and difficulties related to graphics. Last, but certainly not least, the editors thank Ms. Kimberley McGrath, whose continued wit and guidance kept *World of Anatomy and Physiology* a labor of love.

The editors lovingly dedicate this book to Mary Josephine Wilmoth.

A

ABDOMEN

The abdomen refers to the cavity in the mid-portion of the body bounded superiorly by the **diaphragm** that separates it from the thorax. The abdomen contains a number of **organs** including the **pancreas**, stomach, intestines, **liver**, spleen, bladder, and gallbladder.

The abdominal cavity is the largest cavity in the human body. It is somewhat oval in shape, with the long axis of the oval running upward and downward. There is a flaring out of the abdomen from bottom to top. The abdomen more accurately resembles an egg, with the narrower region of the oval pointing down. Variation in the size of the abdomen occurs with age and sex. The abdomen of an adult male tends to be oval but flattened in the front. In the adult female the oval shape tends to be narrower to accommodate the larger pelvic region. In children the bottom portion of the abdominal egg shape tends to be more pointed.

The abdomen sits between the **lungs** and the pelvis. At the top of the abdomen is the dome-like diaphragm. The bottom of the abdomen extends into the bony pelvic girdle. The muscles in this region are referred to as the diaphragm of the pelvis.

Anatomists tend to divide the abdomen into two imaginary regions for the purposes of description. The upper part of the oval is called the abdomen proper. The lower part of the oval is referred to as the pelvis. While there is no barrier between these regions, the abdomen does narrow, and this narrow region forms the boundary.

The abdomen proper is surrounded by muscle and a coating of muscle that is called the **fascia**. The action of the muscles allows this region of the abdomen to change in size and shape. This is particularly evident when someone is panting for breath.

The middle region of the abdomen houses the stomach. The lower region of the abdomen contains the bladder, colon, rectum, some of the small intestine, and some reproductive organs.

Another feature of the abdomen, which is especially evident when the region is surgically exposed, is a glistening membrane that covers many of the organs and the wall of the abdomen. This membrane, called the **peritoneum**, is the largest membrane in the body. The peritoneum is highly lubricated, which assists the packaging of the various organs into the abdominal space and the movement of these over one another during **digestion** and as the abdomen changes shape. Folds of the peritoneum, which are called the mesentery, extend from the abdominal walls to the various organs. The mesentery suspends the organs in place and houses **blood** vessels that supply the various organs with nutrients. **Infection** of the peritoneum can occur. This malady is called peritonitis.

Superficially the abdomen is divided by anatomists into nine regions. In the midline, the most superior region (closet to the thorax) is the epigastric region. The epigastric region is flanked by left and right hypochondriac regions. The navel lies in the center umbilical region. The umbilical region is flanked by left and right lumbar regions. The most inferior midline abdominal region is the hypogastric region. The hypogastric region is flanked by left and right iliac regions.

See also Abdominal aorta; Abdominal veins; Anatomical nomenclature

ABDOMINAL AORTA

The abdominal aorta is a region of the descending aorta, originating superiorly as a continuation of the **thoracic aorta** as it passes through an opening in the **diaphragm**, and terminating inferiorly as the abdominal aorta bifurcates (divides into two structures) into the left and right common **iliac arteries**.

The abdominal aorta is a large-lumened, unpaired arterial vessel that is part of the main trunk of the systemic arterial system. As such, the abdominal aorta supplies oxygenated **blood**, pumped by the left ventricle of the **heart**, to the abdominal and pelvic **organs** and structures via visceral and parietal arterial branches.

The abdominal aorta and its major arterial branches are highly elastic. During systole (heart **muscle contraction**), the aortic and arterial walls expand to accommodate the increased blood flow. Correspondingly, the vessels contract during diastole and elastin fibers assure that this contraction also serves to drive blood through the arterial vessels.

As the thoracic aorta passes through the aortic hiatus (an opening in the diaphragm) it becomes the abdominal aorta. The abdominal aorta ultimately branches into left and right common iliac **arteries**. The common iliac arteries then branch into internal and external iliac arteries to supply oxygenated blood to the organs and tissues of the lower **abdomen**, pelvis, and legs.

Major branches of the abdominal aorta include, ventrally, the celiac branches, and superior and inferior mesenteric arteries. On the dorsal side of the aorta are the lumbar and median sacral branch arteries. Lateral to the aorta are the inferior phrenics, middle supernal, renal, and ovarian or testicular arteries. Because the branches from the abdominal aorta are large, the aorta rapidly decreases in size as it courses downward (inferiorly) through the abdomen.

The celiac trunk divides into three major branches: the left gastric artery to the stomach, the hepatic artery to the lobes of the **liver**, and splenic artery—surrounded by a plexus of nerves—that ultimately terminates in branches entering the hilus of the spleen.

The superior mesenteric artery supplies oxygenated blood to the small intestine below the duodenum and portions of the cecum and colon. There is often a remnant of the umbilical artery, in the form of a fibrous strand that runs between the navel (umbilicus) and the superior mesenteric artery. Branches of the superior mesenteric artery include the inferior pancreaticoduodenal artery, jejunal and ileal branches, illeocolic artery and the right and middle colic arteries.

The inferior mesenteric arteries supply the transverse colon, descending colon, and rectum. Branches of the inferior mesenteric include the left colic artery, the sigmoid arteries (inferior left colic artery and the superior rectal artery).

Middle suprarenal arteries branch from the abdominal aorta to supply the suprarenal glands. Renal arteries branch form the abdominal aorta to supply the **kidneys**. Phrenic branches of the abdominal aorta supply oxygenated blood to the diaphragm.

As a region of the descending aorta, the abdominal aorta arises in the embryo from the dorsal aortas that are located on each side of the **notochord**. At about the end of the first month of development these embryonic dorsal aorta fuse to form the descending aorta. In the embryo, there are three important divisions of arteries branching from the abdominal aorta. A splanchnic group of arteries supplies the gastro-intestinal tract. A urogenital group of arteries supplies organs and structures derived from the intermediate cell mass, including the kidneys, suprarenals, and **gonads**. A dorsal and lateral group of arteries supply arterial branches ot the **spinal cord**, vertebrae, and muscles and skin of the dorsal body wall.

See also Angiology; Blood pressure and hormonal control mechanisms; Fetal circulation; Systemic circulation

ABDOMINAL MUSCLES · *see* MUSCLES OF THE THORAX AND ABDOMEN

ABDOMINAL VEINS

Blood from the **abdomen** and legs is returned to the right atrium of the **heart** by the vein called the inferior vena cava. The inferior vena cava, which is one of the largest blood vessels in the body, runs along the anterior (forward) side of the spine alongside the **abdominal aorta**. The blood that the abdominal aorta sends the down to the abdomen and legs is returned by the inferior vena cava.

At its upper end, about one inch (2.5 cm) above the **diaphragm**, the inferior vena cava empties into the right atrium of the heart along with the superior vena cava, which drains the upper half of the body. At its lower end, at the level of the fifth lumbar **vertebra** (in the lower back), the inferior vena cava is formed by the convergence of the two **veins** that drain the legs and pelvic region, the left and right iliac veins. As the inferior vena cava ascends from this point it receives blood from several veins that drain the abdomen, including the renal veins which drain the **kidneys**, and the hepatic veins.

The hepatic veins are part of the portal circulation, which drains blood from the digestive system. Blood collected from the capillary beds of the spleen, **pancreas**, gallbladder, stomach, and large and small intestines drains first into the portal vein. Unlike other venous blood, this is not returned directly to the right side of the heart for circulation through the **lungs**; it is sent instead to the **liver**, where it is re-dispersed through a second network of tiny blood vessels called sinusoids, which resemble **capillaries**. During its passage through the sinusoids, the blood is purified and filtered. The sinusoids then converge, forming larger and larger vessels. Blood finally leaves the liver via the hepatic veins that, depending on the individual, vary in number from six to 23. After further convergence, the hepatic veins drain into the inferior vena cava, returning the blood collected by the portal circulation to the flow of regular venous blood.

See also Anatomical nomenclature; Circulatory system; Collateral circulation; Vascular system (embryonic development); Vascular system overview

ABDUCTION AND ADDUCTION · *see* ANATOMICAL NOMENCLATURE

ACCLIMATIZATION

Acclimatization is a process whereby the human body becomes adapted to the challenges brought about by a change in altitude.

The decreasing amount of oxygen available as altitude—the vertical distance above the level of the sea—increases can make the body's ability to function more difficult. If the body is not given time to get used to changes in altitude, a syndrome called altitude sickness may arise. Altitude sickness is characterized by a severe **headache, nausea**, physical weakness, and increased respiration rate (as the body attempts to take in enough oxygen to maintain proper oxygenation). With sustained exposure, high altitude and lower air pressure can cause fluid build-up in the **lungs** and the **brain**. These conditions, called high altitude pulmonary **edema** and cerebral edema, can be life-threatening.

All these symptoms can most often be avoided with acclimatization. Given time—typically one to three days at the particular altitude—the body can adapt to the decrease in oxygen concentration at various altitudes. The "staging" of mountain climbers at various altitudes on their way up Mount Everest is an example of acclimatization in action.

A number of changes take place in acclimatization, in order to allow a body to operate with decreased oxygen. The depth of respiration increases. Put another way, the volume of air breathed in to the lungs in a breath is greater. Technically, this adaptation is known as the hypoxic ventilatory response. Second, the pressure in the **arteries** of the lung increases, powering **blood** into regions of the lung that otherwise would not be reached when **breathing** at sea level. Also, more oxygen-carrying red blood cells are made, in order to more efficiently distribute the available oxygen throughout the body. The rate at which the **heart** beats increases initially but then, as acclimatization progresses, decreases, so as to limit the consumption of oxygen. Finally, there is an increase in the amount of an enzyme that functions to promote the release of oxygen from **hemoglobin** (the protein that transports it throughout the body) to body tissues.

Athletes have exploited high altitude acclimatization. Training at high altitude prior to a competition at lower altitudes may imbue the athlete with an increased capacity to carry oxygen.

See also Oxygen transport and exchange

ACETYLCHOLINE · *see* NEUROTRANSMITTERS

ACID-BASE BALANCE

According to one definition, an acid is a substance that can donate, or make available, a hydrogen ion to another molecule. A base is a substance that can accept a hydrogen ion. The balance of acidic and basic compounds in the body is vitally important for the proper functioning of most physiological processes. Examples of these processes are the delivery of oxygen to cells, the use of oxygen by the cells, and the hormonal regulation of metabolic operations.

Waste products of **metabolism**, such as carbon dioxide (CO_2), hydrogen ions (H^+), sulphuric acid, and phosphoric acid will alter the pH—the concentration of hydrogen ions—in body fluids if not 'buffered,' or counteracted, by interaction with more acidic compounds in **organs** like the **lungs** and the **kidneys**.

The higher a person's metabolic rate is, the faster the ability to break down complex components into more basic energy-generating components, the more acid enters and lowers the pH of the **blood**. The body is equipped with mechanisms to cope with the fluctuations in acid production and so to maintain the pH at a fairly constant level. The main buffering system in the body is the bicarbonate/carbonic acid system in the **extracellular fluid**. This is the fluid that is outside of cells. Via reversible chemical reactions, the bicarbonate/carbonic acid system can neutralize acidic or basic compounds. Within cells, in the intracellular fluid, proteins are important in maintaining the acid-base balance. The chemical structure of protein molecules can be such that they can bind acidic or basic compounds. Yet another means of buffering, in both extracellular and intracellular fluids, particularly in the urinary tract, is via a compound called phosphate (PO_4.).

These and other buffering mechanisms can fail. When this occurs, the acid-base balance is not maintained. **Acidosis** is the condition caused by the removal of bicarbonate or by an increase in carbonic acid in the blood. The result is an increased hydrogen ion concentration and, hence, the lowering of the pH of the blood. Diabetes, starvation, or a high **fat** diet can lead to acidosis. Conversely, a condition called **alkalosis** occurs when bicarbonate increases. In alkalosis, the blood pH rises, which can cause vomiting, **nausea** and **headache**.

Athletes can alter their **breathing** just prior to competition in order to create alkalosis, which will aid in the absorption of the **lactic acid** generated by muscle activity, or to minimize alkalosis, which can compromise endurance activities.

See also Cell membrane transport; Diabetes mellitus; Respiration control mechanisms

ACIDOSIS

Acidosis is an abnormal or pathologic condition caused by an imbalance in the acid-base **equilibrium** of the body. The body does not maintain an exactly neutral pH of 7, but rather a **blood** serum ph of 7.3-7.5. Accordingly, a pH of 7.293—although in strict terms an alkaline pH—is for the body an acidic rather than an alkaline state.

Acidosis is indicated by a lowering of blood pH that corresponds to an elevated hydrogen ion concentration in reaction to depletion in the normally counterbalancing alkaline sodium bicarbonate levels found in tissues and **plasma**. The depletion of neutralizing alkaline agents may result from a disruption or failure in the acid basis chemical buffers that maintain pH levels within a small and normal range of variation.

With regard to acid-base imbalance, the opposite end of the disease spectrum is characterized by **alkalosis.**

Many systems can become acidiotic. Diabetic acidosis results from a surplus of ketones. Metabolic acidosis results from excessive levels of by-products of **fat metabolism** termed keto acids.

Acidosis can occur directly, by the addition of acidic agents, failure to remove acidic agents, or by the removal of alkaline sodium bicarbonate. The loss of bicarbonate is usually due to a failure in the renal reuptake mechanism that preserves bicarbonate and protects the body from excessive bicarbonate loss. Such renal tubular acidosis is indicated by high levels bicarbonate in the urine, which, of course, becomes highly alkaline. **Diarrhea** frequently accompanies metabolic acidosis as the loss of excretory fluid results in a severe loss of sodium bicarbonate—lost before it can be reabsorbed by the kidneys—lowers the blood pH.

The **respiratory system** can also suffer acidosis, and this is specifically distinguished from metabolic acidosis by being termed respiratory acidosis. With respiratory acidosis, a lowering of blood pH can result from inadequate ventilation of carbon dioxide (CO_2). In many cases it is not the lack of oxygen that causes damage in obstructed airways, rather it is the acute (sudden) onset of severe acidosis caused by an inability to vent carbon dioxide. Chronic (occurring over a long term) acidosis may result as a consequence of a gradual deterioration in the ability to ventilate carbon dioxide. The effects chronic acidosis may be masked by a compensating action in the **kidneys,** where increased levels of sodium bicarbonate may be retained to neutralize the acidic buildup in the lung. With acute acidosis, however, the slower reacting kidneys do not have time to perform this compensating action.

See also Acid-base balance; Renal system; Respiration control mechanisms

ACQUIRED IMMUNITY · *see* IMMUNOLOGY

ACTION POTENTIAL

An action potential represents a change in electrical potential from the resting potential of the neuronal cell membrane, and is a series of electrical and underlying chemical changes that travel down the length of a neural cell (neuron). The neural impulse is created by the controlled development of action potentials that sweep down the body (axon) of a neural cell.

There are two major control and communication systems in the human body, the **endocrine system** and the nervous system. In many respects, the two systems compliment each other. Although long duration effects are achieved through endocrine hormonal regulation, the nervous system allows nearly immediate control, especially regulation of **homeostatic mechanisms** (e.g., **blood** pressure regulation).

The neuron **cell structure** is specialized so that at one end, there is a flared structure termed the dendrite. At the dendrite, the neuron is able to process chemical signals from other **neurons** and endocrine **hormones.** If the signals received at the dendritic end of the neuron are of a sufficient strength and properly timed, they are transformed into action potentials that are then transmitted in a "one-way" direction (unidirectional propagation) down the axon.

In neural cells, electrical potentials are created by the separation of positive and negative electrical charges that are carried on ions (charged atoms) across the cell membrane. There are a greater number of negatively charged proteins on the inside of the cell, and unequal distribution of cations (positively charged ions) on both sides of the cell membrane. Sodium ions (Na+) are, for example, much more numerous on the outside of the cell than on the inside. The normal distribution of charge represents the resting **membrane potential** (RMP) of a cell. Even in the rest state there is a standing potential across the membrane and, therefore, the membrane is polarized (contains an unequal distribution of charge). The inner cell membrane is negatively charged relative to the outer shell membrane. This potential difference can be measured in millivolts (mv or mvolts). Measurements of the resting potential in a normal cell average about 70mv.

The standing potential is maintained because, although there are both electrical and concentration gradients (a range of high to low concentration) that induce the excess sodium ions to attempt to try to enter the cell, the channels for passage are closed and the membrane remains almost impermeable to sodium ion passage in the rest state.

The situation is reversed with regard to potassium ion (K+) concentration. The concentration of potassium ions is approximately 30 times greater on the inside of the cell than on the outside. The potassium concentration and electrical gradient forces trying to move potassium out of the cell are approximately twice the strength of the sodium ion gradient forces trying to move sodium ions into the cell. Because, however, the membrane is more permeable to potassium passage, the potassium ions leak through he membrane at a greater rate than sodium enters. Accordingly, there is a net loss of positively charges ions from the inner part of the cell membrane, and the inner part of the membrane carries a relatively more negative charge than the outer part of the cell membrane. These differences result in the net RMP of –70mv.

The structure of the cell membrane, and a process termed the **sodium-potassium pump** maintains the neural cell RMP. Driven by an ATPase enzyme, the sodium potassium pump moves three sodium ions from the inside of the cell for every two potassium ions that it brings back in. The ATPase is necessary because this movement or pump of ions is an active process that moves sodium and potassium ions against the standing concentration and electrical gradients. Equivalent to moving water uphill against a gravitational gradient, such action requires the expenditure of energy to drive the appropriate pumping mechanism.

When a neuron is subjected to sufficient electrical, chemical, or in some cases physical or mechanical stimulus that is greater than or equal to a threshold stimulus, there is a rapid movement of ions, and the resting membrane potential changes from –70mv to +30mv. This change of approximately 100mv is an action potential that then travels down the neuron like a wave, altering the RMP as it passes.

The creation of an action potential is an "all or none" event. Accordingly, there are no partial action potentials. The stimulus must be sufficient and properly timed to create an action potential. Only when the stimulus is of sufficient strength will the sodium and potassium ions begin to migrate done their concentration gradients to reach what is termed threshold stimulus and then generate an action potential.

The action potential is characterized by three specialized phases described as depolarization, repolarization, and hyperpolarization. During depolarization, the 100mv electrical potential change occurs. During depolarization, the neuron cannot react to additional stimuli and this inability is termed the absolute refractory period. Also during depolarization, the RMP of –70mv is reestablished. When the RMP becomes more negative than usual, this phase is termed hyperpolarization. As repolarization proceeds, the neuron achieves an increasing ability to respond to stimuli that are greater than the threshold stimulus, and so undergoes a relative refractory period.

The opening of selected channels in the cell membrane allows the rapid movement of ions down their respective electrical and concentration gradients. This movement continues until the change in charge is sufficient to close the respective channels. Because the potassium ion channels in the cell membrane are slower to close than the sodium ion channels, however, there is a continues loss of potassium ion from the inner cell that leads to hyperpolarization.

The sodium-potassium pump then restores and maintains the normal RMP.

In demyelinated nerve fibers, the depolarization induces further depolarization in adjacent areas of the membrane. In myelinated fibers, a process termed salutatory conduction allows transmission of an action potential, despite the insulating effect of the myelin sheath. Because of the sheath, ion movement takes place only at the Nodes of Ranvier. The action potential jumps from node to node along the myelinated axon. Differing types of nerve fibers exhibit different speed of action potential conduction. Larger fibers (also with decreased electrical resistance) exhibit faster transmission than smaller diameter fibers.

The action potential ultimately reaches the presynaptic portion of the neuron, the terminal part of the neuron adjacent to the next **synapse** in the neural pathway. The synapse is the gap or intercellular space between neurons. The arrival of the action potential causes the release of ions and chemicals (**neurotransmitters**) that travel across the synapse and act as the stimulus to create another action potential in the next neuron.

See also Adenosine triphosphate (ATP); Nerve impulses and conduction of impulses; Nervous system overview; Reflexes; Touch, physiology of

ACTIVE TRANSPORT • *see* CELL MEMBRANE TRANSPORT

ADDUCTION • *see* ANATOMICAL NOMENCLATURE

ADENOIDS • *see* TONSILS

ADENOSINE TRIPHOSPHATE (ATP)

Adenosine 5'–triphosphate (ATP) was discovered in 1929 by the German chemist Karl Lohmann. Its chemical structure consists of an adenosine nucleotide structure to which three phosphoryl groups are sequentially attached via cleavable bonds. It is the most common energy "currency" of living cells, functioning as an intermediate which drives "energy requiring" (endergonic) reactions through the cleavage of its phosphate groups. The processes that maintain life require an input of energy in order to proceed and are driven by the exergonic (energy yielding) reactions of nutrient oxidation. This coupling is most often mediated through the syntheses of high-energy intermediates, such as ATP, whose consumption provides energy for the endergonic processes. In biosynthesis, for example, ATP is the immediate product of all processes leading to the chemical storage of energy.

The capacity of ATP to act as the biochemical energy "currency" comes from its high potential to transfer its phosphate groups. The molecule can be cleaved in several different ways. For example transfer of the orthophosphate to alcoholic hydroxyl groups, acid or amide groups and the release of ADP has a free energy of hydrolysis of 29.4 kJ/mol. The **enzymes**, which catalyse these reactions are kinases, which in some cases (e.g., creatine kinase) can also catalyse the hydrolysis of ATP from ADP. Another way to cleave ATP is to remove two pyrophosphate groups (two phosphates) and release adenosine monophosphate (AMP). The free energy of hydrolysis for this reaction is 36.12 kJ/mol and it occurs, for example, in the course of purine synthesis.

The energy stored in ATP is used in the vast majority of central biochemical reactions, including the synthesis of macromolecules such as proteins, fats and **carbohydrates** from their corresponding monomeric subunits. Such endergonic reactions are often driven forward by enzymatic coupling to the hydrolysis of ATP. Many catabolic pathways, including glycolysis, require an "investment" of ATP, which is later re-synthesized. Essentially all anabolic pathways require ATP, either directly or indirectly. In the human body ATP provides energy for the contraction of muscles. The cleavage of ATP to ADP and inorganic phosphate is generally coupled to another process in the cell so that the free energy released by ATP is used to drive another endergonic process. In the actomyosin complex in muscle, the cleavage of ATP provides energy for contraction. Also, the active transport of many substances across membranes depends on a source of ATP. For example in the Na^+/K^+ pump in cell membranes, the energy required for the transport of ions against their concentration gradients is

Three-dimensional micrograph of crystallized adenosine triphosphate (ATP). © M.W. Davidson, National Audubon Society Collection/Photo Researchers, Inc. Reproduced by permission.

provided by ATP. ATP is also released together with acetylcholine at synapses in the **peripheral nervous system** and has been proposed as a modulator of nervous transmission, as it inhibits the release of acetylcholine.

See also Biochemistry; Metabolism; Muscle contraction

ADIPOSE TISSUE

Adipose **tissue** is a specialized type of connective tissue, the tissue that binds various structures together and also provides cushioning and support. Viewed under a microscope, the cells that make up adipose tissue consist almost entirely of **fat**; even the cell nucleus is pushed over to the edge of the cell. This appearance is consistent with the function of adipose tissue as the major storage site for fat. The fat is usually in the form of **triglycerides**.

There are two types of adipose tissue: white and brown. Most adipose tissue is of the white type. White adipose tissue has three functions: heat insulation, cushioning, and as an energy source. The later is the most important. The insulation function is mainly attributable to the adipose tissue found just underneath the skin—the subcutaneous adipose tissue—as this tissue conducts heat only one third as efficiently as other tissues. Lastly, adipose tissue surrounds internal **organs**, and helps protect them from jarring trauma.

Adipose tissue is an important biological energy source because the fat it contains yields more energy per unit weight than either carbohydrate or protein. Thus, not as much adipose tissue needs to be laid down by the body to produce as much energy as the other two materials. Also, adipose tissue is hydrophobic (from the Latin for "water hating"), so the advantages of energy and insulation do not come at a price of carrying around a huge excess of water, which would make mobility difficult.

White adipose tissue is used as a substrate, or starting material, from which energy is generated by a series of bio-

chemical reactions. In contrast, brown adipose tissue yields energy directly, without intervening chemical breakdown reactions. Brown adipose tissue takes its name from its color, which is due to the many **blood** vessels and **mitochondria** (the **ATP** factories of a cell) rich cells that are present in the tissue. This adipose tissue is localized in the body, especially near the thymus and the **abdomen**.

While essential and beneficial as an energy and heat source, too much adipose tissue can be detrimental to health. Normal body content of adipose tissue is 15-20% of body weight in males and 20-25% in females. The condition of obesity, where body fat content exceed these levels can be much higher, and has genetic and lifestyle causative factors.

See also Adenosine triphosphate (ATP); Adolescent growth and development; Fat, body fat measurement; Lipids and lipid metabolism; Metabolism; Mitochondria and cellular energy

ADOLESCENT GROWTH AND DEVELOPMENT

Adolescence is the time between the beginning of sexual maturation (also known as puberty, from the Latin pubertas, meaning adult) and the beginning of adulthood. Adolescence usually spans the years between ages 13 and 19. Adolescence includes physical growth and emotional, psychological, and mental change. Psychological maturation occurs as the child acquires adult-like behavior. During this period, adolescents are expected to become capable of adult behavior and response. Adolescence is a period of many transitions. During the teen years, adolescents experience changes in their physical development at a rate of speed unparalleled since infancy. Physical development includes rapid gains in height and weight. During a one-year growth spurt, boys and girls can gain an average of 4.1 in. (10.4 cm) and 3.5 in. (8.9 cm) in height, respectively. This spurt typically occurs two years earlier about for girls than for boys. Weight gain results from increased muscle development in boys and body fat in girls.

Hormones play two different roles in adolescent development, the organizational role and the activational role. The organizational role is the ability of hormones to generate different patterns of behavior in the male and in the female brain, respectively. Already in the prenatal age, the hormones organize the brain differently. The activational role is the ability to initiate the modifications related to puberty and to differentiate them for male and females. During and just before puberty, the hypothalamus both stops the inhibition upon the factors able to initiate puberty and begins to produce substances that set the puberty in motion. The first signal begins due to higher concentrations of leptin, a protein produced by adipocytes of the fat tissue. The hypothalamus stimulates the hypophysis to secrete hormones able to promote the overall growth of the body and to maturate the gonads, as well as the adrenal cortex and thyroid. Scientists suggest that adrenal cortex maturation is involved in sexual attraction. Hormone concentrations are due to gland activations that are controlled by several mechanisms of feed-back.

The development of primary sex characteristics includes the further maturing of the gonads, the testis in boys, and the ovaries in girls. In both sexes, hormonal regulation of reproduction is regulated by the brain. Until eight weeks of gestation, the brain is organized in a female direction irrespectively with the gender of the fetus. Successively, testosterone, for example, organizes the male brain in patterns of behavior, many of which may not appear until much later. Hypothalamic gonadotropin releasing hormone (GnRH) controls release of both luteinizing hormone (LH) and follicle stimulating hormone (FSH). LH acts primarily on endocrine cells of the gonads. FSH acts primarily on gamete-producing cells. Both sexes produce androgens and estrogens. Testosterone, the main androgen produced in the testis is converted to dihydrotestosterone (DHT) in many tissues. Estradiol, the main estrogen, is made from testosterone by the action of the enzyme aromatase. Both the ovary and testis produce peptide hormones that have feedback effects on the hypophysis. Inhibins are hormones that inhibit FSH secretion. Activins stimulate FSH secretion as well as spermatogenesis, oocyte maturation. Children (both males and females) with a deficiency of GnRH will not mature in the absence of gonadotropin stimulation due to lower levels of androgens and estrogens.

Secondary sex characteristics can be considered traits which give an individual an advantage over its rivals in courtship. During puberty, changing hormonal levels play a role in activating the development of secondary sex characteristics. These include: (1) growth of pubic hair (pubarche); (2) Growth of the breasts in girls (thelarche) (3) menarche (first menstrual period for girls) or penis growth (for boys); (4) voice changes (for boys); (5) growth of underarm hair; (5) facial hair growth (for boys);(6) nighttime ejaculations (nocturnal emissions; "wet dreams" for boys) and (7) increased production of oil, increased sweat gland activity, and the beginning of acne.

Normal growth is categorized in a range used by pediatricians to gauge how a child is growing. The following are some average ranges of weight and height, based on growth charts developed by the Centers for Disease Control and Prevention (CDC): briefly, at 12 years of age a male, should be 54–63.5 in. (1.37-1.6 m) and a female 55–64 in. (1.4-1.6 m) The weight would be 66–130 lb. (29.9-58.9 kg) and 68–136 lb. (30.8-61.7 kg) respectively. At 18-years-old, a male would be 65–74 in. (1.7-1.9 m) tall and a female 60–68.5 in. (1.5-1.7 m). The weight would be 116–202 lb. (52.6-91.6 kg) and 100–178 lb. (45.4-80.7 kg), respectively. Growth not only involves length and weight of a body, but also includes internal growth and development, including the brain. Growth also affects different parts of the body at different rates; the head reaches almost its entire size by age one. Throughout childhood, a child's body becomes more proportional to other parts of his/her body. Growth is complete between the ages of 16 and 18, at which time the growing ends of bones fuse. Although a child may be growing, his/her growth pattern may deviate from the normal. Ultimately, the child should grow to normal height by adulthood. Teens frequently sleep longer. Research suggests that teens actually need more sleep to allow their bodies to conduct the internal work required for such rapid growth. On average, teens need about 9.5 hours of sleep a night.

Teens' brains are not completely developed until late in adolescence. Studies suggest that the connections between neurons affecting emotional, physical, and mental abilities are incomplete. This could explain why some teens seem to be inconsistent in controlling their emotions, impulses, and judgments. The advances in thinking can be divided into several areas, including the development of advanced reasoning skills, as well as formulation and testing of hypotheses, demonstrating agreement or disagreement, questioning, giving examples, and making distinctions and connections. Advanced reasoning skills include the ability to think logically about multiple options. It includes the ability to think hypothetically. Adolescents also develop abstract thinking skills. Abstract thinking includes faith, beliefs and spirituality. Additionally, adolescents develop the ability of "meta-cognition." Meta-cognition is awareness of one's own thinking process, and involves the ability to think about how one is perceived by others. Meta cognition involves of skills that are genuinely transferable and can, therefore, be applied in different settings, resulting in an improved ability to learn.

Teens demonstrate a heightened level of self-consciousness. Teens tend to believe that everyone is as concerned with their thoughts and behaviors as they are. This leads teens to believe that they have an "imaginary audience" of people who are always watching them. Teens tend to believe that no one else has ever experienced similar strong feelings and emotions. Teens tend to exhibit a "justice" orientation. They are quick to point out inconsistencies between adults' words and their actions. They have difficulty seeing shades of gray. They see little room for error.

There are four recognized psychosocial issues that teens normally deal with during their adolescent years. These include: 1) Establishing an identity. This has been called one of the most important tasks of adolescents. Over the course of the adolescent years, teens begin to integrate the opinions of influential others (e.g., parents, other caring adults, friends, etc.) into their own likes and dislikes. 2) Establishing autonomy. Some people assume that autonomy refers to becoming completely independent from others. They equate it with teen "rebellion." Rather than severing relationships, however, establishing autonomy during the teen years actually means becoming an independent and self-governing person within relationships. Autonomy is a necessary achievement if the teen is to become self-sufficient in society. 3) Establishing intimacy. Many people, including teens, equate intimacy with sex. In fact, intimacy and sex are not the same. Intimacy is usually first learned within the context of same-sex friendships, then utilized in romantic relationships. Intimacy refers to close relationships in which people are open, honest, caring, and trusting. Friendships provide the first setting in which young people can practice their social skills with those who are their equals. It is with friends that teens learn how to begin, maintain, and terminate relationships, practice social skills, and become intimate. 4) Becoming comfortable with one's sexuality. The teen years mark the first time that young people are both physically mature enough to reproduce and cognitively advanced enough to think about sexuality. Given this, the teen years are the prime time for the development of sexuality. How teens are educated about and exposed to sexuality will largely determine whether or not they develop a healthy sexual identity. More than half of most high school students report being sexually active. Many experts agree that the mixed messages teens receive about sexuality contribute to problems such as teen pregnancy and sexually transmitted diseases.

Teens begin to spend more time with their friends than their families. It is within friendship groups that teens can develop and practice social skills. Teens are quick to point out to each other behaviors which are acceptable and which are not. It is important to remember that even though teens are spending increased amounts of time with their friends, they still tend to conform to parental ideals when it comes to decisions about values, education, and long-term plans. Teens may become involved in multiple hobbies or clubs. In an attempt to find out where they excel, teens may try many activities. Teens' interests also change quickly. Teens may become more argumentative. Teens may question adults' values and judgments. When teens don't get their way, they may say, "you just don't understand." Teens may begin to interact with parents as people. Even though they may not want to be seen with parents in public, teens may begin to view parents more as people. They may ask more questions about how a parent was when he or she was a teen. They may attempt to interact with adults more as equals.

Teens may be clumsier because of growth spurts. If it seems that a teen's body is all arms and legs, the perception is correct. During this phase of development, body parts don't all grow at the same rate. This can lead to clumsiness as the teen tries to cope with limbs that seem to have grown overnight. Teens can appear gangly and uncoordinated. Teenage girls may become overly sensitive about their weight. This concern arises because of the rapid weight gain associated with puberty. Sixty percent of adolescent girls report that they are trying to lose weight. A small percentage of adolescent girls (one to three percent) become so obsessed with their weight that they develop severe eating disorders such as anorexia nervosa or bulimia. Anorexia nervosa refers to starvation; bulimia refers to binge eating and vomiting. Teens may be concerned because they are not physically developing at the same rate as their peers. Teens may be more developed than their peers ("early-maturers") or less developed than their peers ("late-maturers"). Being out of developmental step with peers is a concern to adolescents because most just want to fit in. Early maturation affects boys and girls differently. Research suggests that early maturing boys tend to be more popular with peers and hold more leadership positions. Adults often assume that early maturing boys are cognitively mature as well. This assumption can lead to false expectations about a young person's ability to take on increased responsibility. Because of their physical appearance, early maturing girls are more likely to experience pressure to become involved in dating relationships with older boys before they are emotionally ready. Early maturing girls tend to suffer more from depression, eating disorders, and anxiety.

Teens may feel awkward about demonstrating affection to the opposite sex parent. As they develop physically, teens are beginning to rethink their interactions with the opposite sex. An adolescent girl who used to hug and kiss her dad when he returned home from work may now shy away. A boy who used to kiss his mother good night may now wave to her on his way up the stairs. Teens may ask more direct questions about sex. At this stage, adolescents are trying to figure out their sexual values. Teens often equate intimacy with sex. Rather than exploring a deep emotional attachment first, teens tend to assume that if they engage in the physical act, the emotional attachment will follow. They may ask questions about how to abstain without becoming embarrassed or about how they will know when the time is right. They may also have specific questions about methods of birth control and protection from sexually transmitted diseases.

See also Gonads and gonadotropic hormone physiology; Growth and development; Hormones and hormone action; Human development (timetables and developmental horizons); Menstruation

ADRENAL GLANDS AND HORMONES

The adrenal glands are a pair of **endocrine glands** that sit atop the **kidneys** and that release their **hormones** directly into the bloodstream. The adrenals are flattened, somewhat triangular bodies that, like other endocrine glands, receive a rich **blood** supply. The phrenic (from the **diaphragm**) and renal (from the kidney) **arteries** send many small branches to the adrenals, while a single large adrenal vein drains blood from the gland.

Each adrenal gland is actually two **organs** in one. The inner portion of the adrenal gland, the adrenal medulla, releases substances called catecholamines, specifically **epinephrine**, adrenaline, norepinephrine, noradrenaline, and dopamine. The outer portion of the adrenal gland, the adrenal cortex, releases steroids, which are hormones derived from cholesterol.

There are three somewhat distinct zones in the adrenal cortex: the outer part, the zona glomerulosa (15% of cortical mass) made up of whorls of cells; the middle part, the zona fasciculata (50% of cortical mass) made up of columns of cells and that are continuous with the whorls; and an innermost area called the zona reticularis (7% of cortical mass), which is separated from the zona fasciculata by venous sinuses.

The cells of the zona glomerulosa secrete steroid hormones known as mineralocorticoids, which affect the fluid balance in the body, principally aldosterone, while the zona fasiculata and zona reticularis secrete glucocarticoids, notably cortisol and the androgen **testosterone**, which are involved in carbohydrate, protein, and **fat metabolism**.

The secretion of the adrenal cortical hormones is controlled by a region of the **brain** called the **hypothalamus**, which releases a corticotropin-releasing hormone. This hormone targets the anterior part of the pituitary gland, situated directly below the hypothalamus. The corticotropin-releasing hormone stimulates the release from the anterior pituitary of adrenocorticotropin (ACTH), which, in turn, enters the blood and targets the adrenal cortex. There, it binds to receptors on the surface of the gland's cells and stimulates them to produce the steroid hormones.

Steroids contain as their basic structure three 6-carbon (hexane) rings and a single 5-carbon (pentane) ring. The adrenal steroids have either 19 or 21 carbon atoms. These important hormones are collectively called corticoids. The 21-carbon steroids include glucocorticoids and mineralocorticoids, while the 19-carbon steroids are the androgens. Over 30 steroid hormones are made by the cortex, but only a few are secreted in physiologically significant amounts. These hormones can be classified into three main classes, glucocorticoids, mineralocorticoids, and corticosterone.

Cortisol (hydrocortisone) is the most important glucocorticoid. Its effect is the opposite that of insulin. It causes the production of the sugar glucose from **amino acids** and **glycogen** stored in the **liver**, called gluconeogenesis, thus increasing blood glucose. Cortisol also decreases the use of glucose in the body (except for the brain, **spinal cord**, and **heart**, and it stimulates the use of fatty acids for energy.

Glucocorticoids also have anti-inflammatory and antiallergenic action, hence they are often used in the treatment of rheumatoid arthritis. The excessive release of glucocorticoids causes Cushing's disease, which is characterized by fatigue and loss of muscle mass due to the excessive conversion of amino acids into glucose. In addition, there is redistribution of body fat to the face, causing the condition known as "moon face." The mineralocorticoids are essential for maintaining the balance of sodium in the blood and body tissues and the volume of the **extracellular fluid** in the body. Aldosterone, the principal mineralocorticoid produced by the zona glomerulosa, enhances the uptake and retention of sodium in cells, as well as the cells' release of potassium. This steroid also causes the tubules of the kidneys to retain sodium, thus maintaining levels of this ion in the blood, while increasing the excretion of potassium into the urine. Simultaneously, aldosterone increases reabsorption of bicarbonate by the kidney, thereby decreasing the acidity of body fluids.

A deficiency of adrenal cortical hormone secretion causes Addison's disease, characterized by fatigue, weakness, skin pigmentation, a craving for salt, extreme sensitivity to stress, and increased vulnerability to **infection.**

The adrenal androgens are weaker than testosterone, the male hormone produced by the testes. However, some of these androgens, including androstenedione, dehydroepiandrosterone (DHEA), and dehydroepiandrosterone sulfate can be converted by other tissues to stronger androgens, such as testosterone. The cortical output of androgens increases dramatically after **puberty**, giving the adrenal gland a major role in the developmental changes in both sexes. The cortex also secretes insignificant amounts of **estrogen.**

The steroid hormones are bound to steroid-binding proteins in the bloodstream, from which they are released at the surface of target cells. From there they move into the nucleus of the cell, where they may either stimulate or inhibit gene activity.

The release of the cortical hormones is controlled by adrenocorticotropic (ACTH) from the anterior pituitary gland. The level of ACTH has a diurnal periodicity, that is, it undergoes a regular, periodic change during the 24-hour time period. ACTH concentration in the blood rises in the early morning, peaks just before awaking, and reaches its lowest level shortly before **sleep.**

Several factors control the release of ACTH from the pituitary, including corticotropin-releasing hormone from the hypothalamus, free cortisol concentration in the **plasma**, stress (e.g., surgery, hypoglycemia, exercise, emotional trauma), and the sleep-wake cycle.

Mineralocorticoid release is also influenced by factors circulating in the blood. The most important of these factors is angiotensin II, the end product of a series of steps starting in the kidney. When the body's blood pressure declines, this change is sensed by a special structure in the kidney called the juxtaglomerular apparatus. In response to this decreased pressure in kidney arterioles the juxtaglomerular apparatus releases an enzyme called renin into the kidney's blood vessels. There, the renin is converted to angiotensin I, which undergoes a further enzymatic change in the bloodstream outside the kidney to angiotensin II. Angiotensin II stimulates the adrenal cortex to release aldosterone, which causes the kidney to retain sodium. The increased concentration of sodium in the blood-filtering tubules of the kidney causes an osmotic movement of water into the blood, thereby increasing the blood pressure.

The adrenal medulla, which makes up 28% of the mass of the adrenal glands, is composed of irregular strands and masses of cells that are separated by venous sinuses. These cells contain many dense vesicles, which contain granules of catecholamines.

The cells of the medulla are modified ganglion (nerve) cells that are in contact with preganglionic fibers of the **sympathetic nervous system**. There are two types of medullary secretory cells, called chromaffin cells: the epinephrine (adrenalin)-secreting cells, which have large, less dense granules, and the norepinephrine (noradrenalin)-secreting cells, which contain smaller, very dense granules that do not fill their vesicles. Most chromaffin cells are the epinephrine-secreting type. These substances are released following stimulation by the acetylcholine-releasing sympathetic nerves that form synapses on the cells. Dopamine, a neurotransmitter, is secreted by a third type of adrenal medullar cell, different from those that secrete the other amines.

The extensive nerve connections of the medulla essentially mean that this part of the adrenal gland is a sympathetic ganglion, that is, a collection of sympathetic nerve cell bodies located outside the **central nervous system**. Unlike normal nerve cells, the cells of the medulla lack axons and instead, have become secretory cells.

The catecholamines released by the medulla include epinephrine, norepinephrine, and dopamine. While not essential to life, they help to prepare the body to respond to short-lived but intense emergencies.

Most of the catecholamine output in the adrenal vein is epinephrine. Epinephrine stimulates the nervous system and also stimulates glycogenolysis (the breakdown of glycogen to glucose) in the liver and in **skeletal muscle**. The free glucose is used for energy production or to maintain the level of glucose in the blood. In addition, it stimulates lipolysis (the breakdown of fats to release energy-rich free fatty acids) and stimulates metabolism in general. Epinephrine also increases the force and rate of heart **muscle contraction**, which results in an increase in cardiac output.

See also Hormones and hormone action

ADRENAL MEDULLAE

The adrenal glands are paired structures that lie on top of each kidney. This places them in a retroperitoneal position in the body. Just like the **kidneys**, the adrenal glands lie outside the **peritoneum** and near the body wall in close contact with the spine and supporting muscles. The adrenal glands are actually two glands in one. The outside, or adrenal cortex, is part of the **endocrine system** and secretes steroid-type **hormones**. The inner cells, the adrenal medulla, are often discussed in **anatomy** texts as part of the **central nervous system** even though they secrete hormonal-type chemicals.

The adrenal medulla, or inner core of the gland, is actually made up of highly modified nervous cells (cromaffin cells) that are part of the sympathetic, or **autonomic nervous system**. The cells are rounded, but slightly columnar and tightly packed. They originate from ectodermal cells. As a unit they function as part of the **sympathetic nervous system** and are often described as a modified sympathetic ganglion. A ganglion is a cluster of nerve cells that form a lump or swelling in the **peripheral nervous system**. These lumps will process signals from the nervous system and integrate specific information to specifically targeted areas of the body. The neuronal message is sent by means of chemicals

The position of the adrenal medullae puts them in proximity to the **spinal cord**. Preganglionic fibers from the sympathetic nervous system leave the thoracic spinal cord (mid-back region) and go through the adrenal cortex until they reach the cells of the adrenal medullae. When these cells are excited by the splanchnic nervous system (a route of nerve fibers that travels from the spinal cord through the ganglia without synapsing) they release a group of important chemicals called catecholamines. They are better known as **epinephrine** (adrenalin) and norepinephrine (noradrenalin). The concentrations of these chemicals when produced and released are about 85% and 15% respectively. The two chemicals function as **neurotransmitters** and include dopamine as well. The adrenal medullae and the sympathetic nervous system function so closely together they are given the title of the sympathoadrenal system.

The chemicals they produce are part of the "fight or flight" reaction the body experiences during periods of high stress. Epinephrine causes the **heart** rate to accelerate. Norepinephrine is a vasoconstrictor that causes the internal flow size of **arteries** and **veins** to become smaller. The result of these two actions is a greater ability of the body to respond to physical activity. This condition is only temporary, lasting only a few seconds to minutes.

See also Adrenal glands and hormones

ADRENALINE • *see* EPINEPHRINE

ADRIAN, EDGAR DOUGLAS (1889-1977)
English neurophysiologist

Lord Edgar Douglas Adrian, noted Cambridge University physiologist, won renown for his research on the functions of the **brain** and the nervous system. With Sir Charles Scott Sherrington, he received the Nobel Prize in physiology or medicine in 1932 in recognition of his work on the role of **neurons** in the stimulation of muscles and sense **organs**. Adrian's research also made possible the development of electroencephalography, or the measurement of electrical activity in the brain.

Born in London, Adrian was the son of Flora Lavinia Barton and Alfred Douglas Adrian. Adrian attended London's Westminster School and in 1908, won a science scholarship that opened the doors of Cambridge University's Trinity College to him. Besides taking courses in other natural sciences, he studied physiology under the direction of the physiologist Keith Lucas. Lucas was researching the reactions of muscles and nerves to electrical stimulation. When Adrian joined in this pursuit, he set his course for a lifelong career investigating the nervous system. He graduated from Trinity College in 1911 with first-class honors in five subjects.

Adrian's work in neurophysiology with Lucas led Adrian into the analysis of the functioning of neurons (nerve cells) in the stimulation of muscles and sense organs. The physiologist Sherrington had already made discoveries in this field, which Adrian was to advance further. His early research with Lucas resulted in his election as a fellow of Trinity College in 1913. Adrian earned his bachelor of medicine degree in 1915 at St. Bartholomew's in London and was able to pursue his interest in the nervous system when he served in the British Army during World War I. He was assigned to the treatment of nerve injuries and disorders of servicemen at the Hospital for Nervous Diseases. The effect of shell shock was a particular area of study. The young doctor's efforts to get assigned to a post in France were unsuccessful, however.

Adrian's career took an unexpected turn when Lucas died in an airplane crash during the war and Adrian was appointed to take charge of his laboratory in 1919, the same year he received his doctor of medicine degree. In the laboratory, he resumed his work on nerve impulses and began using advanced electrical techniques. He was able to amplify by 5,000 times the impulses in a single nerve fiber and single end organ in a frog's muscle. Adrian published his first observations on these nerve stimuli experiments in 1926 and came forth with definitive conclusions in 1928. Impulses that led to the sensation of **pain** were of particular interest to Adrian, and he directed his attention to a study of the brain. He found that the regions of the brain leading to a particular sense organ varied between species of animals. In pigs, which use their snouts to explore their environment, for instance, almost the entire region of the cortex dedicated to touch is taken up with nerve endings of the fibers that lead to the snout. In humans, a large area is taken up with the endings of fibers leading to the hands. Adrian's work cast new light on the nature of the nervous impulse, the action of the neuron and the physical nature of sensation.

In 1929, Adrian was elected Foulerton Professor of the Royal Society. He made a trip to New York, where he worked with Detlev Wulf Bronk on converting electrical impulses to sound. Returning to Cambridge, he continued his investigations of how sensory impulses reach the brain. One of his aims was to develop a practical method of reading the brain's electrical wave patterns. His work laid the foundation for the development of clinical electroencephalography, which could accomplish such brain analysis. The **electroencephalogram** (**EEG**) made it possible to study such conditions as epilepsy and brain **tumors**.

It was announced on October 27, 1932, that Adrian and Sherrington were to share the Nobel Prize in physiology or medicine. The news was greeted enthusiastically throughout the scientific world and hailed particularly by the British press. Adrian was named professor of physiology at Cambridge in 1937 and was appointed to the Medical Research Council in 1939. From 1951 to 1965, he held the post of master of Trinity College and from 1957 to 1959 was also vice chancellor of Cambridge University. During these later years, he also served terms as president of the Royal Society, president of the British Association for the Advancement of Science, and president of the Royal Society of Medicine. He served on committees of the World Health Organization and, in 1962, was elected a trustee of Rockefeller University in New York.

Among the many awards received for his research achievements were the Royal Medal (1934), the British Order of Merit (1942), the Copley Medal from the Royal Society (1946), the Albert Gold Medal of the Royal Society of the Arts (1953), the Harben Medal (1955), the French Legion of Honor (1956), the Medal for Distinguished Merit of the British Medical Association (1958) and the Jephcott Medal of the Royal Society of Medicine (1963). In 1955, he was knighted First Baron Adrian of Cambridge. Adrian did not confine his activities to the laboratory or lecture hall. He and his wife enjoyed mountain climbing. He also enjoyed fencing, sailing and fast bicycle riding. He took a strong interest in the arts, particularly painting. The exhibit of eighty of his works in Cambridge marked the high point of his hobby. When he retired from Trinity in 1965, he continued to live in the college's Neville's Court almost until his death.

See also Nerve impulses and conduction of impulses; Nervous system overview; Nervous system: embryological development; Neural damage and repair; Neural plexuses; Neurology; Neurons; Neurotransmitters

AGING PROCESS

In the past, there has been a tendency to confuse aging with diseases that are frequently associated with old age. It is, however, important to clearly distinguish the biochemical and physiological processes occurring with the onset of years from pathological conditions, for example **cancer, arteriosclerosis** (a generic term for several diseases in which the arterial wall becomes thickened and loses elasticity due to, for example, fatty deposits) and neurological disorders such as **Alzheimer's disease**. It was Fritz Verzar (1886–1979), Swiss gerontologist, who said that "Old age is not an illness; it is a continuation of life with decreasing capacities for adaptation." This view, that aging is a progressive failure of the body's homeostatic adaptive responses, gained wide acceptance only recently.

The obvious characteristics of old age are well known: greying and loss of **hair**, loss of teeth, wrinkling of the skin, decreased muscle mass and increased **fat** deposits. The physiological signs of aging are the gradual deterioration in function and capacity to respond to environmental stress. Thus basic kidney and digestive metabolic rates decrease, as does the ability to maintain a constant internal environment despite changes in temperature, diet and oxygen supply. These manifestations of aging are related to a decrease in the actual number of cells in the body (100,000 **brain** cells are lost each day) and to the disordered functioning of the cells that remain.

The extracellular components of tissues also change with age. Collagen fibres, responsible for the strength of **tendons**, increase in number and change in quality with aging. These changes within the fibres of arterial walls due to aging can be as much responsible for the loss of elasticity as those of the pathological condition culminating in arterioisclerosis. Elastin, another constituent of the intercellular matrix, is responsible for the elasticity of **blood** vessels and skin. In old age it thickens, fragments, and acquires a greater affinity for **calcium**. All these changes can also be associated with arteriosclerosis.

Most cells are not able to divide indefinitely. Several kinds of cells in the body, for example **heart** cells, **skeletal muscle** cells and **neurons** are incapable of replication. Experiments have also proved that many other cell types are limited in the number of cell divisions that they are able to perform. Embryonic fibroblast cells grown outside the body divide only about 50 times and then stop. Other experiments have shown that the number of cell divisions correlates with the age of the cell donor. The finite lifetime of cultured human cells has been observed in many normal cell types including skin, brain, **liver** and **smooth muscle**. No exception has been found to the general rule that normal cells possess a finite capacity to divide. The number of divisions also correlates with the normal lifespan of different species from which the cells are obtained-strong evidence that the cessation of mitosis is a normal, genetically programmed event.

Just as the factors that limit the life of an individual cell are unknown, so are those that restrict the growth or life of a **tissue** or organ. In women, at **menopause** the ovary ceases to function. Certain ovarian cells die long before the rest of the female body. One can speculate that similar mechanisms determine longevity. Some investigators have studied aging from the standpoint of **immunology**. The ability of a human being to develop antibodies is thought to diminish with age. Senescence, according to researchers, results in the older person's immunological system having a "shotgun" rather than a specific, response to a foreign protein. This shotgun response may include an autoimmune reaction that attacks and gradually destroys the individual's own normal tissue and **organs**.

Most gerontologists now believe that aging is a manifestation of the genetic coding within the cells of an organism. As a logical extension of the concept of development that begins from the moment of conception, aging must be part of the ongoing genetic expression of events that guide human organisms through embryological and fetal development, through childhood changes, adolescence and maturity. There are three general hypotheses that have been used as a basis for additional research. The first hypothesis proposes that, as time passes, the information-processing apparatus of cells begins to make more and more errors. As a result, faulty **enzymes** lead to a decline in the functional capability of the cells. Experiments designed to study **protein synthesis** in aging cells have not produced evidence that supports this hypothesis. The second hypothesis argues that many of the genes along the **DNA** molecule are repeated in identical sequences, making the genetic message very redundant. When an active gene is damaged, it may be replaced by one of the copies. Thus, as a person lives, the copies are gradually used up, errors accumulate, and functional losses result. According to this hypothesis, the greater the redundancy, the longer the lifespan. This hypothesis is also not substantiated by much evidence, although recent advances in the human genome analysis may provide more information.

The third hypothesis proposes that aging is a continuation of a normal sequence of genetic signals that regulate development. This idea suggests that aging genes are activated in the proper sequence, which causes a slowing down of biochemical pathways that are expressed in the recognizable changes of aging. There are two cell lines that are known to escape the inevitability of aging and **death**. These are **germ cells** and cancer cells. Perhaps these cell lines share a common genetic mechanism that makes them immortal. **Fertilization** may be a process that resets or reprograms a cell's clock by reshuffling the genetic information. These mechanisms make certain that each member of the species is programmed to die but ensure that the species will survive. Taking this view, one can look at an organism as a carrier of immortal germ cells and death as an effective survival strategy for the species. Since **viruses** can also reshuffle the genetic information of cells by integrating into the DNA, it is possible that viruses can cause cells to become "immortal" and cancerous.

See also Cell cycle and cell division; Growth and development

AGNODICE (4TH CENTURY B.C.)
Greek physician

Agnodice was the first Greek woman licensed to practice medicine in ancient Athens. According to Hyginus, a Latin author of the A.D. first century, Agnodice was a Greek sage who lived five hundred years earlier in the republic of Athens,

then a patriarchal society, where the learning and practice of medicine was an exclusive privilege of men. Determined to study medicine, Agnodice cut her **hair**, wore male clothing, and so disguised became a student under the famous anatomist Herophilus (ca. 335 B.C.).

Agnodice dedicated herself to the study of gynecological diseases, which would to become the focus of her clinical practice as a graduated physician. Always disguised as a man, she treated women, and when needed, revealed to her patients her real gender in order to be accepted by those who felt embarrassed in allowing a man to treat their gynecological or labor-related complications. Her knowledge and ability in the field of gynecology granted her an ever-growing female clientele that started refusing other male doctors. Agnodice's success among female patients aroused envy and suspicion from other local physicians, who accused Agnodice (whom they thought to be a man) of seducing the female clientele of the city. Patients were also accused of feigning illness to secure visits from Agnodice. Taken to court, Agnodice defended herself by lifting her tunic and revealing her true gender as a woman, being thus acquitted from such charges. However, she was now accused of illegal practice of medicine, because the law forbid women to both learn medicine and work as a physician. The wives of Athens came to the court in her support. Further considerations led the judges to lift such legal restrictions against women, therefore recognizing Agnodice's right to practice medicine and paving the way for other female citizens to be accepted as healers. Agnodice's legend was again heralded in the nineteenth century, as women in the Western world sought entrance into traditionally male institutions of medical education.

See also History of anatomy and physiology: The Classical and Medieval periods

AIDS

The advent of AIDS (acquired immunity deficiency syndrome) in early 1981 stunned the scientific community, as many researchers at that time viewed the world to be on the brink of eliminating infectious disease. AIDS, an infectious disease syndrome that suppresses the **immune system**, is caused by the human immune deficiency virus (HIV), part of a group of **viruses** known as retroviruses. The name AIDS was coined in 1982. Victims of AIDS most often die from opportunistic infections that take hold of the body because the immune system is severely impaired.

Following the discovery of AIDS, scientists attempted to identify the virus that causes the disease. In 1983–4, two scientists and their teams reported isolating HIV, the virus that causes AIDS. One was French immunologist Luc Montagnier (1932–), working at the Pasteur Institute in Paris, and the other was American immunologist Robert Gallo (1937–) at the National **Cancer** Institute in Bethesda, Maryland. Both identified HIV as the cause of AIDS and showed the pathogen to be a retrovirus, meaning that its genetic material is **RNA**, instead of **DNA**. Following the discovery, a dispute ensued over who made the initial discovery, but today Gallo and Montagnier are credited as co-discoverers.

Inside its host cell, the HIV retrovirus uses an enzyme called reverse transcriptase to make a DNA copy of its genetic material. The single strand of DNA then replicates and, in double stranded form, integrates into the chromosome of the host cell where it directs synthesis of more viral RNA. The viral RNA in turn directs the synthesis protein capsids and both are assembled into HIV viruses. A large number of viruses emerge from the host cell before it dies. HIV destroys the immune system by invading lymphocytes and macrophages, replicating within them, killing them, and spreading to others.

Scientists believe that HIV originated in the region of sub-Saharan Africa and subsequently spread to Europe and the United States by way of the Caribbean. Since viruses exist that suppress the immune system in monkeys, scientists hypothesize that these viruses mutated to HIV in the bodies of humans who ate the meat of monkeys, and subsequently caused AIDS. A fifteen-year-old male with skin lesions who died in 1969 is the first documented case of AIDS. Unable to determine the cause of **death** at the time, doctors froze some of his tissues, and upon recent examination, the **tissue** was found to be infected with HIV. During the 1960s, doctors often listed **leukemia** as the cause of death in many AIDS patients. After several decades however, the incidence of AIDS was sufficiently widespread to recognize it as a specific disease. Epidemiologists, scientists who study the incidence, cause, and distribution of diseases, turned their attention to AIDS. American scientist James Curran, working with the Centers for Disease Control and Prevention (CDC), sparked an effort to track the occurrence of HIV. First spread in the United States through the homosexual community by male-to-male contact, HIV rapidly expanded through all populations. Presently new HIV infections are increasing more rapidly among heterosexuals, with women accounting for approximately twenty percent of the AIDS cases. The worldwide AIDS epidemic is estimated to have killed more than 6.5 million people, and infected another 29 million. A new **infection** occurs about every fifteen seconds. HIV is not distributed equally throughout the world; most afflicted people live in developing countries. Africa has the largest number of cases, but the fastest rate of new infections is occurring in Southeast Asia and the Indian subcontinent. In the United States, although the disease was concentrated in large cities, it has spread to towns and rural areas. Once the leading cause of death among people aged 25–44 in the Unites States, AIDS is now second to accidents.

HIV is transmitted in bodily fluids. Its main means of transmission from an infected person is through sexual contact, specifically vaginal and anal intercourse, and oral to genital contact. Intravenous drug users that share needles are at high risk of contracting AIDS. An infected mother has a 15-25% chance of passing HIV to her unborn child before and during birth, and an increased risk of transmitting HIV through breast-feeding. Although rare in countries such as the United States where **blood** is screened for HIV, the virus can be transmitted by **transfusions** of infected blood or blood-clotting factors. Another consideration regarding HIV transmis-

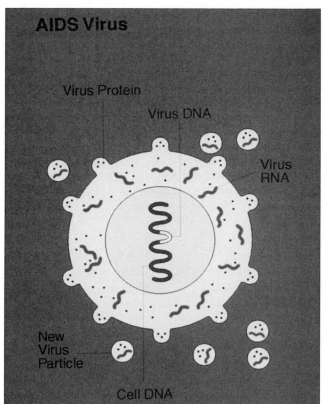

Diagram depicting HIV viral infection of cell and eventual lysis of new viral particles. *Courtesy of National Institutes of Health.*

sion is that a person who has had another sexually transmitted disease is more likely to contract AIDS.

Laboratories use a test for HIV-1 that is called ELISA (Enzyme-Linked-ImmunoSorbent Assay). There is another type of HIV called HIV-2. First developed in 1985 by Robert Gallo and his research team, ELISA is based on the fact that, even though the disease attacks the immune system, B cells begin to produce antibodies to fight the invasion within weeks or months of the infection. The test detects the presence of HIV-1 type antibodies, and reacts with a color change. Weaknesses of the test lie in the fact that it does not diagnose patients who are infectious but have not yet produced HIV-1 antibody or those that are infected with HIV-2. In addition, ELISA may give a false positive result to persons suffering from a disease other than AIDS. Patients that test positive with ELISA are given a second more specialized test to confirm the presence of AIDS. Developed in 1996, this test detects HIV **antigens**, proteins produced by the virus and can therefore identify HIV before the patient's body produces antibodies. In addition, separate tests for HIV-1 and HIV-2 have been developed.

After HIV invades the body, the disease passes through different phases, culminating in AIDS. During the earliest phase, the infected individual may experience general flu-like symptoms such as fever and **headache** within one to three weeks after exposure, and then remains relatively healthy while the virus replicates and the immune system produces antibodies. This stage continues for as long as the body's immune response keeps HIV in check. Progression of the disease is monitored by the declining number of particular antibodies called CD4-T lymphocytes. HIV attacks these immune cells by attaching to their CD4 receptor site. The virus also attacks macrophages, the cells that pass the antigen to helper **T lymphocytes**. The progress of HIV can also be determined by the amount of HIV in the patient's blood. After several months to several years, the disease progresses to the next stage in which the CD4-T cell count declines, and non-life-threatening symptoms such as weakness or swollen lymph glands may appear. The CDC has established a definition for the diagnosis of AIDS in which the CD4 T-cell count is below 200 cells per cubic millimeter of blood or, an opportunistic disease has set in.

Although progress has been made in the treatment of AIDS, a cure is yet to be found. In 1995, scientists developed a potent cocktail of drugs that help stop the progress of HIV. Among other substances, the cocktail combines zidovudine (AZT), didanosine, and a protease inhibitor. AZT and didanosine are nucleosides that are building blocks of DNA. The enzyme, reverse transcriptase, mistakenly incorporates the drugs into the viral chain, thereby stopping DNA synthesis. Used alone, AZT works temporarily until HIV develops immunity to the nucleoside. Proteases are **enzymes** that are needed by HIV to reproduce, and when protease inhibitors are administered, HIV replicates are no longer able to infect cells. In 1995, the Federal Drug Administration approved saquinaviras, the first protease inhibitor to be used in combination with nucleoside drugs such as AZT, followed in 1996 by approval for the protease inhibitors, ritonavir and indinavir to be used alone or in combination with nucleosides. The combination of drugs brings about a greater increase of antibodies and a greater decrease of HIV than either type of drug alone. Although patients improve on a regimen of mixed drugs, they are not cured due to the persistence of inactive virus left in the body. Researchers are looking for ways to flush out the remaining HIV. In the battle against AIDS, researchers are also attempting to develop a vaccine. As an adjunct to the classic method of preparing a vaccine from weakened virus, scientists are attempting to create a vaccine from a single virus protein.

In addition to treatment, the battle against AIDS includes preventing transmission of the disease. Infected individuals pass HIV-laden macrophages and T lymphocytes in their bodily fluids to others. Sexual behaviors and drug-related activities are the most common means of transmission. Commonly, the virus gains entry into the bloodstream by way of small abrasions during sexual intercourse, or direct injection with an infected needle. Preventing HIV transmission among the peoples of the world has resulted in unprecedented public health education and social programs aimed at increasing understanding of the nature and transmission of AIDS, and changing behaviors identified at-risk to spread the disease.

See also Immune system; Immunity; Viruses and responses to viral infection

ALIMENTARY SYSTEM • *see* GASTROINTESTIONAL TRACT

ALKALOSIS

Alkalosis is an acid-base imbalance toward alkalinity, or the base end of the acid-base range. Alkalosis is indicated by an elevated pH that corresponds to a decrease in hydrogen ion concentration. The body does not maintain an exactly neutral pH of 7, but rather a **blood** serum ph of 7.3 to 7.5. Accordingly, a pH of 7.293—although in strict terms an alkaline pH—is for the body an acidic rather than an alkaline state. Alkalosis exists when blood serum pH is greater than pH 7.5 (some clinicians and medical texts set this level at pH 7.45).

With regard to acid-base imbalance, the other extreme of disorder (lower pH) is termed **acidosis**. Alkalosis occurs less frequently that acidosis. An individual may suffer from both an acidosis and an alkalosis at the same time. For example, a metabolic acidosis may occur at the same time as a respiratory alkalosis. Invariably, one disorder is physiologically more dominant than the other. Accordingly, the less dominant imbalance is sometime referred to as a compensating or balancing mechanism.

Alkalosis may result from a loss of acidic compounds, or ingestion of base (alkaline) compounds that then shifts the acid base balance in favor of alkalinity. A loss of potassium and a reduction in potassium ion concentration is also a cause of alkalosis. If large amount of chloride ions are lost from the gastro-intestinal tract, there is also a large loss of potassium ions that results in hypochloremic alkalosis. Direct loss of potassium is termed hypokalemic alkalosis.

Especially in infants and small children prolonged malnutrition, constipation and vomiting may accompany alkalosis. Due to potassium depletion, individuals with alkalosis often complain of severe muscle cramping. The loss of gastric contents or the ingestion of diuretic compounds may also result in alkalosis. An increasingly common cause of alkalosis results from the over ingestion of antacids or bicarbonate of soda in an effort to ease indigestion.

Respiratory alkalosis results from an excessive ventilation of carbon dioxide (CO_2) and a corresponding decrease in CO_2 concentration (as measured by blood gas analysis). Excessive ventilation of carbon dioxide also results from hyperventilation, a form of deep and rapid **breathing**. Treatments for alkalosis resulting from hyperventilation include rebreathing techniques, the simplest of which allows individuals, under proper supervision, to breath into a paper bag for a few seconds to increase carbon dioxide levels.

See also Kidneys; Renal system; Respiration control mechanisms; Respiratory system

ALLERGIES

An allergy is an excessive or hypersensitive response of the **immune system** to harmless substances in the environment. Instead of fighting off a disease-causing foreign substance, the immune system launches a complex series of actions against an irritating substance, referred to as an allergen. The immune response may be accompanied by a number of stressful symptoms, ranging from mild to severe to life threatening. In rare cases, an allergic reaction leads to anaphylactic shock—a condition characterized by a sudden drop in **blood** pressure, difficulty in **breathing**, skin irritation, collapse, and possible **death**.

The immune system may produce several chemical agents that cause allergic reactions. Some of the main immune system substances responsible for the symptoms of allergy are the histamines that are produced after an exposure to an allergen. Along with other treatments and medicines, the use of antihistamines helps to relieve some of the symptoms of allergy by blocking out histamine receptor sites. The study of allergy medicine includes the identification of the different types of allergy, **immunology**, and the diagnosis and treatment of allergy.

The most common cause of allergy is pollens that are responsible for seasonal or allergic rhinitis. The popular name for rhinitis, hay fever, a term used since the 1830s, is inaccurate because the condition is not caused and its symptoms do not include fever. Throughout the world during every season, pollens from grasses, trees, and weeds produce allergic reactions like sneezing, runny **nose**, swollen nasal tissues, headaches, blocked sinuses, and watery, irritated eyes. Of the 46 million allergy sufferers in the United States, about 25 million have rhinitis.

Dust and the house dust mite constitute another major cause of allergies. While the mite itself is too large to be inhaled, its feces are about the size of pollen grains and can lead to allergic rhinitis. Other types of allergy can be traced to the fur of animals and pets, food, drugs, insect bites, and skin contact with chemical substances or odors. In the United States there are about 12 million people who are allergic to a variety of chemicals. In some cases an allergic reaction to an insect sting or a drug reaction can cause sudden death. Serious asthma attacks are associated with seasonal rhinitis and other allergies. About nine million people in the United States suffer from asthma.

Some people are allergic to a wide range of allergens, while others are allergic to only a few or none. The reasons for these differences can be found in the makeup of an individual's immune system. The immune system is the body's defense against substances that it recognizes as dangerous to the body. Lymphocytes, a type of white blood cell, fight **viruses**, **bacteria**, and other **antigens** by producing antibodies. When an allergen first enters the body, the lymphocytes produce an antibody called immunoglobulin E (IgE). The IgE antibodies attach to mast cells, large cells that are found in connective **tissue** and contain histamines along with a number of other chemical substances.

Studies show that allergy sufferers produce an excessive amount of IgE, indicating a hereditary factor for their allergic responses. How individuals adjust over time to allergens in their environments also determines their degree of susceptibility to allergic disorders.

The second time any given allergen enters the body, it becomes attached to the newly formed Y-shaped IgE antibodies. These antibodies, in turn, stimulate the mast cells to discharge its histamines and other anti-allergen substances. There are two types of histamine: H_1 and H_2. H_1 histamines travel to receptor sites located in the nasal passages, **respiratory system**, and skin, dilating smaller blood vessels and constricting airways. The H_2 histamines, which constrict the larger blood vessels, travel to the receptor sites found in the salivary and tear glands and in the stomach's mucosal lining. H_2 histamines play a role in stimulating the release of stomach acid, thus contributing to a seasonal stomach ulcer condition.

The simplest form of treatment is the avoidance of the allergic substance, but that is not always possible. In such cases, desensitization to the allergen is sometimes attempted by exposing the patient to slight amounts of the allergen at regular intervals.

Antihistamines, which are now prescribed and sold over the counter as a rhinitis remedy, were discovered in the 1940s. There are a number of different ones, and they either inhibit the production of histamine or block them at receptor sites. After the administration of antihistamines, IgE receptor sites on the mast cells are blocked, thereby preventing the release of the histamines that cause the allergic reactions. The allergens are still there, but the body's "protective" actions are suspended for the period of time that the antihistamines are active. Antihistamines also constrict the smaller blood vessels and **capillaries**, thereby removing excess fluids. Recent research has identified specific receptor sites on the mast cells for the IgE. This knowledge makes it possible to develop medicines that will be more effective in reducing the symptoms of various allergies.

Corticosteroids are sometimes prescribed to allergy sufferers as anti-inflammatories. Decongestants can also bring relief, but these can be used for a short time only, because their continued use can set up a rebound effect and intensify the allergic reaction.

ALVEOLI

An alveolus (alveoli is plural) is a tiny air sac located within the **lungs**. The exchange of oxygen and carbon dioxide takes place within these sacs.

The basic structure of the **respiratory system** can be envisioned as an upside-down tree. Air is breathed into the trachea, which is the tree trunk, and thus the broadest part of the respiratory tree. The trachea divides into two major tree limbs, the right and left **bronchi**, each of which branches off into multiple smaller bronchi, which course through the **tissue** of the lung. Just as a tree's limbs branch off into ever-smaller branches and twigs, so each bronchus divides into tubes of smaller and smaller diameter, finally ending in the terminal bronchioles. The air sacs of the lung, in which oxygen-carbon dioxide exchange actually takes place, are clustered at the ends of the bronchioles like the leaves of a tree at the ends of the smallest twig-like branches, and arecalled alveoli.

The alveoli are surrounded by tiny **blood** vessels called **capillaries**. When air is inhaled (breathed into the lungs), it ultimately enters the alveoli. Because the alveoli are composed of only a single, thin layer of tissue, oxygen in the inhaled air can cross out of the alveoli and into the capillaries, where it binds with the **hemoglobin** found in red blood cells. Blood containing oxygen is then carried throughout the body, for delivery to every type of tissue and organ system.

Carbon dioxide is one of the body's waste products. Carbon dioxide circulates through the body in the blood, until it reaches the alveolar capillaries. Carbon dioxide crosses out of the capillaries into the alveoli at the same time that oxygen is crossing out of the alveoli and into the capillaries. The carbon dioxide is then breathed out during exhalation.

It is interesting to note that, when comparing the alveoli of various species, the alveoli change in terms of both size and quantity. They are smallest but most numerous in mammals, intermediate in size and number in reptiles, and largest in size but smallest in quantity in amphibians. Humans continue to develop alveoli up until about the age of eight, when the human lung contains the adult number of approximately 300 million alveoli.

See also Respiration control mechanisms

ALZHEIMER'S DISEASE

Alzheimer's disease is a progressive **brain** disease that produces mental deterioration. Its symptoms, which include increasingly poor memory, personality changes, and loss of concentration and judgment, are caused by the physiological **death** of brain cells, and a decrease in the anatomical connections between those cells that do survive. The disease affects approximately four million persons in the United States. Although most victims are over age 65, Alzheimer's disease is not a normal result of aging. Medication can relieve some symptoms but there is no effective treatment or cure. Its cause is unknown.

People have long assumed that physical and mental decline were normal and unavoidable features of old age. Such deterioration was called senility. As recently as the early 1970s, much of the public and many physicians and nurses were not familiar with Alzheimer's disease. It was not until the 1980s that scientists realized that Alzheimer's disease was the most common cause of senility in middle-aged and older people. Since then, the public has become more aware of the disease, and scientists have increasingly focused on associating the anatomical and physiological disease patterns with specific genetic and environmental factors.

In addition to generating a high level of interest in the scientific community, in the United States public awareness and concern regarding Alzheimer's disease increases along with the increasing average age.

The term Alzheimer's disease is less than a century old. A German neurologist, Dr. Alois Alzheimer (1864-1915), was the first person to describe the disease. In 1906, he studied a 51-year-old woman whose personality and mental abilities were obviously deteriorating: she forgot things, became paranoid, and acted strangely. She died approximately four and one-half years after Alzheimer first treated her. Following an autopsy, Alzheimer examined sections of her brain under a microscope. He noted deposits of an unusual substance in her **cerebral cortex** (the outer, wrinkled layer of the brain, where many of the higher brain functions such as memory, speech, and thought originate). The substance Alzheimer saw under the microscope is now known to be a protein called amyloid beta-protein.

Many scientists today believe this protein plays an important role in causing Alzheimer's disease. Others believe it is not the primary cause of the disease, but rather a response to it. Eighty years or so after Alzheimer described the first case of the disease, researchers have found that a small percentage of Alzheimer's disease cases are apparently caused by genetic mutations. Most cases, however, are the result of unknown causes.

Ninety years after Alois Alzheimer first saw the amyloid beta-protein under his microscope, direct microscopic examination of the brain is still the only certain method of diagnosis. Nevertheless, by combining the results of medical histories, physical examinations, laboratory tests, and neurological exams to rule out other causes of dementia, physicians can accurately diagnose 90% or more of cases.

Prescription drugs are useful for treating symptoms like insomnia, anxiety, and depression, but there is no cure or drug that will stop the progressive degeneration produced by the disease. Estimates by the Alzheimer's Association indicate that Alzheimer's disease is the fourth leading cause of death in the United States, after **heart** disease, **cancer**, and **stroke**. It causes the deaths of more than 100,000 adults in the United States each year. As more and more citizens live into their 80s and 90s, the number of Alzheimer's disease cases world wide could reach 12-14 million or more before the middle of the twenty first century.

Alzheimer's disease is in most cases a disease of the elderly. Ten percent of people 65 years of age or older and 40-50% of everyone 85 or older has or will get the disease. Except for a small number of cases of inherited Alzheimer's disease, the disease is rarely seen in persons in their 30s or 40s.

The brains of Alzheimer's disease victims appear shrunken (atrophied) compared to nondiseased brains. The atrophy is particularly apparent in large parts of the neocortex, the outer layer of **gray matter** responsible for higher brain functions such as thought and memory. The **hippocampus** (a part of the brain near the temple) and the area around it are also heavily affected in Alzheimer's disease patients. This area plays an important role in forming memories. Much of the shrinkage of the brain is due to loss of brain cells and decreased numbers of connections, or synapses, between them.

In the surviving brain cells of an Alzheimer's patient, two hallmark features of the disease can be seen under the microscope: neurofibrillary tangles and amyloid plaques. Neurofibrillary tangles are abnormal collections of bunched and twisted fibrils in **neurons**. The fibrils are derived from components of a network of tubules and filaments that provide cells with structure and organization. The protein in neurofibrillary tangles is called tau. Such tangles are not specific to Alzheimer's disease but are also found in a dozen or so other brain diseases. The severity of mental impairment, however, correlates best with loss of connections between brain cells (synaptic loss), followed by neurofibrillary tangles; plaques appear to correlate only slightly or not at all.

Amyloid plaques consist of a core of amyloid beta-protein and other proteins surrounded by neurites. Neurites are the long projections of nerve cell bodies called axons and dendrites. Axons send signals to other cells and dendrites receive them. In addition to amyloid beta-protein, plaques contain other proteins, such as those normally found in **blood** serum.

Scavenger cells called microglia can be found in the center of plaques, and astrocytes, a type of cell that usually helps protect neurons, are found around the outside of the spherical plaques. The microglia may be part of an inflammatory response of the body to the plaques. Some researchers believe this response may actually end up killing brain cells rather than destroying plaques.

Amyloid plaques can be found in the brains of healthy persons in small numbers, but they are greatly increased in the brains of Alzheimer's disease patients. The disease, in fact, is defined by the presence of a large number of plaques in a section of brain **tissue**. The concentrations of many types of **neurotransmitters** (chemical messengers used by brain cells to communicate with one another) are also lower in the brains of Alzheimer's disease victims.

Scientists are studying many factors that may contribute to Alzheimer's disease. These include toxins, metabolic abnormalities, and infectious agents alone and in combination with each other. Some substances being investigated as possibly contributing to the development of Alzheimer's disease include mercury, aluminum, and **viruses** and their particles. Among the known risk factors for Alzheimer's disease are head trauma, age, Down syndrome, and in a small percentage of cases—approximately 10%—gene mutations. Women are more likely than men to suffer from Alzheimer's disease, although the use of **estrogen** in the post-menopausal period may possibly be somewhat protective. Some small studies have also suggested that the use of NSAIDs (nonsteroidal anti-inflammatory drugs) may also provide some protection against the development of Alzheimer's disease.

Familial Alzheimer's disease, an inherited form of the disease, accounts for approximately 10% of cases. Certain genes located on **chromosomes** 1 and 14 have been dubbed presenilins, and appear to be involved in the development of Alzheimer's disease. The pesenilin on chromosome 14 appears to be most frequently responsible for early-onset Alzheimer's disease (about 70%) with a shorter, more rapid course (age of onset, 45 years; duration of disease: 6-7 years). The presenilin on chromosome 1 has been implicated in a slightly later onset, with a somewhat more protracted course (age of onset: 53 years; duration of disease: 11 years).

In the majority of Alzheimer's disease cases, symptoms appear after age 65 and are not linked to a specific mutation. Some scientists suspect that the genes that are mutated in persons with familial Alzheimer's may play some role in the disease in victims without the mutations. While genes alone may cause 10% of all cases, the remaining 90% may be caused by various combinations of genetic and as yet undefined environmental factors. It is possible that most cases result from a genetic predisposition combined in varying ways with other factors.

Recently scientists have discovered that the gene for apolipoprotein E (ApoE), a protein that moves cholesterol in the bloodstream and can bind to amyloid beta-protein and other proteins, can affect a person's risk of developing the disease. There are three forms of the ApoE gene: ApoE2, ApoE3, and ApoE4. These three forms of the gene are not mutations but are normally occurring variations found in many populations. Unlike the genes inherited by the families with familial Alzheimer's disease, it is not a mutation in the ApoE gene that increases the risk factor; rather, it is the variety or type of ApoE gene someone inherits that affects his or her chances of developing Alzheimer's disease.

No one knows how this gene contributes to Alzheimer's disease, but it is known that inheritance of ApoE4 increases the risk and lowers the age of onset of the disease. Inheritance of ApoE2 decreases the risk and increases the age of onset. The ApoE gene seems to be a susceptibility gene.

Researchers are attacking the problem of Alzheimer's disease by studying the basic biology of the disease and by trying to develop drugs they hope will counteract or slow adverse anatomical and physiological disease processes.

See also Brain: intellectual functions; Genetics and developmental genetics; Histology and microanatomy; Nerve impulses and conduction of impulses; Nervous system overview; Neural damage and repair; Neurology

AMINO ACIDS

Amino acids, the building blocks of all protein molecules, are nitrogen-containing organic compounds that consist of at least one acidic carboxyl group (COOH) and one amino group (NH_2). In alpha amino acids that are contained in the proteins found in cells, these two groups are both attached to a carbon atom, which also carries a hydrogen atom, plus a side chain known as the R group. The R group varies from one amino acid to another and gives each amino acid its distinctive properties. Although relatively simple compounds, amino acids can vary widely and to date more than 80 different amino acids have been found in living organisms. Of these 80 amino acids, 22 are considered the precursors of animal proteins.

Proteins are one of the most common types of molecules in living matter. There are countless members of this class of molecules. They have many functions from composing **cell structure** to enabling cell-to-cell communication. One thing that all proteins have in common is that they are composed of amino acids.

Amino acids are amphoteric organic acids that are able to biochemically react with both acids or bases.

Codons along the messenger **RNA** molecule (mRNA), synthesized from a **DNA** template, control the sequence of the insertion of amino acids into the protein chain during the process of translation.

The first few amino acids were discovered in the early 1800s. In 1806, French chemist, Louis-Nicolas Vauquelin, isolated a compound in asparagus that proved to be the amino acid, asparagine. In 1812, William Hyde Wollaston found a substance in urine that he identified as a cystic oxide, and was later named cystine. And in 1820, another French chemist, Henri Braconnot, discovered the first two natural amino acids, glycine and leucine. Several other compounds were discovered toward the end of the 19th century. In 1895, Sven Hedin isolated the compound arginine; in 1896, with the help of his colleague Albrecht Kossel, he discovered histidine. Three years later, in 1899, Edmund Dreschel identified another important amino acid, lysine. Although these scientists were able to determine that these were unique compounds, they were unsure of their exact significance. Scientists were also uncertain of the relationship between amino acids and protein molecules.

In 1899, the German chemist, Emil Fischer, began investigating both questions. Fischer synthesized many of the thirteen amino acids that were already known, and identified three more. More importantly, Fischer, showed how the various amino acids combined with each other inside the protein molecule. The amino group of one amino acid is linked to the acidic carboxyl group of the next by a peptide bond. Fischer suggested that the sequences and patterns formed by the various chains of amino acids helped establish the characteristics of different proteins.

Fischer also developed a method for linking amino acids together, as they were in natural proteins, to form polypeptides. In 1907, he managed to put together a synthetic protein molecule that contained eighteen amino acid units, a molecule so remarkably authentic that, as he demonstrated, digestive **enzymes** attacked it just as they would a natural protein.

Although much was now known about the structure of amino acids, their nutritional significance had yet to be determined. Since the early 1800s, scientists such as Gerardus Mulder, **François Magendie** and William Prout had established the nutritional importance of the proteins themselves. But even here, with few exceptions (Magendie, for instance, had proven that gelatin had almost no nutritional value) the various proteins were considered roughly identical. So, most felt, were their amino acid units.

By the turn of the twentieth century, however, the situation began to change. In 1901, the British biochemist, Frederick Gowland Hopkins, not only discovered the amino acid tryptophan but later also showed that it played an important role in the diet. In one of his feeding experiments, Hopkins demonstrated that the protein in corn, zein, a protein that contains no tryptophan, could not sustain life in laboratory rats if used as the sole protein. Only when the tryptophan-rich

protein casein was added to the diet did the rats once again begin to thrive. Hopkins's experiment suggested that, if proteins were not nutritionally identical (which seemed increasingly evident), perhaps it was the amino acids they contained that made the difference.

At roughly the name time, two Americans, Thomas B. Osborne and Lafayette B. Mendel, reached similar conclusions. Between 1909 and 1928, the two biochemists were investigating the proteins in a great variety of plant seeds. They found that two amino acids in particular, tryptophan and lysine, were essential for normal growth in rats. Moreover, neither of the amino acids could be synthesized by the rats themselves, but had to be present in their diets.

In the 1930s, another American biochemist, William Rose, added the finishing touch to the amino acid story. In 1935, Rose isolated threonine, the last nutritionally important amino acid to be discovered, and, over the next decade or so, determined which amino acids could be synthesized by humans and certain mammals, and which had to be supplied by the diet. Unless all these amino acids were attained through various protein foods, Rose explained, the body would not have the building blocks to form new protein molecules, and the growth and repair of body cells would be impaired.

The amino acids that receive the most research attention are the alpha-amino acids that genes are codes for, and that are used to construct proteins. These amino acids include glycine NH_2CH_2COOH, alanine $CH_3CH (NH_2) COOH$, valine $(CH_3)2CHCH (NH_2)COOH$, leucine $(CH_3)_2CHC H_2CH(NH_2)COOH$, isoleucine $CH_3CH_2CH(CH_3)CH(NH_2) COOH$, methionine $CH_3SCH_2CH_2CH(NH_2)COOH$, phenylalanine $C_6H_5CH_2CH(CH_2)COOH$, proline C_4H_8NCOOH, serine $HOCH_2CH(NH_2)COOH$, threonine $CH_3CH(OH)CH(NH_2) COOH$, cysteine $HSCH_2CH(NH_2)COOH$, asparagine, glutamine $H_2NC(O)(CH_2)2CH(NH_2)COOH$, tyrosine $C_6H_4OH CH_2CHNH_2COOH$, tryptophan $C_8H_6NCH_2CHNH_2COOH$, aspartate $COOHCH_2CH(NH_2)COOH$, glutamate $COOH (CH_2)2CH(NH_2)COOH$, histidine $HOOCCH(NH_2)CH_2C_3 H_3H_2$, lysine $NH_2(CH_2)_4CH(NH_2)COOH$, and arginine $(NH_2)C(NH)HNCH_2CH_2CH_2CH(NH_2)COOH$.

Proteins consist of long chains of amino acids connected by peptide linkages (-CO·NH-). A protein's primary structure refers to the sequence of amino acids in the molecule. The protein's secondary structure is the fixed arrangement of amino acids that results from interactions of amide linkages that are close to each other in the protein chain. The secondary structure is strongly influenced by the nature of the side chains, which tend to force the protein molecule into specific twists and kinks. Side chains also contribute to the protein's tertiary structure, i.e., the way the protein chain is twisted and folded. The twists and folds in the protein chain result from the attractive forces between amino acid side chains that are widely separated from each other within the chain. Some proteins are composed of two of more chains of amino acids. In these cases, each chain is referred to as a subunit. The subunits can be structurally the same, but in many cases differ. The protein's quaternary structure refers to the spatial arrangement of the subunits of the protein, and describes how the subunits pack together to create the overall structure of the protein.

Even small changes in the primary structure of a protein may have a large effect on that protein's properties. A single misplaced amino acid can alter the protein's function. This situation occurs in certain genetic diseases such as **sickle cell anemia**. In that disease, a single glutamic acid molecule has been replaced by a valine molecule in one of the chains of the **hemoglobin** molecule, the protein that carries oxygen in red **blood** cells and gives them their characteristic color. This seemingly small error causes the hemoglobin molecule to be misshapen and the red blood cells to be deformed. Such red blood cells cannot distribute oxygen properly, do not live as long as normal blood cells, and may cause blockages in small blood vessels.

Enzymes are large protein molecules that catalyze a broad spectrum of biochemical reactions. If even one amino acid in the enzyme is changed, the enzyme may lose its catalytic activity.

The amino acid sequence in a particular protein is determined by the protein's **genetic code**. The genetic code resides in specific lengths (called genes) of the polymer doxyribonucleic acid (DNA), which is made up of from 3000 to several million nucleotide units, including the nitrogeneous bases: adenine, guanine, cytosine, and thymine. Although there are only four nitrogenous bases in DNA, the order in which they appear transmits a great deal of information. Starting at one end of the gene, the genetic code is read three nucleotides at a time. Each triplet set of nucleotides corresponds to a specific amino acid.

Occasionally there an error, or mutation, may occur in the genetic code. This mutation may correspond to the substitution of one nucleotide for another or to the deletion of a nucleotide. In the case of a substitution, the result may be that the wrong amino acid is used to build the protein. Such a mistake, as demonstrated by sickle cell anemia, may have grave consequences. In the case of a deletion, the protein may be lose its functionality or may be completely missing.

Amino acids are also the core construction materials for **neurotransmitters** and **hormones**. Neurotransmitters are chemicals that allow nerve cells to communicate with one another and to convey information through the nervous system. Hormones also serve a communication purpose. These chemicals are produced by glands and trigger metabolic processes throughout the body. Plants also produce hormones.

Important neurotransmitters that are created from amino acids include serotonin and gamma-aminobutyric acid. Serotonin ($C_{10}H_{12}N_2O$) is manufactured from tryptophan, and gamma-aminobutyric acid ($H_2N(CH_2)_3COOH$) is made from glutamic acid. Hormones that require amino acids for starting materials include thyroxine (a hormone produced by the thyroid gland), and auxin (a hormone produced by plants). Thyroxine is made from tyrosine, and auxin is constructed from tryptophan.

A class of chemicals important for both neurotransmitter and hormone construction are the catecholamines. The amino acids tyrosine and phenylalanine are the building materials for catecholamines, which are used as source material for both neurotransmitters and for hormones.

Amino acids also play a central role in the **immune system**. Allergic reactions involve the release of histamine, a chemical that triggers inflammation and swelling. Histamine is a close chemical cousin to the amino acid histidine, from which it is manufactured.

Melatonin, the chemical that helps regulate **sleep** cycles, and melanin, the one that determines the color of the skin, are both based on amino acids. Although the names are similar, the activities and component parts of these compounds are quite different. Melatonin uses tryptophan as its main building block, and melanin is formed from tyrosine. An individual's melanin production depends both on genetic and environmental factors.

Proteins in the diet contain amino acids that are used within the body to construct new proteins. Although the body also has the ability to manufacture certain amino acids, other amino acids cannot be manufactured in the body and must be gained through diet. Such amino acids are called the essential dietary amino acids, and include arginine, histidine, isoleucine, leucine, lysine, methionine, phenylalanine, threonine, tryptophan, and valine.

Foods such as meat, fish, and poultry contain all of the essential dietary amino acids. Foods such as fruits, vegetables, grains, and beans contain protein, but they may lack one or more of the essential dietary amino acids. However, they do not all lack the same essential dietary amino acid. For example, corn lacks lysine and tryptophan, but these amino acids can be found in soy beans. Therefore, vegetarians can meet their dietary needs for amino acids as long by eating a variety of foods.

Amino acids are not stockpiled in the body, so it is necessary to obtain a constant supply through diet. A well-balanced diet delivers more protein than most people need. In fact, amino acid and protein supplements are unnecessary for most people, including athletes and other very active individuals. If more amino acids are consumed than the body needs, they will be converted to **fat**, or metabolized and excreted in the urine.

However, it is vital that all essential amino acids be present in the diet if an organism is to remain healthy. Nearly all proteins in the body require all of the essential amino acids in their synthesis. If even one amino acid is missing, the protein cannot be constructed. In cases in which there is an ongoing deficiency of one or more essential amino acids, an individual may develop a condition known as kwashiorkor. Which is characterized by severe weight loss, stunted growth, and swelling in the body's tissues. The situation is made even more grave because the intestines lose their ability to extract nutrients from whatever food is consumed. Children are more strongly affected by kwashiorkor than adults because they are still growing and their protein requirements are higher. Kwashiorkor often accompanies conditions of famine and starvation.

Phenylketonuria (PKU) is an inherited metabolic disorder in which an enzyme (phenylalanine hydroxylase) that is crucial to the appropriate processing of the amino acid phenylalanine, is absent or deficient. Normally, phenylalanine is converted to tyrosine in the body. When phenylalanine cannot be broken down, it accumulates in excess quantities throughout the body, causing mental retardation and other neurological complications. Treatment is usually started during babyhood; delaying such treatment results in a significantly lowered intelligence quotient (IQ) by age one. Because tyrosine is involved in the production of melanin (pigment), people with PKU usually have lighter skin and **hair** than other family members.

PKU is an autosomal recessive disorder, and is caused by mutations in both alleles of the gene responsible for phenylalanine hydroxylase. Understanding of the exact mechanisms of the neurological complications associated with PKU are, however, little understood, and knowledge of the precise genetic mutations responsible for PKU have yet to yield significant advances in treatment or prevention of PKU. Because it is vital to begin diet treatment immediately, most nations in the developed world require all that all infants be tested for the disease within the first week of life.

See also Biochemistry; Inherited diseases; Pharmacogenetics; Protein metabolism; Protein synthesis; Ribonucleic acid (RNA)

AMNIOCENTESIS

Amniocentesis is a procedure used to obtain amniotic fluid for prenatal diagnosis of a fetus. Cells naturally are exfoliated from the surface of the fetus and some of these cells survive for a time in the fluid surrounding the fetus in the amniotic cavity. Soluble biochemical material of clinical significance produced by the fetus may also accumulate in the amniotic fluid. The fluid can be analyzed for these substances directly. During the procedure, a local anesthetic is given and a hollow needle is inserted through the mother's abdominal wall into the amniotic cavity. A small sample of the fluid is then withdrawn with a syringe attached to the needle. In order to insure the safety of the fetus, the procedure is monitored via an ultrasound scan. Viable cells in the fluid are then cultured (grown) *in vitro* in the laboratory. The **chromosomes** of the cultured cells can then be examined. Viewing the chromosomes under a light microscope will reveal if a normal diploid number of chromosomes are present or if extra or fewer chromosomes are present. Additionally, structural chromosomal aberrations can be detected.

Amniocentesis is an elective procedure that can detect the presence of many types of genetic disorders, thus allowing doctors and prospective parents to make important decisions about early treatment and intervention. Down syndrome is a chromosomal disorder characterized by a diversity of physical abnormalities, mental retardation, and shortened life expectancy. It is by far the most common, nonhereditary, genetic birth defect, afflicting about one in every 1,000 babies. Since the risk of bearing a child with a nonhereditary genetic defect such as Down syndrome is directly related to a woman's age, amniocentesis is often recommended for women who will be older than 35 on their due date. Thirty-five

is the recommended age to begin amniocentesis because that is the age at which the risk of carrying a fetus with such a defect roughly equals the risk of miscarriage caused by the procedure, about 1 in 200.

Amniocentesis is ordinarily performed between the 14th and 16th week of **pregnancy**, with results usually available within three weeks. It is possible to perform amniocentesis as early as the 11th week but this is not usually recommended because there appears to be an increased risk of miscarriage when done at this time. The advantage of early amniocentesis is the extra time for decision making if a problem is detected. Potential treatment for the fetus can begin earlier. Also, elective abortions are often safer and less controversial the earlier they are performed.

See also Birth defects and abnormal development; Prenatal growth and development

ANATOMICAL NOMENCLATURE

Over the centuries, anatomists developed a standard nomenclature, or method of naming anatomical structures. Terms such as "up" or "down" obviously have no meaning unless the orientation of the body is clear. When a body is lying on it's back, the thorax and **abdomen** are at the same level. The upright sense of up and down is lost. Further, because anatomical studies and particularly embryological studies were often carried out in animals, the development of the nomenclature relative to **comparative anatomy** had an enormous impact on the development of human anatomical nomenclature. There were obvious difficulties in relating terms from quadrupeds (animals that walk on four legs) who have abdominal and thoracic regions at the same level as opposed to human bipeds in whom an upward and downward orientation might seem more obvious.

In order to standardize nomenclature, anatomical terms relate to the *standard anatomical position*. When the human body is in the standard anatomical position it is upright, erect on two legs, facing frontward, with the arms at the sides each rotated so that the palms of the hands turn forward.

In the standard anatomical position, *superior* means toward the head or the *cranial* end of the body.

The term *inferior* means toward the feet or the *caudal* end of the body.

The frontal surface of the body is the *anterior* or *ventral* surface of the body. Accordingly, the terms "anteriorly" and "ventrally" specify a position closer to—or toward—the frontal surface of the body. The back surface of the body is the *posterior* or *dorsal* surface and the terms "posteriorly" and "dorsally" specify a position closer to—or toward—the posterior surface of the body.

The terms *superficial* and *deep* relate to the distance from the exterior surface of the body. Cavities such as the thoracic cavity have internal and external regions that correspond to deep and superficial relationships in the midsagittal plane.

The bones of the **skull** are fused by sutures that form important anatomical landmarks. Sutures are **joints** that run jaggedly along the interface between the bones. At birth, the

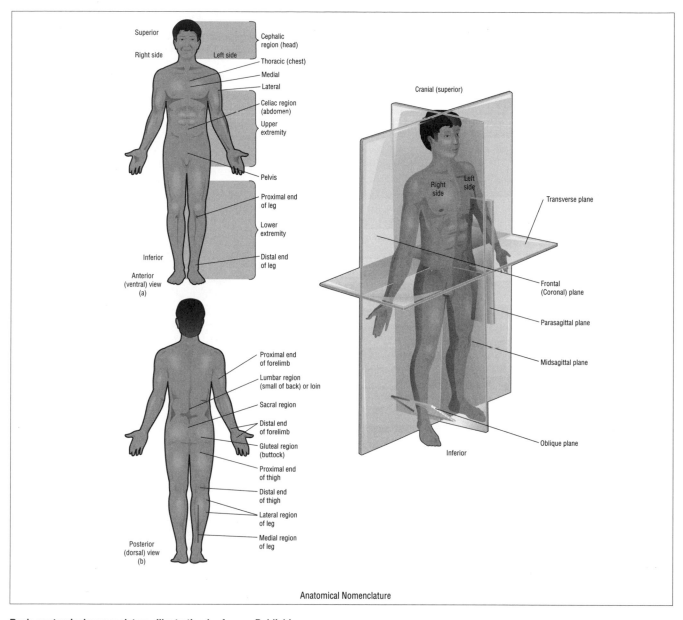

Anatomical Nomenclature

Basic anatomical nomenclature. *Illustration by Argosy Publishing.*

sutures are soft, broad, and cartilaginous. The sutures eventually fuse and become rigid and ossified near the end of **puberty** or early in adulthood.

The sagittal suture unties the parietal bones of the skull along the midline of the body. The suture is used as an anatomical landmark in anatomical nomenclature to establish what are termed *sagittal planes* of the body. The primary sagittal plane is the sagittal plane that runs through the length of the sagittal suture. Planes that are parallel to the sagittal plane, but that are offset from the midsagittal plane are termed *parasagittal planes*. Sagittal planes run anteriorly and posteriorly, are always at right angles to the coronal planes. The *medial plane* or *midsagittal plane* divides the body vertically into superficially symmetrical *right* and *left* halves.

The medial plane also establishes a centerline axis for the body. The terms *medial* and *lateral* relate positions relative to the medial axis. If a structure is medial to another structure, the medial structure is closer to the medial or center axis. If a structure is lateral to another structure, the lateral structure is farther way from the medial axis. For example, the **lungs** are lateral to the **heart**.

The coronal suture unites the frontal bone with the parietal bones. In anatomical nomenclature, the primary *coronal plane* designates the plane that runs through the length of the coronal suture. The primary coronal plane is also termed the *frontal plane* because it divides the body into frontal and back halves.

Planes that divide the body into superior and inferior portions, and that are at right angles to both the sagittal and coronal planes are termed transverse planes. Anatomical planes that are not parallel to sagittal, coronal, or transverse planes are termed oblique planes.

The body is also divided into several regional areas. The most superior area is the *cephalic region* that includes the head. The *thoracic region* is commonly known as the chest region. Although the *celiac region* more specifically refers to the center of the *abdominal region*, celiac is sometimes used to designate a wider area of abdominal structures. At the inferior end of the abdominal region lies the *pelvic region* or *pelvis*. The posterior or dorsal side of the body has its own special regions, named for the underlying vertebrae. From superior to inferior along the midline of the dorsal surface lie the *cervical, thoracic, lumbar* and *sacral* regions. The buttocks is the most prominent feature of the *gluteal region.*

The term *upper limbs* or *upper extremities* refers to the arms. The term *lower limbs* or *lower extremities* refers to the legs.

The *proximal* end of an extremity is at the junction of the extremity (i.e., arm or leg) with the trunk of the body. The *distal* end of an extremity is the point on the extremity farthest away from the trunk (e.g., fingers and toes). Accordingly, if a structure is proximate to another structure it is closer to the trunk (e.g., the elbow is proximate to the wrist). If a structure is distal to another, it is farther from the trunk (e.g., the fingers are distal to the wrist).

Structures may also be described as being medial or lateral to the midline axis of each extremity. Within the upper limbs, the terms radial and ulnar may be used synonymous with lateral and medial. In the lower extremities, the terms fibular and tibial may be used as synonyms for lateral and medial.

Rotations of the extremities may de described as medial rotations (toward the midline) or lateral rotations (away from the midline).

Many structural relationships are described by combined anatomical terms (e.g. the eyes are anterio-medial to the ears).

There are also terms of movement that are standardized by anatomical nomenclature. Starting from the anatomical position, *abduction* indicates the movement of an arm or leg away from the midline or midsagittal plane. *Adduction* indicates movement of an extremity toward the midline.

The opening of the hands into the anatomical position is *supination* of the hands. Rotation so the dorsal side of the hands face forward is termed *pronation.*

The term *flexion* means movement toward the flexor or anterior surface. In contrast, *extension* may be generally regarded as movement toward the extensor or posterior surface. Flexion occurs when the arm brings the hand from the anatomical position toward the shoulder (a curl) or when the arm is raised over the head from the anatomical position. Extension returns the upper arm and or lower to the anatomical position. Because of the embryological rotation of the lower limbs that rotates the primitive dorsal side to the adult form ventral side, flexion occurs as the thigh is raised anteriorly and superiorly toward the anterior portion of the pelvis.

Extension occurs when the thigh is returned to anatomical position. Specifically, due to the embryological rotation, flexion of the lower leg occurs as the foot is raised toward the back of the thigh and extension of the lower leg occurs with the kicking motion that returns the lower leg to anatomical position.

The term *palmar surface* (palm side) is applied to the flexion side of the hand. The term *plantar surface* is applied to the bottom sole of the foot. From the anatomical position, extension occurs when the toes are curled back and the foot arches upward and flexion occurs as the foot is returned to anatomical position.

Rolling motions of the foot are described as *inversion*(rolling with the big toe initially lifting upward) and *eversion* (rolling with the big toe initially moving downward).

See also Embryology; Histology and microanatomy; History of anatomy and physiology: The Classical and Medieval periods; History of anatomy and physiology: The Renaissance and Age of Enlightenment; History of anatomy and physiology: The science of medicine; Skeletal system overview (morphology)

ANATOMICAL PLANES • *see* ANATOMICAL NOMENCLATURE

ANATOMY

Anatomy is the study of the structure of living things. There are three main areas of anatomy: cytology studies the structure of cell; **histology** examines the structure of tissues; and gross anatomy deals with **organs** and organ groupings called systems. **Comparative anatomy**, which strives to identify general structural patterns in families of plants and animals, provided the basis for the classification of species. Human anatomy is a crucial element of the medical curriculum.

Modern anatomy, as a branch of Western science, was founded by the Flemish scientist **Andreas Vesalius**, who in 1543 published *De Humani Corporis Fabrica* (Structure of the human body), one of the great works in the history of science. In addition to correcting numerous misconceptions about the human body, Vesalius's book was the first description of human anatomy that organized the organs into systems. Although initially rejected by many followers of classical anatomical doctrines, Vesalius's systematic conception of anatomy soon became the foundation of anatomical research and education throughout the world; anatomists still use his systematic approach.

Human anatomy divides the body into the following distinct functional systems: cutaneous, muscular, skeletal, circulatory, nervous, digestive, urinary, endocrine, respiratory, and reproductive. This division helps the student understand the organs, their relationships, and the relations of individual organs to the body as a whole.

The cutaneous system consists of the integument—the covering of the body, including the skin, **hair**, and **nails.** The skin is the largest organ in the body, and its most important function is to act as a barrier between the body and the outside world. The skin's minute openings (pores) also provide an outlet for sweat, which regulates the body temperature. Melanin, a dark pigment found in the skin, provides protection from sunburn. The skin also contains oil-producing cells.

Beneath the skin is the muscular system. The muscles enable the body to move and provide power to the hands and fingers. There are two basic types of muscles. Voluntary (skeletal) muscles enable us to perform movements of our own decision, to walk, move our arms, or smile and frown. Involuntary (smooth) muscles are not consciously controlled, and operate on their own. For example, they play an important role in **digestion**. There is a third type of muscle, the **heart** muscle (myocardium), which is involuntary, but is striated, as in skeletal muscles. This life-sustaining muscle pumps **blood** throughout the body constantly, without pause, from the embryonic stage to **death**.

The skeletal system, or the skeleton, is underneath the muscular system. This bony frame provides the support that muscles need in order to function. Of the 206 bones in the human body, the largest is the femur, or thigh bone. The smallest are the tiny ear ossicles, three in each ear, named the hammer (malleus), anvil (incus), and stirrup (stapes). Often included in the skeletal system are the ligaments, which connect bone to bone; the **joints**, which allow the connected bones to move; and the **tendons**, which connect muscle to bone.

The **circulatory system** comprises the heart, **arteries**, **veins**, **capillaries**, blood and blood-forming organs, and the lymphatic sub-system. The four chambers of the heart pump blood throughout the body and the **lungs**. From the heart, the blood circulates through arteries. The blood is distributed through smaller and smaller tubes until it passes into the microscopic capillaries that bathe every cell. The veins collect the "used" blood from the capillaries and return it to the heart.

The nervous system consists of the **brain**, the **spinal cord**, and the sensory organs that provide information to them. For example, our eyes, ears, **nose**, tongue, and skin receive stimuli and send signals to the brain. The brain then makes decisions about any needed action. The brain is an intricate system of complicated cells that allow us to think, read, hear, and enjoy a movie. It also regulates **breathing**, movement, body temperature, and many other functions.

The digestive system is essentially a long tube extending from the mouth to the anus. Food entering the mouth is conducted through the stomach, small intestine, and large intestine, where accessory organs contribute digestive juices to break down the food, extracting the molecules that can be used to nourish the body. The unusable parts of the ingested food are expelled through the anus as fecal matter. The salivary glands (in the mouth), the **liver**, and the **pancreas** are the primary digestive glands.

The urinary system consists of the **kidneys**, the bladder, and the connecting tubules. The kidneys filter water and waste products from the blood and pass them into the bladder. At intervals, the bladder is emptied through the urinary tract, ridding the body of unneeded substances.

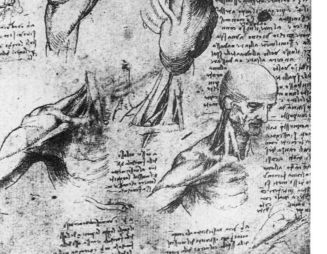

During the Renaissance, an explosion of anatomical studies based upon human dissection, such as this Leonardo da Vinci drawing of the muscular and skeletal structures of the upper body, defined anatomy as the central element of the science of medicine. *Corbis Corporation. Reproduced by permission.*

The **endocrine system** consists of ductless (endocrine) glands that produce **hormones** that regulate various bodily functions. The pancreas secretes insulin to regulate sugar **metabolism**, for example. The pituitary gland in the brain is the principal gland that regulates the others.

The **respiratory system** includes the lungs, the **diaphragm**, and the tubes that connect them to the outside air. Respiration is the process whereby an organism absorbs oxygen from the air and returns carbon dioxide. The diaphragm is the muscle that enables the lungs to work.

Finally, the reproductive system enables **sperm** and egg to unite and the egg to remain in the uterus or womb to develop into a functional human.

See also Anatomical nomenclature; History of anatomy and physiology: The Classical and Medieval periods; History of anatomy and physiology: The Renaissance and Age of Enlightenment; History of anatomy and physiology: The science of medicine

ANESTHESIA AND ANESTHETIC DRUG ACTIONS

Aesthesia is the depression or numbing of nerve pathways in all or part of the nervous system. The effect of anesthesia is the loss of sensation, principally the loss of **pain**. Thus, anesthesia functions to keep a patent free from pain during surgery.

A second hallmark of an anesthetic is its reversibility. The loss of sensation induced by an anesthetic is only temporary; the times vary depending on the anesthetic.

Anesthesia has accompanied surgical procedures for hundreds of years. Ether—until relatively recently a popular anesthetic, until its flammable nature proved too dangerous—was first made in 1540. Injectable anesthetics were in use by the mid-seventeenth century, and nitrous oxide (also commonly called laughing gas) was inhaled as both an anesthetic and for social frivolity by the early years of the nineteenth century.

There are four basic categories of anesthesia; general, regional, local, and sedation. General anesthesia affects the **brain** cells, producing unconsciousness. Regional anesthesia affects a large bundle of nerves to a certain area of the body. Sensation in that area of the body is lost, but consciousness is unaffected. Regional anesthesia is often employed in childbirth, producing a loss of sensation and pain to the lower body during labor and delivery. Local anesthesia produces a loss in sensation in a very specific area of the body. Anesthetizing the gums in dental surgery or the testicular area in a vasectomy are two examples of local anesthesia. Finally, application of a lower concentration of an anesthetic can produce sedation; a sleep-like state. Nitrous oxide, also commonly called laughing gas, is a sedative.

While under general anesthesia, a patient is unaware of his surroundings, is immobile, has no memory of the time they are anesthetized, and is pain-free. Both the brain and the **spinal cord** are affected. General anesthetics can be administered as an inhaled gas (popular examples are nitrous oxide, sevoflurane, desflurane, isoflurane and halothane) or as an injected liquid (barbiturates, propofol, ketamine, etomidate, narcotics, and Valium-like drugs like benzodiazepines).

Patients under general anesthesia must be carefully monitored, as the drugs pass through the brain and other **organs** of the body. **Heart** rate and rhythm, **blood** pressure and respiration rate are examples of monitored parameters.

Different anesthetic drugs have different actions in the body and exert their effects for different periods of time. For example, halothane can cause the heart rate to slow down and the blood pressure to decrease, while desflurane can have the opposite effects.

Anesthetics also vary with respect to their basis of action. A local anesthetic, which is typically applied by injection of the drug just underneath the designated area of skin, blocks nerve impulses by decreasing the permeability of nerve membranes to sodium ions. General anesthetics are thought to act on proteins and membrane channels that are involved in the transfer of information from one neuron to another (**neurotransmitters**). But, still after all these years of use, little is known of the exact molecular basis of action of general anes-

thetic drugs. Recent experiments on goldfish indicate that general anesthetics swell the proteins that form the membrane channels, and that this swelling is what blocks the transfer from neuron to neuron.

See also Brain stem function and reflexes; Central nervous system (CNS); Drug effects on the nervous system

ANGIOLOGY

The study of **blood** vessels, i.e., **veins** and **arteries**, and of lymphatic vessels, as well as the related physiological regulatory mechanisms, is known as angiology. The network of blood vessels constitutes a closed system of ducts that transports blood to the body tissues and from tissues back to the **heart**. This vascular system consists of both the arterial and venous systems. The blood flows out of the heart through the arterial system and returns to the heart through the venous system. Part of the interstitial or extra cellular fluid also penetrates the tiny capillary veins, and is thus carried to the vascular system.

The main cause of blood circulation is the cardiac systole followed by the contraction of the arterial walls, known as diastolic recoil, as well as the compression of the veins by the skeletal muscles during body movement and the negative intra-thoracic pressure caused by inbreathing. The walls of aorta and other large-diameter arteries are distended during cardiac systole and contracted during diastole, thus transmitting the pulsatile movement to a network of progressively thinner arteries. The walls of the great arteries are rich in elastic fibers and contain lesser amounts of smooth muscles than the walls of capillary arteries or arterioles, which have more smooth muscles and lesser quantities of elastic fibers. Therefore, arterioles offer greater resistance to the blood flux than do the much larger arteries. The arterioles also ramify into a network of even thinner vessels, which subdivide into hair-like vessels, or **capillaries**. They are connected to hair-like veins or venules, constituting the capillary bed that permeates tissues, supplying them with micronutrients and oxygen, and removing metabolic waste and carbonic gas excreted by cells. Capillary vessels have single-layer walls with a high permeability. The **anatomy** of capillary vessels varies from one organ to another, and is organized accordingly to specific organ needs. Capillaries may present intercellular clefts between the endothelial cells of the capillary walls, as well as windows or pores, covered by a very thin membrane known as basement membrane. In the glomerular tufts of the kidney, for instance, multiple oval windows or fenestrae (i.e., "windows" in Latin) allows the filtration of a great amount of ionic and small molecular substances by the renal glomeruli. The intestinal villi, have capillaries with medium-size intercellular clefts, which allow several different sizes of small and medium molecules to be absorbed, whereas in the **liver**, the clefts are wide opened, thus permitting the passage of almost all substances from the blood to the hepatic tissues, including **plasma** proteins. However, blood flow in the capillaries is not continuous, and is interrupted every few seconds or minutes, according to oxygen concentrations in the tissues, a phenomenon known as vasomotion.

Cells are kept apart from each other by interstitial spaces, constituted by collagen fiber bundles and proteoglycan filaments, which entrap the **interstitial fluid**. Interstitial fluid is originated from fluid diffusion and filtration from the capillaries, containing the same elements of the plasma, except by the much lower protein concentrations. Capillary pressure forces fluid and its solutes through the capillary pores to the interstitial spaces, whereas the osmotic pressure causes fluid movement back to the capillaries, thus preventing significant loss of blood fluid volume. Although most of the interstitial fluid is reabsorbed by the venules of the capillary bed, the lymphatic system helps the return to the circulation of small amounts of protein and excess of interstitial fluid through a network of lymphatic vessels that eventually drains into the venous system. Lymphatic capillary vessels can also transport larger metabolic substances and proteins that could not be otherwise removed from the interstitial spaces through direct reabsorption into the venous capillaries. Fluid and suspended particles are able to flow into the lymphatic capillaries (where they form the lymph) due to the presence of valves and anchoring filaments that attach the endothelial cells in the lymphatic capillaries to the connective **tissue** around them. Approximately two-thirds of the lymph is originated from the liver and intestines, and the lymphatic system is one important route of absorption of nutrients from the **gastrointestinal tract**, especially the absorption of **lipids**.

Veins and venules have thinner walls than arteries, are very flexible, with relatively little **smooth muscle**. They contain a series of one-way valves to prevent blood reflux due to gravity force and also suffer constriction through the action of adrenergic nerves and chemical agents such as norepinephrine. Veins are responsible for the return of blood into the atrium of the heart.

See also Arteriosclerosis; Blood pressure and hormonal control mechanisms; Coronary circulation; Metabolic waste removal; Pulmonary circulation

ANTERIOR • *see* ANATOMICAL NOMENCLATURE

ANTIANXIETY DRUG ACTIONS • *see*
PSYCHOPHARMACOLOGY

ANTIARRHYTHMIC DRUG ACTIONS • *see*
DRUG TREATMENT OF CARDIOVASCULAR AND VASCULAR DISORDERS

ANTIDEPRESSANT DRUG ACTIONS • *see*
PSYCHOPHARMACOLOGY

ANTIGENS AND ANTIBODIES

Antibodies, or Y-shaped immunoglobulins, are proteins found in the **blood** where they help to fight against foreign substances called antigens. Antigens, which are usually proteins or polysaccharides, stimulate the **immune system** to produce antibodies. The antibodies inactivate the antigen and help to remove it from the body. While antigens can be the source of infections from pathogenic (disease-causing) **bacteria** and **viruses**, organic molecules detrimental to the body from internal or environmental sources also act as antigens.

Once the immune system has created an antibody for an antigen whose attack it has survived, it continues to produce antibodies for subsequent attacks from that antigen. This long-term memory of the immune system provides the basis for the practice of vaccination against disease. The immune system, with its production of antibodies, has the ability to recognize, remember, and destroy well over a million different antigens.

There are several types of simple proteins known as globulins in the blood: alpha, beta, and gamma. Antibodies are gamma globulins produced by beta lymphocytes when antigens enter the body. The gamma globulins are referred to as immunoglobulins. In medical literature, immunoglobulins appear in the abbreviated form as Ig. Each antigen stimulates the production of a specific antibody (Ig).

Antibodies are all in a Y-shape with differences in the upper branch of the Y. These structural differences of **amino acids** in each of the antibodies enable the individual antibody to recognize an antigen. An antigen has on its surface a combining site that the antibody recognizes from the combining sites on the arms of its Y-shaped structure. In response to the antigen that has called it forth, the antibody wraps its two combining sites like a "lock" around the "key" of the antigen combining sites to destroy it.

An antibody's mode of action varies with different types of antigens. With its two-armed Y-shaped structure, the antibody can attack two antigens at the same time with each arm. If the antigen is a toxin produced by pathogenic bacteria that cause an **infection** like diphtheria or tetanus, the binding process of the antibody will nullify the antigen's toxin. When an antibody surrounds a virus, such as one that causes influenza, it prevents it from entering other body cells. Another mode of action by the antibodies is to call forth the assistance of a group of immune agents that operate in what is known as the **plasma** complement system. First, the antibodies will coat infectious bacteria and then white blood cells will complete the job by engulfing the bacteria, destroying them, and then removing them from the body.

There are five different antibody types, each one having a different Y-shaped configuration and function. They are the IgG, A, M, D, and E antibodies.

IgG is the most common type of antibody. It is the chief Ig against microbes. It acts by coating the microbe to hasten its removal by other immune system cells. It gives lifetime or long-standing immunity against infectious diseases. It is highly mobile, passing out of the blood stream and between cells, going from **organs** to the skin where it neutralizes surface bacteria and other invading microorganisms. This mobility allows the antibody to pass through the **placenta** of the mother to her fetus, thus conferring a temporary defense to the unborn child.

After birth, IgG is passed along to the child through the mother's milk, assuming that she nurses the baby. But some of the IgG will still be retained in the baby from the placental transmission until it has time to develop its own antibodies. Placental transfer of antibodies does not occur in horses, pigs, cows, and sheep. They pass their antibodies to their offspring only through their milk.

IgA antibodies are found in body fluids such as tears, **saliva**, and other bodily secretions. It is found in small quantities in the bloodstream and protects other wet mucosal surfaces of the body. While they have basic similarities, each IgA is further differentiated to deal with the specific types of invaders that are present at different openings of the body.

Since IgM is the largest of the antibodies, it is effective against larger microorganisms. Because of its large size (it combines five Y-shaped units), it remains in the bloodstream where it provides an early and diffuse protection against invading antigens, while the more specific and effective IgG antibodies are being produced by the plasma cells.

The ratio of IgM and IgG cells can indicate the various stages of a disease. In an early stage of a disease there are more IgM antibodies. The presence of a greater number of IgG antibodies would indicate a later stage of the disease. IgM antibodies usually form clusters that are in the shape of a star.

IgD antibodies appear to act in conjunction with B and T-cells to help them in location of antigens. Research continues on establishing more precise functions of this antibody.

The antibody responsible for allergic reactions, IgE acts by attaching to cells in the skin called mast cells and basophil cells (mast cells that circulate in the body). In the presence of environmental antigens like pollens, foods, chemicals, and drugs, IgE releases histamines from the mast cells. The histamines cause the nasal inflammation (swollen tissues, running **nose**, sneezing) and the other discomforts of hay fever or other types of allergic responses, such as hives, asthma, and in rare cases, anaphylactic **shock** (a life-threatening condition brought on by an allergy to a drug or insect bite). An explanation for the role of IgE in allergy is that it was an antibody that was useful to early man to prepare the immune system to fight parasites. This function is presently overextended in reacting to environmental antigens.

The presence of antibodies can be detected whenever antigens such as bacteria or red blood cells are found to agglutinate (clump together), or where they precipitate out of solution, or where there has been a stimulation of the plasma complement system. Antibodies are also used in laboratory tests for blood typing when **transfusions** are needed and in a number of different types of clinical tests, such as the Wassermann test for syphilis and tests for typhoid fever and infectious mononucleosis.

By definition, anything that makes the immune system respond to produce antibodies is an antigen. Antigens are living foreign bodies such as viruses, bacteria, and fungi that cause disease and infection. Or they can be dust, chemicals, pollen grains, or food proteins that cause allergic reactions.

Antigens that cause allergic reactions are called allergens. A large percentage of any population, in varying degrees, is allergic to animals, fabrics, drugs, foods, and products for the home and industry. Not all antigens are foreign bodies. They may be produced in the body itself. For example, **cancer** cells are antigens that the body produces. In an attempt to differentiate its "self" from foreign substances, the immune system will reject an organ transplant that is trying to maintain the body or a blood transfusion that is not of the same blood type as itself.

There are some substances such as nylon, plastic, or Teflon that rarely display antigenic properties. For that reason, nonantigenic substances are used for artificial blood vessels, component parts in **heart** pacemakers, and needles for hypodermic syringes. These substances seldom trigger an immune system response, but there are other substances that are highly antigenic and will almost certainly cause an immune system reaction. Practically everyone reacts to certain chemicals, for example, the resin from the **poison** ivy plant, the venoms from insect and reptile bites, solvents, formalin, and asbestos. Viral and bacterial infections also generally trigger an antibody response from the immune system. For most people penicillin is not antigenic, but for some there can be an immunological response that ranges from severe skin rashes to **death**.

Another type of antigen is found in the **tissue** cells of organ transplants. If, for example, a kidney is transplanted, the surface cells of the kidney contain antigens that the new host body will begin to reject. These are called human leukocyte antigens (HLA), and there are four major types of HLA subdivided into further groups. In order to avoid organ rejection, tissue samples are taken to see how well the new organ tissues match for HLA compatibility with the recipient's body. Drugs will also be used to suppress and control the production of helper/suppressor T-cells and the amount of antibodies.

Red blood cells with the ABO antigens pose a problem when the need for blood transfusions arises. Before a transfusion, the blood is tested for type so that a compatible type is used. Type A blood has one kind of antigen and type B another. A person with type AB blood has both the A and B antigen. Type O blood has no antigens. A person with type A blood would require either type A or O for a successful transfusion. Type B and AB would be rejected. Type B blood would be compatible with a B donor or an O donor. Because O has no antigens, it is considered to be the universal donor. Type AB is the universal recipient because its antibodies can accept A, B, AB, or O. One way of getting around the problem of blood types in transfusion came about as a result of World War II. The great need for blood transfusions led to the development of blood plasma, blood in which the red and white cells are removed. Without the red blood cells, blood could be quickly administered to a wounded soldier without the delay of checking for the blood antigen type.

Another antigenic blood condition can affect the life of newborn babies. Rhesus disease (also called erythroblastosis fetalis) is a blood disease caused by the incompatibility of Rh factors between a fetus and a mother's red blood cells. When an Rh negative mother gives birth to an Rh positive baby, any transfer of the baby's blood to the mother will result in the production of antibodies against Rh positive red blood cells. At her next **pregnancy** the mother will then pass those antibodies against Rh positive blood to the fetus. If this fetus is Rh positive, it will suffer from Rh disease. Tests for Rh blood factors are routinely administered during pregnancy.

Western medicine's interest in the practice of vaccination began in the eighteenth century. This practice probably originated with the ancient Chinese and was adopted by Turkish doctors. A British aristocrat, Lady Mary Wortley Montagu (1689–1762), discovered a crude form of vaccination taking place in a lower-class section of the city of Constantinople while she was traveling through Turkey. She described her experience in a letter to a friend. Children who were injected with pus from a smallpox victim did not die from the disease, but instead built up immunity to it. Rejected in England by most doctors who thought the practice was barbarous, smallpox vaccination was adopted by a few English physicians of the period. They demonstrated a high rate of effectiveness in smallpox prevention.

By the end of the eighteenth century, Edward Jenner (1749–1823) improved the effectiveness of vaccination by injecting a subject with cowpox, then later injecting the same subject with smallpox. The experiment showed that immunity against a disease could be achieved by using a vaccine that did not contain the specific pathogen for the disease. In the nineteenth century, Louis Pasteur (1822–1895) proposed the germ theory of disease. He went on to develop a rabies vaccine that was made from the spinal cords of rabid rabbits. Through a series of injections starting from the weakest strain of the disease, Pasteur was able, after 13 injections, to prevent the death of a child who had been bitten by a rabid dog.

There is now greater understanding of the principles of vaccines and the immunizations they bring because of our knowledge of the role played by antibodies and antigens within the immune system. Vaccination provides active immunity because our immune systems have had the time to recognize the invading germ and then to begin production of specific antibodies for the germ. The immune system can continue producing new antibodies whenever the body is attacked again by the same organism or resistance can be bolstered by booster shots of the vaccine.

For research purposes there were repeated efforts to obtain a laboratory specimen of one single antibody in sufficient quantities to further study the mechanisms and applications of antibody production. Success came in 1975 when two British biologists, **César Milstein** and Georges Kohler (1946–) were able to clone immunoglobulin (Ig) cells of a particular type that came from multiple myeloma cells. Multiple myeloma is a rare form of cancer in which white blood cells keep turning out a specific type of Ig antibody at the expense of others, thus making the individual more susceptible to outside infection. By combining the myeloma cell with any selected antibody-producing cell, large numbers of specific monoclonal antibodies can be produced. Researchers have used other animals, such as mice, to produce hybrid antibodies which increase the range of known antibodies.

Monoclonal antibodies are used as drug delivery vehicles in the treatment of specific diseases, and they also act as catalytic agents for protein reactions in various sites of the body. They are also used for diagnosis of different types of diseases and for complex analysis of a wide range of biological substances. There is hope that monoclonal antibodies will be as effective as **enzymes** in chemical and technological processes, and that they currently play a significant role in genetic engineering research.

See also B lymphocytes; Bacteria and responses to bacterial infection; Immunology; T lymphocytes; Viruses and responses to viral infection

ANTIOXIDANTS

Antioxidants are molecules that prevent or slow down the breakdown of other substances by oxygen. In biology, antioxidants are scavengers of small, reactive molecules known as free radicals and include intracellular **enzymes** such as superoxide dismutase (SOD), catalase and glutathione peroxidase. Antioxidants can also be extracellular originating as exogenous cofactors such as **vitamins**. Nutrients functioning as antioxidants include vitamins, for example ascorbic acid (vitamin C), tocopherol (vitamin E) and vitamin A. Trace elements such as the divalent metal ions selenium and zinc also have antioxidant activity as does **uric acid**, an endogenous product of purine **metabolism**. Free radicals are molecules with one or more unpaired electrons, which can react rapidly with other molecules in processes of oxidation. They are the normal products of metabolism and are usually controlled by the antioxidants produced by the body or taken in as nutrients. However, stress, aging, and environmental sources such as polluted air and cigarette smoke can add to the number of free radicals in the body, creating an imbalance. The highly reactive free radicals can damage nucleic acids and have been linked to changes that accompany aging (such as age-related macular degeneration, an important cause of **blindness** in older people) and with disease processes that lead to **cancer**, **heart** disease, and **stroke**.

The last few years have witnessed an explosion of information on the role of oxidative stress in causing a number of serious diseases, and there appears to be a potential therapeutic role for antioxidants in preventing such diseases. For example, recent epidemiological studies have shown that a higher consumption of vitamin E and to a lesser extent B-carotene is associated with a large decrease in the rate of coronary arterial disease. The most effective dose of vitamin E is apparently 400–800 mg/day. Other studies have shown that a diet rich in fruits and vegetables leads to a marked decline in the cancer rate in most **organs** with the exception of **blood**, breast, and prostate. Antioxidants play a major role in this. It is now abundantly clear that toxic free radicals play an important role in carcinogenesis. In several studies high cancer rates were associated with low blood levels of antioxidants particularly vitamin E. Similarly, vitamin C is thought to protect against stomach cancer by scavenging carcinogenic nitrosamines in the stomach.

The **brain** is particularly vulnerable to oxidative stress. Free radicals play an important role in a number of neurological conditions including stroke, Parkinson's disease, **Alzheimer's disease**, epilepsy and schizophrenia. Some other diseases in which oxidative stress and depletion of antioxidant defence mechanisms are prominent features include hepatic cirrhosis, pre-eclampsia, pancreatitis, rheumatoid arthritis, mitochondrial diseases, systemic sclerosis, **malaria**, neonatal oxidative stress, and renal dialysis.

Studies have suggested that the antioxidants occurring naturally in fresh fruits and vegetables are very beneficial and protect against excessive oxidative stress. There is still some question as to whether antioxidants in the form of dietary supplements are equally beneficial. Some scientists assert that regular consumption of such supplements interferes with the body's own production of antioxidants.

See also Aging processes; Neurology; Uric acid

ANTIPSYCHOTIC DRUG ACTIONS • *see* PSYCHOPHARMACOLOGY

ANUS • *see* GASTRO-INTESTINAL TRACT

AORTIC ARCH

The aortic arch describes the large bend in the ascending aorta after it leaves the **heart**. Running behind the **sternum** and the manubrium near, at about the level of the second rib (sternocostal **cartilage**), the ascending aorta makes a sweeping, double twisting bend toward the back (dorsal) surface of the body. The twisting and bending ultimately results in a generalized 180-degree bend or arch (aortic arch) that transforms into the descending aorta. Several important **arteries** that supply oxygenated **blood** to the neck and head have their origin in branches (trunks) off the aortic arch.

During the bending of the aortic arch, normally, three branches of the aorta split off from the trunk of the aortic arch. The three branches—the brachiocephalic trunk aorta, the left common carotid artery, and the left subclavian artery—usually split from the aortic arch as three separate arterial trunks, arising from different positions on the arch. The spacing of the branches may vary. It is not uncommon for one of more of these major arteries to be fused for a time—that is, for two of the branches to split off from a common trunk—or for the number of branches to be increased to four or more if, for example, the right common carotid artery also branch directly from the aortic arch instead of from the brachiocephalic trunks.

The brachiocephalic artery is the largest diameter branch of the aortic arch and normally gives rise to right subclavian and right common **carotid arteries** at about the level where the sternum joins the clavicle (sternoclavicular joint).

The right common carotid artery splits from the brachiocephalic artery. In contrast, the left common carotid artery usually branches directly from the aortic arch. The common carotids then branch into the external and internal carotid arteries that supply blood to the neck and head regions.

Although the right subclavian artery normally arises from the brachiocephalic trunk, the left subclavian artery usually spits directly from the aortic arch. The left and right **subclavian arteries** ultimately provide the arterial path for blood destined for the vertebral arteries, the internal **thoracic arteries**, and other vessels that provide oxygenated blood to the thoracic wall, **spinal cord**, parts of the upper arm (upper limb), neck, **meninges**, and the **brain**.

In the early embryo (4 weeks), blood leaves the developing heart into the trunk of the arterial system (*truncus arteriosis*) and the aortic sac. The ascending aorta and the aortic arch of the ascending aorta develop specifically from the embryonic pharyngeal arterial system. The generation, fusion, and degeneration of embryonic arteries, especially the trunks and branches of the embryonic pharyngeal arch system, are representative of the generalized head-to-tail (craniocaudal) scheme of development found in the early embryo. The ultimate twisting and curved orientation of the ascending aorta, the aortic arch region of the ascending aorta, and of the branches and trunks arising from the aortic arch in the adult are due to the dynamic development of the embryonic pharyngeal arch system.

Essentially, a transitional region between the ascending aorta and the descending aorta, the pressures found in the aortic arch are high, with the highest pressure normally at or near the upper limit of the individual's systolic high pressure (average 120 mm Hg). As branches arise and the arch turns into the descending artery, pressure begin to decrease due to vascular resistance.

Aortic arch syndrome describes a number of disorders, including such disorders as atherosclerosis, that result in a blockage of one or more branches of the aortic arch.

See also Angiology; Blood pressure and hormonal control mechanisms; Ductus arteriosis; Systemic circulation; Vascular system (embryonic development)

APPENDICULAR SKELETON

The appendicular skeletal system is the series of bones that form the arms and legs. These appendages are connected to the axial or central skeleton vertebrae and ribs by a series of bones that are collectively called the girdles. There is strong evolutionary evidence that suggests vertebrates are descendants of a group of fish named the Rhipidistians. Both the fossil and living fish have the same kind of bone structure and sequence of bone types in their fin girdles. It is believed that as these bones became increasingly functional in body weight support that they adapted to a variety of vertebrate functions.

The human pectoral girdle supports the arms. The broad and flat scapula attaches to the back of the body in the thoracic or chest region. Large muscles of the neck and back are anchored to it. In the front (anterior) region of the chest the small and long clavicles (collar bones) attach to the scapula.

This circular base around the upper body provides the strength for arms to circulate and move around the body.

The arms are a series of bones that follow the same pattern found in the legs. The humerus articulates with the scapula and is large and long. It is the point of attachment for the large working muscles of the arm and shoulders. At the elbow two bones, the radius and ulna allow the lower arm to move forward and backwards (ulna) and to rotate (radius). The wrist is a series of small square bones collectively named the carpals. It is easy to appreciate the value of the carpals by just turning your hand in a circle. The palm of the hand is supported by the metacarpals, some long and relatively slender bones that lead to the fingers. Each of the four long fingers is a series of three small **phalanges**. The thumb is the only phalange with only two bones.

The hip or pelvic girdle is actually a fusion of three bones: the ilium (superior), pelvic (anterior), and ischium (posterior). These bones form a very robust structure for bearing the weight of the body. The pelvis, as this structure is called, is attached to the fused vertebrae of the sacrum. A large fossa (deep pit) holds the major leg bone and is called the acetabulum. The pelvis provides additional support for the internal **organs**. In women, the pelvis is wider and flattened for support of the fetus during **pregnancy**.

The major bone of the leg is the large femur, or thigh bone. It is the largest bone of the body which is not surprising since it holds the entire torso upright. The pattern of the leg bones is the same as the arm. Next to the body there is one large long bone followed by another pair of long bones, the tibia and fibula. The tibia is also quite large while the fibula is very reduced and does not have a large role in standing. Following these bones is another series of squarish bones, the tarsals. Their function is similar to the wrist bones. The foot is supported by the metatarsals followed by the phalanges of the toes.

See also Bone histophysiology; Bone injury, breakage, repair, and healing; Bone reabsorbtion; Lower limb structure; Skeletal and muscular systems, embryonic development; Skeletal muscle; Skeletal system overview (morphology); Upper limb structure

APPENDIX

The appendix is a small finger-like appendage found near the juncture of the small and large intestines. Also termed the vermiform appendix, the organ is vestigial and has no apparent function. Vestigial **organs** and structures are those that do not serve an anatomical or physiological purpose and which are considered evolutionary remnants of an ancestral species or nonfunctional remnants of organs and structures created during embryonic development. Despite the fact that the appendix is not essential to good health, appendicitis—an inflammation of the appendix—can be life threatening and usually requires surgery to remove the appendix.

Inflammation of the appendix in the form of appendicitis may cause the appendix to rupture and spill its contents,

including intestinal **bacteria** into the peritoneal cavity. As a result of the introduction of intestinal bacteria into the **peritoneum**, an individual usually develops peritonitis, an inflammation of the peritoneum or peritoneal cavity. The peritoneal cavity is the space separating the visceral and parietal layers of the peritoneum. Peritonitis results in a high fever, **pain**, and may result in **death** if not aggressively treated with antibiotics.

During the course of **digestion** and waste elimination, the appendix may become clogged with fecal matter. Such a blockage prevents the normal drainage from the appendix and may result in acute (rapid onset) inflammation. Other agents, including parasitic worms, may also block appendix drainage and result in appendicitis.

The appendix is located in the lower right portion of the **abdomen**. Inflammation of the appendix is usually accompanied by a painful tenderness in this region, especially in response to palpation (the exerting of pressure by the hand). It is not uncommon for patients to experience pain in other areas, especially when the appendix becomes severely inflamed or ruptures.

See also Bacteria and responses to bacterial infection; Gastrointestinal tract; Infection and resistance; Vestigial structures

ARISTOTLE (384 B.C.-322 B.C.)
Greek philosopher and scientist

While he is highly regarded as a philosopher and father of logic and reasoning, Aristotle is also known for accomplishments in and contributions to other sciences. Throughout his life, he wrote several biological works which laid the foundations for **comparative anatomy**, taxonomy (classification), and **embryology**.

Aristotle was born in the northern Greek village of Stagira. His father was the court physician to the king of Macedonia, and it was at the Macedonian court where Aristotle spent much of his early boyhood. His father died before Aristotle was ten years old, and the young Aristotle was raised by friends of the family.

At age seventeen, Aristotle was sent to the Academy of Plato in Athens where he plunged wholeheartedly into Plato's pursuit of truth and goodness, and soon became Plato's best pupil, earning the nickname "intelligence of the school." Twenty years after Aristotle's arrival, Plato died; Aristotle then left the Academy to travel. His journeys led him through the Greek empire and Asia Minor for twelve years during which he began his research into natural history and biology.

In 342 B.C., Philip II invited Aristotle to return to the Macedonian court and teach his son, Alexander. Aristotle's student later became known in history as Alexander the Great.

After the death of Philip and Alexander's rise to the throne, Aristotle left the court for a brief visit to his hometown; he soon returned to Athens to resume his scientific studies. In 335 B.C., he founded a university called the

Lyceum. He had renounced some of Plato's theories and began his own style of teaching at the newly established school. In the mornings, he would stroll through the Lyceum gardens, discussing problems and theories with his advanced students. Because he walked about while teaching, Athenians nicknamed Aristotle's school the *Peripatetic*, the Greek term meaning, "to walk about." Like their headmaster, Lyceum pupils performed research in nearly every existing field of knowledge. They dissected animals and studied the habits of insects, helping Aristotle to compile data for his classification system.

The school became the basic building block for the great library and museum in the area. Unfortunately, in about the year 323 B.C., the ruling emperor Alexander died, forcing Aristotle to leave Athens due to anti-Macedonian sentiment and accusations of impiety. He went to his mother's homeland of Chalcis where he died a year later.

Aristotle contributed much to the field of biology, especially through his early work on classification. In helping devise a classification system, he established the basic principles of dividing and subdividing plants and animals. At that time, only about a thousand species were known and he was able to group them into simple categories of animals with red **blood** (with backbones), and animals with no red blood (without backbones). Plants were divided into different categories that dealt more with size and appearance.

Aristotle's classification system remained intact for almost 2,000 years, until in the 1500s, scientists recognized that the growth of knowledge called for an expanded system. Modification came slowly, with great debate, and it revealed the complexity of the process. In the late 1700s, the Aristotelian classification system was finally replaced by a much more comprehensive and systematic one developed by Swedish naturalist, Carl Linnaeus.

Aristotle's work on classification was not accidental. He was a painstaking observer, believing nature never created anything without a reason. He was particularly fascinated by sea creatures, often dissecting them and studying their natural habitats. This approach to **anatomy** led Aristotle to look for correlations between **structure and function** and to a belief that each biological part has its own special uses.

Ultimately, he established a teleological approach through constant inquiry about the ultimate purpose of things. This approach persisted in biological thinking well into the twentieth century.

Besides classifying animals and plants in nature, Aristotle also was the first to define and classify the various branches of knowledge. He sorted them into physics, metaphysics, rhetoric, poetics, and logic. In doing so, Aristotle laid the foundation of most of the sciences.

See also History of anatomy and physiology: The Classical and Medieval periods

ARM • *see* UPPER LIMB STRUCTURE

ARTERIES

Arteries are **blood** vessels that transport oxygenated blood from the **heart** to other **organs** and systems throughout the body.

A typical artery contains an elastic arterial wall that can be divided into three principal layers, although the absolute and relative thickness of each layer varies with the type or diameter of artery. The outer layer is termed the tunica adventia, the middle layer is termed the tunica media, and an inner layer is the tunica intima. These layers surround a lumen, or opening, that varies in size with the particular artery, through which blood passes.

Arteries of varying size comprise a greater arterial blood system that includes, in descending diameter, the aorta, major arteries, smaller arteries, arterioles, metaarterioles, and **capillaries**. It is only at the level of the capillary that branches of arteries become thin enough to permit gas and nutrient exchange. As the arterial system progresses toward the smaller diameter capillaries, there is a general and corresponding increase in the number of branches and total area of lumen available for blood flow. As a result, the rate of flow slows as blood approaches the capillary beds. This slowing is an important feature that enables efficient exchange of gases, especially oxygen.

In larger arteries, the outer, middle, and inner endothelial and muscle layers are supported by elastic fibers, and serve to channel the high pressure and high rate of blood flow. A difference in the orientation of cells within the layers (e.g., the outer endothelial cells are oriented longitudinally, while the middle layer **smooth muscle** cells run in a circumference around the lumen) also contributes both strength and elasticity to arterial structure.

The aorta and major arties are highly elastic, and contain walls with high amounts of elastin. During heart systole (contraction of the heart ventricles), the arterial walls expand to accommodate the increased blood flow. Correspondingly, the vessels contract during diastole and this contraction also serves to drive blood through the arterial system.

In the systemic arterial network that supplies oxygenated blood to the body, aortas are regions of the large-lumened singular artery arising from the left ventricle of the heart. Starting with the ascending aorta that arises from the left ventricle, the aortas form the main trunk of the systemic arterial system. Before the ascending aorta curves into the **aortic arch**, right and left coronary arteries branch off to supply the heart with oxygenated blood. Before the aortic arch turns to continue downward (inferiorly) as the descending aorta, it gives rise to a number of important arteries. Branching either directly off of—or from a trunk communicating with the aortic arch—is a brachiocephalic trunk that branches into the right subclavian and right common carotid artery that supply oxygenated blood to the right sight of the head and neck, as well as portions of the right arm.

The aortic arch also gives rise to the left common carotid artery that, along with the right common carotid artery, branches into the external and internal **carotid arteries** to supply oxygenated blood to the head, neck, **brain**.

The left subclavian artery branches from the aortic arch and—with the right subclavian arising from the brachiocephalic trunk—supplies blood to neck, chest (thoracic wall), **central nervous system**, and arms via axillary, brachial, and vertebral arteries.

In the chest (thoracic region), the continuation of the aortic arch—the descending aorta—is specifically referred to as the **thoracic aorta**. The thoracic aorta is the trunk of arterial blood supply to the thoracic region. Parietal branches of arteries derived from the thoracic aorta supply blood to the walls of thoracic organs and cavities. Visceral arterial branches supply blood to interior thoracic organs.

As the thoracic aorta passes through an opening in the **diaphragm** (aortic hiatus) to become the **abdominal aorta**, parietal and visceral branches supply oxygenated blood to abdominal organs and structures. The abdominal aorta ultimately branches into left and right common **iliac arteries** that then branch into internal and external iliac arteries, supplying oxygenated blood to the organs and tissues of the lower **abdomen**, pelvis, and legs.

In the pulmonary arterial system, the pulmonary trunk arises from the right ventricle of the heart to divide into left and right pulmonary arteries that supply deoxygenated (unaerated) blood to the **lungs** (by a number of pulmonary branch arteries) to different regions of lung **tissue**.

See also Angiology; Blood pressure and hormonal control mechanisms; Circulatory system; Gaseous exchange; Vascular system (embryonic development); Vascular system overview

Arteriosclerosis

Arteriosclerosis is a group of arterial diseases in which the arterial opening (lumen) becomes blocked (occluded) or narrowed concurrent with a loss of elasticity in the arterial wall. The three major forms of arteriosclerosis include atherosclerosis, medial arteriosclerosis, and arteriolar sclerosis. The group of arteriosclerotic diseases is most commonly known as hardening of the **arteries**.

Atherosclerosis, the most frequent form of arteriosclerosis, results from the deposition of fatty deposits (atheromas) that form plaques along the inner layer of the arterial vessel wall. As the deposition of plaques continues, the lumen becomes increasing narrow and the deposition of plaque causes a calcification or hardening of the vessel wall. The calcification of the wall results in a loss of elasticity and an ability to dilate to increase **blood** flow when oxygen demand increases. Plaques due to **fat** deposition also provide sites for the deposition of blood **platelets** and blood clot formation. Breakage of these clots usually results in an occlusion or **stroke** to a smaller vessel through which the clot cannot pass.

Although not all of the causes of arteriosclerotic disease are known, there is a familial and genetic predisposition to the disease. Hypertension also accelerates the disease process and,

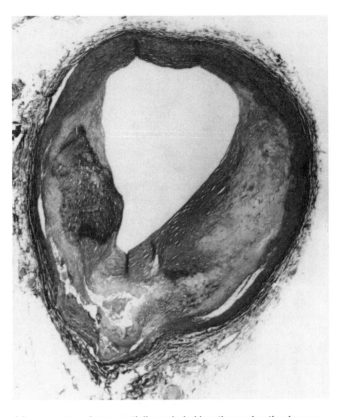

A human artery shown partially occluded by atherosclerotic plaques. *Photograph by Martin M. Rotker. Phototake NYC. Reproduced by permission.*

in turn, the narrowing of the lumen increases pressure with the vessel that potentiates (increases) hypertension.

Arteriosclerotic disease can lead to **heart** disease, aneurysm, stroke, and generalized peripheral vascular disease. Individual organ and regional symptoms (e.g., cold feet, painful throbbing after exercise) may also appear as early symptoms as the result of occlusions in arteries serving those **organs** or regions. Reduced blood flow to organs or regional **tissue** groups may also impede the removal of metabolic waste products from affected sites.

Elevated levels of low-density lipoproteins (LDL) contribute to atherosclerosis. The risk of atherosclerosis is usually measured by determining the proportion of high-density lipoprotein (HDL) to LDL.

The narrowed lumen and plaques contributed to increased vascular friction and resistance to flow. Plaque formation leads to both a disturbance of laminar flow (the smooth flow of **plasma** with the vessel channel) and increased resistance.

Arteriosclerosis is a normal consequence of aging, and arteriosclerotic diseases are the leading causes of morbidity (illness) and mortality in most developed countries, especially the United States. Because plaque formation is slow, the risks and serious consequences resulting from arteriosclerotic diseases generally increase with age. Men suffer arteriosclerotic diseases at far higher rates than women.

Nonatheromatous variations of arteriosclerosis include medial arteriosclerosis (Mönckeberg's arteriosclerosis) and arteriolosclerosis. Medial arteriosclerosis particularly affects the medial layer of arteries and leads to impairment of the medial muscle and elastic fibers. Arteriolar sclerosis affects smaller arteries (arterioles).

See also Angiology; Arteries; Blood pressure and hormonal control mechanisms; Cardiac disease; Circulatory system; Collateral circulation

ARTHROLOGY

Arthrology is a term that refers to the study and classification of **joints** of the body and how the joints contribute to the articulation (movement) of the various parts of the body.

Joints are located between bones. They can be thought of as the functional junction of the bones. For example, the position of the elbow and its ability to assume different positions makes possible a great range of motions in the arms. Without an elbow, an arm would be capable of only a rigid motion originating at the shoulder. With the articulation power of the elbow, however, an arm can be simultaneously lifted vertically and the lower arm can be turned at a right angle to the vertical movement.

Joints also allow for the alteration and expansion of bone during childhood development and growth, and permit the body to link muscular contraction to movement. Using the elbow as an example again, the contraction of the upper arm biceps muscle and the resulting upward movement of the lower arm would be impossible without the presence of the moveable elbow joint.

The classification of joints is based on the type of **tissue** present in the joint and on the degree of movement of the joint. With respect to tissue type, joints can be fibrous (containing fibrous connective tissue), cartilaginous (containing a structure of hyaline or **fibrocartilage**) or synovial (containing lubricating fluid). With respect to movement, joints can be completely immobile (synarthrotic), slightly movable (amphiarthrodic), or totally movable (diarthrodic).

There are several designs of diarthrodic joints in the human body. A ball and socket joint, in which a ball at the end of one bone fits into a cavity at the end of another bone, allows for up-down, side-to-side and rotational movement. Shoulder and hip joints are two examples of an enarthrodial joint. An ellipsoidal joint, such as that between carpals and metacarpals, can move in an up-down and side-to-side fashion. Both gliding (arthrodial) and hinged (ginglymoidal) joints are capable of movement in only one plane (up-down or side-to-side). The elbow and the knee are two examples of a hinged joint. The shoulder controls side-to-side motion of the arm, not the elbow. Another type of diarthrodic joint is also capable of only movement in one plane. But, the movement is rotational rather than up-down or side-to-side. Finally, as exemplified by the thumb, a saddle (or stellar) joint can move in two planes. The thumb joint is important from an evolutionary sense, as it

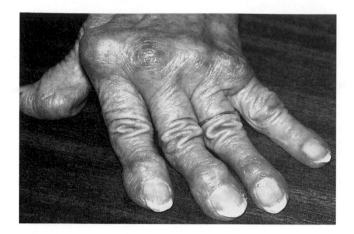

Joint malformations, as shown in this hand, often occur with inflammatory diseases such as rheumatoid arthritis. *Science Photo Library, National Audubon Society Collection/Photo Researchers, Inc. Reproduced by permission.*

allows the thumb to be used to grasp objects. This ability has set humans and primates apart from other life forms.

Diarthrodic joints vary in their construction. Some, like the shoulder, are relatively simple, with a ball meeting a cavity. Other joints can have three or more surfaces that are capable of motion. Still other joints, the spine for example, incorporate cartilaginous discs in the midst of the moveable joints. Some joints are packed closely together, while other joints, the knee for example, are arranged somewhat more loosely, with the bones being tied together via muscles and ligaments. The latter arrangement reduces friction in a joint that is frequently in motion.

See also Skeletal and muscular systems, embryonic development; Skeletal system overview (morphology)

ATP • *see* ADENOSINE TRIPHOSPHATE (ATP)

AUENBRUGGER, JOSEPH LEOPOLD (1722-1809)

Austrian physician

Leopold Auenbrugger invented the method of percussion for the diagnosis of chest diseases. He was born the son of an innkeeper in Graz, Austria. A student of Gerard van Swieten (1700–1772), he received his medical degree from the University of Vienna in 1752. From 1751 to 1762, he practiced medicine at the Spanish Military Hospital in Vienna and there became an expert on chest diseases. After serving as the hospital's chief physician from 1758 to 1762, he practiced medicine privately in Vienna for the rest of his life. A man of varied and cultivated interests, he contributed the libretto for Antonio Salieri's (1750–1825) comic opera *Der Rauchfangkehrer* in

1781. Holy Roman Emperor Joseph II (1741–1790) elevated him to the nobility in 1784.

In 1754, Auenbrugger discovered that sounds made by striking the chest gently but firmly with the fingertips would vary according to the health or condition of the chest. This technique, which he named "percussion," came naturally to him because in his childhood he had learned to ascertain the relative fullness of his father's wine barrels by tapping on them. Combined with "auscultation," listening to the chest, percussion soon became a powerful diagnostic tool for Auenbrugger, capable of determining the specific sites of lesions as well as different types of chest conditions. Diseased chests contain different amounts and thicknesses of liquid. The percussed sound of a healthy chest, dry and full of air, resembles a cloth-covered drum; but a diseased chest, containing various amounts of various thick fluids, sounds more muffled, as increased amounts of liquid dull the sound.

Auenbrugger assiduously correlated these different sounds with their respective thoracic conditions. He classified the sounds as high, low, distinct, indistinct, etc., and published his results in 1761 as *Inventum Novum ex Percussione Thoracis Humani ut Signo Abstrusos Interni Pectoris Morbos Detegendi (A New Discovery Enabling Detection of Diseases of the Chest by the Percussion of the Human Thorax)*. His findings were mostly ignored. Even though his book was translated into French as *Manuel des Pulmoniques ou Traité Complet des Maladies de la Poitrine (Manual of Pulmonics, or a Complete Treatise on the Diseases of the Chest)* by Rozière de la Chassagne, a physician in Montpellier, in 1770, it continued to attract little attention. Even a favorable review by **Albrecht von Haller** (1708–1777) in 1762 could not get Auenbrugger's innovation more widely recognized.

Auenbrugger's 1761 book remained obscure until Baron Jean Nicolas Corvisart des Marets (1755–1821) translated it into French in 1808 as *La Nouvelle Méthode pour Reconnaître les Maladies Internes de Poitrine par la Percussion de cette Cavité (The New Method of Recognizing Internal Diseases of the Chest by the Percussion of this Cavity)*. From that point, the diagnosis of chest diseases became a major concern among French medical researchers and practitioners. Corvisart's student, René-Théophile-Hyacinthe Laennec (1781–1826), invented the stethoscope in 1816. By the time Sir John Forbes (1787–1861) translated Auenbrugger's book into English as *On Percussion of the Chest* in 1824, physicians worldwide had already universally acknowledged percussion as a key diagnostic technique.

Besides chest diseases, Auenbrugger also studied nervous and mental disorders. He published in 1776 a Latin book recommending treatment of certain kinds of mania with camphor, and in 1783 a German book about suicide, which argued that to will one's own death was a genuine disease. The German-speaking world was then experiencing an epidemic of suicides of romantic young men, in the wake of the 1774 publication of *The Sorrows of Young Werther* by Johann Wolfgang von Goethe (1749–1832).

See also Cardiac disease; Lungs; Pathology; Respiratory system

AUTOIMMUNE DISORDERS

Autoimmune diseases are conditions in which a person's **immune system** attacks the body's own cells, causing **tissue** destruction. Autoimmune diseases are classified as either general, in which the autoimmune reaction takes place simultaneously in a number of tissues, or organ specific, in which the autoimmune reaction targets a single organ. Autoimmunity is accepted as the cause of a wide range of disorders, and is suspected to be responsible for many more. Among the most common diseases attributed to autoimmune disorders are rheumatoid arthritis, systemic lupus erythematosis (lupus), multiple sclerosis, myasthenia gravis, pernicious anemia, and scleroderma.

The reason why the immune system become dysfunctional is not well understood. Most researchers agree that a combination of genetic, environmental, and hormonal factors play into autoimmunity. The fact that autoimmune diseases run in families suggests a genetic component. Recent studies have identified an antiphospholipid antibody (APL) that is believed to be a common thread among family members with autoimmune diseases. Among study participants, family members with elevated APL levels showed autoimmune disease, while those with other autoantibodies did not. Family members with elevated APL levels also manifested different forms of autoimmune disease, suggesting that APL may serve as a common trigger for different autoimmune diseases. Further study of the genetic patterns among unrelated family groups with APL suggests that a single genetic defect resulting in APL production may be responsible for several different autoimmune diseases. Current research focuses on finding an established APL inheritance pattern, as well as finding the autoimmune gene responsible for APL production.

To further understand autoimmune disorders, it is helpful to understand the workings of the immune system. The purpose of the immune system is to defend the body against attack by infectious microbes (germs) and foreign objects. When the immune system attacks an invader, it is very specific. A particular immune system cell will only recognize and target one type of invader. To function properly, the immune system must not only develop this specialized knowledge of individual invaders, but it must also learn how to recognize and not destroy cells that belong to the body itself. Every cell carries protein markers on its surface that identifies it in one of two ways: what kind of cell it is (e.g., nerve cell, muscle cell, **blood** cell, etc.) and to whom that cell belongs. These markers are called major histocompatability complexes (MHCs). When functioning properly, cells of the immune system will not attack any other cell with markers identifying it as belonging to the body. Conversely, if the immune system cells do not recognize the cell as "self," they attach themselves to it and put out a signal that the body has been invaded, which in turn stimulates the production of substances such as antibodies that engulf and destroy the foreign particles. In case of autoimmune disorders, the immune system cannot distinguish between self cells and invader cells. As a result, the same destructive operation is carried out on the body's own cells that would normally be carried out on **bacteria, viruses,** and other such harmful entities.

A number of tests can help diagnose autoimmune diseases; however the principle tool used by physicians is antibody testing. Such tests involve measuring the level of antibodies found in the blood and determining if they react with specific **antigens** that would give rise to an autoimmune reaction. An elevated amount of antibodies indicates that a humoral immune reaction is occurring. Elevated antibody levels are also seen in common infections. These must be ruled out as the cause for the increased antibody levels. The antibodies can also be typed by class. There are five classes of antibodies and they can be separated in the laboratory. The class IgG is usually associated with autoimmune diseases. Unfortunately, IgG class antibodies are also the main class of antibody seen in normal immune responses. The most useful antibody tests involve introducing the patient's antibodies to samples of his or her own tissue, if antibodies bind to the tissue it is diagnostic for an autoimmune disorder. Antibodies from a person without an autoimmune disorder would not react to self tissue. The tissues used most frequently in this type of testing are thyroid, stomach, **liver,** and kidney tissue.

Treatment of autoimmune diseases is specific to the disease, and usually concentrates on alleviating symptoms rather than correcting the underlying cause. For example, if a gland involved in an autoimmune reaction is not producing a hormone such as insulin, administration of that hormone is required. Administration of a hormone, however, will restore the function of the gland damaged by the autoimmune disease. The other aspect of treatment is controlling the inflammatory and proliferative nature of the immune response. This is generally accomplished with two types of drugs. Steroid compounds are used to control inflammation. There are many different steroids, each having side effects. The proliferative nature of the immune response is controlled with immunosuppressive drugs. These drugs work by inhibiting the replication of cells and, therefore, also suppress non-immune cells leading to side effects such as anemia. Prognosis depends upon the **pathology** of each autoimmune disease.

See also Antigens and antibodies; Immunity

AUTONOMIC NERVOUS SYSTEM

The autonomic nervous system (ANS) is the specialized component of the nervous system that functions to regulate the activities of cardiac muscle, **smooth muscle, endocrine glands,** and exocrine glands. The ANS functions involuntarily and reflexively in an automatic manner without conscious control.

The ANS achieves its ability to either excite or inhibit activity via a dual innervation of target tissues and **organs.** The ANS achieves this control via two divisions of the ANS, the **sympathetic nervous system** and the **parasympathetic nervous system.**

The ANS is the mediator of visceral reflex arcs. In contrast to the somatic nervous system that always acts to excite

muscles groups, the autonomic nervous systems can act to excite or inhibit innervated **tissue.** The autonomic nervous system also differs from the somatic nervous systems in the types of tissue innervated and controlled. The somatic nervous system regulates **skeletal muscle** tissue, while the ANS services smooth muscle, cardiac muscle, and glandular tissue.

The involuntary ANS is controlled in the **hypothalamus** while the somatic system is regulated by other regions of the **brain** (cortex). In contrast, the somatic nervous system may control motor functions by neural pathways that contain only a single axon that innervates an effector (e.g., a target) muscle. The ANS is comprised of pathways that must contain at least two axons separated by a ganglia (clusters of neural cells outside of the brain and **spinal cord** of the **central nervous system**) that lies in the path between the axons.

ANS reflex arcs are stimulated by input from sensory or visceral receptors. The signals are processed in the hypothalamus (or regions of the spinal cord) and target effector control is then regulated via myelinated preganglionic **neurons** (cranial and spinal nerves that also contain somatic nervous system neurons). Ultimately, the preganglionic neurons terminate in a neural ganglion. Direct effector control is then regulated via unmyelinated postganglionic neurons.

The principal **neurotransmitters** in ANS synapses are acetylcholine and norepinephrine.

See also Muscular innervation; Nerve impulses and conduction of impulses; Nervous system overview; Nervous system, embryological development; Neural plexi; Neurology; Neurons; Neurotransmitters

A-V NODE • *see* HEART, RHYTHM CONTROL, AND IMPULSE CONDUCTION

AVIATION PHYSIOLOGY

Aviation **physiology** deals with the physiological challenges encountered by pilots and passengers when subjected to the environment and stresses of flight.

Human physiology is evolutionarily adapted to be efficient up to about 12,000 ft. (3,658 m) above sea level (the limit of the physiological efficiency zone). Outside of this zone, physiological compensatory mechanisms may not be able to cope with the stresses of altitude.

Military pilots undergo a series of exercises in high altitude simulating hypobaric (low pressure) chambers to simulate the early stages of **hypoxia.** The tests provide evidence of the rapid deterioration of motor skills and critical thinking ability when pilots undertake flight above 10,000 ft. (3,048 m) above sea level without the use of supplemental oxygen. Hypoxia can also lead to hyperventilation as the body attempts to increase **breathing** rates.

Altitude-induced decompression sickness is another common side effect of high altitude exposure in unpressurized or inadequately pressurized aircraft. Although the percentage of oxygen in the atmosphere remains about 21% (the other 79% of the atmosphere is composed of nitrogen and a small amount of trace gases), there is a rapid decline in atmospheric pressure with increasing altitude. Essentially, the decline in pressure reflects the decrease in the absolute number of molecules present in any given volume of air.

Pressure changes can adversely affect the **middle ear**, sinuses, teeth, and **gastrointestinal tract**. Any sinus block (barosinusitis) or occlusions that inhibit equalization of external pressure with pressure within the ear usually results in severe **pain**. In severe cases, rupture of the **tympanic membrane** may occur. Maxillary sinusitis may produce pain that is improperly perceived as a toothache. This is an example of **referred pain**. Pain related to trapped gas in the tooth itself (barondontalgia) may also occur.

Ear block (barotitis media) also causes loss of hearing acuity (the ability to hear sounds across a broad range of pitch and volume). Pilots and passengers may use the **Valsalva maneuver** to counteract the effects of water pressure on the **Eustachian tubes** and to eliminate pressure problems associated with the middle ear. When subjected to pressure, the tubes may collapse or fail to open unless pressurized. Eustachian tubes connect the corresponding left and right middle ears to the back of the **nose** and throat, and function to allow the equalization of pressure in the middle ear air cavity with the outside (ambient) air pressure. The degree of Eustachian tube pressurization can be roughly regulated by the intensity of abdominal, thoracic, neck, and mouth muscular contractions used to increase pressure in the closed airway.

Rapid changes in altitude allow trapped gases to cause pain in **joints** in much the same way—although to a far lesser extent—that the **bends** causes pain in scuba divers. Lowered outside atmospheric pressure creates a strong pressure gradient that permits dissolved nitrogen and other dissolved or "trapped" gases within the body to attempt to "bubble off" or leave the **blood** and tissues in an attempt to move down the concentration gradient toward a region of lower pressure.

Spatial disorientation trainers demonstrate the disorientation and loss of balance (vestibular disorientation) that can be associated with flight at night—or in clouds—where the pilot losses the horizon as a visual reference frame. Balance and the sense of turning depend upon the ability to discriminate changes in the motion of fluids within the semicircular canals of the ear. When turns are gradual, the changes become imperceptible because the fluids are moving at a constant velocity. Accordingly, without visual reference, pilots can often enter into steep turns or dives without noticing any changes. Spatial disorientation chambers allows pilots to learn to "trust their instruments" as opposed to their error-prone sense of balance when flying in IFR (Instrument Flight Rules) conditions.

In addition to vestibular disorientation, spatial disorientation can also lead to motion sickness.

Because of the highly repetitive nature of the active pilot scan of instruments, fatigue is a chronic problem for pilots. Fatigue combined with low oxygen pressures may induce strong and disorienting visual illusions.

Although not often experienced in general aviation, military pilots operate at high speeds and undertake maneuvers that subject them to high "g" (gravitational forces) forces. In a vertical climb, the increased g forces (called positive "g" forces" because they push down on the body) tend to force blood out of the **circle of Willis** supplying arterial blood to the **brain**. The loss of oxygenated blood to the brain eventually causes pilots to lose their field of peripheral vision. Higher forces cause "blackouts" or temporary periods of unconsciousness. Pilots can use special abdominal exercises and "g" suits (essentially adjustable air bladders that can constrict the legs and **abdomen**) to help maintain blood in the upper half of the body when subjected to positive "g" forces.

In a dive, a pilot experiences increased upward "g" forces (termed negative "g" forces) that force blood into the arterial circle of Willis and cerebral **tissue**. The pilot tends to experience a red out. Increased arterial pressures in the brain can lead to **stroke**. Although pilots have the equipment and physical stamina to sustain many positive "g" forces (routinely as high as five to nine times the normal force of gravity) pilots experience red out at about 2–3 negative "g's." For this reason, maneuvers such as loop, rolls, and turns are designed to minimize pilot exposure to negative "g" forces.

See also Aviation physiology; Ear (external, middle and internal ear); Ear: Otic embryological development; Sense organs: Balance and orientation; Space physiology; Swallowing and dysphagia; Underwater physiology

AVICENNA (ABU ALI AL-HUSSAIN IBN ABDULLAH IBN-SINA) (981-1037)
Arabic physician

Avicenna (Ibn-Sina) was a philosopher and physician, born in Afshana, near Boukhara (presently, Pakistan). He wrote an early tenth-century medical encyclopedia, *al-Qanun fi al-Tibb*, known as *Canons of Medicine* in the West.

A precocious child, Avicenna was already well versed in the Al-Quram (the sacred book of Islam) and in the Shariah or Islamic law, at the age of ten. Showing great intelligence, he learned virtually without tutoring in mathematics, physics, medicine, and Eastern and Greek philosophy, the latter recently translated into Arabic by the Afghan sage al-Biruni (A.D. 975–1067). At the age of seventeen, Avicenna cured the prince of Boukhara of a serious illness, and the prince, out of gratitude, gave him free access to his vast royal library. After the fall of this ruler, Avicenna traveled through the Middle East and the western part of the Indian subcontinent, and finally established himself in Djouzdjan (Jurjan) in Afghanistan, under the protection of Abu-Muhammad Chirazi, who offered him a house for the opening of a public school. It was in this city that he first met the astute Abu Raihan al-Biruni, and they entertained the local scholars with their famous philosophical debate.

In Jurjan, Avicenna built a prestigious name as a physician as well as a philosopher, and teacher. He also began writing his famous *Canons of Medicine*. Some years later, the ruler of Hamadan appointed Avicenna as his Great Vizier, but the envy and political intrigue of his rivals forced him to seek protection under the prince of Isphahan in Persia (now Iran), where he made important contributions to the advancement of medicine at the Isphahan School of Medicine. In Isphahan, Avicenna finished his *Cannon of Medicine*. The Canons were later translated into Latin by Gerardo de Cremona in the twelfth century, achieving great popularity for six centuries among the European physicians of the Middle Ages and Renaissance. The Canons cover thousands of topics, such as **pathology** (i.e., the study of disease, its causes, and natural history); pharmacopoeia (description of medicinal drugs and their properties), and **anatomy.**

Among Avicenna's many contributions to the science of medicine, he was a pioneer in the recognition of the contagious nature of tuberculosis and other infectious diseases, the first to describe the symptoms and natural history of meningitis, and developed the hypothesis of infectious diseases being transmitted through water, food and soil contamination. Avicenna described in detail the different parts of the **eye** as it is known today: conjunctive sclera, **cornea**, choroids, iris, **retina**, layer lens, aqueous humor, optic nerve, and optic chiasma. He also described the three aortic valves and their role in avoiding the reflux of **blood** into the **heart.** Avicenna explained both muscular movement and muscle **pain** as originated from the stimuli of the nerves attached to them. He observed that although the **liver**, spleen, and **kidneys** did not contain any nerves inside them, the nerves were embedded in the tissues that covered these **organs.** He introduced the use of medical probe for the lachrymal channel in the treatment of lachrymal fistula. He recommended that physicians and surgeons should study the human body directly, in order to understand anatomy, and should never base their theories in conjectures and presumptions. Avicenna took a special interest in child health and obstetrics, as well as in gynecology. In a similar manner to Hindu traditional medicine, he recognized the interaction between psychology and health. He died from a stomach disease in Hamadan, Persia (now, Iran) at the age of 56. A famous portrait of Avicenna (or Ibn-Sina) hangs in the hall of the Faculty of Medicine at the University of Paris, France.

See also Aortic arch; Eye and ocular fluids; History of anatomy and physiology: The Classical and Medieval periods; Muscle contraction

AXELROD, JULIUS (1912-   )

American biochemist and pharmacologist

Julius Axelrod is a biochemist and pharmacologist whose discoveries relating to the role of **neurotransmitters** in the **sympathetic nervous system** earned him the Nobel Prize in physiology or medicine in 1970, together with Ulf Euler of Sweden and Sir **Bernard Katz** of Great Britain. As Axelrod himself has said, he was a late starter as a distinguished scien-

tist, due to both the humble circumstances of his birth and his coming of age in the Great Depression of the 1930s. He only began real scientific research in 1946, and earned his Ph.D. in 1955. From then on he compensated for lost time and became the first chief of the **pharmacology** section of the National Institute of Mental Health, a branch of the prestigious National Institutes of Health.

Axelrod was born on May 30, 1912, in a tenement house in New York City, the son of Isadore Axelrod, a maker of flower baskets for merchants and grocers, and Molly Leichtling Axelrod. His parents had immigrated to the United States from Polish Galicia in the early years of the century, met and married in New York, and settled in the heavily Jewish area of the Lower East Side of Manhattan. Julius Axelrod attended public elementary and high schools near his home but later recalled that he got his real education in the neighborhood public library, reading voraciously through several books a week, everything from pulp novels to Upton Sinclair and Leo Tolstoy. He studied for a year at New York University, but when his money ran out he transferred to the tuition-free City College of New York, from which he graduated in 1933 with majors in biology and chemistry. He later claimed that he did most of his studying on the long subway rides between his home and the uptown Manhattan campus of City College.

Axelrod applied to several medical schools but was not admitted to any. It has been widely reported, in the *New York Times,* for example, that he failed to get into medical school because of quotas for Jewish applicants. It was difficult to find any work in New York in the depths of the Depression, and Axelrod was fortunate to find employment in 1933 as a laboratory assistant at the New York University Medical School at $25 per month. In 1935, he took a position as chemist at the Laboratory of Industrial Hygiene, a nonprofit organization set up by the New York City Department of Public Health to test vitamin supplements added to foods. He married Sally Taub on August 30, 1938, and they eventually had two sons. Axelrod took night courses and received an M.A. in chemistry from New York University in 1941. In the early 1940s, Axelrod lost the sight of one **eye** in a laboratory accident.

Axelrod later speculated that he might have remained at the Laboratory of Industrial Hygiene for the rest of his working life. The work, he said, was moderately interesting, and the pay adequate. However, in 1946, quite by chance, he received the opportunity to do scientific research and found it exciting. The laboratory received a small grant to study the problem of why some persons taking large quantities of acetanilide, a non-aspirin pain-relieving drug, developed methemoglobinemia, the failure of **hemoglobin** to bind oxygen for delivery throughout the body. Axelrod, who had little experience in such work, consulted Dr. Bernard B. Brodie of Goldwater Memorial Hospital of New York. Brodie was intrigued with the problem and worked closely with Axelrod in finding its solution. He also found Axelrod a place among the research staff at New York University. The two men soon discovered that the body metabolizes acetanilide into a substance with an analgesic effect, and another substance that causes methemoglobinemia. They recommended that the beneficial metabolic product be administered directly, without the use of acetanilide. Related analgesics were investigated in the same manner.

In 1949, Axelrod, Brodie, and several other researchers at Goldwater Hospital were invited to join the National **Heart** Institute of the National Institutes of Health in Bethesda, Maryland. There Axelrod studied the physiology of caffeine absorption and then turned to the sympathomimetic amines, drugs which mimic the actions of the body's sympathetic nervous system in stimulating the body to prepare for strenuous activity. He studied such compounds as amphetamine, mescaline, and ephedrine and discovered a new group of **enzymes**, which allowed these drugs to metabolize in the body. By the mid 1950s, Axelrod decided that he needed a doctorate to advance in his career at the National Institutes of Health. He took a year off to prepare for comprehensive examinations at George Washington University in the District of Columbia, submitted research work he had already done to satisfy the thesis requirements, and received a Ph.D. in pharmacology in 1955, at the age of forty-three. He was then offered the opportunity to create a section in pharmacology within the Laboratory of Clinical Sciences at the National Institute of Mental Health, another branch of the National Institutes of Health. He became chief of the section in pharmacology and held that position until his retirement in 1984.

In 1957, Axelrod began the research, which eventually led to the Nobel Prize. Axelrod, his colleagues, and his students studied the manner in which neurotransmitters, the chemicals that transmit signals from one nerve ending to another across the very small spaces between them, operate in the human body. In the 1940s, the Swedish scientist **Ulf von Euler** had discovered that noradrenaline, or norepinephrine, was the neurotransmitter of the sympathetic nervous system. Axelrod was concerned with the way in which noradrenaline was rapidly deactivated in order to make way for the transmission of later nerve signals. He discovered that this was accomplished in two basic ways. First, he found a new enzyme, which he named catechol-O-methyltransferase (COMT), which was essential to the **metabolism**, and hence the deactivation, of noradrenaline. Second, through a series of experiments on cats, he determined that noradrenaline was reabsorbed by the nerves and stored to be reused later. These seemingly esoteric discoveries in fact had enormous implica-

tions for medical science. Axelrod demonstrated that psychoactive drugs such as antidepressants, amphetamines, and cocaine achieved their effects by inhibiting the normal deactivation or reabsorption of noradrenaline and other neurotransmitters, thus prolonging their impact upon the nervous system or the **brain**. His experiments also pointed the way to many new discoveries in the rapidly growing field of neurobiological research and the chemical treatment of mental and neurological diseases. The 1978, Nobel Prize in physiology or medicine, shared with Ulf von Euler and Bernard Katz, crowned his achievements in this area.

In his later years, Axelrod has worked in many areas of biochemical and pharmacological research, notably in the study of **hormones**. Especially important to the advancement of medical science was his development of many new experimental techniques that could be widely applied in the work of other researchers. He also had a great impact through his training of and assistance to a long line of visiting researchers and postdoctoral students at the National Institutes of Health. He continued his own research at the National Institute of Mental Health following his formal retirement in 1984. Early in 1993, Axelrod had the unusual experience of having his own life saved through a scientific discovery he had made many years before. At the age of eighty, he suffered a massive heart attack. The cardiologists at Georgetown University Medical Center soon determined that several of his coronary **arteries** were almost completely blocked by **blood** clots and that he must have immediate triple coronary artery bypass surgery. The complication was that his blood pressure had fallen so dangerously low that he might not survive the operation. The solution to this crisis was to inject a synthetic form of noradrenaline to stimulate the contractions of his heart and thus raise his blood pressure to a more acceptable level. Axelrod survived the operation and within two months was back at work and attending conferences in foreign countries.

See also Nerve impulses and conduction of impulses; Nervous system overview; Nervous system: embryological development; Neural damage and repair; Neural plexuses; Neurology; Neurons

AXON • *see* NEURONS

B

B LYMPHOCYTES

B lymphocytes, also known as B cells, are one of the five types of white **blood** cells, or **leukocytes**, that circulate throughout the blood. They and T lymphocytes are the most abundant types of white blood cells. B lymphocytes are a vital part of the body's **immune system**. They function to specifically recognize a foreign protein, designated as an antigen, and to aid in destroying the invader.

B lymphocytes are produced and mature in the bone marrow. The mature form of the cell is extremely diverse, with a particular B cell being tailored to recognize just a single antigen. This recognition is via a molecule on the surface of the B cell, called a B cell receptor. There are thousands of copies of the identical receptor scattered over the entire surface of a B cell. Moreover, there are many thousands of B cells, each differing in the structure of this receptor. This diversity is possible because of rearrangement of genetic material to generate genes that encode the receptors. The myriad of receptors are generated even before the body has been exposed to the protein antigen that an individual receptor will recognize. B cells thus are one means by which our immune system has 'primed' itself for a rapid response to invasion.

The surface receptor is the first step in a series of reactions in the body's response to a foreign antigen. The receptor provides a 'lock and key' fit for the target antigen. The antigen is soluble; that is, floating free in the fluid around the B cell. A toxin that has been released from a bacterium is an example of a soluble antigen. The binding of the antigen to the B cell receptor triggers the intake of the bound antigen into the B cell, a process called receptor-mediated endocytosis. Inside the cell the antigen is broken up and the fragments are displayed one the surface of the B cell. These are in turn recognized by a receptor on the surface of a T lymphocyte, which binds to the particular antigen fragment. There follows a series of reactions that causes the B cell to differentiate into a **plasma** cell, which produces and secretes large amounts of an antibody to the original protein antigen.

Plasma cells live in the bone marrow. They have a limited lifetime of 2-12 weeks. Thus, they are the immune system's way of directly addressing an antigen threat. When the threat is gone, the need for plasma cells is also gone. But, B lymphocytes remain, ready to differentiate into the antibody-producing plasma cells when required.

Within the past several years, research has indicated that the deliberate depletion of B cells might aid in thwarting the progression of autoimmune disease—where the body's immune system reacts against the body's own components—and so bring relief from, for example, rheumatoid arthritis. However, as yet the data is inconclusive, and so this promising therapy remains to be proven.

See also Antigens and antibodies; Lymphatic system

BACTERIA AND BACTERIAL INFECTION

Infectious diseases depend on the interplay between the ability of pathogens to invade and/or proliferate in the body and the degree to which the body is able to resist. If the ability of a microorganism to invade, proliferate and cause damage in the body exceeds the body's protective capacities, a disease state occurs. **Infection** refers to the growth of microorganisms in the body of a host. Infection is not synonymous with disease because infection does not always lead to injury, even if the pathogen is potentially virulent (able to cause disease). In a disease state, the host is harmed in some way, whereas infection refers to any situation in which a microorganism is established and growing in a host, whether or not the host is harmed.

The steps of pathogenesis, the progression of a disease state, include entry, colonization, and growth. Pathogens like bacteria use several strategies to establish virulence. The bacteria must usually gain access to host tissues and multiply before damage can be done. In most cases this requires the

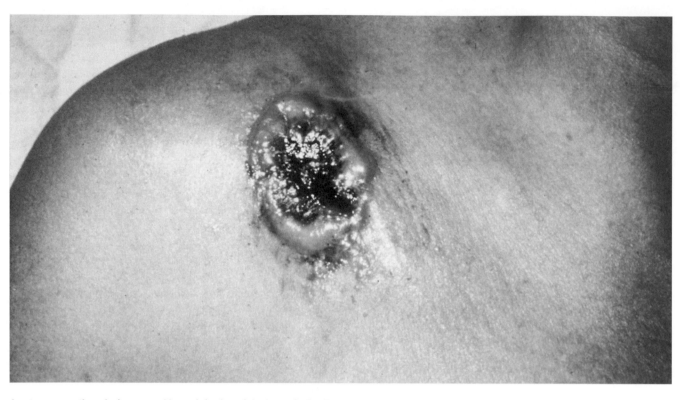

A cutaneous anthrax lesion caused by an infection of the bacteria *Bacillus anthracis*, normally infecting livestock, has been used as an agent of biological terrorism. *NMSB/Custom Medical Stock Photo. Reproduced by permission.*

penetration of the skin, mucous membranes, or intestinal epithelium, surfaces that normally act as microbial barriers. Passage through the skin into subcutaneous layers almost always occurs through wounds and in rare instances pathogens penetrate through unbroken skin.

Most infections begin with the adherence of bacteria to specific cells on the mucous membranes of the respiratory, alimentary or genitourinary tract. Many bacteria possess surface macromolecules that bind to complementary acceptor molecules on the surfaces of certain animal cells, thus promoting specific and firm adherence. Certain of these macromolecules are polysaccharides and form a meshwork of fibres called the glycocalyx. This can be important for fixing bacteria to host cells. Other proteins are specific, e.g. M-proteins on the surface of *Streptococcus pyogenes* which facilitate binding to the respiratory mucosal receptor. Also structures known as fimbrae may be important in the attachment process. For example, the fimbrae of *Neiseria gonorrhoeae* play a key role in the attachment of this organism to the urogenital epithelium where it causes a sexually transmitted disease. Also, it has been shown that fimbriated strains of *Escherichia coli* are much more frequent causes of urinary tract infections than strains lacking fimbrae, showing that these structures can indeed promote the capacity of bacteria to cause infection.

The next stage of infection is invasion, that is the penetration of the epithelium to generate pathogenicity. At the point of entry, usually at small breaks or lesions in the skin or mucosal surfaces, growth is often established in the submu-

cosa. Growth can also be established on intact mucosal surfaces, especially if the normal flora is altered or eliminated. Pathogen growth may also be established at sites distant from the original point of entry. Access to distant, usually interior, sites occurs through the **blood** or lymphatic system.

If a pathogen gains access to tissues by adhesion and invasion, it must then multiply, a process called colonization. Colonization requires that the pathogen bind to specific **tissue** surface receptors and overcome any non-specific or immune host defences. The initial inoculum is rarely sufficient to cause damage. A pathogen must grow within host tissues in order to produce disease. If a pathogen is to grow, it must find appropriate nutrients and environmental conditions in the host. Temperature, pH, and reduction potential are environmental factors that affect pathogen growth, but the availability of microbial nutrients in host tissues is most important. Not all nutrients may be plentiful in different regions. Soluble nutrients such as sugars, **amino acids** and organic acids may often be in short supply and organisms able to utilize complex nutrient sources such as **glycogen** may be favored. Trace elements may also be in short supply and can influence the establishment of a pathogen. For example, iron is thought to have a strong influence on microbial growth. Specific iron binding proteins called transferrin and lactoferrin exist in human cells and transfer iron through the body. Such is the affinity of these proteins for iron, that microbial iron deficiency may be common and administration of a soluble iron salt may greatly increase the virulence of some pathogens. Many bacteria produce iron-chelating compounds known as siderophores, which help them to obtain iron from the environment. Some iron chelators isolated from pathogenic bacteria are so efficient that they can actually remove iron from host iron binding proteins. For example, a siderophore called aerobactin, produced by certain strains of *E. coli* and encoded by the Col V plasmid, readily removes iron bound to transferring.

After initial entry, the organism often remains localized and multiplies, producing a small focus of infection such as a boil, carbuncle or pimple. For example, these commonly arises from *Staphylococcus* infections of the skin. Alternatively, the organism may pass through the lymphatic vessels and be deposited in lymph nodes. If an organism reaches the blood, it will be distributed to distal parts of the body, usually concentrating in the **liver** or spleen. Spread of the pathogen through the blood and lymph systems can result in generalized (systemic) infection of the body, with the organism growing in a variety of tissues. If extensive bacterial growth in tissues occurs, some of the organisms may be shed into the bloodstream, a condition known as bacteremia.

A number of bacteria produce extracellular proteins, which break down host tissues, encourage the spread of the organism and aid the establishment and maintenance of disease. These proteins, which are mostly **enzymes**, are called virulence factors. For example, streptococci, staphylococci and pneumococci produce hyaluronidase, an enzyme which breaks down hyaluronic acid, a host tissue cement. They also produce proteases, nucleases and lipases that depolymerize host proteins, nucleic acids and fats. Clostridia that cause gas gangrene produce collagenase, and κ-toxin, which breaks down the collagen network supporting the tissues.

The ways in which pathogens bring about damage to the host are diverse. Only rarely are symptoms of a disease due simply to the presence of a large number of microorganisms. Although a large mass of bacterial cells can block vessels or **heart** valves or clog the air passages of the **lungs**. In many cases pathogenic bacteria produce toxins that are responsible for host damage. Toxins released extracellularly are called exotoxins, and these may travel from the focus of infection to distant parts of the body and cause damage in regions far removed from the site of microbial growth. The first example of an exotoxin to be discovered was the diptheria toxin produced by *Corynebacterium diptheriae*. Some Gram negative bacteria produce lipopolysaccharides as part of their cell walls, which under some conditions can be toxic. These are called endotoxins and have been studied primarily in the genera *Escherichia, Shigella,* and *Salmonella.*

See also Immune system; Immunology; Infection and resistance

BAER, KARL ERNST VON (1792-1876)
Estonian biologist

Karl Ernst von Baer's most famous discoveries in **anatomy** grew out of his work at the University of Königsberg. For more than a century, scientists had attempted to determine the exact nature and location of the mammalian egg. In 1673, **Regnier de Graaf** had discovered follicles in the ovaries that he thought might be eggs. However, he later found structures even smaller than follicles in the uterus, raising doubts about the role of the follicles themselves. During his research at Königsberg, Baer discovered the mammalian egg by identifying a yellowish spot within the follicle visible only with a microscope. He developed this idea in his 1827 treatise, *De ovi mammalium et hominis genesi (On the Origin of the Mammalian and Human Ovum).*

Another of Baer's accomplishments was his explanation of early embryonic development, a theory that he summarized in his two-volumetextbook *Über die Entwicklunggeschichte der Thiere (On the Development of Animals),*1828-1837. Here, Baer set forth the theory that embryos of all animals begin as similar structures that are both simple and homogeneous, and develop into complex heterogeneous forms. In fact, Baer claimed it is impossible to distinguish among the early embryos of birds, reptiles, and mammals until their later stages of development. Baer set forth the germ-layer theory of development which stated that during the early stages of development, embryos ultimately form four (later shown to be three) distinct layers that eventually differentiate into specific **organs** or body structures. These are the **ectoderm, mesoderm** and **endoderm**. While observing embryos, Baer discovered the extraembryonic membranes, the chorion, amnion, and allantois and determined their functions. He also discovered the presence of a **notochord** in early vertebrate embryos. Although the notochord quickly changes into a spinal column, its early presence indicates the evolutionary connection between vertebrates and other organisms now classified together in the phylum *Chordata*. For his work, Baer was awarded the Copley Medal of the Royal Society in 1876.

One of ten children, Karl Ernst von Baer was born in Piep, Estonia, to parents descended from Prussian nobility. Due to the large size of his family, Baer was sent to live with his childless uncle and aunt until the age of seven. Initially tutored at home, he later spent three years at a school for the children of nobility. Although his father and uncle encouraged him to pursue a military career, from 1810-1814, Baer attended the University of Dorpat in Vienna, Austria and obtained an M.D. degree. From 1814-1817, he studied **comparative anatomy** at Würzburg. In 1817, he was appointed prosector at the University of Königsberg, where in 1819, he accepted an appointment to teach zoology and anatomy, and serve as chief at the new zoological museum that he organized. In 1828, Baer became a member of the St. Petersburg Academy of Sciences, where he taught zoology from 1829-30. Baer then returned to Königsberg until 1834. At that time, he became librarian at St. Petersburg Academy of Sciences, and conducted research in anatomy and zoology. In 1846, Baer was also appointed to the position of Professor of Comparative Anatomy and Physiology at the Medico-Chirugical Academy in St. Petersburg.

Baer also took part in other types of scientific projects. In 1837, he led a scientific expedition to Novaya Zemlya in Arctic Russia, and from 1851-6, studied the fisheries of lake

Peipus and the Baltic and Caspian Seas. He served as inspector of fisheries for the empire from 1851-1852. Baer founded the St. Petersburg Society for Geography and Ethnography and the German Anthropological Society. Baer died in Dorpat, Estonia.

See also Embryology; Embryonic development: Early development, formation, and differentiation

BANTING, FREDERICK G. (1891-1941)
Canadian physiologist

Frederick Grant Banting was a Canadian physician who discovered insulin with the collaboration of Charles H. Best, John James Rickard Macleod, and James Bertram Collip. Insulin is a hormone that is secreted by the **pancreas**, and that regulates the level of glucose (sugar) in the **blood**. The discovery of insulin led to a treatment for diabetes, a disease that at the time of Banting's youth, usually meant **death** within a few years for those who developed it. For his contribution to this work, Banting shared the 1923 Nobel Prize in physiology or medicine with Macleod, the physiologist in whose laboratory the insulin research was carried out. It was the first Nobel Prize to be awarded to a Canadian.

Banting was born near Alliston, Ontario, about 40 mi. (64 km) north of Toronto. He was the youngest of five children of William Thompson Banting, a farmer whose parents had emigrated to Canada from Ireland, and Margaret Grant Banting, the daughter of a miller and the first Canadian of European descent to be born in Alliston. Banting attended local schools, where he was a serious, but average, student, excelling in sports. Following his father's wishes, Banting entered Victoria College of the University of Toronto to prepare for the Methodist ministry. After a year, however, he switched to medicine, and he began medical courses in the fall of 1912. The university accelerated his course of study with the outbreak of World War I. Banting completed his degree in December 1916, and was dispatched to England shortly thereafter, as a lieutenant in the Canadian Army Medical Corps. He was assigned to an orthopedic hospital at Ramsgate. There he gained experience in surgery under the direction of Clarence L. Starr. Later he was sent to France, and in September 1918, Banting's right forearm was severely wounded by shrapnel near Haynscourt, in the Cambrai sector. In 1919, the British Government awarded him the Military Cross for valorous conduct in action.

Banting returned to Toronto after his injury healed and worked as an intern for a year at the Hospital for Sick Children under his wartime mentor Clarence Starr, who was surgeon-in-chief. In 1920 Banting set up his own surgical practice in London, Ontario. He chose London because his fiancée, Edith Roach, was working there as a teacher. The couple had become engaged before Banting went off to war, but they never married. Banting's practice got off to a slow start. In the first month after he posted office hours only one patient visited. To make ends meet, Banting sought an appointment as a

demonstrator in surgery and **anatomy** at the University of Western Ontario, a job that took up only a few hours a week of his time. He also got his first taste of medical research there, working in the laboratory of neurophysiologist Frederick R. Miller.

To prepare for a lecture to his students on the pancreas in October 1920, Banting was reading up on how the body metabolizes **carbohydrates**, including sugar. The pancreas was known to secrete digestive **enzymes** and thus, to play a role in the **metabolism** of sugar. Thirty years earlier the researchers Oscar von Mering and Joseph Minkowski had found that when they surgically removed the pancreas of a dog, the animal developed diabetes. Among the first symptoms were frequent urination and high levels of sugar in the urine, the same symptoms as in human diabetes. But in the intervening years, no one had clarified just how the pancreas prevented diabetes. On October 30, 1920, reading himself to sleep with the November issue of the journal *Surgery, Gynecology, and Obstetrics,* Banting noted the lead article, titled "The Relation of the Islets of Langerhans to Diabetes, With Special Reference to Cases of Pancreatic Lithiasis." This article gave Banting an idea for an experiment that would reveal the role of the pancreas in diabetes.

The article described an autopsy where the main duct of the pancreas was found to be blocked by a stone. The pancreas itself had wasted away, except for certain groups of cells known as the islets of Langerhans. Banting suspected that these islet cells held the key to understanding diabetes. As he later recalled in his Cameron Lecture in Edinburgh in 1928, he was unable to sleep. So at 2:00 a.m., he jotted down notes for an experiment: "Tie off pancreas ducts of dogs. Wait six or eight weeks. Remove and extract."

The next day Banting mentioned his idea to Miller. Miller suggested he talk to J. J. R. Macleod at the University of Toronto, a world-renowned expert on carbohydrate metabolism. Early in November 1920, Banting met with Macleod to propose his idea and to ask Macleod to be his sponsor by providing his laboratory. The elder scientist was not impressed. In fact, Macleod told Banting, other researchers had already spent years in unsuccessful attempts to find the role of pancreatic extracts in diabetes. Although scientists had previously tied off the pancreatic ducts of dogs and observed the organ atrophy, it seemed that no one had tried to prepare an extract. Banting wanted to do this and administer the extract to dogs that had been made diabetic by the removal of the pancreas. This idea intrigued Macleod. After six months of persuasion, Banting finally convinced Macleod to help him. In mid-May of 1921, Banting shut down his practice, which he considered an utter failure, resigned his university position, and moved to Toronto to begin his experiments. Macleod had agreed to provide him for eight weeks with a laboratory, ten dogs, and an assistant who knew how to perform chemical analysis of blood and urine. That assistant was 22-year-old Charles H. Best, who had just finished his undergraduate degree in physiology and **biochemistry**.

Banting and Best set about performing surgery on the dogs, tying off the pancreatic ducts in some and removing the pancreas from others, to render them diabetic. Macleod supervised the research for a month and then left the pair to themselves while he went to his home in Scotland for vacation. Banting and Best spent several weeks working out technical surgical problems. In their first attempts, they tied off the pancreatic ducts of the dogs with catgut, which disintegrated, leaving the ducts open and the pancreas healthy. They solved this problem by switching to silk. Banting also developed a method for removing the pancreas in one procedure, whereas previously it had been done in two operations. Finally on July 30, both a diabetic dog and a dog whose pancreatic ducts had been tied were ready.

Banting removed the shriveled pancreas from the dog whose ducts had been tied. He and Best prepared an extract from it by chopping the pancreas into small pieces, grinding it in a chilled mortar with salt water, and filtering the mixture through cheesecloth. A blood sample from the diabetic dog showed its blood sugar level to be 0.2. Banting and Best injected some of their extract into the dog. An hour later its blood sugar level had dropped to 0.12. After another injection it registered 0.11. This dog died the next day, presumably from an **infection**. But Banting and Best were encouraged by the result and tested their extract on more diabetic dogs. They called the extract "isletin."

During the following months Banting and Best performed additional experiments to confirm and explain their results. They also looked for a more practical way of obtaining their extract than by performing risky duct-tying surgery on dogs and then waiting seven weeks for the pancreas to atrophy. They also wanted a procedure that would provide larger quantities of the substance. First they tried a technique of forcing the pancreas to secrete all of its digestive enzymes except "isletin." This, however, still required surgery. Then Banting recalled that the pancreases of fetal animals contain more islet cells than those of adults. He prepared an extract using pancreases from fetal calves obtained from a local slaughterhouse. It worked, and this new method provided the researchers with a steady supply of the hormone.

In the meantime, Macleod returned from his holiday. He was pleased with Banting's progress, but suggested that still more experiments be done. Banting, who had been working without a salary, threatened to quit unless Macleod gave him a job and more help in the lab. Macleod eventually granted this, enlisting J. B. Collip, a Ph.D. biochemist and an expert on blood chemistry, to obtain a purified version of the pancreatic extract. But combined with Macleod's initial doubts about Banting, the confrontation had driven a rift in their relationship that never healed.

By January 1922, the researchers felt ready to test insulin, a name suggested by Macleod, on a human diabetic. Banting and Best initially injected each other with the extract to be sure it would not cause harm. Their first patient was 14-year-old Leonard Thompson. He had suffered from diabetes for two years, was lethargic and debilitated, and without treatment, it was assumed that he would soon slip into a **coma** and die. The boy's blood sugar level did fall after insulin injections, but the researchers stopped the trial because impurities in the extract seemed to cause complications. Twelve days later, however, they started treatment again, with a more pure

and potent extract. The boy's blood sugar dropped, and so did the level of sugar in his urine. The researchers reported in the *Journal of the Canadian Medical Association* in 1922 that "the boy became brighter, more active, looked better and said he felt stronger." The first human test of insulin was a success: Leonard Thompson lived another 13 years, dying of pneumonia following a motorcycle accident. Combined with a regulated diet, insulin soon became the standard treatment for diabetes. The University of Toronto contracted with Eli Lilly and Company to scale up production of the drug.

In 1923, the Nobel Prize in physiology or medicine was awarded to Banting and Macleod. Banting split his half of the money with Best and was furious that Macleod could claim credit for the discovery. In Banting's view, the idea that started the experiments had been his alone, and the crucial scientific work had been done while Macleod was on vacation. The Nobel Committee, however, wished to acknowledge Macleod's role in directing the research, which also required the collaboration of Best and Collip to carry through. Macleod, for his part, split his share of the award with Collip. The Nobel Prize was followed by numerous additional accolades for Banting. The Canadian parliament awarded him an annuity in 1923. The same year the University of Toronto established a Banting and Best Chair of Medical Research, with Banting its first recipient. This grew into the Banting and Best Department of Medical Research, which Banting headed for the rest of his life. In 1930, the University of Toronto honored him by naming a new medical building the Banting Institute. In 1934, King George V, who was a diabetic, made Banting a knight commander of the British Empire. Banting was elected a fellow of the Royal Society of London in 1935.

Banting achieved fame for the discovery of insulin. He did not, however, contribute much to its refinement and use, perhaps because of his feud with Macleod. Instead, he turned his attention to a range of medical problems far removed from diabetes. He contributed to **cancer** research, studying a type of tumor caused by a virus, known as Rous sarcoma. He studied the physiological problems associated with insufficient secretion of the suprarenal gland, for which he attempted to develop an extract, and he did research on silicosis. He also became involved in the Canadian government's support of medical research as chairman of the Medical Research Committee of the National Research Council, and chairman of the Associate Committee on Aviation Medical Research. In the latter position, he did research on the physiological effects of flying at high altitudes.

Banting married Marion Robertson in 1924. They had one son. The marriage was unhappy almost from the start, however, and the couple divorced in 1932 after a long separation. During the 1920s Banting began to nurture a long-standing interest in art. He collected paintings by Canadian artists and joined the Arts and Letters Club in Toronto. Eventually he began painting himself, under the tutelage of his friend A. Y. Jackson, a well-known landscape artist. Beginning in 1927 Banting and Jackson made trips to northeastern Quebec, Ellesmere Island, and other northern reaches of Canada to sketch and paint.

In 1939 Banting married Henrietta Ball, a technician in his department at the University of Toronto. That same year he volunteered for military service when Canada entered World War II. He was soon promoted to the rank of major, and spent much of the next two years shuttling between Canada and England as a liaison between authorities on wartime medicine in the two countries. He had embarked on such a mission on February 21, 1941, when the airplane carrying him crashed in Newfoundland and he was killed.

See also Carbohydrates; Glucose utilization, transport and insulin; Hormones and hormone action

Bárány, Robert (1876-1936)
Swedish physician

Robert Bárány made significant contributions to the understanding of the vestibular apparatus, part of the inner ear that plays an important role in maintaining balance. He devised ingenious tests to diagnose inner-ear disease, and he investigated the relationship between the vestibular and nervous systems. Because of his ground-breaking research in this area, he is credited with creating a new field of study, otoneurology. Bárány's achievements were recognized in 1914 with the Nobel Prize in physiology or medicine.

Bárány was born in Rohonc (near Vienna), Austria-Hungary (now Austria), the eldest of six children. His father, Ignaz Bárány, was a bank official. His mother, Marie Hock Bárány, was the daughter of a well-known Prague scientist, and it was her intellectual influence that predominated in the family. When Bárány was young, he contracted a form of tuberculosis that left him with a permanent stiffness in his knee, but also first awakened his interest in medicine.

Always a top student, Bárány began attending medical school at the University of Vienna in 1894. In 1900, he received a doctor of medicine degree. He then spent two years studying internal medicine, **neurology**, and psychiatry at clinics in Frankfurt, Heidelberg, and Freiburg. Next, he returned to Vienna, where he received hospital surgical training. Finally, in 1903, he accepted a post at the Ear Clinic, also in Vienna, which was then directed by Adam Politzer. Bárány's association with Politzer, a leading figure in the history of otology (the study of the ear and its diseases), proved to be highly fruitful.

It was the chance observation that clinic patients often became dizzy after having their ears irrigated that led Bárány to develop one test that still bears his name. The Bárány caloric test involves stimulating each of a patient's inner ears separately by syringing one with hot liquid and the other with cold. Normally, this results in rapid, involuntary movements of the eyeballs, termed nystagmus. Bárány demonstrated that the direction of the nystagmus is determined by the temperature of the water and the position of the head. He also showed that the absence or delay of nystagmus indicates a problem with the balance structures of the ear. The test was an emi-

Robert Bárány. *Corbis-Bettmann. Reproduced by permission.*

nently practical technique for diagnosis, since it could easily be performed at a patient's bedside.

Another diagnostic procedure introduced by Bárány was the chair test. The patient is turned in a rotating chair with a specially designed headrest that inclines the head slightly forward. Once again, any deviation from the normal pattern of nystagmus afterward indicates a problem. Yet another of Bárány's inventions during this period was the noise box, a much-used device that effectively isolates the hearing performance of one ear by creating a masking noise in the other.

Bárány's phase of great productivity was to be interrupted by the start of World War I. Bárány was dispatched by the army to the fortress of Przemysl on the border between Poland and Russia, where he served as a medical officer. While there, Bárány continued to study the connection between the vestibular apparatus and the nervous system. He also developed an improved surgical technique for dealing with fresh bullet wounds to the **brain**.

However, in April 1915, the Russians occupied Przemysl, and Bárány was transported along with other prisoners by cattle car to Merv in central Asia. Conditions there were unsanitary and difficult, and Bárány came down with **malaria**. Still, he was relatively fortunate: the medical commander in Merv knew him by reputation and placed him in charge of otolaryngology (the medical specialty concerned with the ear, **nose** and throat) for both Russian natives and Austrian prisoners. The Russians were grateful patients. After Bárány had successfully treated the local mayor and his family, he became a daily dinner guest in their home.

It was while he was still a prisoner of war that Bárány received the news that he had won the Nobel Prize. Thanks to the personal intervention of Prince Carl of Sweden, Bárány was released in 1916. He returned to Vienna that same year, but was bitterly disappointed by the reception he received from colleagues there. They claimed he had inadequately cited their own contributions to his work. These accusations were investigated by the Nobel Prize Committee, which found them groundless. Nevertheless, the attacks prompted Bárány to accept a post as professor at the University of Uppsala in Sweden in 1917, where he remained for the rest of his life. Eventually, he rose to the position of chairman of the department of ear, nose, and throat medicine there.

While at Uppsala, Bárány studied the role of the part of the brain called the **cerebellum** in controlling body movement. He had previously devised another test for disturbances in cerebellar function, known as the pointing test, in which the patient points at a fixed object with the eyes alternately open and closed. Consistent errors while the eyes are closed indicates a possible brain lesion.

Bárány also developed a surgical technique for treating chronic sinusitis. For this, he was awarded the Jubilee Medal of the Swedish Society of Medicine in 1925. Among his numerous other awards were the Belgian Academy of Sciences Prize, the ERB Medal from the German Neurological Society, and the Guyot Prize from the University of Groningen in the Netherlands. He also received honorary degrees from several universities, including the University of Stockholm. Austria issued a stamp in his honor in 1976, to commemorate the 100th anniversary of his birth.

Bárány was described as a quiet and solitary man, fanatically devoted to his work. Yet at home, he also enjoyed music and played the piano well; he particularly liked the music of composer Robert Schumann. And despite having had a stiff knee since childhood, he was an avid mountain hiker and tennis player.

Bárány married 1909. The couple had three children, all of whom went on to become physicians or medical scientists. Bárány's last years were marred by a series of strokes, which resulted in partial paralysis. He was aware of an international meeting that was organized to celebrate his sixtieth birthday, but, sadly, he died in Uppsala only a few days before the occasion on April 8, 1936. Yet his memory has been kept alive with the Bárány medal, first awarded by the University of Uppsala in 1948, honoring deserving scientists for investigations of the vestibular system. In addition, the Bárány Society was established in 1960 to conduct international symposia on vestibular research.

See also Ear (external, middle and internal ear); Ear: Otic embryological development; Equilibrium

BASAL GANGLIA

The basal ganglia has been recognized as an important area of the **brain** since the seventeenth century. Then, the English anatomist **Thomas Willis** published a description of the basal ganglia, although at that time it was referred to as the corpus striatum. This name reflected its striped appearance, due to fibers. It was a structure thought to receive all sensory input and which initiated all body movements.

The term basal ganglia means "basement structures." This reflects the location in the brain, deep beneath the **cerebral cortex**. Basal ganglia are comprised of so-called subcortical nuclei: the caudate, putamen, globus pallidus, nucleus accumbens, substantia nigra, and the subthalamic nucleus. Bearing out the earlier speculations, basal ganglia have an important role in regulating and coordinating movement.

The basal ganglia, along with the **cerebellum**, function to modify movement on an almost continuous basis. Outputs from the basal ganglia are inhibitory, while outputs from the cerebellum are excitatory. It is the balance between these two systems that produces the smooth and coordinated movement of an organism.

The caudate and putamen receive input from another part of the brain called the cerebral cortex. Some information is routed onward to the substantia nigra, which in turn communicates back to the caudate and putamen, and sends information to control head and **eye** motion. As a contributor to these functions, the substantia nigra produces dopamine, a chemical that is critical for normal body movement. Breakdown in dopamine production produces erratic and uncontrolled movement, as is seen in Parkinson's disease.

Information from the caudate and putamen is also routed to the globus pallidus, which in turn communicates in an inhibitory way with the **thalamus**. This inhibition of thalamus function is thought to be necessary for takes such as proper **posture**, where the activity of **reflexes** that would detract from upright control must be reduced.

See also Brain stem function and reflexes; Gray matter and white matter

BASAL METABOLIC RATE • *see* METABOLISM

BEAUMONT, WILLIAM (1785-1853)
American surgeon

American Army surgeon William Beaumont made some of the first *in vivo* (in the living body) observations of the physiology of the human digestive system.

Beaumont, Connecticut-born and the son of a farmer, worked briefly as a school teacher, then studied medicine at St. Albans, Vermont. After an apprenticeship, Beaumont received a license to practice medicine in time to serve as an assistant army surgeon during the War of 1812. Although he left the army in 1815 to start a medical practice in Plattsburgh, New York, he returned in 1820 and remained an Army surgeon, serving at various posts, until 1839.

It was at one of those army posts, Fort Mackinac in northern Michigan, that Beaumont met the patient that was to make both of them famous. The patient was a 19-year-old

William Beaumont. *The Bettmann Archive. Reproduced by permission.*

French Canadian trapper, Alexis St. Martin who, on June 6, 1822, was accidentally shot while visiting the Mackinac branch of the American Fur Company. The bullet wound tore a deep chunk out of the left side of St. Martin's lower chest and, although Beaumont was sent for immediately, everyone assumed that the young man would never survive. Miraculously, he did—although his wound needed to be rebandaged daily for a year—and in time St. Martin recovered virtually all his strength. (He lived to be 82, in fact.) However, St. Martin's bullet hole never fully closed. An inch-wide opening (called a fistula) remained through which Beaumont could insert his finger all the way into the stomach.

About a year later, St. Martin needed a cathartic of rhubarb and sulphur and Beaumont decided to try administering the medicine through the hole in his patient's stomach. To the surgeon's surprise, the cathartic seemed to work exactly as it would have if it had been administered orally, and Beaumont promptly began planning other experiments as well.

Beaumont started by taking small chunks of food, tying them to a string, and inserting them directly into St. Martin's stomach. At varying intervals, he then pulled the food out, and was therefore able to observe, first hand, the results of **digestion**, hour by hour. Later, by using a hand lens, Beaumont began peering into his patient's stomach, and could actually see how the human stomach behaved at various stages of

digestion and under varying circumstances. He was also able to extract and analyze samples of digestive fluids.

Over the next few years, Beaumont conducted well over two hundred carefully detailed experiments and, in 1833, published his findings as *Experiments and Observations on the Gastric Juice and the Physiology of Digestion.* The book provided invaluable information on the digestive process and also suggested to other scientists (including Claude Bernard) that artificial fistulas might be a practical way to learn more about the body. A year after Beaumont's work was published, St. Martin, perhaps tiring of the scrutiny Beaumont had subjected him to, refused to cooperate with further studies and returned to Canada.

Although Beaumont resigned from the army in 1840, went into private practice in St. Louis, Missouri, and stayed out of the laboratory, his one classic work earned him lasting fame as one of America's remarkable pioneer researchers.

See also Digestion; Gastro-intestinal tract; Intestinal motility; Stomach: Histophysiology

BEHRING, EMIL VON (1854-1917)
German bacteriologist

Emil von Behring was one of the founders of the science of **immunology**. His discovery of the diphtheria and tetanus antitoxins paved the way for the prevention of these diseases through the use of immunization. It also opened the door for the specific treatment of such diseases with the injection of immune serum. Behring's stature as a seminal figure in modern medicine was recognized in 1901, when he received the first Nobel Prize in physiology or medicine.

Emil Adolf von Behring was born in Hansdorf, West Prussia (now Germany). He was the eldest son of August Georg Behring, a schoolmaster with thirteen children, and his second wife, Augustine Zech Behring. Although his father planned for him to become a minister, young Behring had an inclination toward medicine. The family's meager circumstances seemed to put this goal out of his reach, however. Then one of Behring's teachers, recognizing great promise, arranged for his admission to the Army Medical College in Berlin, where he was able to obtain a free medical education in exchange for future military service. He received his doctor of medicine degree in 1878, and two years later he passed the state examination that allowed him to practice.

The army promptly sent Behring to Posen (now Poznan, Poland), then to Bonn in 1887, and finally back to Berlin in 1888. His first published papers, which date from this period, dealt with the use of iodoform as an antiseptic. After completing his military service in 1889, Behring became an assistant at the Institute of Hygiene in Berlin, joining a team of researchers headed by **Robert Koch**, a leading light in the new science of bacteriology.

It was while working in Koch's laboratory that Behring began his pioneering investigations of diphtheria and tetanus. Both of these diseases are caused by bacilli (**bacteria**) that do not spread widely through the body, but produce generalized symptoms by excreting toxins. Diphtheria, nicknamed the "strangling angel" because of the way it obstructs **breathing**, was a terrible killer of children in the late nineteenth century. Its toxin had first been detected by others in 1888. Tetanus, likewise, was fatal more often than not. In 1889, the tetanus bacillus was cultivated in its pure state for the first time by Shibasaburo Kitasato, another gifted member of Koch's team.

The next year, Behring and Kitasato jointly published their classic paper, "Ueber das Zustandekommen der Diphtherie-Immunität und der Tetanus-Immunität bei Thieren" ("The Mechanism of Immunity in Animals to Diphtheria and Tetanus"). One week later, Behring alone published another paper dealing with immunity against diphtheria and outlining five ways in which it could be achieved. These reports announced that injections of toxin from diphtheria or tetanus bacilli led animals to produce in their **blood** substances capable of neutralizing the disease **poison**.

Behring and Kitasato dubbed these substances antitoxins. Furthermore, injections of blood serum from an animal that had been given a chance to develop antitoxins to tetanus or diphtheria could confer immunity to the disease on other animals, and even cure animals that were already sick.

The news created a sensation. Several papers confirming and amplifying these results, including some by Behring himself, appeared in rapid succession. In 1893, Behring described a group of human diphtheria patients who were treated with antitoxin. That same year, he was given the title of professor. However, Behring's diphtheria antitoxin did not yield consistent results. It was the bacteriologist **Paul Ehrlich**, another of the talented associates in Koch's lab, who was chiefly responsible for standardizing the antitoxin, thus making it practical for widespread therapeutic use. Working together, Ehrlich and Behring also showed that high-quality antitoxin could be obtained from horses, as well as from the sheep used previously, opening the way for large-scale production of the antitoxin.

In 1894, Behring accepted a position as professor at the University of Halle. A year later, he was named a professor and director of the Institute of Hygiene at the University of Marburg. Thereafter he focused much of his attention on the problem of immunization against tuberculosis. His assumption, unfounded as it turned out, was that different forms of the disease in humans and in cattle were closely related. He tried immunizing calves with a weakened strain of the human tuberculosis bacillus, but the results were disappointing. Although his bovine vaccine was widely used for a time in Germany, Russia, Sweden, and the United States, it was found that the cattle excreted dangerous microorganisms afterward. Nevertheless, Behring's basic idea of using a bacillus from one species to benefit another influenced the development of later vaccines.

Behring did not entirely abandon his work on diphtheria during this period. In 1913, he announced the development of a toxin-antitoxin mixture that resulted in longer-lasting immunity than did antitoxin serum alone. This approach was a forerunner of modern methods of preventing, rather than just

treating, the disease. Today children are routinely and effectively vaccinated against diphtheria and tetanus.

However, the first great drop in diphtheria mortality was due to the antitoxin therapy introduced earlier by Behring, and it is for this contribution that he is primarily remembered. The fall in the diphtheria **death** rate around the turn of the century was one of the sharpest ever recorded for any treatment. In Germany alone, an estimated 45,000 lives per year were saved. It is no wonder, then, that Behring received the 1901 Nobel Prize "for his work on serum therapy, especially its application against diphtheria, by which he... opened a new road in the domain of medical science and thereby placed in the hands of the physician a victorious weapon against illness and deaths." Behring was also elevated to the status of nobility and shared a sizable cash prize from the Paris Academy of Medicine with Émile Roux, the French bacteriologist who was one of the men who had the diphtheria toxin in 1888. In addition, Behring was granted honorary memberships in societies in Italy, Turkey, France, Hungary, and Russia.

There were other, financial rewards as well. From 1901 onward, ill health prevented Behring from giving regular lectures, so he devoted himself to research. A commercial firm in which he had a financial interest built a well-equipped laboratory for his use in Marburg, Germany. Then, in 1914, Behring established his own company to manufacture serums and vaccines. The profits from this venture allowed him to keep a large estate at Marburg, on which he grazed cattle used in experiments. This house was a gathering place of society. Behring also owned a vacation home on the island of Capri in the Mediterranean.

In 1896 Behring married 18-year-old Else Spinola, daughter of the director of a Berlin hospital. They had seven children. Yet despite all outward appearances of personal and professional success, Behring was subject to frequent bouts of serious depression, some of which required sanatorium treatment. In addition, a fractured thigh led to a condition that increasingly impaired his mobility. He was already in a weakened state when he contracted pneumonia in 1917. His body was unable to withstand the added strain, and he died on March 31 in Marburg, Germany.

See also Antigens and antibodies; Immune system; Immunity

BÉKÉSY, GEORG VON (1899-1972)

Georg von Békésy was educated as a physicist, but is best known for his research on the physics of hearing. He was awarded the 1961 Nobel Prize in physiology or medicine for his research on the physical mechanism of hearing, particularly with regard to the changes that take place within the cochlea. His discoveries have been important in the development of surgical procedures and prosthetic devices for the treatment of hearing disorders.

Békésy was born in Budapest, Hungary. His father was Alexander von Békésy, a member of the Hungarian diplomatic service, and his mother was the former Paula Mazaly. As his father was assigned to a variety of diplomatic posts, Georg grew up in a number of cities, including Münich, Constantinople (now Istanbul), and Zürich, as well as his native Budapest. He entered the University of Bern in 1916, where he concentrated in chemistry. Békésy remained at Bern until 1920. After serving briefly in the military, he enrolled at the University of Budapest and was awarded a Ph.D. in physics for his thesis on fluid dynamics in 1923.

Intrigued as a boy by gypsy music, Békésy maintained a lifelong interest in the study of sound. His first job as a communications engineer with the Hungarian Post Office, beginning in 1923, allowed Békésy to begin a serious investigation into the physics of sound, as the postal service was also responsible for the nation's telephone system. As he began to deal with the practical problems of telephone systems, Békésy realized that it would be helpful to better understand the precise physical mechanism by which sound travels through the human ear. By the 1920s, anatomists had a reasonably thorough understanding of the physical structure of the human ear, but physiologists were still disputing the physical process by which sound stimulates the inner ear and causes the **brain** to detect differences in pitch.

Békésy conducted research on the physics of hearing throughout his tenure with the Hungarian Postal Service from 1923 to 1946. During that period, he also worked as a consultant in the central laboratories of Siemens and Halske in Berlin (1926–27) and was employed by the University of Budapest as lecturer (1932–34), special professor (1939–40), and full professor (1940–46). In 1946–47, he spent a year as research professor at the Karolinska Institute in Stockholm. Throughout this time, Békésy designed and carried out an extensive series of experiments intended to determine exactly what happens to sounds that enter the ear. In some of these experiments, he performed very delicate surgery on the inner ear itself. In order to do so, he had to design and build very small tools with which to operate on the less-than-fingernail-sized cochlea. He is reputed, for example, to have designed and built a pair of scissors with blades measuring less than a hundredth of an inch long. Using these tools and various combinations of mirrors, Békésy was able to observe changes that occurred within the cochlea as it received sounds.

In 1947, Békésy accepted an appointment as senior research fellow at Harvard University's Psycho-Acoustic Laboratory, where he had built and carried out experiments using a model cochlea. Békésy's model cochlea consisted of a water-filled plastic tube 11.8 in (30 cm) long. He stimulated one end of the tube with a sound and then placed his forearm along the length of the tube. He found that the sound wave was noticeable at only one particular point on his forearm, and that he could vary that point by changing pitch. From his many experiments, Békésy concluded that sound moves as a traveling wave across the cochlea by way of the basilar membrane. Based on its frequency, a wave produces a maximum vibration at one particular point along the membrane—waves of a higher pitch reach their peak nearer the base of the cochlea than those of a lower pitch. As the wave peaks along the membrane, adjacent cells along the organ of Corti are stimulated, and an auditory message is sent to the brain. It is by this mech-

anism, then, that the brain is able to determine the pitch of the sound that it is "hearing."

In 1961, Békésy became the first physicist to receive the Nobel Prize in physiology or medicine. He left Harvard in 1966 for the University of Hawaii, where he became professor of sensory sciences. As this title suggests, his interests broadened from the physics of sound to the relationship among all senses, especially sound, **taste**, and touch.

Among his many honors are the Denker Prize in Otology (1931), the Leibnitz Medal of the German Academy of Sciences (1937), the Howard Crosby Warren Medal of the Society of Experimental Psychologists (1955), and the Gold Medals of the American Otological Society (1957) and the Acoustical Society of America (1961). Békésy died in Honolulu, Hawaii, in 1972.

See also Ear (external, middle and internal); Hearing (physiology of sound transmission)

BELL, CHARLES (1774-1842)

Scottish surgeon

Charles Bell was a Scottish surgeon and anatomist who pioneered neurophysiological research. Bell's experimental work served as a catalyst to other researchers in **neurology** and led to several important discoveries. Bell is remembered today for giving his name to Bell's palsy after demonstrating that lesions on the seventh cranial nerve (facial nerve) can cause facial paralysis.

Born in Edinburgh, Charles Bell was the son of a clergyman and the younger brother of John Bell, an eminent surgeon and anatomist who first taught him **anatomy**. Charles Bell attended Edinburgh University and, after qualifying, became a surgeon at Edinburgh Royal Infirmary in 1799. In 1804, he moved to London where he lectured and became the owner of an anatomy school. Bell rose to prominence as a surgeon and in 1812, was appointed surgeon at Middlesex Hospital.

In 1821, Bell demonstrated that the seventh cranial nerve was a separate nerve. In 1824, he became Professor of Anatomy and Surgery at the Royal College of Surgeons. He helped found the Middlesex Hospital Medical School in 1828 and in that same year became the Principal of the Medical School at University College, London. In 1831, he was knighted and in 1836, he was appointed Professor of Surgery at the University of Edinburgh. He died in 1842 at Hallow, Worcestershire.

Bell executed meticulous dissection which played an important part in his discovery that facial paralysis, or Bell's palsy, occurs when one of the two facial muscles (right or left) become injured or inflamed causing paralysis to half of the face. This paralysis is usually temporary and is characterized by drooping of the eyelid and corner of the mouth. Bell also discovered the long thoracic nerve (Bell's nerve), which controls a muscle in the chest wall. Perhaps Bell's most important contribution to neurophysiology was his demonstration that nerves consist of separate fibers that are bound together; the fibers serve either sensory or motor functions but not both and transmit in one direction only. His ideas were set forth in his essays "Idea of a New Anatomy of the Brain" (1811) and "The Nervous System of the Human Body" (1830).

Bell sustained a life long interest in anatomy and was also a skilled draftsman. He enthusiastically taught anatomy to artists. He published *Essays on the Anatomy of Expression in Painting* (1806), which described the science of physiognomy. His interest in led him to treat gunshot wounds in the battle of Corunna. He was also instrumental in founding a hospital in Brussels following the battle of Waterloo.

See also Cranial nerves; Nerve impulses and conduction of impulses; Nervous system overview

BENACERRAF, BARUJ (1920-)
Venezuelan American geneticist

Baruj Benacerraf's experience with asthma as a child sparked his interest in the body's immune reactions. He researched immunological hypersensitivity with American chemist, Elvin Kabat at the College of Physician's and Surgeons at Columbia University in 1948. Benacerraf went on to conduct research on immunity at the National Center for Scientific Research in Paris. Benacerraf returned to the United States in 1956, joining the faculty at the New York University School of Medicine. There, Benacerraf began his Nobel Prize-winning research on cells involved in immune reactions. Benacerraf continued these investigations at the National Institute of Allergy and Infectious Disease in Bethesda, Maryland, from 1968 to 1970, and subsequently as chairman of the department of **pathology** at the Harvard University Medical School.

While attempting to produce uniform antibodies in test animals, Benacerraf noticed that some guinea pigs responded to **antigens** by producing antibodies, while others did not. Benacerraf demonstrated that the animals' responses were determined by genes he called immune-response (IR) genes. Other researchers found similar genes in mice, rats, and monkeys. In 1969, Benacerraf confirmed that the IR genes were located within the MHC, (major histocompatibility complex). In the 1940's, American geneticist **George Snell** along with British researcher Peter Gorer had discovered a group of genes that later became known as the MHC. The MHC is the main system by which a mammal distinguishes between self and non-self and determines whether or not the body launches an **immune system** response. The identification of self vs. non-self by the immune system depends on the characteristics of surface molecules on cells. Jean Baptiste Dausset, a French immunologist and hematologist, discovered that humans carry the MHC as well as other animals, and in humans it is called the HLA (human leukocyte antigen) system. Benacerraf shared the 1980 Nobel Prize in physiology or medicine with Snell and Dausset for their work on immunological reactions. In 1980, Benacerraf became president and chief executive

officer of the Dana Farber Cancer Center in Boston, Massachusetts.

Baruj Benacerraf was born in Caracas, Venezuela, the son of a wealthy textile merchant. Benacerraf grew up in France, then attended Columbia University in New York, where he graduated in 1942. Benacerraf married and became a naturalized United States citizen in 1943. Benacerraf obtained a medical degree from the Virginia College of Medicine in 1945, then served a tour of duty with the United States Army Medical Corps stationed in Nancy, France.

See also Immune system; Immunity; Immunology

BENDS

"The bends," or decompression sickness, is a health hazard associated with pressure changes during diving underwater. At the water surface, a column of air weighs 14.7 pounds per square inch and this corresponds to atmospheric pressure or one atmosphere. When divers go underwater, they experience the weight of the water in addition to the atmospheric pressure. A 33-ft. (10-m) column of water weighs 14.7 pounds per square inch. Thus, pressure increases by 1 atmosphere for every 33 ft. of descent underwater.

Nitrogen comprises almost 80% of atmospheric air and normally this gas does not dissolve in body tissues to any great extent. However, during dives, the increased pressures force nitrogen gas into solution in body tissues, especially within **adipose tissue**. Because the **blood** supply to adipose **tissue** is not great and since nitrogen gas diffuses slowly, it takes hours for the equilibration of nitrogen to occur between body tissues and air.

The potential for decompression sickness is related to the rate of ascent following a dive. If decompression is too rapid, ambient pressure decreases rapidly and nitrogen comes out of solution as bubbles in body tissues and fluids. A few small bubbles can probably be tolerated without undue consequences but large numbers of bubbles become painful stimuli, especially in the **joints**, giving rise to the so-called "bends." In severe cases, neurological symptoms arise including **deafness**, impaired vision and even paralysis and occur when bubbles of nitrogen gas are present in blood vessels serving parts of the **brain**.

The treatment of decompression sickness is immediate recompression in a hyperbaric chamber where the greater than atmospheric pressure conditions force the nitrogen gas bubbles back into solution and can often produce a dramatic reduction in symptoms. A slow decompression then follows to allow gradual removal of nitrogen from body fluids.

The correct method of preventing decompression sickness during ascent is carefully spelled out during proper dive-training procedures. Slow ascents in graduated steps as set out in empirically derived decompression tables can reduce the risk to virtually zero. Another method of reducing the incidence of decompression sickness is to substitute helium gas for the nitrogen in the air to be breathed by divers. Helium is about half as soluble in body tissues as nitrogen and is about one seventh the

Scuba divers returning from an underwater archeological site, stop at the 10-ft. (3-m) decompression point to avoid decompression sickness (the bends). © Jonathan Blair/Corbis. Reproduced by permission.

molecular weight of nitrogen. It can thus diffuse much more rapidly in and out of tissues. Pure or elevated concentrations of oxygen are not used in diving because of oxygen toxicity.

See also Adipose tissue; Breathing; Gaseous exchange

BERGSTRÖM, SUNE KARL (1916-)

Swedish biochemist

Sune Karl Bergström is best known for his research on **prostaglandins**. These substances, which were first discovered in the **prostate gland** and seminal vesicles, were found by Bergström and his colleagues to affect circulation, **smooth muscle tissue**, and general **metabolism** in ways that can be medically beneficial. Certain prostaglandins, for example, lower **blood** pressure, while others prevent the formation of ulcers on the stomach lining. For his research, Bergström shared the 1982 Nobel Prize in physiology or medicine with John R. Vane and **Bengt Samuelsson**.

Sune Bergström was born in Stockholm, Sweden, to Sverker and Wera (Wistrand) Bergström. Upon completion of high school he went to work at the Karolinska Institute as an assistant to the biochemist Erik Jorpes. The young Bergström was assigned to do research on the **biochemistry** of fats and steroids. Jorpes was impressed enough with his assistant to sponsor a year-long research fellowship for Bergström in 1938 at the University of London. While there, Bergström focused his research on **bile** acid, a steroid produced by the **liver** which aids in the **digestion** of cholesterol and similar substances.

Bergström had planned to continue his research in Edinburgh the following year thanks to a British Council fellowship, but the fellowship was canceled after World War II broke out. He did, however, receive a Swedish-American Fellowship in 1940, which allowed him to study for two years at Columbia University and to conduct research at the Squibb Institute for Medical Research in New Jersey. At Squibb, Bergström researched the steroid cholesterol, particularly its reaction to chemical combination with oxygen at room temperature, a process called auto-oxidation.

Bergström returned to Sweden in 1942, receiving doctorates in medicine and biochemistry from the Karolinska Institute two years later. He was appointed assistant in the biochemistry department of Karolinska's Medical Nobel Institute. While there, he continued experiments with auto-oxidation, working with linoleic acid, which is found in some vegetable oils. He discovered a particular enzyme was responsible for the oxidation of linoleic acid, and helped attempt to purify the enzyme while working with biochemist Hugo Theorell.

While attending a meeting of Karolinska's Physiological Society in 1945, Bergström met the physiologist **Ulf von Euler**. Von Euler, who was better known as the discoverer of the hormone norepinephrine, had been doing research on prostaglandins. Scientists had observed in the 1930s that seminal fluid used in artificial insemination stimulated contraction and subsequent relaxation in the smooth muscles of the uterus. Von Euler isolated a substance from the seminal fluid of sheep and found it had the same effect in relaxing the smooth muscle of blood vessels. Impressed with Bergström's work on enzyme purification, von Euler gave him some of the extract for further purification.

Bergström began initial experiments but put his work on hold when in 1946 he was named a research fellow at the University of Basel. Returning from Switzerland in 1947, he was appointed professor of physiological chemistry at the University of Lund. His first task was to help revitalize the university's research facilities, which had fallen into disuse during the war. Afterwards, he resumed his research on prostaglandins, assisted by graduate students such as Bengt Samuelsson. Working with new large supplies of sheep seminal fluid, Bergström and his colleagues were able to isolate and purify two prostaglandins by 1957. Bergström was appointed professor of chemistry at Karolinska a year later, and brought his research on prostaglandins and his collaboration with Samuelsson with him. By 1962, six prostaglandins, identified as A through F, had been identified.

Bergström and Samuelsson then worked on determining how prostaglandins are formed. They discovered that prostaglandins are formed from common fatty acids, and further identified specific functions performed by each prostaglandin. Over the next few years, Bergström and Samuelsson surmised that certain prostaglandins could be used to treat high blood pressure, blocked **arteries**, and other circulatory problems by relaxing muscle tissue. These prostaglandins were also shown to prevent ulceration of the stomach lining and to protect against side effects of such drugs as aspirin, long known to irritate the stomach lining. Other prostaglandins could be used to raise blood pressure or stimulate uterine muscle by their contracting effect.

Bergström remained at Karolinska, serving as dean of its medical school from 1963 to 1966 and as rector of the institute from 1969 to 1977. He was chairman of the Nobel Foundation's Board of Directors from 1975 to 1987, and from 1977 to 1982 he served as chairman of the World Health Organization's Advisory Committee on Medical Research. He retired from teaching in 1981, choosing to devote his full time to research at Karolinska.

A modest, reserved man, Bergström's reaction upon learning of his Nobel award was gratitude—first, that his colleagues appreciated his efforts, and second, that his former student Samuelsson had also been named. The book *Nobel Prize Winners* reports him as saying that there is "no greater satisfaction than seeing your students successful." His connection with the Nobel Foundation had led some to wonder whether he might be passed over for a prize of his own. But the *New York Times,* reporting on the 1982 awards, noted that "it was only a matter of time, most scientists agree, before Dr. Bergström's research would be honored by the foundation he directs—for the work was too important to be ignored through any concern over apparent conflicts of interest."

See also Lipids and lipid metabolism; Hormones and hormone action

BICHAT, MARIE FRANÇOIS XAVIER
(1771-1802)
French physician

Marie François Xavier Bichat shifted the primary focus of anatomical study from **organs** and gross structures to tissues and microscopic structures, and thereby was one of the main founders of **histology**. He also furthered descriptive **anatomy** and pathological anatomy. His concentration on tissues and their lesions opened many doors to physiological inquiry.

Bichat was born in Thoirette, France, the son of a physician. He attended the best schools in Lyons, including courses in philosophy at the Séminaire Saint-Irénée. He began his life-long study of anatomy and surgery in 1791 at the Hôtel Dieu in Lyons, then in 1794 became the surgical protege of Pierre Joseph Desault (1744–1795), surgeon-in-chief at the prestigious Hôtel Dieu in Paris. Influenced by Philippe Pinel (1745–1826) and Andreas Bonn (1738–1818) and befriended

Marie François Xaiver Bichat. *The Library of Congress.*

by Napoleon's personal physician, Baron Jean Nicolas Corvisart des Marets (1755–1821), Bichat made amazingly rapid progress. Around 1796, he began giving private lessons in anatomy and physiology. Dead at only 31, Bichat was universally mourned, not only by physicians and scientists, but even by Napoleon. He was so highly regarded that the famous Cathedral of Notre Dame was the site of his funeral.

Bichat performed hundreds of autopsies but did most of his close-level "microscopic" observation of small structures without the aid of a microscope. Concentrating on tissues rather than organs or large structures, he distinguished according to their physical properties twenty-one different kinds of **tissue**: absorbent, animal muscle, arterial, bony, capillary, cartilaginous, cellular, connective, dermoid, epidermoid, exhalation, fibrocartilaginous, fibrous, glandular, medullary, mucous, nervous, organic muscle, serous, synovial, and venous. The list is controversial; some historians count as many as twenty-three in his writings.

Bichat viewed each kind of tissue as essential to life, and believed that disease resulted when tissue was infected or damaged, regardless of its location in the body. Persuaded by the animism of Georg Ernst Stahl (1660–1734), the doctrine that the soul is the body's life force and that disease is a disturbance in the soul, he developed his own theory of vitalism, the idea that bodies, or body parts, have soul-like qualities. Most historians argue that Bichat's vitalism is really just Stahl's animism particularized in tissues rather than generalized over the entire body.

Bichat had a profound effect on medical philosophy. Even those who disagreed with his vitalism, such as the physiologist Claude Bernard (1813–1878), the pathologist Jean Cruveilhier (1791–1874), the positivist philosopher Auguste Comte (1798–1857), the mechanist philosopher Pierre Jean George Cabanis (1757–1808), and the voluntarist philosopher Arthur Schopenhauer (1788–1860) all acknowledged his effect on their thought. Surgery also benefited immensely from Bichat's work. His student and confidant, Philibert Joseph Roux (1780–1854), succeeded Guillaume Dupuytren (1777–1835) as surgeon-in-chief at Hôtel Dieu in Paris. British surgical pioneer Sir Benjamin Collins Brodie (1783–1862) owed much of his professional success to Bichat. Moreover, the world's first lithographic anatomical atlas, *Anatomie de l'homme* [Human Anatomy], that Jules Germain Cloquet (1790–1883) published in five volumes from 1821 to 1831, depended on Bichat for much of its data.

Because he knew that tuberculosis would soon kill him, he began rushing his works through the press in 1799. The first to appear was *Traité des membranes en général et diverses membranes en particulier* [Treatise on Membranes in General and Various Membranes in Particular] in 1800. Later in 1800, he published *Recherches physiologiques sur la vie et la mort* [Physiological Investigations of Life and Death]. During this busy time he also edited the first two volumes of Desault's *Oeuvres chirurgicales* [Surgical Works]; Roux edited the third, and they appeared from 1798 to 1803. His four-volume *Anatomie générale, appliquée à la physiologie et à la médecine* [General Anatomy Applied to Physiology and Medicine] appeared in 1802. When he died, he had completed only three of the projected five volumes of his masterpiece, *Traité d'anatomie descriptive* [Treatise on Descriptive Anatomy]. Matthieu François Régis Buisson (1776–1804) edited volume three and wrote volume four, Roux did volume five, and it was all published between 1801 and 1803.

See also History of anatomy and physiology: The Renaissance and Age of Enlightenment; History of anatomy and physiology: The science of medicine

BILE, BILE DUCTS, AND THE BILIARY SYSTEM

Bile is a bitter fluid, yellow, brown or green in color, that is comprised of water, **electrolytes**, and organic molecules, including bile acids, cholesterol, phospholipids and **bilirubin**. Present in many species, bile has two main functions. First, the bile acids it contains are critical for **digestion** and for the absorption of fats and fat-soluble **vitamins** in the small intestine. Second, many of the body's waste products are secreted into the bile and are then eliminated in the fecal material. Bile ducts and the biliary system participate in these functions.

Humans normally produce 400-800 ml of bile daily. If this production or the flow of bile from the **liver** to the small intestine is impeded, a condition called jaundice can result.

Bile is produced in hepatocytes, which are the main functional cells of the liver. The bile flows into the bile ducts. As bile flows through the bile ducts, it is modified by the addition of secretions from the epithelial cells that line the ducts. The transport of bile from the liver hepatocytes to the small intestine is accomplished via a system of bile ducts that collectively are called the biliary system.

The biliary system consists of canaliculi; spaces between adjacent hepatocyte cells. Further on towards the small intestine each canaliculus feeds into a true bile duct. Small bile ducts, called ductules, join together to form a large and common bile duct, which dumps bile into the duodenum (a part of the small intestine). A powerful muscle known as a sphincter is located at the entry point to the small intestine, and functions to control the flow rate of the bile.

Humans and many other animals have an organ close to the liver, the gallbladder, that can store and concentrate bile when food is scarce, to protect the body from the toxic effects of the bile. Food stimulates the production of a hormone called cholecystokinen, which acts to trigger the secretion of bile.

The main route for eliminating cholesterol from the body is via the bile. Cholesterol, being fatty in composition, cannot dissolve in the water-based fluids that are present in the body. But the bile acids and lecithin in bile allow cholesterol to dissolve into solution. If this mechanism goes awry, cholesterol can precipitate out of the bile solution, forming gallstones. Aside from its function as a waste disposal fluid, bile also helps neutralize stomach acid in the small intestine, providing a more hospitable environment for **enzymes** that break down food. Bile acids function to promote the complete digestion of food, by facilitating the uptake of fat-soluble vitamins (A, D, E, and K) through the wall of the small intestine. Bile acids can also act as **hormones** to control the manufacture of cholesterol.

See also Elimination of waste; Homeostatic mechanisms; Metabolic waste removal

BILIRUBIN

Bilirubin is the waste product that results from the breakdown of **hemoglobin** molecules in red **blood** cells. It is excreted from the body via the bile ducts of the **liver**, typically as the main component of **bile**.

Bilirubin is an end product of the **metabolism** of the heme component of red blood cells. In our bodies, bilirubin serves an important purpose. It is one of the most powerful anti-oxidants produced by the body (**vitamins** E and C are other examples of anti-oxidants). An anti-oxidant functions, as the name implies, to prevent the chemical process of oxidation, which can be destructive to the functioning of cells, tissues and **organs**. In normal quantities, bilirubin aids the body. But, when it accumulates, bilirubin poses problems.

Accumulation of bilirubin can occur in several ways. A liver malfunction can result in the impaired or blocked excretion of bilirubin. Secondly, the compound can be over-produced. Lastly, the normal binding of bilirubin to another molecule in body fluids, albumin, may be impaired, allowing more of the unbound form of bilirubin to be present. It is this unbound form that is problematic.

The build-up of bilirubin in the fatty tissues of the skin turns the tissues yellow. This condition is known as jaundice. Excessive accumulation of bilirubin in newborns may be serious if untreated, causing permanent **brain** damage. In this condition, kernicterus, a decrease in pH causes bilirubin to come out of solution and accumulate in solid form in the brain. The brain malfunction can lead to a characteristic form of crippling called athetoid cerebral palsy. To prevent these catastrophes, bilirubin levels are monitored in newborns. Excessive levels can be treated in two ways. A newborn can be exposed to a "bilirubin light", which emits light of a certain wavelength in order to slightly alter the three-dimensional shape of the bilirubin molecule so that it is able to dissolve back into solution again. The dissolved bilirubin is then able to be excreted form the body more readily. Another treatment procedure called exchange transfusion washes out excess bilirubin by a series of **transfusions**. In each transfusion a small amount of the baby's blood is removed and replaced with an equal volume of adult donor blood (not the mother's, to avoid antigenic reaction) that is the same type as the mother's blood.

See also Bile, bile ducts and the biliary system

BIOCHEMISTRY

Biochemistry seeks to describe the structure, organization, and functions of living matter in molecular terms. Essentially two factors have contributed to the excitement in the field today and have enhanced the impact of research and advances in biochemistry on other life sciences. Firstly, it is now generally accepted that the physical elements of living matter obey the same fundamental laws that govern all matter, both living and non-living. Therefore the full potential of modern chemical and physical theory can be brought in to solve certain biological problems. Secondly, incredibly powerful new research techniques, notably those developing from the fields of biophysics and **molecular biology**, are permitting scientists to ask questions about the basic process of life that could not have been imagined even a few years ago.

Biochemistry now lies at the **heart** of a revolution in the biological sciences and it is nowhere better illustrated than in the remarkable number of Nobel Prizes in chemistry or medicine and physiology that have been won by biochemists in recent years. A typical example is the award of the 1988 Nobel Prize in medicine and physiology, to **Gertrude Elion** and **George Hitchings** of the United States and Sir **James Black** of Great Britain for their leadership in inventing new drugs. Elion and Hitchings developed chemical analogs of nucleic acids and **vitamins** which are now being used to treat leukaemia, bacterial infections, **malaria**, gout, herpes virus infections and **AIDS**. Black developed beta blockers that are now used to reduce the risk of heart attack and to treat diseases such as asthma. These drugs were designed and not discovered through random organic synthesis. Developments in knowl-

edge within certain key areas of biochemistry, such as protein structure and function, nucleic acid synthesis, enzyme mechanisms, receptors and metabolic control, vitamins, and coenzymes all contributed to enable such progress to be made.

Two more recent Nobel Prizes give further evidence for the breadth of the impact of biochemistry. In 1997, the Chemistry Prize was shared by three scientists—the American Paul Boyer and the British J. Walker for their discovery of the "rotary engine" that generates the energy-carrying compound **ATP**, and the Danish J. Skou, for his studies of the "pump" that drives sodium and potassium across membranes. In the same year, the Nobel Prize in medicine and physiology went to **Stanley Prusiner**, for his studies on the prion, the agent thought to be responsible for "mad cow disease" and several similar human conditions.

Biochemistry draws on its major themes from many disciplines. For example from organic chemistry, which describes the properties of biomolecules; from biophysics, which applies the techniques of physics to study the structures of biomolecules; from medical research, which increasingly seeks to understand disease states in molecular terms and also from **nutrition**, microbiology, physiology, cell biology and **genetics**. Biochemistry draws strength from all of these disciplines but is also a distinct discipline, with its own identity. It is distinctive in its emphasis on the structures and relations of biomolecules, particularly **enzymes** and biological catalysis, also on the elucidation of metabolic pathways and their control and on the principle that life processes can, at least on the physical level, be understood through the laws of chemistry. It has its origins as a distinct field of study in the early nineteenth century, with the pioneering work of Freidrich Wöhler. Prior to Wöhler's time it was believed that the substance of living matter was somehow quantitatively different from that of non-living matter and did not behave according to the known laws of physics and chemistry. In 1828 Wöhler showed that urea, a substance of biological origin excreted by humans and many animals as a product of nitrogen **metabolism**, could be synthesized in the laboratory from the inorganic compound ammonium cyanate. As Wöhler phrased it in a letter to a colleague, "I must tell you that I can prepare urea without requiring a kidney or an animal, either man or dog". This was a shocking statement at the time, for it breached the presumed barrier between the living and the nonliving. Later, in 1897, two German brothers, Eduard and Hans Buchner, found that extracts from broken and thoroughly dead cells from yeast, could nevertheless carry out the entire process of fermentation of sugar into ethanol. This discovery opened the door to analysis of biochemical reactions and processes in vitro (Latin "in glass"), meaning in the test tube rather than in vivo, in living matter. In succeeding decades many other metabolic reactions and reaction pathways were reproduced in vitro, allowing identification of reactants and products and of enzymes, or biological catalysts, that promoted each biochemical reaction.

Until 1926, the structures of enzymes (or "ferments") were thought to be far too complex to be described in chemical terms. But in 1926 J.B. Sumner showed that the protein urease, an enzyme from jack beans, could be crystallized like other organic compounds. Although proteins have large and complex structures, they are also organic compounds and their physical structures can be determined by chemical methods.

Today, the study of biochemistry can be broadly divided into three principal areas: (1) the structural chemistry of the components of living matter and the relationships of biological function to chemical structure; (2) metabolism, the totality of chemical reactions that occur in living matter; and (3) the chemistry of processes and substances that store and transmit biological information. The third area is also the province of molecular genetics, a field that seeks to understand heredity and the expression of genetic information in molecular terms.

Biochemistry is having a profound influence in the field of medicine. The molecular mechanisms of many diseases, such as **sickle cell anemia** and numerous errors of metabolism, have been elucidated. Assays of enzyme activity are today indispensable in clinical diagnosis. To cite just one example, **liver** disease is now routinely diagnosed and monitered by measurements of **blood** levels of enzymes called transaminases and of a **hemoglobin** breakdown product called **bilirubin**. **DNA** probes are coming into play in diagnosis of genetic disorders, infectious diseases and cancers. Genetically engineered strains of **bacteria** containing recombinant DNA are producing valuable proteins such as insulin and growth hormone. Furthermore, biochemistry is a basis for the rational design of new drugs. Also the rapid development of powerful biochemical concepts and techniques in recent years has enabled investigators to tackle some of the most challenging and fundamental problems in medicine and physiology. For example in **embryology**, the mechanisms by which the fertilized embryo gives rise to cells as different as muscle, **brain** and liver are being intensively investigated. Also, in **anatomy**, the question of how cells find each other in order to form a complex organ, such as the liver or brain, are being tackled in biochemical terms. The impact of biochemistry is being felt in so many areas of human life, through this kind of research and the discoveries are fuelling the growth of the life sciences as a whole.

See also History of anatomy and physiology: The science of medicine; Human genetics; Molecular biology

BIOFEEDBACK

Biofeedback is a means by which a person can mentally influence a natural physiologic process that may or may not be consciously regulated under normal conditions. This could include lowering **blood** pressure, regulating the **heart** rate, or influencing the skin temperature.

As a form of therapy, biofeedback is a relatively recent development that has begun to gain acceptance among some members of the medical community. Research in the topic began in the 1960s, and by the end of the decade a number of research projects had demonstrated its effectiveness. While some early studies indicated that physiologic processes not usually under conscious control could not be influenced by biofeedback, subsequent research soon disproved this assumption.

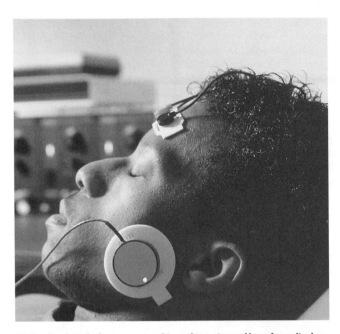

Biofeedback techniques are used to reduce stress. Use of monitoring equipment allows patients to monitor their progress and obtain feedback regarding often complex physiological processes. *Photograph by Will & Deni McIntyre. Photo Researchers, Inc. Reproduced by permission.*

During the 1970s, biofeedback developed a devout following and as a result an entire industry was created for the manufacture and marketing of the instruments used to measure biofeedback alterations and alpha rhythms in the **brain**. Alpha rhythms are electrical waves formed by the brain at the rate of 8–13 cycles per second. Because they are associated with the state of meditation attained by practitioners of yoga or transcendental meditation, they were accepted as the optimal state of biofeedback. Instruments to detect and measure alpha rhythms soon became readily available.

Biofeedback training must begin with an auditory or visual signal to measure the activity of the organ being influenced and to indicate any changes that take place in it. Heart rate, for example, can be signaled by a beeping sound that occurs with each heartbeat. A subject can detect any increase or decrease in the number of heartbeats per minute by the increasing or decreasing rapidity of the beeping. A visual signal also could be devised—for example, spikes in a horizontal line that appear closer together or farther apart as the heartbeat changes. The ultimate objective is to develop a such a high level of consciousness that such changes can be determined without the signal device.

Biofeedback can be separated into three processes or steps. The first involves detecting the biological process being measured and amplifying it so as to be seen or heard. The second step is to convert the electrical signal into an easily understood form from which its alterations can be read. The third step is to make this signal available to the subject as soon as possible after the event being measured has occurred.

Clinical applications for biofeedback include the control of blood pressure for patients with hypertension, relief or control of migraine headaches, easing of muscle cramps, and relief of insomnia.

Biofeedback training begins with the basic control of heart rate or other readily accessible and controllable functions. The subject is provided an auditory or visual signal at first and is gradually weaned from it as he becomes more skillful at the practice. Once the basic skill has been learned he or she can then shift concentration to a specific problem. Ideally, patients will continue to practice biofeedback techniques and so increase their effectiveness over time.

Biofeedback has gained acceptance in the United States as its clinical use has increased. Most major cities have a biofeedback association, and practitioners can be certified by a national certification institute. Certification standards are rigorous to assure that the practitioner has a thorough understanding of **physiology** and psychology to better apply the methodology.

See also Blood pressure and hormonal control mechanisms

BIRTH • *see* PARTURITION

BIRTH, CHANGES AT BIRTH • *see* PARTURITION

BIRTH CONTROL (PHARMACOLOGICAL REGULATION MECHANISMS) • *see* OVARIAN
CYCLE AND HORMONAL REGULATION

BIRTH DEFECTS AND ABNORMAL DEVELOPMENT

Birth defects, or congenital defects result from heredity, environmental influences, or maternal illness. Such defects range from the very minor, such as a dark spot or birthmark that may appear anywhere on the body, to more serious conditions that may result in marked disfigurement, impaired functioning, or decreased lifespan.

A number of factors individually or in combination may cause birth defects. Heredity plays a major role in passing birth defects from one generation to the next. Inherited conditions are passed on when a baby receives a flawed gene from one or both parents. Conditions such as **sickle cell anemia**, color **blindness**, **deafness**, and extra digits on the hands or feet are hereditary. The condition may not appear in every generation, but the defective gene usually is passed on to successive generations.

Low birth weight is the most common birth defect, with one in every 15 babies being born at less than their ideal weight. A baby who weighs 5 lb, 8 oz (2,500 g) at birth has a low birth weight. One who is born weighing 3 lb, 5 oz (1,500 g) has a very low birth weight. Low birth weight may occur if the baby is born before the normal gestation period of 38 weeks has elapsed, in which case the baby is preterm or premature. A low birth weight baby born after a normal gestation

period is called a small-for-date or small-for-gestational-age baby. Premature birth, other than being a birth defect in itself, also may have accompanying defects. A baby born before the 28th week of gestation, for example, may have great difficulty **breathing** because the **lungs** have not developed fully.

Exposure of the mother to chemicals such as mercury or to radiation during the first three months of **pregnancy** may result in an abnormal alteration in the growth or development of the fetus. The mother's diet may also be a factor in her baby's birth defect. A balanced and healthy diet is essential to the proper formation of the fetus because the developing baby receives all of its **nutrition** from the mother.

Prenatal development of the fetus may also be affected by disease that the mother contracts, especially those that occur during the first trimester (three months) of pregnancy. For example, if a pregnant woman catches rubella, the virus crosses the **placenta** and infects the fetus. In the fetus, the virus interferes with normal **metabolism** and cell movement and can cause blindness (from cataracts), deafness, **heart** malformations, and mental retardation. The risk of the fetal damage resulting from maternal rubella **infection** is greatest during the first month of pregnancy (50%) and declines with each succeeding month.

It is especially important that the mother not smoke, consume alcohol or take drugs while she is pregnant. Drinking alcohol heavily can result in fetal alcohol syndrome (FAS), a condition that is physically apparent. FAS newborns have small eyes and a short, upturned **nose** that is broad across the bridge, making the eyes appear farther apart than normal. These babies also are underweight at birth and do not catch up as time passes. They often have some degree of mental retardation and may exhibit behavior problems. A mother who continues to take illicit drugs while pregnant will have a baby who is already addicted. The addiction may not be fatal, but the newborn may undergo severe withdrawal, unless the addiction is revealed and carefully treated. Furthermore, some behavior problems/cognitive deficits are suspected to be associated with fetal drug exposure and addiction.

Some therapeutic drugs taken by pregnant women have also been shown to produce birth defects. The most notorious example is thalidomide, a mild sedative. During the 1950s, women in more than 20 countries who had taken this drug gave birth to more than 7,000 severely deformed babies. The principal defect these children suffered from is a condition called phocomelia, characterized by extremely short limbs often with no fingers or toes.

Approximately one newborn out of every 735 has a form of clubfoot. In the most serious form, known as equinovarus, the foot is twisted inward and downward and the foot itself is cupped or flexed. If both feet are clubbed in this manner the toes point to each other rather than straight ahead. Often the heel cord or Achilles tendon is taut so that the foot cannot be straightened without surgery.

Approximately 7,000 newborns (one of every 930 births) are born with cleft lip and/or cleft palate each year in the United States. Cleft lip and palate describe a condition in which a split remains in the lip and roof of the mouth. During growth *in utero* (in the womb) the lip or palate, which develop from the edges toward the middle, fail to grow together. The condition occurs most often among Asian Americans and certain Native American groups, less frequently among whites, and least often among African Americans.

Approximately 25% of infants born with cleft palate have inherited the trait from one or both parents. The cause for the other 75% remains unknown, but may be a combination of heredity, poor nutrition, use of drugs, or a disease the mother contracted while pregnant. The cleft may involve only the upper lip, may extend into the palate, or may be located on the back of the palate. Surgery is especially important to correct the defect in the palate.

Spina bifida, or open spine, occurs once in 2,000 births in the United States. It occurs when the edges of the spine that should grow around the **spinal cord** do not meet. An open area remains, which can mean that an area of the spinal cord (or the entire spinal cord, in the most severe cases) is unprotected. The mildest form of spina bifida may be so slight that the defect does not have any effect on the child and is discovered by accident, usually when an x ray is taken for another reason. The term spina bifida means the spine is cleft, having an opening or space, in two parts. Hydrocephaly, an abnormally large amount of fluid that dilates the ventricles of the **brain**, often can occur in infants who have spina bifida. The cause of spina bifida is not known, nor is any means of prevention. It can be diagnosed before birth by **amniocentesis** or ultrasound. The risk of having a baby with spina bifida or other associated defects seems to be reduced if a woman takes at least 400 mg of folic acid just before and throughout pregnancy.

Congenital **heart defects** occur in one of every 115 births in the United States. The defect may be so mild that it is not detected for some years or it may be fatal. A baby with a heart defect may be born showing a bluish tinge around the lips and on the fingers. This condition, called cyanosis, is a signal that the body is not receiving enough oxygen. The blue color may disappear shortly after birth, indicating that all is normal, or it may persist, indicating that further testing is needed to determine the nature of the heart defect.

During fetal development, **blood** does not need to flow through the fetal lungs. It receives its oxygen from the mother through the placenta via the umbilical cord. A special shunt that bypasses the lungs is in place during development of the fetus, and normally should close at birth. After birth, blood should begin to circulate through the lungs for the first time, because the newborn baby's lungs are now responsible for delivering oxygen to the blood. Sometimes, however, the shunt does not close properly, and blood is not appropriately circulated through the lungs. When this occurs, surgery is required to close the shunt and restore normal circulation.

If it is undetected at birth, a heart defect may impair the growth of a child. He will be unable to exert the energy that other children do at play because he cannot supply sufficient oxygen to his body. He may become breathless at small amounts of exertion and may squat frequently because it is easier to breathe in that position.

Some minor heart defects may disappear over time as the child grows. A small hole in the wall between the left and right sides of the heart, which causes symptoms by allowing the mixing of oxygen-poor and oxygen-rich blood, for example, may spontaneously close over time. A larger defect will require surgical patching.

Physical defects in newborns are common. They can affect any of the bones or muscles in the body and may or may not be correctable. Among the more common are the presence of extra fingers or toes (polydactyly), which presents no health threat and can be corrected surgically. Similarly, webbed fingers and toes, a genetic disorder, seen in approximately one of every 1,700 to 2,000 births, can be treated surgically to resemble a normal appendage.

A more serious, though relatively rare, condition is called achondroplasia; this term means without **cartilage** formation and refers to the supposed lack of cartilage growth plates near the ends of a child's bones. In fact in the plates are present, but grow poorly. Achondroplasia is a type of dwarfism. This genetic disorder of bone growth is seen in one in 20,000 births and is one of the oldest known birth defects. Ancient Egyptian art shows individuals with this condition. The child who has this condition will be slow at walking and sitting because of his short arms and legs, and this may be interpreted as mental retardation. However, these individuals have normal intelligence.

In addition to physical deformities, certain diseases and syndromes also are passed to the infant through the parents' genetic material. Some of these conditions can be controlled or treated while others are untreatable and fatal.

Sickle cell anemia is an inherited disease of the blood cells that occurs in about one of every 400 African Americans. An individual can be a carrier of sickle cell anemia, in which case he or she has the gene but does not show any active signs of the disease. If two carriers become parents, however, some of their children may have sickle cell anemia.

The disease gets its name because certain red blood cells assume a sickle shape and lodge in small blood vessels. This altered shape is a function of the **hemoglobin** molecule present in red blood cells. Two forms of hemoglobin make up these cells: hemoglobin A (Hb A) and hemoglobin B (Hb B). In individuals with sickle cell anemia, Hb B is instead produced as Hb S, a form of hemoglobin with a rigid, sickle shape that deforms the red blood cell. When the cell becomes wedged in a small blood vessel it prevents the flow of blood through the vessel and can initiate what is called a sickle cell crisis. The lack of blood flow to the tissues being blocked causes **pain** and inflammation of the oxygen-deprived **tissue**.

Tay-Sachs disease affects mainly Jewish people of eastern European origin, the Ashkenazi Jews, and is a condition that is fatal at an early age. A carrier of the disease will have a gene for Tay-Sachs disease and another gene that is normal. If two carriers have children, every pregnancy will have a 25% chance of producing a completely normal child; a 50% chance of producing a child who will carry the trait, but reveal no symptoms; and a 25% chance of producing a child who actually suffers from the disease.

The newborn Tay-Sachs child lacks a blood enzyme called hexosaminidase A, which breaks down certain fats in the brain and nerve cells. When first born, the baby appears totally normal. However, over a short period of time, the brain cells become clogged with fatty deposits, and the child begins to lose functioning. As the disease progresses, the child will no longer be able to smile, crawl, or turn over, and will ultimately become blind and unaware of his surroundings. Usually the child dies by the age of three or four years.

There is currently no cure for Tay-Sachs disease, although carriers can be detected by a simple blood test that measures the amount of hexosaminidase A. A carrier will have half the amount of the enzyme as a normal person, and two carriers can be counseled to explain the probability of producing an offspring with Tay-Sachs disease. Researchers are trying to find a way to provide sufficient levels of the missing enzyme in the newborn, or to find a suitable substitute that could be supplied as the child ages, much like insulin is used to treat diabetes. A more technologically advanced line of research is examining the possibility of transplanting a normal gene to replace the defective one in carriers.

One in every 800–1,000 babies is born with Down syndrome. Down syndrome babies may have eyes that slant upward, small ears that may turn over at the top, a small mouth and nose that also is flattened between the eyes (at the bridge). Mental retardation is present in varying degrees, but most Down syndrome children have only mild to moderate retardation. Down syndrome results when either the egg or the **sperm** that fertilizes it has an extra chromosome. Normally a human has 23 pairs of **chromosomes**, for a total of 46. An extra chromosome, specifically an extra number 21 chromosome, present when the egg is fertilized, leads to a baby with Down's syndrome. If either parent has Down syndrome, the probability of passing the condition on to the offspring is increased. Also, parents who have had one Down's syndrome child and mothers older than 35 years of age are at increased risk of having a Down's syndrome baby.

It should be apparent from this small sample that some birth defects are hereditary, passed from parents to offspring; little can now be done to prevent or cure these conditions, but genetic therapy offers hope that this situation may change in the future. Other birth defects result from infections of the mother during pregnancy, or from maternal consumption of alcohol or drugs, use of tobacco, or exposure to radiation or chemicals during pregnancy. In some cases, these birth defects can be prevented through education or improved prenatal care.

See also Cell differentiation; Embryology; Embryology; Embryonic development: Early development, formation, and differentiation; Gene therapy; Genetic code; Genetic regulation of eukaryotic cells; Genetics and developmental genetics; Human genetics

BLACK, JAMES WHYTE (1924-)

English pharmacologist and physician

Sir James Black was one of the founders of a revolution in the way pharmaceutical companies search for medicines. He developed a method of discovering and evaluating new medicines by studying the basic biological mechanisms that under-

lie disease. His approach led to new, more effective treatments for **heart** ailments, including heart attack, and to the first successful drug to treat ulcers. For his pioneering efforts, Black shared the 1988 Nobel Prize in physiology or medicine with George H. Hitchings and Gertrude Belle Elion of Burroughs Wellcome Co. in the United States.

Black was born in Uddingston, Scotland, to a working class family. His father was a Scottish coal miner who worked his way up to mining engineer. Black was the youngest of four sons. One of his older brothers studied medicine and Black soon followed in his footsteps. At age fifteen, he won a residential scholarship to St. Andrew's University, where he received his medical degree in 1946. He remained as an assistant lecturer from 1946 to 1947 before traveling to Malaysia to serve as a senior lecturer in physiology at the University of Malaya from 1947 to 1950. He returned to Scotland in 1950 and lectured in physiology at Glasgow Veterinary School until 1958. During this time, Black began research on the mechanism of increase in gastric secretions caused by the body's production of histamine. This research formed the basis for his later work on blocking histamine receptors (chemical groups in **plasma** membrane or cell interior that have an affinity for a specific chemical or compound, in this case histamine) to reduce gastric secretions. While in Glasgow, Black also became familiar with the alpha and beta adrenergic receptors, which are responsible for regulating heart beat.

Black joined Imperial Chemical Industries in 1958. There he sought better ways of treating angina pectoris, a painful disease caused by insufficient oxygenation of the heart. The painful episodes suffered by angina patients are accentuated by increased heart rate, which increases the heart's requirement for oxygen. Black's research led him to theorize that a drug that would neutralize the effects of the **hormones** adrenaline and noradrenaline, which mediate heart rate, would relieve the symptoms of angina.

The existence of receptors for these hormones had been understood since 1948, when the biochemist Raymond P. Ahlquist first described their action. Black developed a chemically similar but nonfunctional version of the active hormones that would block one of these receptors, the beta receptor. His first studies were with analogs of isoprenaline, a compound similar to noradrenaline. One of these analogs, known as propanolol, had the desired effect. It constricted heart muscle, stopping angina attacks.

In 1964, Black joined the British subsidiary of Smith Kline & French Laboratories. There he worked on new approaches to treating intestinal ulcers. Black knew from his earlier studies that histamine stimulated the secretion of excess acid that causes ulcers. The antihistamines in use at that time inhibited muscle contractions but not acid secretion. Black attacked the problem using the same strategy that worked in the development of the angina treatment—he sought a chemical that would inhibit histamine receptors, blocking the action of the hormone. Many thousands of compounds were tested. Finally in 1972, a partial histamine receptor antagonist was found, guanylhistamine. Unfortunately, it had serious side effects and clinical tests were halted in 1974.

After further modification to the chemical structure, Black's group introduced cimetidine, a successful ulcer drug.

Black left Smith Kline in 1973, in order to spend four years as head of the department of **pharmacology** at University College in London. Then in 1978, he returned to industry, accepting a post as director of therapeutic research at the Wellcome Research Laboratories in Kent. He remained there until 1984, when the lure of academia led him to King's College of Medicine and Dentistry, where he remains today.

In 1988, Black was honored with the Nobel Prize in physiology or medicine, an award he shared with **George Hitchings** and **Gertrude Elion**, pharmaceutical researchers from Burroughs Wellcome in the United States. Although it is unusual for the prize to go to pharmacologists, the fruits of Black's research, propanolol and cimetidine, are among the most widely-used medications prescribed by physicians today.

Black's success in designing new medicines may be attributed in part to the rational method he employed. Instead of randomly searching for chemicals with a physiological effect, he sought to understand the underlying biological processes and designed drugs that mimic life processes. To test his drugs, he designed bioassays that tested how well his drugs would work in the body.

Black is regarded as a shy man who does not like to publicize his personal life. He is said to enjoy reading beyond his scientific subjects, music, and the arts. Black has received many awards and honorary degrees for his work. He was elected to the Royal Society of London in 1976, and received its Mullard Award in 1978. He received the Albert Lasker Clinical Medicine Award in 1976, and was elected a foreign associate of the U.S. National Academy of Sciences in 1991. He was knighted in 1981.

See also Cardiac cycle; Cardiac disease; Cardiac muscle; Hormones and hormone action; Intestinal histophysiology; Pharmacology

BLADDER • *see* UROLOGY

BLINDNESS AND VISUAL IMPAIRMENTS

Blindness is usually considered as an inability to see or a complete loss of vision, although legally, a blind person may retain some vision. In contrast, visual impairment indicates a loss of vision such that there is an impact on daily living, which usually implies partial loss of vision.

There are many causes of visual impairment or blindness, and all parts of the **eye** (**cornea**, **retina**, lens, optic nerve) can be affected. The causes can be genetic (inherited eye diseases affecting both eyes), accidental (mechanical injury to the eyeball), inflammation of the eye tissues (uveitis), acute or extended exposure to harmful chemicals or radiation (acids, alkali, tobacco smoke, UV radiation), dietary imbalance (lack of vitamin A), medication (corticosteroids), systemic diseases (diabetes, renal failure), or simply an **aging process**.

The majority of visual impairments do not lead to blindness and are related to the refractive power of the lens and cornea. However, they are often troublesome and possibly restrictive in one's choice of job. A large number of people have problems focusing due to a variety of conditions. These can include near-sightedness (myopia), far-sightedness (hyperopia), astigmatism (inability to obtain a sharp focus), presbyopia (difficulty in accommodation), animetropia (unequal vision in each eye), and finally, aniseikonia can develop as a result of surgery, resulting in the images in two eyes being a different size and shape.

Keratoconus, which arises from the thinning of the central stromal layer of the cornea possibly due to abnormalities in collagen **metabolism**, affects the cornea and usually causes some impairment of vision, but can be treated.

Cataracts are the leading cause of blindness in developing countries and result from increased opacity of the lens, which interferes with vision. In developed countries, cataracts are mainly age-related or arise as a diabetic complication. They can also result from an environmental trauma (toxic substance exposure, radiation, mechanical or electrical injury), and a small proportion of cataracts are congenital, resulting from the over-proliferation of lens epithelial cells. Most cataracts can be removed by surgery, although in rare cases, post-operative bacterial infection (endophthalmitis) develops, which can compromise newly restored vision.

The eye tissues are all interconnected and a problem with one can cause a problem with another. The best example is the vitreous, which in addition to the accumulation of **calcium** and cholesterol leading to decreased transparency and subsequent impairment of vision, can shrink, leading to vitreal or retinal detachment. If the macular region is affected, some loss of visual acuity can follow and in any case 'floaters' or 'flushes' appear in the visual field.

Disorders and changes affecting the retina are the leading cause of blindness in developed countries. The abnormalities in the central retina can affect retinal pigment epithelium leading to blurry vision, or can affect the macular region (photoreceptors) leading to color misperception. Color blindness can also originate from the lack of one or more type of cones. Total color blindness (monochromatic vision) is very rare; most commonly various levels of single color deficits are found. The central vision can also be destroyed by hemorrhages of the neovascular vessels developing in the retina as a result of the aging process or diabetic retinopathy.

Irreversible loss of vision occurs due to optic nerve damage resulting from glaucoma. Glaucoma is caused by an increase in the intraocular pressure (IOP), which develops in the aqueous and is transmitted to the back of the eye, damaging the optic nerve and consequently causing severe reduction of visual field and loss of peripheral vision.

In the older population, complete or partial blindness is caused mainly by the aging process. Changes that lead to the destruction of vision are the non-enzymatic modifications in proteins, **lipids**, and **DNA**, which affect their structure, composition, and function. Glycosylation, carbamylation and deamination of the proteins, oxidation of proteins and lipids, UV induced damage to proteins and to DNA are the main culprits.

An accumulation of these changes leads to decreased transparency of the lens (cataracts) and retinal degeneration (age-related macular degeneration - AMD) both resulting in blindness or severe visual impairment. Most of the age-related changes are non-reversible, with the exception of the cataracts that can be surgically treated.

Research into the causes of blindness, especially glaucoma and AMD, is being undertaken by many groups in order to develop preventative measures and new treatment methods.

See also Eye and ocular fluids; Retina and retinal imaging; Vision: histophysiology of the eye; Sense organs: ocular (visual) structures

BLOBEL, GÜNTER (1936-)
German American cell biologist

Günter Blobel was awarded the 1999 Nobel Prize in physiology and medicine for the discovery that proteins have intrinsic signals that govern their transport and localization in the cell. Born May 21, 1936, in Waltersdorf/Silesia, Germany, he now lives in the New York, where he is a Howard Hughes Medical Institute investigator and faculty member at Rockefeller University.

Blobel's research has focused on the process by which newly made proteins are moved across the membranes of eukaryotic cell structures called organelles. Organelles are membrane bound regions within the cell. Accurate delivery of proteins to their target organelles is vital to proper cellular function. Malfunction in this delivery process can produce aberrant (abnormal) cells. The study of protein processing and transport has immediate bearing on diseases such as cystic fibrosis, **Alzheimer's** and **AIDS**.

In the average eukaryotic cell, there are thousands of different types of proteins, and about a billion proteins in total. Their lifetime is finite. Hence, constant replacement of proteins is occurring. The proteins are made by structures called ribosomes. The work of Blobel and his colleagues has shown how the newly made proteins make their way from the ribosome to their final destination.

The journey is determined by a process similar to the use of a postal code of a letter. A protein is made with a portion of its sequence acting as an organelle-specific address. This section is called the signal sequence. The signal sequence, typically about 15 **amino acids** long, is recognized by an organelle's surface.

The importance of the signal sequence has been exemplified by Blobel and his colleagues using an organelle called the endoplasmic reticulum. Binding of the signal sequence to its receptor stimulates the formation of a narrow channel across the endoplasmic reticulum membrane. The protein then passes through the channel to the interior of the endoplasmic reticulum. The channel is so narrow that the protein must pass through it in an unfolded configuration. A number of polypeptide-binding proteins assist in keeping the protein unfolded. After passage of the protein, the channel closes, and the pro-

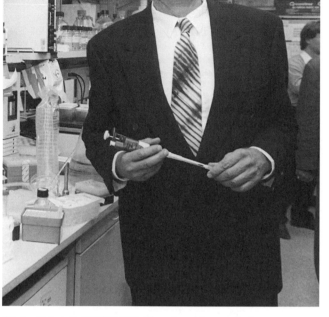

Günter Blobel. © AFP/CORBIS. Reproduced by permission.

tein emerging on the opposite side of the membrane can fold into its preferred configuration.

The membrane pores also have been shown to be capable of opening laterally into the membrane layer they are spanning, in order to facilitate the insertion of new material into the membrane. This is accomplished by the presence of a stop transfer sequence of the translocating protein, which stimulates the lateral opening of the pore. The translocating protein would move into the membrane, with the channel simultaneously closing in both dimensions.

Both of the above processes are unidirectional. Similar mechanisms as the above are thought to exist in the outer and inner membranes of chloroplasts and **mitochondria**.

Current research in Blobel's laboratory also explores the movement of proteins across nuclear pore complexes. These are large protein molecules found in openings in the membrane surrounding the nucleus of a cell. They are exceedingly complex, being comprised of upwards of 100 different proteins. The complexes are about 50-times wider than the pores across the endoplasmic reticulum, because the nucleus has two membrane layers instead of one. The pore complexes are so large that entities like folded up proteins and complexes of nucleic acid and protein can pass through. This so-called nucleocytoplasmic transport is a high-traffic process where 10 to 20 events occur per second. In contrast to the signal sequence-mediated translocation, nucleocytoplasmic transport can be bidirectional. Nucleocytoplasmic transport also differs from the endoplasmic reticulum situation in that there are various targeting factors operating. These factors are referred to as importins, exportins and transportins. Blobel refers to them as karyopherins, or kaps for short.

Work on *Saccharomyces cerevisiae*, Baker's yeast, has identified 14 separate kaps, which are thought to act to position proteins near the mouth of the channels. Currently, the precise mechanisms are being studied.

Another area of study concerns the finding the cytoplasm of cells contains filaments, which collectively are called the cytoskeleton. The cytoskeleton acts as a kind of highway for movement of proteins and **RNA** around the cytoplasm. Whether the nucleus contains an analogous network is as yet unknown. However, protein fibers within the nucleus have been observed by staining nuclei with antibodies against the protein. If the fibers are tracks involved in transport, they could open up a whole new level of biology, which Blobel calls intranuclear transport.

Aside from its clinical implications, Blobel's discoveries are fundamentally important to the emerging technologies of **gene therapy** and nuclear transplantation therapy.

See also Cell membrane transport; Cell structure; Protein metabolism; Protein synthesis

BLOCH, KONRAD EMIL (1912-2000)
German-born American biochemist

Konrad Bloch's investigations of the complex processes by which animal cells produce cholesterol have helped to increase scientific understanding of the **biochemistry** of living organisms. Bloch's research established the vital importance of cholesterol in animal cells, and helped lay the groundwork for further research into treatment of various common diseases. For his contributions to the study of the **metabolism** of cholesterol, he was awarded the 1964 Nobel Prize in physiology or medicine.

Bloch was born in the German town of Neisse (now Nysa, Poland) to Frederich (Fritz) D. Bloch and Hedwig Bloch. Sources list his mother's maiden name variously as Steiner, Steimer, or Striemer. After receiving his early education in local schools, Bloch attended the Technische Hochschule (technical university) in Munich from 1930 to 1934, studying chemistry and chemical engineering. He earned the equivalent of a B.S. in chemical engineering in 1934, the year after Adolf Hitler became chancellor of Germany. As Bloch was Jewish, he moved to Switzerland after graduating and lived there until 1936.

While in Switzerland, Bloch conducted his first published biochemical research. He worked at the Swiss Research Institute in Davos, where he performed experiments involving the biochemistry of phospholipids in tubercle bacilli, the **bacteria** that causes tuberculosis.

In 1936, Bloch emigrated from Switzerland to the United States; he would become a naturalized citizen in 1944.

With financial help provided by the Wallerstein Foundation, he earned his Ph.D. in biochemistry in 1938 at the College of Physicians and Surgeons at Columbia University, and then joined the Columbia faculty. Bloch also accepted a position at Columbia on a research team led by Rudolf Schoenheimer. With his associate David Rittenberg, Schoenheimer had developed a method of using radioisotopes (radioactive forms of atoms) as tracers to chart the path of particular molecules in cells and living organisms. This method was especially useful in studying the biochemistry of cholesterol.

Cholesterol, which is found in all animal cells, contains 27 carbon atoms in each molecule. It plays an essential role in the cell's functioning; it stabilizes cell membrane structures and is the biochemical "parent" of cortisone and some **sex hormones**. It is both ingested in the diet and manufactured by **liver** and intestinal cells. Before Bloch's research, scientists knew little about cholesterol, although there was speculation about a connection between the amount of cholesterol and other fats in the diet and **arteriosclerosis** (a buildup of cholesterol and lipid deposits inside the **arteries**).

While on Schoenheimer's research team, Bloch learned about the use of radioisotopes. He also developed, as he put it, a "lasting interest in intermediary metabolism and the problems of biosynthesis." Intermediary metabolism is the study of the biochemical breakdown of glucose and **fat** molecules and the creation of energy within the cell, which in turn fuels other biochemical processes within the cell.

After Schoenheimer died in 1941, Rittenberg and Bloch continued to conduct research on the biosynthesis of cholesterol. In experiments with rats, they "tagged" acetic acid, a 2-carbon compound, with radioactive carbon and hydrogen isotopes. From their research, they learned that acetate is a major component of cholesterol. This was the beginning of Bloch's work in an area that was to occupy him for many years—the investigation of the complex pattern of steps in the biosynthesis of cholesterol.

Bloch stayed at Columbia until 1946, when he moved to the University of Chicago to take a position as assistant professor of biochemistry. He stayed at Chicago until 1953, becoming an associate professor in 1948 and a full professor in 1950. After a year as a Guggenheim Fellow at the Institute of Organic Chemistry in Zurich, Switzerland, he returned to the United States in 1954 to take a position as Higgins Professor of Biochemistry in the Department of Chemistry at Harvard University. Throughout this period he continued his research into the origin of all 27 carbon atoms in the cholesterol molecule. Using a mutated form of bread mold fungus, Bloch and his associates grew the fungus on a culture that contained acetate marked with radioisotopes. They eventually discovered that the two-carbon molecule of acetate is the origin of all carbon atoms in cholesterol. Bloch's research explained the significance of acetic acid as a building block of cholesterol, and showed that cholesterol is an essential component of all body cells. In fact, Bloch discovered that all steroid-related substances in the human body are derived from cholesterol.

The transformation of acetate into cholesterol takes 36 separate steps. One of those steps involves the conversion of acetate molecules into squalene, a hydrocarbon found plentifully in the livers of sharks. Bloch's research plans involved injecting radioactive acetic acid into dogfish, a type of shark, removing squalene from their livers, and determining if squalene played an intermediate role in the biosynthesis of cholesterol. Accordingly, Bloch traveled to Bermuda to obtain live dogfish from marine biologists. Unfortunately, the dogfish died in captivity, so Bloch returned to Chicago empty-handed. Undaunted, he injected radioactive acetate into rats' livers, and was able to obtain squalene from this source instead. Working with Robert G. Langdon, Bloch succeeded in showing that squalene is one of the steps in the biosynthetic conversion of acetate into cholesterol.

Bloch and his colleagues discovered many of the other steps in the process of converting acetate into cholesterol. **Feodor Lynen**, a scientist at the University of Munich with whom he shared the Nobel Prize, had discovered that the chemically active form of acetate is acetyl coenzyme A. Other researchers, including Bloch, found that acetyl coenzyme A is converted to mevalonic acid. Both Lynen and Bloch, while conducting research separately, discovered that mevalonic acid is converted into chemically active isoprene, a type of hydrocarbon. This in turn is transformed into squalene, squalene is converted into anosterol, and then, eventually, cholesterol is produced.

In 1964, Bloch and his colleague Feodor Lynen, who had independently performed related research, were awarded the Nobel Prize in physiology or medicine "for their discoveries concerning the mechanisms and regulation of cholesterol and fatty acid metabolism." In presenting the award, Swedish biochemist **Sune Bergström** commented, "The importance of the work of Bloch and Lynen lies in the fact that we now know the reactions that have to be studied in relation to inherited and other factors. We can now predict that through further research in this field... we can expect to be able to do individual specific therapy against the diseases that in the developed countries are the most common cause of death." The same year, Block was honored with the Fritzsche Award from the American Chemical Society and the Distinguished Service Award from the University of Chicago School of Medicine. He also received the Centennial Science Award from the University of Notre Dame in Indiana and the Cardano Medal from the Lombardy Academy of Sciences the following year.

Bloch continued to conduct research into the biosynthesis of cholesterol and other substances, including glutathione, a substance used in **protein metabolism**. He also studied the metabolism of olefinic fatty acids. His research determined that these compounds are synthesized in two different ways: one comes into play only in aerobic organisms and requires molecular oxygen, while the other method is used only by anaerobic organisms. Bloch's findings from this research directed him toward the area of comparative and evolutionary biochemistry.

Bloch's work is significant because it contributed to creating "an outline for the chemistry of life," as E.P. Kennedy and F.M. Westheimer of Harvard wrote in *Science*. Moreover, his contributions to an understanding of the biosynthesis of cholesterol have contributed to efforts to comprehend the human body's regulation of cholesterol levels in **blood** and tis-

Konrad Bloch. *AP/Wide World Photos, Inc. Reproduced by permission.*

sue. Bloch served as an editor of the *Journal of Biological Chemistry,* chaired the section on metabolism and research of the National Research Council's Committee on Growth, and was a member of the biochemistry study section of the United States Public Health Service.

See also Biochemistry; Lipids and lipid metabolism

BLOOD

Blood is the fluid that circulates in the blood vessels of the body. Morphologically, it is a mesenchymatous **tissue** derived from the diffuse network of cells forming the embryonic **mesoderm**, that gives rise to blood and also blood vessels, the lymphatic system, and cells of the mononuclear phagocyte system. Blood cells (or hemocytes) are suspended in a fluid matrix, known as **plasma**. The chief non-cellular components of plasma are proteins (albumin and globulins), anions (mainly chloride and bicarbonate) and cations (mainly sodium, with smaller concentrations of potassium, **calcium** and magnesium).

Functionally, blood separates and protects the cells of the body from the external environment. The composition of blood and its **acid-base balance** are regulated within certain narrow limits, thereby ensuring that the body's internal environment is constant. As part of its function, blood mediates the exchange of material between the environment and the cells and tissues of the body. All vertebrates and many invertebrates depend on this exchange and also the transport of materials by the blood for the preservation of life.

Humans, as do other chordates, have a closed vascular system with blood and lymph circulating separately in, respectively, blood vessels and lymphatic vessels. In capillary beds, lymph is constantly produced by filtration through the capillary walls. Lymph flows directly into the venous part of the blood circulation via special lymph vessels (in mammals via the thoracic duct). Vertebrate lymph contains only colorless lymphocytes. In contrast, the blood plasma contains colorless **leukocytes**, thrombocytes, and specialized red cells known as **erythrocytes**. The erythrocytes contain the red, iron containing pigment called **hemoglobin**, which serves to transport oxygen. Oxygenated hemoglobin is light red, whereas deoxygenated hemoglobin is dark red. The ratio of total blood cell volume to blood plasma is expressed as the hematocrit, which is determined by centrifugation of anticoagulated fresh blood in a glass capillary tube. Normal hematocrit values are 42% for women and 46% for men.

Blood has many functions, the primary of which is transport. During respiration, oxygen bound to blood pigments is transported from the respiratory **organs** to the tissues and carbon dioxide is carried from the tissue to the respiratory organs. Blood also serves to transport nutrients absorbed during **digestion** or mobilized from reserve materials, to various sites in the body. In addition, it serves to transport waste products of **metabolism** to the excretory organs for removal from the body. Blood participates in the chemical regulation of the entire organism by the transport of **hormones** from their sites of secretion to their sites of action, and also by transporting other active substances such as **vitamins**.

Blood is important for maintaining the constant physical-chemical **equilibrium** necessary for the metabolic activity of all cells. In this respect, the inorganic ions of the internal milieu surrounding all cells are particularly important. Thus, normal cell function is only possible if the total concentration of all ions in the blood is held constant within certain limits. This constancy is achieved essentially by the regulatory activity of the excretory organs, the total salt concentration of human blood being around 300 mmol/l. In addition, acidic compounds like carbon dioxide and carboxylic acids constantly arise from the metabolic oxidation of food materials. These acidic compounds cannot be allowed to accumulate and decrease the pH of the body fluid, because cells function normally only if they are bathed in medium of a constant pH. This in turn means that the pH of the blood must be controlled within narrow limits.

The acid-base equilibrium of blood is regulated by the activity of the excretory organs (e.g., **kidneys**) and respiratory organs (carbon dioxide excretion by the **lungs**), and especially by the buffering capacity of the plasma hydrogen carbonate, which neutralizes the acids that spill over into the blood from the tissues. The normal pH of human blood is 7.3-7.4.

In mammals, blood also has an important function in maintaining a relatively constant body temperature. Heat pro-

duced by metabolism is transferred to the body surface by the blood where it is lost from the skin by radiation and conduction. Heat loss can be regulated by the contraction or dilation of blood vessels and these processes are under hormonal control.

Another very important function of blood is defense against **infection**. It carries defensive substances, including antibodies and antitoxins against foreign bodies (**antigens**) and toxic substances. Antigens are precipitated or coagulated, then cytosed and digested by colorless blood cells (leukocytes, phagocytes). Antibodies, or immunoglobulins, are special glycoproteins, produced in a major defensive reaction against foreign substances such as the components of **bacteria** and **viruses**, including also their endo and exo-toxins. In addition to antibodies, blood also contains agglutinins, which adsorb to foreign erythrocytes and cause them to clump.

Blood cells known as thrombocytes participate in the closing and sealing of wounds and other perforations in vessel walls by blood staunching or **blood coagulation**. When mammalian blood is shed, it congeals rapidly into a gelatinous clot of enmeshed fibrin threads, which trap blood cells and serum. Modern theories envision a series of reactions leading to the formation of insoluble fibrin from a soluble precursor, fibrinogen, through the mediation of a proteolytic enzyme called thrombin.

Many serum proteins show changes in their concentration during disease states. The determination of serum levels of certain proteins is, therefore, important for diagnostic purposes. The most striking abnormalities are observed in a malignant disorder of the plasma cell system called myeloma and in a disease of the lymphoid system called macroglobulinaemia Waldenström. The former is commonly associated with the presence of large amounts of proteins of the immunoglobulin class in the patient's serum while the latter is characterized by the presence of an increased amounts gamma macroglobulin. Other serum proteins show altered concentrations with inflammations, with tissue destruction or with increased loss of proteins in the urine due to kidney damage. Damage to **liver** cells often leads to impairment of **protein synthesis** and subsequently serum levels of proteins such as albumin and clotting factors decrease.

See also Blood coagulation and blood coagulation tests; Blood pressure and hormonal control mechanisms; Immune system; Immunity; Immunology

BLOOD CELL FORMATION • *see* HEMOPOIESIS

BLOOD COAGULATION AND BLOOD COAGULATION TESTS

Blood coagulation, also called blood clotting, is the term given to a complex series of reactions that results in the formation of a tightly knit meshwork at the site where a blood vessel has been severed. The ability of a body to stem the flow of blood following injury to blood vessel is vital to survival of the organism.

Coagulation is a complex process that involves blood cells called **platelets**, various coagulation factors that are circulating in the blood and blood vessels. When a blood vessel is cut open, the open end shrinks in size to limit the opening. Platelets are recruited to the site to plug the opening. Then, the various enzymatic coagulation factors in the blood **plasma** are activated, leading to the formation of fibrin. Fibrin is a strong, cross-linked protein that becomes an integral part of the clot.

The coagulation process is actually a cascade, where one reaction initiates another reaction, and so on, until the tightly cross-linked fibrin polymer is formed. The cascade of reaction consists of two pathways, the intrinsic pathway and the extrinsic pathway. One or the other can be initiated depending on the extent of an injury. The extrinsic pathway is triggered by **tissue** injury while the intrinsic pathway is not. The two pathways do converge at a common point, where a clotting factor called X is converted to Xa. The pathways are complex. There are over 20 proteins, or clotting factors, which must be activated in a set sequence for clotting to be successful. Other components, such as **calcium** ion, also need to be present for activation (change) of factors

The activated factor Xa, in turn, catalyzes the activation of a protein called prothrombin to form thrombin. Thrombin converts fibrinogen to fibrin. It is at this point that coagulation nears completion. The body tightly controls the concentration of thrombin, as too much circulating thrombin would cause coagulation of blood in areas other than at a cut. This could be lethal. A series of feedback mechanisms is operative throughout the coagulation cascade to put the system in readiness for rapid response to injury.

Another aspect of coagulation is the dissolution of the clot over time, as the injury to the vessel is repaired. Degradation of fibrin clots is due to an enzyme called plasmin. Plasmin results from several activation steps involving several proteins. The degradation pathway is also under tight control. The whole process of coagulation and degradation is called hemostasis. Defects in hemostasis cause bleeding disorders. Examples include hemophilia and von Willebrand Disease.

Coagulation tests can reveal whether the clotting process is proceeding normally or abnormally. The time for blood to clot at room temperature and the physical appearance of the resulting clot is one such test. The number and shape of red blood cells and platelets is determined by light scattering or the change in electrical resistance as blood streams through a small hole. Tests of platelet function, namely aggregation, are visually based tests. Assays also examine the performance of some of the clotting factors, including thrombin. In a test called template bleeding time, a pressure cuff is fitted onto an arm and inflated to restrict blood flow through arm vessels. A small cut is deliberately made and the time for clotting to occur is determined, normally 2–9 minutes. The efficiency of clot formation can be measured in a test tube by lowering a pin coated with platelets into activated whole blood. Fibrin strands interact with the platelets and as a clot forms the pin moves. The pattern of pin movement can be related to the progress of coagulation. Monitoring the pin's motion on a computer can be diagnostic of abnormalities in blood coagulation. The apparatus is called a Thromboelastograph.

See also Homeostatic mechanisms

BLOOD PRESSURE AND HORMONAL CONTROL MECHANISMS

The pressure exerted by the **blood** inside the **arteries** is termed blood pressure. Several factors are accountable for its levels, the **heart** rate, volume and viscosity of blood pumped per beat, force of the heartbeat, elasticity and resistance of vessel walls, and the resistance of the capillary bed (i.e., the network of capillary vessels that permeates tissues). **Capillaries** are minute blood vessels connecting the arterioles to either **veins** or lymphatic vessels. Other factors with a role in blood pressure levels are the balance between potassium and sodium levels, and the action of pressure-controlling **hormones**.

As the heart contracts and relaxes in a pulsatile rhythm, systole, the contraction of the left ventricle that ejects blood into the aorta, distending its walls, exerts a pressure level of 120 mm Hg, whereas diastole (i.e., momentary heart relaxation) has a level pressure of 80 mm Hg, due to the elastic recoil of arterial walls. As the blood flows along the systemic vessels, the pressure gradually falls to almost 0 mm Hg when the blood reaches the end of the cava vein, as is emptied into the right coronary atrium. In the capillary bed, the average pressure is about 17 mm Hg, varying from about 35 mm Hg in the arteriolar ends to 10 mm Hg near the venous ends.

The normal mean blood pressure is about 100 mm Hg. If for some reason the pressure level falls significantly below the mean level, it triggers a cascade of nervous **reflexes** that promotes contraction in the large venous reservoirs, and increases both the rate and force of cardiac contractions as well as induces a general constriction of small arteries (arterioles) throughout the body. Therefore, more blood is made available in the arterial tree. The substance released by the nervous system in the **smooth muscle** cells of the blood vessel walls is norepinephrine, a vasoconstrictor (vessel-constricting chemical).

However, if the causes leading to such low blood pressure persist and are no longer beneficial, other regulatory systems are activated, such as the secretion of pressure-controlling hormones. For instance, the **kidneys** control arterial pressure inducing changes in the volume of extra cellular fluids through the renin-angiotensin system. Renin is an enzyme released by the kidneys when the blood pressure is dangerously low. Renin helps to increase blood pressure through several ways. It promotes the release of angiotensin I, a mild vasoconstrictor, by entering the blood circulation. Angiotensin I is then enzymatically processed to become angiotensin II, a powerful vasoconstrictor that acts mainly on the small arterioles, and in a lesser way on veins. Arteriolar constriction increases the total vascular peripheral resistance, what elevates blood pressure in the arteries and the mild venal constriction helps the return of the blood to the heart. Angiotensin II also inhibits the elimination of sodium and water by the kidneys, thus augmenting the volume of extra cellular fluid. Even small elevations in the volume of extra cellular fluid can induce a blood pressure increase. Whereas the vasoconstriction by angiotensin II lasts for just a few minutes, the elevation of extra cellular fluid volume lasts for several hours or days, and is, therefore, the major effect in pressure elevation by the renin-angiotensin system. A smaller quantity of renin remains in the **renal system** where it elicits other regulatory functions.

Another pressure-controlling hormone, vasopressin, is secreted by the posterior pituitary gland, and also increases water reabsorption by the kidneys, and in turn, causes constriction of the blood vessels, thus elevating blood pressure. ADH (antidiuretic hormone), also secreted by the posterior pituitary, promotes the renal water reabsorption and vasoconstriction, increasing blood pressure. Aldosterone, secreted by the cortex of adrenal glands, is a hormone activated by angiotensin II that also increases sodium reabsorption and the elimination of potassium, which increases blood pressure as well. It also increases extra cellular fluid volume.

Arterial pressure is also regulated by vasodilator substances, such as bradykinin, acetylcholine, mineral ions, and endogenous nitric oxide, carbon dioxide and hydrogen gases. Bradykinin promotes arteriolar dilatation and increased capillary permeability. Mineral ions that induce vasodilatation are potassium and magnesium.

High intake of salt in the diet causes an increase in the volume of extra cellular fluid that ultimately leads to an increase in blood pressure above normal levels (i.e., hypertension). Although some cases of chronic high blood pressure are due to hereditary traits, a sodium-rich diet through childhood and young adulthood may also lead to chronic hypertension during later life. Hypertension, in turn, may lead to blood vessel ruptures in the **brain**, or strokes (cerebral infarct). Hypertension causes progressive destruction of the kidneys through successive ruptures of vessels in this organ, which leads to renal failure, an increased level of urea in the blood (uremia), and ultimately **death**.

Blood pressure measurements are usually done in millimeters of mercury (mm Hg) with a mercury manometer used for this purpose. The mercury manometer measures the force of blood against any unit area of the vessel wall. However, it is only useful for measuring stable pressures. When it is necessary to monitor unstable blood pressure, oscillating rapidly, electronic pressure transducers are utilized. These instruments convert pressure into electrical signals, that are recorded at high speed.

See also Angiology; Cardiac cycle; Cardiac muscle; Hormones and hormone action

BLUSHING

Blushing is a general term applied to a temporary erythema (redness) of the skin, especially the upper thorax, neck, and facial areas. The discoloration results from a brief infusion of **blood** following a rapid dilation of blood vessels in the affected area.

Erythema may occur as a result of **infection**, inflammation, or trauma to the skin. Erythema associated with blushing usually has an initiating emotional stimulus. Often blushing is

a characteristic response to a set of environmental circumstances. Although the emotional key that triggers the blushing response (e.g., embarrassment at public speaking, flattering comments, etc.) may vary from individual to individual, the physiological alterations in skin circulation remain essentially the same. In all forms of blushing, there is an increase in the diameter of the **capillaries** (the smallest diameter blood vessels that bridge the arterial and venous system and that are the sites of **vascular exchange**) underlying the affected skin area.

The flow of blood in the integumentary system (skin) is generally regulated by a number of neural and humoral mechanisms. Normally the blood flow through integumentary cutaneous **tissue** represents less than five percent of the total circulatory flow. With blushing and other infiltrations of cutaneous tissue, the portion of blood dedicated to cutaneous tissue may increase ten to twenty times the normal levels.

As part of the sympathetic response to stress, although visceral arterioles may constrict for a brief time, cutaneous vessels may dilate and cause blushing. Sympathetic pathways to integumentary glands cause the release of bradykinin (one of several peptides that act to regulate blood vessels and **smooth muscle**) that dilates the skin after compression. Vasodilators such as bradykinin expand capillary diameters and allow the lumen of the capillaries to become infused and engorged with blood.

Blushing also results in a localized elevation of temperature in the infused skin tissue. This reddening is often noticed on cold days as the skin of the usually exposed face and neck attempts to counter external cold temperature by increasing blood flow to the capillaries in the face and neck. A number of sensory receptors, including thermoreceptors, especially those found in the **hypothalamus** and **spinal cord**, regulate the dilation of arterioles in cutaneous tissue. These receptors provide a vital adaptation mechanism to allow the skin to respond to extremes of external heat and cold.

In contrast to the **physiology** of blushing, vasoconstrictor biochemicals force capillary vessels to contract and produce a pale skin tone in the overlying skin.

See also Autonomic nervous system; Blood pressure and hormonal control mechanisms; Fight or flight reflex; Inflammation of tissues; Integument

BONE HISTOPHYSIOLOGY

Bone is an organ composed of dense, hard, and flexible connective **tissue** that makes up the skeletal system of the human body. The extracellular matrix of bone is made up of several components. Protein, collagen in particular, provides bone with its hardness and flexibility. Minerals known as hydroxyapatite are composed primarily of **calcium** and phosphate with minuscule amounts of magnesium, sodium, potassium, and carbonate and confer hardness in bone. Proteoglycans make up the ground substance in bone and may help to mineralize the osseous tissue. Water is also found in the matrix of bone; some of it is bound to other molecules and some of it is free.

Four types of bone cells help to form and maintain the extracellular matrix of bone. The first type is unspecialized cells, called osteoprogenitor cells that eventually differentiate into osteoblasts and other connective tissue. Osteoblasts are the second type that is found on the outer surfaces of bone. Osteoblasts function in the formation and deposition of bone by secreting the materials that make up the matrix. Osteoblasts can mature to become the third type of cell known as osteocytes that are buried within the bone. They no longer secrete the matrix, but rather help to maintain the osseous tissue. Osteocytes can be found in cavities called lacunae. Processes of the osteocytes extend into tiny canals called canaliculi, allowing osteocytes to communicate with one another. Finally, osteoclasts reabsorb bone by the secretion of proteolytic **enzymes** and acids. Osteoclasts originate in the bone marrow, where they are derived from monocytes. Together, the bone cells and extracellular matrix form the osseous tissue known as bone.

Bone is composed of two types of osseous tissue, spongy cancellous tissue and compact tissue. Compact tissue surrounds cancellous tissue and is enclosed in a membranous periosteum. Compact (cortical) bone is made up of an intricate network of canals within the osseous tissue. Haversian canals, also known as central canals, run the long axis of the bone and contain **blood** vessels and nerves. Volkmann's or perforating canals run parallel to and connect the Haversian canals. Blood vessels and nerves can also be found in the Volkmann's canals. Surrounding the central canals are concentric (circular) tubes of matrix called concentric lamellae. Spaces between the concentric lamellae are called lacunae and house the osteocytes. Branching out from the lacunae are the canaliculi that connect the lacunae to each other and to the Haversian canals. Together these structures form a complex system known as the Haversian system, or osteon, that enables the bone to receive oxygen and nutrients while ridding itself of wastes. It also facilitates transport between cells. Interstitial lamellae are pieces of old osteons that were broken down during bone remodeling and fill the spaces between the individual osteons.

Cancellous bone consists of an irregular, porous meshwork of bone called trabeculae. The spaces between the plates of trabeculae are filled with bone marrow. The trabeculae consist of extracellular matrix as well as osteocytes that are found in lacunae. Like compact bone, the osteocytes have projections that extend into canaliculi. However, unlike compact bone, the osteocytes are not buried deep; they receive nutrients directly from the blood circulating in the spaces containing bone marrow. Bone marrow forms the precursors to the cells found in the blood.

Calcium, the most abundant mineral in the body, is stored in the bones and teeth and is present in all of the body's cells and fluids. Calcium is necessary for the conduction of nerve impulses, contraction of muscle, clotting of blood, function of enzymes, beating of the **heart**, and respiration. Even small changes in blood calcium level can be deadly. Because 99% of the calcium in the body is stored in bone, bone plays a major role in maintaining calcium homeostasis in the blood.

The calcium levels in the bone are constantly changing to compensate for the need of calcium in the blood and other parts of the body. When blood calcium levels increase, calcitonin, a hormone that decreases blood calcium levels, is secreted by the thyroid gland. Calcitonin works to increase the activity of bone and also functions in its deposition. Additionally, calcitonin acts to slow down the rate at which bone is reabsorbed by inhibiting the activity of osteoclasts. On the other hand, when blood calcium levels are low, parathyroid hormone is released from the parathyroid gland and increases the activity and number of osteoclasts. Osteoclasts reabsorb bone, making calcium available to the blood. Parathyroid hormone also acts to facilitate the recovery of calcium from the urine and stimulates formation of calcitriol to further increase blood calcium levels.

As people age, the skeletal system is faced with two adverse effects. The first effect is demineralization of bone, the loss of calcium and other minerals, which usually begins between the age of 30–35 in females and 50–60 years of age in males. Women can lose 40% of the bone minerals in their body and men as much as 55% by the time they reach the age of 80. Secondly, a decreased rate in **protein synthesis** decreases the amount of collagen available to provide bone with hardness and tensile strength. In both conditions, bones become brittle and can fracture easily leading to conditions such as **osteoporosis**, osteopenia, osteoarthritis, and osteomalacia.

See also Bone injury, breakage, repair, and healing; Bone reabsorbtion; Hormones and hormone action; Skeletal system overview (morphology)

BONE INJURY, BREAKAGE, REPAIR, AND HEALING

Bone is a living tissue that is encased in a rigid, stress-bearing frame. Both individually and together as the skeleton of vertebrate creatures, bone provides the scaffolding upon which the body is constructed.

As bone is rigid, bone is subject to partial or total breakage and other types of damage. But bone has the capacity to repair such damage. The healing process is a dynamic balance between bone deposition, resorption and remodeling.

Bone damage takes several forms. A fracture, where bone is disrupted, is one class of bone damage. A fracture can be due to injury (traumatic fracture) or due to disease (pathologic or spontaneous fracture). Fractures can cause the broken part of the bone to pierce through the skin (open fracture), or the skin can remain intact (closed fracture).

Fractures can be further classified with respect to the pattern of injury. A transverse fracture occurs across the long axis of a bone. An oblique fracture runs more along the long axis. In a leg bone, for example, an oblique fracture would be longer than a transverse fracture. A spiral fracture runs more along a long axis, but the fracture is not in a straight line. Another type of fracture, called a fissured fracture, is an incomplete break in the plane of a bone's long axis. A greenstick fracture is also an incomplete break that occurs on only

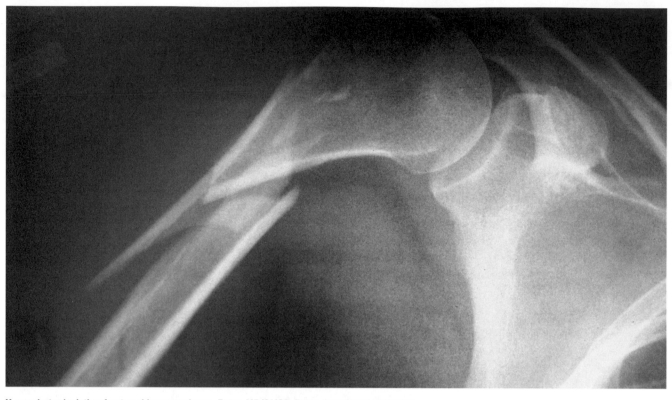

X-ray photo depicting fractured humerus bone. *Bates MD/CMSP. Reproduced by permission.*

one side of a bone. Finally, a comminuted fracture is a shattering of a bone into many pieces. This is the worst type of fracture, and the bone may not fully heal.

Breakage of bone can also occur because of disease, rather than because of physical force. A common example is **osteoporosis**, where too much of a compound called hydroxyapatite is removed from the bone. **Calcium** loss causes bones to become brittle and more easily subject to breakage. Osteoporosis can happen in the elderly or in younger people with a dysfunctional parathyroid or thyroid.

Repair and healing of bones is promoted by immobilizing the damaged region, as with a cast. The damaged areas of the bone are positioned near one another so that the body's repair mechanisms can act. Upon injury, bleeding of the bone occurs, because the inside of bone is rich in **blood** vessels. A blood clot forms and specialized cells called osteoclasts and phagocytic cells are recruited to the site to clear away bone fragments and blood clot debris. Fibroblasts invade the area between the edges of the broken bone and lay down a material called **fibrocartilage**. Then, other cells called osteoblasts convert the fibrocartilage to bone.

Surgical assistance in healing bone takes the form of a bone graft, where bone is transplanted from another part of the person's own body (autograft) or from another member of the same species, living or dead (allograft). Both types of bone grafts are commonly used in spine surgery.

Modern technology is enhancing the body's ability to repair damaged bone and enhance the success of grafts. Living bone substitutes that can attach to and 'fill in' the fracture, similar to how plastic wood can fill in chips and cracks in wood. In another approach, a cocktail of high concentrations of natural growth factors (**platelets**, white blood cells) and proteins are delivered to the site of bone damage. The cocktail acts as a kind of glue, enabling bone grafting to be more successful.

See also Appendicular skeleton; Blood coagulation and blood coagulation tests; Skeletal system overview

BONE MARROW • *see* BONE HISTOPHYSIOLOGY

BONE REABSORPTION

Bone reabsorption is a process whereby established bone is reabsorbed into its constituent parts. This process occurs throughout life, and is a normal part of growth and aging.

The process of bone or osteo-reabsorption is also called bone turnover. Specialized cells known as osteoclasts are responsible for the reabsorption. Other cells, known as osteoblasts, are responsible for laying down new bone.

Osteoclasts are located in notches or indentations, called Howship's lacunae, that are found at the surface of bones. The cells themselves contain **ATP** producing bodies called **mitochondria** and lysosomes. The lysosomes can be thought of as bags. They contain **enzymes**, in particular an enzyme known as acid phosphatase, that acts to hydrolyze, or break down, the bone collagen. Osteoclasts also have a ruffled appearing border. It is believed that the ruffled border promotes interaction of the osteoclast with the bone, giving the collagen-dissolving acid phosphatase the time and contact exposure needed to act on the bone. The dissolution of the bone requires energy, which is supplied by the ATP produced by the mitochondria of the osteoclasts.

As humans grow, there is more bone formation than reabsorption—in other words more osteoblast activity than osteoclast activity—hence bones become bigger and stronger. Once growth has stopped, at around 25 years of age, there should be a balance of bone formation and reabsorption. However, over 40 years of age, the rate of reabsorption can become greater than the rate of bone formation. If this imbalance is pronounced, the bone mass in the body is reduced as the bones are eaten away, and bones can become brittle and break easily. This condition is called **osteoporosis**.

The trigger for the imbalance of osteoclast and osteoblast activities is not known. It is known that the loss of osteoblast activity is also coincident with a loss of **calcium**. Also, exercise, or rather the lack of it, is important, as if bones are not stressed, bone reabsorption increases.

See also Bone histophysiology; Ossification

BORDET, JULES (1870-1961)
Belgian physician

Jules Bordet was a pioneer in the field of **immunology**. It was his research that made clear the exact manner by which serums and antiserums act to destroy **bacteria** and foreign **blood** cells in the body, thus explaining how human and animal bodies defend themselves against the invasion of foreign elements. Bordet was also responsible for developing complement fixation tests, which made possible the early detection of many disease-causing bacteria in human and animal blood. For his various discoveries in the field of immunology, Bordet was awarded the Nobel Prize in physiology or medicine in 1919.

Jules Jean Baptiste Vincent Bordet was born in Soignies, Belgium, a small town situated twenty-three miles southwest of Brussels. He was the second son of Charles Bordet, a schoolteacher, and Célestine Vandenabeele Bordet. The family moved to Brussels in 1874, when his father received an appointment to the École Moyenne, a primary school. Jules and his older brother Charles attended this school and then received their secondary education at the Athénée Royal of Brussels. It was at this time that Bordet became interested in chemistry and began working in a small laboratory that he constructed at home. He entered the medical program at the Free University of Brussels at the age of sixteen, receiving his doctorate of medicine in 1892. Bordet began his research career while still in medical school, and in 1892 published a paper on the adaptation of **viruses** to vaccinated organisms in the *Annales de l'Institut Pasteur* of Paris. For

this work, the Belgian government awarded him a scholarship to the Pasteur Institute, and from 1894 to 1901, Bordet stayed in Paris at the laboratory of the Ukrainian-born scientist Elie Metchnikoff. In 1899, Bordet married Marthe Levoz; they eventually had two daughters, and a son who also became a medical scientist.

During his seven years at the Pasteur Institute, Bordet made most of the basic discoveries that led to his Nobel Prize of 1919. Soon after his arrival at the Institute, he began work on a problem in immunology. In 1894, Richard Pfeiffer, a German scientist, had discovered that when cholera bacteria was injected into the **peritoneum** of a guinea pig immunized against the **infection**, the pig would rapidly die. This bacteriolysis, Bordet discovered, did not occur when the bacteria was injected into a non-immunized guinea pig, but did so when the same animal received the antiserum from an immunized animal. Moreover, the bacteriolysis did not take place when the bacteria and the antiserum were mixed in a test tube unless fresh antiserum was used. However, when Bordet heated the antiserum to 131°F (55°C), it lost its power to kill bacteria. Finding that he could restore the bacteriolytic power of the antiserum if he added a little fresh serum from a non-immunized animal, Bordet concluded that the bacteria-killing phenomenon was due to the combined action of two distinct substances: an antibody in the antiserum, which specifically acted against a particular kind of bacterium; and a non-specific substance, sensitive to heat, found in all animal serums, which Bordet called "alexine" (later named "complement").

In a series of experiments conducted later, Bordet also learned that injecting red blood cells from one animal species (rabbit cells in the initial experiments) into another species (guinea pigs) caused the serum of the second species to quickly destroy the red cells of the first. And although the serum lost its power to kill the red cells when heated to 55 degrees centigrade, its potency was restored when alexine (or complement) was added. It became apparent to Bordet that hemolytic (red cell destroying) serums acted exactly as bacteriolytic serums; thus, he had uncovered the basic mechanism by which animal bodies defend or immunize themselves against the invasion of foreign elements. Eventually, Bordet and his colleagues found a way to implement their discoveries. They determined that alexine was bound or fixed to red blood cells or to bacteria during the immunizing process. When red cells were added to a normal serum mixed with a specific form of bacteria in a test tube, the bacteria remained active while the red cells were destroyed through the fixation of alexine. However, when serum containing the antibody specific to the bacteria was destroyed, the alexine and the solution separated into a layer of clear serum overlaying the intact red cells. Hence, it was possible to visually determine the presence of bacteria in a patient's blood serum. This process became known as a complement fixation test. Bordet and his associates applied these findings to various other infections, like typhoid fever, carbuncle, and hog cholera. August von Wasserman eventually used a form of the test (later known as the Wasserman test) to determine the presence of syphilis bacteria in the human blood.

Already famous by the age of thirty-one, Bordet accepted the directorship of the newly created Anti-rabies and Bacteriological Institute in Brussels in 1901; two years later, the organization was renamed the Pasteur Institute of Brussels. From 1901, Bordet was obliged to divide his time between his research and the administration of the Institute. In 1907, he also began teaching following his appointment as professor of bacteriology in the faculty of medicine at the Free University of Brussels, a position that he held until 1935. Despite his other activities, he continued his research in immunology and bacteriology. In 1906, Bordet and Octave Gengou succeeded in isolating the bacillus that causes whooping cough in children and later developed a vaccine against the disease. Between 1901 and 1920, Bordet conducted important studies on the coagulation of blood. When research became impossible because of the German occupation of Belgium during World War I, Bordet devoted himself to the writing of *Traité de l'immunité dans les maladies infectieuses* (1920), a classic book in the field of immunology. He was in the United States to raise money for new medical facilities for the war-damaged Free University of Brussels when he received word that he had been awarded the Nobel Prize. After 1920, he became interested in bacteriophage, the family of viruses that kill many types of bacteria, publishing several articles on the subject. In 1940, Bordet retired from the directorship of the Pasteur Institute of Brussels and was succeeded by his son, Paul. Bordet himself continued to take an active interest in the work of the Institute despite his failing eyesight and a second German occupation of Belgium during World War II. Many scientists, friends, and former students gathered in a celebration of his eightieth birthday at the great hall of the Free University of Brussels in 1950. He died in Brussels in 1961.

See also Immune system; Immunity; Immunology

BOVERI, THEODOR HEINRICH (1862-1915)

German zoologist

In exploring fundamental issues in heredity and development, Theodor Boveri carried out intriguing studies of the role of the nucleus and **chromosomes** in development. Boveri, a pioneer in the field of cytology, developed the theory of the genetic continuity of the chromosomes and analyzed the development of spermatozoa and ova.

Boveri was born in Bamberg, Bavaria (now Germany) and earned his M.D. (1885) from the University of Munich in 1885. As a medical student, Boveri's special interest was **anatomy**. He studied zoology with Richard Hertwig (1850-1937). In 1887, Boveri was appointed lecturer in zoology and **comparative anatomy** at Munich. From 1885 to 1893, Boveri carried out cytological research, including his landmark studies of the chromosomes, at the Zoological Institute in Munich. In 1893, Boveri became professor of zoology and comparative anatomy at the University of Wurzburg, where he remained until his death.

In 1887, Boveri described the development of an unfertilized egg, including the formation of polar bodies (small cells that result from the division of an unfertilized egg). His next major report described finger-shaped structures that appeared in the nuclei of eggs of the roundworm *Ascaris* during early cleavage stages. Boveri concluded that these entities were chromosomes. At the time, most scientists thought that chromosomes were part of the nucleus that only appeared during mitosis (**cell division**). Boveri's work with roundworm eggs was instrumental in proving that chromosomes are separate, continuous entities within the nucleus of a cell. In his next important publication Boveri provided convincing evidence that during **fertilization** the ovum and **sperm** cell contribute equal numbers of chromosomes to the zygote. Boveri's studies clarified ideas about the nature of cell division and fertilization previously suggested by the Belgian embryologist Edouard van Beneden (1846-1910).

In 1876, Van Beneden discovered the subcellular structure for which Boveri coined the term "centrosome." Boveri and van Beneden independently proved that this structure does not disappear at the end of the process known as mitosis (cell division). Boveri proved that the centrosome is the division center for a dividing egg cell. In most cells, arrays of microtubules grow from the centrosome, a special area of the cytoplasm near the nuclear envelope to the **plasma** membrane. Just before cell division, the centrosome divides and the two halves become the poles of the mitotic spindle, which plays a key role in separating the chromosomes and other cell components into the daughter cells during mitosis.

Theodor Boveri was the first to perform experiments in which the enucleated egg of one species is fertilized with sperm from a different species. This process is known as merogony. These experiments suggested that chromosomes are influenced by the cytoplasm surrounding the nucleus.

In an intriguing series of experiments on sea urchin embryos with abnormal numbers of chromosomes, Boveri investigated the behavior of the chromosomes in germ cell formation and embryological development. Boveri used this system to test the hypothesis that all the chromosomes were essentially identical. He demonstrated that abnormal combinations of chromosomes led to aberrations in the development of doubly fertilized eggs. Although Boveri could not identify the role of individual chromosomes, his observations proved that the chromosomes were not functionally equivalent. Boveri's experiments on double-fertilized eggs inspired Walter S. Sutton's (1877-1916) studies of the chromosomes of grasshoppers.

Thus, Boveri's work, along with that of Sutton, Nettie M. Stevens's (1861-1912), and Edmund B. Wilson (1856-1939) supported the idea that the hereditary factors must be on the chromosomes and that chromosomes were morphologically and functionally distinct entities. Both Boveri and Sutton argued that, when taken together, cytological observations and Mendelian breeding studies supported the hypothesis that the chromosomes carried the Mendelian factors. This concept came to be known as the Boveri-Sutton theory. Although the Boveri-Sutton theory seemed to explain many aspects of both cytology and heredity, it did not gain immediate and universal approval. As the author of *The Cell in Development and*

Heredity, and mentor of many prominent scientists, Wilson was instrumental in converting several skeptical scientists into advocates of the chromosome theory.

See also Cell cycle and cell division; Chromosomes; Genetics and developmental genetics

BOWELS AND BOWEL MOVEMENTS • *see* ELIMINATION OF WASTE

BRACHIAL PLEXUS

The brachial plexus is a neural plexus (a grouping and branching of nerves) located deep in the neck, shoulder, and maxilla region that is responsible for the proper innervation and control of the muscles of the shoulder, upper chest, and arms (upper limbs). Because of the complexities of branching nerve roots, trunks, and cords of the brachial plexus, injuries to the brachial plexus region often cause loss or impairments of function at distant muscle groups.

The nerves forming the brachial plexus come from spinal nerves, specifically the last four cervical and first thoracic spinal nerve. The cervical nerves are designated C5, C6, C7, and C8. Although there are only seven cervical vertebrae, there are eight cervical nerves. The thoracic spinal nerve is designated as T1. Spinal nerves result from the unification of dorsal and ventral spinal roots in the intervertebral foramen. The spinal nerves then divide into anterior and posterior primary rami. The spinal verves comprising the brachial plexus are from the anterior rami divisions.

The spinal nerves C5 and C6 fuse to form the upper trunk of the brachial plexus. Spinal nerve C7 becomes a middle trunk of the brachial plexus. Spinal nerves C8 and T1 join to form a lower trunk of the brachial plexus.

At about the level of the clavicle, the trunks of the brachial plexus divide to form anterior and posterior nerves. The anterior nerve divisions from the upper and middle trunks form the lateral cord. The anterior branch nerves from the lower trunk form a medial cord. The posterior divisions band together to form the posterior cord of the brachial plexus. As the cords continue, they come to lie lateral, medial, and posterior to the axillary artery, and it is this anatomical relationship from which each cord derives its name.

The brachial plexus has many branches from its root, trunk, and cord regions.

Above the level of the clavicle are the supraclavicular branches that branch directly from the spinal nerve roots. These nerves include the long thoracic nerve (also known as the nerve of Bell) that forms from C5, C6, and C7 spinal nerves. The long thoracic nerve innervates the serratus anterior muscle. Injuries to this nerve may result in an inability to push an object because of the loss of function to the agonist muscles in the scapular region. Also branching from the brachial plexus in the supraclavicular region are the dorsal scapular nerve, and nerves to the subclavius.

The medial brachial cutaneous nerve, and the median pectoral nerve have their origins in the inferior trunk of the brachial plexus.

The thorcodorsal nerve and middle subscapular nerves branch from the posterior cord. The axillary nerve and the upper and lower subscapular nerves also trace back to the posterior cord. The axillary nerve ultimately continues on as the radial nerve.

The medial root of the median nerve comes from the medial cord of the brachial plexus. The medial nerve ultimately continues down the arm as the ulnar nerve.

Injury to the axillary nerve branching from the brachial plexus can result in a paralysis of the deltoid and a loss of sensation in the skin in the scapular area. Injuries to the deltoid region can also result in the ability to abduct the arm.

Injury to the median nerve of the brachial plexus can cause a loss of flexion of the fingers. This loss of flexion results in a loss of the critical ability to oppose the thumb with individual fingers. Median nerve impairment can also result in a loss of range of motion of the arm. Individuals who sustain median nerve injury causing loss of index finger flexion may develop an index finger that "points" or remains extended. Because the median nerve ultimately passes through the carpal tunnel of the wrist, injuries or inflammation of the wrist (e.g., **carpal tunnel syndrome**) can result in **pain** and loss of feeling well away from the wrist itself.

Injuries to the radial nerve that branches from the brachial plexus, injuries commonly associated with injuries to the humerus bone of the arm, often develop into an excessive or permanent flexion of the wrist or flaccid wrist drop.

Impairment of the ulnar nerve derived from the brachial plexus can result in paralysis or wrist extension. Severe ulnar nerve injuries can result in the hand taking on a claw-like appearance with the fingers spread out and unable to flex.

A protective neurovascular sheath made up of **tascia** protects the brachial plexus. Microscopic and **imaging** examinations establish that the **connective tissues** invest and help partition the brachial plexus.

See also Brachial system; Motor functions and controls; Nerve impulses and conduction of impulses; Nervous system overview; Nervous system, embryological development; Neural damage and repair; Neural plexi; Neurology; Upper limb structure

BRACHIAL SYSTEM

Arteries are the conduits for the delivery of oxygenated **blood** to tissues and cells. The brachial arterial system delivers blood from the **heart** to the arms and associate upper limb musculature. Each arm contains a brachial artery, the major artery of the brachial system. Each brachial artery runs from the portion of the arm located just below the shoulder down to just below (distal to) the bend of the elbow. There, it branches into the radial and ulnar arteries. The radial artery continues the delivery of blood to the hands.

Along the brachial artery are branching arteries, which carry the blood to all regions of the arms. The arteria profunda brachii (also known as the superior profunda artery) is large. It branches off just below the border of the shoulder and the arm and follows the route of the radial nerve laterally to the sides of the arm. Two other branches of this artery, the middle collateral branch and the radial collateral branch, run to the meaty back portion of the upper arm.

Another branch of the brachial artery is called the nutrient artery. It branches off about the middle of the arm. The superior ulnar collateral artery (also called the inferior profunda artery) branches off from the brachial artery a bit below the middle of the arm. This artery is smaller in diameter. It runs down the arm from the elbow. The inferior ulnar collateral artery (also known as the anastomotica magna artery) branches off just above the elbow. Smaller sub-branches radiate upwards and downwards from the inferior artery. Finally, three to four muscular branches (also known as rami musculares) radiate outward into the various muscles groups along the length of the arm.

The brachial artery is one of the arteries used to measure a **pulse**. When the biceps muscle is flexed, a groove can be felt just below the armpit on the inside of the arm. It is at this location that the pulse of blood through the brachial artery can be detected.

Injury to the elbow can impinge on the brachial artery. For this reason, it is important to quickly treat an elbow injury, especially blunt or crushing trauma, to ensure that blood flow to the lower arm is not being compromised. Interruption of blood flow even for a short while can lead to cell **death (necrosis)** and loss of function of tissues.

See also Pulse and pulse pressure; Upper limb structure

BRAIN • *see* CEREBRAL MORPHOLOGY

BRAIN HEMISPHERES • *see* CEREBRAL HEMISPHERES

BRAIN: INTELLECTUAL FUNCTIONS

The **cerebral cortex**, or **gray matter**, is the most external layer of the brain of vertebrates, constituting the largest portion of the nervous system. It is connected to the **thalamus**, forming the thalamocortical system, which, except by some sensory olfactory pathways to the cortex, constitutes the major two-way unit between sensory perception and intellectual functions in humans. The cortex is divided into several cortical areas, each responsible for separate functions, such as planning of complex movements, memory, personality, elaboration of thoughts, word formation, language understanding, motor coordination, visual processing of words, spatial orientation, and body spatial coordination.

Sensations received from peripheral sense **organs** by the primary motor and sensory cortical areas are detected as specific sensations (visual, somatic, auditory). The secondary sensory areas are activated in the process of recognition of these signals, and begin the analysis of sensory signals, such as the interpretation of color, shape, and texture. As for somatic sensations, the secondary motor areas recognize patterns of motor activity. However, large areas of the cortex, termed association areas, have a more flexible role than the rigid primary and secondary areas mentioned so far. The association areas of the cortex receive and simultaneously analyze multiple sensations received from several regions of the brain, such as the motor and sensory cortical areas, and subcortical areas such as the limbic system, responsible for animal behavior, emotions and motivation.

The association areas are identified as two major areas, the parieto-occiptotemporal, and the prefrontal association areas. The parieto-occiptotemporal area occupies the large parietal and occipital cortical space, bordered by the somatosensory, auditory, and visual cortices. The second association area is the anterior section of the cortex, bordered by the olfactory lobe, the limbic, and the motor areas. The parieto-occiptotemporal cortex is divided in the following sub areas, in accordance with their respective functional roles: analysis of special body coordinates; language understanding (also known as Wernicke's area); reading; naming of objects. These functions are essential for the learning process and intelligent activity. The prefrontal cortex works in close association with the motor cortex and the parieto-occiptotemporal association area, from which it receives pre-analyzed sensory information. From these data, it processes the planning of sequential and complex patterns of motor-coordinated activity and the elaboration of thoughts. The prefrontal area contains a special region termed Broca's area, where a neural circuitry, responsible for word formation, works in close relationship with the Wernicke's area. Another region of association activity, the limbic association area, is localized in the limbic system (the source of emotional drives, behavior, and motivation) and stimulates the other association areas of the brain.

The brain is divided in two large lobes, interconnected by a bundle of nerves, the corpus callosum. It is now known that in approximately 95% of all people, the area of the cortex in the left hemisphere can be up to 50% larger than in the right hemisphere, even at birth. Both Wernicke's and the Broca's areas are usually much more developed in the left hemisphere, which gave origin to the theory of left hemisphere dominance. The motor area for hand coordination is also dominant in nine of out ten persons, accounting for the predominance of right-handedness among the population. However, the nature of such hemisphere dominance is associated with the linguistic abilities, which is just one of the many aspects linked to intellectual activity. The non-dominant hemisphere is responsible for other important forms of intelligence as well. Moreover, the interpretative and motor areas of the brain do receive sensory information from both the right and the left hemispheres through the corpus callosum. Studies show that the non-dominant hemisphere plays an important role in musical understanding, composition and learning, perception of spatial relations, perception of visual and other esthetical patterns, understanding of connotations in verbal speeches, perception of voice intonation, identification of other's emotions and mood, and body language.

Memory is another fundamental aspect of brain intellectual function, because the learning process and the ability to elaborate abstract thoughts into an ordered and coherent sequence depend upon the memory-processing centers of the brain. Memory-processing centers include the prefrontal cortex for short-term memory, and the intermediate and long-term memory centers of the **hippocampus**. The hippocampus is an elongated portion of the cerebral cortex belonging to the limbic system, which is responsible, among other functions, by the storage of verbal and symbolic types of memory into both intermediate and long-term memory.

See also Central nervous system (CNS); Gray matter and white matter; Hypothalamus

BRAIN STEM FUNCTION AND REFLEXES

The **brain** stem is an extension of the **spinal cord** into the cranial cavity consisting of the medulla, **pons**, and mesencephalon. The brain stem contains the motor and sensory nuclei that control sensorial and motor functions of the face and the head, whereas the spinal cord controls the movements of the rest of the body below the neck. The brain stem also accounts for many other involuntary and autonomic functions, such as the control of respiration, **eye** movement, **equilibrium**, gastrointestinal function, cardiovascular system, and several stereotyped body movements as well.

Higher brain neural centers (i.e., motor cortical areas, **cerebellum, hypothalamus,** and limbic system) send signals to the brain stem that trigger or change specific control functions throughout the body.

The brain stem's reticular nuclei and vestibular nuclei support the body against gravity by controlling the degree of **muscle contraction** through the excitatory-inhibitory interplay between pontine and medullary reticular nuclei. The pontine nuclei transmit excitatory signals to the spinal cord, through the pontine reticulospinal tract that ends on the medial anterior motor **neurons**, which excite the axial skeletal muscles of the **vertebral column** and the limbs to contract, thus supporting the body against gravity. Conversely, the medullary reticular system sends inhibitory signals to the anterior motor neurons through the medullary reticulospinal tract, thus counterbalancing the excitatory signals and avoiding excessive muscle tension.

These two systems are modulated (i.e., excited or inhibited) through signals received from the cortical motor areas and other related centers. However, children born without cerebral structures above the mesencephalic region are still able to perform several stereotyped movements, such as suckling, yawning, stretching, crying, following objects with the eyes and head, sitting, and sucking their own fingers. This shows that such stereotyped motor functions are controlled by the brain stem.

Neural areas of the medullary, pontine and mesencephalic structures of the brain stem also control the **auto-** nomic nervous system and the involuntary functions of the body, such as **heart** rate, respiratory rate (involuntary functions), gastrointestinal movements (peristaltic motility), **blood** pressure, gastrointestinal gland secretion, body heat control, and urinary bladder contraction (autonomic functions).

The autonomic centers of the brain stem are controlled by the hypothalamus and by signals received from the cerebrum as well. The posterior hypothalamus, for instance, activates the medullary cardiovascular control areas to increase blood pressure, whereas other hypothalamic centers may control body heat, modulate the levels of salivation, stimulate urinary bladder emptying, and gastrointestinal activity (secretion of gastric acidy, motility) through the stimulation of specific neural center in the brain stem.

Other autonomic control functions include the regulation of the pancreatic secretion reflex, perspiration, gallbladder release of **bile**, increase of blood flow to active muscles involved in rapid motor activity, hepatic glycolysis, augmented glucose concentration in the blood, and increased rate of **blood coagulation**. These **reflexes** are activated through the mediation of the **sympathetic nervous system**. For example, as a response to a rage signal elicited mainly by the hypothalamus. The signals are then transmitted through the reticular centers of the brain stem to the spinal cord, which causes a massive sympathetic discharge that triggers the flight-or-fight reflex, followed by a mobilization of the autonomous reflexes already mentioned. This pathway of excitatory signals from the sympathetic system to the brain stem areas and the spinal cord is known as sympathetic alarm reaction and it is activated through mental and/or physical stress as well.

See also Central nervous system

BRAIN, VASCULAR SUPPLY • *see* CIRCLE OF WILLIS

BRAIN WAVES • *see* ELECTROENCEPHALOGRAPH (EEG)

BREASTS

Breasts are located on the anterior aspect of the chest and are composed of glandular **tissue, adipose tissue,** and ligamentous tissue. In women, breasts function to produce and secrete milk after childbirth, a process called **lactation**. The accessory reproductive glands, or **mammary glands**, are actually modified **sweat glands** responsible for lactation in women; they are present in men, but tend to be underdeveloped.

Supportive fibrous tissue, called Cooper's ligaments, is found throughout the breasts. These ligaments partially cover the lobes of glandular tissue, helping to bind the breasts together and give them shape. The nipples are externally located and serve as a passage for the milk to the environment. Each nipple is innervated by many nerves and contains **smooth muscle**. The nerves cause the nipples to be sensitive to external stimuli while the muscle fibers allow the nipples to become erect. A pigmented area called areola surrounds each nipple. The areola have small, raised bumps that contain oil-producing glands, called Montgomery's glands that help to lubricate the nipples during breast-feeding.

The breasts also contain a network of lymphatic vessels that rid the body of foreign materials and waste products. The lymphatic vessels in the breasts connect to the axillary nodes in the armpits and the internal mammary nodes behind the **sternum**. The fluid in the lymphatic vessels, called lymph, consists of lymphocytes that function as part of the body's **immune system**.

Many **hormones** are responsible for the growth, development, and function of the mammary glands. **Estrogen** encourages the proliferation of the mammary glands and **progesterone** is necessary for the development of the mammary glands. Besides the **sex hormones**, cortisol, insulin, growth hormone, and thyroid hormones also have an active role in the growth of breasts during **puberty**. Two pituitary hormones are necessary for the production and secretion of milk following **pregnancy**. Prolactin stimulates production of milk, while **oxytocin** causes contraction of the smooth muscle in the ducts that release milk from the breast in response to suckling.

Mammary tissue begins to develop in human embryos even before the development of the distinguishing reproductive **organs** of males and females. Human embryos, like all mammals, have similar developmental patterns regardless of whether they become male or female. All human embryos are morphologically female until the sixth week of development when gender is determined. If the Y chromosome is present, **testosterone** is released, resulting in the development of male characteristics. During childhood, the breasts of females and males are underdeveloped and very much the same. Occasionally breast development does occur in men, a condition known as gynecomastia. This condition is often due to a change in hormone levels, but does have other causes.

Before puberty, the female's breasts remain underdeveloped. This is because the ovaries have not yet begun to secrete the sex hormones estrogen and progesterone. Once puberty begins, breast **growth and development** result from hormonal influence and an increase in **fat** deposition. Additionally, the ducts of the mammary glands begin to grow and branch out. During **menstruation**, the breasts may swell and become sensitive to the touch; however, this does not occur in all women. During pregnancy, the breasts tend to increase in size due to the enlargement of the ducts and glandular tissue. The nipples also increase in size and the areola become darker. After **menopause**, hormonal activity decreases, the mammary glands are replaced with fat, and breasts begin to sag under the influence of gravity and deteriorating musculo-ligamental support.

Mammography is the use of x-ray imagery specially designed for examining breasts. Mammography screens for any abnormalities in the breast tissue and is a diagnostic tool for the early detection of breast **cancer**. Two types of mammography exist; they are classified based on the invasiveness of the procedure. Screening mammography is used to detect breast cancer while lumps are still too small to be felt. Usually women receive screenings once they reach the age of 40.

Diagnostic mammography is a more in depth procedure utilized once an abnormality has been determined.

Most lumps found in the breasts are noncancerous cysts filled with fluid or benign **tumors** called fibroadenomas. The most common type of breast cancer, ductal carcinoma, occurs in the milk ducts of the mammary glands. A carcinoma is a malignant tumor found in epithelial cells. Another carcinoma of the breast, lobular carcinoma, grows in the lobes of the mammary glands. Other types of breast cancer include inflammatory carcinoma characterized by red, swollen breasts with a rash. Paget's disease is a cancer of the nipple and areola. Most types of breast cancer are infiltrating, meaning they spread to other tissue.

See also Hormones and hormone action; Reproductive system and organs (female)

BREATHING

Breathing provides the oxygen needed for the conversion of foodstuff to the energy used to support and maintain the various functions of life. The intake of oxygen is facilitated by the **lungs**.

Breathing is vital for life. Humans can live only a very short time without an infusion of oxygen. Without oxygen, irreversible alterations in the **brain** set in after only about 4–5 minutes. Reflecting this importance to life, breathing is an automatic function, and is controlled by nerve cells in the brain stem. Signals from the brain are sent down the **spinal cord** via the phrenic nerve to two sets of muscles that control the operation of the the **diaphragm**, a sheet-like muscle located at the bottom of the chest cavity, that moves down to force air into the lungs and up to force air out of the lungs.

Air can be breathed into the lungs through the **nose** and the mouth. When entering through the nose, air is warmed, moistened and cleaned on its travels through passages called sinuses. The air then passes through the throat, down the windpipe and into the lungs. As this is happening the chest expands downward and horizontally outwards. This allows the lungs to act like bellows, expanding to receive the incoming air. Once in the lungs, the air goes through a maze of smaller and smaller tubes—bronchi, secondary **bronchi** and bronchioles—until it reaches the tiny sacs called **alveoli**. The alveoli, 300 million in each lung, are the site where oxygen from the air enters the **blood**, and where carbon dioxide from the body enters the air, to be expelled from the lungs. These gases are able to diffuse across the thin walls of the alveoli.

The efficiency of breathing can be affected by several factors. As people age, and generally begin to assume a poorer, stooped **posture**, the expansion of the chest is more restricted. But even young people who adopt a poor posture can compromise their breathing efficiency. Gender also places a role; men generally breathe more fully and lower in their **abdomen**, whereas women tend to breathe higher in their thorax. Presence of a fetus pressing on the diaphragm and ribcage can lessen lung expansion in pregnant women. Respiratory infections or medical conditions such as asthma can also compro-

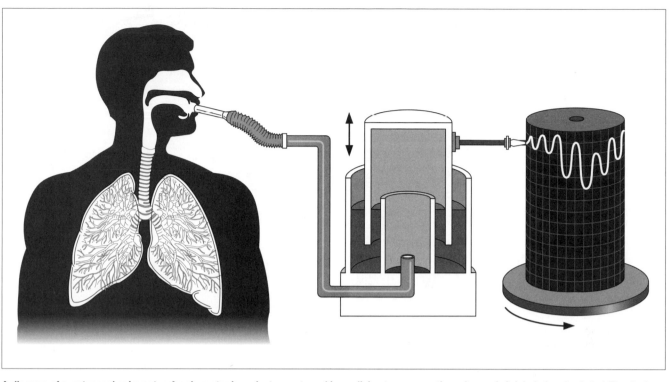

A diagram of a water seal spirometer. A spirometer is an instrument used in medicine to measure the volume of air inhaled and exhaled. The device is considered an essential tool in the detection of chronic obstructive pulmonary disease. *Illustration by Hans & Cassidy.*

mise breathing. Colds, flu and pneumonia can make inhalation more difficult and impair gas exchange in the lungs. Smoking has devastating effects on the breathing process. The toxic components of tobacco smoke can kill the many tiny hairlike **cilia** that line the airways, leading to the accumulation and stagnation of **mucus**, germs, and inhaled debris.

See also Acidosis; Gaseous exchange; Oxygen transport and exchange; Respiratory system

BROCA, PIERRE PAUL (1824-1880)

French anthropologist and anatomist

Paul Broca is best known for his role in the discovery of specialized functions in different areas of the **brain**. In 1861, he was able to show, using post-mortem analysis of patients who had lost the ability to speak, that such loss was associated with damage to a specific area of the brain. The area, located toward the front of the brain's left hemisphere, became known as Broca's convolution. Along with its importance to the understanding of human physiology, Broca's findings addressed questions concerning the evolution of language.

Broca, the son of a Huguenot doctor, was born near Bordeaux, France. After studying mathematics and physical science at the local university, he entered medical school at the University of Paris in 1841. He received his M.D. in 1849. Though trained as a pathologist, anatomist and surgeon, Broca's interests were not limited to the medical profession.

His versatility and tireless dedication to science permitted him to make significant contributions to the fields of **anatomy** and anthropology.

The application of his expertise in anatomy outside the field of medicine began in 1847, as a member of a commission charged with reporting on archaeological excavations of a cemetery. The project permitted Broca to combine his anatomical and mathematical skills with his interests in anthropology.

The discovery in 1856, of Neanderthal man once again drew Broca into anthropology. Controversy surrounded the interpretation of Neanderthal. It was clearly a human **skull**, but more primitive and apelike than a modern skull and the soil stratum in which it was found indicated a very early date. Neanderthal's implications for evolutionary theory demanded thorough examination of the evidence to determine decisively whether it was simply a congenitally deformed *Homo sapiens* or a primitive human form. Both as an early supporter of Charles Darwin and as an expert in human anatomy, Broca supported the latter view. Broca's view eventually prevailed, though not until the discovery of the much more primitive Java Man (then known as *Pithecanthropus*, but later *Homo erectus*).

All animals living in groups communicate with one another. Non-human primates have the most complex communication system other than human language. They use a wide range of gestures, facial expressions, postures and vocalizations, but are limited in the variety of expressions and are unable to generate new signals under changing circumstances. Humans alone possess the capacity for language rather than

Paul Pierre Broca. © Hulton-Deutsch Collection/Corbis. Reproduced by permission.

relying on a body language vocabulary. Language permits humans to generate an infinite number of messages and ultimately allows the transmission of information—the learned and shared patterns of behavior characteristic of human social groups, which anthropologists call culture—from generation to generation. The development of language spurred human evolution by permitting new ways of social interaction, organization and thought.

Given the importance assigned to human speech in human evolution, scientists began to look for the physical preconditions of speech. The fact that apes have the minimal parts necessary for speech indicated that the shape and arrangement of the vocal apparatus was insufficient for the development of speech. The vocalizations produced by other animals are involuntary and incapable of conscious alteration. However, human speech requires codifying thought and transmitting it in patterned strings of sound. The area of the brain isolated by Broca sends the code to another part of the brain, which controls the muscles of the face, jaw, tongue, palate and **larynx** setting the speech apparatus in motion. This area and a companion area which controls the understanding of language, known as Wernicke's area, are detectable in early fossil skulls of the genus *Homo*. The brain of *Homo* was evolving toward the use of language, although the vocal chamber was still inadequate to articulate speech. Broca discovered one piece in the puzzle of human communication and speech which permits the transmission of culture.

Equally important, Broca contributed to the development of physical anthropology, one of the four subfields of anthropology. Craniology, the scientific measurement of the skull, was a major focus of physical anthropology during this period. Mistakenly considering contemporary human groups as if they were living fossils, anthropologists became interested in the nature of human variability and attempted to explain the varying levels of technological development observed worldwide by looking for a correspondence between cultural level and physical characteristics. Broca furthered these studies by inventing at least twenty-seven instruments for making measurements of the human body, and by developing standardized techniques of measurement.

See also Brain: Intellectual functions; Larynx and vocal cords

BRONCHI

Bronchi are the respiratory passageways between the trachea and the lobes of the lung. After the left and right primary bronchi (also termed principal bronchi) arise from the trachea, they decrease in diameter as they ramify (create multiple extensions) and enter the lobes of the lung as lobar and segmental bronchi. The multiple branching of the bronchi is referred to as a bronchial tree.

Corresponding to the lobes of the lung they serve, the right principal bronchus divides into a right superior lobe broncus, a middle lobe bronchus, and right inferior lobe bronchus. The left principal bronchus divides into the left superior principal bronchus and the left inferior lobe bronchus.

In the embryo, the bronchi develop from the caudal end of the tracheal tube as primordial main stem bronchi. As ultimately reflected in adult **anatomy**, the left main bronchus assumes a more transverse course than the right main bronchus. Splanchnopleuric **mesoderm** differentiates into the fibro-elastic and muscular structure of the developing bronchial tree. **Blood** vessels serving both bronchial and lung **tissue** also develop from this mesodermal layer. Vagal nerves invest the developing lobes of the lung surrounding the bronchi serving each lung lobe bud. Pulmonary **arteries** and **veins** also develop within the bronchial tree to form a bronchopulmonary segment—a functional unit of lung—wherein a single segmental bronchus and accompanying pulmonary artery serve a defined region of the lung. Blood returns to the **heart** via pulmonary veins that lie in the connective tissue between the bronchopulmonary segments.

Bronchopulmonary segments, surrounded by connective tissue, form the functionally independent respiratory units of the lung. This redundancy is important, because, if one segment or lobe of the lung is punctured, or otherwise becomes non-functional, other branches of the bronchopulmonary tree are not compromised.

In the adult, there are six main bronchopulmonary segments within the right superior lobe of the lung (one apical, two posterior, three anterior). Within the right middle lobe, there are nine bronchopulmonary segments (four lateral, five medial). The inferior lobe of the right lung usually contains far

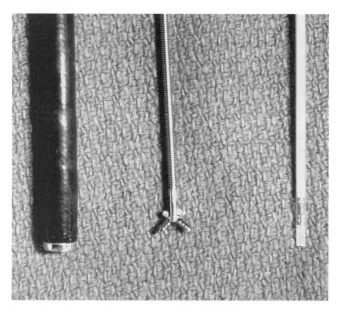

A number of instruments can be used to dilate the airway to allow direct examination of the bronchi (bronchoscopy). *Photograph by Michael English, M.D. Custom Medical Stock Photo. Reproduced by permission.*

more bronchopulmonary segments than the other right lobes combined. The inferior lobe contains 40 bronchopulmonary segments investing all regions of lobe.

On the left side of the lung, the left superior lobe contains 15 bronchopulmonary segments, the same number as the superior and middle lobes of the right lung. As with the right side the left inferior lobe contains 40 bronchopulmonary segments.

The lining of the bronchial tree consists of a layer of pseudostratified ciliated epithelium interspersed with goblet cells. The **cilia** beat so that they move debris and mucous upwards. This broncho-tracheal elevator is an important component in the removal of debris and the resistance to infections (e.g., anthrax **infection**) that can become severe if sufficient **bacteria**, debris, or spores are able to invade the terminal portions of the bronchopulmonary tree.

It possible to view the interior structure and mucosa of individual bronchi (a broncus) via fiberoptic bronchoscopic examination. A bronchoscope is introduced through the naso-oral cavity and trachea. It is also possible to remove bronchial tissue samples for pathological examination. Bronchoscopy also allows for the visualization and removal of foreign objects that become lodged in the bronchial tree. Inflammation of one or more bronchi results in bronchitis. Acute onset bronchitis, with varying degrees of severity, is often associated with upper respiratory infections.

Chronic inflammation of the bronchi may result from a number of pathological conditions, including low-level exposure to allergens (allergic reaction inducing substances), smoking, and a variety of chemical pollutants and toxins. Whether stimulated by immune response or chemical irritation, inflammation in the bronchial tree results in increased mucosal production that blocks the airway and results in bron-

chospasm, coughing, and the expectoration (spitting) of mucosal fluids.

A congenital abnormality of the **alveoli** may result in an increased diameter (dilation) in the terminal portions of the bronchial tree. Infections such as pneumonia and flu (influenza) can also produce chronic dilation of the terminal portions of the bronchial tree (brochiectasis).

See also Anatomical nomenclature; Breathing; Respiration control mechanisms; Respiratory system embryological development; Respiratory system

BROWN, MICHAEL S. (1941-)
American physician

Michael S. Brown, a **genetics** professor and director of the Center for Genetic Diseases at the University of Texas Southwestern Medical School, is one of America's foremost experts on cholesterol **metabolism** in the human body. In the 1970s, Brown and **Joseph Goldstein** investigated familial hypercholesterolemia, a dangerous inherited disorder that causes elevated levels of cholesterol in the **blood**. Their research led them to the discovery of a protein in the membranes of a cell, called the LDL receptor, which plays a central role in the body's ability to lower cholesterol levels. For this discovery and their subsequent research on the LDL receptor, Brown and Goldstein shared the 1985 Nobel Prize in physiology or medicine.

Brown was born in New York City on April 13, 1941, to Harvey and Evelyn Katz Brown. He attended the University of Pennsylvania as an undergraduate, receiving his bachelor's degree in 1962. Following his graduation, Brown enrolled in the medical school at the University of Pennsylvania, where he was awarded the Frederick Packard Prize in Internal Medicine for his research. He earned his M.D. in 1966 and served as an intern and a resident at Massachusetts General Hospital in Boston. It was during his residency that he met Joseph Goldstein, his future research partner, who was also on the staff at Massachusetts General.

In 1968, Brown was made a clinical associate at the National Institutes of Health (NIH) in Bethesda, Maryland. He was assigned to the **biochemistry** lab, where he worked with Earl Stadtman, head of the laboratory for the National **Heart**, Lung, and Blood Institute. While at NIH, Brown focused his research on gastroenterology, particularly on the role of **enzymes** in digestive chemistry. In 1971, while studying a particular enzyme involved in the production of cholesterol, Brown was offered a position as an assistant professor at the University of Texas Southwestern Medical School in Dallas. He accepted, and Goldstein, who had also served at NIH in Bethesda, joined the Texas Southwestern faculty a year later. At this time the two began a collaboration which was to distinguish them as pioneers in genetics.

In Dallas during the 1970s, Brown and Goldstein examined skin samples from people who suffered from hypercholesterolemia, specifically those rare patients whose condition was homozygous, meaning that they had not just one defective gene but two. In these cases, patients often exhibited extremely high levels of low-density lipoprotein, LDL, even during childhood. LDL carries cholesterol to the cells, and in excessive quantities can clog **arteries** and encourage heart disease. Brown and Goldstein discovered that the cells of these patients were missing a crucial protein, called a receptor, which binds to LDL and regulates its level in the body. Without the protein, the body can not break down LDL, and it accumulates in the blood. Brown and Goldstein's breakthrough was the discovery and isolation of this LDL receptor protein.

Brown and Goldstein not only identified the LDL receptor, they also located the gene responsible for its production. By sequencing and cloning the gene, they were able to localize the gene mutations responsible for familial hypercholesterolemia, as well as other inherited conditions involving cholesterol metabolism. Their findings also led to possible drug therapies for people with cholesterol disorders. By administering a combination of drugs which would inhibit the liver's ability to synthesize cholesterol, Brown and Goldstein increased their patients' need for cholesterol from outside sources. The patients' bodies subsequently produced more LDL receptors, and their cholesterol levels fell sharply. They also found that a **liver** transplant can correct genetic deficiencies in the production or expression of LDL receptors. In later research, Brown and Goldstein engineered a mouse which, because of its abnormally high numbers of LDL receptors, could eat a high-fat diet and yet show no significant rise in LDL.

In a series of experiments, Brown and Goldstein were ultimately able to define and analyze each step in the path of cholesterol through the body, from production to dissolution. They also demonstrated a mechanism by which a low-fat diet and regular exercise can decrease cholesterol levels. Brown and Goldstein's work had significant implications not only for genetic defects, but also for **nutrition** and fitness. In addition, the team's research methods contributed to a greater understanding of cell receptors in general, serving as a model for research on over 20 other receptors.

In addition to the Nobel Prize, Brown has received several honorary degrees and a number of awards for his research, including the Pfizer Award from the American Chemical Society in 1976, the Albert Lasker Medical Research Award in 1985, and the National Medal of Science in 1988. He has been a member of the National Academy of Sciences since 1980. He was appointed Paul J. Thomas Professor of Genetics and director of the Center for Genetic Diseases at the University of Texas Southwestern Medical School, positions he has held since 1977.

See also Lipids and lipid metabolism

BURNET, FRANK MACFARLANE (1899-1985)
Australian immunologist and virologist

While working at the University of Melbourne's Walter and Eliza Hall Institute for Medical Research in the 1920s, Frank

Macfarlane Burnet became interested in the study of **viruses** and bacteriophage (viruses that attack **bacteria**). That interest eventually led to two major and related accomplishments. The first of these was the development of a method for cultivating viruses in chicken embryos, an important technological step forward in the science of virology. The second accomplishment was the development of a theory that explains how an organism's body is able to distinguish between its own cells and those of another organism. For this research, Burnet was awarded a share of the 1960 Nobel Prize in physiology or medicine (with Peter Brian Medawar).

Burnet was born in Traralgon, Victoria, Australia. His father was Frank Burnet, manager of the local bank in Traralgon, and his mother was the former Hadassah Pollock MacKay. As a child, Burnet developed an interest in nature, particularly in birds, butterflies, and beetles. He carried over that interest when he entered Geelong College in Geelong, Victoria, where he majored in biology and medicine.

In 1917, Burnet continued his education at Ormond College of the University of Melbourne, from which he received his bachelor of science degree in 1922 and then, a year later, his M.D. degree. Burnet then took concurrent positions as resident pathologist at the Royal Melbourne Hospital and as researcher at the University of Melbourne's Hall Institute for Medical Research. In 1926, Burnet received a Beit fellowship that permitted him to spend a year in residence at the Lister Institute of Preventive Medicine in London. The work on viruses and bacteriophage that he carried out at Lister also earned him a Ph.D. from the University of London in 1927. At the conclusion of his studies in England in 1928, Burnet returned to Australia, where he became assistant director of the Hall Institute. He maintained his association with the institute for the next thirty-seven years, becoming director there in 1944. In the same year, he was appointed professor of experimental medicine at the University of Melbourne.

Burnet's early research covered a somewhat diverse variety of topics in virology. For example, he worked on the classification of viruses and bacteriophage, on the occurrence of psittacosis in Australian parrots, and on the epidemiology of herpes and poliomyelitis. His first major contribution to virology came, however, during his year as a Rockefeller fellow at London's National Institute for Medical Research from 1932 to 1933. There he developed a method for cultivating viruses in chicken embryos. The Burnet technique was an important breakthrough for virologists since viruses had been notoriously difficult to culture and maintain in the laboratory.

Over time, Burnet's work on viruses and bacteriophage led him to a different, but related, field of research, the vertebrate **immune system**. The fundamental question he attacked is one that had troubled biologists for years: how an organism's body can tell the difference between "self" and "notself." An organism's immune system is a crucial part of its internal hardware. It provides a mechanism for fighting off invasions by potentially harmful—and sometimes fatal—foreign organisms (**antigens**) such as bacteria, viruses, and fungi. The immune system is so efficient that it even recognizes and fights back against harmless invaders such as pollen and dust, resulting in allergic reactions.

Burnet was attracted to two aspects of the phenomenon of immunity. First, he wondered how an organism's body distinguishes between foreign invaders and components of its own body, the "self" versus "not-self" problem. That distinction is obviously critical, since if the body fails to recognize that difference, it may begin to attack its own cells and actually destroy itself. This phenomenon does, in fact, occur in some cases of **autoimmune disorders**.

The second question on which Burnet worked was how the immune system develops. The question is complicated by the fact that a healthy immune system is normally able to recognize and respond to an apparently endless variety of antigens, producing a specific chemical (antibody) to combat each antigen it encounters. According to one theory, these antibodies are present in an organism's body from birth, prior to birth, or an early age. A second theory suggested that antibodies are produced "on the spot" as they are needed and in response to an attack by an antigen.

For more than two decades, Burnet worked on resolving these questions about the immune system. He eventually developed a complete and coherent explanation of the way the system develops in the embryo and beyond, how it develops the ability to recognize its own cells as distinct from foreign cells, and how it carries with it from the very earliest stages the templates from which antibodies are produced. For this work, Burnet was awarded a share of the 1960 Nobel Prize in physiology or medicine. Among the other honors he received were the Royal Medal and the Copley Medal of the Royal Society (1947 and 1959, respectively) and the Order of Merit in 1958. He was elected a fellow of the Royal Society in 1947 and knighted by King George V in 1951.

Burnet retired from the Hall Institute in 1965, but continued his research activities. His late work was in the area of autoimmune disorders, **cancer**, and aging. He died of cancer in Melbourne 1985. Burnet was a prolific writer, primarily of books on science and medicine, during his lifetime.

See also Antigens and antibodies; Immune system; Immunity; Immunology

BURNS

Burns can be of various origins—thermal, electrical, chemical, and mechanical—and can cause varying degrees of damage to the body. All burns involve injury to the surface of the skin and to the underlying layers of **tissue**.

The skin is composed of several layers. The outermost visible layer is the **epidermis**. This layer sloughs off continuously. Thus, damage to the epidermis is not long lasting. Immediately underneath the epidermis is the **dermis**, which is composed of connective tissue, **blood** vessels, nerve endings, **hair** follicles, and sweat and oil glands. This myriad of functions makes damage to the dermis potentially more damaging and longer lasting. The lowest layer of skin is the hypodermis, or basement membrane. Damage to this layer, the foundation of the skin, constitutes a serious full-thickness injury.

The injuries caused by the various types of burns are classified according to the amount of the body that is affected and of the depth of injury. The extent of the damage is described either as a first, second or third-degree burn. A first-degree burn is limited to the epidermis, the uppermost portion of the skin. The skin becomes moist and red and is painful for a short period of time. A sunburn is an example of a first-degree burn. A second-degree, or partial-thickness, burn penetrates further into the skin, affecting both the epidermis and dermis. The dermis may be only slightly harmed or can be very damaged, depending on the severity of the burn. Both conditions are extremely painful and healing is more prolonged; up to a month. Scarring is possible with second-degree burns. A third-degree, or full-thickness, burn is the most severe type of burn. All the tissue layers are affected, requiring skin grafting to restore the integrity of the skin. Hence, recovery from a third-degree burn can be prolonged. Paradoxically, given the extent of damage, third-degree burns are initially not as painful as other burns, because nerve endings are also destroyed.

Thermal burns can be caused in several ways. Open flame typically produces a deep, second or third-degree burn. Scalding caused by hot liquids is not usually as deep as the damage caused by flame, but severe damage can result. Contact with a hot object, such as the burner of a stove, causes contact burns. The body's reflexive response determines the extent of burn damage, from minor to major. A final category of thermal burn is termed flash injury. Flash burns result from an explosion of something like a gasoline tank. The extent of injury is quite variable as is the affected portion of the body, depending on the force of the explosion and the proximity of the person to the blast.

Strong acids, alkalis, or other corrosive material causes chemical burns. The conversion of chemical energy to thermal energy is the damaging reaction. This damage persists as long as the chemical agent remains in contact with the skin.

Electrical burns are produced when an electrical current travels from one contact point on the body to another point. The current is converted to heat as it moves through the body, usually in a path along blood vessels or nerve cells. Muscle and bone may also be used as pathways. So, electrical burn damage can involve various parts of the body, such as the **heart** and the lens of the **eye**, and can occur deep within the body.

Finally, the mechanical type of burn is a friction injury, and is typically caused by ropes and carpets. Many sports-related burn injuries of the mechanical type.

Dealing with burns depends on the type and extent of the burn. Treatment can range from minor efforts—application of a soothing suave in the case of sunburn—to extensive fluid replacement and skin grafts for severe burns. More severe burns may also damage various cells, which function in the immune response. The damage to the **immune system** leaves the person much more susceptible to **infection**. Thus, treatment can also involve vaccination, the administration of antibiotics, or the isolation of the person in an environment free from disease causing microorganisms.

See also Electrolytes and electrolyte balance; Osmotic equilibria between intercellular and extracellular fluids

C

CALCITONIN • *see* PARATHYROID HORMONES

CALCIUM AND PHOSPHATE METABOLISM

The average adult human body contains 3.3 lb. (1.5 kg) of calcium, almost 99% of which is stored in bones, 0.1% is in the extra cellular fluid, and approximately 1% is in the cells. The usual **plasma** concentration of calcium is kept rather constant due to the regulatory action of the parathyroid gland. When the dietary calcium absorption by the intestine is not enough to keep the plasma concentration at normal levels, the parathyroid triggers calcium reabsorbtion from the bones to avoid a significant depletion. Such fine control is important because calcium takes part in several physiological functions, such as the contraction of cardiac, skeletal, and smooth muscles, synaptic signal transmission (i.e., communication between neural cells), **blood coagulation**, regulation of intra cellular **metabolism**, and bone formation. Even a slight shortage of calcium (hypocalcemia) for instance, may trigger progressive neural excitement. When calcium levels fall below 50% of normal, the excitement of peripheral nerves induces muscular spasms, whereas an above-the-normal calcium plasma concentration, or hypercalcemia, leads to nervous system depression, decreased cardiac activity, and loss of contractility of **smooth muscle** of the **gastrointestinal tract**.

Like calcium, phosphate is necessary for the mineralization of bones, and it is also found in all body cells, where it takes part in the formation of different compounds such as nucleic acids, coenzymes, and other substances indispensable for cellular biological processes. Phosphate present in the diet is easily absorbed by the intestine into the **blood**, whereas only about 35% of dietary calcium is absorbed. Calcium absorption can be negatively affected when the pH of the duodenum is alkaline (i.e., has low acidity). The average calcium daily elimination is about 1 gram, of which 100 mg are found in the urine, and the remaining is excreted in the feces. Therefore, a chronic ineffective intestinal absorption or a diet poor in calcium may lead to a premature and progressive reabsorbtion of calcium from the bones, and consequently, **osteoporosis**.

The active form of vitamin D (dihydroxycolecalciferol), acts on the calcium/phosphate metabolism in three different ways: promoting intestinal absorption of calcium and enhancing phosphate absorption, promoting calcium and phosphate deposition in the skeleton, and inhibiting the excretion of calcium and phosphate in the urine. Calcium and phosphate are the main components of the bone salts deposited in the organic bone matrix to form the hydroxyapatite crystals. The other mineral salts in the composition of hydroxyapatite are sodium, magnesium, potassium, and carbonate ions. In the absence of vitamin D, the organic bone matrix continues to be formed; however, as the necessary mineral salts are not deposited.

See also Bone hystophysiology; Bone reabsorbtion; Muscle contraction

CANCER

Cancer may be as old as humankind. During the fifteenth century, what might now be considered a cancer-like growth was referred to as a scirrus, or scar. Environmental substances have long been associated with the disease. In 1775, Sir Percivall Pott connected frequent occurrences of scrotal cancer among chimney sweeps with their continual exposure to flue dust. Until the 1700s, Europeans treated cancer with crude methods like cauterization and bloodletting. Although cancer research has been continuous for centuries, the most important conclusions have been drawn in the twentieth century.

Cancer is a group of many diseases in which certain cells within the body lose their ability to regulate **cell division**. The cancerous cell multiplies uncontrollably, causing other normal cells to be crowded out and destroyed. If this growth takes place in a vital organ, malfunctions and **death** can result.

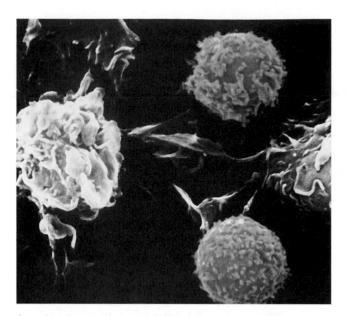

Scanning electron micrograph (SEM) of dividing cancer cells.
Photograph by Nina Lampen. Science Source/Photo Researchers, Inc.
Reproduced by permission.

The causes of cancer are diverse and the cure rates associated with different types of cancer vary. Scientists have long assumed that cancer is linked to changes in the genetic material of the cell. It has been noted that the **chromosomes** of cancerous cells show abnormalities that may include deletions or translocations of certain genes. Many agents known to cause cancer (carcinogens) are also proven mutagens, or substances that cause atypical genetic changes. Ultraviolet radiation, x rays, **viruses**, and some chemicals are carcinogens that cause genetic mutations.

These initial observations triggered the search for other causes of cancer. In 1950, scientists demonstrated that nucleic acids were important hereditary chemicals. Frank L. Horsfall, Jr. (1906–1971), a clinician and virologist, was interested in finding the causes of cancer. He knew that once cancerous cells appeared, these cells would produce more cells with similar cancerous properties. In cell culture, the cancerous alteration was passed on even when the carcinogen was removed. This clue lead Horsfall to believe that the changes in the cell must have been a result of changes in **DNA**. He also saw similarities in animal cancers and changes in virus-infected bacterial cells. Horsfall's discovery that all cancer is attributed to changes in the nucleic acid of the cell provided a unifying concept for studying cancer.

Nearly 50 years before Horsfall's observations, American physician **Peyton Rous** began research in **pathology** at what is now Rockefeller University. Rous was interested in the **physiology** of cancer within mammals and birds and discovered the first virus-induced cancer. This connective **tissue** cancer in chickens, called Rous sarcoma, causes enlargement of the **liver** and is fatal. Rous's initial experiment included grafting sarcoma tumor cells from diseased hens into healthy hens. He noted that the healthy hens soon contracted the dis-

ease. Even filtered fluid extracts from diseased hens were contagious. Rous hypothesized that a virus may be the cause of the cancer. Unfortunately, other scientists were not able to duplicate Rous's experiments using different bird species and his work was ignored for several decades. Soon after Horsfall and other researchers, established a better understanding of viruses and cancer, Rous's work was finally recognized when he received the Nobel Prize in 1966. Rous's viral theory of cancer engendered an entirely new perspective. It became accepted that some cancers are infectious, that is, able to spread from individual to individual via certain viruses. Today more than 24 potentially oncogenic (cancer-causing) viruses are known. Several of these, including the Epstein-Barr and hepatitis viruses, cause other diseases in addition to cancer.

Because of our better understanding of the causes of cancer, various therapies have recently been introduced in an effort to improve cure rates for many cancers. **Chemotherapy** employs numerous chemical medications to attack and kill cancerous cells. Unfortunately, many of these compounds are toxic to the human body and cause severe side effects. Radiotherapy uses radiation to shrink cancerous growths prior to or in place of surgical removal. Although radiation is site directed from an exterior source, or in some cases of breast cancer an implanted device, it also kills normal cells in massive numbers. Radiation therapy is frequently used in combination with chemotherapy. Immunotherapy is yet another useful cancer treatment. This type of treatment is relatively new and appears to have great potential. Immunotherapy uses several different methods to bolster the patient's own **immune system**. The body is then better able to recognize and destroy the cancerous cells. Finally, herbal therapy has experienced a renewed interest by some in the medical community. Several herbs have been shown to contain cancer-inhibiting compounds under certain conditions.

The twentieth century's final decade heralded enormous advances in **genetics** research, including the ability to pinpoint genes and the role they play in disease. Scientists have identified mutations in three primary types of genes that can cause cancer. Mutations in tumor suppressor gene inhibit their ability to prevent the reproduction of damaged cells and tumor growth. Some oncogenes, a mutated form of protooncogenes (before oncogenes), produce a dominant protein that encourages cell growth. DNA repair genes correct genetic errors; when they cannot perform their function due to a mutation, cancer occurs. Perhaps the most frequent genetic mutation associated with cancer occurs in the p53 gene and its encoded protein. Located on the short arm of the human chromosome 17, this tumor suppressor gene is often lost or mutated in **tumors**, leading to abnormal cell proliferation.

A growing number of genes have been identified as playing a major role in many types of cancer. A gene on chromosome 3 is thought to be a tumor suppressor gene associated with up to two-thirds of kidney cancer cases. A gene on the upper part of chromosome 2 causes colon cancer by fostering mutations in other genes. A mutation in a gene on chromosome 9 causes basal cell carcinoma, the most common type of skin cancer. A dysfunction in the p16 gene prevents it from producing a protein, leading to benign pituitary tumors, which can still cause vision problems and disrupt hormone balances.

While it is still uncertain whether many of these genetic mutations are inherited, the result of environmental influences, or a combination of both, researchers are closing in on many familial cancer-causing genetic mutations, including forms of colon and breast cancer. In 1995 the identification of the "breast cancer genes" BRCA1 and BRCA2 represented the first "susceptibility" genes found for a prevalent cancer. BRCA1, found on chromosomes 17, is responsible for a prevalent form of hereditary breast cancer. BRCA2 is found on chromosome 13 and appears to cause as many cases of breast cancer as BRCA-1, including rare forms of male breast cancer. Originally thought to raise a women's chances of getting breast cancer by 85%, follow-up research has indicated that women with either one of these mutations has an increased risk of closer to 50%. Nevertheless, these findings, as well as similar findings for other cancers, certainly will lead to tests that can identify individuals with genes that may dispose them to cancer. Understanding the genetic causes of cancer will also lead to **gene therapy** as a viable approach to curing and helping to prevent cancer.

Even with the confirmation of a genetic link to cancer, it may be many more decades before cancer is fully understood. Some forms of cancer are extremely rare and deadly, while others are common and may boast a cure rate of 90% or better. Every day potential carcinogens are discovered and novel therapies tested.

See also Stem cells; Tumors and tumerous growth

CAPILLARIES

Capillaries are the microscopic **blood** vessels branching from the arterioles and merging into venules. Despite the fact that there are approximately 40 billion capillaries in the body, they hold only 5% of total blood volume. There are two reasons for this. First, the size of the capillaries is only 5–10nm in diameter. Second, at any give time only a fraction (25%) of capillaries are fully filled with blood, especially in tissues at rest, as blood flow in microvessels is dependent on the metabolic activity of the **tissue** and is regulated at the sites of their origin by the precapillary sphincter muscles.

Capillaries are essential for the delivery of oxygen to the tissues and the exchange of nutrients between the blood and **interstitial fluid** surrounding the cells. This function is well supported by the **anatomy** of the vessels. The thin walls of the capillaries are composed of a single layer of endothelial cells. As a result, gasses such as CO_2 and O_2 can diffuse through their walls, as can lipid soluble substances. In contrast, an exchange of lipid-insoluble substances occurs via transcytosis, which involves formation of pinocytotic vesicles at one side of the endothelial cell, their transport across the cells, and release of contents from the other side of the cell.

Capillaries also play an important role in regulating the relative volume of the blood and interstitial fluid by allowing a bulk flow through their walls. This exchange of water and solutes occurs in response to the pressure gradient across the capillary wall.

Based on the structure of the endothelial cells, there are three types of capillaries. Continuous capillaries are a tube developed by the endothelial cells with no intercellular or intracellular gaps (**brain, retina**), or small intercellular gaps. In contrast, fenestrated endothelium has pores of 70–100nm in size which allow some substances to pass through. Finally, there are discontinuous capillaries (or sinusoids) that are the largest capillaries and have little or no basal membrane, and large intercellular pores and fenestrations.

Capillaries form functional units known as capillary beds and these are not uniformly distributed among the different tissues. Sites of high metabolic activity (**liver, kidneys**) contain numerous capillaries, while sites with little metabolic activity (such as the lens of the **eye**) are capillary-free.

See also Vascular system overview; Oxygen transport and exchange; Fluid transport

CARBOHYDRATES

Carbohydrates are present in every plant or animal cell, and make up the largest portion, in terms of mass, of organic compounds present on Earth. Together with fats and proteins, they are the organic nutrients of humans and animals. Carbohydrates are a large class of naturally occurring polyhdroxycarbonyl compounds which have the general molecular formula $(C)_n(H2O)_n$. They were originally characterized as hydrated forms of carbon and the name has been retained, although it is chemically inaccurate. Today, other compounds having different elemental compositions are also categorized as carbohydrates, e.g., aldonic acids, uronic acids, deoxysugars, amino sugars and mucopolysaccharides.

The carbohydrates are subdivided on the basis of their molecular size. Monosaccharides, or simple sugars, cannot be further hydrolysed. Glucose, galactose and fructose are three common monosaccharides of physiological importance which share the same molecular formula: $C_6H_{12}O_6$. Because of their six carbon atoms they are termed hexoses. Even though they have the same molecular formulae, the atoms are arranged differently in the monosaccharide units of each of the three sugars, making them into distinct substances. Substances having identical molecular formulae but different structural formulae are known as structural isomers. Glucose is also called "blood sugar" as it is present at concentrations of up to 0.1 % in **blood**. It is the most important animal monosaccharide because it is the immediate source of energy for cellular respiration.

Oligosaccharides are made up of two to ten monosaccharides, linked by α- or β-glycoside bonds. Common disaccharides, carbohydrates with two sugar units, include sucrose (table sugar) made up of glucose and fructose, lactose (milk sugar) made up of glucose and galactose, and maltose (a product of starch **digestion**) composed of two glucose units. The polysaccharides are a large group of carbohydrates composed of ten or more monosaccharides linked together to branched or unbranched chains. They commonly contain hundreds of thousands of monosacchride units and have very high molecular weights. The polysaccharide **glycogen** is the reserve carbohydrate in animals. Excess glucose from **metabolism** is polymerised through α-1,4 glycosidic linkages to form long chains. At branching points, the units are attached with α-1,6 links. The molecular weight of glycogen ranges from 1 to 16 million, which means a maximum of 10^5 glucose units. It is most abundant in the **liver** (up to 10 %) and the muscles (up to 1%) and serves as a strorage substance, which is subject to continuous synthesis and degradation. For longer-term storage, excess carbohydrate is converted to and stored as **fat**. There is some evidence that intense exercise and a high-carbohydrate diet ("carbo-loading") can increase the reserves of glycogen in the muscles and thus may help sportsmen like marathon runners work their muscles somewhat longer and harder than otherwise.

See also Biochemistry; Fat, body fat measurements; Protein metabolism; Protein synthesis; Sports physiology

CARBON DIOXIDE LEVELS AND VENTILATION • *see* RESPIRATION CONTROL MECHANISMS

CARDIAC ARRHYTHMIAS • *see* HEART, RHYTHM CONTROL AND IMPULSE CONDUCTION

CARDIAC CYCLE

The cardiac cycle describes the coordinated and rhythmic series of muscular contractions associated with the normal **heart** beat.

The cardiac cycle can be subdivided into two major phases, the systolic phase and the diastolic phase. Systole occurs when the ventricles of the heart contract. Accordingly, systole results in the highest pressures within the systemic and pulmonary circulatory systems. Diastole is the period between ventricular contractions when the right and left ventricles relax and fill.

The cardiac cycle cannot be described as a linear series of events associated with the flow of **blood** through the four chambers. One can not accurately describe the cardiac cycle by simply tracing the path of blood from the right atrium, into the right ventricle, into the **pulmonary circulation**, the venous pulmonary return to the left atrium, and finally the ejection into the aorta and **systemic circulation** by the contraction of the left ventricle. In reality, the cardiac cycle is a coordinated series of events that take place simultaneously on both the right pulmonary circuit and left systemic circuit of the heart.

The cardiac cycle begins with a period of rapid ventricular filling. The right atrium fills with deoxygenated blood from the superior vena cava, the inferior vena cava, and the coronary venous return (e.g., the coronary sinus and smaller coronary **veins**). At the same time, the pulmonary veins return oxygenated blood from the **lungs** to the left atrium. During the early diastolic phase of the cardiac cycle, both ventricles relax and fill from their respective atrial sources. The atrio-ventricular valves (the tricuspid valve is located between the right atrium and right ventricle; the mitral valve is between the left atrium and left ventricle) open and allow blood to flow from the atria into the ventricles.

The flow of blood through the atrio-ventricular valves is unidirectional and as volume related pressure increases within the ventricles, the atrioventricular valves close to prevent backflow from the ventricles into the atria.

At the onset of the systolic phase, specialized cardiac muscle fibers within the sino-atrial node (**S-A node**) contract and send an electrical signal propagated throughout the heart. In a sweeping fashion, the right atrium contracts and forces the final volume of blood into the right ventricle. The left atrium contracts and contributes the final 20% of volume to the left ventricle.

The S-A node signal is delayed by the atrioventricular node to allow the full contraction of the atria that allows the ventricles to reach their maximum volume. A sweeping right to left wave of ventricular contraction then pumps blood into the pulmonary and systemic circulatory systems. The semilunar valves that separate the right ventricle from the pulmonary artery and the left ventricle from the aorta open shortly after the ventricles begin to contract. The opening of the semilunar valves ends a brief period of isometric (constant volume) ventricular contraction and initiates a period of rapid ventricular ejection.

As muscle fibers contract, they lose their ability to contract forcefully (i.e., the greatest force of muscular contraction in the ventricle occurs earlier in the contraction phase and decreases as contraction proceeds). When ventricular pressures fall below their respective attached arterial pressures, the semilunar pulmonary and aortic valves close. At the end of systole, the semilunar valves shut to prevent the backflow of blood into the ventricles.

After emptying, both ventricles collapse to undergo a period of repolarization and refilling. The receptivity of the ventricles to filling corresponding lowers atrial pressures and allows them to fill from their respective venous sources. At the outset, ventricular pressures remain greater than atrial pressures and the atrioventricular valves remain closed. Because the volume of blood in the ventricle is once again static—closed off by both the atrio-ventricular and semilunar valves—this period is described as isometric (same volume) relaxation.

The cardiac cycle is complete with the onset of another period of rapid ventricular filling that takes place when atrial pressures exceed ventricular pressures and the atrio-ventricular valves open to allow rapid filling.

It is the opening and snapping shut of the atrio-ventricular and semilunar pulmonary and aortic valves that creates the familiar pattern of sound associated with the cardiac cycle. Because the right to left contractions of the atria and ventricles are sweeping, the opening and closings of the right side and left side valves are separated by a short interval. The first heart sound results from the closure of the atrio-ventricular valves.

The second heart sound results from the closure of the semilunar valves.

The electrical events associated with cardiac cycle are measured with the **electrocardiogram (EKG)**.

See also Coronary circulation; Heart defects; Heart, embryonic development and changes at birth; Heart, rhythm control and impulse conduction; Nerve impulses and conduction of impulses; Nervous system overview; Neurons; Purkinje system

CARDIAC DISEASE

Throughout history, diseases of the **heart** have captured the concern and interest of investigators. Ancient Greek and Roman physicians observed the serious and often fatal consequences of heart disease. But effective treatment for heart disease was limited to rest and painkillers until the eighteenth-century discovery of the therapeutic properties of the foxglove plant, whose dried leaf is still used to make the medicine digitalis.

While the heart was once considered a part of the body that could never be improved surgically, the twentieth century has seen a revolution in surgical treatment for heart disease. Blocked coronary **arteries** can be bypassed using new **tissue** and failing hearts can be transplanted. Yet heart disease remains the primary cause of **death** in the United States. Preventive health measures, such as improved diet and regular exercise, have become fundamental tools in the battle against heart disease.

In the twentieth century, physicians acquired the tools to prevent heart disease in some cases and treat it effectively in many others. Major medical advances, such as the development of antibiotic therapy in the 1940s, have dramatically reduced heart disease due to syphilis and rheumatic fever. Developments in surgery, new drugs, diagnostic skill, and increasing knowledge about preventive medicine have also greatly reduced deaths from heart disease. Between 1980 and 1990, the death rate from heart disease dropped 26.7% in the United States, according to the American Heart Association (AHA).

Congenital heart disease, the atrial-septal defect in particular, was first described in 1900 by George Gibson of Edinburgh. This problem, which occurs when there is an opening in the wall (or septum) between two atria, can cause the right ventricle to be overwhelmed with **blood**, a condition that eventually leads to heart failure. In some cases, however, the holes are small and do not cause problems. For years, there was little physicians could do to help children with the problem. The development of successful surgical procedures to repair atrial septal defects was one of a multitude of dramatic modern advances in heart surgery.

New types of surgery were made possible with a series of technological advances. In 1934, the American John H. Gibbon developed a machine that allowed the heart to stop beating during surgery while the blood was oxygenated outside the body. Gibbon spent nearly 20 years testing the machine on animals. In 1953, Gibbon became the first surgeon to operate on an open heart when he repaired an 18-year-old girl's atrial septal defect.

With new access to the heart, the treatment of heart disease changed dramatically. The development of the first electric pacemaker in 1950 enabled doctors to correct many arrhythmias (abnormal heart rhythms) and numerous types of heart block (a conduction disorder of the heart's electrical signals). In 1992, a total of 113,000 pacemakers were implanted in the United States, according to the American Heart Association.

A significant change in the treatment of coronary heart disease was the development of coronary bypass surgery in 1967 in the United States. The surgery uses blood vessels taken from elsewhere in the body, often the leg, to pass around diseased tissue. By 2001, over half a million coronary artery bypass grafts were performed yearly in the United States, according to the AHA.

Another commonly performed procedure for individuals with coronary heart disease is angioplasty, during which narrowed arteries are stretched to enable blood to flow more easily. The surgery involves threading a tube through the body and stretching the artery by using a plastic balloon that is inflated when the tube is in the coronary artery. A total of 399,000 angioplasty procedures were performed in 1992, according to the AHA.

The most dramatic change in treatment of heart disease was the development of methods to replace the most damaged hearts with healthy human hearts or even animal hearts. The first successful human heart transplant was performed by South African surgeon Christiaan Barnard in 1967. The patient, however, died in 18 days. Though many surgeons tried the operation, success was limited, most patients dying after days or months, until the early 1980s, when effective drugs were developed to fight organ rejection. In 1993, a total of 2,300 heart transplants were performed in the United States, where the one-year survival rate is 81.6%, according to the AHA.

Physician William Osler observed in 1910 that certain types of people were most likely to develop coronary artery disease, particularly individuals who were "keen and ambitious." Contemporary efforts to prevent heart disease focus on identifying types of behavior and activity that increase the risk of heart disease and on encouraging individuals to adopt healthier lifestyles.

The massive body of evidence linking various types of risk factors to heart disease derives from a series of ambitious twentieth-century studies of heart disease in large groups of people over a long period of time. One of the best known of these efforts is the Framingham study, which has traced thousands of residents since 1949. This and other studies have led to findings that individuals are at a greater risk of heart disease if they have high levels of certain types of cholesterol in the blood, if they smoke cigarettes, if they are obese, if they have high blood pressure, and if they are male. Blood cholesterol, a fat-like substance found in all human and animal tissue, is a primary focus of efforts to prevent heart disease. High cholesterol levels are shaped, in part, by diet and can be lowered. Experts suggest limiting consumption of foods high in satu-

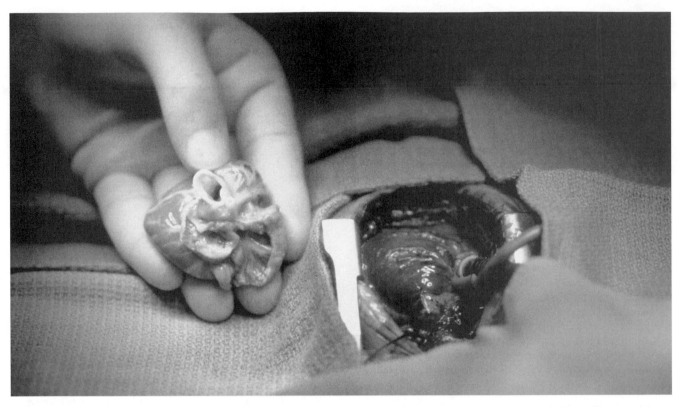

A comparison of hearts during transplant surgery shows the diseased heart (held in the hand) beside the implanted healthy heart. *Photograph by Alexander Tsiaras. National Audubon Society Collection/Photo Researchers, Inc. Reproduced by permission.*

rated fats, such as cream, meat, and cheese. Such a diet reduces the risk of high levels of low-density lipoprotein, or LDL, the type of cholesterol which increases the risk of heart disease.

Exercise has also been promoted as a protection against heart disease. Numerous studies have shown that individuals who do not exercise are more likely to develop coronary heart disease. Exercise reduces blood pressure and eases blood flow through the heart. In addition, people who exercise are less likely to be overweight.

Individuals who burn more calories are also more likely to have higher levels of what has been called the "good" cholesterol—high-density lipoprotein, or HDL. This type of cholesterol is believed to reduce the risk of heart disease. Other activities that boost the level of HDL include maintaining average weight and not smoking cigarettes.

The last few decades of the twentieth century have also seen the introduction of numerous drugs that prolong life and activity for individuals with heart disease. Beta-blockers are used to treat angina, high blood pressure, and arrhythmia. They are also given to individuals who have had heart attacks. These drugs block the neurohormone norepinephrine from stimulating the **organs** of the body. This makes the heart beat more slowly and slows the dilation of certain blood vessels.

Another important class of drugs for the treatment of heart disease is the vasodilators, that cause blood vessels to dilate, or increase in diameter. These drugs, including the so-called ACE inhibitors, are used to ease the symptoms of angina by easing the work of the heart, to forestall complete congestive heart failure, and to prolong life in people who have had heart attacks.

A third important type of drug reduces cholesterol in the blood. The process by which these drugs eliminate cholesterol from the blood varies, but several work by preventing the reabsorbtion of **bile** salts by the body. Bile salts play a role in **digestion**, and they contain cholesterol.

Diagnostic advances have also made a difference in the treatment of heart disease. Cardiac catheterization enables doctors to see how the heart works without surgery. The process, which was first explored in humans in 1936, involves sending a tube through an existing blood vessel and filling the tube with a contrast material that can be tracked as it circulates through the heart. In 1992, a total of 1,084,000 of these procedures were performed to diagnose heart problems.

Though knowledge about the treatment and prevention of heart disease has expanded dramatically, heart disease remains an immense threat. A total of 925,000 Americans die each year of cardiovascular disease, a general category including heart disease, **stroke**, and high blood pressure, all of which are linked. The biggest killer is coronary heart disease, claiming almost 500,000 U.S. deaths each year.

Genetic therapy for heart disease is considered a fertile area for progress. In 1994, surgeons performed a procedure on a woman who had a genetic defect that prevented her **liver** from removing adequate amounts of LDL cholesterol. She had suffered a heart attack at age 16. The procedure, which took place in Michigan, involved the insertion of genetically modified cells in her liver, to enable the organ to remove LDL cholesterol properly. With the new cells, her heart should no longer be threatened by high levels of cholesterol.

Though much is known about risk factors for heart disease, new theories will continue to be tested in the future. For example, various studies have shown that individuals who eat large amounts of fish (especially containing particular oils called omega-3 fatty acids) or who consume vitamin E have a lower than average rate of coronary heart disease. Ambitious studies are under way to confirm or deny this information.

Researchers are also looking carefully at women and heart disease, a topic that has been overshadowed by research concerning men and heart disease in the past. **Estrogen** supplements are known to reduce the risk of heart disease in women, but extensive studies have not been conducted. The National Institutes of Health (NIH) is currently funding a long-term study of 27,000 women who will either take estrogen, **progesterone**, or a placebo. These women will then be evaluated for heart disease and other indicators. Such studies will help document what may and may not be helpful to women seeking protection from heart disease.

In 1998, the American Heart Association published its third list of what it considers to be the most promising research areas in heart disease. These included: **gene therapy** which could potentially encourage the growth of new blood vessels to and from the heart, thus bypassing diseases vessels; the discovery of new "super aspirins" which seem to have even greater protective effects for both heart attack and stroke; more data to support the association between inflammation and heart attacks; better techniques for early detection of obstructed vessels in the heart (using magnetic resonance imaging, or MRI); hope that damaged left ventricular muscle can regain better functioning, if a mechanical device called a left ventricular assist device (LVAD) takes over the work of the left ventricle for a time; further evidence that tobacco is a crucial risk factor in the development of heart disease, as evidenced by research which showed that as few as 10 cigarettes a day shortens life; more research supporting the importance of diet and exercise on levels of cholesterol in the blood; efforts to encourage people to seek treatment more quickly when a heart attack is suspected; the association between non-responsiveness to nitric oxide, and the development of high blood pressure.

See also Arteriosclerosis; Blood pressure and hormonal control mechanisms; Cardiac cycle; Cardiac muscle; Circulatory system; Collateral circulation; Electrocardiograms (ECG); Heart defects; Heart: embryonic development and changes at birth; Heart: rhythm control and impulse conduction; Lipids and lipid metabolism

CARDIAC MUSCLE

Cardiac muscle is found only in the **heart**. The middle layer of the heart, the myocardium, is composed entirely of cardiac muscle. The myocardium is composed of interconnected bundles of branching muscle fibers arranged in a spiral around the heart. This spiral shape is due to the complex twisting of the heart during development.

Functionally and structurally, cardiac muscle shares characteristics of both skeletal and **smooth muscle**. Like **skeletal muscle**, it contains an abundance of **mitochondria** and myoglobin, because of the tremendous energy requirements of the heart. In addition, like skeletal muscle, it has a striated appearance. This is due to a regular banding pattern made up of thin and thick filaments.

The individual cardiac muscle cells are joined end to end by intercalated disks. These disks contain two types of specialized junctions. One, the desmosomes, acts like a spot rivet to hold the cells together. These types of adhesion junctions are found in great quantities in tissues that are subject to considerable mechanical stress. The other type of junction is called a gap junction. Its function is to allow the passage of the **action potential** (the electrochemical propagation of a nerve signal) to spread from one cell to another.

Cardiac muscle is capable of generating a spontaneous electrochemical signal without stimulation from the nerves. This self-excitable capacity is shared with smooth muscle. The action potential then spreads to adjoining cells. This allows contraction of the cells as a single coordinated unit. These groups of interconnected cells that function mechanically and electronically together are referred to as a functional syncytium. Two other histological properties of cardiac muscle are a central nuclei (often football shaped) and a high cytoplasm to nucleus ratio.

See also Cardiac cycle; Electrocardiogram (ECG); Muscles of the thorax and abdomen; Muscular innervation; Muscular system overview

CARLSSON, ARVID (1923-   )
Swedish biochemist

Arvid Carlsson, along with **Paul Greengard** and **Eric Kandel**, received the Nobel Prize in physiology or medicine in 2000 for work concerning signal transduction in the nervous system. The three scientists have enhanced the understanding of normal **brain** function by discovering slow synaptic transmission, one type of signal transduction between nerve cells. In addition, their work establishes a basis for the evaluation of receptor genes and the genetic control of signal transmission.

The human brain has billions of nerve cells, or **neurons**. The junction between neurons is the **synapse**, the precise point of the transmission of information, or signal transduction, from the brain. **Neurotransmitters** are formed from chemical compounds, or precursors, and are released by the presynaptic cell into the synapse. There the neurotransmitters combine with protein molecules (receptors) from the postsynaptic cell to regulate electrical signal transmission between neurons.

Arvid Carlsson received the Nobel Prize for his discovery that dopamine is a neurotransmitter. Paul Greengard then discovered how dopamine and other neurotransmitters interact with the nervous system, and Eric Kandel determined that the synapse is the location of short-term and long-term memory in the sea slug. The research of the three Nobel laureates has led to a better understanding of how the brain functions and how neurological and psychiatric diseases can be caused by disturbances in signal transduction. Because of the findings of Carlsson, and the further research of Greengard and Kandel, scientists are able to understand the molecular mechanisms involved in certain neuropsychiatric diseases.

Arvid Carlsson's groundbreaking research in the 1950's established the relationship between neurotransmitters and diseases of the **central nervous system** when he discovered that a depleted neurotransmitter causes Parkinson's disease, a debilitating brain condition named after the English doctor who first described it in 1817. Carlsson moreover established that Parkinson's and other diseases caused by neurotransmitter dysregulation could be treated with drugs that regulate the governing neurotransmitters.

Born in Uppsala, Sweden, Arvid Carlsson received his medical degree from the University of Lund, Sweden, in 1951. He became professor of **pharmacology** at the University of Göteborg, Sweden, in 1959, and Professor Emeritus in 1989. A member of the Swedish Academy of Sciences and a foreign affiliate of the U.S. National Academy of Sciences, Professor Carlsson has authored several hundred peer-reviewed journal articles and received almost twenty prestigious awards before winning the Nobel Prize in 2000.

While conducting research at the University of Lund on how neurons transmit signals, Carlsson discovered that dopamine is a neurotransmitter. Scientists had previously believed that dopamine was only a precursor of noradrenaline, another neurotransmitter. Carlsson, however, was able to measure levels of dopamine in different brain **tissue**. Finding that dopamine existed in areas where there was no noradrenaline, he concluded that dopamine itself was a transmitter. He also discovered that dopamine was especially concentrated in the **basal ganglia**, the portion of the brain that controls motor behavior.

When Carlsson conducted several experiments with animals to deplete them of some of their neurotransmitters, he observed that the animals lost the ability for spontaneous movement. After he administered a precursor of dopamine called levodopa to the animals, they regained normal motor function. Carlsson then made a comparable experiment using the precursor of another transmitter, serotonin, but the animals did not regain their motor ability.

Recognizing the similarity between the impaired motor functioning in his animals and the symptoms (involuntary, uncontrollable tremors and increasing muscle rigidity) of Parkinson's disease in humans, Carlsson concluded that patients with Parkinson's must have a dopamine deficiency. His work led to the drug L-dopa being developed and used to treat Parkinson's, the first disease treated by drugs that replaced abnormally functioning neurotransmitters.

Carlsson's research led to an important discovery about schizophrenia, a debilitating mental illness. He found that the reuptake of dopamine by the receptors in the basal ganglia is much higher than normal in schizophrenics. Drugs that block dopamine receptors, therefore, are now prescribed for schizophrenia.

Carlsson also discovered that the dysregulation of another neurotransmitter, serotonin, causes clinical depression and several other behavioral disorders, including OCD (Obsessive-Compulsive Disorder), obesity, violent and aggressive behaviors, and suicide, one of the ten leading causes of **death** in the United States. His work led to the development of anti-depressive drugs called selective serotonin reuptake inhibitors (SSRIs, such as Prozac and Paxil) to treat the various illnesses caused by low levels of serotonin.

The drugs that emerged from Carlsson's research, however, are not cure-alls. L-dopa helps temporarily to alleviate some symptoms of Parkinson's, but it does not cure the disease; moreover, people who take L-dopa often suffer several unpleasant side effects, including **nausea** and cardiac irregularities. Patients who use SSRIs also complain of complications, including unwanted weight gain and some loss of sexual functioning. Nevertheless, Carlsson's research led to the understanding and treatment of serious neurological and psychiatric diseases.

See also Nerve impulses and conduction of impulses; Nervous system overview

CAROTID ARTERIES

Arising from the brachiocephalic trunk of the **aortic arch**, and the aortic arch itself, the common carotid **arteries** ultimately branch into the external and internal carotid arteries that provide the oxygenated **blood** supply to the neck and head. The right and left common carotid arteries, the right and left external carotid arteries, and the right and left internal carotid arteries form what anatomists term the carotid arterial system.

The common carotid arteries normally vary in origin. Although the right common carotid artery arises from the brachiocephalic artery, the left common carotid artery normally splits directly off the aortic arch. Normally the left common carotid branches from the highest point of the aortic arch near the junction of the neck and thorax. The common carotids then branch into the external and internal carotid arteries.

The respective external carotid arteries arising from the common carotid arteries supply blood to branches that provide blood to the pharyngeal and facial regions, superior thyroid, neck, and the **skull**. The internal carotids that arise from the common carotids supply blood to the **brain**, the orbit of the **eye**, the **middle ear**, and the hypophysis.

In the early embryo (4 weeks), primitive carotid arteries branching from the developing pharyngeal arterial system and a pair of dorsal aortae provide early circulation to the developing brain. The generation, fusion, and degeneration of embryonic arteries that accompany the development of the pharyngeal arch system into the ascending aorta and aortic arch region of the ascending aorta—and of carotid arteries derived from them—are typical of a head-to-tail (craniocaudal) scheme of development. This dynamic development usually explains the varied and often tortuous (twisting and turning) origin and course of vessels derived from the pharyngeal branch system around adult **organs** (e.g., the esophagus and trachea).

At about the level of the thyroid **cartilage** in the adult, the right and left common carotid arteries divide into the right and left sets of internal and external carotid arteries. Where this division takes place, the common carotids expand in diameter (dilate). Anatomists commonly refer to the enlargement of a vessel or chamber as a sinus. Accordingly, the carotid sinus refers to the location where the dilation occurs (dilatation of the vessel). Because the carotid arteries are dilated, the walls of the vessels are stretched thinner than in adjacent areas.

At the carotid sinus, the carotid system is most vulnerable to external pressure and compression.

Sufficient pressure on the carotid arteries can cause fainting or a more severe and prolonged lack of consciousness (carotid sinus syncope). The carotids are most sensitive to pressure at the point at which they split (bifurcate). The physiological mechanism for syncopal episodes involves the activation (excitation) of pressure receptors (baroreceptors) that may lead to a rapid and dramatic decrease in arterial pressure of 20-25 mm Hg (mercury) or more. It is this sudden loss or arterial pressure—and concurrent loss of oxygenated blood supply to the brain—that causes the fainting (syncopal) episode. If sufficient or prolonged pressure is applied, it is possible to cause a temporary cardiac arrest a few seconds in duration. In individuals with healthy hearts, the **heart** usually begins to beat again due the internal (intrinsic) contractility of cardiac muscle. Prolonged compression of the carotids may be fatal.

Within what is termed the carotid triangle (an area of the anterior neck defined by muscles—specifically by the sternocleidomastoid, and portions of the omohyoid, stylohyoid, and digastric muscles) gentle pressure is often applied to the external carotid artery to determine the presence of a **pulse**. The pulsations of the heart are easily felt because within the carotid triangle the external carotid is essentially only covered by skin and superficial **fascia**. According the external carotid is easily felt (palpated) and is highly vulnerable to external pressure and injury.

See also Anatomical nomenclature; Angiology; Blood pressure and hormonal control mechanisms; Circle of Willis; Ductus arteriosis; Pulse and pulse pressure; Systemic circulation; Vascular system (embryonic development)

CARPAL TUNNEL SYNDROME

Carpal tunnel syndrome results from compression and irritation of the median nerve where it passes through the wrist. In

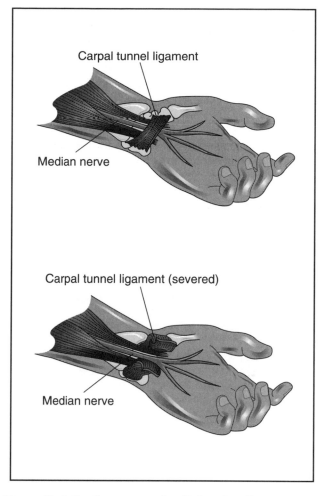

Diagram illustrating the compressed and inflamed median nerve typical of carpal tunnel syndrome. *Illustration by Electronic Illustrators Group.*

the end, the median nerve is responsible for both sensation and movement. When the median nerve is compressed, an individual's hand will feel as if it has "gone to sleep." The individual will experience numbness, tingling, and a prickly pin like sensation over the palm surface of the hand, and the individual may begin to experience muscle weakness, making it difficult to open jars and hold objects with the affected hand. Eventually, the muscles of the hand served by the median nerve may begin to atrophy, or grow noticeably smaller.

Compression of the median nerve in the wrist can occur during a number of different conditions, particularly conditions that lead to changes in fluid accumulation throughout the body. Because the area of the wrist through which the median nerve passes is very narrow, any swelling in the area will lead to pressure on the median nerve, which will interfere with the nerve's ability to function normally. **Pregnancy**, obesity, arthritis, certain thyroid conditions, diabetes, and certain pituitary abnormalities all predispose to carpal tunnel syndrome. Furthermore, overuse syndrome, in which an individual's job requires repeated strong wrist motions (in particular, motions which bend the wrist inward toward the forearm) can also predispose to carpal tunnel syndrome.

Research conducted by the American Academy of Orthopedic Surgeons has found that advanced carpal tunnel syndrome can be prevented in many cases. By doing an uncomplicated set of wrist exercises consistently before work, during breaks, and after work, pressure on the median nerves that leads to carpal tunnel syndrome can be avoided. The exercises are simple, involving mild flexion and extension of the wrists. By stretching the associated **tendons**, trauma from repetitive exertion is made less likely by significantly lowering pressure within carpal tunnels. People most likely to benefit from such exercise are those who use computers and other electronic keyboard devices daily. Women are known to experience carpal tunnel syndrome more frequently than do men.

Carpal tunnel syndrome is initially treated by splinting, which prevents the wrist from flexing inward into the position that exacerbates median nerve compression. When carpal tunnel syndrome is more advanced, injection of steroids into the wrist to decrease inflammation may be necessary. The most severe cases of carpal tunnel syndrome may require surgery to decrease the compression of the median nerve and restore its normal function.

An often-underestimated disorder, carpal tunnel syndrome affects significant numbers of workers. In some years, according to federal labor statistics, carpal tunnel syndrome exceeded lower back **pain** in its contribution to the duration of work absences. One estimate reports that as many as 5–10 workers per 10,000 will miss work for some length of time each year due to work-related carpal tunnel syndrome. Additionally, the affliction is not limited to those whose jobs involve long hours of typing. International epidemiological data indicate that the highest rates of the disorder also include occupations such as meat-packers, automobile and other assembly workers, and poultry processors. Also from these studies, strong evidence is presented which positively correlates carpal tunnel syndrome with multiple risk factors, rather than a single factor alone. It is believed that the risk of developing carpal tunnel syndrome is far greater when continual repetition of action is combined with increased force of the action, wrist vibration, and overall poor **posture**.

See also Neural damage and repair; Posture and locomotion

CARREL, ALEXIS (1873-1944)

French physician

Alexis Carrel was an innovative surgeon whose experiments with the transplantation and repair of body **organs** led to advances in the field of surgery and the art of **tissue** culture. An original and creative thinker, Carrel was the first to develop a successful technique for suturing **blood** vessels together. For his work with blood vessel suturing and the **transplantation of organs** in animals, he received the 1912 Nobel Prize in medicine and physiology. Carrel's work with tissue culture also contributed significantly to the understand-

ing of **viruses** and the preparation of vaccines. A member of the Rockefeller Institute for Medical Research for thirty-three years, Carrel was the first scientist working in the United States to receive the Nobel Prize in medicine and physiology.

Carrel was born in Sainte-Foy-les-Lyon, a suburb of Lyons, France. He was the oldest of three children. His mother, Anne-Marie Ricard, was the daughter of a linen merchant. His father, Alexis Carrel Billiard, was a textile manufacturer. Carrel dropped his baptismal names, Marie Joseph Auguste, and became known as Alexis Carrel upon his father's death when the boy was five years old. As a child, Carrel attended Jesuit schools. Before studying medicine, he earned two baccalaureate degrees, one in letters (1889) and one in science (1890). In 1891, Carrel began medical studies at the University of Lyons. For the next nine years, Carrel gained both academic knowledge and practical experience working in local hospitals. He served one year as an army surgeon with the Alpine Chasseurs, France's mountain troops. He also studied under Leo Testut, a famous anatomist. As an apprentice in Testut's laboratory, Carrel showed great talent at dissection and surgery. In 1900, he received his medical degree but continued on at the University of Lyons teaching medicine and conducting experiments in the hope of eventually receiving a permanent faculty position there.

In 1894, the president of France bled to death after being fatally wounded by an assassin in Lyons. If doctors had known how to repair his damaged artery, his life may have been saved, but such surgical repair of blood vessels had never been done successfully. It is said that this tragic event captured Carrel's attention and prompted him to try and find a way to sew severed blood vessels back together. Carrel first taught himself how to sew with a small needle and very fine silk thread. He practiced on paper until he was satisfied with his expertise, then developed steps to reduce the risk of **infection** and maintain the flow of blood through the repaired vessels. Through his careful choice of materials and long practice at various techniques, Carrel found a way to suture blood vessels. He first published a description of his success in a French medical journal in 1902.

Blood transfusion and organ transplantation seemed within reach to Carrel, now that he had mastered the ability to suture blood vessels. In experiments with dogs, he performed successful kidney transplants. His bold investigations began to attract attention not only from other medical scientists but from the public as well. His work was reviewed in both medical journals and popular newspapers such as the *New York Herald*. In the era of Ford, Edison, and the Wright Brothers, the public was easily able to imagine how work in a scientific laboratory could lead to major changes in daily life. Human organ transplantation and other revolutions in surgery did not seem far off.

In 1906, the opportunity to work in a world-class laboratory came to Carrel. The new Rockefeller Institute for Medical Research (now named Rockefeller University) in New York City offered him a position. Devoted entirely to medical research, rather than teaching or patient care, the Rockefeller Institute was the first institution of its kind in the United States. Carrel would remain at the institute until 1939.

At the Rockefeller Institute, Carrel continued to improve his methods of blood-vessel surgery. He knew that mastering those techniques would allow for great advances in the treatment of disorders of the **circulatory system** and wounds. It also made direct blood **transfusions** possible at a time when scientists did not know how to prevent blood from clotting. Without this knowledge, blood could not be stored or transported. In the *Journal of the American Medical Association* in 1910, Carrel described connecting an artery from the arm of a father to the leg of an infant in order to treat the infant's intestinal bleeding. Although the experiment was a success, the discovery of anticoagulants soon made such direct transfer unnecessary. For his pioneering efforts, Carrel won the Nobel Prize in 1912.

Carrel's success with tissue cultures through animal experiments led him to wonder whether human tissues and even whole organs, might be kept alive artificially in the laboratory. If so, lab-raised organs might eventually be used as substitutes for diseased parts of the body. The art of keeping cells and tissue alive, and even growing, outside of the body is known as tissue culture. Successfully culturing tissue requires great technical skill. Carrel was particularly interested in perfusion—a procedure of artificially pumping blood through an organ to keep it viable. Carrel's work with tissue culture contributed greatly to the understanding of normal and abnormal cell life. His techniques helped lay the groundwork for the study of viruses and the preparation of vaccines for polio, measles, and other diseases. Carrel's discoveries, in turn, built upon the successes of, among others, Ross G. Harrison, a contemporary anatomist at Yale who worked with frog tissue cultures and transplants.

When World War I began, Carrel was in France. The French government called him to service with the army, assigning him to run a special hospital near the front lines for the study and prompt treatment of severely infected wounds. There, Madame Carrel, his wife of less than one year and a trained surgical nurse, assisted him. In collaboration with biochemist Henry D. Dakin, Carrel developed an elaborate method of cleansing deep wounds to prevent infection. The method was especially effective in preventing gangrene, and was credited with saving thousands of lives and limbs. The Carrel-Dakin method, however, was too complicated for widespread use, and has since been replaced by the use of antibiotic drugs.

After an honorable discharge in 1919, Carrel returned to the Rockefeller Institute in New York City. He resumed his work in tissue culture, and began an investigation into the causes of **cancer**. In one experiment, he built a huge mouse colony to test his theories about the relationship between **nutrition** and cancer. But the experiment produced inconclusive results, and the Institute ceased funding it after 1933. Nevertheless, Carrel's tissue culture research was successful enough to earn him the Nordhoff-Jung Cancer Prize in 1931 for his contribution to the study of malignant **tumors.**

In the early 1930s, Carrel returned again to the challenge of keeping organs alive outside the body. With the engineering expertise of aviator Charles A. Lindbergh, Carrel designed a special sterilizing glass pump that could be used to circulate nutrient fluid around large organs kept in the lab. This perfusion pump, a so-called artificial **heart**, was germ-free and was successful in keeping animal organs alive for several days or weeks, but this was not considered long enough for practical application in surgery. Still, the experiment laid the groundwork for future developments in heart-lung machines and other devices. To describe the use of the perfusion pump, Carrel and Lindbergh jointly published *The Culture of Organs* in 1938. Lindbergh was a frequent sight at the Rockefeller Institute for several years, and the Lindberghs and the Carrels became close friends socially. They appear together on the July 1, 1935, cover of *Time* magazine with their "mechanical heart."

Carrel's mystical bent, publicly revealed after his visit to Lourdes as a young man, was displayed again in 1935. That year Carrel published *Man, the Unknown,* a work written upon the recommendation of a loose-knit group of intellectuals that he often dined with at the Century Club. In *Man, the Unknown,* Carrel posed highly philosophical questions about mankind, and theorized that mankind could reach perfection through selective reproduction and the leadership of an intellectual aristocracy. The book, a worldwide best seller and translated into nineteen languages, brought Carrel international attention. Carrel's speculation about the need for a council of superior individuals to guide the future of mankind was seen by many as anti-democratic. Others thought that it was inappropriate for a renowned scientist to lecture on fields outside his own.

Unfortunately, one of those who disliked Carrel's habit of discussing issues outside the realm of medicine was the new director of the Rockefeller Institute. Herbert S. Gasser had replaced Carrel's friend and mentor, Simon Flexner, in 1935. Suddenly Carrel found himself approaching the mandatory age of retirement with a director who had no desire to bend the rules and keep him aboard. In 1939, Carrel retired and his laboratories, along with the Division of Experimental Surgery, were closed.

Carrel's retirement coincided with the beginning of World War II in September 1939. Carrel and his wife were in France at the time and Carrel immediately approached the French Ministry of Public Health and offered to organize a field laboratory, much like the one he had run during World War I. When the government was slow to respond, Carrel grew frustrated. In May 1940, he returned to New York alone. As his steamship was crossing the Atlantic, Hitler invaded France.

Carrel made the difficult return to war-torn Europe as soon as he was able, arriving in France via Spain in February 1941. Paris was under the control of the Vichy government, a puppet administration installed by the German military command. Although Carrel declined to serve as director of public health in the Vichy government, he stayed in Paris to direct the Foundation for the Study of Human Problems. The Foundation, supported by the Vichy government and the German military command, brought young scientists, physicians, lawyers, and engineers together to study economics, political science, and nutrition. When the Allied forces reoccupied France in August 1944, the newly restored French government immediately suspended Carrel from his directorship of the Foundation and accused him of collaborating with the Germans. Mercifully, perhaps, a serious heart attack forestalled any further prosecution. Attended by French and American physicians, and nursed by his wife, Carrel died of heart failure in Paris in 1944. After his death, his body was buried in St. Yves chapel near his home on the island of Saint Gildas, Cotes-du-Nord.

Carrel's reputation remains that of a brilliant, yet temperamental man. His motivations for his involvement with the Nazi-dominated Vichy government remain the subject of debate. Yet there is no question that his achievements ushered in a new era in medical science. His pioneering techniques paved the way for successful organ transplants and modern heart surgery, including grafting procedures and bypasses.

See also Circulatory system; Transplantation of organs

CARTILAGE

Cartilage is connective **tissue** that lacks **blood** vessels, and functions to bind structures together or provide support. Cartilage cushions **organs** and articulating (flexing) **joints** in the body. In an embryo, some of the precursors to bones are made of cartilage. Ribs, ears and the **nose** are also made of cartilage.

Typical of connective tissue, cartilage is composed of glycosaminoglycans, proteoglycans, and glycoproteins. Cells called chondroblasts produce these compounds. Additionally, with the exception of cartilage found in articulating joints, cartilage is covered with a connective tissue later called the perichondrium.

There are three mains types of cartilage: **hyaline cartilage**, **elastic cartilage**, and **fibrocartilage**. Hyaline cartilage gets its name from its translucent (hyaline) appearance. The collagen and elastin fibers cannot be seen. It is the most common and best-studied form of cartilage. The elastic cartilage is found in the ear, auditory canals of the outer ear, and in the **larynx**. It is essentially hyaline cartilage enriched in fibers made out of elastin. These fibers are visible under microscopic magnification. The presence of more elastin confers a more elastic quality to the cartilage relative to hyaline cartilage. Fibrocartilage represents an intermediate between regular connective tissue and hyaline cartilage. It is located in some ligaments, the public region of the body and in the intervertebral discs of the spine.

Cartilage damage is infamous as the source of knee **pain** for many people. In the knee, cartilage comprises the meniscus. Tears can occur in the meniscus due to injury (often sports-related) and age-related deterioration. Because the majority of the meniscus does not have a blood supply, the normal healing process that occurs elsewhere in the body does not occur in the knee. The torn meniscus causes abnormal, often painful movement of the knee joint. Surgical intervention is usually necessary to effect repair.

See also Arthrology; Cartilaginous joints

CARTILAGINOUS JOINTS

Joints are the connections, or articulations, between bones. Joints may be classified either structurally, by what material links the bones, or functionally, by how much movement is allowed in the joint. In common usage, the word "joints" generally refers to the freely moving joints with joint cavities. These are called **synovial joints**. However, some joints allow little or no movement between the bones. These joints may consist either of fibrous connective **tissue** or **cartilage**. Fibrous and cartilaginous joints may eventually turn to bone, or ossify, over time.

There are two types of cartilaginous joints. Primary cartilaginous joints are also called synchondroses. In these joints, which are always immovable, a plate of smooth, pearly hyaline cartridge connects the bones. Examples of these joints are the epiphyseal or growth plates at the ends of the long bones. These serve as bone-forming centers during growth, and ossify in adults.

The other class of cartilaginous joint consists of the secondary cartilaginous joints, or symphyses. In these joints, the surfaces of the bonds to be joined are each covered with a smooth layer of **hyaline cartilage**. The two hyaline cartilage layers are themselves joined by **fibrocartilage**, which contains thick strands of collagen fibers. The result is a sort of sandwich of cartilaginous material between the bones. The cartilaginous pad can be compressed and stretched, allowing limited movement at these joints.

There are a few especially important examples of symphyses in the human body, such as the pads between the vertebrae. These intervertebral discs have tough fibrocartilage around their rims and less dense, fluid-soaked centers. The discs serve as shock absorbers and provide flexibility in the spine.

The **pubic symphysis** joint in the lower pelvis is usually immovable. However, during **pregnancy**, hormone changes influence its structure and make it more flexible. This allows it to stretch so that the baby's head can pass through the birth canal.

See also Arthrology; Costochondral cartilage; Fibrous joints; Joints and synovial membranes

CAUDAD • *see* ANATOMICAL NOMENCLATURE

CELL CYCLE AND CELL DIVISION

The series of stages that a cell undergoes while progressing to division is known as cell cycle. In order for an organism to grow and develop, the organism's cells must be able to duplicate themselves. Three basic events must take place to achieve this duplication: the **deoxyribonucleic acid (DNA)**, which makes up the individual **chromosomes** within the cell's nucleus must be duplicated; the two sets of DNA must be packaged up into two separate nuclei; and the cell's cytoplasm must divide itself to create two separate cells, each complete with its own nucleus. The two new cells, products of the single original cell, are known as daughter cells.

Although prokaryotes (e.g. **bacteria**, non-nucleated unicellular organisms) divide through binary fission, eukaryotes (including, of course, human cells) undergo a more complex process of cell division because DNA is packed in several chromosomes located inside a cell nucleus. In eukaryotes, cell division may take two different paths, in accordance with the cell type involved. Mitosis is a cellular division resulting in two identical nuclei that takes place in somatic cells. Sex cells or gametes (ovum and spermatozoids) divide by meiosis. The process of meiosis results in four nuclei, each containing half of the original number of chromosomes. Both prokaryotes and eukaryotes undergo a final process, known as cytoplasmatic division, which divides the parental cell in new daughter cells.

Mitosis is the process during which two complete, identical sets of chromosomes are produced from one original set. This allows a cell to divide during another process called cytokinesis, thus creating two completely identical daughter cells.

During much of a cell's life, the DNA within the nucleus is not actually organized into the discrete units known as chromosomes. Instead, the DNA exists loosely within the nucleus, in a form called chromatin. Prior to the major events of mitosis, the DNA must replicate itself, so that each cell has twice as much DNA as previously.

Cells undergoing division are also termed competent cells. When a cell is not progressing to mitosis, it remains in phase G0 ("G" zero). Therefore, the cell cycle is divided into two major phases: interphase and mitosis. Interphase includes the phases (or stages) G1, S and G2 whereas mitosis is subdivided into prophase, metaphase, anaphase and telophase.

Interphase is a phase of cell growth and metabolic activity, without cell nuclear division, comprised of several stages or phases. During Gap 1 or G1 the cell resumes protein and **RNA** synthesis, which was interrupted during previous mitosis, thus allowing the growth and maturation of young cells to accomplish their physiologic function. Immediately following is a variable length pause for DNA checking and repair before cell cycle transition to phase S during which there is synthesis or semi-conservative replication or synthesis of DNA. During Gap 2 or G2, there is increased RNA and **protein synthesis**, followed by a second pause for proofreading and eventual repairs in the newly synthesized DNA sequences before transition to mitosis.

The cell cycle starts in G1, with the active synthesis of RNA and proteins, which are necessary for young cells to grow and mature. The time G1 lasts, varies greatly among eukaryotic cells of different species and from one **tissue** to another in the same organism. Tissues that require fast cellular renovation, such as mucosa and endometrial epithelia, have shorter G1 periods than those tissues that do not require frequent renovation or repair, such as muscles or **connective tissues**.

The first stage of mitosis is called prophase. During prophase, the DNA organizes or condenses itself into the specific units known as chromosomes. Chromosomes appear as double-stranded structures. Each strand is a replica of the other and is called a chromatid. The two chromatids of a chromosome are joined at a special region, the centromere. Structures called centrioles position themselves across from each other, at either end of the cell. The nuclear membrane then disappears.

During the stage of mitosis called metaphase, the chromosomes line themselves up along the midline of the cell. Fibers called spindles attach themselves to the centromere of each chromosome.

During the third stage of mitosis, called anaphase, spindle fibers will pull the chromosomes apart at their centromere (chromosomes have two complementary halves, similar to the two nonidentical but complementary halves of a zipper). One arm of each chromosome will migrate toward each centriole, pulled by the spindle fibers.

During the final stage of mitosis, telophase, the chromosomes decondense, becoming unorganized chromatin again. A nuclear membrane forms around each daughter set of chromosomes, and the spindle fibers disappear. Sometime during telophase, the cytoplasm and cytoplasmic membrane of the cell split into two (cytokinesis), each containing one set of chromosomes residing within its nucleus.

Cells are mainly induced into proliferation by growth factors or **hormones** that occupy specific receptors on the surface of the cell membrane, being also known as extra-cellular ligands. Examples of growth factors are as such: epidermal growth factor (EGF), fibroblastic growth factor (FGF), platelet-derived growth factor (PDGF), insulin-like growth factor (IGF), or by hormones. PDGF and FGF act by regulating the phase G2 of the cell cycle and during mitosis. After mitosis, they act again stimulating the daughter cells to grow, thus leading them from G0 to G1. Therefore, FGF and PDGF are also termed competence factors, whereas EGF and IGF are termed progression factors, because they keep the process of cellular progression to mitosis going on. Growth factors are also classified (along with other molecules that promote the cell cycle) as pro-mitotic signals. Hormones are also pro-mitotic signals. For example, thyrotrophic hormone, one of the hormones produced by the pituitary gland, induces the proliferation of thyroid gland's cells. Another pituitary hormone, known as growth hormone or somatotrophic hormone (STH), is responsible by body growth during childhood and early adolescence, inducing the lengthening of the long bones and protein synthesis. Estrogens are hormones that do not occupy a membrane receptor, but instead, penetrate the cell and the nucleus, binding directly to specific sites in the DNA, thus inducing the cell cycle.

Anti-mitotic signals may have several different origins, such as cell-to-cell adhesion, factors of adhesion to the extracellular matrix, or soluble factor such as TGF beta (tumor growth factor beta), which inhibits abnormal cell proliferation, proteins p53, p16, p21, APC, pRb, etc. These molecules are the products of a class of genes called tumor suppressor genes. Oncogenes, until recently also known as proto-oncogenes, synthesize proteins that enhance the stimuli started by growth factors, amplifying the mitotic signal to the nucleus, and/or promoting the accomplishment of a necessary step of the cell

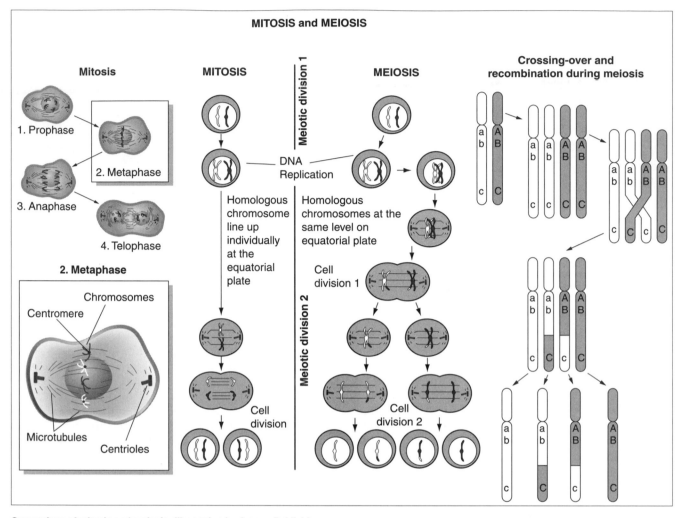

Comparison of mitosis and meiosis. *Illustration by Argosy Publishing.*

cycle. When each phase of the cell cycle is completed, the proteins involved in that phase are degraded, so that once the next phase starts, the cell is unable to go back to the previous one. Next to the end of phase G1, the cycle is paused by tumor suppressor gene products, to allow verification and repair of DNA damage. When DNA damage is not repairable, these genes stimulate other intra-cellular pathways that induce the cell into suicide or apoptosis (also known as programmed cell **death**). To the end of phase G2, before the transition to mitosis, the cycle is paused again for a new verification and "decision:" either mitosis or apoptosis.

Along each pro-mitotic and anti-mitotic intra-cellular signaling pathway, as well as along the apoptotic pathways, several gene products (proteins and **enzymes**) are involved in an orderly sequence of activation and inactivation, forming complex webs of signal transmission and signal amplification to the nucleus. The general goal of such cascades of signals is to achieve the orderly progression of each phase of the cell cycle.

Mitosis always creates two genetically identical cells from the original cell. In mitosis, the total amount of DNA

doubles briefly, so that the subsequent daughter cells will ultimately have the exact amount of DNA initially present in the original cell. Mitosis is the process by which all of the cells of the body divide and therefore reproduce. The only cells of the body which do not duplicate through mitosis are the sex cells (egg and **sperm** cells). These cells undergo a slightly different type of cell division called meiosis, which allows each sex cell produced to contain half of its original amount of DNA, in anticipation of doubling it again when an egg and a sperm unite during the course of conception.

Meiosis, also known as reduction division, consists of two successive cell divisions in diploid cells. The two cell divisions are similar to mitosis, but differ in that the chromosomes are duplicated only once, not twice. The end result of meiosis is four haploid daughter cells. Because meiosis only occurs in the sex **organs** (**gonads**), the daughter cells are the gametes (spermatozoa or ova), which contain hereditary material. By halving the number of chromosomes in the sex cells, meiosis assures that the fusion of maternal and paternal gametes at **fertilization** will result in offspring with the same chromosome number as the parents. In other words, meiosis compensates for chromosomes doubling at fertilization. The two successive nuclear divisions are termed as meiosis I and meiosis II. Each is further divided into four phases (prophase, metaphase, anaphase, and telophase) with an intermediate phase (interphase) preceding each nuclear division.

The events that take place during meiosis are similar in many ways to the process of mitosis, in which one cell divides to form two clones (exact copies) of itself. It is important to note that the purpose and final products of mitosis and meiosis are very different.

Meiosis I is preceded by an interphase period in which the DNA replicates (makes an exact duplicate of itself), resulting in two exact copies of each chromosome that are firmly attached at one point, the centromere. Each copy is a sister chromatid, and the pair are still considered as only one chromosome. The first phase of meiosis I, prophase I, begins as the chromosomes come together in homologous pairs in a process known as synapsis. Homologous chromosomes, or homologues, consist of two chromosomes that carry genetic information for the same traits, although that information may hold different messages (e.g., when two chromosomes carry a message for **eye** color, but one codes for blue eyes while the other codes for brown). The fertilized eggs (zygotes) of all sexually reproducing organisms receive their chromosomes in pairs, one from the mother and one from the father. During synapsis, adjacent chromatids from homologous chromosomes "cross over" one another at random points and join at spots called chiasmata. These connections hold the pair together as a tetrad (a set of four chromatids, two from each homologue). At the chiasmata, the connected chromatids randomly exchange bits of genetic information so that each contains a mixture of maternal and paternal genes. This "shuffling" of the DNA produces a tetrad, in which each of the chromatids is different from the others, and a gamete that is different from others produced by the same parent. Crossing over does, in fact, explain why each person is a unique individual, different even from those in the immediate family. Prophase I is also marked by the appearance of spindle fibers (strands of microtubules) extending from the poles or ends of the cell as the nuclear membrane disappears. These spindle fibers attach to the chromosomes during metaphase I as the tetrads line up along the middle or equator of the cell. A spindle fiber from one pole attaches to one chromosome while a fiber from the opposite pole attaches to its homologue. Anaphase I is characterized by the separation of the homologues, as chromosomes are drawn to the opposite poles. The sister chromatids are still intact, but the homologous chromosomes are pulled apart at the chiasmata. Telophase I begins as the chromosomes reach the poles and a nuclear membrane forms around each set. Cytokinesis occurs as the cytoplasm and organelles are divided in half and the one parent cell is split into two new daughter cells. Each daughter cell is now haploid (n), meaning it has half the number of chromosomes of the original parent cell (which is diploid-2n). These chromosomes in the daughter cells still exist as sister chromatids, but there is only one chromosome from each original homologous pair.

The phases of meiosis II are similar to those of meiosis I, but there are some important differences. The time between the two nuclear divisions (interphase II) lacks replication of DNA (as in interphase I). As the two daughter cells produced in meiosis I enter meiosis II, their chromosomes are in the form of sister chromatids. No crossing over occurs in prophase II because there are no homologues to **synapse**. During metaphase II, the spindle fibers from the opposite poles attach to the sister chromatids (instead of the homologues as before). The chromatids are then pulled apart during anaphase II. As the centromeres separate, the two single chromosomes are drawn to the opposite poles. The end result of meiosis II is that by the end of telophase II, there are four haploid daughter cells (in the sperm or ova) with each chromosome now represented by a single copy. The distribution of chromatids during meiosis is a matter of chance, which results in the concept of the law of independent assortment in **genetics**.

The events of meiosis are controlled by a protein enzyme complex known collectively as maturation promoting factor (MPF). These enzymes interact with one another and with cell organelles to cause the breakdown and reconstruction of the nuclear membrane, the formation of the spindle fibers, and the final division of the cell itself. MPF appears to work in a cycle, with the proteins slowly accumulating during interphase, and then rapidly degrading during the later stages of meiosis. In effect, the rate of synthesis of these proteins controls the frequency and rate of meiosis in all sexually reproducing organisms from the simplest to the most complex.

Meiosis occurs in humans, giving rise to the haploid gametes, the sperm and egg cells. In males, the process of gamete production is known as **spermatogenesis**, where each dividing cell in the testes produces four functional sperm cells, all approximately the same size. Each is propelled by a primitive but highly efficient flagellum (tail). In contrast, in females, **oogenesis** produces only one surviving egg cell from each original parent cell. During cytokinesis, the cytoplasm and organelles are concentrated into only one of the four daughter cells-the one which will eventually become the female ovum or egg. The other three smaller cells, called polar bodies, die and are reabsorbed shortly after formation. The concentration of cytoplasm and organelles into the oocyte greatly enhances the ability of the zygote (produced at fertilization from the unification of the mature ovum with a spermatozoa)to undergo reapid cell division.

The control of cell division is a complex process and is a topic of much scientific research. Cell division is stimulated by certain kinds of chemical compounds. Molecules called cytokines are secreted by some cells to stimulate others to begin cell division. Contact with adjacent cells can also control cell division. The phenomenon of contact inhibition is a process where the physical contact between neighboring cells prevents cell division from occurring. When contact is interrupted, however, cell division is stimulated to close the gap between cells. Cell division is a major mechanism by which organisms grow, tissues and organs maintain themselves, and wound healing occurs.

Cancer is a form of uncontrolled cell division. The cell cycle is highly regulated by several enzymes, proteins, and cytokines in each of its phases, in order to ensure that the resulting daughter cells receive the appropriate amount of genetic information originally present in the parental cell. In the case of somatic cells, each of the two daughter cells must contain an exact copy of the original genome present in the parental cell. Cell cycle controls also regulate when and to what extent the cells of a given tissue must proliferate, in order to avoid abnormal cell proliferation that could lead to dysplasia or tumor development. Therefore, when one or more of such controls are lost or inhibited, abnormal overgrowth will occur and may lead to impairment of function and disease.

See also Cell structure; Cell theory; Genetic regulation of eukaryotic cells

CELL DIFFERENTIATION

Cells differ from each other in morphology (structure), and this difference is a reflection of physiological activities and biochemical functions that are ultimately under the control of genes. The differences that can be seen grossly, observed in the microscope, and detected by biochemical and molecular procedures together comprise what is known as differentiation.

Differentiation is associated with **embryology**. The undifferentiated cells of a zygote, morula, and blastula give rise to progressively more differentiated cells until the adult forms, which is a mosaic of many highly differentiated cells. Ordinarily, differentiated cells have lost the competence to give rise any other kind of cell. For example, muscle cells never give rise to organ cells, and vise versa. Moreover, contained within the mosaic of terminally differentiated cells are a number of **stem cells**. A stem cell is a less than fully differentiated cell that has retained the competence to give rise to another stem cell and a cell that will become fully differentiated. Consider the skin—it would rapidly be lost because of abrasion and the wear and tear of use were it not for replacement by cells from the basal layer of **epidermis**. Differentiation of the skin stem cell progeny gives rise to postmitotic keratinized protective cells. **Blood** is a **tissue** type that must be continually replaced. It is not surprising, therefore, to note that new blood cells develop from stem cells, which like skin stem cells, at division give rise to both more stem cells and cells which will differentiate as blood.

Differentiation resulting from selective gene action of a genome (the entire genetic complement of an organism) held in common by all cells has been a tenet of modern genetic biology. It is the business of a cell to produce all of the proteins and **enzymes** held in common by most cells. The commonly produced gene products are sometimes referred to as housekeeping proteins. However, the adult fly, frog or human are comprised of a great diversity of differentiated cells. The differentiated cells produce, in addition to the housekeeping gene products, tissue-specific proteins. A unique portion of the genome of differentiated cells is activated and this accounts for differentiation. While all other cells have these gene sequences, they are silent except in the specific cell type under consideration. Thus, cell types differ from each other not because of genomes, nor because of the activity of housekeeping genes, but by the activation of tissue-specific genes, which convey cell specificity. Gene regulation that permits differentiation is the result of promoters and enhancers (which occur in **DNA** on either side of specific genes) and regulatory proteins which bind to the promoters and enhancers that in turn, enhance or inhibit gene expression.

See also Cell cycle and cell division; Cell structure; Cell theory

CELL MEMBRANE TRANSPORT

The cell is bound by an outer membrane that, in accord with the fluid mosaic model, is comprised of a phospholipid lipid bilayer with proteins—molecules that also act as receptor sites—interspersed within the phospholipid bilayer. Varieties of channels exist within the membrane. There are a number of internal cellular membranes that partially partition the intercellular matrix, and that ultimately become continuous with the nuclear membrane.

There are three principal mechanisms of outer cellular membrane transport (i.e., means by which molecules can pass through the boundary cellular membrane). The transport mechanisms are passive, or gradient diffusion, facilitated diffusion, and active transport.

Diffusion is a process in which the random motions of molecules or other particles result in a net movement from a region of high concentration to a region of lower concentration. A familiar example of diffusion is the dissemination of floral perfumes from a bouquet to all parts of the motionless air of a room. The rate of flow of the diffusing substance is proportional to the concentration gradient for a given direction of diffusion. Thus, if the concentration of the diffusing substance is very high at the source, and is diffusing in a direction where little or none is found, the diffusion rate will be maximized. Several substances may diffuse more or less independently and simultaneously within a space or volume of liquid. Because lightweight molecules have higher average speeds than heavy molecules at the same temperature, they also tend to diffuse more rapidly. Molecules of the same weight move more rapidly at higher temperatures, increasing the rate of diffusion as the temperature rises.

Driven by concentration gradients, diffusion in the cell usually takes place through channels or pores lined by proteins. Size and electrical charge may inhibit or prohibit the passage of certain molecules or **electrolytes** (e.g., sodium, potassium, etc.).

Osmosis describes diffusion of water across cell membranes. Although water is a polar molecule (i.e., has overall partially positive and negative charges separated by its molecular structure), transmembrane proteins form hydrophilic (water loving) channels to through which water molecules may move.

Facilitated diffusion is the diffusion of a substance not moving against a concentration gradient (i.e., from a region of low concentration to high concentration) but which require the assistance of other molecules. These are not considered to be energetic reactions (i.e., energy in the form of use of **adenosine triphosphate** molecules [**ATP**] is not required). The facilitation or assistance—usually in physically turning or orienting a molecule so that it may more easily pass through a membrane—may be by other molecules undergoing their own random motion.

Transmembrane proteins establish pores through which ions and some small hydrophilic molecules are able to pass by diffusion. The channels open and close according to the physiological needs and state of the cell. Because they open and close transmembrane proteins are termed "gated" proteins. Control of the opening and closing mechanism may be via mechanical, electrical, or other types of membrane changes that may occur as various molecules bind to cell receptor sites.

Active transport is movement of molecules across a cell membrane or membrane of a cell organelle, from a region of low concentration to a region of high concentration. Since these molecules are being moved against a concentration gradient, cellular energy is required for active transport. Active transport allows a cell to maintain conditions different from the surrounding environment.

There are two main types of active transport; movement directly across the cell membrane with assistance from transport proteins, and endocytosis, the engulfing of materials into a cell using the processes of pinocytosis, phagocytosis, or receptor-mediated endocytosis.

Transport proteins found within the phospholipid bilayer of the cell membrane can move substances directly across the cell membrane, molecule by molecule. The **sodium-potassium pump**, which is found in many cells and helps nerve cells to pass their signals in the form of electrical impulses, is a well-studied example of active transport using transport proteins. The transport proteins that are an essential part of the sodium-potassium pump maintain a higher concentration of potassium ions inside the cells compared to outside, and a higher concentration of sodium ions outside of cells compared to inside. In order to carry the ions across the cell membrane and against the concentration gradient, the transport proteins have very specific shapes that only fit or bond well with sodium and potassium ions. Because the transport of these ions is against the concentration gradient, it requires a significant amount of energy.

Endocytosis is an infolding and then pinching in of the cell membrane so that materials are engulfed into a vacuole or vesicle within the cell. Pinocytosis is the process in which cells engulf liquids. The liquids may or may not contain dissolved materials. Phagocytosis is the process in which the materials that are taken into the cell are solid particles. With receptor-mediated endocytosis the substances that are to be transported into the cell first bind to specific sites or receptor proteins on the outside of the cell. The substances can then be engulfed into the cell. As the materials are being carried into the cell, the cell membrane pinches in forming a vacuole or other vesicle. The materials can then be used inside the cell. Because all types of endocytosis use energy, they are considered active transport.

See also Action potential; Cell cycle and cell division; Cell structure

CELL STRUCTURE

The cell is the basic unit of life. The **cell theory**, set forth in the 1850s, states that cells are the fundamental units of life, because a cell is the simplest unit capable of independent existence. All living things are made of cells. Cells are composed up of 90% water. The rest of the present molecules are 50% protein, 15% carbohydrate, 15% nucleic acid, 10% **lipids** and 10% others.

Two types of cells are recognized in living things. Prokaryotes (literally, "before the nucleus") are cells that have no distinct nucleus that is found to float around the cell. Most prokaryotic organisms are single-celled, such as **bacteria** and algae. Eucaryotic (literally, "true nucleus") organisms, on the other hand, have a distinct nucleus and a highly organized internal structure. The nucleus directs all cellular activities by controlling the synthesis of proteins. The nucleus contains encoded instructions for the synthesis of proteins in a helical double-stranded molecule called **deoxyribonucleic acid** (**DNA**). The nucleus, therefore, directs all cellular activities and thus, all body processes.

Prokaryotes can grow rapidly and are widespread in the biosphere. Some of them can double size, mass and number in 20 minutes. They are highly versatile and can survive in a wide variety of nutrients and environments. These cells of these organisms have a rigid cell wall immediately inside which is a cell membrane, which play a critical role in the transfer of chemical to and from the cells. Under a microscope, grainy dark spots distinctly apparent inside the cells are the ribosome's, which are the sites of important chemical reactions. Ribosomes do not have a membrane. Ribosome's fall into two separate units while not synthesizing protein. The cytoplasm is the fluid material inside the cell while the bubble-like regions inside the cell are the storage granules. While sharing many common features, prokaryotes have considerable structural and biochemical diversity.

Eucaryotic cells, on the other hand are 1,000-10,000 times larger than prokaryotic cells. All cells of plants, animals, and fungi belong to this family. Multi-cellular organisms contain a vast array of highly specialized cells. Plants, for example, contain root cells, leaf cells, and **stem cells**. Humans have skin cells, nerve cells, and sex cells. Each kind of cell is structured to perform a highly specialized function. In order to meet the many different specialized functions, these cells exist in many different forms. They act in a cooperative manner in multi-cellular organisms. However, many important microbial species are also eucaryotes.

Eucaryotic cells have a substantial degree of spatial organization and differentiation. They have a number of membrane-enclosed domains called organelles inside the cells. Besides the nuclear membrane, which clearly differentiates the nucleus and the ribosome, eukaryotes have a complex convoluted membrane system called the endoplasmic reticulum from the cell wall into the cell. The endoplasmic reticulum is the transport system for molecules needed for certain changes and specific destinations, instead of molecules that float freely in the cytoplasm. There are two types, rough and smooth. Rough ones have ribosomes attached to it, as mentioned

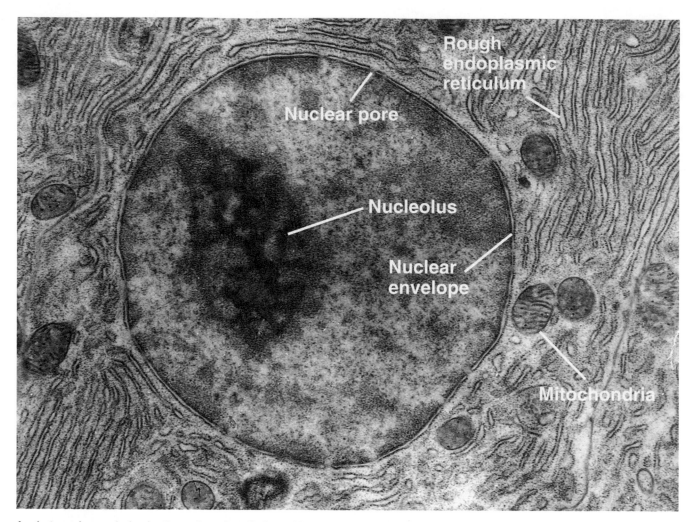

An electron micrograph showing the nucleus of a cell, along with organelles including mitochondria, the site of ATP production. © Don W. Fawcett, National Audubon Society Collection/ Photo Researchers, Inc. Reproduced by permission.

before, and smooth ones do not. The **mitochondria** are highly specialized organelles, which catalyze reactions whose products are major energy supplies to the cell. Mitochondria are the sites of aerobic respiration, and generally are the major energy production center in eucaryotes. Mitochondria have two membranes, an inner membrane and an outer membrane. The reticulations, or many foldings of the inner membrane, serve to increase the surface area of membrane on which membrane-bound reactions can take place. The golgi complex, lysosomes, and vacuoles are the remaining organelles that isolate chemical reactions and chemicals compounds from the cytoplasm. The golgi changes molecules and divides them into small membrane contained sacs called vesicles. These sacs can be sent to various locations in the cell. The lysosome is the digestive system in the cell. It breaks down molecules into their base components digestive **enzymes**. This demonstrates one of the reasons for having all parts of a cell compartmentalized; the cell could not use the destructive enzymes if they were not sealed off from the rest of the cell. Chloroplasts are the site of photosynthesis in eucaryotic cells.

They are disk-like structures composed of a single membrane surrounding a fluid containing stacks of membranous disks. Because of their green color, chloroplast are the only organelles that can be easily seen with a light microscope.

See also Cell cycle and cell division; Cell differentiation; Cell membrane transport

CELL THEORY

Modern cell theory asserts that all living things are made of basic structural units termed cells. Moreover, living cells arise only from other living cells. Cells also function as the basic physiological unit of living things. Cells, as a basic structural and functional unit, mirror and support the processes apparent in the whole organism.

The term cell was first used by the English scientist **Robert Hooke**, who, in the mid-seventeenth century, used the term to describe the structure of cork. Shortly thereafter, the

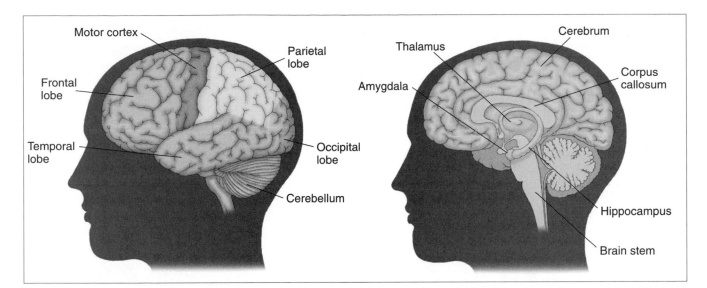

The anatomy of the human brain, exterior view showing the lobes (left) and the interior sections (right). *Illustration by Frank Forney. Reproduced by permission.*

Dutch scientist **Anton van Leeuwenhoek** made the first recorded observations of bacterial cells (termed "animalcules" by Leeuwenhoek) in scrapings taken from his own teeth. In the 1830's, German botanist **Matthias Schleiden** served that plants were composed of cells, and within a year German zoologist **Theodor Schwann** extended Schwann's assertions to assert that animals were also composed of fundamental cellular units or cells.

Schwann was the first to assert a modern form of cell theory. Schwann asserted that both plants and animals are composed entirely of cells or the fluid products of cells. Schwann also asserted that cells not have structural integrity but also a separate internal **physiology**. In multi-cellular organism, however, this internal cellular physiology functioning in a supportive and coordinated role with the cell comprising the organism as a whole.

In the mid-nineteenth century, the German physician Rudolph Virchow's (1821–1902) embryological work demonstrated that living cells could arise only from other living cells (biogenesis), and not from inanimate matter (abiogenesis). Virchow's assertion, *"Omnis cellula e cellula"* (All cells from cells) was the basis upon which French scientist Louis Pasteur (1822–1895) was later to fully extend cell theory—and in the process quash centuries of speculation regarding spontaneous generation (the generation of modern living organisms from inanimate matter) and provide a tremendous boost to evolutionary theory by demonstrating that living organisms arise only from living organisms.

See also Cell cycle and cell division; Cell differentiation; Cell membrane transport; Cell structure; Histology and microanatomy; History of anatomy and physiology: The Classical and medieval periods; History of anatomy and physiology: The Renaissance and age of Enlightenment; History of anatomy and physiology: The science of medicine

CENTRAL NERVOUS SYSTEM (CNS)

The central nervous system is comprised of the **brain** and the **spinal cord**. The brain receives sensory information from the nerves that pass through the spinal cord and from other nerves as well, such as those from sensory **organs** involved in sight and **smell**. Once received, the brain processes the sensory signals and initiates responses.

The spinal cord is a principle conduit of sensory information to and from the brain. Information flows to the central nervous system from the **peripheral nervous system**, which senses signals from the environment outside the body (sensory-somatic nervous system) and from the internal environment (**autonomic nervous system**). The brain's responses to incoming information flow through the spinal cord nerve network to the various effector organs and **tissue** regions where the target responsive action will take place.

Both the spinal cord and the brain are made up of structures of nerve cells called **neurons**. The long main body extension of a neuron is termed an axon. Depending on the type of nerve, the axons may be coated with a material called myelin. Both the brain and spinal cord components of the central nervous system contain bundles of cell bodies (out of which axons grow) and branched regions of nerve cells that are called dendrites. Between the axon of one cell body and the dendrite of another nerve cell is an intervening region called the **synapse**. In the spinal cord of humans, the myelin-coated axons are on the surface and the axon-dendrite network is on the inside. In the brain this arrangement is reversed.

Another important component of the central nervous system are the **meninges**. The meninges are three sheets or layers of connective tissue that cover all of the spinal cord and the brain. Going from the outside in the layers are called the dura matter, arachnoid and the pia matter. Infections of the

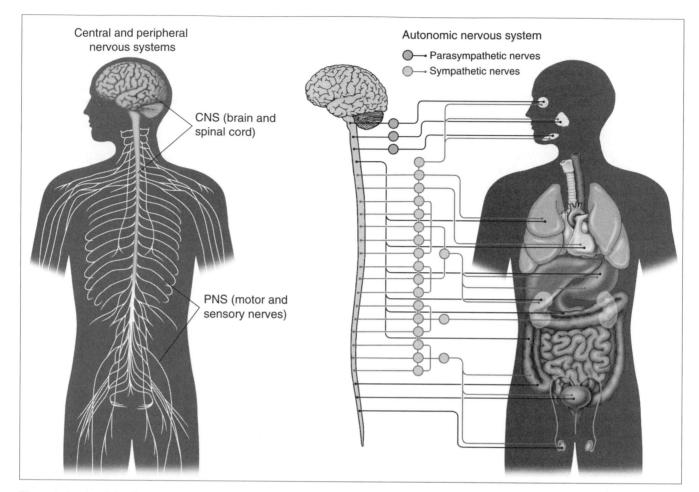

The central and peripheral nervous systems (left) and the autonomic nervous system (right). *Illustration by Frank Forney. Reproduced by permission.*

meninges are called meningitis. Bacterial, viral, and proto-zoan meningitis is serious and requires prompt medical atten-tion. Between the arachnoid and the pia matter is a fluid called the **cerebrospinal fluid**. Bacterial infections of the cere-brospinal fluid can occur and are life-threatening.

Along the length of the spinal cord are positioned thirty-one pairs of nerves. These are known as mixed spinal nerves, as they convey sensory information to the brain and response information back from the brain. Spinal nerve roots emerge from the spinal cord that lies within the spinal canal. Both doral and ventral roots fuse in the intervertebral foramen to create a spinal nerve. Although there are only seven cervical **vertebra**, there are eight cervical nerves. Cervical nerves one through seven (C1-C7) emerge above (superior to) the corre-sponding cervical vertebrae. The last vervical nerve (C8) emerges below (inferior to) the last cervical vertebrae from that point downward the spinal nerves exit below the corre-sponding vertebrae for which they are named.

The brain itself is an enlargement of one end of the spinal cord. There are three regions that comprise the brain: the forebrain (also known as the prosencephalon), the mid-brain (mesencephalon), and the hindbrain (rhombencephalon).

The hindbrain consists of the **medulla oblongata, pons** and **cerebellum**. Functions associated with this region include **breathing** and coordinated body movements. The midbrain is involved in body movement. The so-called pleasure center is located here, which has been implicated in the development of addictive behaviors. The forebrain contains the intellectual centers. Regions of the forebrain control attributes of person-ality, vision, speech, and hearing.

Diseases that affect the nerves of the central nervous system include rabies, polio, and sub-acute sclerosing pan-encephalitis. Such diseases affect movement and can lead to mental incapacitation. The brain is also susceptible to disease, including toxoplasmosis and the development of empty region due to prions. Such diseases cause a wasting away of body function and mental ability. Brian damage can be so compro-mised as to be lethal.

See also Autonomic nervous system; Brain stem function and reflexes; Gray matter and white matter; Nerve impulses and conduction of impulses; Nervous system overview; Nervous system, embryological development; Neural damage and repair

CEPHLAD • *see* ANATOMICAL NOMENCLATURE

CEREBELLUM

The cerebellum, which is Latin for "little brain," is located in the posterior region of the **skull**. In the human **brain**, the cerebellum occupies a place partially tucked under the forebrain's **cerebral hemispheres**. The cerebellum is extensively folded into parallel folds, or folia, giving it an appearance of having irregular pleats. The cerebellum possesses right and left hemispheres that are connected with the **spinal cord** and forebrain. Between the two hemispheres is a worm-like structure known as the vermis. The two hemispheres are each divided into three lobes, the paleocerebellum (anterior), the neocerebellum (mid-lobe), and the archicerebellum (posterior). Each of the cerebellum's hemispheres connect with spinal cord nerves on the same side of the body, but with the opposite cerebral hemisphere.

The cerebellum's specialized function in the human brain is to maintain **posture** and balance, and to carry out coordinated movement, by processing signals that are transmitted from the cerebral cortex's motor area to the spinal cord and then to muscle groups, creating movements. The cerebellum also receives muscle and joint signals. It compares these with the cortex's signals, and makes adjustments as necessary to achieve the coordinated movement intended. Some evidence exists that the cerebellum can store a sequence of instructions for movements that are repeated frequently, and for repetitious skilled movements that are learned by rote. In some studies of the brain's responses to language-related tasks, researchers have found that, as tasks became more complex, several sites in the cerebellum are activated along with areas of the forebrain that process many types of information. Thus, some scientists suggest that the cerebellum may play a role in cognitive learning.

Another important function played by the cerebellum is its role in the reticular activating system, a widespread network of nerve cells that are the means by which humans maintain consciousness. The reticular activating system is also involved in the brain's ability to focus attention, blocking out some distractions that originate both within and outside the body.

See also Brain stem function and reflexes; Brain, intellectual functions; Cerebral cortex; Cerebral morphology; Motor functions and controls

CEREBRAL CORTEX

The cerebral cortex is a simple structure of the **brain** with complex functions. The neocortex, as it is called by neuroanatomists, is a thin covering of **gray matter** on the outer surface of the entire cerebrum. The layer is about two to three mm thick and primarily composed of two types of **neurons**. The surface area of the cortex of the average adult brain is around 964 in.^{2} (2,500 cm^2). The reason for this enormous surface area compared to brain size is the result of the infolding of the surface of the brain. The hills (gyri) and valleys (sulci)

that are characteristic of the brain provide the greatly increased surface. In some places the fissures or large sulci are very deep and provide a sort of map for dividing the cortex into distinct regions with specific functions.

The cortext contains about 50-100 billion neurons. There are two major neuronal types of cells distributed throughout six layers of the thin cortex. One type is the stellate cells, with round bodies and short dendrites (extensions that receive stimuli) that extend in all directions. These cells receive sensory signals and process the information in a local area. The other main cell type, the pyramidal cells, are elongate and triangular in cross-section. One end of the cell points toward the brain surface. The dendrites are thick and many are horizontal and pass into the white matter just below. They are responsible for sending signals out of the cerebrum.

There are five anatomical and functional regions in the cortex. Four are clearly visible on the surface and are named for the cranial bones that overlie them. The fifth lies deep to the lateral sulcus. It is called the insula, and its function is not yet clear. The frontal lobe is named for its place under and just behind the frontal bones and is the location for voluntary motor control, imagination, emotions, judgment, and aggressive behavior. The parietal lobe is largely responsible for integrating sensory information except hearing, **smell**, and vision. The occipital lobe receives and processes visual signals. The temporal lobe is a long horizontal lobe that is separated from the parietal by a long, deep, lateral sulcus. The functions are many and include hearing, learning, visual recognition (different from straight vision), and many types of emotional behavior. Memory is also a major function of this region.

The cerebral cortex covers both hemispheres of the brain and is joined by the corpus callosum. It not incorrect to say that the cerebral cortex is the main processing center of the brain. Without many of the higher brain functions associated with the cerebral cortex, many higher functions of the human brain would be impossible.

See also Brain: Intellectual functions

CEREBRAL HEMISPHERES

The cerebral hemispheres are literally the two halves of the **brain**. They are identified on the outside or external surface by the folded hills (gyri) and valleys (sulci). The deep longitudinal fissure runs down the sagital (middle) plane of the brain and separates the two halves of the cerebrum. It clearly shows the separation between the two hemispheres. The hemispheres make up about 83% of the volume of the human brain. They are connected to one another by a thick bundle of nerve fibers, the corpus callosum.

Different lobes have been identified in the hemispheres that are named for the cranial bones that overlie them. The neuroanatomy of the hemispheres is simple in that the majority of the tissues are myelinated white matter covered by a thin layer of **gray matter**, the neocortex. However, the function of the regions of the hemispheres is complicated. Regions of the hemispheres are responsible for almost all higher functions of the brain such as reasoning and emotions, as well as motor skills. Sensory information is both received (afferent) and sent out (efferent) to various regions of the body. Many of the **cranial nerves** leave the brain at the base or ventral side of the hemispheres.

One study conducted recently showed that there are differences between male and female hemispheres in the number of neuronal densities. They are higher in males than in females. However, the **neurons** in females tend to be larger than those in males. Whether or how this leads to any difference in brain functioning is not known. This difference, however, has been identified in the cortex and upper white matter. This may suggest that these neuronal patterns can account for the differences in female's right-handedness, language advantage, and bilateral (both sides) brain activation patterns. In other words, these results may explain why women tend to use both sides of the brain during certain tasks while males use a dominant side.

The **anatomy** of the cerebral hemispheres has been well mapped. Most of the sulci and gyri have been identified and named. Regions tested for visual, auditory (hearing), gustatory (**taste**), and other sensory functions have been fairly well located. However, research on the brain's hemispheres remains difficult. Researchers believe they have identified many regions responsible for certain functions, but there are still many questions. Because it is not possible to experiment on human brain **tissue** as it is with mice or other traditional laboratory animals, much of the information researchers have is gained from work done on other mammals.

See also Brain: Intellectual functions; Central nervous system (CNS); Cerebellum; Cerebral cortex; Cerebral morphology

CEREBRAL MORPHOLOGY

Cerebral morphology is the general description of the structures and features of the **brain.**

As an anatomical organ, the brain is an enlargement of the superior end of the **spinal cord.** The brain itself is divided into three major anatomical regions, the prosencephalon (forebrain), mesencephalon (midbrain) and the rhombencephalon (hindbrain).

The prosencephalon is divided into the diencephalon and the telencephalon (also known as the cerebrum). The cerebrum contains the two large bilateral hemispherical **cerebral cortex** that are responsible for the intellectual functions, personality, speech, and higher level sensory processing (e.g., coordinating and interpreting vision and hearing). The mesencephalon region serves as a bridge between higher and lower brain functions, and contains a number of centers associated with regions that create strong drives to certain behaviors. The rhombencephalon consisting of the **medulla oblongata, pons,** and **cerebellum** is an area largely devoted to lower brain functions, including autonomic functions involved in the regulation of **breathing** and general body coordination.

The brain contains a number of internal cavities (ventricles) that are continuous with the spinal canal and that contain cerebralspinal fluid. The ventricles are connected via cerebral aqueducts. The brain is covered by three layers of **meninges** (dura matter, arachnoid matter, and pia mater that dip into the many folds and fissures.

Within the hindbrain, the medulla oblongata is a conically shaped structure that lies between the spinal cord and the pons. A median fissure (deep convoluted fold) separates swellings (pyramids) on the surface of the medulla.

The pons (also known as the metencephalon) is located on the anterior surface of the cerebellum and is continuous with the superior portion of the medulla oblongata. The pons contains large tracts of transverse fibers that serve to connect the left and right **cerebral hemispheres.**

The cerebellum lies superior and posterior to the pons at the back base of the head. The cerebellum consists of left and right hemispheres connected by the vermis. Specialized tracts (peduncles) of neural **tissue** also connect the cerebellum with the midbrain, pons, and medulla. The surface of the cerebral hemispheres (the cortex) is highly convoluted into many folds and fissures.

The midbrain serves to connect the forebrain region to the hindbrain region. Within the midbrain a narrow aqueduct connects ventricles in the forebrain to the hindbrain. There are four distinguishable surface swellings (colliculi) on the midbrain. The midbrain also contains a highly vascularized mass of neural tissue named the red nucleus that is reddish in color (a result of the vascularization) compared to other brain structures and landmarks.

Although not visible from an exterior inspection of the brain, the diencephalon contains a dorsal **thalamus** (with a large posterior swelling termed the pulvinar) and a ventral **hypothalamus** that forms a border of the third ventricle of the brain. In this third ventral region lie a number of important structures, including the optic chiasma (the region where the ophthalmic nerves cross), and infundibulum.

Obscuring the diencephalon are the two large, well-developed and highly convoluted cerebral hemispheres that comprise the cerebrum. The cerebrum is the largest of the regions of the brain. The two large cerebral hemispheres are separated by a wide fissure (longitudinal fissure)that is connected by a large tract of white matter termed the corpus callosum. Within each cerebral hemisphere lies a lateral ventricle. The cerebral hemispheres run under the frontal, parietal, and occipital bones of the **skull.** The **gray matter** cortex is highly convoluted into folds (gyri) and the covering meninges dip deeply into the narrow gaps between the folds (sulci). The divisions of the superficial **anatomy** of the brain use the gyri and sulci as anatomical landmarks to define particular lobes of the cerebral hemispheres. As a rule, the lobes are named according to the particular bone of the skull that covers them. Accordingly, there are left and right frontal lobes, parietal lobes, an occipital lobe, and temporal lobes.

In a reversal of the pattern found within the spinal cord, the cerebral hemispheres have white matter tracts on the inside of the hemispheres and gray matter on the outside or cortex regions. Masses of gray matter that are present within the interior white matter are termed **basal ganglia** or basal nuclei.

See also Anatomical nomenclature; Brain stem function and reflexes; Brain: Intellectual functions; Nervous system overview; Nervous system, embryological development; Neural damage and repair; Neural plexi; Neurology; Neurons

CEREBROSPINAL FLUID

The entire surface of the **central nervous system** is bathed by a clear, colorless fluid called cerebrospinal fluid (CSF). The CSF is contained within a system of fluid-filled cavities called ventricles. It serves to both cushion the **brain** and **spinal cord**, and to bathe the central nervous system (CNS) in a dynamic, watery milieu. Its main function is to provide buoyancy and protection for the CNS. The total CSF volume in humans is approximately 140 ml. About 23 ml of this total is contained within the ventricular system and the remainder is distributed between the subarachnoid space surrounding the brain and spinal cord and the interstitial space surrounding the individual elements of the CNS. The lining of the ventricular system, known as the ependymal lining, is freely permeable to the CSF and also to large molecules. Thus the ventricles, subarachnoid space and interstitial space in the brain are anatomically continuous. However, the vascular compartment of the brain is separated from the CSF spaces by specialized capillary endothelium, which restricts the passage of molecules larger than 20 angstroms in diameter. This layer of ependyma forms the anatomical basis of the blood-brain barrier.

The average rate of CSF production in humans is approximately 0.3-0.4 ml/min. Thus the total CSF volume is renewed every 5-7 hours. About 70% of CSF is produced in the choroid plexus, and the remainder is formed from the metabolic activity of the brain and spinal chord parenchyma. Production of the CSF by the choroids plexus begins as **blood** is filtered through the fenestrations of the choroidal **capillaries**. This protein rich ultrafiltrate enters the choroid plexus stroma and moves by bulk flow into the clefts between the choroids epithelial cells. At this point, several processes determine which constituents of the ultrafiltrate enter the CSF. Sodium is moved into the ventricle in exchange for potassium via a sodium-potassium-adenosine triphospahte (Na+-K+-ATPase) pump, which operates within the epithelial cell. Chloride and bicarbonate ions move passively into the CSF as a result of carbonic anhydrase activity within the epithelial cell. Protein enters the ventricular system by two possible mechanisms: pinocytosis and through small pores. The osmotic gradient established by active sodium secretion appears to be responsible for the passive migration of water across the choroidal epithelium into the ventricles. Extrachoroidal production of CSF is also known and perhaps as much as 30% of CSF formation occurs within the CNS parenchyma.

See also Nervous system overview; Neurology

CERVICAL SPINE • *see* VERTEBRAL COLUMN

CERVIX • *see* REPRODUCTIVE SYSTEM AND ORGANS (FEMALE)

CHAIN, ERNST BORIS (1906-1979)
German-born English biochemist

Ernst Chain was instrumental in the creation of penicillin, the first antibiotic drug. Although the Scottish bacteriologist **Alexander Fleming** discovered the *penicillium notatum* mold in 1928, it was Chain who, together with **Howard Florey**, isolated the breakthrough substance that has saved countless victims of infections. For their work, Chain, Florey, and Fleming were awarded the Nobel Prize in physiology or medicine in 1945.

Chain was born in Berlin to Michael Chain and Margarete Eisner Chain. His father was a Russian immigrant who became a chemical engineer and built a successful chemical plant. The death of Michael Chain in 1919, coupled with the collapse of the post-World War I German economy, depleted the family's income so much that Margarete Chain had to open up her home as a guesthouse.

One of Chain's primary interests during his youth was music, and for a while it seemed that he would embark on a career as a concert pianist. He gave a number of recitals and for a while served as music critic for a Berlin newspaper. A cousin, whose brother-in-law had been a failed conductor, gradually convinced Chain that a career in science would be more rewarding than one in music. Although he took lessons in conducting, Chain graduated from Friedrich-Wilhelm University in 1930 with a degree in chemistry and physiology.

Chain began work at the Charite Hospital in Berlin while also conducting research at the Kaiser Wilhelm Institute for Physical Chemistry and Electrochemistry. But the increasing pressures of life in Germany, including the growing strength of the Nazi party, convinced Chain that, as a Jew, he could not expect a notable professional future in Germany. Therefore, when Hitler came to power in January 1933, Chain decided to leave. Like many others, he mistakenly believed the Nazis would soon be ousted. His mother and sister chose not to leave, and both died in concentration camps.

Chain arrived in England in April 1933, and soon acquired a position at University College Hospital Medical School. He stayed there briefly and then went to Cambridge to work under the biochemist Frederick Gowland Hopkins. Chain spent much of his time at Cambridge conducting research on **enzymes**. In 1935, Howard Florey became head of the Sir William Dunn School of Pathology at Oxford. Florey, an Australian-born pathologist, wanted a top-notch biochemist to help him with his research, and asked Hopkins for advice. Without hesitation, Hopkins suggested Chain.

Florey was actively engaged in research on the bacteriolytic substance lysozyme, which had been identified by Fleming in his quest to eradicate **infection**. Chain came across

Fleming's reports on the penicillin mold and was immediately intrigued. He and Florey both saw great potential in the further investigation of penicillin. With the help of a Rockefeller Foundation grant, the two scientists assembled a research team and set to work on isolating the active ingredient in *penicillium notatum.*

Fleming, who had been unable to identify the antibacterial agent in the mold, had used the mold broth itself in his experiments to kill infections. Assisted in their research by fellow scientist Norman Heatley, Chain and Florey began their work by growing large quantities of the mold in the Oxford laboratory. Once there were adequate supplies of the mold, Chain began the tedious process of isolating the "miracle" substance. Succeeding after several months in isolating small amounts of a powder that he obtained by freeze-drying the mold broth, Chain was ready for the first practical test. His experiments with laboratory mice were successful, and it was decided that more of the substance should be produced to try on humans. To do this, the scientists needed to ferment massive quantities of mold broth; it took 125 gal. (118.3 l) of the broth to make enough penicillin powder for one tablet. By 1941, Chain and his colleagues had finally gathered enough penicillin to conduct experiments with patients. The first two of eight patients died from complications unrelated to their infections, but the remaining six, who had been on the verge of death, were completely cured.

One potential use for penicillin was the treatment of wounded soldiers, an increasingly significant issue during the Second World War. However, for penicillin to be widely effective, the researchers needed to devise a way to mass-produce the substance. Florey and Heatley went to the United States in 1941 to enlist the aid of the government and of pharmaceutical houses. New ways were found to yield more and stronger penicillin from mold broth, and by 1943, the drug went into regular medical use for Allied troops. After the war, penicillin was made available for civilian use. The ethics of whether to make penicillin research universally available posed a particularly difficult problem for the scientific community during the war years. While some believed that the research should not be shared with the enemy, others felt that no one should be denied the benefits of penicillin. This added layers of political intrigue to the scientific pursuits of Chain and his colleagues. Even after the war, Chain experienced firsthand the results of this dilemma. As chairman of the World Health Organization in the late 1940s, Chain had gone to Czechoslovakia to supervise the operation of penicillin plants established there by the United Nations. He remained there until his work was done, even though the Communist coup occurred shortly after his arrival. When Chain applied for a visa to visit the United States in 1951, his request was denied by the State Department. Though no reason was given, many believed his stay in Czechoslovakia, however apolitical, was a major factor.

After the war, Chain tried to convince his colleagues that penicillin and other antibiotic research should be expanded, and he pushed for more state-of-the-art facilities at Oxford. Little came of his efforts, however, and when the Italian State Institute of Public Health in Rome offered him the opportunity to organize a biochemical and microbiological

department along with a pilot plant, Chain decided to leave Oxford.

Under Chain's direction, the facilities at the State Institute became known internationally as a center for advanced research. While in Rome, Chain worked to develop new strains of penicillin and to find more efficient ways to produce the drug. Work done by a number of scientists, with Chain's guidance, yielded isolation of the basic penicillin molecule in 1958, and hundreds of new penicillin strains were soon synthesized.

In 1963, Chain was persuaded to return to England. The University of London had just established the Wolfson Laboratories at the Imperial College of Science and Technology, and Chain was asked to direct them. Through his hard work the Wolfson Laboratories earned a reputation as a first-rate research center.

In 1948, Chain had married Anne Beloff, a fellow biochemist, and in the following years she assisted him with his research. She had received her Ph.D. from Oxford and had worked at Harvard in the 1940s. The couple had three children.

Chain retired from Imperial College in 1973, but continued to lecture. He cautioned against allowing the then-new field of **molecular biology** to downplay the importance of **biochemistry** to medical research. He still played the piano, for which he had always found time even during his busiest research years. Over the years, Chain also became increasingly active in Jewish affairs. He served on the Board of Governors of the Weizmann Institute in Israel, and was an outspoken supporter of the importance of providing Jewish education for young Jewish children in England and abroad—all three of his children received part of their education in Israel.

In addition to the Nobel Prize, Chain received the Berzelius Medal in 1946, and was made a commander of the Legion d'Honneur in 1947. In 1954, he was awarded the **Paul Ehrlich** Centenary Prize. Chain was knighted by Queen Elizabeth II in 1969. Chain died of heart failure at age 73.

See also Bacteria and responses to bacterial infection

CHEMOTHERAPY

Chemotherapy is the treatment of a disease or condition with chemicals that have a specific effect on its cause, such as a microorganism or **cancer** cell. The first modern therapeutic chemical was derived from a synthetic dye. The sulfonamide drugs developed in the 1930s, penicillin and other antibiotics of the 1940s, **hormones** in the 1950s, and more recent drugs that interfere with cancer cell **metabolism** and reproduction have all been part of the chemotherapeutic arsenal.

The first drug to treat widespread **bacteria** was developed in the mid-1930s by the German physician-chemist **Gerhard Domagk**. In 1932, he discovered that a dye named prontosil killed streptococcus bacteria, and it was quickly used medically on both streptococcus and staphylococcus. One of the first patients cured with it was Domagk's own daughter. In 1936, the Swiss biochemist Daniele Bovet, working at the

Pasteur Institute in Paris, showed that only a part of prontosil was active, a sulfonamide radical long known to chemists. Because it was much less expensive to produce, sulfonamide soon became the basis for several widely used "sulfa drugs" that revolutionized the treatment of formerly fatal diseases. These included pneumonia, meningitis, and puerperal ("childbed") fever. For his work, Domagk received the 1939 Nobel prize in physiology or medicine. Though largely replaced by antibiotics, sulfa drugs are still commonly used against urinary tract infections, Hanson disease (leprosy), and **malaria**, and for burn treatment.

At the same time, the next breakthrough in chemotherapy, penicillin, was in the wings. In 1928, the British bacteriologist **Alexander Fleming** noticed that a mold on an uncovered laboratory dish of staphylococcus destroyed the bacteria. He identified the mold as *Penicillium notatum,* which was related to ordinary bread mold. Fleming named the mold's active substance penicillin, but was unable to isolate it.

In 1939, the American microbiologist René Jules Dubos (1901–1982) isolated from a soil microorganism an antibacterial substance that he named tyrothricin. This led to wide interest in penicillin, which was isolated in 1941 by two biochemists at Oxford University, **Howard Florey** and **Ernst Chain.**

The term antibiotic was coined by American microbiologist Selman Abraham Waksman, who discovered the first antibiotic that was effective on gram-negative bacteria. Isolating it from a Streptomyces fungus that he had studied for decades, Waksman named his antibiotic streptomycin. Though streptomycin occasionally resulted in unwanted side effects, it paved the way for the discovery of other antibiotics. The first of the tetracyclines was discovered in 1948 by the American botanist Benjamin Minge Duggar. Working with *Streptomyces aureofaciens* at the Lederle division of the American Cyanamid Co., Duggar discovered chlortetracycline (Aureomycin).

The first effective chemotherapeutic agent against **viruses** was acyclovir, produced in the early 1950s by the American biochemists **George Hitchings** and Gertrude Bell Elion for the treatment of herpes. Today's antiviral drugs are being used to inhibit the reproductive cycle of both **DNA** and **RNA** viruses. For example, two drugs are used against the influenza A virus, Amantadine and Rimantadine, and the **AIDS** treatment drug AZT inhibits the reproduction of the human immunodeficiency virus (HIV).

Cancer treatment scientists began trying various chemical compounds for use as cancer treatments as early as the mid-nineteenth century. But the first effective treatments were the **sex hormones**, first used in 1945, estrogens for prostate cancer and both estrogens and androgens to treat breast cancer. In 1946, the American scientist Cornelius Rhoads developed the first drug especially for cancer treatment. It was an alkylating compound, derived from the chemical warfare agent nitrogen mustard, which binds with chemical groups in the cell's DNA, keeping it from reproducing. Alkylating compounds are still important in cancer treatment.

In the next twenty years, scientists developed a series of useful antineoplastic (anti-cancer) drugs, and, in 1954, the

forerunner of the National Cancer Institute was established in Bethesda, MD. Leading the research efforts were the so-called "4-H Club" of cancer chemotherapy: the Americans **Charles Huggins**, who worked with hormones; George Hitchings (1905–1998), purines and pyrimidines to interfere with cell metabolism; Charles Heidelberger, fluorinated compounds; and British scientist Alexander Haddow (1907–1976), who worked with various substances. The first widely used drug was 6-Mercaptopurine, synthesized by Elion and Hitchings in 1952.

Chemotherapy is used alone, in combination, and along with radiation and/or surgery, with varying success rates, depending on the type of cancer and whether it is localized or has spread to other parts of the body. They are also used after treatment to keep the cancer from recurring (adjuvant therapy). Since many of the drugs have severe side effects, their value must always be weighed against the serious short-and long-term effects, particularly in children, whose bodies are still growing and developing.

In addition to the male and female sex hormones androgen, **estrogen**, and progestins, scientists also use the hormone somatostatin, which inhibits production of growth hormone and growth factors. They also use substances that inhibit the action of the body's own hormones. An example is Tamoxifen, used against breast cancer. Normally the body's own estrogen causes growth of breast tissues, including the cancer. The drug binds to cell receptors instead, causing reduction of **tissue** and cancer cell size.

Forms of the B-vitamin folic acid were found to be useful in disrupting cancer cell metabolism by the American scientist Seymour Farber (1912–) in 1948. Today they are used on **leukemia**, breast cancer, and other cancers.

Plant alkaloids have long been used as medicines, such as colchicine from the autumn crocus. Cancer therapy drugs include vincristine and vinblastine, derived from the pink periwinkle by American Irving Johnson (1925–). They prevent mitosis (division) in cancer cells. VP-16 and VM-16 are derived from the roots and rhizomes of the may apple or mandrake plant, and are used to treat various cancers. Taxol, which is derived from the bark of several species of yew trees, was discovered in 1978, and is used for treatment of ovarian and breast cancer.

Another class of naturally occurring substances are anthracyclines, which scientists consider to be extremely useful against breast, lung, thyroid, stomach, and other cancers.

Certain antibiotics are also effective against cancer cells by binding to DNA and inhibiting RNA and **protein synthesis**. Actinomycin D, derived from Streptomyces, was discovered by **Selman Waksman** and first used in 1965 by American researcher Seymour Farber. It is now used against cancer of female reproductive **organs, brain tumors**, and other cancers.

A form of the metal platinum called cisplatin stops cancer cells' division and disrupts their growth pattern. Newer treatments that are biological or based on proteins or genetic material and can target specific cells are also being developed. Monoclonal antibodies are genetically engineered copies of proteins used by the **immune system** to fight disease.

Rituximab was the first moncoclonal antibody approved for use in cancer, and more are under development. Interferons are proteins released by cells when invaded by a virus. Interferons serve to alert the body's immune system of an impending attack, thus causing the production of other proteins that fight off disease. Interferons are being studied for treating a number of cancers, including a form of skin cancer called multiple myeloma. A third group of drugs are called anti-sense drugs, which affect specific genes within cells. Made of genetic material that binds with and neutralizes messenger-RNA, anti-sense drugs halt the production of proteins within the cancer cell.

Genetically engineered cancer vaccines are also being tested against several virus-related cancers, including **liver**, cervix, **nose** and throat, kidney, lung, and prostate cancers. The primary goal of genetically engineered vaccines is to trigger the body's immune system to produce more cells that will react to and kill cancer cells. One approach involves isolating white **blood** cells that will kill cancer and then to find certain **antigens**, or proteins, that can be taken from these cells and injected into the patient to spur on the immune system. A "vaccine gene gun" has also been developed to inject DNA directly into the tumor cell. An RNA cancer vaccine is also being tested. Unlike most vaccines, which have been primarily tailored for specific patients and cancers, the RNA cancer vaccine is designed to treat a broad number of cancers in many patients.

As research into cancer treatment continues, new cancer-fighting drugs will continue to become part of the medical armamentarium. Many of these drugs will come from the burgeoning biotechnology industry and promise to have fewer side effects than traditional chemotherapy and radiation.

See also Pharmacogenetics; Pharmacology

CHEWING · *see* MASTICATION

CHILD GROWTH AND DEVELOPMENT

Child growth includes the period between toddlerhood to preadolescence. In the post-toddler period, **nutrition** continues to be critically important. The child's coordination, language, ability to think, and social skills advance rapidly. In the preschool years (approximately ages four and five, and sometimes six), socialization and preparation for schooling take on greater importance. From age four onward, early childhood programs are more likely to be associated with education and preschools, but health and nutrition remain key components of what young children need. In the early primary school, a period of transition into school and the world at large (roughly ages six to eight), depending on the degree of synchronicity between home and school, this transition can be relatively easy or extremely difficult. In preadolescence, the period of childhood just before the onset of **puberty**, (10–12 in girls and 11–13 in boys) children feel disorganized, and their growth is

rapid and uneven. They are not quite adolescents yet because their sexual maturity has not fully completed, and they are often referred to as tweens, meaning between the stages of childhood and teen years. Children at these ages try to meet the expectations of both parents and friends.

At three years of age, children can use short sentences, follow simple instructions, and often repeat words they overhear in conversations. At four years of age, the child can understand most sentences, understand physical relationships (on, in, under), uses sentences that are four to five (or more) words long, can say his/her name, age, and gender, and use pronouns. Strangers can understand the child's spoken language. According to Brown's Stages, a framework designed to understand and predict the path that normal expressive language development, children at 36–42 months are expected to have a "mean length of utterance measured in morphemes" of about 2.75 morphemes. This corresponds to the Stage III language development. Stage IV and V, that should be reached at 40–46 months and 42–52 months respectively, are characterized by improvements in speaking as well as using articles, regular past tense (-ed endings), third person regular and irregular, auxiliary verbs, contractible copulas, and contractible auxiliary verbs.

By the end of the fourth year, the child asks abstract (why?) questions, and understands concepts of same versus different. By age five, children should be able to retell a story in their own words and use more than five words in a sentence. By the time a child is six years, improvements consist of answering "What would happen if..." questions, understanding the "opposite of", telling left from right, using all pronoun forms correctly, and using sentences that are close to simple adult sentences in terms of formulation.

As children enter their school years, they become increasingly independent, spending much of their days outside the home while in school and with peers. As the child progresses in school, comprehension, communication patterns, and usage of language will become more sophisticated. Usually, children will understand more vocabulary words and concepts than they may be able to express. Children at this time should be able to engage in narrative discourse and share ideas and opinions in clear speech. Further development depends on family, social and educational environment.

Many believe that **genetics** is the most significant factor in a child's potential to succeed in the academic world. However, genetics is not the only determining component of a child's potential to succeed. Environment and experience both play a significant role in the learning process.

A child's **brain** develops exponentially during the earliest years of their young life. Brain cells are formed during the first two years and, in the early years a child develops the basic brain and physiological structures upon which later growth and learning are dependent. All interactions between babies and their parents encourage the development and growth of synapses. Even the earliest experiences can have the most significant effect on the young developing brain. In addition, stimulation of a child's senses affects the structure and organization of neural pathways in the brain during the formative period The journey to learning begins with the initial step when a parent teaches their infant to adapt to their new surroundings, by teaching them to eat, learn, and respond to stimuli. Evidence from the fields of **physiology**, nutrition, psychology, education, and other fields continues to accumulate to indicate that the early years are critical to all of later life. The early years are critical in the formation of intelligence, personality, and social behavior, and the effects of early neglect can be cumulative. **Molecular biology** offers new understandings of the way the nervous system functions, the ways in which the brain develops, and the impact of the environment on that development. By age six, most of these connections are already organized and able to provide opportunities for perceptual and motor experiences at an early age, favorably affecting various learning abilities in later life. The environment affects not only the number of brain cells and the number of connections, but also the ways in which they are wired. In fact, the brain uses its experience with the world to refine the way it functions. Early experiences are important in shaping the way the brain works. Again, there is evidence of the negative impact of stress during the early years on brain function for developing a variety of cognitive, behavioral, and emotional difficulties.

A child's growth is a continuous process of a gradual sequencing from one stage of physical development to another. Growth pattern is largely determined by genetics and under hormonal control. It is a very complex process, and requires the coordinated action of several **hormones. Human growth hormone** (hGH), produced in the body by the pituitary gland plays a significant role in the building of cells, tissues, **blood** vessels, muscles, bones, and **organs**. Human growth hormone is responsible for cell growth through the teens and rebuilding and repairing cells throughout our entire life cycle. hGH is produced in abundance in the preadolescence and adolescence stages. In adolescents it stimulates linear growth and the aging of the bones. In addition, hGH stimulates the intracellular transport of **amino acids** and causes nitrogen retention, a supposed marker of protein anabolism. The levels are four times higher in the child through adolescent years than in more advanced ages. The major role of hGH in stimulating body growth is to stimulate the **liver** and other tissues to secrete insulin-like growth factor-1 (IGF-1). IGF-1 stimulates proliferation of chondrocytes (**cartilage** cells), resulting in bone growth. Growth hormone does seem to have a direct effect on bone growth in stimulating differentiation of chondrocytes. IGF-1 also appears to be the key player in muscle growth. It stimulates both the differentiation and proliferation of myoblasts. It also stimulates amino acid uptake and **protein synthesis** in muscle and other tissues.

Normal growth, supported by good nutrition, adequate **sleep**, and regular exercise is one of the best overall indicators of good health. A significantly malnourished child may be pushed off the "natural" genetically determined growth curve. Severe hyponutrition, enough to affect a child's growth rate, is uncommon today in the United States and other developed countries unless the child has an associated chronic illness or disorder. Extra food or greater than recommended amounts of **vitamins**, minerals, or other nutrients will not increase the ultimate height to which a child will grow.

See also Adolescent growth and development; Growth and development; Nutrition and nutrient transport to cells; Toddler growth and development

CHROMOSOMES

Chromosomes are microscopic units containing organized genetic information, located in the nuclei of diploid and haploid cells (e.g., human somatic and sex cells), and are also present in one-cell non-nucleated organisms (unicellular microorganisms) such as **bacteria**, which do not have an organized nucleus. The sum-total of genetic information contained in different chromosomes of a given individual or species are generically referred to as the genome.

In humans, chromosomes are structurally made of roughly equal amounts of proteins and **DNA**. Each chromosome contains a double-strand DNA molecule, arranged as a double helix, and tightly coiled and neatly packed by a family of proteins called histones. DNA strands are comprised of linked nucleotides. Each nucleotide has a sugar (deoxyribose), a nitrogenous base, plus one to three phosphate groups. Each nucleotide is linked to adjacent nucleotides in the same DNA strand by phosphodiester bonds. Phosphodiester is another sugar, made of sugar-phosphate. Nucleotides of one DNA strand link to their complementary nucleotide on the opposite DNA strand by hydrogen bonds, thus forming a pair of nucleotides, known as a base pair, or nucleotide base.

Chromosomes contain the genes, or segments of DNA, that encode for proteins of an individual. Genes contain up to thousands of sequences of these base pairs. What distinguishes one gene from another is the sequence of nucleotides that code for the synthesis of a specific protein or portion of a protein. Some proteins are necessary for the structure of cells and tissues. Others, like **enzymes**, a class of active (catalyst) proteins, promote essential biochemical reactions, such as **digestion**, energy generation for cellular activity, or **metabolism** of toxic compounds. Some genes produce several slightly different versions of a given protein through a process of alternate transcription of bases pairs segments known as codons. When a chromosome is structurally faulty, or if a cell contains an abnormal number of chromosomes, the types and amounts of the proteins encoded by the genes are altered. Changes to proteins often result in serious mental and physical defects and disease.

Within the chromosomes, the DNA is tightly coiled around proteins (e.g., histones) allowing huge DNA molecules to occupy a small space within the nucleus of the cell. When a cell is not dividing, the chromosomes are invisible within the cell's nucleus. Just prior to **cell division**, the chromosomes uncoil and begin to replicate. As they uncoil, the individual chromosomes take on a distinctive appearance that allows physicians and scientists to classify the chromosomes by size and shape.

Numbers of autosomal chromosomes differ in cells of different species; but are usually the same in every cell of a given species. **Sex determination** cells (mature ovum and

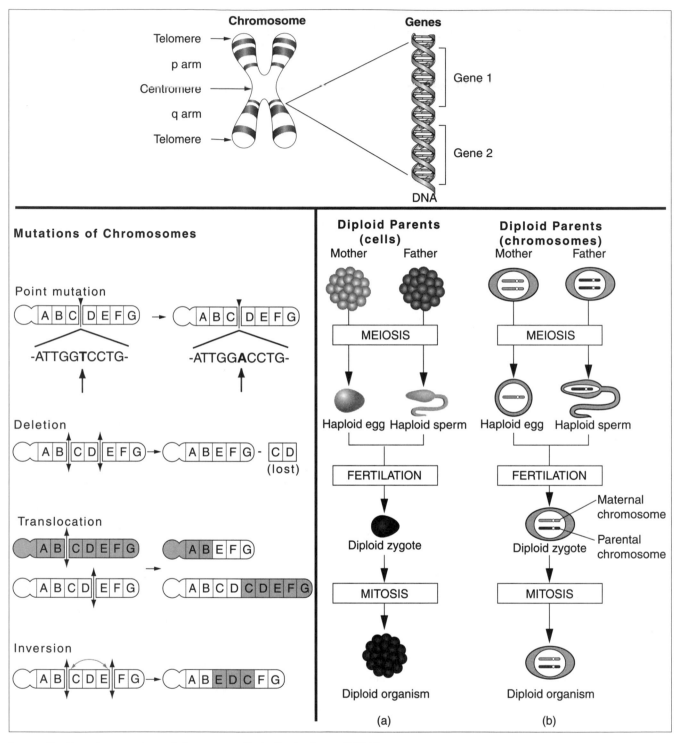

A generalized homologous autosomal chromosome. *Illustration by Argosy Publishing.*

sperm) are an exception, where the number of chromosomes is halved. Chromosomes also differ in size. For instance, the smallest human chromosome, the sex chromosome Y, contains 50 million base pairs (bp), whereas the largest one, chromosome 1, contains 250 million base pairs. All 3 billion base pairs in the human genome are stored in 46 chromosomes.

Human genetic information is therefore stored in 23 pairs of chromosomes (totaling 46), 23 inherited from the mother, and 23 from the father. Two of these chromosomes are sex chromosomes (chromosomes X and Y). The remaining 44 are autosomes (in 22 autosomal pairs), meaning that they are not sex chromosomes and are present in all somatic cells (i.e., any other body cell that is not a germinal cell for spermatozoa in males or an ovum in females). Sex chromosomes specify the offspring gender: normal females have two X chromosomes and normal males have one X and one Y chromosome. These chromosomes can be studied by constructing a karyotype, or organized depiction, of the chromosomes.

Each set of 23 chromosomes constitutes one allele, containing gene copies inherited from one of the progenitors. The other allele is complementary or homologous, meaning that they contain copies of the same genes and on the same positions, but originated from the other progenitor. As an example, every normal child inherits one set of copies of gene BRCA1, located on chromosome 13, from the mother and another set of BRCA1 from the father, located on the other allelic chromosome 13. Allele is a Greek-derived word that means "one of a pair," or any one of a series of genes having the same locus (position) on homologous chromosomes.

The first chromosome observations were made under light microscopes, revealing rod-shaped structures, in varied sizes and conformations commonly J-, or V-shaped in eukaryotic cells and ring-shaped chromosome in bacteria. Staining reveals a pattern of light and dark bands. Today, those bands are known to correspond to regional variations in the amounts of the two nucleotide base pairs: Adenine-Thymine (A-T or T-A) in contrast with amounts of Guanine-Cytosine (G-C or C-G).

In humans, two types of cell division exist. In mitosis, cells divide to produce two identical daughter cells. Each daughter cell has exactly the same number of chromosomes. This preservation of chromosome number is accomplished through the replication of the entire set of chromosomes just prior to mitosis.

Two kinds of chromosome number defects can occur in humans: aneuploidy, an abnormal number of chromosomes, and polyploidy, more than two complete sets of chromosomes. Most alterations in chromosome number occur during meiosis. During normal meiosis, chromosomes are distributed evenly among the four daughter cells. Sometimes, however, an uneven number of chromosomes are distributed to the daughter cells.

Genetic abnormalities and diseases occur if chromosomes or portions of chromosomes are missing, duplicated or broken. Abnormalities and diseases may also occur if a specific gene is transferred from one chromosome to another (translocation), or there is a duplication or inversion of a segment of a chromosome. Down syndrome, for instance, is caused by trisomy in chromosome 21, the presence of a third copy of chromosome 21. Some structural chromosomal abnormalities have been implicated in certain cancers. For instance, myelogenous **leukemia** is a **cancer** of the white **blood** cells. Researchers have found that the cancerous cells contain a translocation of chromosome 22, in which a broken segment switches places with the tip of chromosome 9.

In non-dividing cells, it is not possible to distinguish morphological details of individual chromosomes, because they remain elongated and entangled to each other. However, when a cell is dividing, i.e., undergoing mitosis, chromosomes become highly condensed and each individual chromosome occupies a well-defined spatial location.

Karyotype analysis was the first genetic screening utilized by geneticists to assess inherited abnormalities, like additional copies of a chromosome or a missing copy, as well as DNA content and gender of the individual. With the development of new molecular screening techniques and the growing number of identified individual genes, detection of other more subtle chromosomal mutations is now possible (e.g., determinations of gene mutations, levels of gene expression, etc). Such data allow scientists to better understand disease causation and to develop new therapies and medicines for those diseases.

In mitosis, cells divide to produce two identical daughter cells. Each daughter cell has exactly the same number of chromosomes. This preservation of chromosome number is accomplished through the replication of the entire set of chromosomes just prior to mitosis.

Sex cells, such as eggs and sperm, undergo a different type of cell division called meiosis. Because sex cells each contribute half of a zygote's genetic material, sex cells must carry only half the full complement of chromosomes. This reduction in the number of chromosomes within sex cells is accomplished during two rounds of cell division, called meiosis I and meiosis II. Prior to meiosis I, the chromosomes replicate, and chromosome pairs are distributed to daughter cells. During meiosis II, however, these daughter cells divide without a prior replication of chromosomes. Mistakes can occur during either meiosis I and meiosis II. Chromosome pairs can be separated during meiosis I, for instance, or fail to separate during meiosis II.

Meiosis produces four daughter cells, each with half of the normal number of chromosomes. These sex cells are called haploid cells (meaning half the number). Non-sex cells in humans are called diploid (meaning double the number) since they contain the full number of normal chromosomes.

Most alterations in chromosome number occur during meiosis. When an egg or sperm that has undergone faulty meiosis and has an abnormal number of chromosomes unites with a normal egg or sperm during conception, the zygote formed will have an abnormal number of chromosomes. If the zygote survives and develops into a fetus, the chromosomal abnormality is transmitted to all of its cells. The child that is born will have symptoms related to the presence of an extra chromosome or absence of a chromosome.

See also Cell cycle and cell division; Cell differentiation; Genetic code; Genetic regulation of eukaryotic cells; Genetics and developmental genetics; Human genetics; Molecular biology; Ribonucleic acid (RNA)

CILIA AND CILIATED EPITHELIAL CELLS

Cilia, which are specialized arrangements of microtubules, have two general functions. They propel certain unicellular organisms, such as aramecium, through the water. In multicellular organisms, if cilia extend from stationary cells that are part of a **tissue** layer, they move fluid over the surface of the tissue.

Cilia line the respiratory epithelium of the **nose, pharynx**, and trachea. The tracheal cilia sweep dust, pollen and other particulate matter up to the pharynx so that it may be swallowed. Elements in cigarette smoke reduce this motion, allowing particles into the **lungs**. Defects in the respiratory cilia may cause bronchitis and sinusitis.

The oviduct is also lined with cilia. Those in the funnel-like opening of the oviduct attract the newly ovulated egg by drawing in fluid from the body cavity. Cilia farther along in the oviduct help to move the egg toward the uterus. The cilia of the rete testis and ductulus efferens of the testis move newly formed **sperm** from the seminiferous tubules to the vas deferens. Ependymal cells, which line the fluid-filled ventricles of the **brain**, are also ciliated.

Cilia are present in large numbers, as many as 100 per cell. They are about.25 um in diameter and about 2–20 um long. They work like the oars of a crew team, alternating power and recovery strokes in synchrony.

Each cilium is an extension of the cell, and is covered by the cell's **plasma** membrane. The core of the cilium is called the axoneme. It consists of microtubules, which in turn are composed of the protein tubulin, arranged in a characteristic pattern. Nine pairs of microtubules, called doublets, form a ring. Two single microtubules are located in the center of the ring. The doublets are connected to the center of the axoneme by spokes that end near the central microtubules. Each doublet has two "arms" made of the protein dynein. These arms reach toward the neighboring doublet.

Dynein is able to hydrolize **ATP**, allowing the dynein arms to slide one doublet past the other. Because the microtubules are connected at their base by a basal body, however, the amount of sliding is limited by the fixed position of their bases, causing the sliding motion to be translated into bending of the axoneme.

The basal bodies, also made of tubulin, are self-replicating. A newly replicated basal body can move away from its original site and generate a new cilium.

See also Cell structure; Reproductive system and organs (female); Reproductive system and organs (male); Respiratory system

CIRCLE OF WILLIS

A large portion of the oxygen needed by the **brain** is supplied via a set of vessels located at the base of the brain that, together, form a structure generally known as the arterial circle of Willis. Principally formed by junctures and communicating vessels (anastomosis) among the basilar artery—formed by the fusion of the left and right vertebral arteries—and the left and right internal **carotid arteries**, the circle of Willis aids in assuring a continuous supply of oxygenated **blood** to all portions of the brain.

Although physiological studies utilizing radioisotopes and other traceable markers establish that the majority of the blood originally passing through the left vertebral and left internal carotid **arteries** normally supply the left side of the

brain (with a similar situation found on the right with the right vertebral and right internal carotid arteries), the circle of Willis allows a communication through which oxygenated blood may pass from the left to right side if one side becomes blocked (occluded). Accordingly, oxygenated blood from either vertebral artery or either internal carotid may be able to supply vital oxygen to either cerebral hemisphere.

The circle of Willis is composed of the right and left internal carotid arteries joined by the anterior communicating artery. The basilar artery (formed by the fusion of the vertebral arteries) divides into left and right posterior cerebral arteries that are connected (anastomsed) to the corresponding left or right internal carotid artery via the respective left or right posterior communicating artery. A number of arteries that supply the brain originate at the circle of Willis including the anterior cerebral arteries that originate from the anterior communicating artery.

In the embryo, the components of the circle of Willis develop from the embryonic dorsal aortae and the embryonic intersegmental arteries.

Insufficient arterial pressure within the arterial circle of Willis can cause fainting (syncope) or a more severe loss of consciousness. A continuous supply highly oxygenated blood is critical to brain **tissue** function and a decrease in pressure or oxygenation (percentage of oxygen content) can cause tissue damage within minutes. Depending on a number of other physiological factors (e.g., temperature, etc.) brain damage or **death** may occur within two to ten minutes of severe oxygen deprivation. The circle of Willis provides the communications that allow multiple paths for oxygenated blood to supply the brain if any of the principal suppliers of oxygenated blood (i.e., the vertebral and internal carotid arteries) are constricted by physical pressure, occluded by disease, or interrupted by injury. This redundancy of blood supply is termed **collateral circulation**.

As an additional safety measure, the size of the communicating arteries and the arteries branching from the circle of Willis are able to change in response to increased blood flow that accompanies occlusion or interruption of blood supply to a another component of the circle.

See also Anatomical nomenclature; Angiology; Blood pressure and hormonal control mechanisms; Cranial circulation; Pulse and pulse pressure; Systemic circulation; Vascular system (embryonic development)

CIRCULATORY SYSTEM

The human circulatory system is termed the cardiovascular system, from the Greek word *kardia*, meaning **heart**, and the Latin *vasculum*, meaning small vessel. The basic components of the cardiovascular system are the heart, the **blood** vessels, and the blood. The work done by the cardiovascular system is astounding. Each year, the heart pumps more than 1,848 gal. (7,000 l) of blood through a closed system of about 62,100 mi. (100,000 km) of blood vessels. This is more than twice the distance around the equator of the Earth. As blood circulates around the body, it picks up oxygen from the **lungs**, nutrients from the small intestine, and **hormones** from the **endocrine glands**, and delivers these to the cells. Blood then picks up carbon dioxide and cellular wastes from cells and delivers these to the lungs and **kidneys**, where they are excreted. Substances pass out of blood vessels to the cells through the interstitial or **tissue** fluid which surrounds cells.

The adult heart is a hollow cone-shaped muscular organ located in the center of the chest cavity. The lower tip of the heart tilts toward the left. The heart is about the size of a clenched fist and weighs approximately 10.5 oz (300 g). Remarkably, the heart beats more than 100,000 times a day and close to 2.5 billion times in the average lifetime. A triple-layered sac, the **pericardium**, surrounds, protects, and anchors the heart. A liquid pericardial fluid located in the space between two of the layers, reduces friction when the heart moves.

The heart is divided into four chambers. A partition or septum divides it into a left and right side. Each side is further divided into an upper and lower chamber. The upper chambers, atria (singular atrium), are thin-walled. They receive blood entering the heart, and pump it to the ventricles, the lower heart chambers. The walls of the ventricles are thicker and contain more cardiac muscle than the walls of the atria, enabling the ventricles to pump blood out to the lungs and the rest of the body. The left and right sides of the heart function as two separate pumps. The right atrium receives oxygen-poor blood from the body from a major vein, the vena cava, and delivers it to the right ventricle. The right ventricle, in turn, pumps the blood to the lungs via the pulmonary artery. The left atrium receives the oxygen-rich blood from the lungs from the pulmonary **veins**, and delivers it to the left ventricle. The left ventricle then pumps it into the aorta, a major artery that leads to all parts of the body. The wall of the left ventricle is thicker than the wall of the right ventricle, making it a more powerful pump able to push blood through its longer trip around the body.

One-way valves in the heart keep blood flowing in the right direction and prevent backflow. The valves open and close in response to pressure changes in the heart. Atrioventricular (AV) valves are located between the atria and ventricles. Semilunar (SL) valves lie between the ventricles and the major **arteries** into which they pump blood. The "lub-dup" sounds that the physician hears through the stethoscope occur when the heart valves close. The AV valves produce the "lub" sound upon closing, while the SL valves cause the "dup" sound. People with a heart murmur have a defective heart valve that allows the backflow of blood.

The rate and rhythm of the heartbeat are carefully regulated. The heart continues to beat even when disconnected from the nervous system. This is evident during heart transplants when donor hearts keep beating outside the body. The explanation lies in a small mass of contractile cells, the sino-atrial (SA) node or pacemaker, located in the wall of the right atrium. The SA node sends out electrical impulses that set up a wave of contraction that spreads across the atria. The wave reaches the atrio-ventricular (AV) node, another small mass of contractile cells. The AV node is located in the septum

between the left and right ventricle. The AV node, in turn, transmits impulses to all parts of the ventricles. The bundle of His, specialized fibers, conducts the impulses from the AV node to the ventricles. The impulses stimulate the ventricles to contract. An electrocardiogram, **ECG** or EKG, is a record of the electric impulses from the pacemaker that direct each heartbeat. The SA node and conduction system provide the primary heart controls. In patients with disorganized electrical activity in the heart, surgeons implant an artificial pacemaker that serves to regulate the heart rhythm. In addition to self-regulation by the heart, the **autonomic nervous system** and hormones also affect its rate.

The heart cycle refers to the events associated with a single heartbeat. The cycle involves systole, the contraction phase, and diastole, the relaxation phase. In the heart, the two atria contract while the two ventricles relax. Then, the two ventricles contract while the two atria relax. The heart cycle consists of a systole and diastole of both the atria and ventricles. At the end of a heartbeat all four chambers rest. The rate of heartbeat averages about 75 beats per minute, and each **cardiac cycle** takes about 0.8 seconds.

The blood vessels of the body make up a closed system of tubes that carry blood from the heart to tissues all over the body and then back to the heart. Arteries carry blood away from the heart, while veins carry blood toward the heart. **Capillaries** connect small arteries (arterioles) and small veins (venules). Large arteries leave the heart and branch into smaller ones that reach out to various parts of the body. These divide still further into smaller vessels called arterioles that penetrate the body tissues. Within the tissues, the arterioles branch into a network of microscopic capillaries. Substances move in and out of the capillary walls as the blood exchanges materials with the cells. Before leaving the tissues, capillaries unite into venules, which are small veins. The venules merge to form larger and larger veins that eventually return blood to the heart. The two main circulation routes in the body are the **pulmonary circulation**, to and from the lungs, and the **systemic circulation**, to and from all parts of the body. Subdivisions of the systemic system include the **coronary circulation**, for the heart, the cerebral circulation, for the **brain**, and the renal circulation, for the kidneys. In addition, the hepatic portal circulation passes blood directly from the digestive tract to the **liver**.

The walls of arteries, veins, and capillaries differ in structure. In all three, the vessel wall surrounds a hollow center through which the blood flows. The walls of both arteries and veins are composed of three coats. The inner coat is lined with a simple squamous endothelium, a single flat layer of cells. The thick middle coat is composed of **smooth muscle** that can change the size of the vessel when it contracts or relaxes, and of stretchable fibers that provide elasticity. The outer coat is composed of elastic fibers and collagen. The difference between veins and arteries lies in the thickness of the wall of the vessel. The inner and middle coats of veins are very thin compared to arteries. The thick walls of arteries make them elastic and capable of contracting. The repeated expansion and recoil of arteries when the heart beats creates the **pulse**. A person can feel the pulse in arteries near the body surface, such as the radial artery in the wrist. The walls of veins

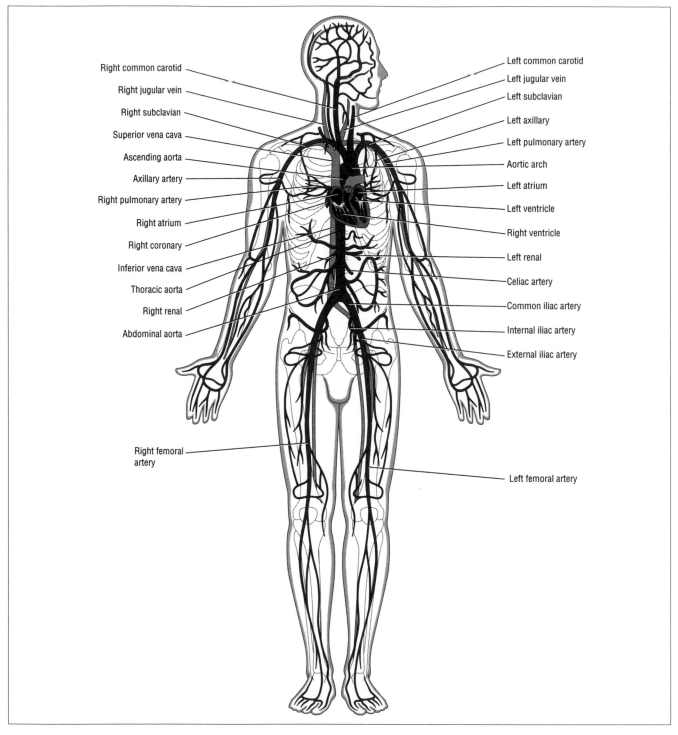

Right common carotid
Right jugular vein
Right subclavian
Superior vena cava
Ascending aorta
Axillary artery
Right pulmonary artery
Right atrium
Right coronary
Inferior vena cava
Thoracic aorta
Right renal
Abdominal aorta

Left common carotid
Left jugular vein
Left subclavian
Left axillary
Left pulmonary artery
Aortic arch
Left atrium
Left ventricle
Right ventricle
Left renal
Celiac artery
Common iliac artery
Internal iliac artery
External iliac artery

Right femoral artery

Left femoral artery

The circulatory system. *Illustration by Argosy Publishing.*

are more flexible than artery walls and they change shape when muscles press against them. Blood returning to the heart in veins is under low pressure often flowing against gravity. One-way valves in the walks of veins keep blood flowing in one direction. Skeletal muscles also help blood return to the heart by squeezing the veins as they contract. Varicose veins develop when veins lose their elasticity and become stretched. Faulty valves allow blood to sink back thereby pushing the vein wall outward. The walls of capillaries are only one cell thick. Of all the blood vessels, only capillaries have walls thin enough to allow the exchange of materials between cells and the blood. Their extensive branching provides a sufficient surface area to pick up and deliver substances to all cells in the body.

Blood pressure is the pressure of blood against the wall of a blood vessel. Blood pressure originates when the ventricles contract during the heartbeat. In a healthy young adult male, blood pressure in the aorta during systole is about 120 mm Hg, and approximately 80 mm Hg during diastole. The sphygmomanometer is an instrument that measures blood pressure. A combination of nervous carbon and hormones help regulate blood pressure around a normal range in the body. In addition, there are local controls that direct blood to tissues according to their need. For example, during exercise, reduced oxygen and increased carbon dioxide stimulate blood flow to the muscles.

Blood is liquid connective tissue. It transports oxygen from the lungs and delivers it to cells. It picks up carbon dioxide from the cells and brings it to the lungs. It carries nutrients from the digestive system and hormones from the endocrine glands to the cells. It takes heat and waste products away from cells. The blood helps regulate the body's base-acid balance (pH), temperature, and water content. It protects the body by clotting and by fighting disease through the **immune system**.

The structure of blood is that it is heavier and stickier than water, has a temperature in the body of about 100.4°F (38°C), and a pH of about 7.4. Blood makes up approximately 8% of the total body weight. A male of average weight has about 1.5 gal. (5-6 l) of blood in his body, while a female has about 1.2 gal. (4-5 l). Blood is composed of a liquid portion (the **plasma**), and blood cells.

Plasma is composed of about 91.5% water that acts as a solvent, heat conductor, and suspending medium for the blood cells. The rest of the plasma includes plasma proteins produced by the liver, such as albumins, that help maintain water balance, globulins, that help fight disease, and fibrinogen, that aids in blood clotting. The plasma carries nutrients, hormones, **enzymes**, cellular waste products, some oxygen, and carbon dioxide. Inorganic salts, also carried in the plasma, help maintain osmotic pressure. Plasma leaks out of the capillaries to form the **interstitial fluid** (tissue fluid) that surrounds the body cells and keeps them moist, and supplied with nutrients.

The cells in the blood are **erythrocytes** (red blood cells), **leukocytes** (white blood cells), and thrombocytes (**platelets**). More than 99% of all the blood cells are erythrocytes, or red blood cells. Red blood cells look like flexible biconcave discs about 8 nm in diameter that are capable of squeezing through narrow capillaries. Erythrocytes lack a nucleus and therefore are unable to reproduce. **Antigens**, specialized proteins on the surface of erythrocytes, determine the ABO and Rh blood types. Erythrocytes contain **hemoglobin**, a red pigment that carries oxygen, and each red cell has about 280 million hemoglobin molecules. An iron ion in hemoglobin combines reversibly with one oxygen molecule, enabling it to pick up, carry and drop off oxygen. Erythrocytes are formed in red bone marrow, and live about 120 days. When they are worn out, the liver and spleen destroy them and recycle their breakdown products. Anemia is a blood disorder characterized by too few red blood cells.

Leukocytes are white blood cells. They are larger than red blood cells, contain a nucleus, and do not have hemoglobin. Leukocytes fight disease organisms by destroying them or by producing antibodies. Lymphocytes are a type of leukocyte that bring about immune reactions involving antibodies. Monocytes are large leukocytes that ingest **bacteria** and get rid of dead matter. Most leukocytes are able to squeeze through the capillary walls and migrate to an infected part of the body. Formed in the white/yellow bone marrow, a leukocyte's life ranges from hours to years depending on how it functions during an **infection**. In **leukemia**, a malignancy of bone marrow tissue, abnormal leukocytes are produced in an uncontrolled manner. They crowd out the bone marrow cells, interrupt normal blood cell production, and cause internal bleeding. Treatment for acute leukemia includes blood **transfusions**, anticancer drugs, and, in some cases, radiation.

Thrombocytes or platelets bring about clotting of the blood. Clotting stops the bleeding when the circulatory system is damaged. When tissues are injured, platelets disintegrate and release the substance thromboplastin. Working with **calcium** ions and two plasma proteins, fibrinogen and prothrombin, thromboplastin converts prothrombin to thrombin. Thrombin then changes soluble fibrinogen into insoluble fibrin. Finally, fibrin forms a clot.

The lymphatic system is an open transport system that works in conjunction with the circulatory system. Lymphatic vessels collect intercellular fluid (tissue fluid), kill foreign organisms, and return it to the circulatory system. The lymphatic system also prevents tissue fluid from accumulating in the tissue spaces. Lymph capillaries pick up the intercellular fluid, now called lymph, and carry it into larger and larger lymph vessels. Inside the lymph vessels, lymph passes through lymph nodes, where lymphocytes attack **viruses** and bacteria. The lymphatic system transports lymph to the large brachiocephalic veins below the collarbone where it is re-enters the circulatory system. Lymph moves through the lymphatic system by the squeezing action of nearby muscles, for there is no pump in this system. Lymph vessels are equipped with one-way valves that prevent backflow. The spleen, an organ of the lymphatic system, removes old blood cells, bacteria, and foreign particles from the blood.

See also Blood coagulation and blood coagulation tests; Blood pressure and hormonal control mechanisms; Cardiac muscle; Carotid arteries; Collateral circulation; Coronary circulation; Elimination of waste; Fetal circulation; Homeostatic mechanisms; Pulmonary circulation; Renal system

CLAUDE, ALBERT (1899-1983)
Belgian-born American cell biologist

Biologist Albert Claude received the Nobel Prize in 1974 for his discoveries concerning the fine structure of the cell. His early work described the nature of **mitochondria** as the power-house of the cell, paving the way for much groundbreaking research by others. In addition, he demonstrated that the interior of cells were not merely an arbitrary mass of substances, but rather a highly organized space delineated by the net-like endoplasmic reticulum, a formation that he was the first to recognize.

Born in Longlier, Belgium (now Luxembourg), Albert Claude chose to become a United States citizen at age 43. Though he maintained dual citizenship, his decision was the logical outcome of a growing research career in the U.S., a place of opportunity for an individual who began life with what seemed like limited prospects. Claude's father, Florentin Joseph Claude, was a baker. His mother, Marie-Glaudicine Wautriquant, and his father evidently provided the right attitude for young Claude, for he overcame the constraints of a limited education and income to gain acceptance into the University of Liege. This was possible because of his service in World War I, in which he won the Interallied Medal along with veteran status. The university admitted him under a special program designed for war veterans.

Claude earned an M.D. degree in 1928 under his continuing government scholarship, and attended the Kaiser Wilhelm Institute in Berlin for further study. He relocated to the United States in 1929 to join the staff of the Rockefeller Institute in New York City, home to much of the great biomedical research and discoveries of the early twentieth century. There, Claude studied the tumor agent of Rous sarcoma, a virus of chickens. Though Claude had not been invited to join the Institute, the director, Simon Flexner, one of the country's leading medical educators, approved his hiring.

In the laboratory of James B. Murphy, Claude began earnest work on isolating the originating factor of the sarcoma, a malignant **plasma**, first discovered in 1911 at Rockefeller Institute by **Peyton Rous**, that was a type of soft-tissue **cancer** in chickens. Only recently had microbiologists first suggested that cancers might be caused by newly discovered agents known as **viruses**. But it was not until 1932 that Rous' work was vindicated by the discovery of transmissible wild rabbit cancers that were proven to be viral in nature.

Diligently pursuing the new field of virology, Claude developed a technique using a high-speed centrifuge to spin fractionated (broken-up) cells infected with viruses in an attempt to isolate their agents. Though his primitive machine was constructed from meat grinders and sieves, Claude was able to fractionate various components of cells that had never been separated before, paving the way for new understanding of their varying functions. Though he never succeeded in fully isolating the virus within the cell mixture (a development that came years later by other investigators), his discoveries nevertheless became crucial to the study of cell biology.

The Rous virus is among those now known as a **ribonucleic acid** (**RNA**) virus, that is, its genetic material is derived from RNA rather than the more common **deoxyribonucleic acid** (**DNA**). Claude was surprised to find that it was not only virus-infected cells that showed a high RNA content, but also healthy cells. By the early 1940s, Claude joined forces with biochemists George Hogeboom and Rollin Hotchkiss in an attempt to determine the origins of this cellular RNA.

Claude, Hogeboom, and Hotchkiss found a variety of different "granules" in the cells that they determined were

mitochondria, which were first discovered in 1897. However, the purpose of these often-abundant cell components, especially in the **liver** cells, still remained unknown. Claude found that the mitochondria were not the source of the cells' RNA, but they did harbor certain **enzymes** that seemed to be involved in the cells' energy **metabolism**, a process dimly understood at the time. Claude and his colleagues, in fact, proved in 1945 that mitochondria are the "powerhouses" of all cells, from **bacteria** to liver, from plants to fungi to animals. The RNA, it turned out, was concentrated in other cell particles that fellow researcher **George Palade** discovered and called microsomes. Later renamed ribosomes, these particles were shown to be the centers of protein production in all cells of every type of living thing. In 1974 Claude, Palade, and a third researcher, Christian R. Duvé, shared the Nobel Prize in physiology or medicine.

By the early 1940s, Claude had significantly perfected ultracentrifugation (the process of separating cell particles) and was seeking other new technologies with which to probe the cell. In 1942, he became convinced that the newly developed electron microscope would be useful in furthering his studies and secured the use of the device at the Interchemical Corporation, home to the only electron microscope in New York City which was used primarily for metallurgical purposes.

The cells that Claude and his associate, Keith Porter, observed under the microscope showed the presence of a "lace-work" structure that was eventually proven to be the major structural feature of the interior of all but bacterial cells. This lace-work structure was also responsible in part for providing the shape of cells as well as the location for many granular cell components, including ribosomes. The discovery of this endoplasmic reticulum (derived from the Latin word for "fishnet") altered biologists' view of cells as simply bags of "stuff" to highly organized biological units.

In 1948, Claude returned to his native Belgium and for a time gave up active research to become an administrator at the Université Libre de Bruxelles, where he spent the next twenty years developing a significant cancer research center. During the same period he headed the Institut **Jules Bordet**, where he resumed research on the fine structure of cells.

In 1972, the Rockefeller University (formerly Institute) awarded Claude emeritus standing. Other honors accrued over the span of his career include the Medal of the Belgian Academy of Medicine, the Louisa G. Horowitz Prize of Columbia University, and the **Paul Ehrlich** and Ludwig Darmstaedter Prize of Frankfurt. In addition, Claude was a full member of the Belgian and French academies of science and an honorary member of the American Academy of Arts and Sciences. Other honors included the Order of the Palmes Académiques of France, the Grand Cordon of the Order of Léopold II, and the Prix Fonds National de la Recherche Scientifique from Belgium.

See also Cell cycle and cell division; Cell differentiation; Cell membrane transport; Cell structure; Cell theory; Mitochondria and cellular energy; Molecular biology

CLITORIS · *see* GENITALIA (FEMALE EXTERNAL)

CLOTTING · *see* BLOOD COAGULATION AND BLOOD
COAGULATION TESTS

CNS · *see* CENTRAL NERVOUS SYSTEM (CNS)

COCCYX · *see* VERTEBRAL COLUMN

COCHLEA · *see* EAR (EXTERNAL, MIDDLE, AND INTERNAL)

COHEN, STANLEY (1922-)
American biochemist

A pioneer in the study of growth factors—the nutrients that differentiate the development of cells—Stanley Cohen is best known for isolating nerve growth factor (NGF), the first known growth factor, and for subsequently discovering and fully identifying the epidermal growth factor (EGF). Cohen shared the 1986 Nobel Prize in physiology or medicine with his colleague, Italian American neurobiologist **Rita Levi-Montalcini**, who first discovered NGF. Research on NGF has led to better understanding of such degenerative disorders as **cancer** and **Alzheimer's disease**, while studies concerning EGF have proved useful in exploring alternative burn treatments and skin transplants.

Cohen was born in Brooklyn, New York, in 1922 to Russian immigrant parents. Though his father earned only a modest living as a tailor, both parents, Louis and Fannie (Feitel) Cohen, ensured that their four children received quality educations. As a child, Cohen was stricken with polio, imparting him with a permanent limp. His illness, however, influenced him to pursue intellectual interests. While a student at James Madison High School he earnestly studied science as well as classical music, learning to play the clarinet. Cohen entered Brooklyn College to study chemistry and zoology, graduating in 1943 with a B.A. Following his undergraduate studies, Cohen received a scholarship to Oberlin College in Ohio, where he earned an M.A. in zoology in 1945. He then attended the University of Michigan on a teaching fellowship in **biochemistry**, earning his Ph.D. in 1948.

From 1948 until 1952, Cohen worked at the University of Colorado School of Medicine in Denver, holding a research and teaching position in the Department of Biochemistry and Pediatrics. There Cohen earned the respect of his peers for his collaborative studies with pediatrician Harry H. Gordon on the metabolic functions of creatinine (a chemical found in **blood**, muscle **tissue**, and urine) in newborn infants. Cohen moved to St. Louis, Missouri, in 1952 to work as a postdoctoral fellow in the radiology department at Washington University. The following year, he was asked to become a research associate in the laboratory of renowned zoologist Viktor Hamburger, who was conducting studies on growth processes. Levi-Montalcini, who had been researching nerve cell growth in chicken embryos that had been injected with the tumor cells of male

mice, had just returned from Rio de Janeiro, where she had conducted successful tissue culture experiments that definitively proved the existence of NGF. Working at the lab in St. Louis, Levi-Montalcini relied on Cohen's expertise in biochemistry to isolate and analyze NGF.

The collaboration between Levi-Montalcini and Cohen combined two similar personalities. Both scientists have been characterized by their unassuming manners despite their obvious intellectual abilities and perceptive intuitions. Describing her early recollections of Cohen, Levi-Montalcini wrote in her autobiography *In Praise of Imperfection*, "I had been immediately struck by Stan's absorbed expression, total disregard for appearances—as evidenced by his motley attire—and modesty.... He never mentioned his competence and extraordinary intuition which always guided him with infallible precision in the right direction." Between the years 1953 and 1959, Cohen and Levi-Montalcini conducted intense research, both enthusiastically pursuing their findings concerning NGF.

By 1956 Cohen had succeeded in extracting NGF from a mouse tumor; however, this proved to be a difficult substance to work with. Upon the suggestion of biochemist **Arthur Kornberg**, Cohen added snake venom to the extract, hoping to break down the nucleic acids that made the extract too gelatinous. Fortuitously, the snake venom produced more nerve growth activity than the tumor extract itself, and Cohen was able to proceed more rapidly with his studies. In 1958 he discovered that an abundant source of NGF could be found in the salivary glands of male mice—glands not unlike the venom sacs of snakes. Cohen's biochemical advances enabled Levi-Montalcini to study the neurological effects of NGF in rodents.

At a time when Levi-Montalcini and Cohen were advancing rapidly in their collaborative research, funding for Hamburger's laboratory could no longer support Cohen. Before leaving Washington University, Cohen was able to purify NGF as well as produce an antibody for it; however, its complete chemical structure was not fully determined until 1970 when researchers at Washington University completed analysis of NGF's two identical chains of **amino acids**. Before departing St. Louis, Cohen also observed an unusual occurrence in newborn rodents that had been injected with unpurified salivary NGF. Unlike control mice, whose eyes opened on the thirteenth or fourteenth day, those injected with the unpurified NGF opened their eyes on the seventh day; they also sprouted teeth earlier than did the control group.

Cohen left Washington University in 1959 to join a research group at Vanderbilt University in Nashville, Tennessee; there he continued his work with growth factors, focusing on identifying the unknown factor in unpurified NGF that had caused the mice to open their eyes earlier than normal. By 1962, Cohen had extracted the contaminant in these samples of NGF and was able to purify a second substance, a protein that promoted skin cell and **cornea** growth that he called epidermal growth factor, or EGF. This protein has found widespread use in treating severe **burns**; a solution rich in EGF can promote the speedy healing of burned skin, while a skin graft soaked in EGF will quickly bond with damaged tissue. Cohen also isolated the protein which acted as a receptor for EGF—an important step toward understanding the transmission of signals that stimulate normal and abnormal cell growth—that has been particularly crucial in studying cancer development. Cohen was successful in fully identifying the amino acid sequence of EGF by 1972.

Despite his significant contributions, Cohen has never managed a large laboratory, and for many years his work went unacknowledged. He remarked in *Science* that while the scientific community took little notice of his early studies on growth factors, this anonymity proved beneficial. "People left you alone and you weren't competing with the world," he recalled. "The disadvantage was that you had to convince people that what you were working with was real." Cohen's work has subsequently gained wide recognition, and he has received numerous awards in addition to the Nobel, including the Alfred P. Sloan Award in 1982, as well as both the National Medal of Science and the Albert Lasker Award in 1986.

See also Cancer; Cell cycle and cell division; Cell differentiation; Cell membrane transport; Cell structure; Cell theory

COLLATERAL CIRCULATION

When the primary **blood** supply to some part of the body is blocked, any circulation to that part that continues through smaller, parallel vessels is called collateral circulation. Because collateral circulation is sometimes able to compensate for lost primary circulation, it is also sometimes called compensatory circulation.

Collateral circulation is made possible by anastomoses—bridges or interconnections between **arteries**. Anastomoses allow blood to detour around blockages, and vary in size from microscopic to large. Arteries connected by anastomoses are collateral arteries. Tissues whose primary arteries are connected to more (or larger) collateral arteries are better able to survive blockage of their primary blood supply. The arteries of the stomach walls, for example, are interconnected by so many anastomoses that a surgeon can tie off several of them without worrying that part of the stomach will die for lack of blood. In contrast, the arteries of the **heart**, **brain**, **liver**, and **kidneys** have few anastamoses, so an arterial blockage in one of these **organs** almost always kills part of the organ. In the brain, an arterial blockage with resultant **tissue death** is called a **stroke**; in the heart, a heart attack.

In surgery, arteries backed up by lots of collateral circulation can be tied off without consequence (as in the stomach). Others cannot be tied off (or tied off downstream of a certain point) without injuring tissue or even killing the patient. Surgeons can sometimes increase collateral circulation to some part of the body by cutting nerves that cause the **smooth muscle** tissue lining the collateral arteries to contract. These permanently relaxed arteries then carry more collateral circulation.

Collateral arteries have the power to enlarge over time if unusually large amounts of blood try to push through them. Thus, if an artery supplying part of the heart becomes gradually blocked over time by ischemic heart disease, even the

small anastomoses available in heart muscle may be able to widen enough to take over the blocked artery's job. However, these collateral vessels are subject to the same disease process that blocked the primary artery, and may become blocked themselves.

See also Circulatory system; Vascular system overview

COLON • *see* GASTRO-INTESTINAL TRACT

COMA

A coma is a deep state of unconsciousness, characterized by an inability to respond to stimuli. There are grades or levels of coma corresponding to an afflicted individual's ability to respond to stimuli. A complete or irreversible coma is usually characterized by an inability to respond to stimuli or, conversely, a mild coma is often characterized by a less complete sensory unreceptivity in patients.

In addition to coma, there are other stages of consciousness including lethargy and stupor that reflect a greater degree of **brain** activity and awareness. With some injuries and disease processes, it is possible for patient to move between varying states of consciousness, excepting irreversible coma. Accordingly, the ability to define and determine irreversible coma and brain **death** are important considerations in medical ethics and in the practical treatment of patients.

In addition to an inability to respond to stimuli, there are a number of indications of coma. With irreversible coma, there is a changed **electroencephalogram** (i.e., an **EEG** that shows a lack of electrical activity found in the normally functioning brain). Patients in a comatose state are unable speak or respond to normally painful stimuli such as pinpricks. In many cases, there is also an inability to breath without mechanical assistance.

Coma can be induced by a number of factors, most commonly traumatic brain injury. Diabetic coma may also result from a catastrophic **acidosis** induced by **diabetes mellitus**. Hepatic comas are those comas brought on by damage to cerebral cells resulting from metabolic defects produced by damage or degeneration (cirrhosis) of hepatic (**liver**) cells. Acute alcohol or drug intoxication may also lead to coma.

See also Brain stem function and reflexes; Brain: Intellectual functions; Reflexes; Reticular activating system (RAS); Sleep

COMPARATIVE ANATOMY

There are many forms of evidence for **evolution**. One of the strongest forms of evidence is comparative **anatomy**; comparing structural similarities of organisms to determine their evolutionary relationships. Organisms with similar anatomical features are assumed to be relatively closely related evolutionarily, and they are assumed to share a common ancestor.

As a result of the study of evolutionary relationships, anatomical similarities and differences are important factors in determining and establishing classification of organisms.

Some organisms have anatomical structures that are very similar in embryological development and form, but very different in function. These are called **homologous structures**. Since these structures are so similar, they indicate an evolutionary relationship and a common ancestor of the species that possess them. A clear example of homologous structures is the forelimb of mammals. When examined closely, the forelimbs of humans, whales, dogs, and bats all are very similar in structure. Each possesses the same number of bones, arranged in almost the same way. While they have different external features and they function in different ways, the embryological development and anatomical similarities in form are striking. By comparing the anatomy of these organisms, scientists have determined that they share a common evolutionary ancestor and in an evolutionary sense, they are relatively closely related.

Other organisms have anatomical structures that function in very similar ways, however, morphologically and developmentally these structures are very different. These are called analogous structures. Since these structures are so different, even though they have the same function, they do not indicate an evolutionary relationship nor that two species share a common ancestor. For example, the wings of a bird and dragonfly both serve the same function; they help the organism to fly. However, when comparing the anatomy of these wings, they are very different. The bird wing has bones inside and is covered with feathers, while the dragonfly wing is missing both of these structures. They are analogous structures. Thus, by comparing the anatomy of these organisms, scientists have determined that birds and dragonflies do not share a common evolutionary ancestor, or that, in an evolutionary sense, they are closely related. Analogous structures are evidence that these organisms evolved along separate lines.

Vestigial structures are anatomical features that are still present in an organism (although often reduced in size) even though they no longer serve a function. When comparing anatomy of two organisms, presence of a structure in one and a related, although vestigial structure in the other is evidence that the organisms share a common evolutionary ancestor and that, in an evolutionary sense, they are relatively closely related. Whales, which evolved from land mammals, have vestigial hind leg bones in their bodies. While they no longer use these bones in their marine habitat, they do indicate that whales share an evolutionary relationship with land mammals. Humans have more than 100 vestigial structures in their bodies.

Comparative anatomy is an important tool that helps determine evolutionary relationships between organisms and whether or not they share common ancestors. However, it is also important evidence for evolution. Anatomical similarities between organisms support the idea that these organisms evolved from a common ancestor. Thus, the fact that all vertebrates have four limbs and gill pouches at some part of their development indicates that evolutionary changes have occurred over time resulting in the diversity we have today.

See also Evolution and evolutionary mechanisms; Structure and function

COMPUTER ASSISTED TOMOGRAPHY SCANNING (CAT SCAN) • *see* IMAGING

CONES • *see* VISION: HISTOPHYSIOLOGY OF THE EYE

CONGESTIVE HEART FAILURE • *see* CARDIAC DISEASE

CONNECTIVE TISSUES

Connective **tissue** is found throughout the body and includes **fat**, **cartilage**, bone and **blood**. The main functions of the different types of connective tissue include providing support, filling in spaces between **organs**, protecting organs and aiding in the transport of materials around the body.

Connective tissue is composed of living cells and protein fibers suspended in a gel-like material called matrix. Depending on the type of connective tissue, the fibers are collagen fibers, reticular fibers, elastin fibers, or a combination of two or more types. The type and arrangement of the fibers gives each type of connective tissue its particular properties.

Of the three types of protein fibers in connective tissue, collagen is by far the most abundant, and accounts for almost one third of the total body weight of humans. Under the microscope, collagen looks like a rope with three individual protein fibers twined around each other. Collagen is extremely strong, but has little flexibility. Reticular fibers are composed of very small collagen fibers, but are shorter than collagen fibers, and they form a net-like supporting structure that gives shape to various organs. Elastin fibers have elastic properties and can stretch and be compressed, imparting flexibility in the connective tissues where they are found.

Two main types of fibrous connective tissue are found in the body: dense and loose. In dense connective tissue, almost all the space between the cells is filled by large numbers of protein fibers. In loose connective tissue, there are fewer fibers between the cells which imparts a more open, loose structure.

Dense connective tissue contains large numbers of collagen fibers, and so it is exceptionally tough. Dense regular connective tissue has parallel bundles of collagen fibers and forms **tendons** that attach muscles to bone and ligaments that bind bone to bone. Dense irregular connective tissue, with less orderly arranged collagen fibers, forms the tough lower layer of the skin known as the **dermis**, and encapsulates delicate organs such as the **kidneys** and the spleen.

Loose connective tissue has fewer collagen fibers than dense connective tissue, and therefore is not as tough. Loose connective tissue (also known as areolar connective tissue) is widely distributed throughout the body and provides the loose packing material between glands, muscles, and nerves.

Two other connective tissues with fibers are **adipose tissue** and reticular tissue. Adipose tissue is composed of specialized fat cells and has few fibers; this tissue functions as an insulator, a protector of delicate organs and as a site of energy storage. Reticular connective tissue is composed mostly of reticular fibers that form a net-like web, which forms the internal framework of organs like the **liver**, lymph nodes, and bone marrow.

Cartilage is composed of cartilage cells, and collagen fibers or a combination of collagen and elastin fibers. An interesting characteristic of cartilage is that when it is compressed it immediately springs back into shape. **Hyaline cartilage** is rigid yet flexible, due to evenly spaced collagen fibers. Hyaline cartilage is found at the ends of the ribs, around the trachea (windpipe), and at the ends of long bones that form **joints**. Hyaline cartilage forms the entire skeleton of the embryo, which is gradually replaced by bone as the newborn grows. **Fibrocartilage** contains densely packed regularly arranged collagen fibers, which impact great strength to this connective tissue. Fibrocartilage is found between the bones of the vertebrae as discs that act as a cushion. **Elastic cartilage** contains elastin fibers and is thus more flexible that either hyaline cartilage or fibrocartilage. Elastic cartilage is found in the pinnas of the **external ear**.

Bone is composed of bone cells (osteocytes), suspended in a matrix consisting of collagen fibers and minerals. The mineral portion imparts great strength and rigidity to bone. Osteocytes are located in depressions called lacunae connected by canals called Haversian canals.

Two types of bone form the mammalian skeleton: cancellous bone and compact bone. Cancellous bone is more lattice-like than compact bone, and does not contain as many collagen fibers in its matrix. Cancellous bone is lightweight, yet strong, and is found in the **skull**, the **sternum** and ribs, the pelvis and the growing ends of the long bones. Compact bone is densely packed with fibers, and forms the outer shell of all bones and the shafts of the long bones of the arms and legs. Compact bone is heavier than cancellous bone, and provides great strength and support.

Blood is a liquid connective tissue composed of a fluid matrix and blood cells. The blood cells include white blood cells, which function in the **immune system**, and red blood cells, which transport oxygen and carbon dioxide. The fluid part of the blood (the **plasma**) transports **hormones**, nutrients, and waste products, and plays a role in **temperature regulation**.

See also Fibrous joints; Joints and synovial membranes; Skeletal and muscular systems: embryonic development; Skeletal system overview (morphology)

CORI, CARL FERDINAND (1896-1984)
Austro-Hungarian-born American biochemist

Carl Ferdinand Cori and his wife, biochemist Gerty T. Cori, were prominent researchers in **physiology, pharmacology**, and biology. Their most important work involved carbohydrate **metabolism** (especially in **tumors**), phosphate processes in the muscles, the process of glucose-glycogen interconversion, and the action of insulin. The Coris shared the 1947 Nobel Prize in physiology or medicine (along with the Argentine physiologist **Bernardo Houssay**) "for their discovery of the course of the catalytic conversion of glycogen."

Cori was born in Prague (Czech Republic), which was then part of the Austro-Hungarian Empire. His parents were Carl Isidor Cori, a professor of zoology at the German University of Prague, and Maria Lippich Cori. When Cori was still young, the family moved to Trieste, Italy, where his father had been appointed director of the Marine Biology Station. Cori studied at the Gymnasium in Trieste from 1906 to 1914, and then returned to Prague and began medical studies at the German University. His studies, however, were interrupted by World War I. Serving in the Austrian army, Cori worked in hospitals for infectious diseases on the Italian front.

It was during his first term at the University of Prague that Cori met his wife, who was also a medical student. Described as redheaded and vivacious, Gerty Theresa Radnitz was the daughter of a Prague businessman and a lifelong resident of that city. As medical students, the Coris coauthored their first scientific publication; ultimately, they would publish over two hundred research articles together. They were married on August 5, 1920, shortly after receiving their medical doctorates.

From 1920 to 1922, Cori served first as a researcher at the First Medical Clinic in Vienna, Austria, and then in the same capacity at Austria's University of Graz. During this time, his wife worked as an assistant at a children's hospital in Vienna. In 1922, Cori accepted a position at the New York State Institute for the Study of Malignant Diseases in Buffalo. **Gerty Cori** joined him soon thereafter, and they continued their research together. During this period, the Coris were studying carbohydrate metabolism, particularly in tumor cells. They also researched the effects of the surgical removal of the ovaries on the incidence of tumors.

The Coris became American citizens in 1928, and in 1931 they accepted positions at the medical school of Washington University in St. Louis, Missouri, where Cori was to remain until 1966. Their research on carbohydrate metabolism now centered on glucose, or "blood sugar," the energy source for animal life. They developed methods to analyze the relationship of glucose to **glycogen**, the starchlike form in which glucose is stored in the **liver** and muscles. In the 1930s, the Coris performed groundbreaking research on the biochemical processes involved in the interconversion of glucose to glycogen, a process now called the Cori cycle. This interconversion is responsible for maintaining the **blood** sugar at a constant level.

In 1936, the Coris isolated glucose–1-phosphate, now known as the Cori ester, which is involved in the formation and breakdown of glycogen. The Coris also analyzed the function of insulin, a hormone in the **pancreas** that is vital to the body's processing of glucose. In 1938, the Coris analyzed the conversion of glucose–1-phosphate to glucose–6-phosphate. Then, in 1943, they isolated phosphorylase, an enzyme important to the glucose-glycogen interconversion, in crystalline form. The

Coris were able in 1944 to synthesize glycogen in a test tube, the first such synthesis of a high molecular substance.

Cori was appointed professor of **biochemistry** at Washington University in 1944, and two years later he became chairman of the department. In 1947, he and his wife were awarded the Nobel Prize in physiology or medicine for their research on the relationship between glucose and glycogen. This led to many comparisons in the press between the Coris and the first husband-and-wife team to win the Nobel Prize, Pierre and Marie Curie.

In addition to sharing the Nobel Prize in 1947, Cori received numerous awards and honors, including the Isaac Adler Prize from Harvard University in 1944, the Midwest Award of the American Chemical Society in 1945, and the Harry M. Lasker Award of the American Society for the Control of Cancer in 1946. He also received the Squibb Award of the Society of Endocrinologists, which was bestowed on him along with his wife in 1947, and the Willard Gibbs Medal of the American Chemical Society in 1948. Cori received honorary degrees from Cambridge, Yale, and other universities, and was a member of various scientific societies.

In the same year they won the Nobel Prize, his wife was diagnosed with myelosclerosis, a disease of the blood. She died ten years later, on October 26, 1957, of complications from the disease, and Cori suffered the loss of both wife and scientific partner. The couple had one child, a son. In 1966, after retiring from Washington University, Cori served as a visiting professor at the Harvard University School of Medicine. Cori died in Cambridge, Massachusetts.

See also Carbohydrates; Glucose utilization, transport and insulin

Cori, Gerty T. (1896-1957)
Austro-Hungarian-born American biochemist

Gerty T. Cori made significant contributions in two major areas of **biochemistry**, which increased understanding of how the body stores and uses sugars and other **carbohydrates**. For much of her early scientific career, Cori performed pioneering work on sugar **metabolism** (how sugars supply energy to the body), in collaboration with her husband, Carl Ferdinand Cori. For this work they shared the 1947 Nobel Prize in physiology or medicine with **Bernardo A. Houssay**, who had also carried out fundamental studies in the same field. Cori's later work focused on a class of diseases called **glycogen** storage disorders. She demonstrated that these illnesses are caused by disruptions in sugar metabolism. Both phases of Gerty Cori's work illustrated for other scientists the importance of studying **enzymes** (special proteins that permit specific biochemical reactions to take place) for understanding normal metabolism and disease processes.

Gerty Theresa Radnitz was the first of three girls born to Otto and Martha Neustadt Radnitz. She was born in Prague, then part of the Austro-Hungarian Empire, in 1896. Otto was a manager of sugar refineries. It is not known if his work

helped shape his eldest daughter's early interest in chemistry and later choice of scientific focus. However, her maternal uncle, a professor of pediatrics, did encourage her to pursue her interests in science. Gerty was first taught by tutors at home, then enrolled in a private girls' school. At that time, girls were not expected to attend a university. In order to follow her dream of becoming a chemist, Gerty first studied at the Tetschen *Realgymnasium*. She then had to pass a special entrance exam (*matura*) that tested her knowledge of Latin, literature, history, mathematics, physics, and chemistry.

In 1914, Gerty Radnitz entered the medical school of the German University of Prague (Ferdinand University). There she met a fellow classmate, Carl Ferdinand Cori, who shared her interest in doing scientific research. Together they studied human complement, a substance in **blood** that plays a key role in immune responses by combining with antibodies. This was the first of a lifelong series of collaborations. In 1920, they both graduated and received their M.D. degrees.

Shortly after graduating, the couple moved to Vienna and married. Carl worked at the University of Vienna's clinic and the University of Graz's **pharmacology** department, while Gerty took a position as an assistant at the Karolinen Children's Hospital. Some of her young patients suffered from a disease called congenital myxedema, in which deposits form under the skin and cause swelling, thickening, and paleness in the face. The disease is associated with severe dysfunction of the thyroid gland, located at the base of the neck, which helps to control many body processes, including growth. Gerty's particular research interest was in how the thyroid influenced body **temperature regulation**.

In the early 1920s, Europe was in the midst of great social and economic unrest in the wake of World War I, and in some regions, food was scarce. Faced with these conditions, the Coris saw little hope there for advancing their scientific careers. In 1922, Carl moved to the United States to take a position as biochemist at the New York State Institute for the Study of Malignant Diseases (later the Roswell Park Memorial Institute). Gerty joined him in Buffalo a few months later, becoming an assistant pathologist at the institute.

Colleagues cautioned Gerty and Carl against working together, arguing that collaboration would hurt Carl's career. However, Gerty's duties as an assistant pathologist allowed her some free time, which she used to begin studies of carbohydrate metabolism jointly with her husband. This work, studying how the body **burns** and stores sugars, was to become the mainstream of their collaborative research. During their years in Buffalo, the Coris jointly published a number of papers on sugar metabolism that reshaped the thinking of other scientists about this topic. In 1928, Gerty and **Carl Cori** became naturalized citizens of the United States.

In 1931, the Coris moved to St. Louis, Missouri, where Gerty took a position as research associate at Washington University School of Medicine; Carl was a professor there, first of pharmacology and later of biochemistry. The Coris' only child, a son, was born in 1936. Gerty become a research associate professor of biochemistry in 1943 and in 1947, a full professor of biochemistry. During the 1930s and 1940s, the Coris continued their work on sugar metabolism. Their laboratory gained an international reputation as an important center of biochemical breakthroughs. No less than five Nobel laureates spent parts of their careers in the Coris' lab working with them on various problems.

For their pivotal studies in elucidating the nature of sugar metabolism, the Cori's were awarded the Nobel Prize in physiology or medicine in 1947. They shared this honor with Argentine physiologist Bernardo A. Houssay, who discovered how the pituitary gland functions in carbohydrate metabolism. Gerty Cori was only the third woman to receive a Nobel Prize in science. Previously, only Marie Curie and Iréne Joloit-Curie had been awarded such an honor. As with the previous two women winners, Cori was a co-recipient of the prize with her husband.

In the 1920s, when the Coris began to study carbohydrate metabolism, it was generally believed that the sugar called glucose (a type of carbohydrate) was formed from another carbohydrate, glycogen, by the addition of water molecules (a process known as hydrolysis). Glucose circulates in the blood and is used by the body's cells in virtually all cellular processes that require energy. Glycogen is a natural polymer (a large molecule made up of many similar smaller molecules) formed by joining together large numbers of individual sugar molecules for storage in the body. Glycogen allows the body to function normally on a continual basis, by providing a store from which glucose can be broken down and released as needed.

Hydrolysis is a chemical process that does not require enzymes. If, as was believed to be the case in the 1920s, glycogen were broken down to glucose by simple hydrolysis, carbohydrate metabolism would be a very simple, straightforward process. However, in the course of their work, the Coris discovered a chemical compound, glucose–1-phosphate, made up of glucose and a phosphate group (one phosphorus atom combined with three oxygen atoms—sometimes known as the Cori ester) that is derived from glycogen by the action of an enzyme, phosphorylase. Their finding of this intermediate compound, and of the enzymatic conversion of glycogen to glucose, was the basis for the later understanding of sugar metabolism and storage in the body. The Coris' studies opened up research on how carbohydrates are used, stored, and converted in the body.

Cori had been interested in **hormones** (chemicals released by one **tissue** or organ and acting on another) since her early thyroid research in Vienna. The discovery of the hormone insulin in 1921 stimulated her to examine its role on sugar metabolism. Insulin's capacity to control diabetes lent great clinical importance to these investigations. In 1924, Gerty and Carl wrote about their comparison of sugar levels in the blood of both **arteries** and **veins** under the influence of insulin. At the same time, inspired by earlier work by other scientists (and in an attempt to appease their employer), the Coris examined why **tumors** used large amounts of glucose.

Their studies on glucose use in tumors convinced the Coris that much basic research on carbohydrate metabolism remained to be done. They began this task by examining the rate of absorption of various sugars from the intestine. They also measured levels of several products of sugar metabolism, particularly **lactic acid** and glycogen. The former compound results when sugar combines with oxygen in the body.

The Coris measured how insulin affects the conversion of sugar into lactic acid and glycogen in both the muscles and **liver**. From these studies, they proposed a cycle (called the Cori cycle in their honor) that linked glucose with glycogen and lactic acid. Their proposed cycle had four major steps: (1) blood glucose becomes muscle glycogen, (2) muscle glycogen becomes blood lactic acid, (3) blood lactic acid becomes liver glycogen, and (4) liver glycogen becomes blood glucose. Their original proposed cycle has had to be modified in the face of subsequent research, a good deal of which was carried out by the Coris themselves. For example, scientists learned that glucose and lactic acid can be directly inter-converted, without having to be made into glycogen. Nonetheless, the Coris' suggestion generated much excitement among carbohydrate metabolism researchers. As the Coris' work continued, they unraveled more steps of the complex process of carbohydrate metabolism. They found a second intermediate compound, glucose–6-phosphate, that is formed from glucose–1-phosphate. (The two compounds differ in where the phosphate group is attached to the sugar.) They also found the enzyme that accomplishes this conversion, phosphoglucomutase.

By the early 1940s, the Coris had a fairly complete picture of carbohydrate metabolism. They knew how glycogen became glucose. Rather than the simple non-enzymatic hydrolysis reaction that, twenty years earlier, had been believed to be responsible, the Coris' studies painted a more elegant, if more complicated picture. Glycogen becomes glucose–1-phosphate through the action of one enzyme (phosphorylase). Glucose–1-phosphate becomes glucose–6-phosphate through the action of another enzyme (phosphoglucomutase). Glucose–6-phosphate becomes glucose, and glucose becomes lactic acid, each step in turn mediated by one specific enzyme. The Coris' work changed the way scientists thought about reactions in the human body, and it suggested that there existed specific, enzyme-driven reactions for many of the biochemical conversions that constitute life.

In her later years, Cori turned her attention to a group of inherited childhood diseases known collectively as glycogen storage disorders. She determined the structure of the highly branched glycogen molecule in 1952. Building on her earlier work on glycogen and its biological conversions via enzymes, she found that diseases of glycogen storage fell into two general groups, one involving too much glycogen, the other, abnormal glycogen. She showed that both types of diseases originated in the enzymes that control glycogen metabolism. This work alerted other workers in biomedicine that understanding the structure and roles of enzymes could be critical to understanding diseases. Here again, Cori's studies opened up new fields of study to other scientists. In the course of her later studies, Cori was instrumental in the discovery of a number of other chemical intermediate compounds and enzymes that play key roles in biological processes.

At the time of her death, Cori's influence on the field of biochemistry was enormous. She had made important discoveries and prompted a wealth of new research, receiving for her contributions, in addition to the Nobel Prize, the prestigious Garvan Medal for women chemists of the American Chemical Society as well as membership in the National Academy of Sciences. As the approaches and methods that she helped pioneer continue to result in increased scientific understanding, the importance of her work only grows greater.

See also Carbohydrates; Glucose utilization, transport and insulin

CORMACK, ALLAN M. (1924-1998)
South African-born American physicist

Allan M. Cormack was a physicist whose theoretical analysis and experiments in nuclear and particle physics, computer tomography, and math led to his invention of a mathematical technique for computer-assisted x-ray tomography, which revolutionized noninvasive medical diagnosis.

Computerized axial tomography, otherwise known as the CAT scan, is a process by which x rays can be concentrated on specific sections of the human body at a variety of angles. Once this information is analyzed by a computer, it is combined to reproduce images of internal structures previously unviewable by medical technology. It is considered the most revolutionary development in the field of radiography since the discovery of the x ray by Wilhelm Conrad Röntgen in 1895.

Cormack was the first to analyze the possibility of such an examination of a biological system in 1963 and 1964, and to develop the equations needed for computer-assisted x-ray reconstruction of pictures of the human **brain** and body. In 1979, he was awarded the Nobel Prize in physiology or medicine, along with **Godfrey Hounsfield**, a British engineer who, independently of Cormack, developed the first commercially successful CAT scanning devices.

Allan MacLeod Cormack was born in Johannesburg, South Africa, the son of George and Amelia (MacLeod) Cormack, a civil service engineer and a teacher, respectively. The pair had emigrated from Scotland to South Africa prior to World War I, and after young Cormack's father died in 1936, the family settled in Cape Town, where Cormack attended the Rondebosch Boys High School. Here he developed a keen interest in astronomy, physics, mathematics, tennis, debating and acting. Although his first love was astronomy, upon enrolling at the University of Cape Town, Cormack chose the field of engineering, intending to obtain a degree that would allow him to earn a good living. However, within two years he changed his major to physics and completed a baccalaureate of science in 1944 and a Master of Science degree in the field of crystallography in 1945.

During the years that followed, Cormack pursued graduate studies in the field of theoretical physics at Cambridge University in England. Working as a research student in the university's Cavendish Laboratory, he studied radioactive helium under the tutelage of Otto Robert Frisch. He also attended lectures on quantum physics given by Nobel Prize winner Paul Dirac. While his study of physics did not lead him to a career in astronomy as he had anticipated, he remained avidly interested in that field. However, his studies of physics did lead him to meet Barbara Jeanne Seavey, an American whom he met in the lecture hall and married on January 6, 1950. With little money, he returned to South Africa from Cambridge to resume his position as a lecturer in physics at the University of Cape Town, where he would remain until 1956. It was during this period that he was asked to serve a six-month service as resident medical physicist in the radiology department at the Groote Schuur Hospital in Cape Town, where he supervised the use of radioisotopes, as well as the calibration of film badges used to measure hospital workers' exposure to radiation.

At Groote Schuur, Cormack witnessed how radiation was being used in the diagnosis and treatment of **cancer** patients. Baffled by deficiencies in the technology used for such procedures, the experience helped plant the seeds of theoretical analysis that would lead Cormack to develop the CAT scan equations and techniques that would earn him the Nobel Prize in 1979. "I asked myself," he told the *New York Times* following the announcement of the prize, "how can you give a dose of radiation if you don't know the material through which it has to pass?"

This simple question led Cormack to a series of experiments and analyses, the results of which were two papers published separately between 1963 and 1964 in the *Journal of Applied Physics*. By this time, Cormack was also conducting theoretical physics research in Boston on subatomic particles, following a 1956 Harvard University sabbatical as a Research Fellow, where he worked in the cyclotron laboratory under director Andreas Koehler.

Following a brief return to Cape Town in 1957, Cormack returned to the United States, accepting a post as assistant professor of physics at Tufts University in Medford, Massachusetts. Between 1956 and 1964, most of his research in connection with the development of computerized axial tomography was conducted on his own time. Indeed, neither of his two *Journal of Applied Physics* papers met with significant response, despite the fact that they proved the feasibility of his method for producing images of heretofore unviewable or barely viewable cross sections of the human body.

Cormack was naturalized as a citizen of the United States in 1966, and continued his academic career and his research in particle physics at Tufts. He was eventually promoted to associate, and then full professor of physics, serving as chairman of the physics department from 1968 to 1976. Meanwhile, Hounsfield was independently coming to conclusions similar to Cormack's, and developed the first CAT scanner as early as 1972.

In 1979, Cormack and Hounsfield were awarded the Nobel Prize in physiology or medicine for their joint, but independent development of CAT scan theory and technology. At the time, their selection as recipients of the prize was considered highly unusual. Unlike previous Nobel recipients, neither Cormack nor Hounsfield held a doctorate in medicine or

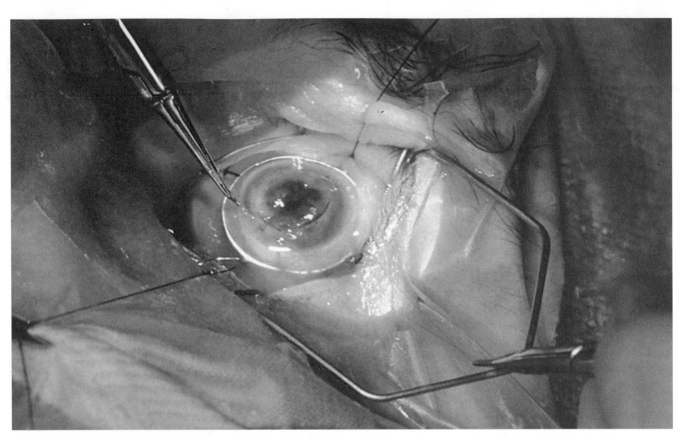

Surgeons implant a donor cornea during a corneal transplant. *Photograph by Chet Szymecki. © Chet Szymecki/Phototake. Reproduced by permission.*

science; further, their discovery was awarded the prize only after the Nobel Assembly vetoed the first choice of the selection committee, reportedly due to a split between factions, with one side favoring discoveries in basic science and the other, those in applied science. Finally, it was highly unusual that the two men had never met or worked together, yet had worked on the same invention concurrently. The story of their simultaneous research is a classic example of independent scientific discovery.

In 1980, Tufts appointed Cormack to university professor, its highest professorial rank, and awarded him an honorary doctoral degree. In 1990, as one of several scientists receiving the National Medal of Science, Cormack was recognized by President George Bush. Bush was quoted in the *New York Times,* lauding the group of scientists as "real life pioneers who press the very limits of their fields." Cormack is a member of the National Academy of Science and the American Academy of Arts and Sciences, and is a fellow of the American Physical Society.

Cormack, who died in 1998, was remembered by his friends as a man with a potent sense of humor, with passionate interests in tennis, swimming, sailing, rock climbing, music, and, as always, astronomy.

See also Imaging

CORNEA AND CORNEAL TRANSPLANT

The cornea is the clear, transparent, anterior covering of the **eye**. The cornea provides almost 70% of the eye's refractive power, and is composed almost entirely of a special type of collagen. The cornea is composed of five layers: the epithelium, Bowman's membrane, the stroma, Descment's membrane, and endothelium. Together, these layers are approximately 5ml (500 microns) thick. The cornea normally contains no **blood** vessels, but because it contains nerve endings, damage to the cornea can be painful.

Corneal transplant is used when vision is lost in an eye because the cornea has been damaged by disease or traumatic injury. Some of the disease conditions that might require corneal transplant include the bulging outward of the cornea (keratoconus), a malfunction of the inner layer of the cornea (Fuchs' dystrophy), and painful swelling of the cornea (pseudophakic bullous keratopathy). Some of these conditions cause cloudiness of the cornea; others alter its natural curvature, which can also reduce the quality of vision.

Injury to the cornea can occur because of chemical **burns**, mechanical trauma, or **infection** by **viruses**, **bacteria**, fungi, or protozoa. Herpes virus is one of the more common infections leading to corneal transplant.

In a corneal transplant, also known as keratoplasty, a patient's damaged cornea is replaced by the cornea from the eye of a human cadaver. This is the single most common type of human transplant surgery and has the highest success rate. Eye banks acquire and store eyes from donor individuals largely to supply the need for transplant corneas.

During a corneal transplant, a disc of **tissue** is removed from the center of the eye and replaced by a corresponding disc from a donor eye. The circular incision is made using an instrument called a trephine. In one form of corneal transplant (penetrating keratoplasty), the disc removed is the entire thickness of the cornea and so is the replacement disc. Over 90% of all corneal transplants in the United States are of this type. In lamellar keratoplasty, on the other hand, only the outer layer of the cornea is removed and replaced.

Surgery is only utilized when damage to the cornea is too severe to be treated with corrective lenses. Refractive surgeries such as LASIK (laser assisted *in situ* keratomileusis), reshape the cornea with a laser without invading the adjacent cell layers. Refractive surgeries offer treatment for mild corneal refractive errors that cause common visual problems, such as myopia (nearsightedness), hyperopia (farsightedness), and astigmatism (distorted vision due to irregular corneal shape). Occasionally, corneal transplant is combined with other types of eye surgery (such as cataract removal). Over 40,000 corneal transplants are performed in the United States each year.

See also Sense organs: Ocular (visual) structures; Vision: histophysiology of the eye

CORONAL PLANE · *see* ANATOMICAL NOMENCLATURE

CORONARY CIRCULATION

Coronary circulation refers to the specialized set of **arteries** and **veins** that supply and drain **blood** form the **heart**. The heart does not derive oxygen and nutrients from the oxygenated blood returned by the pulmonary veins to the left atrium. In fact, the heart does not derive any significant exchanges from the blood that passes through any of its four chambers. Because the heart is a dynamic muscular organ, with especially high oxygen demands, the coronary circulation, a specialized sub-circulation of the **systemic circulation**, ensures delivery of oxygen and nutrients to cardiac **tissue** as well as the pick-up of carbon dioxide and metabolic waste from cardiac tissue.

Two primary coronary arteries, the right and left coronary arteries, branch from the aorta not far form the juncture of the aorta and the left ventricle. The left coronary artery divides into two major branches, the anterior interventricular branch and the circumflex branch. The anterior interventricular branch courses through the anterior interventricular sulcus. As it run through the sulcus it gives off several branches that supply the interventricular septum and the atrio-ventricular bundle (A-V bundle) that is important in electrical conduction within the heart. The circumflex branch courses posteriorly in coronary sulcus.

The right coronary artery courses through the coronary sulcus and gives off a right marginal branch and a posterior interventricular branch. Oxygenated blood traveling through the right coronary artery supplies the sino-atrial and atrio-ventricular nodes. The posterior interventricular branch has a number of fusions (anastomoses) with the anterior interventricular artery branching from the left coronary artery.

As the major arteries further ramify and subdivide, they assure an adequate arteriole and capillary bed for the cardiac muscles and tissues comprising the heart.

The cardiac veins removing blood from heart tissue following gaseous and nutrient exchange drain into the coronary sinus, a wide vessel located between the left and right atrium is what is termed the coronary sulcus. The blood in the coronary sulcus drains directly into the right atrium, directly mixing with the venous return from the inferior vena cava that returns venous blood from the systemic **circulatory system** before passing through the right atrio-ventricular valve into the right ventricle to be pumped into the pulmonary arterial system.

The great cardiac vein runs from the apex of the heart (situated at the inferior, lowest left tip of the heart in a standing body) and after receiving contributions from many branch veins returning blood from the left and right ventricular tissue, the great cardiac vein drains into the coronary sulcus. Other major veins include the small cardiac vein, the middle cardiac vein, the posterior vein of the left ventricle, and the oblique vein of the left atrium. A few smaller veins, including the anterior cardiac veins, do not feed into the coronary sinus, but directly return blood to the right atrium.

See also Anatomical nomenclature; Angiology; Aortic arch; Cardiac disease; Ductus arteriosis; Heart defects; Heart, embryonic development and changes at birth; Heart, rhythm control and impulse conduction; Vascular exchange

CORPUS CALLOSUM · *see* CEREBRAL HEMISPHERES

COSTOCHONDRAL CARTILAGE

Costochondral **cartilage** is the material that separates the bones within the rib cage. "Costo" means rib, "chondr" indicates cartilage. Also referred to as costal cartilage, it helps secure the ribs to the **sternum**.

The costal cartilages consist of 12 pairs of **hyaline cartilage**. The first seven pairs connect directly to the sternum, the next three are associated with the lower border of the preceding rib, and the final two end in the abdominal wall. Like the ribs, the costal cartilages vary in length, width and direction. The cartilage associated with the first rib is short and helps position the first costochondral joint close to the sternum. Cartilage length increases inferior (downward) toward the center of the rib cage, then decreases from ribs seven to 12. The width and intervals between each cartilage diminishes from the first to the last. Direction of each paired cartilage is haphazard, the first descends slightly, the second is horizontal and the third slants upward. The remaining costal cartilages are angular.

Inflammation of the costochondral cartilage is a common condition marked by persistent chest wall **pain**. Symptoms are often mistaken for those associated with a **heart** attack.

See also Elastic Cartilage; Fibrocartilage; Hyaline cartilage; Synovial joints; Tempromandibular joint; Tendons and tendon reflex

COUGH REFLEX

The cough reflex is a coordinated neural and muscular response to the irritation of the **respiratory system**. As a reflex action, coughing does not require conscious direction or control. Coughing is a nociceptive reflex, designed to protect the body from injury.

Coughing is initiated by the sensory **neurons** lining the respiratory passages. Afferent neural impulses produced in respiratory structures travel via the vagal nerve to the medulla of the **brain** and result in the appropriate muscular excitations to produce an acute expulsion of air designed to clear the respiratory passages. Irritation of the respiratory passages may result from contact with a foreign body, mechanical obstruction of the airway, or by an excessive buildup of fluid that obstructs a portion of an airway.

The coordinate cough reflex involves the inspiration of air and the subsequent closure of the epiglottis and the **vocal cords** that traps the air in the **lungs**, as the abdominal muscles, **diaphragm**, and intercostal muscles (muscles between the ribs) involved in normal expiration violently contract. The forceful contraction of these muscles causes the pressure in the lungs to rise until the pressure suddenly and explosively forces open the epiglottis and vocal cords. Air is then expelled at speed ranging up to 100 miles per hour through the **bronchi** and trachea.

The transmission of neural **reflexes** in reflex arcs is based upon the same mechanisms of electrical and chemical mechanisms of transmission as all other forms of neural impulses. Within the neural cell body (axon) the impulse travels electrically as an **action potential** and at the **synapse** (the intercommunicating gap or space between neurons) the neural signal is transmitted through the release, diffusion, and binding of **neurotransmitters**. Reflex impulses are limited to the same conduction speeds as any other form of neural impulse and are the result of simplified neural connections as opposed to a special form of neural transmission. Accordingly, cough **reflexes** and the muscles involved in the cough reflex are subject to reflex rebound and reflex fatigue.

The cough reflex is vital to the maintenance of respiratory integrity, the free-flow of air necessary for life. Although irritation of any of the respiratory structures may produce a cough, certain structures are more sensitive then others. The smaller terminal bronchial structures and lung **alveoli** are, for example, easily subject to irritation by corrosive gases and chemicals, and low levels of such substances will usually induce a cough reflex to expel the irritating agent.

See also Action potential; Brain stem function and reflexes; Nerve impulses and conduction of impulses; Nervous system overview; Neurology; Sneeze reflex; Yawn reflex

COURNAND, ANDRÉ F. (1895-1988)
French-born American physician

André F. Cournand shared the 1956 Nobel Prize in physiology or medicine with German surgeon **Werner Forssmann** and American physiologist Dickinson Woodruff Richards, Jr. for pioneering work in the field of cardiac and pulmonary physiology. Cournand helped develop the technique of cardiac catheterization, which was used at the time to obtain **blood** samples from the **heart** for determining cardiac abnormalities.

André Frédéric Cournand was born in Paris in 1895. His father, Jules Cournand, and his grandfather were both dentists. Cournand writes in his autobiography, *From Roots to Late Budding: The Intellectual Adventures of a Medical Scientist,* that his decision to study the sciences and medicine stemmed from his father's regrets of his own choice of dentistry over medicine. At age 15, young André began to accompany his parents to the salon of a physician friend where many internationally known scientists met and discussed issues of their day. Cournand's mother, Marguerite Weber Cournand, loved literature and learning and encouraged in her son a deep interest in philosophy and art, which Cournand maintained even while pursuing his medical studies and research.

In 1913, Cournand received his bachelor's degree from the University of Paris-Sorbonne, where he also began his medical studies in 1914. But in that year the first World War broke out, and many medical professors enlisted in the army. In the spring of 1915, Cournand decided to postpone his studies. In July of that year he joined a surgical unit that provided emergency care on the front lines. By 1916, he was trained as an auxiliary battalion surgeon and was serving in the trenches. He didn't return to medical school until 1919. After serving as an intern, he received his M.D. in 1930.

Cournand had decided to specialize in upper respiratory diseases and, delaying his entry into private practice, pursued further training in the United States. He joined a residency program at the Tuberculosis Service of the Columbia University College of Physicians and Surgeons at Bellevue Hospital in New York City. He stayed at Columbia for the remainder of his career, rising from his initial position as investigator to a full professor in 1951. He became a naturalized citizen of the United States in 1941.

At Bellevue, Cournand began what would become a long collaboration with Dickinson W. Richards. Together, they investigated the theories of a Harvard physiologist, Lawrence J. Henderson, who had postulated that the heart, **lungs**, and **circulatory system** are a functional unit designed to transport respiratory gases from the atmosphere to the tissues in the body and back out again.

In order to study respiratory gases and their concentrations in the blood as it passed through the heart, samples of blood from the heart had to be obtained. At this time, there was no established technique for this task. Catheters—flexible tubes intended to introduce and remove fluids from organs—had been used for the past 100 years, but only in animal experiments. The safety of catheter use in humans was doubtful. But Cournand was aware that in 1929 a German scientist, Werner Forssmann, had dramatically demonstrated the safety of cardiac catheterization by performing it on himself. He had inserted a catheter into one of his arm **veins** and then threaded it into his right atrium. Cournand became convinced of the safety of catheterization after speaking with one of his professors in Paris who had also performed a type of catheterization on himself, and subsequently scores of others, without any problems.

The Bellevue team experimented on animals for four years, working to standardize the procedure and perfect the equipment they were convinced was necessary for their studies of the cardiac system. When at last cardiac catheterization was used to obtain a sample of mixed venous blood in humans, what could previously be only vaguely determined by clinical observation could be physiologically described. Cardiac catheterization not only allows for samples of mixed venous blood to be collected, but it also measures blood pressure in various parts of the cardiac circulatory system—the right atrium, the ventricles, and the arteries—and measures total blood flow and gas concentrations. In short, the functions of the heart and lungs can be fully specified through cardiac catheterization.

During World War II, Cournand led a team of physicians investigating the use of cardiac catheterization on patients suffering from severe circulatory **shock** resulting from traumatic injury. Obtaining physiological measurements of cardiac output in these patients helped identify the cause of shock—a fall in cardiac output and return. As a result of these findings, it was determined that the best treatment for shock was a blood transfusion rather than simply replacing **plasma**, which had previously been used and was found to cause anemia.

After the war, Cournand applied the technique of cardiac catheterization to patients with heart and pulmonary diseases. The team continually worked to improve the technique and was able, at this time, to obtain simultaneous readings of blood pressure in the right ventricle and the pulmonary artery. This allowed for greater diagnostic accuracy of congenital defects as well as evaluations of treatment. Eventually these investigations led to increased understanding of acquired heart diseases and the relation between diseases of the lungs and cardiac function, thus opening up the field of pulmonary heart diseases.

Cournand began to be recognized for his research in the mid–1940s, when he was invited to speak at and lead various conferences. In 1949, he won the Lasker Award, and in 1952, he was invited by the National Institutes of Health to screen grant applications for the Lung, Heart and Kidney Study Section. Cournand's increasing recognition culminated in the fall of 1956, when he was awarded the Nobel Prize. In 1958, he was elected to the National Academy of Science.

During his years of research, Cournand remained interested and involved in the arts. While still in Paris, he had become a follower of the modern art movement and was

friends with such painters as Jacques Lipschitz and Robert Delaunay and such writers as Andre Breton. Cournand retired in 1964 and devoted the years until his death to the study of the social and ethical implications of modern science. He died in Great Barrington, Massachusetts.

See also Cardiac cycle; Vascular exchange

CPR (CARDIOPULMONARY RESUSCITATION)

Cardiopulmonary resuscitation (CPR) is a procedure to support and maintain **breathing** and circulation for a person who has stopped breathing (respiratory arrest) and/or whose **heart** has stopped (cardiac arrest).

CPR is performed to restore and maintain breathing and circulation, and to provide oxygen and **blood** flow to the heart, **brain**, and other vital **organs**. CPR should be performed if a person is unconscious and not breathing. Respiratory and cardiac arrest can be caused by allergic reactions, an ineffective heartbeat, asphyxiation, breathing passages that are blocked, choking, **drowning**, drug reactions or overdoses, electric **shock**, exposure to cold, severe shock, or trauma. CPR can be performed by trained bystanders or healthcare professionals on infants, children, and adults. It should always be performed by the person on the scene who is most experienced in CPR.

CPR is part of the emergency cardiac care system designed to save lives. Many deaths can be prevented by prompt recognition of the problem and notification of the emergency medical system (EMS), followed by early CPR, defibrillation (which delivers a brief electric shock to the heart in attempt to get the heart to beat normally), and advanced cardiac life support measures.

CPR must be performed within four to six minutes after cessation of breathing so as to prevent brain damage or **death**. It is a two-part procedure that involves rescue breathing and external chest compressions. To provide oxygen to a person's **lungs**, the rescuer administers mouth-to-mouth breaths, then helps circulate blood through the heart to vital organs by external chest compressions. Mouth-to-mouth breathing and external chest compression should be performed together, but if the rescuer is not strong enough to do both, the external chest compressions should be done. This is more effective than no resuscitation attempt, as is CPR that is performed poorly.

When performed by a bystander, CPR is designed to support and maintain breathing and circulation until emergency medical personnel arrive and take over. When performed by healthcare personnel, it is used in conjunction with other basic and advanced life support measures.

According to the American Heart Association, early CPR and defibrillation combined with early advanced emergency care can increase survival rates for people with ventricular fibrillation by as much as 40%. CPR by bystanders may prolong life during deadly ventricular fibrillation, giving emergency medical service personnel time to arrive.

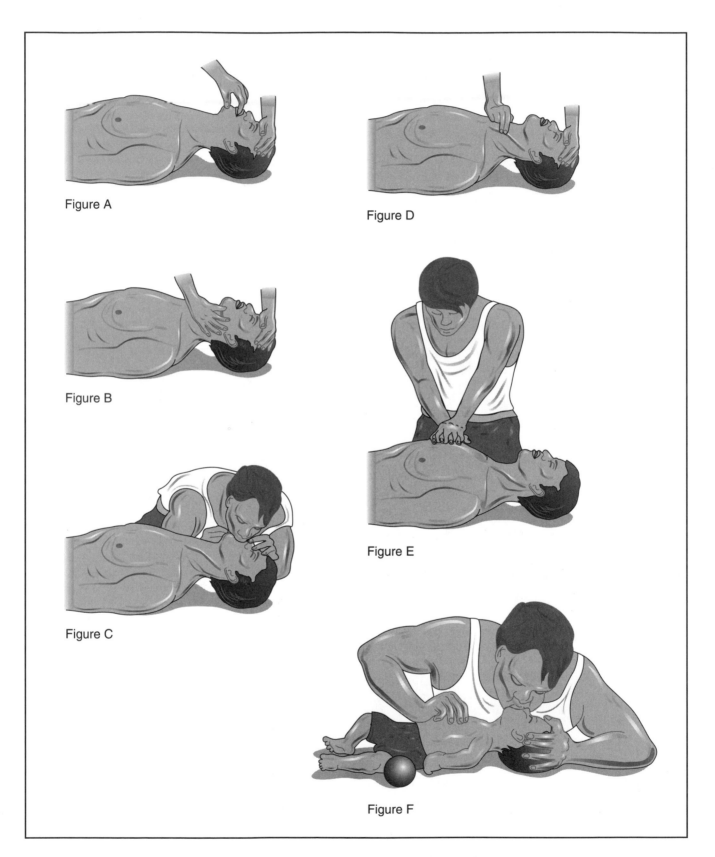

Diagram showing steps in cardiopulmonary resuscitation (CPR), which uses chest compressions and rescue breathing to provide emergency blood supply and oxygen to vital organs. *Illustration by Electronic Illustrators Group.*

However, many CPR attempts are not ultimately successful in restoring a person to a good quality of life. Often, there is brain damage even if the heart starts beating again. CPR is therefore not generally recommended for the chronically or terminally ill, or frail elderly. For these people, it represents a traumatic and not a peaceful end of life.

Each year, CPR helps save thousands of lives in the United States. More than five million Americans annually receive training in CPR through American Heart Association and American Red Cross courses. In addition to courses taught by instructors, the American Heart Association also has an interactive video called Learning System, which is available at more than 500 healthcare institutions. Both organizations teach CPR the same way, but use different terms. These organizations recommend that family members or other people who live with people who are at risk for respiratory or cardiac arrest be trained in CPR. A hand-held device called a CPR Prompt is available to walk people trained in CPR through the procedure, using American Heart Association guidelines. CPR has been practiced for more than 40 years.

The first step is to call the emergency medical system for help by telephoning 911; then to begin CPR, following these steps:

- The rescuer opens a person's airway by placing the head face up, with the forehead tilted back and the chin lifted. The rescuer checks again for breathing (three to five seconds), then begins rescue breathing (mouth-to-mouth artificial respiration), pinching the nostrils shut while holding the chin in the other hand. The rescuer's mouth is placed against the unconscious person's mouth with the lips making a tight seal, then gently exhales for about one to one and a half seconds. The rescuer breaks away for a moment and then repeats. The person's head is repositioned after each mouth-to-mouth breath.

- After two breaths, the rescuer checks the unconscious person's **pulse** by moving the hand that was under the person's chin to the artery in the neck (carotid artery). If the unconscious person has a heartbeat, the rescuer continues rescue breathing until help arrives or the person begins breathing without assistance. If the unconscious person is breathing, the rescuer turns the person onto his or her side.

- If there is no heartbeat, the rescuer performs chest compressions. The rescuer kneels next to the unconscious person, placing the heel of one hand in the spot on the lower chest where the two halves of the rib cage come together. The rescuer puts one hand on top of the other on the person's chest and interlocks the fingers. The arms are straightened, the rescuer's shoulders are positioned directly above the hands on the unconscious person's chest. The hands are pressed down, using only the palms, so that the person's breastbone sinks in about 1½-2 in. (3.8–5 cm). The rescuer releases pressure without removing the hands, then repeats about 15 times per 10-15 second intervals.

- The rescuer tilts the unconscious person's head and returns to rescue breathing for one or two quick breaths. Then breathing and chest compressions are alternated for one minute before checking for a pulse. If the rescuer finds signs of a heartbeat and breathing, CPR is stopped. If the unconscious person is breathing but has no pulse, the chest compressions are continued. If the unconscious person has a pulse but is not breathing, rescue breathing is continued.

- For children over the age of eight, the rescuer performs CPR exactly as for an adult.

To perform CPR on an infant or child under the age of eight, the procedures outlined above are followed with these differences:

- The rescuer administers CPR for one minute, then calls for help.

- The rescuer makes a seal around the child's mouth or infant's **nose** and mouth to give gentle breaths. The rescuer delivers 20 rescue breaths per minute, taking 1½-2 seconds for each breath.

- Chest compressions are given with only one hand for a child and with two or three fingers for an infant. The breastbone is depressed only 1–1½ in. (2.5–3.8 cm) for a child and ½–1 in. for an infant, the rescuer gives at least 100 chest compressions per minute.

Some new ways of performing CPR have been tried. Active compression-decompression resuscitation, abdominal compression done in between chest compressions, and chest compression using a pneumatic vest have all been tested but none are currently recommended for routine use.

The active compression-decompression device was developed to improve blood flow from the heart, but clinical studies have found no significant difference in survival between standard and active compression-decompression CPR. Interposed abdominal counterpulsation, which requires two or more rescuers, one compressing the chest and the other compressing the **abdomen**, was developed to improve pressure and therefore blood flow. It has been shown in a small study to improve survival but more data is needed. A pneumatic vest, which circles the chest of an unconscious person and compresses it, increases pressure within the chest during external chest compression. The vest has been shown to improve survival in a preliminary study but more data is necessary for a full assessment.

If a person suddenly becomes unconscious, a rescuer should call out for help from other bystanders, and then determine if the unconscious person is responsive by gently shaking the shoulder and shouting a question. Upon receiving no answer, the rescuer should call the emergency medical system. The rescuer should check to see whether the unconscious person is breathing by kneeling near the person's shoulders, looking at the person's chest, and placing a cheek next to the unconscious person's mouth. The rescuer should look for signs of breathing in the chest and abdomen, and listen and feel for signs of breathing through the person's lips. If no signs of breathing are present after three to five seconds, CPR should be started.

Emergency medical care is always necessary after successful CPR. Once a person's breathing and heartbeat have been restored, the rescuer should make the person comfortable and stay there until emergency medical personnel arrive. The rescuer can continue to reassure the person that help is coming and talk positively until professionals arrive and take over.

CPR can cause injury to a person's ribs, **liver**, lungs, and heart. However, these risks must be accepted if CPR is necessary to save the person's life.

In many cases, successful CPR results in restoration of consciousness and life. Barring other injuries, a revived person usually returns to normal functions within a few hours of being revived.

See also Cardiac cycle; Cardiac disease; Cardiac muscle; Gaseous exchange; Heart, rhythm control and impulse conduction; Respiration control mechanisms

CRANIAL NERVES

Cranial nerves contain sensory and motor neuron pathways that connect directly to the **brain**. Twelve pairs of cranial nerves pass through the openings in the **skull** to innervate regions of the head and neck. One of the pairs, the vagus nerves, also extend caudally to innervate **respiratory system bronchi** and digestive system **organs** including the stomach and intestines.

In embryological descending order, the twelve cranial nerves, always designated by Roman numerals, are:

- I. Olfactory
- II. Optic
- III. Oculomotor
- IV. Trochlear
- V. Trigeminal
- VI. Abducens
- VII. Facial
- VIII. Auditory (also termed the Vestibulocochlear)
- IX. Glossopharyngeal
- X. Vagus
- XI. Spinal Accessory
- XII. Hypoglossal.

There is a classic mnemonic (a saying where the first letter of each word stands for each of the cranial nerves) used to remember the twelve cranial nerves: "On Old Olympus Towering Tops A Finn And German Viewed Some Hops."

In addition to purely sensory and motor nerves, cranial nerves also contain sensory proprioceptive fibers in their motor neural pathways.

The olfactory nerves are sensory nerves associated with **smell**. Chemical olfactory receptors in the epithelium of the **nose** send signals to the brain via olfactory nerves that pass through the cribiform plate. The olfactory nerves terminate in the olfactory bulb of the brain. Via special neural tracts, olfac-

tory signals are ultimately processed by the **thalamus** and frontal lobe.

Optic nerves are sensory nerves related to vision. The left and right optic nerves originate in the respective optic disk region of the **retina** and pass through, or cross over, in the optic chiasma leading to the thalamus. Ultimately, optic nerve signals are processed by the visual cortex that is located in the occipital lobe and by specialized optic reflex centers located in the midbrain region.

Oculomotor nerves are mixed sensory and motor nerves that innervate and help control some of the intrinsic and extrinsic **eye** muscles. Parasympathetic fibers of the oculomotor nerves serve to control the contractions of the iris that, in turn regulate dilation and constriction of the pupil. Fibers of the oculomotor nerve also help to control muscles that shape the lens of the eye, and are thus important components in adapting visual focus to near and distant objects.

The trochlear cranial nerves are mixed sensory and motor nerves that innervate the superior oblique muscles of the eye.

The trigeminal nerves are the major sensory nerves of the face. The trigeminals derive their name from the fact that each has three major branches, the ophthalmic, maxillary, and mandibular nerves. The ophthalmic and maxillary nerves are sensory nerves. The mandibular nerve is a mixed nerve containing both sensory and motor tracts. The trigeminal nerve is important to the control of chewing (**mastication**).

The abducens cranial nerves provide a motor pathway that innervates the lateral rectus muscles and thereby regulates abduction of the eye.

The facial nerves are mixed sensory and motor cranial nerves that service the facial area. Specific parasympathetic tracts also innervate lacrimal and salivary glands. Facial cranial nerves also carry sensory **taste** related signals from the frontal (or anterior) regions of the tongue.

The acoustic nerves (or vestibulocochlear)—via vestibular and cohlear sensory branches—provide a neural pathway to relay sound and pressure induced stimuli (sound waves are pressure waves) from the inner ear to the bain. The acoustic nerves also provide critical information needed to maintain balance.

The glossopharyngeal cranial nerves provide motor pathways important in the control of the muscles responsible for **swallowing**. Parasympathetic fibers also service the parotid gland, and sensory tracts relay taste related stimuli from back or posterior regions of the tongue. Glossopharyngeal cranial nerves also provide neural pathways that play a role in the regulation of respiration and **blood** pressure and respiration.

The vagus cranial nerves provide parasympathetic innervation to the **heart**, bronchi, stomach, intestines, and other thoracic and abdominal visceral parasympathetic effectors.

Spinal accessory cranial nerves join with spinal nerves to innervate motor pathways of the trapezius and sternocleidomastoid muscles.

The hypoglossal cranial nerves provide motor pathways to muscles related to swallowing.

There are many neurological tests used to evaluate the function of cranial nerves. In addition, specific symptoms relate the to loss or impairment of function in particular senses or regions. For example, the loss of taste from the anterior region of the tongue indicates a pathological condition impairing normal function of a branch of the facial nerve.

In the embryo, the sensory ganglia of cranial nerves derive from neural crest cells that split off from neural folds during the closing of the neural tube. The olfactory, optic, and acoustic nerves develop from specialized neural ectodermal **tissue**. Inner ear receptors derive from special thickenings of **ectoderm** termed placodes. Such placodal thickenings in the pharyngeal arches also help form the facial, glossopharyngeal, and vagus nerves. Extensions, or outgrowths of the axons from the mantle of the basal plate help form the motor neural elements of the trigeminal, facial, and glossopharyngeal nerves.

See also Blood pressure and hormonal control mechanisms; Deafness; Ear (external, middle and internal ear); Face, nose, and palate embryonic development; Hearing (physiology of sound transmission); Nervous system overview; Nervous system: embryological development; Neurology; Sense organs: balance and orientation; Sense organs: embryonic development; Sense organs: ocular (visual) structures; Sense organs: olfactory (sense of smell) structures; Sense organs: otic (hearing) structures; Spinal nerves and rami; Taste, physiology of gustatory structures; Vision: histophysiology of the eye

CRANIUM

The cranium, a portion of the **skull** commonly referred to as the braincase, describes the hard, skeletal structure of the head and face. The bones of the cranium possess distinctive features that allow easy identification of their position and orientation in the body. The human adult mandible, or lower jaw, is not generally considered by anatomists to be a part of the cranium.

The adult human cranium consists of 22 bones: 14 facial bones and 8 cranial bones. The cranial bones are the parietals, located at the top and sides of the skull, two temporal bones above the ears, the ethmoid around the **nose**, the sphenoid at the base of the skull, the frontal over the forehead, and the occipital. The occipital bone at the back of the skull forms a complex joint with the first **vertebra** of the neck (atlas) that permits nodding and rotation of the head.

At birth, the cranial bones are still forming and skull **joints** are soft and flexible. This allows the baby's head to compress and easily pass through the birth canal. The so-called "soft spots" on an infant's skull are termed as fontanels, areas where the bones of the cranium do not meet. Cranial bones fuse together in dovetailed, immovable joints called sutures, which make the skull rigid. By age two, the bones become mostly fused, but full fusion often does not occur until late in **puberty**.

Rare conditions such as Osteitis Deformans and Acromegaly can cause the bones of the skull to grow larger than normal. Falls, blows to the head and others accidents can cause the skull bones to crack and break. Skull fractures can be serious due to the proximity to neural **tissue** in the **brain** or

the subsequent swelling of membranes (e.g., meningeal membranes) covering the **central nervous system** tissue.

See also Meninges; Ossification; Osteology

CRICK, FRANCIS (1916-   )
English molecular biologist

Francis Crick is one half of the famous pair of molecular biologists who unraveled the mystery of the structure of **DNA** (**deoxyribonucleic acid**), the carrier of genetic information, thus ushering in the modern era of **molecular biology**. Since this fundamental discovery, Crick has made significant contributions to the understanding of the **genetic code** and gene action, as well as the understanding of molecular neurobiology. In Horace Judson's book *The Eighth Day of Creation*, Nobel laureate Jacques Lucien Monod is quoted as saying, "No one man created molecular biology. But Francis Crick dominates intellectually the whole field. He knows the most and understands the most." Crick shared the Nobel Prize in medicine in 1962 with **James Watson** and Maurice Wilkins for the elucidation of the structure of DNA.

The eldest of two sons, Francis Harry Compton Crick was born to Harry Crick and Anne Elizabeth Wilkins in Northampton, England. His father and uncle ran a shoe and boot factory. Crick attended grammar school in Northampton, and was an enthusiastic experimental scientist at an early age, producing the customary number of youthful chemical explosions. As a schoolboy, he won a prize for collecting wildflowers. In his autobiography, *What Mad Pursuit*, Crick describes how, along with his brother, he "was mad about tennis," but not much interested in other sports and games. At the age of fourteen, he obtained a scholarship to Mill Hill School in North London. Four years later, at eighteen, he entered University College, London. At the time of his matriculation, his parents had moved from Northampton to Mill Hill, and this allowed Crick to live at home while attending university. Crick obtained a second-class honors degree in physics, with additional work in mathematics, in three years. In his autobiography, Crick writes of his education in a rather light-hearted way. Crick states that his background in physics and mathematics was sound, but quite classical, while he says that he learned and understood very little in the field of chemistry. Like many of the physicists who became the first molecular biologists and who began their careers around the end of World War II, Crick read and was impressed by Erwin Schrödinger's book *What Is Life?*, but later recognized its limitations in its neglect of chemistry.

Following his undergraduate studies, Crick conducted research on the viscosity of water under pressure at high temperatures, under the direction of Edward Neville da Costa Andrade, at University College. It was during this period that he was helped financially by his uncle, Arthur Crick. In 1940, Crick was given a civilian job at the Admiralty, eventually working on the design of mines used to destroy shipping. Early in the year, Crick married, then later welcomed a son

Francis Crick. *The Library of Congress.*

Nobel Prize laureates), but eventually to work on the structure of DNA with Watson.

In 1947, Crick was divorced, then married an art student whom he had met during the war. Their marriage coincided with the start of Crick's Ph.D. thesis work on the x-ray diffraction of proteins. X-ray diffraction is a technique for studying the crystalline structure of molecules, permitting investigators to determine elements of three-dimensional structure. In this technique, x rays are directed at a compound, and the subsequent scattering of the x-ray beam reflects the molecule's configuration on a photographic plate.

In 1941 the Cavendish Laboratory where Crick worked was under the direction of physicist Sir William Lawrence Bragg, who had originated the x-ray diffraction technique forty years before. Perutz had come to the Cavendish to apply Bragg's methods to large molecules, particularly proteins. In 1951, Crick was joined at the Cavendish by James Watson, a visiting American who had been trained by Italian physician Salvador Edward Luria and was a member of the Phage Group, a group of physicists who studied bacterial **viruses** (known as bacteriophage, or simply phages). Like his phage colleagues, Watson was interested in discovering the fundamental substance of genes and thought that unraveling the structure of DNA was the most promising solution. The informal partnership between Crick and Watson developed, according to Crick, because of their similar "youthful arrogance" and similar thought processes. It was also clear that their experiences complemented one another. By the time of their first meeting, Crick had taught himself a great deal about x-ray diffraction and protein structure, while Watson had become well informed about phage and bacterial **genetics**.

Both Crick and Watson were aware of the work of biochemists Maurice Wilkins and Rosalind Franklin at King's College, London, who were using x-ray diffraction to study the structure of DNA. Crick, in particular, urged the London group to build models, much as American chemist Linus Pauling had done to solve the problem of the alpha helix of proteins. Pauling, the father of the concept of the chemical bond, had demonstrated that proteins had a three-dimensional structure and were not simply linear strings of **amino acids**. Wilkins and Franklin, working independently, preferred a more deliberate experimental approach over the theoretical, model-building scheme used by Pauling and advocated by Crick. Thus, finding the King's College group unresponsive to their suggestions, Crick and Watson devoted portions of a two-year period discussing and arguing about the problem. In early 1953, they began to build models of DNA.

Using Franklin's x-ray diffraction data and a great deal of trial and error, they produced a model of the DNA molecule that conformed both to the London group's findings and to the data of Austrian-born American biochemist Erwin Chargaff. In 1950, Chargaff had demonstrated that the relative amounts of the four nucleotides, or bases, that make up DNA conformed to certain rules, one of which was that the amount of adenine (A) was always equal to the amount of thymine (T), and the amount of guanine (G) was always equal to the amount of cytosine (C). Such a relationship suggests pairings of A and T,

who was born during an air raid on London. By the end of the war, Crick was assigned to scientific intelligence at the British Admiralty Headquarters in Whitehall to design weapons.

Realizing that he would need additional education to satisfy his desire to do fundamental research, Crick decided to work toward an advanced degree. Crick became fascinated with two areas of biology, particularly, as he describes it in his autobiography, "the borderline between the living and the nonliving, and the workings of the brain." He chose the former area as his field of study, despite the fact that he knew little about either subject. After preliminary inquiries at University College, Crick settled on a program at the Strangeways Laboratory in Cambridge under the direction of Arthur Hughes in 1947, to work on the physical properties of cytoplasm in cultured chick fibroblast cells. Two years later, he joined the Medical Research Council Unit at the Cavendish Laboratory, ostensibly to work on protein structure with British chemists Max Perutz and John Kendrew (both future and G and C, and refutes the idea that DNA is nothing more than a tetranucleotide, that is, a simple molecule consisting of all four bases.

During the spring and summer of 1953, Crick and Watson wrote four papers about the structure and the supposed function of DNA, the first of which appeared in the journal *Nature* on April 25. This paper was accompanied by papers by Wilkins, Franklin, and their colleagues, presenting experimental evidence that supported the Watson-Crick model. Watson won the coin toss that placed his name first in the authorship, thus forever institutionalizing this fundamental scientific accomplishment as "Watson-Crick."

The first paper contains one of the most remarkable sentences in scientific writing: "It has not escaped our notice that the specific pairing we have postulated immediately suggests a possible copying mechanism for the genetic material." This conservative statement (it has been described as "coy" by some observers) was followed by a more speculative paper in *Nature* about a month later that more clearly argued for the fundamental biological importance of DNA. Both papers were discussed at the 1953 Cold Spring Harbor Symposium, and the reaction of the developing community of molecular biologists was enthusiastic. Within a year, the Watson-Crick model began to generate a broad spectrum of important research in genetics.

Over the next several years, Crick began to examine the relationship between DNA and the genetic code. One of his first efforts was a collaboration with Vernon Ingram, which led to Ingram's 1956 demonstration that sickle cell **hemoglobin** differed from normal hemoglobin by a single amino acid. Ingram's research presented evidence that a molecular genetic disease, caused by a Mendelian mutation, could be connected to a DNA-protein relationship. The importance of this work to Crick's thinking about the function of DNA cannot be underestimated. It established the first function of "the genetic substance" in determining the specificity of proteins.

About this time, South African-born English geneticist and molecular biologist Sydney Brenner joined Crick at the Cavendish Laboratory. They began to work on the coding problem, that is, how the sequence of DNA bases would specify the amino acid sequence in a protein. This work was first presented in 1957, in a paper given by Crick to the Symposium of the Society for Experimental Biology and entitled "On Protein Synthesis." Judson states in *The Eighth Day of Creation* that "the paper permanently altered the logic of biology." While the events of the transcription of DNA and the synthesis of protein were not clearly understood, this paper succinctly states "The Sequence Hypothesis... assumes that the specificity of a piece of nucleic acid is expressed solely by the sequence of its bases, and that this sequence is a (simple) code for the amino acid sequence of a particular protein." Further, Crick articulated what he termed "The Central Dogma" of molecular biology, "that once 'information' has passed into protein, it cannot get out again. In more detail, the transfer of information from nucleic acid to nucleic acid, or from nucleic acid to protein may be possible, but transfer from protein to protein, or from protein to nucleic acid is impossi-

ble." In this important theoretical paper, Crick establishes not only the basis of the genetic code but predicts the mechanism for **protein synthesis**. The first step, transcription, would be the transfer of information in DNA to **ribonucleic acid (RNA)**, and the second step, translation, would be the transfer of information from RNA to protein. Hence, the genetic message is transcribed to a messenger, and that message is eventually translated into action in the synthesis of a protein. Crick is credited with developing the term "codon" as it applies to the set of three bases that code for one specific amino acid. These codons are used as "signs" to guide protein synthesis within the cell.

A few years later, American geneticist Marshall Warren Nirenberg and others discovered that the nucleic acid sequence U-U-U (polyuracil) encodes for the amino acid phenylalanine, and thus began the construction of the DNA/RNA dictionary. By 1966, the DNA triplet code for twenty amino acids had been worked out by Nirenberg and others, along with details of protein synthesis and an elegant example of the control of protein synthesis by French geneticist François Jacob, Arthur Pardée, and French biochemist Jacques Lucien Monod. Brenner and Crick themselves turned to problems in developmental biology in the 1960s, eventually studying the structure and possible function of histones, the class of proteins associated with **chromosomes**.

In 1976, while on sabbatical from the Cavendish, Crick was offered a permanent position at the Salk Institute for Biological Studies in La Jolla, California. He accepted an endowed chair as Kieckhefer Professor and has been at the Salk Institute ever since. At the Salk Institute, Crick began to study the workings of the **brain**, a subject that he had been interested in from the beginning of his scientific career. While his primary interest was consciousness, he attempted to approach this subject through the study of vision. He published several speculative papers on the mechanisms of dreams and of attention, but, as he stated in his autobiography, "I have yet to produce any theory that is both novel and also explains many disconnected experimental facts in a convincing way."

During his career as an energetic theorist of modern biology, Francis Crick has accumulated, refined, and synthesized the experimental work of others, and has brought his unusual insights to fundamental problems in science.

See also Chromosome structure and morphology; DNA structure; DNA transcription and translation; Double helix

CRYPTS OF LIEBERKUHN • *see* INTESTINAL HISTOPHYSIOLOGY

CYTOLOGY • *see* ANATOMY

D

DALE, HENRY HALLETT (1875-1968)
English physiologist

Henry Hallett Dale was a British physiologist who devoted his scientific career to the study of how chemicals in the body regulate physiological functions. Although his work had many facets, the most significant was his collaborative effort with German pharmacologist **Otto Loewi**. In 1936 Dale and Loewi were jointly awarded the Nobel Prize in physiology or medicine for research demonstrating that nerve cells communicate with one another primarily by the exchange of chemical transmitters. In addition to his scientific work, Dale was a prominent figure in science and medicine in England at critical junctures in that nation's history. He was knighted in 1932.

Born London, Dale was the second son of seven children born to Charles Dale, a London businessman, and his wife, Frances Hallett Dale. After graduating from Tollington Park College, London, and the Leys School, Cambridge, Dale entered Trinity College at Cambridge University in 1894. His academic skills gained him first honors in the natural sciences and the Coutts-Trotter studentship at Trinity College. Dale's predecessor in the studentship was Ernest Rutherford, the physicist and chemist who would go on to win the Nobel Prize in chemistry in 1908.

Dale left Cambridge in 1900 to finish his clinical work in medicine at St. Bartholomew's Hospital in London. He received his bachelor's degree in 1903, and his medical doctorate in 1909. During this time, he also was awarded the George Henry Lewes studentship, which allowed him to pursue further physiological research. Later, Dale also received the Sharpey studentship in physiology at University College, London. Dale used these opportunities for research from 1902 to 1904, studying with Ernest Henry Starling and William Maddock Bayliss at University College. Starling and Bayliss identified secretin—a substance secreted by the small intestine—as the first hormone, and Dale collaborated with the pair in further studies on the impact of secretin on cells in the **pan-**creas. Dale's work with Starling and Bayliss instilled in him the idea that physiological functions could be affected by chemicals such as **hormones**. It was also in this laboratory that Dale first met Otto Loewi, who at the time was visiting University College from Germany. Dale and Loewi would go on to become lifelong friends, collaborators, and co-recipients of the 1936 Nobel Prize.

In 1904, Dale spent three months working in the laboratory of the chemist **Paul Ehrlich** in Germany. Members of Ehrlich's laboratory were studying the relationship between the chemical structure of biological molecules and their effect on immunological responses, research that would garner for Ehrlich the 1908 Nobel Prize in physiology or medicine. As did the experience at Starling's laboratory in London, Ehrlich's research introduced Dale to the potential impact that chemicals can have on mediating biological and physiological processes.

After Dale returned to Starling's London laboratory, he was recommended to chemical manufacturer Henry Wellcome for a position with London's Wellcome Physiological Research Laboratories, a commercial laboratory. Established in the 1890s to produce an antitoxin for diphtheria, the laboratories, by the first decade of the 1900s, had begun to promote and pursue basic scientific research. Against the advice of colleagues who distrusted the commercial nature of the laboratory, Dale accepted the post. He reasoned that it would provide him the stability that would allow him to marry. The post it also provided a well-equipped laboratory, freedom from teaching and administrative duties, and the intellectual freedom to pursue his own course of research.

Once Dale had settled at Wellcome, the company suggested that he consider examining the therapeutic properties of ergot, a fungus being used by obstetricians to induce and promote labor. For the next decade, Dale devoted his research efforts to studying the properties of the drug. Although he failed at the stated purpose of his research—articulating the properties of ergot—accidental findings turned out to be of great significance, leading, for instance, to his discovery of the phenomenon of adrenaline (or **epinephrine**) reversal, in which the normally excitatory effects of these drugs are neutralized.

Dale's research on the effects of ergot also introduced him to ongoing efforts to study the **central nervous system**. T. R. Elliott, Dale's friend and colleague at Cambridge, postulated that epinephrine (a neurotransmitter, or substance that transmits nerve impulses) when applied alone could produce an effect similar to stimulating the sympathetic branch of the **autonomic nervous system**. The autonomic nervous system is responsible for involuntary physiological functions, such as **breathing** and **digestion**. This system's sympathetic branch affects such functions as increasing **heart** rate in response to fear and opening **arteries** to increased **blood** flow during exercise. Dale built on Elliott's research and showed, with the chemist George Barger, that epinephrine is one chemical in a class of such chemicals that has "sympathomimetic" properties.

Dale's serendipitous accomplishments drew the attention of Henry Wellcome, and Dale was promoted in 1906 to the directorship of the Wellcome Laboratories. After this promotion, Dale began to apply what he had learned while a student in the laboratories of Starling and Ehrlich. Dale understood that there are a number of active components in ergot. Wanting to understand the chemical mechanisms that underlie physiological functions, Dale began studies of the chemicals that operate in the posterior pituitary lobe of the **brain**. It is this area of the brain where ergot has its effects, since this area is responsible for inducing contractions of the uterine muscles.

Dale resigned from the Wellcome Laboratories in 1914, and joined the scientific staff of the Medical Research Committee; after 1920 this group came to be known as the Medical Research Council. The onset of World War I placed new demands on Dale's administrative and scientific skills. He joined the war effort by engaging in physiological studies of shock, dysentery, gangrene, and the effects of inadequate diet.

After the war, the Medical Research Council evolved to become the National Institute for Medical Research, and Dale served as the organization's first director from 1928 until 1942. Although he continued to perform physiological research, administrative and public duties for the Medical Research Council and the National Institute for Medical Research limited the time and energy that he could devote to the laboratory. His research efforts during the 1920s continued the work he began during the war—studying how histamine contributes to the swelling of **tissue** after traumatic shock. Dale demonstrated that histamine leads to the loss of **plasma** fluid into the tissues and produces swelling. This could lead to more serious problems, including decreased blood circulation, shock, and ultimately **death**.

Dale's study of histamine also contributed to his subsequent work on the nervous system. Histamine, like the neurotransmitter acetylcholine, dilates vascular tissue in the human body. Dale had long known from his work in Starling's laboratory that acetylcholine increases the diameter of vascular tissue. The question remaining for Dale was how this chemical produces the physiological effect.

In 1927, Dale collaborated with H. W. Dudley to isolate acetylcholine from the spleen of an ox and a horse. Having

isolated the crucial compound, Dale sought to understand how and where acetylcholine plays its role in vasodilatation, or the widening of the cavities of blood vessels. Over the next decade, Dale worked with colleagues at the National Institute for Medical Research and concluded that acetylcholine serves as a neurotransmitter and that this is the chemical mediator involved in the transmission of nerve impulses. Dale's findings disproved the proposition of John Carew Eccles and other neurophysiologists who maintained that nerve cells communicate with one another via an electrical mechanism. Dale demonstrated that a chemical process and not an electrical one was the underlying mechanism for nerve transmission. A similar conclusion had been reached by Otto Loewi: As early as 1921, Loewi suggested that a chemical mediator was responsible for the conduction of nerve impulses; it would be Dale who would identify the mediator.

For their work, Dale and Loewi were jointly awarded the 1936 Nobel Prize in physiology or medicine. During the 1930s, Dale continued collaborative research with G. L. Brown, W. Feldberg, J. H. Gaddum, and M. Vogt at the National Institute for Medical Research. Their efforts produced more evidence that acetylcholine is a neurotransmitter involved in nerve impulses.

By the 1940s, Dale was devoting much of his time to administrative duties in various organizations. During World War II, he served as chair of the Scientific Advisory Committee to the War Cabinet. Having been elected a fellow of the Royal Society in 1914, he served as secretary from 1925 to 1935, and as president from 1940 to 1945. His many other public affiliations included serving as president of various organizations, such as the Royal Institution of Great Britain during the mid–1940s, the British Association for the Advancement of Science in 1947, the Royal Society of Medicine from 1948 to 1950, and the British Council during the 1950s.

Other distinctions bestowed upon Dale include receiving the Copley Medal from the Royal Society in 1937 and being knighted with the Grand Cross Order of the British Empire in 1943. He also garnered the Order of Merit in 1944. Since 1959 the Society for Endocrinology has awarded the Dale Medal for the kind of excellence in research exemplified by Dale; and since 1961 the Wellcome Trust he chaired from 1938 until 1960 has endowed the Henry Dale professorship with the Royal Society.

In later years Dale worked with Thorvald Madsen of Copenhagen directing an international campaign to standardize drugs and vaccines. The 1925 conference of the Health Organization of the League of Nations adopted such standards for insulin and pituitary products largely because of Dale's efforts. He repeated these efforts to see into law the Therapeutic Substances Act in England. His other political activities included promoting both the peaceful use of nuclear energy and the value of scientific research.

See also Nerve impulses and conduction of impulses; Nervous system overview; Neurons; Neurotrans-mitters

DALTON, JOHN CALL (1825-1889)

American physiologist

John Call Dalton was one of the first modern American experimental physiologists. While choosing a career path unique to an American physician of the eighteenth century, Dalton's experimental research led to important contributions in **anatomy**, physiology, and medical education.

Dalton was born in Chehnsford, Massachusetts, and after his primary education, attended Harvard University, obtaining his medical degree in 1847. After graduation and serving as a house surgeon at Massachusetts General Hospital for almost four years, Dalton was awarded a prize by the American Medical Association for his essay, "Corpus Leuteum." Dalton then diverted his efforts to physiology and experimentation, traveling to Paris to study with Claude Bernard (1813–1878), widely regarded to be the scientist who defined the principals of experimental medicine. In Paris, Dalton continued his stdies of the anatomy of the **placenta**, and began research on the physiology of the **cerebellum** and intestinal **digestion**. Years later, Dalton would confirm Bernard's experimental results on the sugar-making functions of the **liver**.

Dalton returned from France to accept an appointment as professor of physiology at the University of Buffalo in 1853. While at Buffalo, Dalton was the first in America to illustrate the concepts of physiology using live experimentation on animals. He remained a staunch advocate of live animal research throughout his career. Dalton left Buffalo for a professorship at the Vermont Medical College in 1854, followed by a position as physiology chairman at the Long Island College Hospital in Brooklyn, New York, where he served until 1859. In that same year, Dalton published his physiology textbook, *A Treatise on Human Physiology,* that served as a standard in medical education through seven editions.

During the Civil War, Dalton briefly returned to surgery, serving first in Washington as surgeon of volunteers for the 7th New York regiment, then holding other key medical corps posts until 1864. In 1865, Dalton returned to New York to become professor of physiology and miscropical anatomy at the College of Physicians and Surgeons, where he also served as president of the faculty until his death.

Other contributions by Dalton to the scientific literature of physiology include *A Treatise on Physiology and Hygiene for Schools, Families, and Colleges*(1868), *The Experimental Method of Medicine*(1882), *Doctrines of the Circulation* (1884), and *Topographical Anatomy of the Brain* (1885). Dalton was elected to the National Academy of Sciences in 1864.

See also Histology and microanatomy; History of anatomy and physiology: The science of medicine

DAM, HENRICK (1895-1976)

Danish chemist

Henrik Dam is best known for his discovery of vitamin K, which gives **blood** the ability to clot, or coagulate. The discovery of vitamin K dramatically reduced the number of deaths by bleeding in newborns, and during surgery. For the discovery, Dam received the 1943 Nobel Prize in medicine or physiology. Edward A. Doisy, the American biochemist who isolated and synthetically produced vitamin K, shared this prize with Dam.

Carl Peter Henrik Dam was born in Copenhagen, Denmark. His interest in science was shaped at least in part by his background. His father, Emil Dam, was a pharmaceutical chemist who wrote a history of pharmacies in Denmark. His mother, Emilie Peterson Dam, was a schoolteacher. He attended the Polytechnic Institute in Copenhagen, from which he received his master of science degree in 1920. He was associated with the Royal School of Agriculture and Veterinary Medicine in Copenhagen for the next three years, after which he spent five years as an assistant at the University of Copenhagen's physiological laboratory. He became assistant professor of **biochemistry** in 1928, and associate professor in 1929 (a post he held until 1941).

During these years Dam studied microchemistry under Fritz Pregl in Austria (1925) at the University of Graz, and collaborated with biochemist Rudolf Schoenheimer in Freiburg, Germany (on a Rockefeller Fellowship) from 1932 to 1933. He was awarded a doctorate in biochemistry by the University of Copenhagen in 1934. Afterwards, he worked with the Swiss chemist Paul Xarrer at the University of Zurich in 1935. Dam specialized in **nutrition**, which became his area of expertise.

It was while Dam was studying in Copenhagen that he became interested in what would become the vitamin K factor. In the late 1920s, he began experimenting with hens in an attempt to discover how the animals synthesized cholesterol. Providing them with a synthetic diet, Dam discovered that they developed internal bleeding in the form of hemorrhages under the skin—lesions similar to those found in the disease scurvy. He added lemon juice to the diet (citrus fruits, high in vitamin C, had been found by the eighteenth century Scottish physician James Lind to cure scurvy in sailors), but the supplement did little to reverse the hens' condition.

After experimenting with a variety of food additives, Dam came to the conclusion that some vitamin must exist to give blood the ability to clot—and that this vitamin was what was missing from his synthetic hen diet. He made his findings known in 1934, naming the vitamin "K" from the German word *Koagulation*. Dam's continued research, along with the work of Doisy and other biochemists, led to the isolation of vitamin K and its synthetic production.

Dam's discovery led not only to the Nobel Prize but also the Christian Bohr Award in Denmark in 1939. Dam came to the United States in 1940 for a series of lectures in the U.S. and Canada under the auspices of the American-Scandinavian Foundation. During his visit Nazi Germany invaded Denmark. Dam chose not to return to his native country and accepted a position as senior research associate at the University of Rochester's Strong Memorial Hospital. Because of the war, the Nobel Prize Committee decided to present the awards in New York in 1943. The Nobel recipients of that year, including Dam, were the first to be awarded their prize in the United

States. In 1945, Dam became an associate member of the Rockefeller Institute for Medical Research.

After Denmark was liberated, Dam returned in 1946 to accept the position of head of the biology department at the Polytechnic Institute (the position had been awarded to him in absentia in 1941). He returned to the U.S. in 1949 for a three-month lecture tour, this time to discuss vitamin E. In 1956, he was named head of the Danish Public Research Institute. He was a member of numerous organizations including the American Institute of Nutrition, the Society for Experimental Biology and Medicine, the Royal Danish Academy of Science, the Société Chimique of Zurich, and the American Botanical Society. During his career he published more than one hundred articles in scientific journals on vitamin K, vitamin E, cholesterol, and a variety of other topics. Dam married Inger Olsen in 1925. His primary form of recreation was travel. After he returned to Denmark, he pointedly criticized the American hospital system, saying it was hurt by too much emphasis on the business of running hospitals. He died in Copenhagen at the age of eighty-one.

See also Blood coagulation and blood coagulation tests; Vitamins

DAUSSET, JEAN (1916-   )
French immunologist

Jean Dausset fostered breakthroughs that led to successful organ transplantation. A specialist in **blood** diseases, Dausset discovered the existence of a human biological system that determines whether a body will accept or reject foreign substances, called the major histocompatability complex (MHC). This made it possible for surgeons to type cells to determine whether a prospective donor and patient are a compatible match. His research earned Dausset the Nobel Prize in physiology or medicine in 1980.

Jean Baptiste Gabriel Dausset was born in Toulouse, France, to Henri Dausset and the former Elizabeth Brullard. Dausset followed in the footsteps of his father, a physician and radiologist, by embarking upon a medical career. His studies at the University of Paris were interrupted by the onset of World War II, when he was drafted into military service. Following the fall of France to Germany in 1940, he left his occupied country to fight with the Free French forces in North Africa. Before he left Paris, Dausset gave his identification papers to a Jewish colleague at the Pasteur Institute, who thus survived Nazi persecution by impersonating Dausset.

While in North Africa, Dausset served in the blood transfusion unit of the Free French Army, where he developed his interest in transfusion reactions. Participating in his country's liberation in 1944, Dausset attained the rank of second lieutenant before leaving the service in 1945. After receiving his medical degree from the University of Paris, he was appointed administrator of the National Blood Transfusion Center, a post he held until 1963. In 1948, Dausset became a graduate student at Harvard Medical School, studying hema-

tology and Immunology. His studies paid off in 1951, when he discovered that the type-O blood group, which had been assumed safe for **transfusions** with all other blood types, was in some cases hazardous to the recipients with blood types other than O. Dausset found that persons with type-O blood who had received vaccines against diphtheria or tetanus often developed strong antibodies in their blood system. Since this blood had been conditioned to fight foreign substances, it often reacted with the properties of other blood types—in essence attacking the **antigens** in the recipient's blood. Often when this fortified type-O blood was transfused into other blood types, the reaction would cause dangerous clotting.

Dausset's work at this time also involved research on the disease agranulocytosis, an illness characterized by acute fevers that was often triggered by drug hypersensitivity. He found that those suffering agranulocytosis had an unusually low white blood cell (leukocyte) count. Dausset also found similarly low leukocyte counts in patients who had numerous transfusions. In a 1952 paper titled "Presence of a Leucoagglutinate in the Serum of a Case of Agranulocytosis," he theorized that the patients' systems had created antibodies which reacted against the antigens in foreign white blood cells while simultaneously distinguishing—and not attacking—its own white blood cells. Dausset called these developed antibodies human leukocyte antigens (HLA), and found them to be crucial in determining whether or not blood, **tissue**, or an organ could be transferred from one person to another. The paper built upon work done with the immune systems of mice and, in particular, the discovery of an antigen similar to human HLA called H–2. H–2 had been discovered by American scientist **George Snell** in the 1940s, following research on skin grafts with the animals.

From his knowledge of HLA, Dausset established the existence of a histocompatibility system in humans, known as the major histocompatibility complex (MHC). This genetically controlled system is the human body's method for keeping out material that it deems harmful. It could be used to determine whether a body would accept or reject the tissue of donor. Through typing for HLA, doctors would be able to classify tissue systems as they had blood. It also enabled doctors to determine whether a transplant would be successful by the use of blood tests, instead of the previous system (which was time-consuming and difficult) of taking skin grafts. Both **heart** and kidney transplantation were thus considered far less risky to perform. As a means of persuading surgeons of the importance of screening donors for their HLA prior to transplantation, Dausset helped found the Transplantation Society in 1966 and served as secretary for the organization's first four years.

To obtain greater understanding of the MHC, more research had to be done on human histocompatibility. Dausset, who became professor of hematology at the University of Paris in 1958, and professor of immunohematology ten years later, helped organize a series of conferences at which scientists compared their techniques and discoveries regarding the genetic complex. At a 1965 conference held in Prague, Czechoslovakia, Dausset contributed to a report which revealed that HLA are related to one another and that a genetic system exists which controls what specific antigens appear in human cells.

In order to find out if the genetic laws which had been discovered were valid for the entire human race, and not just certain groups, Dausset and his colleagues, in the late–1960s, conducted anthropological research on 54 widely dispersed racial and ethnic populations. Obtaining blood samples from these groups, the scientists found the genetic laws regarding the MHC were valid for all peoples. The investigation also had the benefit of, as Dausset stated in *Current Biography,* "drawing a biological map of humanity before the modern mixture of population obscures their differences."

After becoming chief biologist for the Paris General Hospital System and cochair of the Institute for Research into Blood Diseases during the early 1960s, Dausset began to investigate the relationship between certain HLAs and diseases. Other scientists followed his lead and found that particular types of histocompatibility antigens are linked to some kinds of arthritis, juvenile diabetes, multiple sclerosis, and perhaps a great many other diseases as well. He theorized that people prone to these diseases could be identified by the presence of these HLAs in their systems. Further study of the HLA factors could result in treatment or prevention of certain diseases in people whose immune systems are identified as susceptible to these types of illnesses.

After teaching at the University of Paris for many years, Dausset was appointed as head of the facility's immunology department in 1977. The same year, he was selected as professor of experimental medicine at the College de France. He also was elected to the Academy of Science at the Institute of France and to the French Academy of Medicine. Dausset's extensive research on HLA and the human **immune system** earned him the 1980 Nobel Prize in physiology or medicine. He shared the prize with Snell, whose work pioneered antigen study in the 1940s, and with **Baruj Benacerraf,** who had built upon Dausset's work.

Dausset and his wife have two children. A man with a strong interest in modern art, Dausset was once co-owner of a Parisian art gallery, La Galerie du Dragon, which specialized in Impressionist paintings.

See also Immunity; Transplantation of organs

DEAFNESS

Deafness is a loss of hearing. This loss can be temporary or permanent and can be due to a variety of reasons.

Damage to hearing is often mechanical. In particular, the eardrum is vulnerable to damage, because of its location at the border between the outer and **middle ear.** Failure to equalize the pressure build up between the external air and the inner part of the ear can cause the eardrum to rupture. As well, the eardrum can be mechanically ruptured, as by a loud noise or, most commonly, by an over-vigorous attempt to clean out the ear with a swab. Such conductive hearing loss can be repaired surgically. Head injuries can damage either the nerves that convey the auditory signals to the **brain** or the auditory cortex, which is the region of the brain that receives and processes the auditory signals.

Infection in the middle ear, such as the viral infections of measles, mumps and meningitis, can produce a build up of fluid in the middle ear. The resulting loss of hearing, which is called sensori-neural hearing loss, can be temporary or permanent. Ototoxic drugs, that is, drugs capable of damaging hearing, can cause deafness. This is an undesirable side effect of the drug. Some of the drugs are life saving, so a loss of hearing is viewed as a price to pay for a life saved. But other ototoxic drugs are widely used for a variety of ailments. A well-known example is aspirin. If taken in large quantities, aspirin can affect hearing.

Sudden Sensorineural Hearing Loss (SSHL) can occur instantaneously. In almost all cases only one ear is affected. Most people will recover the lost hearing. But in 15% of people, recovery does not occur.

Deafness also has a genetic origin. In one type of deafness, non-syndromic autosomal recessive deafness, over 20 locations on human's **chromosomes** are known to be involved. One location is particularly important. The gene at this location, GJB2, encodes a protein that functions to help recycle potassium across cell membranes. The malfunction of the protein may disrupt membrane function. For reasons yet to be determined, this can inhibit hearing. Other genetic mutations cause improper formation of structures called **cilia** in the cochlea of the inner ear. The cilia normally move in response to sound vibrations, which helps convert the sound waves to electrical signals. Impaired cilia could stop the sound information from reaching the brain.

See also Cilia and ciliated epithelial cells; Ear (external, middle and inner)

DEATH AND DYING

Death is defined as the cessation of all vital functions of the body including the heartbeat, **brain** activity (including the brain stem), and **breathing.** Death comes in many forms, whether it is expected after a diagnosis of terminal illness or an unexpected accident or medical condition.

As of 2001, the two leading causes of death for both men and women in the United States were **heart** disease and **cancer.** Accidental death was a distant third followed by **stroke,** chronic lung disorders, pneumonia, suicide, cirrhosis, **diabetes mellitus,** and murder. The order of these causes of death varies among persons of different age, ethnicity, and gender.

In an age of organ transplantation, identifying the moment of death may now have ramifications that involve another life. Its definition thereby takes on supreme legal importance. It is largely due to the need for transplant **organs** that death has been so precisely defined.

The official signs of death include the following:

- No pupil reaction to light.
- No heartbeat (absence of QRS complex).
- No response of the eyes to caloric (warm or cold) stimulation.

- No jaw reflex (the jaw will react like the knee if hit with a reflex hammer).
- No gag reflex (touching the back of the throat induces gagging).
- No response to **pain**.
- No breathing.
- A body temperature above 86°F (30°C) in a cold-water **drowning** victim, which eliminates the possibility of resuscitation.
- No other cause for the above conditions, such as a head injury.
- No drugs present in the body that could cause apparent death.
- All of the above conditions for 12 hours.
- All of the above for six hours and a flat-line **electroencephalogram** (brain wave study).
- No **blood** circulating to the brain, as demonstrated by angiography.

Only recently has there been concerted public effort to address the care of the dying in an effort to improve their comfort and lessen their alienation from those still living. Hospice care represents one of the most effective advances made in this direction. There has also been a liberalization of the use of narcotics and other drugs for symptomatic relief and improvement in the quality of life for the dying.

Autopsy after death is a way to precisely determine a cause of death. The word autopsy is derived from Greek meaning to see with one's own eyes. A pathologist extensively examines a body and submits a detailed report. Although an autopsy can do nothing for an individual after death, it may provide benefit for the family and, in some cases, medical science. Hereditary disorders and diseases may be confirmed. This knowledge could be used to prevent illness in other family members. Information culled from an autopsy can be used to further medical research. The link between smoking and lung cancer was confirmed from data gathered through autopsy. Early information about **AIDS** was also compiled through autopsy reports.

See also Brain stem function and reflexes; Homeostatic mechanisms; Hypoxia; Transplantation of organs

DEGLUTITION · *see* SWALLOWING AND DISPHAGIA

DENDRITE · *see* NEURONS

DEOXYRIBONUCLEIC ACID (DNA)

Deoxyribonucleic acid (DNA) is a double-stranded, helical molecule that forms the molecular basis for heredity.

For replication (duplication) to occur, DNA must first unwind, or "unzip," itself to allow the genetic information-encoding bases to become accessible. The base pairing within

DNA is specific and complementary and, consequently, when the molecule unwinds, two complimentary strands are temporarily produced, each of which acts as a template for a new strand. At the onset of replication, a replication fork is created as the DNA molecule separates at a small region. The enzyme DNA polymerase then adds complementary nucleotides to each side of the freshly separated strands. The DNA polymerase adds nucleotides only to one end of the DNA. As a result, one strand (the leading strand) is replicated continuously, while the other strand (the lagging strand) is replicated discontinuously, in short bursts. Each of these small sections is finally joined to its neighbor by the action of another enzyme, DNA ligase, to yield a complete strand. The whole process gives rise to two completely new and identical daughter strands of DNA.

The discovery of the double-helix molecular structure of deoxyribonucleic acid (DNA) in 1953 was one of the major scientific events of the twentieth century, and marked the culmination of an intense search involving many scientists. Ultimately, credit for the discovery, and the 1962 Nobel Prize in the physiology or medicine category, went to **James Dewey Watson**, who at the time of the discovery was an American postdoctoral student from Indiana University; and **Francis Harry Compton Crick**, a researcher at the Cavendish Laboratory in Cambridge University, England. Their work, conducted at Cavendish Laboratory, significantly impacted the emerging field of **molecular biology**.

The double helix refers to DNA's spiral staircase structure, consisting of two right-handed helical polynucleotide chains coiled around a central axis. Genes, specific regions of DNA, contain the instructions for synthesizing every protein. Because life cannot exist without proteins, and the differences in proteins and protein **enzymes** control biological reactions, the discovery of DNA's structure unveiled one of the fundamental secrets of life: **protein synthesis**. In fact, the central dogma of molecular biology is that DNA is used to build **ribonucleic acid**, which is used to build proteins, which in turn play a role in building DNA and **RNA**.

Prior to Watson and Crick's discovery, it had long been known that DNA contained four kinds of nucleotides, which are the building blocks of nucleic acids, such as DNA and RNA. A nucleotide contains a five-carbon sugar called deoxyribose, a phosphate group, and one of four nitrogen-containing bases: adenine (A), guanine (G), thymine(T), and cytosine (C). Thymine and cytosine are smaller, single-ringed structures called pyrimidines; adenine and guanine are larger, double-ringed structures called purines. Watson and Crick drew upon this and other scientific knowledge in concluding that DNA's structure possessed two nucleotide strands twisted into a double helix, with bases arranged in pairs such as A T, T A, G C, C G. Along the entire length of DNA, the double-ringed adenine and guanine nucleotide bases were probably paired with the single-ringed thymine and cytosine bases. Using paper cutouts of the nucleotides, Watson and Crick shuffled and reshuffled combinations. Later, they used wires and metal to create their model of the twisting nucleotide strands that form the double-helix structure. According to Watson and Crick's model, the diameter of the double helix

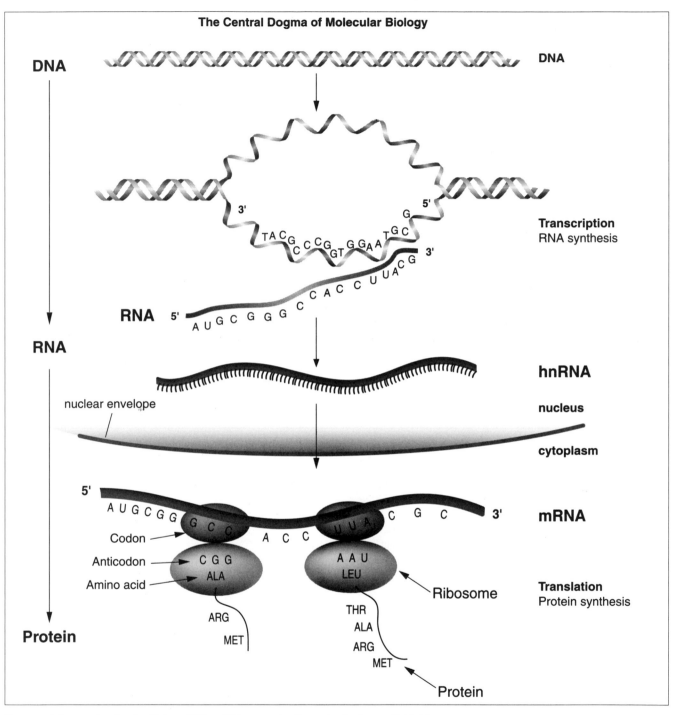

The central dogma of molecular biology, DNA to RNA to protein. *Illustration by Argosy Publishing.*

measures 2.0 nanometers (nm). Each turn of the helix is 3.4 nm long, with 10 bases in each chain making up a turn.

Before Watson and Crick's discovery, no one knew how hereditary material was duplicated prior to **cell division**. Using their model, it is now understood that enzymes can cause a region of a DNA molecule to unwind one nucleotide strand from the other, exposing bases that are then available to become paired up with free nucleotides stockpiled in cells. A half-old, half-new DNA strand is created in a process that is called semiconservative replication.

When free nucleotides pair up with exposed bases, they follow a base-pairing rule which requires that A always pairs with T, and G always with C. This rule is constant in DNA for all living things, but the order in which one base follows another in a nucleotide strand differs from species to species. Thus, Watson and Crick's double-helix model accounts for both the sameness and the immense variety of life.

Even though the long-term survival of species may be improved by genetic changes, the survival of individuals requires genetic stability. To achieve this, each living cell has a complement of enzymes whose function is to repair errors or damage in the DNA. Such lesions can arise spontaneously from normal cellular processes, (e.g., errors in replication) or can be generated by external factors such as chemicals or radiation. The altered portion of the DNA is recognized, usually because the DNA becomes distorted, and the damaged bases of sequences removed by nucleases. The correct sequence is then re-synthesized by a polymerase and the ends joined to the original DNA by a ligase. There are a number of enzyme systems that repair DNA damage and nearly all rely on the existence of two copies of the genetic information, one on each strand of the DNA double helix. Thus if the sequence on one strand is accidentally changed, the complementary strand still holds the correct information and can be used as a template to correct the alteration.

Watson and Crick's discovery of the double helix would not have been possible without significant prior discoveries. In his 1968 book, *The Double Helix, A Personal Account of the Discovery of the Structure of DNA*, Watson wrote that the race to unveil the mystery of DNA was chiefly a matter of five people: Maurice Wilkins, Rosalind Franklin, Linus Pauling, Crick, and Watson. Wilkins, an Irish biophysicist who shared the 1962 Nobel Prize in physiology or medicine with Crick and Watson, extracted DNA gel fibers and analyzed them, using x-ray diffraction. The diffraction showed a helical molecular structure, and Crick and Watson used that information in constructing their double-helix model. Franklin, working in Wilkins' laboratory, between 1950 and 1953, produced improved x-ray data using purified DNA samples, confirming that each helix turn is 3.4 nm. Although her work suggested DNA might have a helix structure, Franklin did not postulate a definite model. Pauling, an American chemist and twice Nobel laureate, in 1951 discovered the three-dimensional shape of the protein collagen. Pauling discovered that each collagen polypeptide or amino acid chain twists helically, and that the helical shape is held by hydrogen bonds. With Pauling's discovery, scientists worldwide began racing to discover the structure of other biological molecules, including the DNA molecule.

DNA fingerprinting, also called DNA profiling or genetic profiling, applies a test to determine the unique DNA sequence that each person carries for the purpose of identification. In the mid 1980s, Sir Alec Jeffreys at the University of Leicester coined the term DNA fingerprint and envisioned its powerful use. A single **hair**, a drop of **blood**, **semen**, or other body fluid can reveal the identity of a person. DNA fingerprinting is used for identifying people, studying populations, and forensic investigations.

Chromatin is a network of deoxyribonucleic acid (DNA) and nucleoproteins that constitutes a chromosome. Chromatin can only be found in a cell with a nucleus, and is therefore not present in a prokaryotic cell. The DNA within a eukaryotic cell can be as long as 12 cm (4.7 in). Due to its length, the DNA must be arranged and organized in order to fit within the small area of a cell nucleus. To accomplish this task, the DNA is bound, through electrostatic forces, with nucleoproteins called histones and nonhistones. The assemblage of DNA with the nucleoproteins is called a nucleosome, which is the fundamental structural unit of chromatin and represents 1.8 turns of DNA wound around a core particle of another histone protein. It is the nucleosomes, along with the DNA material between nucleosomes (linker segment), that gives DNA the characteristic beads-on-a string appearance, with the nucleosomes representing the bead and the linker segment of DNA representing the string.

When chromatin is isolated, it appears to be composed as smooth fibers. While the highest level of chromatin organization is not well understood, scientists have found that chromatin fibers are divided into functional groups, called domains. The domains are grouped and arranged into loops called solenoids. In cells that are dividing, the solenoids are further condensed into chromatids; an identical pair of chromatids comprise the recognizable shape of a chromosome.

There are two types of chromatin: heterochromatin and euchromatin. Heterochromatin is chromatin in condensed form, is seen as dense patches and is transcriptionally inactive while euchromatin is seen as delicate, thread-like structures that are abundant in active transcription cells.

See also Biochemistry; Chromosomes; Genetic code; Genetic regulation of eukaryotic cells; Genetics and developmental genetics; Molecular biology

DEPRESSANTS AND STIMULANTS OF THE CENTRAL NERVOUS SYSTEM

The **central nervous system** (CNS), (i.e., the **spinal cord**, and the cortical and subcortical areas of the **brain**), processes different kinds of sensory information through highly organized neuronal structures that function as specific information-processing areas in the brain in a modular fashion. For instance, the visual, olfactory, motion, and somatic areas are well-defined functional structures, specialized in processing specific types of sensory stimuli, whereas the association areas are able to deal with multiple pre-analyzed sensations, which are received from these other areas through a close interchange with the limbic system. In this way, memory and association areas match the inputting information with stored responses, mainly through the interplay of impulses between the cortical and the limbic system. Therefore, the CNS is a complex system of interlinked structures that requires intercommunication among distant **neurons** (i.e., nervous cells) and peripheral nerves, in order to integrate and regulate the activity of several modular cerebral structures.

Information is transmitted from one neuron to the next under the form of nerve impulses known as synapses. There are two major types of synapses: the electrical synapses and the chemical synapses. Electrical synapses usually occur through the opening of tubular protein structures known as gap junctions in the cells that conduct electricity (ionic solutions) from one neuron to the next. However, these structures are more common in nerves connected to visceral smooth muscles, and are not the main form of signal transmission in the CNS. The prevailing form of signal transmission in the CNS are chemical synapses, through the mediation of chemical substances that occupy specific receptors in the nerve cells, known as **neurotransmitters**. Some neurotransmitters are **hormones** originated from different glands, such as the adrenal, pituitary, pineal, and the **liver**. Other neurotransmitters, such as serotonin and dopamine, or yet, **amino acids**, such as leucine and methionine enkephalins as well as other neuropeptides, are synthesized by the neurons. Among the more than forty substances presently identified as neurotransmitters, the best understood are acetylcholine, histamine, GABA (gamma-aminobutyric acid), glycine, glutamate, serotonin, norepinephrine, and dopamine. These neurochemical substances occur in three different broad types. The first one comprehends those transmitter substances that trigger synaptic transmission, such as serotonin, dopamine, and norepinephrine. The second type are the families of regulatory substances that modulate the action of the first ones, either by competing with a specific neurotransmitter for the same receptors or yet, by binding to the neurotransmitter before it reaches the receptor (inhibitory action), or by facilitating the synthesis of a given neurotransmitter (precursors). Some substances also prevent the decay of a given neurotransmitter, making them available to be reused again in the brain. The third type of neurochemical substances, known as secondary or second messengers, bind to a given primary neurotransmitter to either enhance its action or allow for its activation. Different synaptic pathways seem to be regulated by some specific neurotransmitter systems, such as those that are primarily activated by dopamine and its secondary messengers, whereas some other pathways may be independently activated by several substances, such as glutamate, norepinephrine, dopamine, and serotonin.

Behavioral responses and personality traits are the result of and are affected by the overall activity of neurotransmitters and hormonal interactions. Such interactions are accountable for the following effects upon behavioral response: long-term neurotransmitter baselines that account for differential behavioral-response patterns, changes in a baseline of neurotransmitter activity in response to external stimuli (i.e., social or environmental changes), daily behavioral and mood cycles, such as the circadian rhythm, or monthly cycles such as those involved in mood changes due to hormonal fluctuation, and short-term changes in response to specific stimuli, such as danger, pleasure, or **pain**.

Some neurotransmitters may have both an excitatory and inhibitory effect in different parts of the nervous system. Acetylcholine, for instance, is synthesized mainly by the large pyramidal cells of the motor cortex, although it is also produced in many brain areas, as well as by the motor neurons that innervate **skeletal muscle**, among others. Although acetylcholine usually has an excitatory effect on the nervous system, in the peripheral parasympathetic nerve endings it shows an inhibitory action, such as the inhibition of the **heart**

by the vagus nerve. Norepinephrine helps to control overall neural activity and mood such as increased wakefulness. However, in some areas of the brain norepinephrine also inhibits receptors, preventing their activation by other transmitter substances. Its action in the **sympathetic nervous system** activates some **organs** and inhibits others as well. Serotonin and norepinephrine, two excitatory transmitters of the prefrontal lobes, are directly involved with the regulation of mood and the sense of well-being, appropriate sex drive, contentment, and psychomotor balance. Low levels of either or both these substances are related to depression, loss of appetite, grief, unhappiness, and episodes of suicidal despair. Drugs that inhibit the production of serotonin and norepinephrine in the brain, such as reserpine and lithium compounds can induce depression. On the other hand, compounds that either prevent the destruction of these two transmitters or prevent their reuptake by the nerve endings usually are effective in treating persons with depression. However, an excess of these two neurotransmitters also may cause psychological imbalance, such as increased excitability or even psychotic behavior, due to their excitatory effects. Dopamine has an inhibitory effect on several portions of the prefrontal lobes and other related areas, such as medial and anterior areas of the limbic systems, which are related with the behavioral control centers. Stressful and dangerous situations increase dopamine and adrenaline levels, triggering the fight-or-flight responses. Excess of dopamine production is also associated with schizophrenia and paranoia, and can be induced by drugs that augment dopamine levels, such as L-dopa, used to treat Parkinson's disease.

Neuropeptides usually have a slow but much longer action in the nervous system. Moreover, they are a thousand or more times more potent than the smaller neurotransmitter molecules, such as serotonin, dopamine, **epinephrine**, glutamate, and norepinephrine. They induce prolonged changes in activation and deactivation of specific genes in the nerve cells as well as prolonged changes in the amount of excitatory and inhibitory receptors in neurons, used by other neurotransmitters, which last for days or even months. Examples of slow-action neuropeptides are: leucine and methionine enkephalins, nerve growth factor, neurotensin, and several pituitary, hypothalamic, and other hormones.

Some biochemical substances found in diet have an excitatory effect in the nervous system, such as caffeine (from coffee), theobromine (from tea), and theophylline (from cocoa). Conversely, anesthetic drugs, painkillers, and muscle relaxants do decrease synaptic transmission in many points of the nervous system, depending on the molecular characteristic of each drug.

See also Anesthesia and anesthetic drug actions; Cerebral cortex; Drug effects in the nervous system; Fight or flight reflex; Hypothalamus; Nerve impulses and conduction of impulses; Peripheral nervous system

DERMIS

The middle distinct layer of skin is called the dermis or corium. Anatomically, this layer can be divided into papillary (upper) and reticular (lower) layers. The papillary layer is thrown in folds that produce skin ridges found on the hands and feet. The thickness of the dermis can range from 0.04 in. (1 mm) on the eyelids to 0.16–0.2 in. (4-5 mm) thick on the back. The dermis functions in to maintain **temperature regulation** (thermoregulation) and supplies the avascular **epidermis** with nutrients. It contains sensory nerve endings for **pain**, structures called Meissner's corpuscles (tactile receptors) and a rich **blood** supply. The sebaceous (oil) and shorter **hair** follicles originate in this layer. The dermis is composed of connective **tissue**, cellular elements and ground substance.

The connective tissue of the dermis consists of collagen, elastic, and reticular fibers. These fibers contribute to the elasticity and strength of the skin. Under the electron microscope, the collagenous fibers are observed to be composed of thin, non-branching fibrils held together by cementing ground substance. These fibrils are composed of covalently cross-linked and overlapping units called tropocollagen molecules. Collagenous fibers are responsible for one fourth of man's overall protein mass. The elastic fibers are seen to be thinner than collagen fibers and are entwined among them. They are composed of the protein elastin. Reticular fibers are thought to be immature collagen fibers since their chemical and physical properties are similar. Reticular fibers are sparse in normal skin but abundant in pathological conditions of the skin associated with syphilis, sarcomas and lymphomas.

The cellular elements consist of fibroblasts, histiocytes, mast cells, polymorphonuclear **leukocytes**, eosinophic leukocyte, and lymphocytes. Fibroblasts form collagen fibers and may be the creating base (progenitor) of all other **connective tissues**. Histiocytes can (under pathological conditions) form macrophages that phagocytize **bacteria** and other foreign matter. Mast cells play a role in the inflammatory response to injury of the skin as do the lymphocytes. Polymorphnuclear leukocytes and eosinophilic leukocytes occur quite commonly with various dermatosis, especially those with an allergic source (etiology).

The ground substance of the dermis is a gel-like amorphous (shapeless or shape changing) matrix of physiological importance because it contains proteins, mucopolysaccharides, soluble collagens, **enzymes**, immune chemicals, metabolites and many other substances.

See also Cell differentiation; Ectoderm; Epithelial tissues and epithelium; Histology and microanatomy

DESCARTES, RENÉ (1596-1650)

French mathematician, philosopher, scientist, and soldier

René Descartes, often known by his Latin name, Renatus Cartesius, from which the adjective "Cartesian" is derived, was

René Descartes. *The Library of Congress.*

the prime mover behind the mechanistic conception of the human body. He was born the son of a wealthy magistrate in La Haye, Touraine, France. From 1606 to 1614, he studied logic, mathematics, scholastic theology, Aristotelian philosophy, and the classics at the Jesuit school of La Flèche in Anjou. In 1616, he took a law degree from the University of Poitiers, but never practiced law. Dissatisfied with his education in France, he enrolled in the Dutch military academy at Breda in 1618. There he studied mathematics and mechanics under Isaac Beeckman (1588–1637), whom he acknowledged as his mentor.

Descartes served in the Thirty Years War, first in the Dutch army, then in the Bavarian army. Because of his chronic ill health, he did no fighting, but contributed his mathematical and engineering skills to the war effort. After 1627, he spent hardly any time in his native France. He lived mostly in the Netherlands from 1629 to 1649, then moved at the request of Swedish Queen Christina (1626–1689) to Stockholm, where he died shortly after arriving.

Writing sometimes in French, sometimes in Latin, and sometimes translating one into the other, Descartes published *Discourse on the Method of Rightly Conducting the Reason and Seeking Truth in the Sciences, Dioptrics, Meterology,* and *Geometry* in 1637; *Meditations on First Philosophy* with six sets of objections and replies in 1641; *Principles of Philosophy* in 1644; and *Passions of the Soul* in 1649. His posthumous works include *Treatise on Man and on the Formation of the Fetus* in 1664 and *Rules for the Direction of the Mind* in 1701. In the 1630s, he planned a treatise called *The World,* but, recalling what the Roman Catholic Church had recently done to Nicolaus Copernicus (1473–1543), Giordano Bruno (1548–1600), and Galileo Galilei (1564–1642), he abandoned it out of fear.

Descartes saw all knowledge as a tree. His most fundamental fact, *cogito ergo sum (literally, "I think, therefore, I am")*, his absolute certainty that he existed as a thinking thing, was his basis of metaphysics, the roots. Physics was the trunk. Medicine, mechanics, and morals were the branches.

This view of knowledge committed him to a strict metaphysical dualism between *res cogitans* ("thinking substance") and *res extensa* ("extended substance"). This dualism brought into sharp focus a philosophical problem which was implicit in Plato's (ca. 427 B.C.–ca. 347 B.C.) *Phaedo* and in the spirit/flesh (*pneuma/sarx*) or word/flesh (*logos/sarx*) dualism in the Gospel of John, namely, how can any causal relationship exist between mind and body? This problem, typically just called "the mind/body problem," has intrigued philosophers ever since, has never been solved, and was not even sufficiently argued until Gilbert Ryle (1900–1976) published *The Concept of Mind* in 1949. For Ryle, the Cartesian mind was "the ghost in the machine."

Cartesian dualism immediately wrought a crisis in Continental Rationalism, prompting Baruch Spinoza (1632–1677) to attempt a monistic solution and Gottfried Wilhelm Freiherr von Leibniz (1646–1716) to attempt a pluralistic solution. The rigid natural difference between body and soul entailed separate sets of rules for each. The laws of physics and especially mechanics governed the body but did not affect the soul; while principles of psychology and individual free will determined the nature of the soul and affected the body only in an occult way, as Descartes believed, through the "animal spirits" in the pineal gland. His studies of the pineal gland and his dissections of sheep brains inspired later research in cerebral localization.

Descartes propounded iatromechanism, the belief that physiological and medical processes were reducible to problems of physics and mechanics. Opposed to this school was the iatrochemism of Franciscus de le Boë Sylvius (1614–1672), who claimed that these processes were essentially chemical. Descartes accepted the mechanistic explanation of **blood** circulation proposed by **William Harvey** (1578–1657) in 1628. The Cartesian view that only humans have souls, and that all other animals are merely reactive automatons that feel no **pain** and cannot think, justified vivisection and other cruelty to animals in the name of science.

Descartes believed in divine foreknowledge, but not in predetermination, as he explained with a story in a famous letter to Princess Elisabeth of Bohemia. A king sent two of his subjects, mortal enemies of each other, on simultaneous errands, one to travel from point A to point B, the other from point B to point A. He knew that they would meet and fight. He caused them to meet, but did not force them to fight. The free will of the two enemies was fully preserved even though the orchestrator of their actions knew in advance what would happen.

Descartes discovered analytic geometry, the system of solving problems in plane geometry by graphing algebraic equations as "Cartesian coordinates" on a grid defined by a horizontal x-axis and a vertical y-axis. This standard and indispensable part of mathematics is sometimes called either "coordinate geometry" or "Cartesian geometry." In set theory and formal logic, the "Cartesian product" of any two sets is a third set whose elements are all the pairs such that the first half of each pair is an element of the first set and the second half of each pair is an element of the second set.

See also Brain: Intellectual functions; Comparative anatomy; Embryonic development: Early development, formation, and differentiation; Epithalamus

DEVELOPMENTAL ANATOMY • *see* EMBRYOLOGY

DEVELOPMENTAL HORIZONS • *see* HUMAN DEVELOPMENT (TIMETABLES AND DEVELOPMENTAL HORIZONS)

DIABETES MELLITUS

Diabetes mellitus is a disease of glucose **metabolism** that is defined by **blood** glucose levels. There are three types of diabetes: insulin dependent diabetes mellitus, IDDM (type I), non-insulin dependent diabetes mellitus, NIDDM (type II), and gestational diabetes (onset occurs in women during **pregnancy**).

In IDDM, the **pancreas** does not make enough insulin. Insulin is a hormone that signals **fat**, muscle, and **liver** cells (target cells) to take glucose from the blood **plasma** into the interior on the cells to be used for cellular respiration. Individuals with IDDM must take insulin to manage their blood glucose levels. Studies with twins have shown that concordance for IDDM (the occurrence of IDDM in both individuals) is minimal, therefore the inheritance factor for IDDM is suspected to be minimal. Researchers believe that both environmental factors and viral **antigens** are mostly responsible for IDDM.

NIDDM is associated with obesity and aging, and has a concordance in twin studies of 80-90%, suggesting a strong genetic component. In this type of diabetes, the glycoprotein receptor sites for insulin on the outer surface of the membrane of target cells do not respond to the presence of insulin and so glucose is not taken up into the cells. Blood glucose levels remain high leading to hyperglycemia and, if untreated, eventually, glucose toxicity and **death**. If diagnosed in the early stages of the disease, NIDDM can be managed with diet and exercise.

Genetic studies using Admixture Linkage Analysis have found that NIDDM is genetically complex, involving multiple genes, and multiple gene-environment interactions. The fact that NIDDM was not recognized 100 years ago suggests environmental factors may have created the disease. Further indications of an environmental component to the disease can be found in a four-fold increase in NIDDM in Navajo and Pueblo children since 1990. It is hoped that on-going linkage analyses will find primarily related diabetes genes or modifying genes such as obesity-related genes.

Ethnic groups such a Native North Americans and Hispanics show much higher incidences of diabetes than other ethnic groups. This suggests a genetic susceptibility to NIDDM that may have evolved as a survival mechanism when these populations historically experienced cycles of feast and famine. There may have been a selective advantage in having a genotype that conserved blood sugar during times of famine. A human genome-wide search has revealed a major susceptibility locus for NIDDM on the distal part of the long arm of chromosome 2 that has been called NIDDM1.

See also Carbohydrates; Glucose utilization, transport and insulin

DIAPHRAGM

In general anatomical terms, a diaphragm is a membrane-like mixture of membrane and muscle that forms a partition that separates two chambers. Although most commonly associated with the musculo-membranous diaphragm that separates the thorax and abdomen—a structure that also plays an important role in the mechanics of respiration—there are other important anatomical diaphragms including the muscular pelvic diaphragm, and urogenital diaphragm. In addition to describing a structure that imparts a physical separation, the term diaphragm is often applied to structures that establish an adjustable opening.

The musculo-membranous diagram that separates the thoracic cavity from the abdominal cavity is a major component in the mechanical movement of air during the respiration cycle. Movements of the respiratory diaphragm create pressure differences that drive inspiration and expiration.

The thoraco-abdominal diaphragm attaches to the lower ribs, **sternum**, and vertebral processes that make up the spine. As the boundary between the thorax and **abdomen**, the diaphragm contains an opening (hiatus) through which the **blood** vessels (e.g., the aorta and vena cava), nerves and the esophagus pass.

During inspiration, the muscles of the diaphragm and associated muscles attached to the diaphragm contract in such a manner that the diagram flattens from its natural convex or dome shape. The flattening enlarges the chest (thoracic) cavity and causes a reduction of pressure relative to the ambient (normal, existing) atmospheric pressure. The differences in pressure provide a pressure gradient that draws air into the **lungs**. Because of increased thoracic volume, the lungs are able to expand during inspiration. Conversely, as the diaphragm returns to it convex state, it reduces thoracic cavity volume and imparts a force that pushes air out of the lungs during expiration.

The diaphragm is composed strictly of **skeletal muscle**. There are no **smooth muscle** fibers. Accordingly, the diaphragm does not contract spontaneously and is under nervous control. Just as **breathing** can be controlled voluntarily or unconsciously, control of the diaphragm may be voluntary or be directed by

lower **brain** stem structures that respond to changing levels of oxygen, carbon dioxide, and acids in the blood associated with the respiratory cycle. The phrenic nerve (without sympathetic or parasympathetic nerves) innervates the diaphragm.

See also Cough reflex; CPR (Cardiopulmonary resuscitation); Muscles of the thorax and abdomen; Respiration control mechanisms; Respiratory system embryological development; Respiratory system

DIARRHEA

Diarrhea is a liquid stool or bowel movement that results from the movement of fecal material through the intestines in such a rapid manner that there is insufficient time to properly absorb water. In addition to excessive water, diarrhea waste contains abnormally high amounts of **electrolytes**. According to World Health Organization statistics, diarrhea—resulting from a number of causative agents—is the leading cause of dehydration and **death** in children.

If excessive amounts of the electrolytes are lost as a result of diarrhea, the loss interrupts the normal **acid-base balance** (e.g., excessive potassium loss induces **acidosis**).

Diarrhea may be acute (rapid onset, short duration) or chronic (occur over a long period of time). Many infectious **bacteria** such as Shigella, Salmonella, and Campylobacter may induce diarrhea. In many areas of the world with poor water quality, parasites such as Giardia and Cryptosporidium are leading causes of diarrhea. Viral infections, especially rotavirus infections, also cause diarrhea.

Diarrhea-inducing agents generally produce an irritation of the **gastrointestinal tract**, specifically the gastrointestinal mucosa. Other leading causes of diarrhea include ingestion of spoiled food laden with harmful bacteria (food poisoning). Diarrhea may also result from noninfectious causes including overeating, milk intolerance, or a number of bowel and/or intestinal disorders. Diarrhea may also occur as an unintended side effect of medical treatment with antibiotics or other drugs.

Any increase in peristalsis—the wave-like, directional contraction of longitudinal and **smooth muscle** in lining the intestine that helps food and waste move through the intestines—may result in diarrhea. Accordingly, emotional disturbances that result in increased peristalsis may induce acute diarrhea.

Types and the extent of diarrhea are classified by whether the diarrhea is acute or chronic, the volume of the fecal material expelled, the frequency of expulsion, the water content of the stools and whether **mucus** or **blood** are present in the stool.

When diarrhea occurs without vomiting (the expulsion of food and water through the mouth) it is often possible to rehydrate individuals suffering from diarrhea while attempting to treat the actual cause of the diarrhea. Children are often rehydrated with osmotically balanced liquids such as Pedialyte. The World Heath Organization uses Ricelyte, or it's own osmotically balanced rehydration formula to replenish lost water and needed electrolytes in dehydrated patients.

See also Alimentary system; Alkalosis; Digestion; Electrolytes and electrolyte balance; Elimination of waste; Intestinal histophysiology; Intestinal motility

DIASTOLE AND DIASTOLIC PRESSURE ●
see BLOOD PRESSURE AND HORMONAL CONTROL MECHANISMS

DIENCEPHALON ● *see* CEREBRAL MORPHOLOGY

DIFFUSION ● *see* CELL MEMBRANE TRANSPORT

DIGESTION

Digestion is the physiological process by which food is broken down, mechanically and chemically, into particles small enough to pass through the walls of the intestinal tract and into the **blood**. Once in the bloodstream, these tiny particles can then be distributed throughout the body and used for nourishment. The breaking-down process takes place in almost all parts of the digestive tract, beginning at the mouth and ending, on average, more than fifteen feet later, at the anus.

The digestive process starts as soon as food, taken into the mouth, begins to be chewed into smaller pieces. While being chewed, the food is mixed with **saliva** that contains the enzyme ptyalin, the first of many **enzymes** that will help convert complex and indigestible food molecules into smaller and easier-to-absorb ones. When in the mouth, the activity of ptyalin is already at work, converting some of the complex starches into simple sugars.

After chewing, the food is swallowed and passes through the esophagus, then into the stomach, where the breaking-down process goes into high gear. A strong churning motion causes the food to be thoroughly mixed by the stomach's potent digestive juices, which contain both hydrochloric acid and the enzyme pepsin. The foods are all dissolved into a thick liquid called chyme, but while the protein foods are partly digested, the other nutrients are basically unchanged. Then, in the small intestine, the digestive process is completed by a combination of pancreatic juices (containing the enzymes trypsin, amylase and lipase), intestinal juices and **bile**. (The bile, stored in the gall bladder, works primarily on digestion and absorption of fat.) Thoroughly digested, the nutrient molecules can now by absorbed by blood and lymph vessels in the small intestine's walls and carried into the circulation. Undigestible food particles pass into the large intestine where some water and minerals are absorbed and bacterial action turns the rest into feces, which are eventually eliminated as waste products.

The digestive process is a highly complicated one and, until fairly recently, only vaguely understood. In ancient times, for instance, the early philosophers could see that various foods entered the body and, while some of the food remained in the system presumably to provide nourishment, the rest emerged later in a completely changed form. How did they believe all this was accomplished? The Hippocratic philosophers, in the fourth and fifth centuries B.C., concluded that food was converted through body heat first into liquid form, then into what they called the "four humors"—blood, phlegm, yellow bile and black bile—which could then be absorbed into the body.

A few centuries later, around 280 B.C., the philosopher **Erasistratus** pointed out that, after food was eaten, a great deal of activity soon took place in the stomach and intestines. To him, therefore, digestion was almost certainly a mechanical process. Somehow food was ground down in the stomach to liquid form, then carried to the **liver** where it was transformed into blood. But the famous second century physician **Galen**, although he agreed that the liver was the major organ of digestion, decided that the Hippocratics were right: animal heat must be the guiding force behind digestion. And Galen's views prevailed for hundreds of years.

In the seventeenth century, however, scientists began taking another look at the human body. A number of them, such as **Franciscus Sylvius**, a Dutch physician, believed that most bodily processes could be explained in purely chemical terms. Sylvius studied digestive juices, including saliva, and correctly concluded that digestion was a form of fermentation.

Other scientists of the time tended to see the human body as a kind of machine. To Giovanni Borelli, an Italian mathematician and a contemporary of Sylvius, the stomach was simply a binding device that could mechanically break food down into tiny particles. Erasistratus had been right all along, he concluded.

During the eighteenth century, several scientists made tremendous progress in settling the debate over whether digestion was a mechanical or chemical process. Through his experiments on animal tissues, the Swiss physiologist **Albrecht von Haller** discovered that bile was the key element in the body's digestion of fats. The debate was essentially settled by the French physiologist René de Reaumur (1683–1757) in 1752. De Reaumur fed a hawk two small metal cylinders filled with meat and covered at their ends by gauze. Unable to digest the cylinders, the hawk eventually regurgitated them, and de Reaumur deduced that the meat could only have been dissolved by chemicals present in internal fluids.

Before long, a number of eighteenth-century scientists were able to study the role of digestive juices. The Italian physiologist **Lazzaro Spallanzani** performed experiments similar to de Reaumur's and coined the term "gastric juices." Spallanzani discovered that these juices helped inhibit ad prevent the putrefaction of food within the body. He also helped to further define the important role that saliva plays in the digestive process. Additionally, in 1822, the German chemist Leopold Gmelin (1788–1853) continued to study the chemical processes involved in food digestion by investigating stomach and pancreatic juices. And in 1842, William Prout proved that the most potent acid in gastric juices was actually hydrochloric acid. The discoveries made by these scientists validated the proposition that chemical processes governed the body's digestive system.

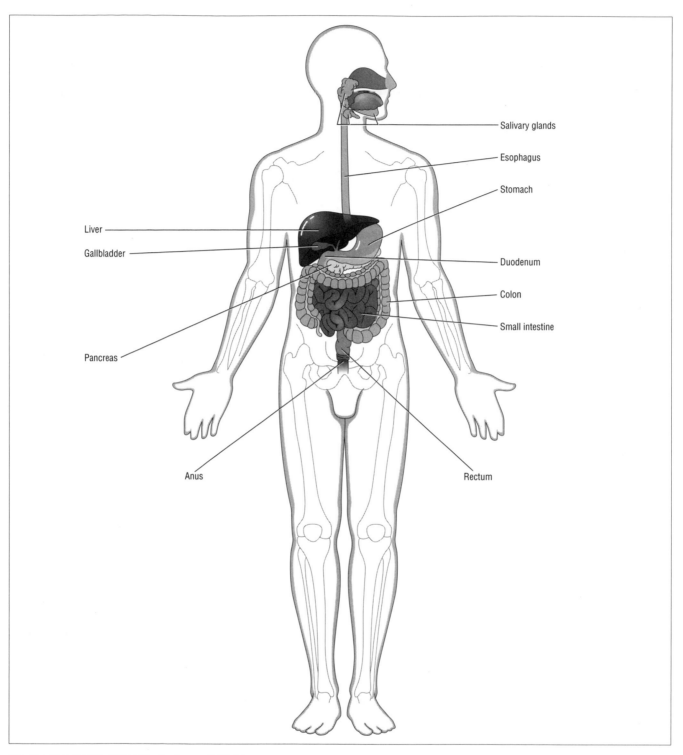

The digestive system. *Illustration by Argosy Publishing.*

A particularly notable contribution to the understanding of digestion was made by the American army surgeon, **William Beaumont**. Beaumont, who joined the army in 1812, was sent, a few years later, to a frontier post in Michigan. While there, be treated a young French-Canadian, Alexis St.

Martin, who had been accidentally shot in the side. Although St. Martin recovered, his bullet wound never fully closed. He retained an inch-wide opening (or fistula) in his side that led to his stomach. Through this opening, Beaumont could not only observe changes in the stomach under varying condi-

•

tions, but could remove samples of gastric juices. Beginning in 1825, then, the army surgeon conducted over two hundred experiments and, by so doing, provided the medical world with a great deal of previously unknown information about gastric **physiology** and the digestive process in a living human being.

Inspired by Beaumont's work, the French physiologist Claude Bernard began, in the mid-1800s, to create artificial fistulas in laboratory animals. Through these openings, Bernard made a number of important discoveries, among them that the small intestine, rather than the stomach, was the major site of digestion and that pancreatic juices were important digestive agents, particularly where **fat** molecules were concerned.

A few years later, in the 1870s, Willy Kuhne's research clarified the role of the intestines in the absorption of digested foods. And, toward the end of the nineteenth century, the Russian physiologist, Ivan Petrovich Pavlov—by experimenting on living dogs—worked out the nervous mechanism that controls the secretion of gastric juices by the digestive glands.

See also Alimentary system; Gastrointestinal embryological development; Gastro-intestinal tract

DIGESTIVE SYSTEM • *see* GASTRO-INTESTINAL TRACT

DIURETICS AND ANTIDIURETIC HORMONES

Diuresis, or the elimination of water, salt and other solutes as well as several metabolites in the form of urine by the **kidneys**, is the main process of **blood** or arterial pressure control. This long-term regulation process acts in two ways: 1), when the levels of **extracellular fluid** rises, and consequently the arterial pressure and blood volume increases, the kidney is stimulated to eliminate water and salt, thus returning the blood pressure to its normal levels; and 2), when the arterial pressure is bellow normal levels, diuresis is decreased, allowing the accumulation of extracellular fluid and blood volume in order to reach the normal pressure levels.

Regulation of body fluid volume involves the alternate excretion and reabsorption by the renal tubules of different concentrations of solutes and water in the urine, such as potassium and sodium and other **electrolytes**. Five different **hormones** control specific tubular reabsorption of different mineral salts and water, therefore interfering with diuresis. Aldosterone, for instance, is a mineralcorticoid hormone that increases sodium tubular reabsorption and augments potassium elimination. Therefore, increased levels of aldosterone leads to higher concentrations of sodium in the extracellular fluid, causing water retention and decreased diuresis. When low blood pressure and/or low extracellular volumes occur, the hormone angiotensin II stimulates aldosterone secretion and constricts arterioles, what ultimately induces tubular reab-

sorption. Moreover, angiotensin II directly promotes sodium reabsorption through the activation of the ATPase pump on the epithelial cells of the renal tubular membranes. Another hormone termed ADH (antidiuretic hormone) regulates body loss of water by increasing its tubular reabsorption to prevent dehydration. ADH also controls the concentrations of solutes in the urine, such as the excretion of normal amounts of sodium in smaller volumes of urine. When increased **plasma** volume occurs, the atrial natriuretic peptide, a diuretic hormone secreted by a group of cells of the cardiac atria, inhibits sodium and water reabsorption by the renal tubules. The amounts of **calcium** and magnesium eliminated in the urine is regulated by the parathyroid hormone, that promotes the tubular reabsorption of these two minerals while inhibits phosphate reabsorption. Modulation of calcium levels in the plasma is crucial to cardiac, muscular, and neural functions. A low level of calcium ions (hypocalcemia) increases neuromuscular excitability and may even cause spasms, when such levels are too low. Conversely, hypercalcemia (abnormally high calcium plasmatic concentrations) may lead to neuromuscular depression and cardiac arrhythmias. Therefore, the balanced interaction of both diuretic and antidiuretic hormones, as well as the other control systems acting upon renal function, are crucial to body homeostasis, affecting a variety of vital functions.

See also Blood pressure and hormonal control mechanisms; Calcium and phosphate metabolism; Renal system

DNA • *see* DEOXYRIBONUCLEIC ACID (DNA)

DOHERTY, PETER C. (1940-)
Australian immunobiologist and molecular biologist

Peter Doherty and co-recipient **Rolf Zinkernagel** received the 1996 Nobel Prize in physiology or medicine for the discoveries concerning the specificity of the cell mediated immune defense system. From Doherty's research, a clearer understanding of the association between **tissue** types and the susceptibility to certain diseases has emerged.

Peter Doherty was born in Queensland, Australia on October 15, 1940. Doherty was educated at Queensland and later at Edinburgh, where he received his Ph.D. in 1970. He worked as a research fellow in microbiology at the John Curtin School of Medical Research at Australian National University in Canberra (1973-75) and was head of the Department of Experimental Pathology from 1982-1988. In the interim, he worked at the Wistar Institute in Philadelphia from 1975 to 1982. In 1988 he returned to the United States to become chairman of the Immunology Department at St. Jude Children's Research Hospital in Memphis, Tennessee.

The work for which Doherty ultimately received the Nobel Prize was carried out at the John Curtin School of Medical Research in Canberra, Australia. His post-doctoral research, begun in 1972, concerned the operation of the **immune**

system, specifically why the system had evolved to identify cells from another member of the same species as foreign.

At the time of Doherty's research, it was known that the immune system could distinguish the body's own material (self) from what was foreign (non-self). It was also known that there were various mechanisms the immune system utilized to accomplish the recognition and elimination of foreign material. Humoral or antibody-mediated immunity involves the production of antibodies by B cells. An antibody is produced in specific recognition of an antigen. Interaction of an antibody with an antigen stimulates white **blood** cells called phagocytes to engulf the invader. The second known mechanism of the immune system dealt with the response to **viruses**. Since viruses multiplied inside the bodies own cells, it was necessary to identify and eliminate them before this multiplication occurred. This was accomplished by a cell called a killer (or cytotoxic) T cell. The T cell is part of cellular or cell-mediated immunity. It was also known that **antigens** called major histocompatibility (MHC) antigens had a significant role in controlling the immune response to transplants.

But, it was not clear at that time how exactly the immune system dealt with invaders, or why certain types of transplanted tissue was rejected by the body. At the time of Doherty's Nobel Prize winning research, the prevailing view was that the presence of bacterial or viral invaders were alone sufficient to invoke an immune response.

At the Curtin School, Doherty and Zinkernagel, a Swiss immunologist, worked on the immune responses to viral infections in mice, focused on elucidating the cause of the **brain** cell destruction in mice associated with lymphocytic choriomeningitis, a meningitis-causing virus. As expected, they observed that infected cells were attacked by the mouse's own **T lymphocytes**. But, unexpectedly, when virus-infected cells were mixed with T cells from other infected mice, they found that the T cells destroyed the virus-infected cells only in a genetically identical strain of mice. No immune response occurred when the infected cells were from another strain of mice. Their work led to the discovery that genes did not mediate the immune response, as was thought at the time, but that the T cells did. They determined the reason was because the T cells needed to recognize two surface features of infected cells before destroying them; the infectious virus and the MHC self antigens. The MHC antigens identify a cell as being from one's own body.

The significance of the discovery was that the immune system can recognize a third state, in addition to self and non-self, a state termed "altered self." When a virus infects a cell and the cell displays viral antigens on its surface, it has become an altered self. The body has developed a means of recognizing an altered cell and in eliminating it. This is the biological role for the MHC system. The recognition of so-called alloantigens (e.g., MHC antigens which differ from the bodies' own MHC antigens) allows the body to recognize the altered self cells.

This discovery the mechanism used by the immune system at the cellular level to distinguish infected molecules from normal ones has aided transplantation research and in finding treatments which lessen the effects of autoimmune reactions in inflammatory disease, such as rheumatic conditions, multiple sclerosis, and diabetes. It has been shown subsequent to Doherty's research that T cells are conditioned in the thymus to recognize self MHC. This must occur to prevent autoimmune activation (a process where the T cells react against their body's own cells). The thymus selects for those T cells, which will recognize foreign antigens complexed with self-MHC proteins. Doherty's work also has relevance with respect to **cancer**. A cell's antigens can be altered not just in virus infections, but in certain cancers. A properly functioning immune system recognizes and destroys such cells before they proliferate. Knowledge of an individual's antigenic profile may make more accurate prediction of disease likelihood a possibility.

DOISY, EDWARD ADELBERT (1893-1986)
American biochemist

Edward Adelbert Doisy was an acclaimed biochemist whose contributions to research involved studying how chemical substances affected the body. In addition to research on antibiotics, insulin, and female **hormones**, he is remembered for his successful isolation of vitamin K, a substance that encourages **blood** clotting. Because he was able to synthesize this substance, many thousands of lives are saved each year. For this research, Doisy shared the 1943 Nobel Prize in physiology or medicine with Danish scientist **Henrik Dam**.

Doisy, one of two children, was born in Hume, Illinois, to Edward Perez Doisy, a traveling salesman, and Ada (Alley) Doisy. His parents, while themselves having little in the way of higher education, encouraged him to attend college. Doisy received his baccalaureate degree in 1914 from the University of Illinois at Champaign, and then obtained his master's in 1916. The advent of World War I interrupted his schooling for two years, during which time he served in the Army. After the war, Doisy received his Ph.D. from Harvard University Medical School in 1920. Beginning in 1919, he rapidly rose through the academic ranks, achieving the position of associate professor of **biochemistry** in the Washington University School of Medicine, St. Louis. He left this position in 1923 to go to the St. Louis University School of Medicine, and a year later he was appointed to the chair of biochemistry, where he engaged in research and teaching. He also was named the biochemist for St. Mary's Hospital. Doisy held these positions until his retirement in 1965.

For 12 years—from 1922 until 1934—Doisy worked with biologist Edgar Allen to study the ovarian systems of rats and mice. During this time he participated in research that isolated the first crystalline of a female steroidal hormone, now called oestrone. He later isolated two other related products, oestriol and oestradiol–17β. When Doisy administered these in tiny quantities to female mice or rats whose ovaries had been removed, the creatures acted as if they still had ovaries. Many women have benefitted from this research, as these compounds and their derivatives have been used to treat several hormone-related problems, including menopausal symptoms.

Doisy, in 1936, turned from this line of research to trying to isolate an antihemorrhagic factor that had been identified by Danish researcher Henrik Dam. Dam had discovered a chemical in the blood of chicks that decreased hemorrhaging; he called this substance *Koagulations Vitamine,* or vitamin K. Using Dam's work as a springboard, Doisy and his co-workers spent three years researching this new vitamin. They discovered that the vitamin had two distinct forms, called K1 and K2, and successfully isolated each—K1 from alfalfa, K2 (which differs in a side chain) from rotten fish. Alter Doisy had isolated these two compounds he successfully determined their structures, and was able to synthesize the extremely delicate vitamin K1.

Synthesizing vitamin K enabled large quantities of it to be produced relatively inexpensively. It has since been used to treat hemorrhages that would previously have been fatal, especially in newborns and other individuals who lack natural defenses; it is estimated that the use of vitamin K saves almost five thousand lives each year in the United States alone. For these research advances, Doisy shared the 1943 Nobel Prize in physiology or medicine with Dam. Some of this research was funded by the University of St. Louis and some of the funds were contributed by the pharmaceutical manufacturer Parke-Davis and Co.—a financial arrangement that Doisy saw as a model for future industry-university research relations.

Over the course of his career, most of Doisy's research focused on how various chemical substances worked in the human body. In addition to vitamin K, his team studying the effects of certain antibiotics, sodium, potassium, chloride, and phosphorus. He also developed a high-potency form of insulin, for use in treating diabetes.

Doisy was made St. Louis University's distinguished service professor in 1951, and later was named emeritus professor of biochemistry. As a sign of his contributions, the university's department of biochemistry was named in his honor in 1965, and he was made its emeritus director. Because of his prominence and his loyalties to the University, there are numerous plaques and buildings bearing his name.

Doisy's contributions to science are recognized by the numerous honorary awards he held and the scientific societies to which he belonged. He was member of the League of Nations Committee on the Standardization of Sex Hormones from 1932 to 1935, and in 1938, was elected to the National Academy of Sciences. In 1941, he was honored with the Willard Gibbs Medal of the American Chemical Society, which is perhaps the highest distinction in chemical science. He served as both the vice president and then president, from 1943 to 1945, of the American Society of Biological Chemists, and was the 29th president of the Endocrine Society in 1949.

See also Blood coagulation and blood coagulation tests; Hormones and hormone action; Vitamins

DOMAGK, GERHARD (1895-1964)
German biochemist

Gerhard Domagk was a biochemist who discovered sulfonamide therapy (sulfa drugs) for bacterial infections. Prior to his work, only a few chemical compounds had been found effective against these infections, and most of these had serious side effects. Domagk was awarded the Nobel Prize in physiology or medicine in 1939 for this discovery, but the German government forced him to decline it. In 1947, he was awarded the Nobel Prize Medal. In presenting this award, Nanna Svartz of the Royal Caroline Institute said that Domagk's discovery "meant nothing less than a revolution in medicine." The introduction of sulfonamide therapy prior to World War II undoubtedly saved many thousands of lives.

Domagk was born in Lagow, Brandenburg, Germany, to Paul and Martha Reiner Domagk. His father was assistant headmaster of a school, and he sent his son to a grade school that specialized in the sciences. Domagk enrolled in the University of Kiel as a medical student in 1914. His studies, however, were almost immediately interrupted by World War I. He enlisted in the German Army, fought at Flanders, and was transferred to the eastern front in December of 1914, where he was wounded. He was then transferred to the medical corps. He served in several hospitals, and his experience attempting to treat wounds and infectious diseases with the inadequate tools of the time undoubtedly influenced the direction of his later research.

Domagk resumed his studies at the University of Kiel following the war and earned his medical degree in 1921. In 1924, he took up the post of lecturer of pathological **anatomy** at the University at Greifswald, and in 1925, he moved on to a similar post at the University at Münster. In 1927, Domagk took a leave of absence from the university, which reshaped his career. He left to work in the laboratories of a company called I. G. Farbenindustrie, where he would remain for the rest of his professional life.

Domagk's career was profoundly influenced by the work of **Paul Ehrlich**. In 1907, Ehrlich had discovered arsphenamine, a compound specifically developed to be toxic to trypanosomes, and in 1909 this drug had been found to be effective against the bacterium that causes syphilis. Ehrlich's work had stimulated a number of searches for other antibacterials, and Domagk systematically continued this work at I. G. Farbenindustrie.

Domagk investigated thousands of chemicals for their potential as antibacterials. He would first test them against bacterial cultures in the test tube, then find the doses tolerated by animals such as mice, and lastly determine if compounds that worked in the test tube also worked against **bacteria** in living animals. For five years Domagk searched in vain for a "magic bullet" that would be toxic to bacteria and not to animals. His success illustrates Pasteur's dictum that chance favors the prepared mind. Methodically checking thousands of compounds for antibacterial activity, Domagk found in 1932 that a red leather dye showed a small effect on bacteria in the test tube. Developed by others at the company, the compound was called prontosil rubrum, and it proved non-toxic to mice.

Domagk's original experiment to determine the effectiveness of prontosil rubrum was straightforward. He injected twenty-six mice with a culture of hemolytic streptococcal bacteria. Fourteen mice served as controls, receiving no therapy, and all died within four days, as expected from previous experiments with untreated animals. The remaining twelve mice were injected with a single dose of prontosil rubrum an hour and a half after being infected with the bacteria. All twelve survived in good condition. In 1932, I. G. Farbenindustries began clinical testing of prontosil rubrum. For reasons that are unknown, however, Domagk delayed publishing the results of his experiment for three years. But it is clear that he understood its implications. During this time his daughter contracted a streptococcal **infection** from a needle prick and failed to respond to traditional therapies. As she lay near death, Domagk injected her with prontosil rubrum, and she subsequently recovered.

There was some initial skepticism when Domagk first published his experimental results, but rapid replication of his findings led to widespread acceptance of the value of prontosil rubrum therapy. Throughout Europe, hospitals treated a variety of illnesses—including pneumonia, meningitis, **blood** poisoning, and gonorrhea—with prontosil rubrum and closely allied compounds. Subsequent laboratory studies have shown that it is only a part of the prontosil rubrum molecule, the sulfonamide group itself, that is responsible for its effect on bacteria. Moreover, the compound does not kill bacteria but interferes with their **metabolism** and therefore with their ability to reproduce.

Although the importance of his work was widely recognized by physicians and fellow scientists, the world of politics obstructed formal acknowledgement of his discovery. Carl von Ossietzky, a German pacifist incarcerated in a prison camp, had been awarded the Nobel Peace Prize in 1936, and Hitler had declared that no German citizen could accept a Nobel Prize. When he was awarded the prize in 1939, Domagk notified the German government and was promptly arrested. He was soon released but was forced to decline the prize. He was awarded the Nobel Medal after the war, although the prize money had reverted to the foundation.

During the late 1930s and throughout World War II, Domagk continued to investigate other compounds for their antibacterial effects. He concentrated considerable effort on anti-tubercular drugs, recognizing the problem of increasing resistance to streptomycin. His work resulted in some drugs of limited use against tuberculosis, though the class of compounds he studied proved to be somewhat toxic. Domagk retired in 1958, but remained active in research. He spent the last few years of his career attempting without success to find an anti-cancer drug.

In addition to the Nobel Prize, Domagk received numerous other accolades. In 1959, he was elected to the Royal Society of London. He was awarded medals by both Spain and Japan, and several German universities conferred honorary doctorates upon him.

See also Bacteria and responses to bacterial infection

DONDERS, FRANS CORNELIS (1818-1889)
Dutch physician

Frans Donders was a Dutch physician who is credited with the invention of phonocardiography (the recording of **heart** sounds) using instrumentation designed for auscultation (listening for sounds produced by the body).

Manual auscultation, also termed immediate auscultation or direct auscultation, is performed by listening directly to sounds with the unaided ear. Indirect or immediate auscultation (most often performed with a stethoscope) utilizes some mechanism or device to amplify sounds produced with the body.

Donders was born in 1818 near Tilburg, in the Netherlands. He attended medical school at the University of Utrecht and completed his thesis at Leiden. Ultimately, Donders returned to Utrecht to become a professor of physiology.

Although well known for his work in ophthalmology, Donders's research interests also included studies in metabolism—especially respiration related processes. During his studies of respiration, Donders extended his studies of auscultation to include heart sounds. Donders attempted to correlate the sounds with known cardiac physiological processes and established heart rhythms.

In a landmark 1868 paper, Donders described the influence of heart rates upon the variation in the length (duration) of systole (the contraction of the ventricles) and diastole in the **cardiac cycle**. Donders listened to the sounds produced during the cardiac cycle with the aid of a stethoscope, and recorded his observations on a rotating cylinder so that the time interval between events could later be precisely determined through calculations derived from the fixed rotational speed of the cylinder (also calibrated with a musical metronome).

Based upon an averaging of many observations, Donders was able to affix average durations to systole and, perhaps more importantly, noticed that the duration was not significantly altered by a moderate change in heart rate. The duration of systole (approximately 0.31 seconds) did not vary under moderate stress that raised a patient's heart rate from a resting 72 beats per minute to 100 beats per minute. Donders was able to record variations in the duration of systole related to very slow heart rates (bradycardia) and very high heart rates (tachycardia). Donders's contribution spurred advancement in cardiac diagnosis.

Using animals, Donders was able to investigate the influence of the vagal nerve ligation (cutting) on heart and respiration rates. Donders was also able to discern specific events with in the cardiac cycle now identified with specific peaks on electrocardiogram recordings. In fact, Nobel Prize winner **Willem Einthoven**, one of the key developers of electrocardiography, spent time training in Donders's research lab.

Donders' research in ophthalmology (i.e., the study of the **eye**) was also notable. He identified in 1858 the cause of hypermetropia (i.e., farsightedness) as being a shortening of the ocular globe, which led to the convergence of the light rays reflected by the ocular lens behind the **retina**. Later, in 1862, he explained the blurred vision associated with astigmatism as due to irregularities on the surfaces of both **cornea** and lens, which caused diffusion of light rays. Donders published in 1864 these findings in his "On the Anomalies of Accommodation and Refraction." Among other works and papers written by Donders, the Ophthalmologic Archives founded by Albrecht von Graefe (1828-1870), also contains: "Laboratory Works on Physiology of the Utrecht Military Suprior School" (1849-1857); "Movements of the Eye" (1849). His studies of optics, physiology and **pathology** of the eye not only brought new lights upon the causes of sight impairments, but Donders also established in 1864 the practice of prescribing eyeglasses with specific fitting lens for different vision problems.

Donders was named a member of the Royal Netherlands Academy of Sciences. After suffering a stroke at the age of 68, he died three years later in 1889.

See also Biofeedback; Blood pressure and hormonal control mechanisms; Breathing; Cardiac muscle; Electrocardiograms (ECG); Heart: rhythm control and impulse conduction

DOPAMINE • *see* NEUROTRANSMITTERS

DORSAL • *see* ANATOMICAL NOMENCLATURE

DROWNING

Drowning refers to **death** that is caused by fluid obstruction of the airway, usually by immersion in water. The main physiological consequence of immersion in water is hypoxemia; a lowered level of oxygen in the **blood**.

Drowning is the culmination of a series of physiological events. Initially, an individual will gasp in a panic reaction. Water may be inhaled into the **lungs** (aspiration). As the body responds to the emergency in a "fight or flight" reaction, the **breathing** rate will increase to such an extent that hyperventilation can occur. Many people will attempt to hold their breath to bring themselves under control. However, if relief from drowning is not found within a minute or so, breathing will necessarily have to commence. A further physiologic response to the presence of water is the spasming of the adductor muscles—the muscles that close the **vocal cords**. This phenomenon is termed laryngospasm. Laryngospasm can also occur when someone tries to eat and breathe at the same time, and is euphemistically described as having something "go down the wrong way." This closure occurs when something other than air enters the trachea. While closure of the windpipe does prevent further entry of water, breathing becomes very difficult. Thus, a cycle of spasm followed by forced opening of the trachea, further entry of water and spasming begins.

Entry of water into the lungs can produce several effects. If the water is salt water, the increased sodium and chloride concentration of the salt water, relative to the fluids in

the body, can cause fluids to be sucked into the lungs, in an effort to equalize ionic concentrations. The result can be drowning in body fluids, which is termed pulmonary **edema**. When fresh water enters the lungs, large volumes can pass from the lungs to the **heart**. This causes the destruction of red blood cells (hemolysis) and release of potassium. The elevated potassium level can produce cardiac arrest (stoppage of the heart) or injury to the **central nervous system**.

Somewhat paradoxically, in many drowning victims the majority of water enters the lungs during the last few moments of life. Death is most typically due to larygospasm and the resulting blockage of air. As the victim becomes unconscious and the spasm reflex abates, water is able to pour into the lungs. Thus drowning can be "wet" or "dry" with respect to the water content of the lungs.

See also Bronchi; Central nervous system (CNS); Fight or flight reflex; Underwater physiology

DRUG ADDICTION

Addiction can be defined as a state where an organism behaves compulsively, even if the consequences of the behavior do not benefit the organism. Drug addiction is a multi-pronged process. Lifestyle has a role, but the biology of an individual is of prominent importance. The pleasure afforded by various drugs and the craving for these sensations as the effects of a drug begin to dissipate has a physiological basis.

Drugs like morphine, cocaine and heroin work by entering a "reward system" in the **brain**. As the brain develops a tolerance for the pleasurable sensations evoked by a drug, more of the drug is necessary to elicit those sensations. The term reward system refers to the tendency of humans and other creatures to perform tasks that are rewarding. Pleasurable feelings provide positive reinforcement that encourages repeating of the behavior. Studies in rats have shown that the ventral tegmental area of the brain is implicated in reward behavior. The **neurons** found in this region contain a neurotransmitter (a substance that functions to pass signal from one neuron to the next) called dopamine. In drug addiction, dopamine is released, triggering the sought-after pleasurable sensations.

Transmission of impulses from neuron to neuron depends upon the rapid release and rapid reabsorption of **neurotransmitters** such as dopamine. Drugs such as cocaine and amphetamine change the flow of neurotransmitters by slowing down the reabsorption of dopamine, in the case of cocaine, or increasing the release of dopamine, in the case of amphetamine. Both drugs act to increases the concentration of dopamine in the synapse—the space between the neurons. Upon repeated stimulation, the dopamine receptors become less sensitive to dopamine, producing an effect known as tolerance.

In more recent years, another component of the brain called the glutamate receptor has been implicated as another focus of addiction. The glutamate receptor is a site that influences the activity of glutamate neurotransmitters. Glutamate is an important neurotransmitter, accounting for upwards of 40 percent of all nerve signals in the brain. Studies with rats have demonstrated that blocking the transmission of glutamate prevents the development of increased sensitivity to drugs such as cocaine. The stimulation of glutamate transmission elicits sensations that require the repeated presence of glutamate.

For both neurotransmitters, a hallmark of addiction is loss of control in limiting the intake of the addictive substance. Recent research indicates that the reward pathway involving dopamine and glutamate can become even more important in the craving associated with addiction than the pleasurable reward itself. Thus, drug addiction can lead to a state where a craving for the drug is overpowering, even though the pleasurable effects produced by the drug might no longer be produced.

A new research avenue being pursued views addictive drugs as foreign invaders. Thus, similar to the body's response to a bacterial invader, the production and administration of antibodies to the particular drug may be a means of preventing the drug from reaching its target in the brain.

See also Depressants and stimulants of the central nervous system; Psychopharmacology

DRUG EFFECTS ON THE NERVOUS SYSTEM

The business of the nervous system is to transmit information from one cell to another. Although this happens in different locations and with different **neurotransmitters**, the basic process is common to all cell-to-cell transmission. First, neurotransmitter molecules must be synthesized. Then, they must be packaged in synaptic vesicles. At the appropriate time, they must exit the cell by exocytosis, cross the synaptic cleft and bind and activate receptors on the post-synaptic neuron. The neurotransmitter molecules must then be either degraded or taken back into the presynaptic neuron, a process known as reuptake. Psychoactive drugs exert their effects by interfering with one or more of these steps.

Some of these drugs, known as stimulants, increase synaptic activity. Amphetamines, for example, increase the release from presynaptic cells of the group of neurotransmitters known as catecholamines. These include **epinephrine**, norepinephrine, and dopamine. The presence of amphetamines can intensify the effect of nerve impulses that occur, and can actually cause the release of catecholamines in the absence of a nerve impulse. Amphetamines also block the reuptake of norepinephrine and dopamine.

Caffeine is a competitive inhibitor of the neuromodulator adenosine, whose function is to inhibit the release of excitatory catecholamine transmitters. Caffeine competes with adenosine for presynaptic receptors, making it unable to perform its inhibitory function.

Nicotine activates certain acetylcholine receptors at neuromuscular junctions and in the **central nervous system**. It also activates receptors in the autonomic ganglia of the **sympathetic nervous system**, resulting in an increase in **heart** rate and a release of epinephrine from the adrenal medulla.

Cocaine blocks the reuptake of catecholamines from synapses. Used short term, this causes heightened alertness and euphoria, as would be expected from increased synaptic activity. Prolonged use, however, produces tolerance, increased use, sleeplessness, and behavior that can resemble schizophrenia.

The hallucinogens LSD and psilocybin are agonists of the serotonin receptors, meaning that they mimic the effects of serotonin. How they produce heightened awareness of sensory stimuli and a dreamlike state is not fully understood.

The cannabinoids, including marijuana and hashish, mimic endogenous molecules named anandamides. Receptors for these molecules are found in the substantia nigra, **hippocampus**, cerebellar cortex, and **cerebral cortex**. Cannabinoids produce hallucinations and relaxation.

The so-called depressants decrease activity in the nervous system. Alcohol, for example, increases the number of GABA receptors on post-synaptic membranes. Because GABA is an inhibitory neurotransmitter, more GABA receptors allow more GABA molecules to affect post-synaptic **neurons**, decreasing their rate of firing.

Opioids, including heroin and morphine, resemble endogenous molecules called **endorphins** that are involved in the control of **pain**. The binding of these molecules to presynaptic receptors decreases the release of acetylcholine, norepinephrine and dopamine. Opioid receptors are present throughout the **brain**, but are especially concentrated in parts of the brain that transmit and modulate information about pain, including the periaquaductal gray region and raphe nuclei of the brain stem and the dorsal horn of the **spinal cord**.

In certain cases, two drugs interact so that one intensifies or negates the effect of the other. Alcohol and barbiturates are both depressant drugs. Taken together, they have a synergistic effect. Doses of these two drugs that would be tolerated if taken separately become lethal if taken together. The drug nalaxone, on the other hand, is an antagonist of the opioids, and prevents the opioids from exerting their full effect. For this reason, it can be used as an antidote for overdoses of morphine.

See also Anesthesia and anesthetic drug actions; Depressants and stimulants of the central nervous system; Drug interactions; Pharmacology

DRUG INTERACTIONS

How well a prescribed or over-the-counter (OTC) medication works in the human body often depends on a patient's diet and any other drugs they consume. Patients prescribed multiple therapies need to be extremely aware of potential drug-drug interactions. Mixing medicines can sometimes cause an adverse effect not typically associated with the medication or an increase or decrease in the drug's action. For example, some antidepressants can prevent some hypertension drugs from properly lowering **blood** pressure. Certain antifungal medications can change the way cholesterol-lowering drugs metabolize in the body.

Drug interactions are typically either pharmacokinetic or pharmacodynamic in nature. Pharmacokinetic interactions involve one drug's ability to alter a secondary medication's absorption, distribution, **metabolism** and and/or excretion. Pharmacodynamic interactions relate to the concentrations of both interacting drugs and their response on organ systems and receptor sites.

Drug-drug interactions can result in serious side effects. Many drugs contain powerful ingredients that, when combined, can speed up the **heart** rate, cause a rapid drop in blood pressure, or create a build-up of deadly toxins that damage the **liver** and heart. Every year, a number of patients die as a direct result of the unintended mixing of medications. These serious consequences may result from the drugs' chemical-physical incompatibility, a change in the rate or quantity of the drug absorbed by the body, or an alteration of the body's receptors to bind to the medication.

Adverse interactions can also occur when mixing medications with certain foods and beverages. While taking oral medication at mealtime or with a snack is often recommended, some foods can alter the body's ability to absorb numerous drugs. Food can slow the body's absorption of antibiotics such as penicillin, tetracycline, and erythromycin. Grapefruit juice, for example, can prevent the body from properly absorbing certain blood pressure-lowering and cardiac medications.

See also Biochemistry; Drug effects on the nervous system

DRUG TREATMENT OF CARDIOVASCULAR AND VASCULAR DISORDERS

There are several lifestyle changes that are the first line of defense against cardiovascular disease. These include adopting a diet low in saturated **fat** and cholesterol, quitting smoking, and beginning a regular exercise program. If these fail, or are insufficient, drug treatment is often the next treatment option.

There are several categories of drugs used in the prevention and treatment of cardiovascular disease. The regimen chosen for a particular patient depends on the nature of their cardiovascular problem and any other medical conditions they may have.

Often, treatment for elevated serum cholesterol is administered as a preventative, before a patient ever exhibits symptoms of **heart** disease. The ability to reduce cholesterol via drug therapy is an important adjunct to lifestyle changes, and can be of great benefit when such changes by themselves are insufficient. However, such a step must be carefully considered, because the medication is taken long term, and may have significant side effects.

One of the most effective categories of cholesterol medications is a class of drugs called statins. They block the action of an enzyme called HMG-COA reductase, which causes the

liver to produce less cholesterol itself and absorb more low-density lipoprotein (commonly known as "bad" cholesterol) from the bloodstream.

High **blood** pressure, or hypertension, is another cardiovascular risk factor that often responds to drug treatment. As with elevated cholesterol, lifestyle changes are generally tried first. Many patients with hypertension respond to a low-salt diet, in addition to the low-fat diet and exercise recommendations generally cited for maintaining a healthy heart.

Diuretics, or "water pills," were the earliest drug treatment for hypertension, and have been administered for more than 40 years. They cause the kidney to excrete more water and sodium, reducing the volume of the blood. High blood pressure may also be treated with beta blockers, which slow down the heart to reduce the blood volume it pumps out. Another class of drugs for hypertension are the vasodilators, which relax the blood vessels, or prevent them from constricting, so the blood can pass through more easily. These include the angiotension-converting enzyme or ACE inhibitors, the alpha blockers, and the **calcium** channel blockers.

Some of the oldest vasodilators, the nitrate drugs, may be used to relieve angina. Angina is a common cardiovascular problem, caused by **arteriosclerosis** or arterial spasm. Nitrates, administered in a quickly-absorbed form by means of a tablet placed under the tongue, can give fast relief.

A type of cardiovascular drug with an even longer pedigree is digitalis and its derivatives. Digitalis comes from the foxglove plant, which had been used in herbal medicine for hundreds of years. It is prescribed for a weakened heart, and strengthens its contractions by regulating the amount of calcium and electrical activity in the heart muscle.

Drugs work to correct irregular heartbeats are called antiarrhythmics. These are powerful medications, and must be carefully administered, since an overdose or skipped dose can be dangerous. Antiarrythmics may interact with each other and with other drugs, so they are often first given in a hospital. Regular follow-up is also essential, since a patient's response to the drugs may change over time.

Another important class of drugs prescribed for heart patients includes anticoagulants, antiplatelets, and thrombolytics. These drugs are commonly called "blood thinners." However, they work not by actually diluting the blood but by inhibiting its ability to clot. Anticoagulants are commonly used for patients who with certain arrhythmias, phlebitis, or who have had valve replacements. Aspirin is often recommended to help prevent heart attacks, or for patients who are recovering from a heart attack. It works by interfering with the blood component called **platelets**, which are instrumental in blood clotting.

Thrombolytics, or "clotbusters," are a relatively new class of drugs that can dissolve clots that have already formed, but before they do any permanent damage. Thrombolytics are administered by emergency medical personnel during the course of a heart attack or **stroke**, and can significantly reduce the likelihood of disability or **death**.

See also Angiology; Blood coagulation and blood coagulation tests; Blood pressure and hormonal control mechanisms;

Cardiac disease; Cardiac muscle; Heart, rhythm control and impulse conduction

Ductus arteriosis

The ductus arteriosis (also spelled ductus arteriosus) is a shunt—a diversionary channel or flow—that allows **blood** to cross between the pulmonary artery and aorta. The shunt of blood from the pulmonary artery to the aorta through the ductus arteriosis, also a shunt of blood from the pulmonary to the **systemic circulation**, is normal and necessary during fetal development. At birth, changes in the pressures brought about by respiration effectively reverse the shunt through the ductus arteriosis and ultimately result in a closure of the shunt. The ductus arteriosis usually closes completely within eight weeks to six months of birth. In a normal adult, the closed ductus arteriosis becomes the ligamentum arteriosum that runs between the **aortic arch** and the left pulmonary artery.

In a small number of cases—about 1 in 2,200 births—the ductus arteriosis remains open (patent) and a patent ductus arteriosis (PDA) develops.

The **lungs** of the developing fetus are filled with amniotic fluid and are non-functional. While the fetus is in the womb, the mother supplies oxygen to the fetal **circulatory system** via the **placenta**. The fluid in the fetal lungs increases pressures within the pulmonary system and provides resistance to the infusion or flow of blood from the pulmonary artery. The open ductus arteriosis vessel allows blood to flow from the pulmonary artery directly into the aorta. The shunt is mechanical in that it allows blood to follow the path of lowest pressures (i.e., least resistance) through the fetal circulatory system.

At birth, the lungs expand with air and the pulmonary resistance drops dramatically so that the path of least resistance for blood flow becomes through the pulmonary artery to the lungs. This effectively does away with the need for the ductus arteriosis. In fact, because systemic pressures are higher than pulmonary pressures (due in part to the ligation or tying off of the umbilical cord that severs the flow of systemic blood to the large and diffusing vascular bed of the placenta), the higher systemic pressures actually lead to a reversal of blood flow within the shut so that a small amount of blood from the aorta crosses over to the pulmonary artery. As long as the ductus arteriosis is not large, the shunt of already oxygenated blood back to the lungs does no harm and may actually help establish full lung function. In most cases, the diminished blood flow through the ductus arteriosis allows the vessel to close and fuse.

Increased blood flow into the pulmonary artery resulting from a large ductus arteriosis that remains open following birth (i.e., where the shunt in blood flow is reversed to a an aorta to pulmonary artery shunt) can eventually lead to increased pulmonary hypertension (elevated pressure levels) associated with Eisenmenger syndrome or congestive **heart** failure.

A patent ductus arteriosis is usually associated with a distinct murmur that can be heard during auscultation (listening for heart sounds with a stethoscope). A definitive diagnosis of PDA can be made via an electrocardiogram or echocardiogram. Blood oxygen saturation tests can also establish the size and/or severity of a PDA. Newborn girls show a much higher rate of PDA than do boys, but the cause of this difference is still under study. Premature infants also show evidence of PDA in much higher numbers than do babies delivered closer to full term.

See also Angiology; Birth defects and abnormal development; Fetal circulation; Heart, embryonic development and changes at birth; Pulmonary circulation; Respiratory system embryological development

Duvé, Christian de (1917-   )
Belgian biochemist and cell biologist

Christian René de Duvé's groundbreaking studies of cellular **structure and function** earned him the 1974 Nobel Prize in physiology or medicine (shared with **Albert Claude** and **George Palade**). However, he did much more than discover the two key cellular organelles—lysosomes and peroxisomes—for which the Swedish Academy honored him. His work, along with that of his fellow recipients, established an entirely new field, cell biology. De Duvé introduced techniques that have enabled other scientists to better study cellular **anatomy** and physiology. De Duvé's research has also been of great value in helping clarify the causes of and treatments for a number of diseases.

De Duvé's parents, Alphonse and Madeleine (Pungs) de Duvé, had fled Belgium after its invasion by the German army in World War I, escaping to safety in England. There, in Thames-Ditton, Christian René de Duvé was born. De Duvé returned with his parents to Belgium in 1920, where they settled in Antwerp. (De Duvé later became a Belgian citizen.) As a child, de Duvé journeyed throughout Europe, picking up three foreign languages in the process, and in 1934, enrolled in the Catholic University of Louvain, where he received an education in the "ancient humanities." Deciding to become a physician, he entered the medical school of the university.

Finding the pace of medical training relaxed, and gravitating toward research, de Duvé joined J. P. Bouckaert's group. Here he studied physiology, concentrating on the hormone insulin and its effects on uptake of the sugar glucose. De Duvé's experiences in Bouckaert's laboratory convinced him to pursue a research career when he graduated with an M.D. in 1941. World War II disrupted his plans, and de Duvé ended up in a prison camp. He managed to escape and subsequently returned to Louvain to resume his investigations of insulin. Although his access to experimental supplies and equipment was limited, he was able to read extensively from the early literature on the subject. On September 30, 1943, he married Janine Herman, the couple eventually had four children. Even before obtaining his Ph.D. from the Catholic University of

Louvain in 1945, de Duvé published several works, including a four hundred page book on glucose, insulin, and diabetes. The dissertation topic for his *Agrégé de l'Enseignement Supérieur* was also insulin. De Duvé then obtained an M.Sc. degree in chemistry in 1946.

After graduation, de Duvé decided that he needed a thorough grounding in biochemical approaches to pursue his research interests. He studied with Hugo Theorell at the Medical Nobel Institute in Stockholm for eighteen months, then spent six months with Carl Ferdinand Cori, **Gerty Cori**, and **Earl Sutherland** at Washington University School of Medicine in St. Louis. Thus, in his early postdoctoral years he worked closely with no less than four future Nobel Prize winners. It is not surprising that, after this hectic period, de Duvé was happy to return to Louvain in 1947 to take up a faculty post at his alma mater teaching physiological chemistry at the medical school. In 1951, de Duvé was appointed full professor of **biochemistry**. As he began his faculty career, de Duvé's research was still targeted at unraveling the mechanism of action of the anti-diabetic hormone, insulin. While he was not successful at his primary effort (indeed, the answer to de Duvé's first research question was to elude investigators for more than thirty years), his early experiments opened new avenues of research.

As a consequence of investigating how insulin works in the human body, de Duvé and his students also studied the **enzymes** involved in carbohydrate **metabolism** in the **liver**. It was these studies that proved pivotal for de Duvé's eventual rise to scientific fame. In his first efforts, he had tried to purify a particular liver enzyme, glucose–6-phosphatase, that he believed blocked the effect of insulin on liver cells. Many enzymes would solidify and precipitate out of solution when exposed to an electric field. Most could then be redissolved in a relatively pure form given the right set of conditions, but glucose–6-phosphate stubbornly remained a solid precipitate. The failure of this electrical separation method led de Duvé to try a different technique, separating components of the cell by spinning them in a centrifuge, a machine that rotates at high speed. De Duvé assumed that particular enzymes are associated with particular parts of the cell. These parts, called cellular organelles (little **organs**) can be seen in the microscope as variously shaped and sized grains and particles within the body of cells. It had long been recognized that there existed several discrete types of these organelles, though little was known about their structures or functions at the time.

The basic principles of centrifugation for separating cell parts had been known for many years. First cells are ground up (homogenized) and the resultant slurry placed in a narrow tube. The tube is placed in a centrifuge, and the artificial gravity that is set up by rotation will separate material by weight. Heavier fragments and particles will be driven to the bottom of the tube while lighter materials will layer out on top. At the time de Duvé began his work, centrifugation could be used to gather roughly four different fractions of cellular debris. This division proved to be too crude for his research, because he needed to separate out various cellular organelles more selectively.

For this reason, de Duvé turned to a technique developed some years earlier by fellow-Belgian Albert Claude while working at the Rockefeller Institute for Medical Research. In the more common centrifugation technique, the cells of interest were first vigorously homogenized in a blender before being centrifuged. In Claude's technique of differential centrifugation, however, cells were treated much more gently, being merely ground up slightly by hand prior to being spun to separate various components.

When de Duvé used this differential centrifugal fractionation technique on liver cells, he did indeed get better separation of cell organelles, and was able to isolate certain enzymes to certain cell fractions. One of his first findings was that his target enzyme, glucose–6-phosphatase, was associated with microsomes, cellular organelles which had been, until that time, considered by cell biologists to be quite uninteresting. De Duvé's work showed that they were the site of key cellular metabolic events. Further, this was the first time a particular enzyme had been clearly associated with a particular organelle.

De Duvé was also studying an enzyme called acid phosphatase that acts in cells to remove phosphate groups (chemical clusters made up of one phosphorus and three oxygen atoms) from sugar molecules under acidic conditions. The differential centrifugation technique isolated acid phosphatase to a particular cellular fraction, but measurements of enzyme activity showed much lower levels than expected. De Duvé was puzzled. What had happened to the enzyme? He and his students observed that if the cell fraction that initially showed this low level of enzyme were allowed to sit in the refrigerator for several days, the enzyme activity increased to expected levels. This phenomenon became known as enzyme latency.

De Duvé believed he had a solution to the latency mystery. He reasoned that perhaps the early, gentle hand-grinding of differential centrifugation did not damage the cellular organelles as much as did the more traditional mechanical grinding. What if, he wondered, some enzymes were not freely exposed in the cells' interiors, but instead were enclosed *within* protective membranes of organelles. If these organelles were not then broken apart by the gentle grinding, the enzyme might still lie trapped within the organelles in the particular cell fraction after centrifugation. If so, it would be isolated from the chemicals used to measure enzyme activity. This would explain the low initial enzyme activity, and why over time, as the organelles' membranes gradually deteriorated, enzyme activity would increase.

De Duvé realized that his ideas had powerful implications for cellular research. By carefully observing what enzymes were expressed in what fractions and under what conditions, de Duvé's students were able to separate various enzymes and associate them with particular cellular organelles. By performing successive grinding and fractionations, and by using compounds such as detergents to break up membranes, de Duvé's group began making sense out of the complex world that exists within cells.

De Duvé's research built on the work of other scientists. Previous research had clarified some of the roles of various enzymes. But de Duvé came to realize that there existed a group of several enzymes, in addition to acid phosphatase, whose primary functions all related to breaking down certain classes of molecules. These enzymes were always expressed in the same cellular fraction, and showed the same latency. Putting this information together, de Duvé realized that he had found an organelle devoted to cellular **digestion**. It made sense, he reasoned, that these enzymes should be sequestered away from other cell components. They functioned best in a different environment, expressing their activity fully only under acidic conditions (the main cell interior is neutral). Moreover, these enzymes could damage many other cellular components if set loose in the cells' interiors. With this research, de Duvé identified lysosomes and elucidated their pivotal role in cellular digestive and metabolic processes. Later research in de Duvé's laboratory showed that lysosomes play critical roles in a number of disease processes as well.

De Duvé eventually uncovered more associations between enzymes and organelles. The enzyme monoamine oxidase, for example, behaved very similarly to the enzymes of the lysosome, but de Duvé's careful and meticulous investigations revealed minor differences in when and where it appeared. He eventually showed that monoamine oxidase was associated with a separate cellular organelle, the peroxisome. Further investigation led to more discoveries about this previously unknown organelle. It was discovered that peroxisomes contain enzymes that use oxygen to break up certain types of molecules. They are vital to neutralizing many toxic substances, such as alcohol, and play key roles in sugar metabolism.

Recognizing the power of the technique that he had used in these early experiments, de Duvé pioneered its use to answer questions of both basic biological interest and immense medical application. His group discovered that certain diseases result from cells' inability to properly digest their own waste products. For example, a group of illnesses known collectively as disorders of **glycogen** storage result from mal-

functioning lysosomal enzymes. Tay Sachs disease, a congenital neurological disorder that usually kills its victims by age five, results from the accumulation of a component of the cell membrane that is not adequately metabolized due to a defective lysosomal enzyme.

In 1962, de Duvé joined the Rockefeller Institute (now Rockefeller University) while keeping his appointment at Louvain. In subsequent years, working with numerous research groups at both institutions, he has studied inflammatory diseases such as arthritis and **arteriosclerosis**, genetic diseases, immune dysfunctions, tropical maladies, and cancers. This work has led, in some cases, to the creation of new drugs used in combatting some of these conditions. In 1971, de Duvé formed the International Institute of Cellular and Molecular Pathology, affiliated with the University at Louvain. Research at the institute focuses on incorporating the findings from basic cellular research into practical applications.

De Duvé's work has won him the respect of his colleagues. Workers throughout the broad field of cellular biology recognize their debt to his pioneering studies. He helped found the American Society for Cell Biology. He has received awards and honors from many countries, including more than a dozen honorary degrees. In 1974, de Duvé, along with Albert Claude and George Palade, both also of the Rockefeller Institute, received the Nobel Prize in physiology or medicine, and were credited with creating the discipline of scientific investigation that became known as cell biology. De Duvé was elected a foreign associate of the United States National Academy of Sciences in 1975, and has been acclaimed by Belgian, French, and British biochemical societies. He has also served as a member of numerous prestigious biomedical and health-related organizations around the globe.

See also Cell cycle and cell division; Cell differentiation; Cell membrane transport; Cell structure; Cell theory; Enzymes and coenzymes

E

EAR (EXTERNAL, MIDDLE, AND INTERNAL)

The ear is the means by which the sounds of the world reach us. The external ear that is evident on humans and other creatures is only a part of the ear. The middle ear and the inner ear have different structures and contributions to the overall ability of the ear to detect sound.

The external ear consists of a hole in the side of the **skull**, known as the auricle or pinna, and of the ear canal. The auricle helps focus the incoming sound waves. The hole leads into the auditory canal, a roughly cylinder-shaped, small diameter canal that is about 0.98 in. (2.5 cm) long. Towards the inner end, the canal widens slightly and ends at the eardrum. The ear canal can be thought of as a shaped tube with a resonating column of air inside it, having open and closed ends, similar to the construction of an organ pipe.

This analogy is apt, for the ear canal enhances the sound vibrations that have traveled in from the outside. The canal can resonate, or vibrate, typically at frequencies that the ear hears most sharply. The vibration increases the wavelength of the sound waves traveling down the canal. The amplified waves eventually contact the ear drum, which is positioned at the inner end of the canal, and marks the boundary between the outer ear and the middle ear.

The ear drum is a membrane. It is capable of vibration, which occurs when the sound waves contact it. The vibrational energy of the ear drum is converted to mechanical vibrations in the solid materials of the middle ear. These solid materials are three bones: the malleus, incus and stapes. The bones form a system of levers that are linked together and are driven by the eardrum. The outer malleus pushes on the incus, which in turn pushes on the stapes. This further amplifies the sound vibrations, typically 2–3 fold. Muscles are positioned around the bones, the smallest muscles in the body, and dampen down the mechanical vibrations if they become too pronounced. They are a form of safety device, restricting movement of one or more of the bones. This protects against the creation of too great a vibration from a very loud sound. Muscle movement is triggered when sound exceeds a certain level.

The stapes bone is in close proximity to a structure called the oval window, which defines the boundary between the middle and inner ears. The oval window is another membrane—15 to 30 times smaller than the eardrum—that covers an opening at the base of the inner ear structure known as the cochlea. Mechanical vibration from the stapes is transferred to the oval window, which further amplifies the sound by 15 to 30 times.

By the time a sound wave has reached the cochlea, it has undergone amplification in the ear canal, the middle ear triplet bone arrangement, and at the oval window. The amplification of the original signal can be upwards of 800 times by this point.

The cochlea contains three fluid-filled regions. The vestibular and tympanic canals contain perilymph; a liquid almost identical to spinal fluid. The third region, the cochlea duct, contains endolymph; a fluid similar to that found within cells. These regions are separated by thin membranes (Reissener's membrane between the vestibular canal and cochlea duct, and the basilar membrane between the tympanic canal and the cochlear duct). Reissener's membrane is only two cell walls thick.

Any rupture of these delicate structures impairs the function of the cochlea, which is to convert the incoming mechanical sound pressure to hydraulic pressure. The cochlear membranes vibrate with the incoming mechanical sound energy. Their vibration is converted to hydraulic pressure (akin to waves in water) that is then transmitted along the entire structure of the cochlea. The waves can be of different heights. The different wave shapes determine which nerve fibres positioned around the basilar membrane will send a signal to the **brain**. This enables sounds of different frequencies to be distinguished and such information routed to the brain for interpretation.

See also Cranial nerves; Deafness; Ear: Otic embryonic development

EAR, OTIC EMBRYOLOGICAL DEVELOPMENT

The embryological development of the ear is best understood by describing separately the individual development of the three major anatomical features that comprise the **ear**. Accordingly, although in the fetus, the development of the external ear, middle ear, and internal ear overlap in many areas and stages, the overall development of the ear may be studied by an individual study of these three structure.

The external ear develops from the upper portion of the first external pharyngeal groove and becomes an area of highly modified skin. Comprising **ectoderm**, the cells in this area, termed the meatal plate, continue to divide until they come in contact with cells destined to form the middle ear structure. The cells at this juncture will ultimately form the **tympanic membrane**. As development proceeds, the meatal plate thickens and becomes plug-like in form (the meatal plug). Eventually, a cavity forms within the meatal plug and bony **tissue** forms within the plug ring to form the external ear (external auditory meatus). The features of the auricle of the adult ear develop from six regionalized swellings (hillocks) that original surround the external pharyngeal groove.

Interestingly, a region of the external ear is the only site of ectodermal cell origin that is innervated by a branch of the Vagal nerve. Anatomists assert that this innervation is an evolutionary vestige of the highly developed lateral-line system found in sharks. In sharks, the lateral-line system enables the shark to discern minute pressure changes in the water caused by moving objects at great distances, and in an important and effective sensory organ used in hunting prey.

The auditory tubes (**Eustachian tubes**) develop from an area of the developing **pharynx** that is composed of endodermal cells. The cells form a tubotympanic recess and ultimately form the Eustachian tubes and other middle ear structures, including the tympanic cavity. At about the end of the first month of development, the cells surrounding the tubotympanic recess form a primitive middle ear cavity. Some of the cells thicken and elongate into the primitive auditory tubes. Eventually the cells in this area come in contact with the otic capsule. An area of cells chondrifies to become the malleus, incus, and handle of the malleus of the middle ear. A separate area of the otic capsule differentiates to form the stapes.

The developing handle of the malleus comes in contact with the developing chorda tympani nerve that ultimately joins other nerves in the eardrum. These structures become wedged between the ectoderm of the outer ear and **endoderm** of the middle ear to eventually lie in a thin layer of **mesoderm** that separates the developing structures.

The malleus and incus become suspended by folds of membrane. These and other folds partially divide the middle ear into a complex series of passages. The otic capsule eventually becomes cartilaginous (condrifies) and then becomes bony (ossifies) as part of the temporal bone. Portions of the convoluted labyrinth of folds become part of the inner ear structure.

The inner ear develops from a thickening of ectoderm near the hindbrain, termed the otic placode. The plate like placode forms an otic pit. When the pit is separated from the surface by further development of surrounding cells, an otic vesicle forms. The otic vesicle is surrounded with a layer of mesoderm to form an otic capsule.

Somewhat like a belt being tightened, the otic capsule becomes contracted about its middle region to form upper and lower chambers. The upper area forms the utriculosaccular area, and the lower chamber the cochlear area. Semicircular canals develop in the utriculosaccular area. Specialized neural cells develop near ampullae that ultimately contribute to sensory feeling associated with balance. Ultimately the utriculosaccular area divides into the utricle and saccule that remain connected by an utriculosaccular duct. The lower cochlear region of the otic capsule grows into the spiral shape found in the adult utriculosaccular area.

Specialized neuroreceptor cells develop within the cochlea to form the spiral organ of Corti. The cells associate with neural ganglia (the spiral ganglion). Eventually regions of the otic vesicle form the membranous labyrinth that becomes filled with lymph fluid. In the adult, the cochlea senses disturbances in the lymphatic fluid caused by sound waves. Hair-like projections respond to change son fluid motion and cause associated nerves to fire electrical signals to the **brain**. The development of the cochlea is usually complete by the end of the second trimester of development.

Although a newborn's ear looks like an adult ear, anatomists and physiologists continue research into determining exactly when the ear completes development and takes on full adult-like capacity. Small differences may lead to discoveries or insights into developmental abnormalities or delays in the development of balance and walking.

See also Cell differentiation; Embryonic development, early development, formation, and differentiation; Nervous system, embryological development; Sense organs: Balance and orientation; Sense organs: Otic (hearing) structures

EATING · *see* MASTICATION

ECCLES, JOHN C. (1903-1997)
Australian neurophysiologist

John Carew Eccles was a neurophysiologist whose research explained how nerve cells communicate with one another. He demonstrated that when a nerve cell is stimulated, it releases a chemical that binds to the membrane of neighboring cells and activates them in turn. He further demonstrated that by the same mechanism a nerve cell can also inhibit the electrical activity of nearby nerve cells. For this research, Eccles shared the 1963 Nobel Prize in physiology or medicine with **Alan Lloyd Hodgkin** and **Andrew Huxley**.

Born in Melbourne, Australia, Eccles was the son of William James and Mary Carew Eccles. Both of his parents were teachers, and they taught him at home until he entered Melbourne High School in 1915. In 1919, Eccles began medical studies at Melbourne University, where he participated in athletics and graduated in 1925 with the highest academic honors. Eccles's academic excellence was rewarded with a Rhodes Scholarship, which allowed him to pursue a graduate degree in England at Oxford University. In September 1925, Eccles began studies at Magdalen College, Oxford. As he had done at Melbourne, Eccles excelled academically, receiving high honors for science and named a Christopher Welch Scholar. In 1927, he received appointment as a junior research fellow at Exeter College, Oxford.

Even before leaving Melbourne for Oxford, Eccles had decided that he wanted to study the **brain** and the nervous system, and he was determined to work on these subjects with Charles Scott Sherrington. Sherrington, who would win the Nobel Prize in 1932, was then the world's leading neurophysiologist; his research had virtually founded the field of cellular neurophysiology. The following year, after becoming a junior fellow, Eccles realized his goal and became one of Sherrington's research assistants. Although Sherrington was then nearly seventy years old, Eccles collaborated with him on some of his most important research. Together, they studied the factors responsible for inhibiting a neuron, or a nerve cell. They also explored what they termed the "motor unit"—a nerve cell, which coordinates the actions of many muscle fibers. Sherrington and Eccles conducted their research without the benefit of the electronic devices that would later be developed to measure a nerve cell's electrical activity. For this work on neural excitation and inhibition, Eccles was awarded his doctorate in 1929.

Eccles remained at Exeter after receiving his doctorate, serving as a Staines Medical Fellow from 1932 to 1934. During this period, he also held posts at Magdalen College as tutor and demonstrator in physiology. The research that Eccles had begun in Sherrington's laboratory continued, but instead of describing the process of neural inhibition, Eccles became increasingly interested in explaining the process that underlies inhibition. He and other neurophysiologists believed that the transmission of electrical impulses was responsible for neural inhibition. Bernhard Katz and Paul Fatt later demonstrated, however, that it was a chemical mechanism and not a wholly electrical phenomenon which was primarily responsible for inhibiting nerve cells.

In 1937, Eccles returned to Australia to assume the directorship of the Kanematsu Memorial Institute for Pathology in Sydney. During the late 1930s and early 1940s, the Kanematsu Institute, under his guidance, became an important center for the study of neurophysiology. With Katz, Stephen Kuffler, and others, he undertook research on the activity of nerve and muscle cells in cats and frogs, studying how nerve cells communicate with muscle or motor cells. His team proposed that the binding of a chemical (now known to be the neurotransmitter acetylcholine) by the muscle cell led to a depolarization, or a loss of electrical charge, in the muscle cell. This depolarization, Eccles believed, occurred because charged ions in the muscle cell were released into the exterior of the cell when the chemical substance released by the nerve cell was bound to the muscle cell.

During World War II, Eccles served as a medical consultant to the Australian army, where he studied vision, hearing, and other medical problems faced by pilots. Returning to full-time research and teaching in 1944, Eccles became professor of physiology at the University of Otago in Dunedin, New Zealand. At Otago, Eccles continued the research that had been interrupted by the war, but now he attempted to describe in greater detail the neural transmission event, using very fine electrodes made of glass. This research continued into the early 1950s, and it convinced Eccles that transmission from nerve cell to nerve cell or nerve cell to muscle cell occurred by a chemical mechanism, not an electrical mechanism as he had thought earlier.

In 1952, Eccles left Otago for the Australian National University in Canberra. Here, along with Fatt and J. S. Coombs, he studied the inhibitory process in postsynaptic cells, which are the nerve or muscle cells that are affected by nerve cells. They were able to establish that whether nerve and muscle cells were excited or inhibited was controlled by pores in the membrane of the cells, through which ions could enter or leave. By the late 1950s and early 1960s, Eccles had turned his attention to higher neural processes, pursuing research on neural pathways and the cellular organization of the brain.

In 1966, Eccles turned sixty-three and university policy at the Australian National University required him to retire. Wanting to continue his research career, he accepted an invitation from the American Medical Association to become the director of its Institute for Biomedical Research in Chicago. He left that institution in 1968 to become professor of physiology and medicine and the Buswell Research Fellow at the State University of New York in Buffalo. The university constructed a laboratory for him where he could continue his research on transmission in nerves. Even at a late stage in his career, Eccles's work suggested important relationships between the excitation and inhibition of nerves and the storing and processing of information by the brain.

In 1975, he retired from SUNY with the title of Professor Emeritus, and subsequently moved to Switzerland. During the final period of his career, Eccles focused on a variety of fundamental problems relating to consciousness and identity, conducting research in areas where physiology, psychology, and philosophy intersect. He died at his home in Contra, Switzerland.

Eccles received a considerable number of scientific distinctions. His memberships included the Royal Society of London, the Royal Society of New Zealand, and the American Academy of Arts and Sciences. He was awarded the Gotch Memorial Prize in 1927, and the Rolleston Memorial Prize in 1932. The Royal College of Physicians presented him with their Baly Medal in 1961, the Royal Society gave him their Royal Medal in 1962, and the German Academy awarded him the Cothenius Medal in 1963. Also in 1963, he shared the Nobel Prize in physiology and medicine with Alan Hodgkin and Andrew Huxley. He was knighted in 1958.

See also Biochemistry; Nerve impulses and conduction of impulses; Nervous system overview; Neurology; Neurons; Neurotransmitters

ECG · *see* ELECTROCARDIOGRAM (ECG)

ECTODERM

Ectoderm is one of three principal germinal layers of cells that are formed in early in embryonic development. Ectoderm comprises the outermost germinal layer from which the nervous system, eyes, ears, **epidermis**, integumentary elements (glands, **hair**, and **nails**) develop. Membranes derived from ectoderm are in contact with **endoderm** derived structures at membranes of the mouth and anus.

In the embryonic disk ectoderm and endoderm sandwich **mesoderm**, the third primitive germinal layer. When the embryonic disk ultimately folds into a tube the basic "tube within a tube" plan of development becomes evident. A core endodermal tube establishes a primitive digestive pathway bounded by an oral orifice and an anal orifice. Around that innermost tube is an outer tube comprised of ectoderm. The ectoderm serves as a protective layer and the layer from which the nervous system and sense **organs** develop. Mesodermal cells fill the space between the inner (endodermal) and outer (ectodermal) tube. Mesodermal cells ultimately contribute to the muscles, organs, and other internal body structures.

About a week following **fertilization**, the human embryonic blastocyst is embedded in the endometrium of the uterus. The blastocyst is a proliferating ball of cells with a cavity termed a blastocoele. At one pole of the blastocyst there is a thickened mass of cells termed the inner cell mass. The inner cell mass contains communicating slit-like openings that form the amniotic cavity and the embryonic disk.

Ectoderm lies on the dorsal side of the embryonic disk. At the cellular or histological level, ectoderm is comprised of pseudostratified columnar cells. On the ventral side of the disk lies the endodermal germ layer. Initially there are only two germ layers but by 18 days following fertilization a thickening occurs in the ectoderm to form a primitive streak. The walls of the primitive groove continue to thicken and, at the anterior (cephalic) end of the groove, expand into a primitive knot also known as Henson's node. Anterior to the primitive knot, primitive ectoderm and endoderm are in direct contact with each other to form a prochordal plate. The primitive streak also establishes the general head-to tail (cephalo-caudal) axis for subsequent development.

Starting at the primitive knot, cells from the ectodermal layer migrate into the primitive groove and invaginate into the space between the ectoderm and endoderm to form mesodermal cells. Other cells, derived from other **fetal membranes** also contribute cells to the mesodermal layers.

As ectodermal cells stream invaginate to form a trilaminar embryonic disk, a head process (ultimately to become the **notochord**) forms in the middle mesodermal layer.

See also Embryonic development: early development, formation, and differentiation; Fetal membranes and fluids; Implantation of the embryo; Limb buds and limb embryological development; Nervous system: embryological development; Organizer experiment; Sexual reproduction; Skeletal and muscular systems: Embryonic development; Somite

EDELMAN, GERALD M. (1929-)
American biochemist

For his "discoveries concerning the chemical structure of antibodies," Gerald M. Edelman and his associate **Rodney Porter** received the 1972 Nobel Prize in physiology or medicine. During a lecture Edelman gave upon acceptance of the prize, he stated that **immunology** "provokes unusual ideas, some of which are not easily come upon through other fields of study.... For this reason, immunology will have a great impact on other branches of biology and medicine." He was to prove his own prediction correct by using his discoveries to draw conclusions not only about the **immune system** but about the nature of consciousness as well.

Born in New York City to Edward Edelman, a physician, and Anna Freedman Edelman, Gerald Maurice Edelman attended New York City public schools through high school. After graduating, he entered Ursinus College, in Collegeville, Pennsylvania, where he received his B.S. in chemistry in 1950. Four years later, he earned an M.D. degree from the University of Pennsylvania's Medical School, spending a year as medical house officer at Massachusetts General Hospital.

In 1955 Edelman joined the United States Army Medical Corps, practicing general medicine while stationed at a hospital in Paris. There, Edelman benefited from the heady atmosphere surrounding the Sorbonne, where future Nobel laureates Jacques Lucien Monod and François Jacob were originating a new study, **molecular biology**. Following his 1957 discharge from the Army, Edelman returned to New York City to take a position at Rockefeller University studying under Henry Kunkel. Kunkel, with whom Edelman would conduct his Ph.D. research, and who was examining the unique flexibility of antibodies at the time.

Antibodies are produced in response to **infection** in order to work against diseases in diverse ways. They form a class of large **blood** proteins called globulins—more specifically, immunoglobulins—made in the body's lymph tissues. Each immunoglobulin is specifically directed to recognize and incapacitate one antigen, the chemical signal of an infection. Yet they all share a very similar structure.

Through the 1960s and 1970s, a debate raged between two schools of scientists to explain the situation whereby antibodies share so many characteristics yet are able to perform many different functions. In one camp, George Wells Beadle and Edward Lawrie Tatum argued that despite the remarkable diversity displayed by each antibody, each immunoglobulin, must be coded for by a single gene. This has been referred to as the "one gene, one protein" theory. But, argued the opposing camp, led by the Australian physician, Sir Frank Macfarlane Burnet, if each antibody required its own code within the **deoxyribonucleic acid (DNA)**, the body's master plan of protein structure, the immune system alone would take up all the possible codes offered by the human DNA.

Both camps generated theories, but Edelman eventually disagreed with both sides of the debate, offering a third possibility for antibody synthesis in 1967. Though not recognized at the time because of its radical nature, the theory he and his associate, Joseph Gally, proposed would later be confirmed as essentially correct. It depended on the vast diversity that can come from chance in a system as complex as the living organism. Each time a cell divided, they theorized, tiny errors in the transcription—or reading of the code—could occur, yielding slightly different proteins upon each misreading. Edelman and Gally proposed that the human body turns the advantage of this variability in immunoglobulins to its own ends. Many strains of **antigens** when introduced into the body modify the shape of the various immunoglobulins in order to prevent the recurrence of disease. This is why many illnesses provide for their own cure—why humans can only get chicken pox once, for instance.

But the proof of their theory would require advances in the state of biochemical techniques. Research in the 1950s and 1960s was hampered by the difficulty in isolating immunoglobulins. The molecules themselves are comparatively large, too large to be investigated by the chemical means then available. Edelman and Rodney Porter, with whom Edelman was to be honored with the Nobel Prize, sought methods of breaking immunoglobulins into smaller units that could more easily be studied. Their hope was that these fragments would retain enough of their properties to provide insight into the functioning of the whole.

Porter became the first to split an immunoglobulin, obtaining an "active fragment" from rabbit blood as early as 1950. Porter believed the immunoglobulin to be one long continuous molecule made up of 1,300 amino acids—the building blocks of proteins. But Edelman could not accept this conclusion, noting that even insulin, with its 51 **amino acids**, was made up of two shorter strings of amino acid chains working as a unit. His doctoral thesis investigated several methods of splitting immunoglobulin molecules, and, after receiving his Ph.D. in 1960 he remained at Rockefeller as a faculty member, continuing his research.

Porter's method of splitting the molecules used **enzymes** that acted as chemical knives, breaking apart amino acids. In 1961 Edelman and his colleague, M. D. Poulik succeeded in splitting IgG—one of the most studied varieties of immunoglobulin in the blood—into two components by using a method known as "reductive cleavage." The technique allowed them to divide IgG into what are known as light and heavy chains. Data from their experiments and from those of the Czech researcher, Frantisek Franek, established the intricate nature of the antibody's "active sight." The sight occurs at the folding of the two chains, which forms a unique pocket to trap the antigen. Porter combined these findings with his, and, in 1962, announced that the basic structure of IgG had been determined. Their experiments set off a flurry of research into the nature of antibodies in the 1960s. Information was shared throughout the scientific community in a series of informal meetings referred to as "Antibody Workshops," taking place across the globe. Edelman and Porter dominated the discussions, and their work led the way to a wave of discoveries.

Still, a key drawback to research remained. In any naturally obtained immunoglobulin sample a mixture of ever so slightly different molecules would reduce the overall purity. Based on a crucial finding by Kunkel in the 1950s, Porter and Edelman concentrated their study on myelomas, cancers of the immunoglobulin-producing cells, exploiting the unique nature of these cancers. Kunkel had determined that since all the cells produced by these cancerous myelomas were descended from a common ancestor they would produce a homogeneous series of antibodies. A pure sample could be isolated for experimentation. Porter and Edelman studied the amino acid sequence in subsections of different myelomas, and in 1965, as Edelman would later describe it: "Mad as we were, [we] started on the whole molecule." The project, completed in 1969, determined the order of all 1,300 amino acids present in the protein, the longest sequence determined at that time.

Throughout the 1970s, Edelman continued his research, expanding it to include other substances that stimulate the immune system, but by the end of the decade the principle he and Poulik uncovered led him to conceive a radical theory of how the **brain** works. Just as the structurally limited immune system must deal with myriad invading organisms, the brain must process vastly complex sensory data with a theoretically limited number of switches, or **neurons**.

Rather than an incoming sensory signal triggering a predetermined pathway through the nervous system, Edelman theorized that it leads to a selection from among several choices. That is, rather than seeing the nervous system as a relatively fixed biological structure, Edelman envisioned it as a fluid system based on three interrelated stages of functioning.

In the formation of the nervous system, cells receiving signals from others surrounding them fan out like spreading ivy—not to predetermined locations, but rather to regions determined by the concert of these local signals. The signals regulate the ultimate position of each cell by controlling the production of a cellular glue in the form of cell-adhesion molecules. They anchor neighboring groups of cells together. Once established, these cellular connections are fixed, but the exact pattern is different for each individual.

The second feature of Edelman's theory allows for an individual response to any incoming signal. A specific pattern of neurons must be made to recognize the face of one's grandmother, for instance, but the pattern is different in every brain. While the vast complexity of these connections allows for some of the variability in the brain, it is in the third feature of the theory that Edelman made the connection to immunology. The neural networks are linked to each other in layers. An incoming signal passes through and between these sheets in a specific pathway. The pathway, in this theory, ultimately determines what the brain experiences, but just as the immune system modifies itself with each new incoming virus, Edelman theorized that the brain modifies itself in response to each new incoming signal. In this way, Edelman sees all the systems of the body being guided in one unified process, a process that depends on organization but that accommodates the world's natural randomness.

Dr. Edelman has received honorary degrees from a number of universities, including the University of

Pennsylvania, Ursinus College, Williams College, and others. Besides his Nobel Prize, his other academic awards include the Spenser Morris Award, the Eli Lilly Prize of the American Chemical Society, Albert Einstein Commemorative Award, California Institute of Technology's Buchman Memorial Award, and the Rabbi Shai Schaknai Memorial Prize.

A member of many academic organizations, including New York and National Academy of Sciences, American Society of Cell Biologists, Genetics Society, American Academy of Arts and Sciences, and the American Philosophical Society, Dr. Edelman is also one of the few international members of the Academy of Sciences, Institute of France. In 1974 he became a Vincent Astor Distinguished Professor, serving on the board of governors of the Weizmann Institute of Science and is also a trustee of the Salk Institute for Biological Studies. Dr. Edelman married Maxine Morrison on June 11, 1950; the couple had two sons and one daughter.

See also Antigens and antibodies; Immunity

EDEMA

An edema is an accumulation of **extracellular fluid** (also known as **interstitial fluid**) that causes swelling of **tissue**.

There are a number of physiological abnormalities that can result in a rise in interstitial fluid pressure and edema.

Low concentration of proteins in **blood plasma** decreases osmotic pressure that normally acts to prevent the leakage of fluid from blood vessels into the extracelluar tissue space (i.e., the spaces between tissue cells). In cases of trauma or inflammation, there is usually an increase in the permeability of the capillary wall that results in greater fluid passage into tissue spaces.

In addition to infusions of fluid that result in edema, poor drainage of lymphatic fluid inhibits elimination of extracelluar fluid. The tropical disease elephantiasis, a result of blockage of lymphatic drainage by infecting nematodes, produces gigantic swellings in the extremities and scrotum. In extreme cases, scrotal enlargement is so profound a wheelbarrow or other device must be used to carry the scrotum.

Edemas are often characterized by the edematous system or region. For example, **brain** edemas result from increased fluid in brain tissue or ventricles.

Edemas in the respiratory systems can be life threatening. In cases of inflammation of tissue in the respiratory tract, edema may act to block or obstruct the airway. True pulmonary edema result from an accumulation of fluid in the tissue and alveolar spaces in the lung.

Congestive **heart** failure can result from edema in cardiac tissue and/or the pulmonary system.

One form of edema that plagues mountain climbers and other working at high altitude is termed High Altitude Pulmonary Edema (HAPE) is characterized by a build-up of excessive fluid in the **lungs**. HAPE sufferers have trouble **breathing** and fatigue easily. The pulmonary edema results from dramatically increased blood flow through the lungs in an effort to compensate for the low pressure of oxygen at alti-

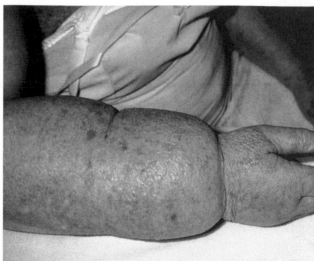

Lymphedema of the arm, caused by an accumulation of lymphatic and extracellular fluid, along with impaired lymphatic drainage. *Photograph by Dr. P. Marazzi. National Audubon Society Collection/Photo Researchers, Inc. Reproduced by permission.*

tude is a result of greatly increased blood flow through the lungs and increased blood pressure.

Diuretics work to counteract some edema by increasing the excretion of salts in the urine. Conversely, ingestion of sodium, usually ingested in the form of table salt (sodium chloride, NaCl) acts to increase fluid retention.

In many cases, edema in the extremities is relieved by elevation that allows gravity to assist in the drainage of blood and lymphatic fluids.

See also Acclimatization; Breathing; Fluid transport; Homeostatic mechanisms; Inflammation of tissues; Lymphatic system; Malaria and the physiology of parasitic inflections; Osmotic equilibria between intercellular and extracellular fluids; Renal system

EDWARDS, ROBERT G. (1925-   )
British physiologist

Robert Geoffrey Edwards and Patrick Christopher Steptoe (1913-1988) pioneered *in vitro* **fertilization** (IVF), making the birth of the first "test-tube baby" possible in 1978. By quickly transferring the oocyte (the egg prior to maturation) to an optimal cultural medium, Edwards was able to replicate the conditions necessary for an egg and **sperm** to survive outside the womb.

Edwards was born in 1925, the son of Samuel and Margaret Edwards. He attended the University of Wales from 1948 to 1951, and the University of Edinburgh from 1951 to 1957. He then worked for a year as research fellow at the California Institute of Technology (Cal Tech) before joining the staff at the National Institute of Medical Research in Mill Hill, England, in 1958.

Edwards took a position at the University of Glasgow in 1962, but moved to Cambridge University the following year. In 1965, he served as visiting scientist at Johns Hopkins University, and in 1966 at the University of North Carolina. Later, he returned to Cambridge, where he became Ford Foundation reader (instructor) in physiology in 1969, a position he held until 1985. While at Cambridge in 1968, Edwards met P. C. Steptoe, with whom he began an important collaboration.

By analyzing the conditions necessary for an egg and sperm to survive outside the womb, Edwards was able to develop an appropriate medium, calling it "a magic culture fluid," in which to achieve fertilization. In 1971, he and Steptoe performed their first attempt to implant a fertilized egg in a patient. They were not successful, however, until the birth of Louise Brown, dubbed the first "test-tube" baby, in July 1978. The IVF method developed by Edwards and Steptoe soon gained wide acceptance, and proved successful in dealing with a number of types of infertility.

After serving as visiting scientist at the Free University of Brussels in 1984, Edwards became professor of human reproduction at Cambridge in 1985. He remained in that position until his retirement in 1989, after which he became a professor emeritus. Together with Steptoe, he established the Bourne Hallam Clinic, and served as its scientific director from 1988 to 1991.

Edwards is Extraordinary Fellow of Churchill College, and served as chair of the European Society of Human Reproduction and Embryology from 1984 to 1986. He is a member and honorary member of several other professional societies, and has received awards from around the world. His publications include *A Matter of Life* (1980), written with Steptoe, and several other works.

See also Infertility (female); Oogenesis; Sexual reproduction

EEG · *see* ELECTROENCEPHALOGRAPH (EEG)

EHRLICH, PAUL (1854-1915)

German immunologist

Paul Ehrlich's pioneering experiments with cells and body **tissue** revealed the fundamental principles of the **immune system**, and established the legitimacy of chemotherapy—the use of chemical drugs to treat disease. His discovery of a drug that cured syphilis saved many lives and demonstrated the potential of systematic drug research. Erlich's studies of dye reactions in **blood** cells helped establish hematology—the scientific field concerned with blood and blood-forming organs—as a recognized discipline. Many of the new terms he coined as a way to describe his innovative research, including "chemotherapy," are still in use. From 1877 to 1914, Ehrlich published 232 papers and books, won numerous awards, and received five honorary degrees. In 1908, Ehrlich received the Nobel Prize in physiology or medicine. Along with **Robert**

Koch, Nobel Prize winner in 1905, and **Emil von Behring**, Nobel Prize winner in 1901, he is considered one of the "Big Three" in early modern medicine. Ehrlich, Koch, and von Behring were contemporaries and frequent collaborators.

Ehrlich was born in Strehlen, Silesia, once a part of Germany, but now a part of Poland known as Strzelin. He was the fourth child after three sisters in a Jewish family. His father, Ismar Ehrlich, and mother, Rosa Weigert, were both innkeepers. As a boy, Ehrlich was influenced by several relatives who studied science. His paternal grandfather, Heimann Ehrlich, made a living as a liquor merchant but kept a private laboratory and gave lectures on science to the citizens of Strehlen. Karl Weigert, cousin of Ehrlich's mother, became a well-known pathologist. Ehrlich, who was close friends with Weigert, often joined his cousin in his lab, where he learned how to stain cells with dye in order to see them better under the microscope. Ehrlich's research into the dye reactions of cells continued during his time as a university student. He studied science and medicine at the universities of Breslau, Strasbourg, Freiburg, and Leipzig. Although Ehrlich conducted most of his course work at Breslau, he submitted his final dissertation to the University of Leipzig, which awarded him a medical degree in 1878.

Ehrlich's 1878 doctoral thesis, "Contributions to the Theory and Practice of Histological Staining," suggests that even at this early stage in his career he recognized the depth of possibility and discovery in his chosen research field. In his experiments with many dyes, Ehrlich had learned how to manipulate chemicals in order to obtain specific effects: Methylene blue dye, for example, stained nerve cells without discoloring the tissue around them. These experiments with dye reactions formed the backbone of Ehrlich's career and led to two important contributions to science. First, improvements in staining permitted scientists to examine cells—healthy or unhealthy—and microorganisms, including those that caused disease. Ehrlich's work ushered in a new era of medical diagnosis and **histology** (the study of cells), which alone would have guaranteed Ehrlich a place in scientific history. Secondly, and more significantly from a scientific standpoint, Ehrlich's early experiments revealed that certain cells have an affinity to certain dyes. To Ehrlich, it was clear that chemical and physical reactions were taking place in the stained tissue. He theorized that chemical reactions governed all biological life processes. If this were true, Ehrlich reasoned, then chemicals could perhaps be used to heal diseased cells and to attack harmful microorganisms. Ehrlich began studying the chemical structure of the dyes he used and postulated theories for what chemical reactions might be taking place in the body in the presence of dyes and other chemical agents. These efforts would eventually lead Ehrlich to study the immune system.

Upon Ehrlich's graduation, medical clinic director Friedrich von Frerichs immediately offered the young scientist a position as head physician at the Charite Hospital in Berlin. Frerichs recognized that Ehrlich, with his penchant for strong cigars and mineral water, was a unique talent, one that should be excused from clinical work and be allowed to pursue his research uninterrupted. The late nineteenth century was a time when infectious diseases like cholera and typhoid fever were incurable and fatal. Syphilis, a sexually transmitted disease caused by a then unidentified microorganism, was an epidemic, as was tuberculosis, another disease whose cause had yet to be named. To treat human disease, medical scientists knew they needed a better understanding of harmful microorganisms.

At the Charite Hospital, Ehrlich studied blood cells under the microscope. Although blood cells can be found in a perplexing multiplicity of forms, Ehrlich was with his dyes able to begin identifying them. His systematic cataloging of the cells laid the groundwork for what would become the field of hematology. Ehrlich also furthered his understanding of chemistry by meeting with professionals from the chemical industry. These contacts gave him information about the structure and preparation of new chemicals and kept him supplied with new dyes and chemicals.

Ehrlich's slow and steady work with stains resulted in a sudden and spectacular achievement. On March 24, 1882, Ehrlich had heard Robert Koch announce to the Berlin Physiological Society that he had identified the bacillus causing tuberculosis under the microscope. Koch's method of staining the bacillus for study, however, was less than ideal. Ehrlich immediately began experimenting and was soon able to show Koch an improved method of staining the tubercle bacillus. The technique has since remained in use.

In March, 1885, Frerichs committed suicide and Ehrlich suddenly found himself without a mentor. Von Frerichs's successor as director of Charite Hospital, Karl Gerhardt, was far less impressed with Ehrlich and forced him to focus on clinical work rather than research. Though complying, Ehrlich was highly dissatisfied with the change. Two years later, Ehrlich resigned from the Charite Hospital, ostensibly because he wished to relocate to a dry climate to cure himself of tuberculosis. The mild case of the disease, which Ehrlich had diagnosed using his staining techniques, was almost certainly contracted from cultures in his lab. In September of 1888, Ehrlich and his wife embarked on an extended journey to southern Europe and Egypt and returned to Berlin in the spring of 1889 with Ehrlich cured.

In Berlin, Ehrlich set up a small private laboratory with financial help from his father-in-law, and in 1890 he was honored with an appointment as Extraordinary Professor at the University of Berlin. In 1891, Ehrlich accepted Robert Koch's invitation to join him at the Institute for Infectious Diseases, newly created for Koch by the Prussian government. At the institute, Koch began his immunological research by demonstrating that mice fed or injected with the toxins ricin and abrin developed antitoxins. He also proved that antibodies were passed from mother to offspring through breast milk. Ehrlich joined forces with Koch and von Behring to find a cure for diphtheria, a deadly childhood disease. Although von Behring had identified the antibodies to diphtheria, he still faced great difficulties transforming the discovery into a potent yet safe cure for humans. Using blood drawn from horses and goats infected with the disease, the scientists worked together to concentrate and purify an effective antitoxin. Ehrlich's particular contribution to the cure was his method of measuring an effective dose.

The commercialization of a diphtheria antitoxin began in 1892 and was manufactured by Höchst Chemical Works. Royalties from the drug profits promised to make Ehrlich and von Behring wealthy men. But Ehrlich—possibly at von Behring's urging—accepted a government position in 1885 to monitor the production of the diphtheria serum. Conflict-of-interest clauses obligated Ehrlich to withdraw from his profit-sharing agreement. Forced to stand by as the diphtheria antitoxin made von Behring a wealthy man, he and von Behring quarreled and eventually parted. Although it is unclear whether or not bitterness over the royalty agreement sparked the quarrel, it certainly couldn't have helped a relationship that was often tumultuous. Although the two scientists continued to exchange news in letters, both scientific and personal, the two scientists never met again.

In June of 1896, the Prussian government invited Ehrlich to direct its newly created Royal Institute for Serum Research and Testing in Steglitz, a suburb of Berlin. For the first time, Ehrlich had his own institute. In 1896, Ehrlich was invited by Franz Adickes, the mayor of Frankfurt, and by Friedrich Althoff, the Prussian Minister of Educational and Medical Affairs, to move his research to Frankfurt. Ehrlich accepted and the Royal Institute for Experimental Therapy opened on November 8, 1899. Ehrlich was to remain as its director until his death sixteen years later. The years in Frankfurt would prove to be among Ehrlich's most productive.

In his speech at the opening of the Institute for Experimental Therapy, Ehrlich seized the opportunity to describe in detail his "side-chain theory" of how antibodies worked. "Side-chain" is the name given to the appendages on benzene molecules that allow it to react with other chemicals. Ehrlich believed all molecules had similar side-chains that allowed them to link with molecules, nutrients, infectious toxins and other substances. Although Ehrlich's hypothesis was incorrect, his efforts to prove it led to a host of new discoveries and guided much of his future research.

The move to Frankfurt marked the dawn of **chemotherapy** as Ehrlich erected various chemical agents against a host of dangerous microorganisms. In 1903, scientists had discovered that the cause of sleeping sickness, a deadly disease prevalent in Africa, was a species of trypanosomes (parasitic protozoans). With help from Japanese scientist Kiyoshi Shiga, Ehrlich worked to find a dye that destroyed trypanosomes in infected mice. In 1904, he discovered such a dye, which was dubbed "trypan red."

Success with trypan red spurred Ehrlich to begin testing other chemicals against disease. To conduct his methodical and painstaking experiments with an enormous range of chemicals, Ehrlich relied heavily on his assistants. To direct their work, he made up a series of instructions on colored cards in the evening and handed them out each morning. Although such a management strategy did not endear him to his lab associates—and did not allow them opportunity for their own research—Ehrlich's approach was often successful. In one famous instance, Ehrlich ordered his staff to disregard the accepted notion of the chemical structure of atoxyl and to instead proceed in their work based on his specifications of the chemical. Two of the three medical scientists working with

Ehrlich were appalled at his scientific heresy and ended their employment at the laboratory. Ehrlich's hypothesis concerning atoxyl turned out to have been correct and would eventually lead to the discovery of a chemical cure for syphilis.

In September of 1906, Ehrlich's laboratory became a division of the new Georg Speyer Haus for Chemotherapeutical Research. The research institute, endowed by the wealthy widow of Georg Speyer for the exclusive purpose of continuing Ehrlich's work in chemotherapy, was built next to Ehrlich's existing laboratory. In a speech at the opening of the new institute, Ehrlich used the phrase "magic bullets" to illustrate his hope of finding chemical compounds that would enter the body, attack only the offending microorganisms or malignant cells, and leave healthy tissue untouched. In 1908, Ehrlich's work on immunity, particularly his contribution to the diphtheria antitoxin, was honored with the Nobel Prize in physiology or medicine. He shared the prize with Russian bacteriologist **Elie Metchnikoff.**

By the time Ehrlich's lab formally joined the Speyer Haus, he had already tested over 300 chemical compounds against trypanosomes and the syphilis spirochete (distinguished as slender and spirally undulating **bacteria**). With each test given a laboratory number, Ehrlich was testing compounds numbering in the nine hundreds before realizing that "compound 606" was a highly potent drug effective against relapsing fever and syphilis. Due to an assistant's error, the potential of compound 606 had been overlooked for nearly two years until Ehrlich's associate, Sahashiro Hata, experimented with it again. On June 10, 1909, Ehrlich and Hata filed a patent for 606 for its use against relapsing fever.

The first favorable results of 606 against syphilis were announced at the Congress for Internal Medicine held at Wiesbaden in April 1910. Although Ehrlich emphasized he was reporting only preliminary results, news of a cure for the devastating and widespread disease swept through the European and American medical communities and Ehrlich was besieged with requests for the drug. Physicians and victims of the disease clamored at his doors. Ehrlich, painfully aware that mishandled dosages could blind or even kill patients, begged physicians to wait until he could test 606 on ten or twenty thousand more patients. But there was no halting the demand and the Georg Speyer Haus ultimately manufactured and distributed 65,000 units of 606 to physicians all over the globe free of charge. Eventually, the large-scale production of 606, under the commercial name "Salvarsan," was taken over by Höchst Chemical Works. The next four years, although largely triumphant, were also filled with reports of patients' deaths and maiming at the hands of doctors who failed to administer Salvarsan properly.

In 1913, in an address to the International Medical Congress in London, Ehrlich cited trypan red and Salvarsan as examples of the power of chemotherapy and described his vision of chemotherapy's future. The City of Frankfurt honored Ehrlich by renaming the street in front of the Georg Speyer Haus "Paul Ehrlichstrasse." Yet in 1914, Ehrlich was forced to defend himself against claims made by a Frankfurt newspaper, *Die Wahrheit* (The Truth), that Ehrlich was testing Salvarsan on prostitutes against their will, that the drug was a

fraud, and that Ehrlich's motivation for promoting it was personal monetary gain. In June 1914, Frankfurt city authorities took action against the newspaper and Ehrlich testified in court as an expert witness. Ehrlich's name was finally cleared and the newspaper's publisher sentenced to a year in jail, but the trial left Ehrlich deeply depressed. In December, 1914, he suffered a mild stroke.

Ehrlich's health failed to improve and the start of World War I had further discouraged him. Afflicted with **arteriosclerosis**, a disease of the **arteries**, his health deteriorated rapidly. He died in Bad Homburg, Prussia (now Germany), on August 20, 1915, after a second stroke. Ehrlich was buried in Frankfurt. Following the German Nazi era—during which time Ehrlich's widow and daughters were persecuted as Jews before fleeing the country and the sign marking Paul Ehrlichstrasse was torn down—Frankfurt once again honored its famous resident. The Institute for Experimental Therapy changed its name to the Paul Ehrlich Institute and began offering the biennial Paul Ehrlich Prize in one of Ehrlich's fields of research as a memorial to its founder.

See also Immunity; Immunology; Pharmacology

EIJKMAN, CHRISTIAAN (1858-1930)
Dutch physician

Christiaan Eijkman was a pioneer in the study of diseases that result from deficiencies in a patient's diet. His major contribution was the discovery that the lack of some vital substance in food caused a disease in chickens similar to beriberi in man. For this work, which helped lead to the concept of **vitamins**, he received the Nobel Prize in physiology or medicine in 1929.

Eijkman was born in Nijkerk, Netherlands. He was the seventh child of a schoolmaster father, also named Christiaan Eijkman, and Johanna Alida Pool Eijkman. Theirs was a family of academically gifted sons whose professional careers would encompass the fields of chemistry, linguistics, and radiology. Soon after Eijkman's birth, the family moved to Zaandam, where he received his early education. In 1875, he began training as a military medical officer at the University of Amsterdam. There, his own ability quickly made itself apparent; he received a medical degree with high honors in 1883.

That same year, Eijkman was dispatched by the army to the Dutch East Indies (now Indonesia), where he served on the islands of Java and Sumatra until a severe case of **malaria** forced him to return home in 1885. This proved to be a blessing in disguise, however, for Eijkman used his recuperation time to study the new science of bacteriology under one of the field's founders, **Robert Koch**, in Berlin. A year later, Eijkman was strong enough to return to the East Indies as part of a Dutch government mission to investigate the disease beriberi. He owed his spot on the research team to personal contacts he had made while working in Koch's laboratory.

Christian Eijkman. *Corbis-Bettmann. Reproduced by permission.*

The name beriberi comes from the Sinhalese word for "extreme weakness." The disease, characterized by impairment of the nerves, **heart**, and digestive system, can cause such symptoms as paralysis, numbness, swelling, and difficulty **breathing**. At the time, the illness was spreading rapidly in Asia, and many European doctors were convinced that the epidemic was bacterial in origin. Eijkman, too, erroneously believed that his search for the source of beriberi would ultimately lead to a microorganism.

In 1888, Eijkman was appointed head of the Javanese Medical School and director of a bacteriological laboratory in Batavia (now Djakarta), where he made a chance observation that would change the course of medicine. An illness, very much like beriberi, suddenly broke out among the laboratory chickens, then just as mysteriously went away. Eijkman learned that an attendant had, for a short time, been feeding the birds cooked white rice from the hospital kitchen. The disappearance of the birds' symptoms coincided with the arrival of a new cook, who refused to allow hospital rice to be used for this purpose. Eijkman soon found that he could produce the disease at will by feeding the chickens hulled and polished rice, and he could just as readily cure it with a diet of whole rice. Eijkman's observations were the starting point for a line of inquiry that led others to an understanding of and treatment for beriberi. Later researchers traced the disease to a lack of thiamine (vitamin B1), a vitamin found in the hulls of unpolished rice. Ironically, Eijkman failed to grasp the true meaning

of his findings. He hypothesized that the rice hulls contained a substance that neutralized a toxin carried in or produced by the rice grains. Nevertheless, Eijkman had shown that not every illness could be explained by the then-revolutionary germ theory of disease, and it is for this achievement that he is primarily remembered today.

Eijkman's other important work in the Dutch East Indies was his study of the physiology of people living in the region. He disproved many widely held notions about the effects of tropical life on Europeans, which in the past had led to several unnecessary precautions. For example, he demonstrated that expected differences in **metabolism**, respiration, perspiration, and **temperature regulation** between Europeans and natives of the tropics did not, in fact, exist. In 1896, Eijkman once again returned to the Netherlands on sick leave. Two years later, he assumed the post of professor of public health and forensic medicine at the University of Utrecht, where he conducted fermentation tests by examining water for signs of bacterial pollution produced by human and animal defecation. As a lecturer, Eijkman was known for his clear demonstrations, perhaps the result of decades of hands-on experience in the lab.

Eijkman did not confine his interests to the university, however. He was actively involved in such issues as the public water supply, housing, school hygiene, and physical education. As a member of his nation's Health Council and Health Commission, he fought against alcoholism and tuberculosis. In recognition of Eijkman's many contributions to society, the Dutch government conferred several orders of knighthood upon him. He was also made a member of the Netherlands' Royal Academy of Sciences in 1907. But the 1929 Nobel Prize, which he shared with Sir Frederick Gowland Hopkins of Great Britain, the discoverer of growth-stimulating vitamins, was his crowning honor. Sadly, though, the seventy-one-year-old Eijkman, who had retired a year earlier, was by this time too sick to travel to Stockholm to receive his prize in person.

Eijkman's first wife died in 1886, soon after his initial return from the East Indies, and just three years after their marriage. Eijkman subsequently remarried 1888, in Batavia. Their son grew to become a physician. Eijkman himself died on November 5, 1930, in Utrecht, Netherlands, after a protracted illness.

EINTHOVEN, WILLEM (1860-1927)
Dutch physician

Although trained in medicine, Willem Einthoven was always interested in physics, and his greatest contributions to science involve the application of physical principles to the development of new instruments and techniques in physiological studies. One such instrument, the string galvanometer, made possible the first valid and reliable electrocardiogram, thereby providing physicians with one of their most valuable tools for the study of cardiovascular disorders. For his invention of the string galvanometer, Einthoven was awarded the Nobel Prize in physiology or medicine in 1924.

Einthoven was born in Semarang, Java, in what was then the Dutch East Indies and is now Indonesia. His father, Jacob Einthoven, was a physician in Semarang. When Jacob died in 1866, his wife, Louise M. M. C. de Vogel, returned to her native Holland with her six children, Willem included. The family settled in Utrecht, where young Willem attended local grammar and high schools. Upon graduation from high school in 1879, he enrolled in the medical program at the University of Utrecht. Six years later, Einthoven received his Ph.D. in medicine, having written his doctoral thesis on the use of color differentiation techniques in spectroscopic analysis. He was immediately offered an appointment as professor of physiology at the University of Leiden, a job he actually began after passing his final state medical examinations on February 24, 1886. Einthoven would remain at his post at the University of Leiden for the next forty-two years until his death in 1927. Also in 1886, Einthoven was married to Frédérique Jeanne Louise de Vogel, a cousin, with whom he would father four children: a son and three daughters.

Perhaps the most significant feature of Einthoven's career is the way he made use of his interest in, and knowledge of, physics in his study of physiological problems. The research for which he is best known involved the detection of the association between electrical currents and the beating human **heart**. Physiologists in the 1880s knew that each contraction of the heart muscle is accompanied by electrical changes in the body, but no precise quantitative data existed for this phenomenon. At that time, the only equipment available to measure electrical charges in the body was not sensitive enough to detect the minute changes in potential difference—the amount of energy released—associated with a heartbeat. The most commonly used device, a capillary electrometer, made use of the rise and fall of a thin column of mercury in a glass tube. Unfortunately, the measurement process of such an instrument took place too slowly to determine actual changes in potential difference resulting from muscular contractions.

Around 1903, Einthoven invented an improved method for measuring such changes: the string galvanometer. Einthoven's new instrument consisted of a very thin quartz wire suspended in a magnetic field. An electric current, even one as small as those associated with muscular contraction, caused a deflection of the wire. By focusing a moving picture camera on the wire, Einthoven could obtain a visual record of the movement of the wire as it was displaced by electrical currents from the heart.

As a result of his research, Einthoven was able to detect and identify a number of different kinds of electrical waves associated with a beating heart, waves that he originally labeled as P, Q, R, S, and T waves. He was eventually able to show that some of these waves result from contractions and electrical changes in the atria and others from contractions and electrical changes in the ventricles of the heart.

Einthoven published a complete description of his string galvanometer in 1909 as *Die Konstruktion des Sitengalvanometers*. In this work, he outlined a method for using the galvanometer to record heart action using three combinations of electrode placement: right hand to left hand, right hand to

Willem Einthoven. *Archive Photos, Inc. Reproduced by permission.*

left foot, and left hand to left foot. Such arrangements of the electrodes could be used, he showed, to locate the position of the heart and to detect any abnormalities in its function.

In *Nobel Prize Winners,* Einthoven's biographer points out that the invention of the string electrode "revolutionized the study of heart disease." For his accomplishment, Einthoven was awarded the Nobel Prize for physiology or medicine in 1924. Interestingly enough, the basic principles of electrocardiography, while first developed by Einthoven, were also derived independently a short time later by the English physicians Sir Thomas Lewis, Sir William Ogler, and James Herrick.

Einthoven continued to refine, develop, and extend the applications of his string galvanometer throughout the rest of his career. For example, later in his life he modified the device so that it could be used to receive long-distance radiotelegraph signals and to measure changes in electric potential in nerves. He was also very popular as a lecturer and made a number of trips to Europe and the United States to talk about his work. Among the many honors Einthoven received was his election as an honorary member of the Physiological Society in 1924, and his induction into England's prestigious Royal Society two years later in 1926.

Einthoven died in Leiden, Netherlands in 1927. His obituary in the periodical *Nature* spoke of the "grace, beauty, and simplicity of his character." Although he left few students behind, Einthoven's impact on the development of electrocardiography was profound.

See also Cardiac cycle; Cardiac muscle; Electrocardiograms (ECG); Heart: rhythm control and impulse conduction

ELASTIC CARTILAGE

Elastic **cartilage** is found in regions of the body where support, resiliency and flexibility are important. Such places include the lobe of the ear, the eustachian tube of the middle **ear**, the epiglottis and in parts of the **larynx**.

Elastic cartilage is similar to another type of cartilage called **hyaline cartilage**. Both types of cartilage contain cells called chondrocytes that are widely spaced in a fluid-filled matrix known as the lacunae (also called ground substance). The lacunae contain a gelatinous substance called chondroitin sulfate. Elastic cartilage differs markedly from hyaline cartilage as the lacunae are also enriched in elastic fibers that run through the matrix in all directions. These fibers compact the spatial arrangement of the chondrocytes, so that they are closer together than those found in hyaline cartilage.

Under microscopic magnification, elastic cartilage appears as chondrocytes dispersed in a network of fibers. Staining for elastin in particular will reveal the extensive network of dense fibrous bundles. This appearance is the best identifier to differentiate elastic cartilage from hyaline cartilage.

Like all cartilage, elastic cartilage exhibits properties that place it between connective **tissue** and bone. In other words, it is tough but flexible. The large water content of elastic and other types of cartilage—about 80% water by weight—allows the cartilage to be compressed and then to regain its shape.

The elasticity of this type of cartilage are well suited to the places where it is found. For example, the epiglottis must be tough, but flexible enough to be capable of closure if necessary. Likewise, the Eustachian tube must be able to withstand increased air pressures. Having a tube constructed of a material that can expand and contract in response to changing internal pressures is vital to the intended function.

Biomedical researchers are taking advantage of the physicochemical properties of elastic cartilage. One advantageous feature is that elastic cartridge does not stimulate much of an immune response from the body. Also, the malleable nature of the cartilage allows it to be packed into a configuration that, in preliminary studies, shows potential as an implant or through direct injection into tissue. Researchers hope that engineered elastic cartilage may be used for reconstructive surgery, such as on facial defects.

See also Cartilaginous joints; Fibrocartilage

ELECTROCARDIOGRAM (ECG)

An electrocardiogram (ECG) is a graphic representation of electrical activity in the **heart** that conveys information related to cardiac rhythms and the pumping of **blood** through the chambers of the heart. Electrocardiograms are also termed EKGs. The electrical activity measured relates to changes in electrical potentials during the **cardiac cycle**.

ECG tracing can provide information related to heart ventricle size, position, and capacity to pump blood. This information is vital to clinicians attempting to assess damage to cardiac **tissue** and function. Variance in electrical conduction patterns can also provide information regarding conductive patterns in the heart and can be related to abnormalities in electrically conductive concentrations of salt and mineral ions.

The heart produces weak electrical currents that can be detected by electrodes that are attached by leads to strategic monitoring locations on the surface of the body. The ECG is a passive test and the electrodes are often cleaned and the area of attachment covered with a conductive gel to improve reception. In order to provide the graphical tracing caused by the movement of a stylus pen over specially lined moving paper, the current received from the heart is amplified thousands of times.

Normal tracings or waves are divided into components termed, in order that they occur, the P, Q, R, S, and T waves. The QRS waves are often described as the QRS complex. The P wave corresponds with the start of atrial contraction and the QRS complex corresponds to the initiation of ventricular contractions. The T wave corresponds to the refractory period and recovery period following ventricular contraction.

In the normal healthy individual, there are approximately 60–100 P waves (corresponding to heartbeats) per minute. Bradycardia describes a rate of less that 60 P waves per minute and tachycardia is a condition where there are in excess of 100 P waves per minute in a resting subject. When there is a great variability (usually greater than ten percent) in P waves from one tracing to the next—or within one tracing—this indicates an arrhythmia in sinus rhythm.

Careful measurement of the heights of waves and intervals between particular wave components can often yield important diagnostic insights into abnormal ventricular rhythms, ventricular hypertrophy, myocardial infarction, myocarditis, and other pathological conditions. For example, enlargements in the atrium (hypertrophy) or in abnormal atrial rhythm are often reflected in P wave abnormalities. Conduction abnormalities (e.g., bundle branch blocks) often result in abnormal QRS complex patterns.

Holter monitors are portable ECG units that can be used to monitor long-term cardiac function. The patient carries the portable unit during normal activities.

See also Cardiac disease; Electroencephalo-graph (EEG); Heart: rhythm control and impulse conduction; Heart; S-A node

ELECTROENCEPHALOGRAM (EEG)

An electroencephalogram, usually abbreviated EEG, is a medical test that records electrical activity in the **brain**. During the test, the brain's spontaneous electrical signals are traced onto

A woman shown during an electroencephalogram, a test that records and measures the electrical activity of the brain. *Photograph by Catherine Pouedras. Science Photo Library, National Audubon Society Collection/Photo Researchers, Inc. Reproduced by permission.*

paper. The electroencephalograph is the machine that amplifies and records the electrical signals from the brain. The electroencephalogram is the paper strip the machine produces. The EEG changes with disease or brain disorder, such as epilepsy, so it can be a useful diagnostic tool, but usually must be accompanied by other diagnostic tests to be definitive.

To perform an EEG, electrodes (wires designed to detect electrical signals) are placed on the **cranium** by attaching the wire with a special adhesive. The electrodes are placed in pairs so that the difference in electric potential between them can be measured. The wires are connected to the electroencephalograph, where the signal is amplified and directed into pens that record the waves on a moving paper chart. The tracing appears as a series of peaks and troughs drawn as lines by the recording pens.

Basic alpha waves, which originate in the cortex, can be recorded if the subject closes his eyes and puts his brain "at rest" as much as possible. Of course, the brain is never still, so some brain activity is going on and is recorded in waves of about 6–12 per second, with an average of about 10 per second. The voltage of these waves is from 5–100 microvolts. A microvolt is one-one millionth of a volt. Thus, a considerable amount of amplification is required to raise the voltage to a discernable level.

The rate of the waves, that is, the number that occur per second, appears to be a better diagnostic indicator than does the amplitude, or strength. Changes in the rate indicating a slowing or speeding up are significant, and unconsciousness occurs at either extreme. **Sleep**, stupor, and deep **anesthesia** are associated with slow waves and grand mal seizures cause an elevated rate of brain waves. The only time the EEG line is

straight and without any wave indication is at **death**. A person who is brain dead has a straight, flat EEG line.

The rates of alpha waves are intermediate compared with other waves recorded on the EEG. Faster waves, 14–50 waves per second, are lower in voltage than alpha waves are called beta waves. Very slow waves, averaging 0.5–5 per second, are delta waves. The slowest brain waves are associated with an area of localized brain damage such as may occur from a **stroke** or blow on the head.

The individual at rest and generating a fairly steady pattern of alpha waves can be distracted by a sound or touch. The alpha waves then flatten somewhat, that is, their voltage is less and their pattern becomes more irregular when the individual's attention is focused. Any difficult mental effort such as multiplying two four-digit numbers will decrease the amplitude of the waves, and any pronounced emotional excitement will flatten the pattern. The brain wave pattern will change to one of very slow waves, about three per second, in deep sleep.

Though the basic EEG pattern remains a standard one from person to person, each individual has his own unique EEG pattern. The same individual given two separate EEG tests weeks or months apart will generate the same alpha wave pattern, assuming the conditions of the tests are the same. Identical twins will both have the same pattern. One twin will virtually match the second twin to the extent that the two tracings appear to be from the same individual on two separate occasions.

Though the EEG is a useful diagnostic tool, its use in brain research is limited. The electrodes detect the activity of only a few **neurons** in the cortex out of the billions that are present. Electrode placement is standardized so the EEG can be interpreted by any trained neurologist. Also, the electrical activity being measured is from the surface of the cortex and not from the deeper areas of the brain.

Neuronal connections are established early in life and remain intact throughout one's lifetime. An interruption of those connections because of a stroke or accident results in their permanent loss. Sometimes, with great effort, alternative pathways or connections can be established to restore function to that area, but the original connection will remain lost.

The electroencephalogram is a means to assess the degree of damage to the brain in cases of trauma, or to measure the potential for seizure activity. It is used also in sleep studies to determine whether an individual has a sleep disorder and to study brain wave patterns during dreaming or upon sudden awakening.

The EEG is also a useful second-level diagnostic tool to follow-up a computerized tomogram (CT) scan to assist in finding the exact location of a damaged area in the brain. The EEG is one of a battery of brain tests available and is seldom used alone to make a diagnosis. The EEG tracing can detect an abnormality but cannot distinguish between, for example, a tumor and a thrombosis (site of deposit of a **blood** clot in an artery).

Although persons with frequent seizures are more likely to have an abnormal EEG than are those who have infrequent seizures, EEGs cannot be solely used to diagnose epilepsy. Approximately 10% of epilepsy patients will have a normal EEG. A normal EEG, therefore, does not eliminate brain damage or seizure potential, nor does an abnormal tracing indicate that a person has epilepsy. Something as simple as visual stimulation or rapid **breathing** (hyperventilation) may initiate abnormal electrical patterns in some patients.

See also Brain stem function and reflexes; Brain: Intellectual functions

ELECTROLYTES AND ELECTROLYTE BALANCE

Electrolytes are molecules, classified as salts, which are distributed in the fluids in the body. These salts are ionic; the sum of all the charged atomic constituents (electrons and protons) is not zero, but is either a positive charge or a negative charge. Positively charged ions are called cations and negatively charged ions are called anions. The most prominent examples of electrolytes found in the human body are potassium, sodium, **calcium**, and chloride.

Body fluids such as **blood**, **plasma** and the so-called **interstitial fluid** that flows between cells are much like seawater in composition, having a high concentration of sodium chloride ($NaCl$). The cation in $NaCl$ is Na^+ and the anion is Cl^-.

In a given compartment, the number of positively charged molecules will be balanced by the number of negatively charged molecules, so that the fluid will be electrically neutral. Preservation of neutrality is very important to the continued proper functioning of many parts of the body. For

example, elevated potassium levels can result in an irregular heartbeat, while decreased levels of potassium can cause paralysis. Excessive levels of sodium in the fluid circulating outside of body cells can cause fluid retention and swollen **joints**.

Potassium and sodium are referred to as micronutrients. Potassium is important in, as examples, maintaining pH balance, transfer of nutrients across cell membranes, maintaining proper water balance, and in the proper contraction of various muscle groups. These functions depend on being able to pass electrical impulses across cell membranes. Electrolytes help to maintain the voltage across membranes, because they can conduct electricity.

The physical effects of electrolyte imbalance are due to the phenomena of hypertonicity (when the electrolyte levels in the blood exceed those inside of the cells) and hypotonicity (when the electrolyte levels of the blood is lower than that inside the cells). In hypertonic conditions, water flows out of cells, to try to dilute the blood concentration. The water loss affects cell shape and function and can be lethal to the cells. In hypotonic conditions, water flows into the cells, again to try and dilute the electrolyte concentrations. The cells swell and can burst.

Intravenous solutions given to patients in hospitals are isotonic, the electrolyte concentrations are the same as in the blood.

The **kidneys** maintain electrolyte concentrations. Sometimes the kidneys require assistance to maintain electrolyte levels, as during heavy exercise when a lot of fluid can be lost from the body as sweat. Fluid replacement can restore the electrolyte concentration.

See also Action potential; Cell membrane transport; Cell structure; Sodium-potassium pump

ELIMINATION OF WASTE

The conversion of foodstuff to energy and various body processes produces excess compounds that cannot be used by the body. They must be removed or else the body will become toxic. The process of excretion involves finding and removing waste materials produced by the body.

The main **organs** in the human body responsible for excretion are the **lungs**, **kidneys** and the skin. The lungs dissipate waste gases that are carried to them in **blood** via **veins**. The gases diffuse across the membrane of the many **alveoli** in the lung and are subsequently passed out of the body during exhalation. The skin is used to remove dead cells, by sloughing off, and sweat, by evaporation. Liquid waste is removed from the body through the kidneys.

The pair of kidneys are small, about 3.9 in. (10 cm) long, and are shaped like beans. Blood is routed through the kidneys, which filter out unwanted water, minerals, and urea from the blood, forming a liquid called urine. The urine is funneled through two tubes called catheters to the bladder. The bladder is a storage facility. When full, muscle contractions force the urine out of the bladder and out of the body through the urethra. The 1–2.11 gal. (1–2 l) of urine produced by the kidneys each day is expelled in this manner, a process commonly referred to as urination.

The **liver** also participates in the elimination of wastes. Principally, the liver eliminates **bilirubin**, a toxic breakdown product of **hemoglobin** generated as red blood cells die. The bilirubin is secreted into **bile** and the bile is routed to the small intestine, to be eliminated via the urine or feces.

Feces (also called excreta or stool) is a solid waste product of **digestion**. The solid forms as waste material is moved through the large intestine by muscular contractions called peristalsis. In humans, approximately 8.8 oz. (250 g) of feces are produced each day. Typically the solid waste is comprised of digestive secretions, **enzymes**, fats, cell debris, **electrolytes**, water, and some protein. Also, up to 20% of the weight of the solid can be **bacteria**.

If the elimination of liquid or solid wastes are disrupted, the consequences for the health of the organism can be serious. For example, if kidneys are diseased and not functioning properly, the buildup of waste can necessitate a treatment known as dialysis. In dialysis, the patient's blood is pumped out of the body and through a dialysis machine. The machine acts as a filter instead of the kidneys. The cleansed blood is then pumped back into the patient. If dialysis fails to correct the problem, a kidney transplant may be necessary. Often the donor of a kidney is a sibling or some other close relation, whose genetic make-up is similar to the patient's.

See also Metabolic waste removal

ELION, GERTRUDE BELLE (1918-1999)
American biochemist

Gertrude Belle Elion's innovative approach to drug discovery furthered the understanding of cellular **metabolism** and led to the development of medications for **leukemia**, gout, herpes, **malaria**, and immunity disorders. Her name appears on 45 patents, and azidothymidine (AZT), the first drug approved for the treatment of **AIDS**, came out of her lab shortly after her 1983 retirement. Known for her intellectual brilliance and attention to detail, she was one of the few women who held a top post at a major pharmaceutical company. Elion worked at Wellcome Research Laboratories for nearly five decades. Her work, with colleague George H. Hitchings, was recognized with the Nobel Prize in physiology or medicine in 1988. The two developed a wide variety of drugs, including acyclovir (Zovirax) for herpes; azathioprine (Imuran) to help prevent rejection; allopurinol (Zyloprim) for gout; pyrimethamine (Daraprim) for malaria; and trimethoprim (part of Septra) for bacterial infections. Her Nobel award was notable for several reasons: few winners have been women, few have lacked a doctorate degree, and few have been industrial researchers.

Elion was born in New York City, the first of two children of Robert Elion and Bertha Cohen. Robert, a dentist, immigrated to the United States from Lithuania as a small boy. Bertha came to the United States from Russia at the age of 14.

Elion, an excellent student, graduated from high school at the age of 15. During her senior year, she had witnessed the painful **death** of her grandfather from stomach **cancer** and vowed to become a cancer researcher. Graduating at the height of the Great Depression after her parents lost everything in the stock market crash, she was able to attend college only because Hunter College in New York City offered free tuition to students with good grades. In college, she majored in chemistry because that seemed the best route to her goal.

Elion graduated Phi Beta Kappa from Hunter College in 1937 with a B.A. at the age of 19. Despite her outstanding academic record, her early efforts to find a job as a chemist failed. One laboratory after another told her that they had never hired a woman chemist. Her self-confidence shaken, Elion began secretarial school. That lasted only six weeks, until she landed a one-semester stint teaching **biochemistry** to nursing students and then took a position in a friend's laboratory. With the money she earned from these jobs, Elion began graduate school. To afford tuition, she continued to live with her parents and to work as a substitute science teacher in the public schools. In 1941, she graduated summa cum laude from New York University with an M.S. in chemistry.

Upon her graduation, Elion again faced difficulties finding work appropriate to her experience and abilities. The only job available to her was as a quality control chemist in a food laboratory, checking the color of mayonnaise and the acidity of pickles for the Quaker Maid Company. After a year and a half, she was finally offered a job as a research chemist at Johnson & Johnson. Unfortunately, her division closed six months after she arrived. The company offered Elion a new job testing the tensile strength of sutures, but she declined.

The start of World War II ushered in a new era of opportunity for Elion, as it did for many women of her generation. As men left their jobs to fight the war, women were encouraged to join the workforce. "It was only when men weren't available that women were invited into the lab," Elion told the *Washington Post*.

The war created an opening for Elion in the research lab of biochemist George Herbert Hitchings at Wellcome Research Laboratories in Tuckahoe, New York, a subsidiary of Burroughs Wellcome Company, a British firm. When they met, Elion was 26 years old and Hitchings was 39. Their working relationship began on June 14, 1944, and lasted for the rest of their careers. Each time Hitchings was promoted, Elion filled the spot he had just vacated, until she became head of the Department of Experimental Therapy in 1967, where she was to remain until her retirement 16 years later. Hitchings became vice president for research. Over the years, they have written many scientific papers together.

Settled in her job and thrilled by the breakthroughs occurring in the field of biochemistry, Elion took steps to earn a Ph.D., the so-called "union card" that all serious scientists are expected to have as evidence that they are capable of doing independent research. Only one school offered night classes in chemistry, the Brooklyn Polytechnic Institute (now Polytechnic University), so that's where Elion enrolled. Attending classes meant taking the train from Tuckahoe into Grand Central Station and transferring to the subway to

Brooklyn. Although the hour-and-a-half commute each way was exhausting, Elion persevered for two years, until the school accused her of not being a serious student and pressed her to attend full time. Forced to choose between school and her job, Elion had no choice but to continue working. Her relinquishment of the Ph.D. haunted her, until her lab developed its first successful drug, 6-mercaptopurine (6MP).

In the 1940s, Elion and Hitchings employed a novel approach to fighting the agents of disease. By studying the biochemistry of cancer cells, and of harmful **bacteria** and **viruses**, they hoped to understand the differences between the metabolism of abnormal and normal cells. In particular, they wondered whether there were differences in how the disease-causing cells used nucleic acids, the chemicals involved in the replication of **DNA**, to stay alive and to grow. Any differences they discovered might serve as a target point for a drug that could destroy the abnormal cells without harming healthy, normal cells. By disrupting one crucial link in a cell's biochemistry, the cell itself would be damaged. In this manner, cancers and harmful bacteria might be eradicated.

Elion's work focused on purines, one of two main categories of nucleic acids. Their strategy, for which Elion and Hitchings would be honored by the Nobel Prize 40 years later, steered a radical middle course between chemists who randomly screened compounds to find effective drugs and scientists who engaged in basic cellular research without a thought of drug therapy. The difficulties of such an approach were immense. Very little was known about nucleic acid biosynthesis. Discovery of the double helical structure of DNA still lay ahead, and many of the instruments and methods that make **molecular biology** possible had not yet been invented. But Elion and her colleagues persisted with the tools at hand and their own ingenuity. By observing the microbiological results of various experiments, they could make knowledgeable deductions about the biochemistry involved. To the same ends, they worked with various species of lab animals and examined varying responses. Still, the lack of advanced instrumentation and computerization made for slow and tedious work. Elion told *Scientific American*, "if we were starting now, we would probably do what we did in ten years."

By 1951, as a senior research chemist, Elion discovered the first effective compound against childhood leukemia. The compound, 6-mercaptopurine (6MP) (trade name Purinethol), interfered with the synthesis of leukemia cells. In clinical trials run by the Sloan-Kettering Institute (now the Memorial Sloan-Kettering Cancer Center), it increased life expectancy from a few months to a year. The compound was approved by the Food and Drug Administration (F.D.A.) in 1953. Eventually 6MP, used in combination with other drugs and radiation treatment, made leukemia one of the most curable of cancers.

In the next two decades, the potency of 6MP prompted Elion and other scientists to look for more uses for the drug. Robert Schwartz, at Tufts Medical School in Boston, and Roy Calne, at Harvard Medical School, successfully used 6MP to suppress the immune systems in dogs with trans-

planted **kidneys**. Motivated by Schwartz and Calne's work, Elion and Hitchings began searching for other immunosuppressants. They carefully studied the drug's course of action in the body, an endeavor known as pharmacokinetics. This additional work with 6MP led to the discovery of the derivative azathioprine (Imuran), that prevents rejection of transplanted human kidneys and treats rheumatoid arthritis. Other experiments in Elion's lab intended to improve 6MP's effectiveness led to the discovery of allopurinol (Zyloprim) for gout, a disease in which excess **uric acid** builds up in the **joints**. Allopurinol was approved by the F.D.A. in 1966. In the 1950s, Elion and Hitchings's lab also discovered pyrimethamine (Daraprim and Fansidar) a treatment for malaria, and trimethoprim (Bactrim and Septra) for urinary and respiratory tract infections. Trimethoprim is also used to treat Pneumocystis carinii pneumonia, the leading killer of people with AIDS.

In 1968, Elion heard that a compound called adenine arabinoside appeared to have an effect against DNA viruses. This compound was similar in structure to a chemical in her own lab, 2,6-diaminopurine. Although her own lab was not equipped to screen antiviral compounds, she immediately began synthesizing new compounds to send to a Wellcome research lab in Britain for testing. In 1969, she received notice by telegram that one of the compounds was effective against herpes simplex viruses. Further derivatives of that compound yielded acyclovir (Zovirax), an effective drug against herpes, shingles, and chicken pox. An exhibit of the success of acyclovir, presented in 1978 at the Interscience Conference on Microbial Agents and Chemotherapy, demonstrated to other scientists that it was possible to find drugs that exploited the differences between viral and cellular **enzymes**. Acyclovir (Zovirax), approved by the F.D.A. in 1982, became one of Burroughs Wellcome's most profitable drugs. In 1984 at Wellcome Research Laboratories, researchers trained by Elion and Hitchings developed azidothymidine (AZT), the first drug used to treat AIDS.

Although Elion retired in 1983, she continued at Wellcome Research Laboratories as scientist emeritus and kept an office there as a consultant. She also accepted a position as a research professor of medicine and **pharmacology** at Duke University, where she worked with a third-year medical student each year on a research project. After her retirement, Elion served as president of the American Association for Cancer Research and as a member of the National Cancer Advisory Board, among other positions. Hitchings, who retired in 1975, also remained active at Wellcome Research Laboratories; he died at age 92 almost exactly one year before Elion.

In 1988, Elion and Hitchings shared the Nobel Prize in physiology or medicine with Sir **James Black**, a British biochemist. Although Elion had been honored for her work before, beginning with the prestigious Garvan Medal of the American Chemical Society in 1968, a host of tributes followed the Nobel Prize. Although she never completed her doctoral work, she received 25 honorary doctorates from universities including Duke, Brown, and the University of North Carolina. She was elected to the National Inventors'

Hall of Fame, the National Academy of Sciences, the American Academy of Arts and Sciences, the National Inventors Hall of Fame, the National Engineering and Science Hall of Fame, the Royal Society, the Institute of Medicine and the National Women's Hall of Fame. She served as past president of the American Association for Cancer Research and as a Presidential Appointee on the National Cancer Advisory Board. Elion maintained that it was important to keep such awards in perspective. "The Nobel Prize is fine, but the drugs I've developed are rewards in themselves," she told the *New York Times Magazine.*

Once engaged to a man who died before they could marry, Elion dismissed thoughts of marriage to anyone else. Elion has said that she never found it necessary to have women role models. "I never considered that I was a woman and then a scientist," Elion told the *Washington Post.* "My role models didn't have to be women—they could be scientists." Her interests were photography, travel, and music, especially opera.

See also Antigens and antibodies; Autoimmune disorders; Immune system; Immunity; Immunology; T lymphocytes; Transplantation of organs

EMBOLISM

An embolism results from the sudden blockage (occlusion) of an artery. In most cases, the blockage results from a **blood** clot traveling in the blood stream. When clots break away form their site of formation, they travel through the **circulatory system** until they reach a vessel, usually a small arteriole or capillary with a small enough diameter to prohibit further passage.

A broken clot obstructing circulation is termed an embolus. Clots that develop with blood vessels are termed thrombic clots or a thrombus. Once the clot breaks free, it is properly termed an emboli. In addition to blood clots, other material including air bubbles, clumps of **fat**, and other plaque matter.

Clots that break free from the left side of the **heart**, aorta, or large **arteries** usually form an embolism in a small artery or arteriole. Clots that originate in the right side of the heart, or in the venous system, flow to the **lungs** and form a pulmonary embolism.

For short periods, a functioning lung on the opposite side helps to compensate for any blockage in the **pulmonary circulation** until the clot can be removed.

Initial thrombus formation can be induced by trauma or disease processes that disturb the endothelium lining the blood vessel. Clots can also form when blood pressure and flow decrease to levels that do not permit the efficient removal of protoclots or microclots normally formed by thrombin and other coagulant agents. At sufficient flow and blood pressure, microclots are normally filtered out by the **liver**, but in cases of diminished flow, microclots and circulating coagulants may lead to clotting within blood vessels. The danger of clotting

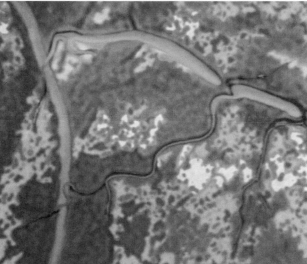

Microphotograph of pulmonary embolism resulting from a clot. Lack of oxygen to the tissue normally supplied by the blocked (occluded) vessel results is tissue death and necrosis. *Custom Medical Stock Photo. Reproduced by permission.*

due to reduced flow, especially in the legs, is often increased in post-surgical patients who are immobilized for long periods.

The ultimate danger in an embolism is the deprivation of oxygenated blood to **tissue** normally served by the blocked vessel. The different tissues in the body have varying abilities to withstand oxygen deprivation. In some cases, **collateral circulation** can develop to compensate for blocked vessels and prevent significant tissue **death** and **necrosis**. In high oxygen demand tissue, such as neural tissue, arterial blockage usually leads to cell death and severe tissue compromise. In addition to the initial site of blockage, once an embolus forms it provides a site for future clotting. Accordingly, it is not unusual for an embolism to increase in size and further impair blood flow as time passes.

See also Angiology; Blood coagulation and blood coagulation tests; Stroke; Transfusions

EMBRYO TRANSFER

Embryo transfer involves the insertion of fertilized early-stage embryos into the female reproductive system. The technique does not alter the genetic composition of the embryo, which was determined at **fertilization.**

Developments in reproductive technology are occurring at a rapid rate in human biology. *In vitro* fertilization, embryo culture, preservation of embryos by freezing (cryopreservation), and cloning technology all yield embryos that are produced outside of the female reproductive system.

There would be no need for embryo transfer if mammalian embryos could be cultured to maturity in the laboratory. While cell culture *in vitro* has made remarkable advances, embryos can be sustained in culture for only a brief period, usually up to the blastocyst stage, or approximately five to six days. Accordingly, their survival is dependent upon transfer to the hospitable and nurturing environment of the uterus of a mother or foster mother.

Embryo transfer started well over a century ago. In one of the first experiments involving embryo transfer, initially performed in 1890 by Walter Heape of Cambridge University, the uterus of one variety of rabbit (a Belgian hare doe) was used for the nourishment, growth and complete fetal development, of another variety of rabbit (an Angora). Heape referred to the Belgian hare doe as a "foster-mother." Since that time, foster-mothering of this type has been extended not only to domestic animals, but also exotic and endangered animal species. Eventually, embryo transfer techniques were applied to humans, as *in vitro* fertilization became available to bypass certain infertility problems.

Transfer can be accomplished by direct non-surgical insertion of the embryo into the female reproductive tract (via the vagina and cervical opening of the uterus), or surgically. A common procedure is to utilize an injection catheter that in some ways functions as a hypodermic syringe. The injection apparatus consists of a thin hollow tube that contains a plunger. At the distal end of the tube is connected a plastic straw containing the embryo and a small drop of culture medium. The apparatus is inserted into the uterus, via the vagina, and the embryo is released within the uterus by gentle pressure on the plunger. Alternatively, surgical embryo transfer is an option in some cases. The upper end of the Fallopian (ovarian) tube is located and the embryo is placed within that site.

See also Embryology; Embryonic development: early development, formation, and differentiation; Implantation of the embryo; In vitro and in vivo fertilization; Infertility (female); Infertility (male)

EMBRYOLOGY

The study of embryology, the science that deals with the formation and development of the embryo and fetus, can be traced back to the ancient Greek philosophers. Originally, embryology was part of the field known as "generation," a term that also encompassed studies of reproduction, development and differentiation, regeneration of parts, and **genetics**. Generation described the means by which new animals or plants came into existence. The ancients believed that new organisms could arise through **sexual reproduction**, asexual reproduction, or spontaneous generation. As early as the sixth century B.C., Greek physicians and philosophers suggested using the developing chick egg as a way of investigating embryology.

Aristotle described the two historically important models of development known as preformation and epigenesis. According to preformationist theories, an embryo or miniature individual preexists in either the mother's egg or the father's **semen** and begins to grow when properly stimulated. Some preformationists believed that all the embryos that would ever

develop had been formed by God at the Creation. Aristotle actually favored the theory of epigenesis, which assumes that the embryo begins as an undifferentiated mass and that new parts are added during development. Aristotle thought that the female parent contributed only unorganized matter to the embryo. He argued that semen from the male parent provided the "form," or soul, that guided development and that the first part of the new organism to be formed was the **heart**.

Aristotle's theory of epigenetic development dominated the science of embryology until the work of physiologist **William Harvey** raised doubts about many aspects of classical theories. In his studies of embryology, as in his research on the circulation of the **blood**, Harvey was inspired by the work of his teacher, **Girolamo Fabrici**. Some historians think that Fabrici should be considered the founder of modern embryology because of the importance of his embryological texts: *On the Formed Fetus* and *On the Development of the Egg and the Chick*. Harvey's *On the Generation of Animals* was not published until 1651, but it was the result of many years of research. Although Harvey began these investigations in order to provide experimental proof for Aristotle's theory of epigenesis, his observations proved that many aspects of Aristotle's theory of generation were wrong.

Aristotle believed that the embryo essentially formed by coagulation in the uterus immediately after mating when the form-building principle of the male acted on the material substance provided by the female. Using deer that had mated, Harvey dissected the uterus and searched for the embryo. He was unable to find any signs of a developing embryo in the uterus until about six or seven weeks after mating had taken place. In addition to his experiments on deer, Harvey carried out systematic studies of the developing chick egg. His observations convinced him that generation proceeded by epigenesis, that is, the gradual addition of parts. Nevertheless, many of Harvey's followers rejected epigenesis and turned to theories of preformation.

Naturalists who favored preformationist theories of generation were inspired by the new mechanical philosophy and by the microscope, a device that allowed them to see the embryo at earlier stages of development. Some naturalists produced very unreliable observations of early embryos, but **Marcello Malpighi** and Jan Swammerdam (1637–1680), two pioneers of microscopy, provided observations that seemed to support preformation. Based on Swammerdam's studies of insects and amphibians, naturalists suggested that embryos preexisted within each other like a nest of boxes. However, given such a theory, only one parent can serve as the source of the sequence of preformed individuals. At the time, the egg of many species was well known, but when the microscope revealed the existence of "little animals" in male semen, some naturalists argued that the preformed individuals must be present in the **sperm**.

Respected scientists of the time, including **Albrecht von Haller**, Charles Bonnet (1720–1793), **Lazzaro Spallanzani**, and René Antoine Ferchault de Reaumur (1683–1757), supported preformation. Bonnet's studies of parthenogenesis in aphids were regarded as strong support of ovist preformationism. Thus, some naturalists argued that the whole human race had

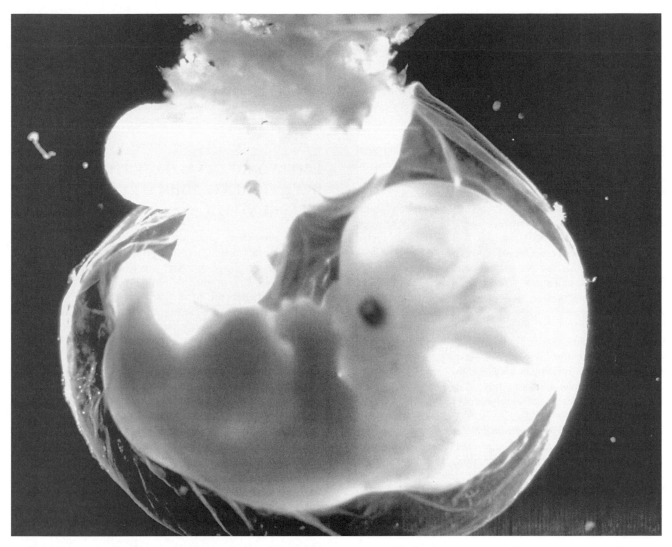

A human embryo at six weeks of development showing a prominent head associated with the normal rapid rate of brain tissue growth. *Photograph by Petit Fromat/Nestle. National Audubon Society Collection/Photo Researchers, Inc. Reproduced by permission.*

preexisted in the ovaries of Eve, while others reported seeing homunculi (tiny people) inside spermatozoa. Other eighteenth century naturalists rejected both ovist and spermist preformationist views. One of the most influential was Casper Friedrich Wolff (1733–1794), who published a landmark article in the history of embryology, "Theory of Generation," in 1759. Wolff argued that the **organs** of the body did not exist at the beginning of gestation, but formed from some originally undifferentiated material through a series of steps. Naturalists who became involved in the movement known as nature philosophy found Wolff's ideas very attractive. During the nineteenth century, **cell theory**, the discovery of the mammalian ovum by **Karl Ernst von Baer**, and the establishment of experimental embryology by Wilhelm Roux (1850–1924) and Hans Driesch (1867–1941) transformed philosophical arguments about the nature of embryological development.

Approximately a century ago, careful observations were made of a number of developing organisms. By this time,

there was a workable cell theory and good microscopes were available. Causal analysis allowed, for example, an understanding of the formation of the three principal germ layers (**ectoderm**, mesodernm and **endoderm**). More specifically it allowed embryologists to describe developmental processes in terms movements of those germ layers (e.g., that the dorsal ectoderm of all vertebrate embryos rolls up into a tube (neural tube) to form the **central nervous system**). It was hypothesized that the certain cells (e.g., the underlying chordamesoderm cells of the gastrula) signaled the ectoderm to become neural. The process was referred to as induction. Other embryonic organs also seemed to appear as a result of induction. Chemical embryology sought to characterize the nature of inducing signals. Now, modern molecular embryology seeks to examine on the level of the gene what controls differentiation of specific **tissue** and cell types of a developing organism.

With regard to genetics, modern molecular biologists have long sought to discover the mechanisms by which a fertilized egg, the recipient of genetic information from its parents, is able to give rise to the various types of cells that comprise an adult organism. The mechanisms of transcription and translation, the essential processes of **protein synthesis**, are much more completely understood than are the mechanisms of **cell differentiation** that regulate the synthesis of specific proteins in specific tissues at certain times during embryonic development.

Beyond the continued study of **birth defects**, modern clinical embryologists also study the interplay between genetic and environmental factors in such increasingly critical fields as cryoembryonics (processes by which frozen embryos are stored for later implantation and development), *in vitro* **fertilization**, and a number of other fertility related areas. Embryologists are also concerned with the similarity between embryonic cells, and dedifferentiated cells often associated with various cancers. Many tumor type cells, for example, lose some of their specialized nature and character, including the ability to regulate cell growth based upon cell density (e.g., contact inhibition), and so in some ways appear similar to embryonic cells.

Strictly defined, an embryo is an organism in the early stages of development. In humans, this term is genrally applied to development during the second to seventh weeks post fertilization. Embryogenesis is the process of embryonic development. Critical areas of study related to embryogeneis include the mechanisms of embryonic induction (a process whereby the development of cells is affected by nearby developing cells), and research on embryonic **stem cells**.

Human embryos are defined as developing humans during the first eight weeks after conception. It is, at best, often difficult to discriminate a human embryo nearing the end of the eighth week from a developing human during the ninth week after conception. Instead of attempts to define embryological processes according to developmental stages or horizons, emphasis in the last few decades, especially among molecular embryologists turned toward the mechanisms of development from a zygote to a multicellular organism. In the particular case of humans, development does not even stop at birth. Note that teeth continue to develop and sex glands with sexual differentiation mature long after birth. For a number of years, many embryologists have referred to their discipline as developmental biology to escape from the need to confine their studies to earlier stages. Accordingly, embryology in the modern sense is the study of the life history of an animal and human embryology considers developmental aspects of life as a whole and not just the first eight weeks.

The causes of developmental abnormalities (congenital malformations) in humans becomes more understandable with a consideration of embryology. The human embryo is extraordinarily vulnerable to drugs, **viruses**, and radiation during the first several months of development when many critical organ systems are developing.

See also Amniocentesis; Birth defects and abnormal development; Embryo transfer; Embryonic development: early development, formation, and differentiation; Fetal membranes and fluids; History of anatomy and physiology: The Classical and Medieval periods; History of anatomy and physiology: The Renaissance and Age of Enlightenment; History of anatomy and physiology: The science of medicine; Human genetics; Implantation of the embryo; *In vitro* and *in vivo* fertilization; Limb buds and limb embryological development; Mesoderm; Nervous system: embryological development; Notochord; Organizer experiment; Skeletal and muscular systems: embryonic development; Somite; Stem cells

EMBRYONIC DEVELOPMENT: EARLY DEVELOPMENT, FORMATION, AND DIFFERENTIATION

Human **pregnancy** is a very inefficient process. In fact, beyond the fact that normally it produces only a single fetus, 50% of the fertilized ovocytes are lost before implantation, and a further 15% are lost within the 10 weeks of embryonal development. Because of this, several events crucial for the conceptus survival take place at this time. Early development can be grouped into four periods: (1) Early preimplantation embryo (until 8 cells) where a symmetric and asynchronic cellular division of toti-potential cells takes place. (2) Late preimplantation embryo characterized by cellular differentiation and embryo hatching. (3) Implantation, where trophoblast invasion of the maternal deciduas allows the survival by means of establishment of maternal-fetal connections.

Although **fertilization** results in the union of maternal and paternal gametes, zygotic gene activity is not required until the blastula stage. In the egg, there exist maternally derived genes which govern embryogenesis through cleavage to the blastula stage. Maternal effect genes encode gene products (**RNA** or protein) that are required in early development prior to zygotic transcription. By day 13–15 after fertilization, embryonic development is apparent. The embryo specific cell population starts to develop and, their specific spatial-temporal organization produces a recognizable embryo. During the first weeks of development (after fertilization) cleavage leads to the formation of the blastomeres. Each blastomere within the zona pellucida becomes smaller with each subsequent division.

As the zygote division reaches sixteen cells, the conceptus becomes a morula (mulberry shaped) and, successively, a compact morula. It leaves the fallopian tube and enters the uterine cavity three to four days after fertilization. After compaction, fluid collects in the internal cavity and the cavity is called the "blastocyst cavity." Some cells become segregated, and these inner cells are called the "inner cell mass" (that will lead to the formation of epiblast and hypoblast). Some of these cells will form the embryo. Embryonic genome activation starts as the conceptus develops to the morula. The outer layer of cells is named the trophoblast and they will form the fetal **placenta**. The blastocyst hatches from the zona pellucida around the sixth day after fertilization, as it enters the uterus. The trophoblast cells secrete an enzyme which erodes the endometrium in order to prepare the implantation site for the blastocyst. The inner cell mass divides, forming a two-layered

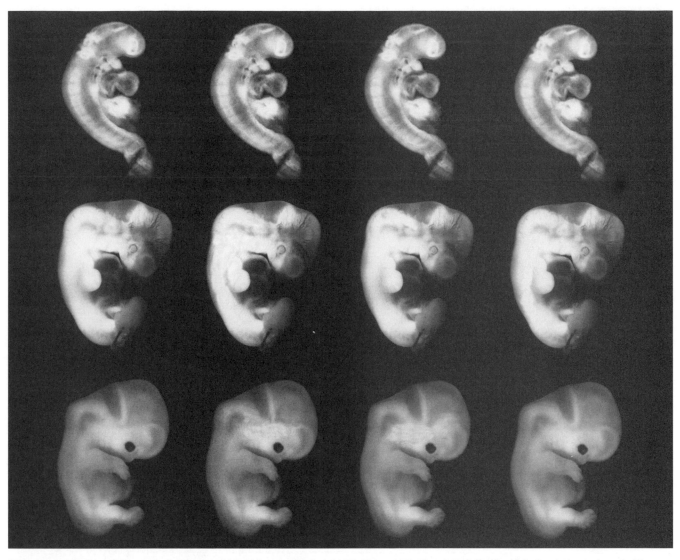

Side views of 29-day-old (5 mm; top), 36.5-day-old (middle), and 44-day-old (13-15 mm; bottom) embryos (heads are in the uppermost area). After eight weeks, the embryo is known as a fetus. © Dr. G. Moscoso/Science Photo Library/Photo Researchers, Inc. Reproduced by permission.

disc. The top layer of cells will become the embryo and amniotic cavity, while the lower cells will become the primitive yolk sac.

The extraembrional events of this period are: (1) Fluid begins to accumulate between cells of the epiblast on day eight. As a result, some epiblast cells are displaced toward the embryonic pole and form the amniotic membrane. These cells are called amnioblasts and the cavity is the amnion. As the amnion grows in volume, amnioblasts migrate along the cytotrophoblastic roof and contribute to growth of the amniotic membrane. (2) Hypoblast cells begin to migrate ventrally along the cytotrophoblast and form a thin membrane called the exocoelomic or Heuser's membrane made of extraembryonic **endoderm** to enclose the exocoelomic cavity or primitive (primary) yolk sac. (3) Cells from the hypoblast come to lie between the primitive yolk sac cytotrophoblast. These cells differentiate into a loose connective **tissue** called the extraem-

bryonic **mesoderm**. Cavities form within the extraembryonic mesoderm and they coalesce to form the extraembryonic coelom or chorionic cavity. (4) Mesoderm next to the trophoblast is now called extraembryonic somatopleuric mesoderm, or extraembryonic somatopleure. Mesoderm next to the primitive yolk sac is now called extraembryonic splanchnopleuric mesoderm or extraembryonic splanchnopleure.

By 14 days after fertilization, the embryo is attached by a connecting stalk (the future umbilical cord), to the developing placenta. Gastrulation continues with the formation of the endoderm and mesoderm by day 17, which develop from the primitive streak, changing the two-layered disc into a three-layered disc. The cells in the central part of the mesoderm release a chemical causing a dramatic change in the size of the cells in the top layer (**ectoderm**) of the flat disc-shaped embryo. The ectoderm grows rapidly over the next few days forming a thickened area. The three layers of the will eventually give rise to: (1) endoderm, that will form the lining of **lungs**, tongue, **tonsils**, urethra, and associated glands, bladder and digestive tract, (2) mesoderm, that will form the muscles, bones, lymphatic tissue, spleen, **blood** cells, **heart**, lungs, and reproductive and excretory systems, and (3) ectoderm, that will form the skin, **nails**, **hair**, lens of **eye**, lining of the internal and external **ear**, **nose**, sinuses, mouth, anus, tooth enamel, pituitary gland, **mammary glands**, and all parts of the nervous system. Gastrulation is completed in the third week (days 15–21).

In the further stages, specific signals, most likely derived from the embryoblast, are needed for embryogenesis to begin. Such signals must be very specific for the initiation of embryogenesis, since if a failure occurs, a blighted ovum will form and lead to spontaneous abortion. Development of embryonic tissues and **organs** is regulated by local interactions between the cells and their extracellular matrix, which take a prevalent role in the further steps of morphogenesis and differentiation. Proliferation and differentiation are part of a complex ontogenetic process that allows the conceptus to grow and to develop physiologic functions. As a rule, growth by cell multiplication is rapid during early development, and then tapers off until it is confined to only a few sites.

Differentiation, the formation of different types of cells and tissues, goes hand in hand with cell growth. Cell migration is among the most important events in early embryonic stages, which allow different types of cells to have intimate contact and to produce tissue induction. Ontogenesis also includes cell maturation and cell apoptosis (programmed cell death) in order to produce mature tissues and organs. Growth Factors (GF) have a major role in determining the different phases of ontogenesis. The next major event for further development of the embryo is the formation of the **circulatory system**, which allows for blood to flow across the body and the **brain**.

See also Cell differentiation; Ear, otic embryological development; Embryology; Eye: Ocular embryological development; Face, nose, and palate embryonic development; Gastrointestinal embryological development; Human development (timetables and developmental horizons); Integumentary system, embryonic development; Limb buds and limb embryological development; Nervous system, embryological development; Renal system, embryological development; Respiratory system embryological development; Skeletal and muscular systems, embryonic development; Urogenital system, embryonic development; Vascular system (embryonic development)

ENDOCRINE SYSTEM AND GLANDS

The endocrine system controls and regulates body **metabolism** and other activities by the production of **hormones**. Hormones are a class of molecules, varying in actual biochemical composition, that act as messenger molecules. Hormones may be steroids, protein peptides, or **lipids**. Regardless of their unique chemical structure, hormones, are secreted by endocrine glands and in some circumstances, specialized tissues.

Just as there are agonist muscle groups that work in opposition to one another to coordinate movements, a number of hormones exist as pairs of hormones that stimulate opposite responses in target tissues. Antagonistic hormones include such chemicals as glucagons and insulin that act to respectively increase or decrease **blood** levels of glucose, and thus affect the availability of glucose to cells to use in cellular metabolism.

Endocrine glands release their homes into the blood stream for distribution to target tissues. Target cells have receptor sites (hormone receptors) on their membranes that specifically bind specific hormone sequences or structures.

Major human endocrine glands include the pituitary gland, **hypothalamus**, thyroid, and **pancreas**.

The pituitary gland is the major and controlling gland of the endocrine system. Often referred to as the master gland of the endocrine system, the structure of pituitary gland, along with the hormones it secretes combine to form the primary site in the body for the coordination of nervous system and endocrine system function. The pituitary gland, also known as the hypophysis or hypophysis cerbri, produces a number of hormones that control hormone production in other endocrine glands.

The pituitary gland is situated in the sella turcica, a small indentation or recess located near the base of the **brain**. About the size of a pea, the pituitary gland is connected to the hypothalamus by a slender pituitary stalk (hypophyseal stalk). This connection is not trivial. The posterior pituitary is really an extension of the hypothalamus, and its **tissue** type differs from the anterior pituitary gland. In addition to an anatomical connection via the hypophyseal-hypothalamic tract, the hypothalamus regulates pituitary function—therefore provides a bridge between the nervous system and the endocrine system—through the hypophyseal portal system. The hypophyseal portal system is a localized collection blood vessels that allow regulating agents secreted by the hypothalamus to travel directly to the pituitary gland.

Neural input (e.g., information gathered by the nervous system) can be interpreted by the hypothalamus and translated into pituitary control, and therefore affect endocrine response. The connection of the pituitary gland to the hypothalamus establishes a clear chain of control: nervous signals in the hypothalamus cause changes in pituitary hormone secretion that, in turn, control the production of hormone in other endocrine glands.

The pituitary gland has two distinct anatomical and physiological regions that produce different hormones. The front (anterior) lobe, termed the adenohypophysis, produces different hormones than the posterior lobe (neurohypophysis). The adenohypophysis secretes seven different hormones. Growth hormone (GH), thyroid-stimulating hormone (TSH), adrenocorticotropic hormone (ACTH), the gonadotropins (LH and FSH), prolactin (PRL), and melanocyte-stimulating hormone (MSH). The neurohypophysis produces **oxytocin** and antidiuretic hormone (ADH).

Growth hormone (GH), or somatotropin, is a hormone that acts directly on tissues to stimulate overall body growth. GH acts to stimulate growth by increasing **protein synthesis** and

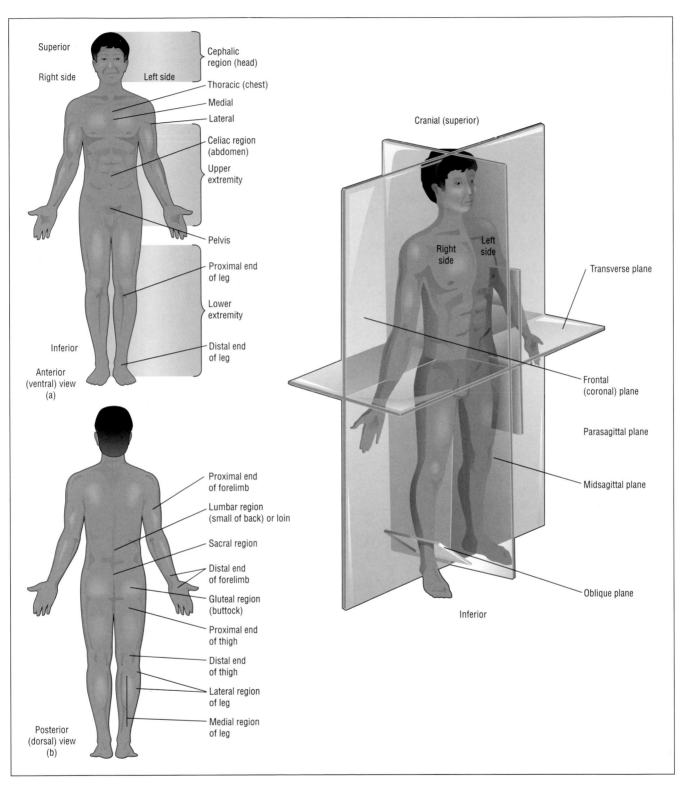

Basic anatomical nomenclature. See entry, "Anatomical nomenclature," page 20. *Illustration by Argosy Publishing.*

A comparison of hearts during transplant surgery shows the diseased heart (held in the hand) beside the implanted healthy heart. See entry, "Cardiac disease," page 83. *Photograph by Alexander Tsiaras. National Audubon Society Collection/Photo Researchers, Inc. Reproduced by permission.*

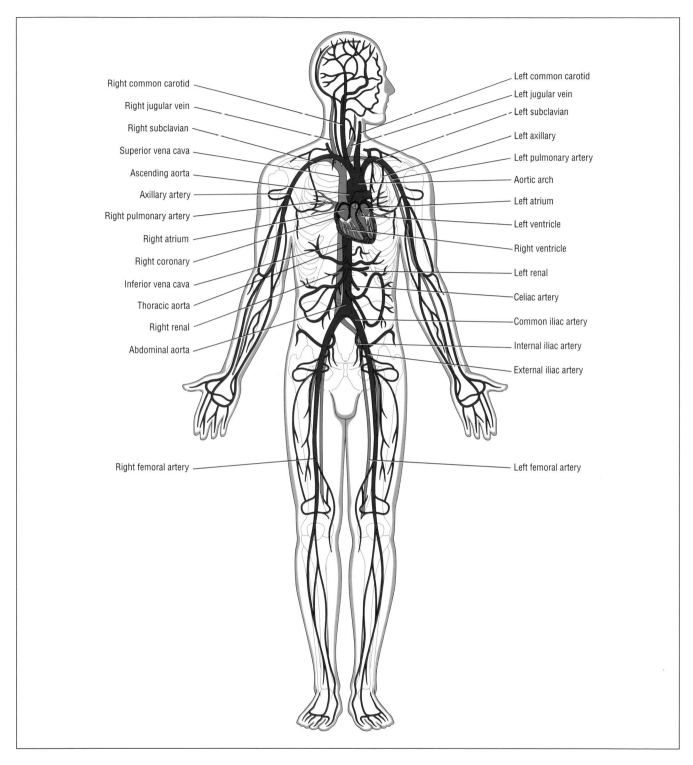

Right common carotid

Right jugular vein

Right subclavian

Superior vena cava

Ascending aorta

Axillary artery

Right pulmonary artery

Right atrium

Right coronary

Inferior vena cava

Thoracic aorta

Right renal

Abdominal aorta

Right femoral artery

Left common carotid

Left jugular vein

Left subclavian

Left axillary

Left pulmonary artery

Aortic arch

Left atrium

Left ventricle

Right ventricle

Left renal

Celiac artery

Common iliac artery

Internal iliac artery

External iliac artery

Left femoral artery

The circulatory system. See entry, "Circulatory system," page 108. *Illustration by Argosy Publishing.*

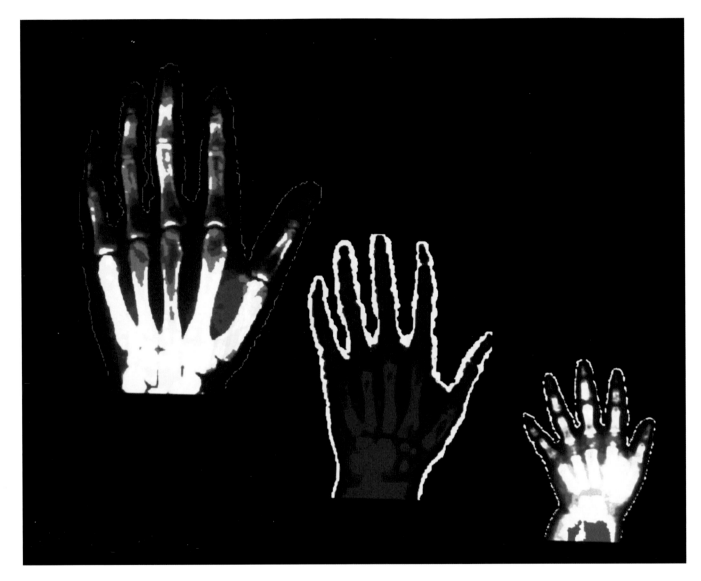

A computer-enhanced x-ray image depicting the hand development of a male at two (far right), six, and nineteen years (far left) of age. See entry, "Growth and development," page 246. *X-ray by Scott Camazine. © Scott Camazine, Photo Researchers, Inc. Reproduced by Permission.*

Healthy lung tissue (left) contrast with the diseased lung tissue (right) damaged by cigarette smoking. See entry, "Lungs," page 364. *Photograph by A. Glauberman. National Audubon Society Collection/Photo Researchers, Inc. Reproduced by permission.*

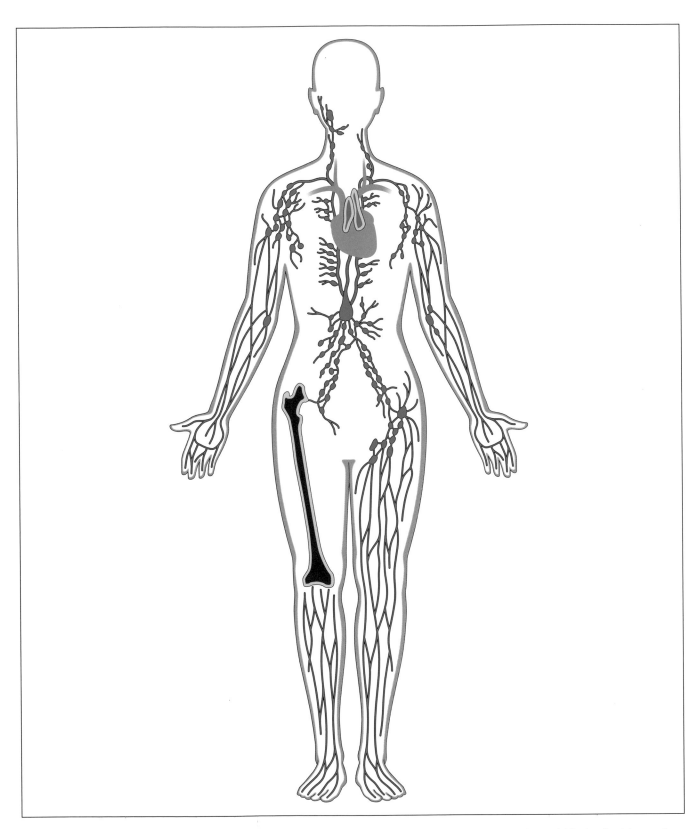

The lymphatic system: lymph nodes (pale green); thymus (deep blue); and one of the bones rich in bone marrow (essential in the circulatory and lymphatic processes)—the femur—shown in purple. See entry, "Lymphatic system," page 366. *Illustration by Argosy Publishing.*

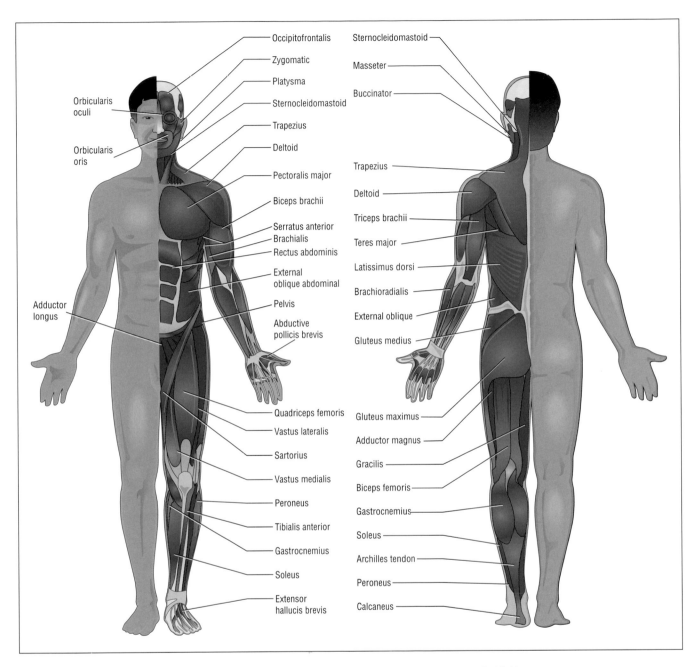

Occipitofrontalis

Zygomatic

Platysma

Sternocleidomastoid

Orbicularis oculi

Trapezius

Orbicularis oris

Deltoid

Pectoralis major

Biceps brachii

Serratus anterior
Brachialis
Rectus abdominis

External oblique abdominal

Adductor longus

Pelvis

Abductive pollicis brevis

Quadriceps femoris

Vastus lateralis

Sartorius

Vastus medialis

Peroneus

Tibialis anterior

Gastrocnemius

Soleus

Extensor hallucis brevis

Sternocleidomastoid

Masseter

Buccinator

Trapezius

Deltoid

Triceps brachii

Teres major

Latissimus dorsi

Brachioradialis

External oblique

Gluteus medius

Gluteus maximus

Adductor magnus

Gracilis

Biceps femoris

Gastrocnemius

Soleus

Archilles tendon

Peroneus

Calcaneus

The superficial muscle system. See entry, "Muscular system overview," page 406. *Illustration by Argosy Publishing.*

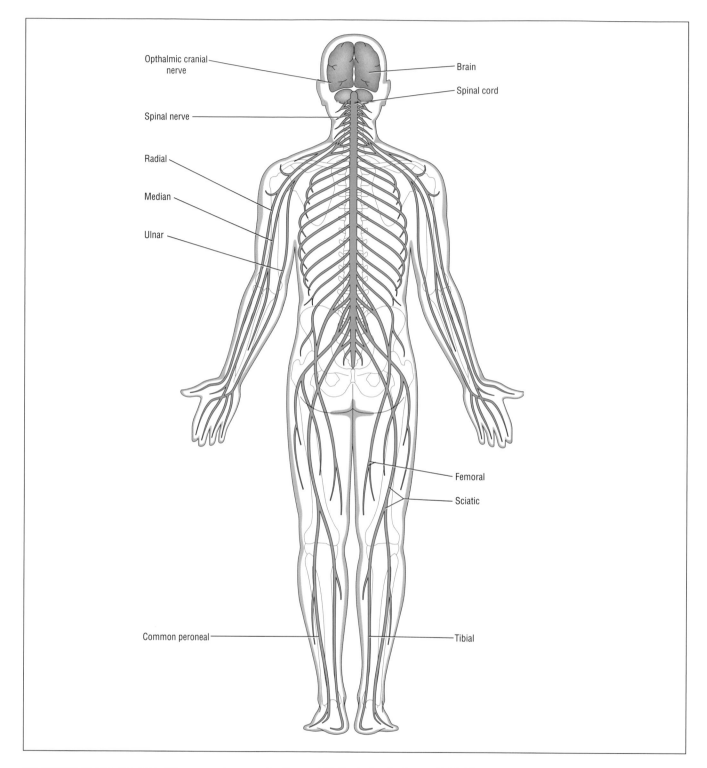

Opthalmic cranial nerve

Brain

Spinal cord

Spinal nerve

Radial

Median

Ulnar

Femoral

Sciatic

Common peroneal

Tibial

The nervous system. See entry, "Nervous system overview," page 415. *Illustration by Argosy Publishing.*

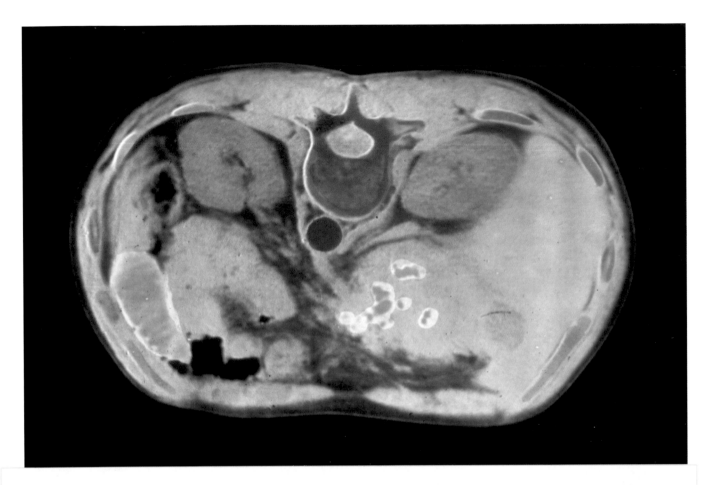

Computed tomography (CT) x-ray scan through the human abdomen. Visible structures include the liver (right), spleen (left), abdominal aorta (center), vertebral column, spinal cord, and kidneys (upper left and upper right). See entry, "Organs and organ systems," page 429. *Photo Researchers, Inc. Reproduced by permission.*

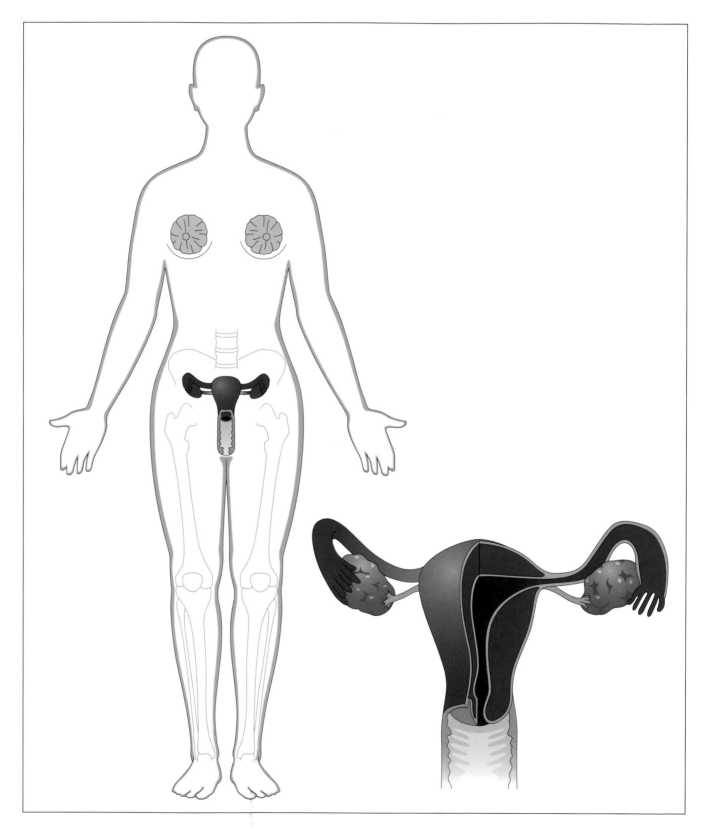

The female reproductive system: uterus (shown in red with the Fallopian tubes); ovaries (blue); vagina (pink with a yellow lining); breasts (apricot). Inset shows detail of ovaries, uterus, Fallopian tubes, and cervix (turquoise). See entry, "Reproductive system (female)," page 480. *Illustration by Argosy Publishing.*

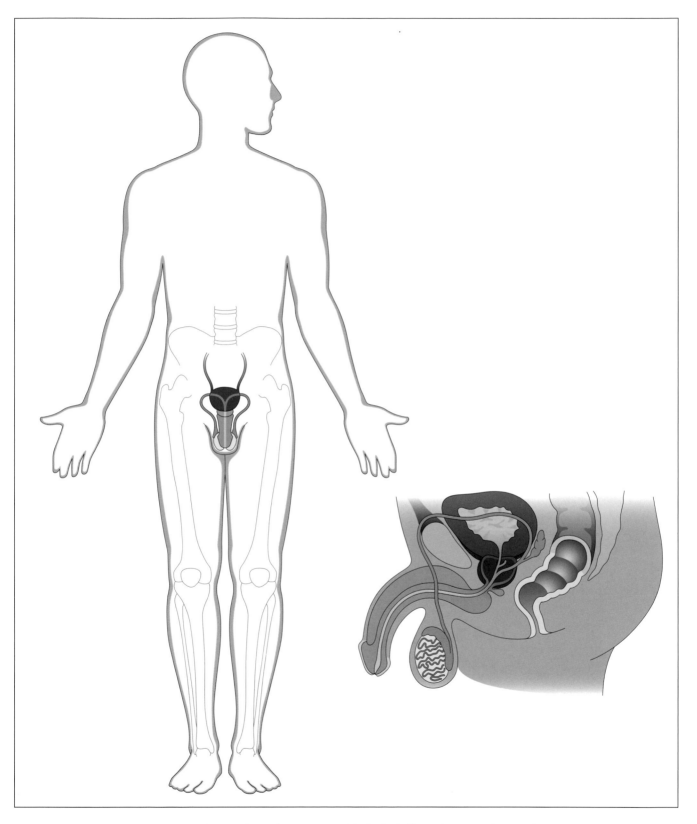

The male reproductive system: penis (pink); testes (yellow); prostate gland (in full-body illustration, shown in peach/apricot, and in the inset as dark blue gland between the bladder and the penis); vas deferens (apricot, in insert); epididymis (yellow-green). See entry, "Reproductive system (male)," page 482. *Illustration by Argosy Publishing.*

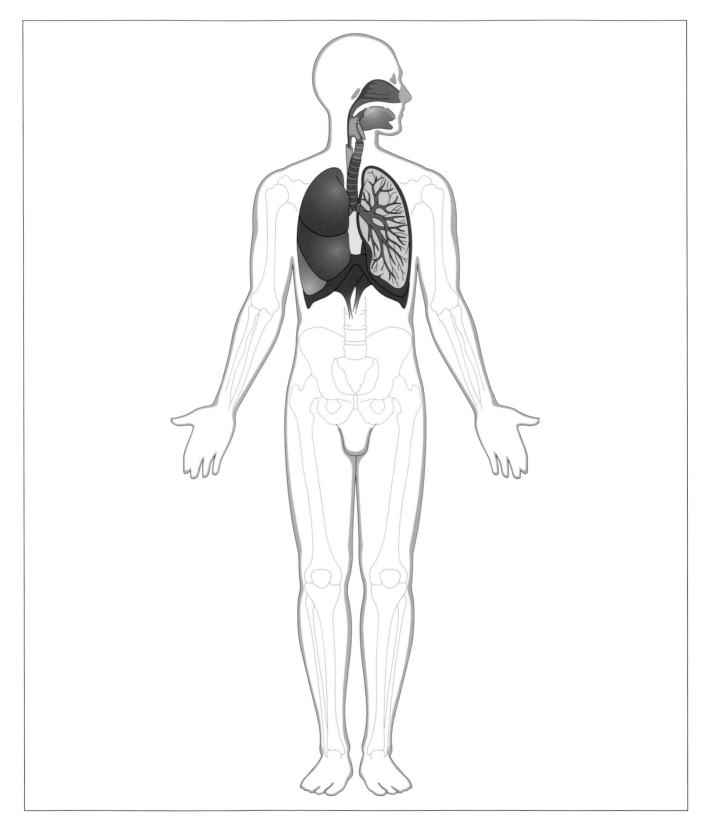

The respiratory system: pharynx (orange); larynx (green ridged tube); esophagus (involved in digestion, not breathing, is shown as smooth green tube); trachea (purple); lungs (deep blue). See entry, "Respiratory system," page 486. *Illustration by Argosy Publishing.*

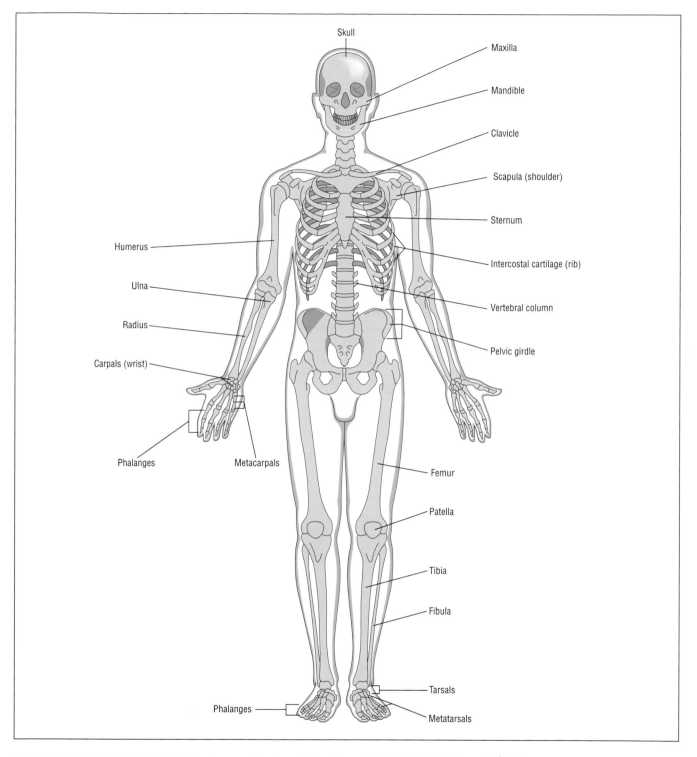

The skeletal system. See entry, "Skeletal system overview (morphology)," page 519. *Illustration by Argosy Publishing.*

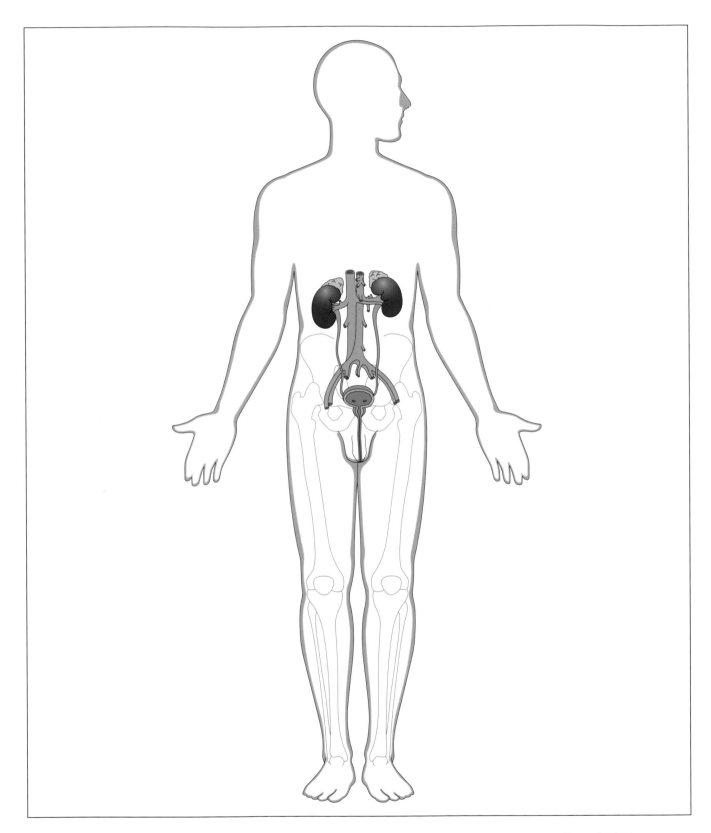

The urogenital system: kidneys (purple); ureters (green); bladder (blue-green). See entry, "Urogenital system," page 570. *Illustration by Argosy Publishing.*

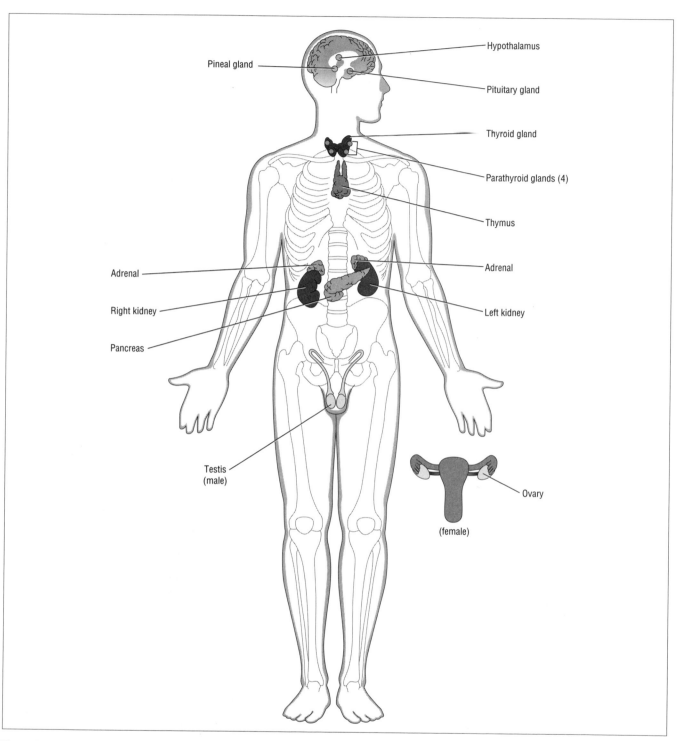

The endocrine system. *Illustration by Argosy Publishing.*

acts metabolically to shunt glucose from **ATP** synthesis pathways, while at the same time promoting **fat** usage. GH stimulates the secretion of somatomedin hormones. Abnormalities in the pituitary gland that result in the overproduction or underproduction of GH can result in gigantism (generalized enlarge-ment), acromegaly (enlargements of the extremities), or—in cases of underproduction—pituitary dwarfism.

Thyroid-stimulating hormone (TSH) stimulates thyroid growth and the secretion of thyroid hormones, including thyroxine and triiodothyronine.

Pituitary secretion of adrenocorticotropic hormone (ACTH) stimulates adenocortical growth (growth of cortex of the adrenal glands locate on the top (apex) of the **kidneys**) and the subsequent secretion of corticosteroids. There is also a feedback mechanism involved in ACTH production. Cortisol (a hormone also important in controlling the inflammatory response) is the principal corticosteroid produced by the adrenal glands. As levels of cortisol increase in the bloodstream, they act to inhibit further production of CTH by the anterior pituitary.

The pituitary gland is also an important regulator of the reproductive system through the secretion of the gonadotropins LH (lutenizing hormone) and FSH (**follicle stimulating hormone**). In the male, LH is usually referred to as interstitial cell stimulating hormone (ICSH). In the female, LH acts to stimulate ovulation and the formation of the corpus luteum on an ovarian follicle. In the male, the action of LH or ICSH acts to stimulate the production of **testosterone**. In the female, FSH stimulates **estrogen** secretion and supports the growth and maturation of the ovarian follicle. In the male, the primary role of FSH is to stimulate **spermatogenesis** (the formation of **sperm** cells).

Pituitary stimulation of melanocyte stimulating hormone (MSH) contributes to the regulation of pigmentation of the skin by stimulating **melanocytes** to produce melanin. Some anatomists actually divide the pituitary into three lobes and argue that the intermediate lobe (pars intermedia) is the area that secretes MSH. Other anatomists argue that the region described as the intermediate lobe of the pituitary gland is too poorly vascularized to be anatomically of functionally distinguishable from the anterior pituitary.

Prolactin (a lactotrophic hormone secreted by anterior pituitary gland) is a multifunctional hormone that stimulates a number of different cell types and cell sites in addition to its primary functions of stimulating milk production by the **mammary glands** and stimulating **progesterone** production by the corpeus luteum located the ovarian follicle. Receptor sites for PRL are found on many cell types scattered throughout the body.

Oxytocin is actually synthesized in the hypothalamus and is stored in the neurohypophysis. Oxytocin stimulated by the neurohypophysis stimulates the contraction of **smooth muscle** in the uterus during birth. Oxytocin also stimulates the ejection of milk by the mammary glands and the release of oxytocin is stimulated when infants suckle.

Antidiuretic hormone (ADH), also termed vasopressin, is also stored in the posterior pituitary gland. When released ADH acts on the renal systems to promote water retention. ADH also stimulates smooth muscle contractions in the smooth muscles lining blood vessels and the digestive tract. Because they act on the smooth muscle of blood vessels, vasopressors are important in the hormonal regulation of blood pressure.

The thyroid gland produces iodine-requiring hormones that regulate metabolism including **temperature regulation** and weight control. In addition to pancreatic role in producing digestive **enzymes**, the special cells with the islet of Langerhans of the pancreas secrete glucagons and insulin to regulate blood sugar levels. Differing regions of the adrenal glands secrete **epinephrine** and corticosteroids. The pineal gland plays a role in metabolic regulation.

In addition to producing sex cells (sperm and ova), the male testes and the female ovaries secrete the **sex hormones** testosterone, estrogen, and progesterone in sex-specific amounts for males and females.

See also Adrenal glands and hormones; Blood pressure and hormonal control mechanisms; Cerebral morphology; Diuretics and antidiuretic hormones; Follicle stimulating hormone (FSH) and luteinizing hormone (LH); Glycogen and glycolysis; Gonads and gonadotropic hormone physiology; Hormones and hormone action; Inflammation of tissues; Oxytocin and oxytocic hormones; Parturition; Pregnancy, maternal physiological and anatomical changes

ENDOCYTOSIS · *see* CELL MEMBRANE TRANSPORT

ENDODERM

Endoderm is one of three principal germinal layers of cells that are formed in early in embryonic development. Endoderm is also referred to as entoderm.

The other principal germinal layers (e.g., layers of cells from which body tissues develop) are **ectoderm** and **mesoderm**. Endoderm comprises the innermost germinal layer from which the epithelium of the digestive system and most of the **respiratory system** are derived. Endodermal cells also develop into parts of the urethra and bladder.

Membranes derived from endoderm are in contact with ectoderm derived membranes at the mouth and anus.

In the embryonic disk ectoderm and endoderm border a third intermediate mesodermal layer. When the embryonic disk ultimately folds into a tube the basic "tube within a tube" plan of development becomes evident. A core endodermal tube establishes a primitive digestive pathway bounded by an oral orifice and an anal orifice. Around that innermost tube is an outer tube comprised of ectoderm. The ectoderm serves as a protective layer and the layer from which the nervous system and sense **organs** develop. Mesodermal cells fill the space between the inner (endodermal) and outer (ectodermal) tube. Mesodermal cells ultimately contribute to the muscles, organs, and other internal body structures.

About a week following **fertilization**, the human embryonic blastocyst is embedded in the endometrium of the uterus. The blastocyst is a proliferating ball of cells with a cavity termed a blastocoele. At one pole of the blastocyst there is a thickened mass of cells termed the inner cell mass. The inner cell mass contains communicating slit-like openings that form the amniotic cavity and the embryonic disk.

Endoderm lies on the ventral side of the embryonic disk. At the cellular or histological level, endoderm is comprised of columnar cells. The ectodermal layer lies on the dorsal side of the disk. Initially there are only two germ layers, but by 18 days post fertilization a thickening occurs in the ectoderm to form a primitive streak. The walls of the primitive groove continue to thicken and, at the anterior (cephalic) end of the groove, expand into a primitive knot also known as Henson's node. Anterior to the primitive knot, primitive ectoderm and endoderm are in direct contact with each other to form a prochordal plate. The primitive streak also establishes the general head-to tail (cephalo-caudal) axis for subsequent development. Starting at the primitive knot, cells from the ectodermal layer migrate into the primitive groove and invaginate into the space between the ectoderm and endoderm to form mesodermal cells. Other cells, derived from other **fetal membranes** also contribute cells to the mesodermal layers. As ectodermal cells stream invaginate to form a trilaminar embryonic disk, a head process (ultimately to become the **notochord**) forms in the middle mesodermal layer.

Endodermal cells ultimately provide the cells that form the tissues comprising the majority of the epithelial lining of the alimentary canal. Endoderm also provides the linings of glands—except for ectodermal derived salivary glands—that open into the alimentary canal. The **liver** and **pancreas** also develop from endoderm. Other organs and systems that contain critical components derived from endoderm include elements of the thyroid gland, parathyroid glands, thymus, **pharynx**, trachea **alveoli**, bladder, prostate and urethra.

See also Embryonic development: Early development, formation, and differentiation; Fetal membranes and fluids; Implantation of the embryo; Limb buds and limb embryological development; Nervous system: Embryological development; Organizer experiment; Sexual reproduction; Skeletal and muscular systems: Embryonic development; Somite

ENDOMETRICAL CYCLE AND MENSTRUATION • *see* MENSTRUATION

ENDORPHINS

Endorphins are naturally produced polypeptides that are produced in the **brain**. Their physiological effect is similar to the analgesia (**pain** relief) of morphine. Chemically, they are classified as opioid-peptides and have 16–31 **amino acids** in their polypeptide chain. There are three main families of opioid-peptides including the endorphins, the enkephalins and the dynorphins. To date, four groups of endorphins (alpha, beta, gamma and sigma) have been identified. Beta-endorphins are the most potent endogenous opioid that affects both physiological and mental processes. The effects range from **central nervous system** and **peripheral nervous system** analgesia and pain modulation to effects on neuro-endocrine control of reproduction, stress, spontaneous behaviour and motivation. It has been found that prolonged physical exercise can increase production and release of endorphins that result in a sense of euphoria that is commonly referred to as "runner's high."

Endorphins are the body's natural way of toning down specific pain responses by binding with opiate receptors in the brain. Although endorphins act as **neurotransmitters** and neurohormones, their role in physiological processes is still not completely understood. One effect of endorphins binding to receptors in the brain may lead to an inhibitory neural response (due to activation of an inhibitory G-protein) or may lead to an excitatory response (hyperpolarization of the neuron) to the point that less neurotransmitter is released into the synaptic gap, resulting in decreased excitation of the neuron. Endorphins may also act to prevent the release of substance P that has been found to transmit pain impulses to the brain.

The search for endorphins began in the 1970s with the attempts to isolate opiate binding sites in the brain. Vincent Dole of Rockerfeller University did the early work in this area in 1970. In 1973, at Johns Hopkins University, Solomon Snyder and his graduate student Candace Pert isolated opiate receptors on nervous **tissue**. With the discovery of the receptor sites, the search for the molecules that bound to these sites intensified. In 1975, Choh Hao Li of the University of California at Berkeley isolated endorphins from a pituitary hormone that he had originally isolated in 1960. He named the biochemical endorphin that means "the morphine within." Endorphin research achieved worldwide recognition in 1977 with the awarding of the Nobel Prize in physiology or medicine to **Roger Guillemin** and **Andrew Schally** for their work on peptide hormone production of the brain.

See also Drug effects on the nervous system; Nerve impulses and conduction of impulses

ENERGETICS • *see* METABOLISM

ENZYMES AND COENZYMES

Enzymes are protein catalysts, the latter being substances that change the speed of chemical reactions. About 90% of cellular proteins are enzymes and some structural proteins are also enzymes. A protein is classified as an enzyme if it is known to catalyse a reaction, although it is always possible that a given protein, which is currently not classified as an enzyme, may catalyse an as yet un-recognized reaction. The Nomenclature Committee of IUBMB (International Union of Biochemistry and Molecular Biology) lists thousands of enzymes, which are classified according to the reactions they catalyse. The Nomenclature Committee publishes an enzyme catalogue, abbreviated as EC, in which each enzyme receives a four-part number. The first part indicates the general category of the reaction it catalyses as one of the following: Oxidoreductase, Transferase, Hydrolase, Lyase, Isomerase, Ligase (synthetase). The second and third parts of the EC number indicate the subgroup and sub-sub-group to which it belongs, and the fourth part the number is assigned arbitrarily, in consecutive order.

The most striking difference between enzymes and chemical catalysts is the specificity of enzymes, with regard to their substrates and to the reactions catalysed. Enzymes distinguish stereoisomers absolutely, due to the three dimensional nature of the binding between enzyme and substrate. Stereoisomers are molecules with the same molecular formula, but having two different configurations that are mirror images of each other. The lock and key model suggests that enzymes, like a complicated lock, can only be fitted by a substrate (key) of precisely the correct shape. A second model, the induced fit model, hypothesizes that when the substrate binds to the enzyme, the confirmation of the latter changes in such a way as to make the fit more complete. In either case, the difference between two stereoisomers of a substrate is analogous to the difference between right and left hands; the enzyme-binding site corresponds to a glove fitting only one of the two hands. The amino acid residues that take part in catalysis and their immediate neighbours constitute the active site of the enzyme. The substrate-binding site is usually close to the active site or may incorporate it. If the substrate is large, the binding site may extend beyond the active site.

Coenzymes can be thought of as any catalytically active, low molecular weight component of an enzyme. This definition includes coenzymes, which are covalently bound to the enzyme as a prosthetic group. In this case the coenzyme and the apoenzyme (or enzyme protein) together make up the holoenzyme. Strictly speaking, the coenzyme enters the reaction stoichiometrically, in that it reacts sequentially with two enzyme proteins, and thus catalyses substrate turnover. An example is nicotinamide adenine dinucleotide (NAD), the working group of dehydrogenases and reductases. It first forms an active complex with a dehydrogenase, for example during glycolysis, and accepts the hydrogen removed from the substrate. Reduced NADH then dissociates from the enzyme I and associates with a reductase (enzyme II), donating the hydrogen to the substrate of this enzyme. Since the coenzyme acts as a second substrate, it can be called a co-substrate. It must be able to react reversibly with the apoenzymes of two different enzymes. Flavin, heme and pyridoxal phosphate are examples of coenzymes while metals are considered inorganic complements of enzyme reactions and are not termed coenzymes but cofactors. Many coenzymes are synthesized from **vitamins**.

Most enzymes are strictly intracellular, but some are excreted into body cavities, for example the digestive enzymes. Proteolytic enzymes, those that degrade other proteins, are secreted in the form of inactive precursors, which are only activated after they are safely out of the cell. Proteolytic enzymes that are not secreted by a cell are sequestered into a special organelle, the lysosome, and degrade intracellular proteins that have come to the end of their lifespan.

Within eukaryotic cells, there is a considerable amount of compartmentation of enzymes (and reactions) within organelles. When a cell is fractionated, these enzymes can be used as markers for the various cellular fractions (**mitochondria**, cytosol, liposomes etc). Some enzymes are specific for particular **organs** or tissues, and their presence in the **blood** can be used as a diagnostic test for damage to a **tissue** of origin. Enzymes are also applied in diagnostics as reagents for the specific detection of other metabolites, such as glucose, **ATP**, **lactic acid** etc., which are diagnostically significant.

Some enzymes may also have some therapeutic value. The proteases, in particular, may be applied to relieve poor **digestion** and problems of **blood coagulation**. Pathologically, increased levels of protease, on the other hand, may be treated with protease inhibitors from, for example, bovine organs, which do not provoke an immune response in humans.

See also Biochemistry; Cell structure; Metabolism

EPIDERMIS

The epidermis is the outermost layer of skin on the body. As such, it is the first barrier to the entry of infectious organisms. The epidermal **tissue** layer also functions to seal in moisture.

The epidermis is composed of keratinocyte cell types. There are five layers, or strata, making up the epidermis. The deepest innermost layer is the stratum germinativum. Next comes the stratum spinosum, then the stratum granulosum. The fourth layer is the stratum lucidum. Finally, the uppermost layer is the stratum corneum.

The keratinocytes that make up the stratum corneum are continually sloughing off from the surface of the skin. This process is termed desquamation. New keratinocytes migrate up through the various layers to reach the stratum corneum, and then migrate through this outer layer to the external surface, a process that takes about three weeks. The skin cells visible on the surface are actually dead. This continual loss and replacement of the epidermal surface cells provides the body with a means to tolerate abrasions without causing long-term damage to the skin's generative surface.

This function occupies some 95% of the cells in the epidermis. The other 5% of the cells, located in the stratum germinativum, are involved in the production of a substance called melanin. The amount and distribution of melanin determines skin coloration. The darker the skin, the more melanin is present. The melanin-producing cells of the epidermis, in response to ultraviolet radiation, can increase the production of the compound, in order to help protect the skin from the harsh ultraviolet rays of the sun. This is the reason skin "tans" in the summer. It is also the melanin producing cells that are affected by skin **cancer**.

Another molecule found in the epidermis is keratin, a fibrous protein that is used as a building block for **hair**. Keratin's properties give the epidermis its firmness, elasticity, and strength. Keratin is produced in the stratum lucidum. The surface of the epidermis also houses the exit sites of **sweat glands**. These and other glands exude a slightly acidic oil, which protects the skin against some **bacteria**.

See also Histology and microanatomy; Infection and resistance; Necrosis

EPINEPHRINE

Epinephrine is a hormone that is produced by the medulla of the adrenal glands that is an important component in the regulation of the sympathetic branch of the **autonomic nervous system**. Epinephrine is also known as adrenaline in some publications, especially those using British English.

Both norepinephrine (a neurotransmitter) and epinephrine are derived from the amino acid tyrosine.

In general, epinephrine causes and elevation of **blood** pressure, **heart** rate, and oxygen consumption, all important components in the fight or flight response.

Within the **liver**, epinephrine causes the release of stored **glycogen** sugars and results in an elevation in blood glucose. Within intact liver cells, epinephrine stimulates enzyme activity (e.g., the activity of glycogen phosphorylase) that cause the breakdown of liver glycogen to blood glucose. At the **tissue** level, epinephrine also acts to increase the rate of **metabolism** of glycogen.

American physiologist **Earl Sutherland** and others, in a series of experiments conducted in the 1950s, demonstrated *in vitro* (in the laboratory) the effects of epinephrine on liver tissue. Research on epinephrine led to the identification of the biologic role of cyclic AMP (adenylic acid) and adenylate cyclase.

There are specific disorders that can lead to diminished production or deficits of epinephrine (e.g., Addison's disease of the adrenal glands). Other adrenal irregularities and **tumors** may result in the overproduction of epinephrine.

Epinephrine can also be produced synthetically and can be given to patients as a vasoconstrictor to constrict blood vessels and reduce blood loss. Epinephrine is used to counter the effects of anaphylactic **shock** (a severe allergic reaction) during which blood pressure falls to low levels and there is a loss of blood flow to vital **organs**. Epinephrine is also used as a cardiac stimulant.

See also Adrenal medullae; Blood coagulation and blood coagulation tests; Blood pressure and hormonal control mechanisms; Drug treatment of cardiovascular and vascular disorders; Glycogen and glycolysis; Liver

EPIPHYSIS • *see* EPITHALAMUS

EPITHALAMUS

The epithalamus, a part of the diencephalon, is located superior and posterior to the **thalamus** and contains the pineal gland (also called the pineal body or epiphysis). The epithalamus also serves as boundary of the third ventricle. Physiological studies have established that the epithalamus plays a regulatory role in the sleep-wake cycle and in controlling mood.

The pineal gland attaches via a stalk to the posterior boundary wall of the third ventricle of the cerebrum. The babenular and posterior commissures are located on the superior and inferior aspects of the base of the pineal stalk.

Controversy regarding the physiological function of the pineal gland is longstanding. The pineal gland was called the "third eye" in medieval writings and René Descarte once pro-

posed a theory of mind and body that asserted that the pineal gland was the "seat of the soul. Some bizarre pseudoscientific beliefs still assert that the pineal gland is a mysterious psycho-metaphysical link.

The role of the pineal gland is, however, more terrestrial and tangible. Although the pineal body plays an import role in birds and other species, it was long considered vestigial in humans. Physiological studies indicate, however, that the pineal body serves as an endocrine gland and that pineal gland stimulation is linked to the production of the hormone melatonin. High levels of melatonin and serotonin are found in the pineal gland. Synthesis and release of melatonin, a derivative of the amino acid tryptophan, is inhibited by light stimulation of the **retina** and stimulated by darkness. The production of melatonin is also related to a daily circadian rhythm during which melatonin levels rise and fall. The highest levels of production generally decrease in the darkest, early morning hours. The gland is most active during **puberty** and decreasing levels of melatonin are produced with age. Ongoing research is dedicated to attempting to more clearly define the causative actions of melatonin in what appear to be at least associative links to **sleep**, aging, and disease processes.

The pineal gland, in conjunction with the **hypothalamus**, may influence the production of gonadatropins and thus play a regulatory role in the onset and course of puberty.

The pineal body does not possess nerve cells but is served by sympathetic ganglia. Two types of cells comprise the pineal gland, pinealocytes and glial cells and with age there is increasing calcification of the pineal gland and the deposition of calcified material called **brain** sand.

Functionally, the epithalamus serves as a linkage point for limbic system components located in the forebrain. Other structural elements of the epithalamus include the habenula—also called the right and left habenular nuclei (nerve cells), sites studies indicate are important in the regulation of food and water intake—and the stria medullaris (a bundle of fibers terminating in the right and left habenular nuclei. The habena is a linkage point in many olfactory reflex pathways.

See also Adolescent growth and development; Cerebral morphology; Gonads and gonadotropic hormone physiology; Hormones and hormone action

EPITHELIAL TISSUE AND EPITHELIUM

Epithelial **tissue** is the tissue that covers surface areas of the body. There are three different types of epithelial tissue. Taken together they are called the epithelium. In any given region of the body, the epithelial tissue is arranged in certain ways, with varying layers of cells. The aggregate is collectively referred to as the epithelium.

The three types of epithelial tissue are: covering and lining tissue, glandular tissue, and sensory tissue. The covering and lining epithelia are associated with, as in these three examples, the skin, the lining of the gastro-intestinal tract, and the lining of the respiratory tract. Their location determines the function of these cells. Again using the above examples,

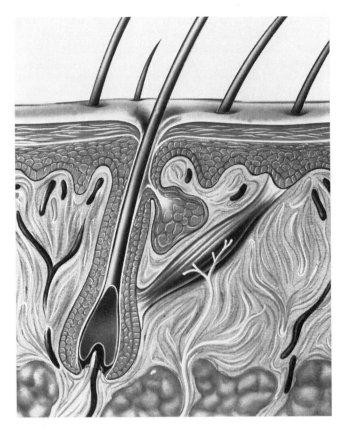

Diagram of a section through human skin, showing the surface protective layer of dead epithelial cells, along with the living epithelium below. *Illustration © 1996 by SPL/John Bavosi. Custom Medical Stock Photo, Inc. Reproduced by permission.*

the skin epithelia is a protective layer, the epithelial tissue of the gastro-intestinal tract is concerned with the absorption of nutrients, and the epithelial tissue of the lung **alveoli** functions in the diffusion of gases across the lung-blood barrier.

Epithilial cells have several characteristics. They are contiguous; that is, the cells are attached to one another. The cells rest on a material called the basement membrane, which consists of a basal lamina produced by the epithelial cells and a fiber-like reticular layer. All epithelial cells have a so-called apical side, which is oriented toward the outside world (be that the lung or the intestinal tract or the true outside world, in the case of the skin) and a so-called basal side, which is oriented towards the basement membrane. As well, all epithelial cells lack **blood** vessels. Finally, epithelial cells are composed largely of a material called keratin.

The basement membrane aids epithelial cells in orienting themselves in relation to other tissues. After an injury to the cells, such as an abrasion, the basement membrane acts as a scaffolding upon which new cells can attach themselves during healing.

Epithilium acts as an interface between the outside world and the rest of the body. For this function, the epithelium needs to be a permeability barrier, capable of letting certain molecules through while excluding other molecules from entry. This function is accomplished by the presence of proteins at the meeting point between adjacent epithelial cells. The proteins span the cell, from its basal surface to its apical surface. These regions are called junctional complexes. Several types of junctional complexes exist. Anchoring junctions facilitate the tight association of adjacent epithelial cells, allowing cells such as skin cells to resist stretching forces that might otherwise pull the cells apart. Tight junctions plug up the space between adjacent cells, so those molecules that are absorbed across a layer of epithelial cells must pass through the cells. This provides a mechanism whereby epithelial tissue can control absorption. Also, gap junctions do allow the passage of molecules, but can do so in a regulated way. This allows cells, such as muscle cells and nervous tissue, to communicate and coordinate their activity.

Epithilial cells are classified according to several criteria: cell shape (squamous, cuboidal, columnar), the number of layers of cells in the epithelium (simple, stratified, pseudostratified), and the function of the cells (microvilli, **cilia**). Their varied shape, arrangement and function makes possible a number of critical activities, including absorption of molecules (in a non-specific or specific manner), secretion of substances such as tears and **hormones** from glands, supporting and maintaining form within the body, as in **cartilage**, bones, **tendons** and ligaments, and in the functioning of sensory cells.

See also Cilia and ciliated epithelial cells; Gaseous exchange

EQUILIBRIUM

Physiological equilibrium—the balance between processes and reactions—extends from the level of the cell membrane to the whole organism.

A well-known example of physiological equilibrium is the Donnan equilibrium. Named after its discoverer, British chemist Frederick George Donnan, this equilibrium is characterized by an unequal distribution of diffusible ions (ions that are capable of moving across a membrane) between two solutions of ions separated by a membrane. The inequality results because the membrane does not allow the passage of one of the ionic species. A result of the inequality of ions is an electrical potential across the membrane. Thus, current flow across the membrane is possible, which can power various enzyme activities and reactions within the membrane. Not surprisingly, the Donnan equilibrium is of fundamental importance in the functioning of cellular processes.

Equilibrium is also the basis of the physiological property of homeostasis (from two Greek words meaning to remain the same), a physiological resistance to change whereby automatically occurring mechanisms act to maintain a constant rate of concentration in the **blood** of certain molecules and ions that are essential to life and to maintain other facets of the body, such as temperature, at specified levels. Homeostasis has been known for a long time. In 1865, Claude Bernard noted, in his book *Introduction to Experimental Medicine,* that the "constancy of the internal milieu was the essential condition to a free life."

As humans begin to venture into space, it is becoming clear that environments different from that on Earth can alter physiological equilibrium. For example, the reduced force on muscles because of the reduced gravity of spaceflight produces atrophy (shrinkage) in muscles and weakens bone, because of an increased loss of **calcium**. Additionally, the equilibrium of gas exchange in the lung can be upset.

See also Acclimatization; Homeostatic mechanisms

ERASISTRATUS (304 B.C.-250 B.C.)
Greek physician and anatomist

Erasistratus, considered the father of physiology, was born on the island of Chios in ancient Greece, to a medical family. His father and brother were doctors, and his mother was the sister of a doctor. He studied medicine in Athens and then, around 280 B.C., enrolled in the University of Cos, a center of the medical school of Praxagoras. Erasistratus then moved to Alexandria, where he taught and practiced medicine, continuing the work of **Herophilus**. In his later years, he retired from medical practice and joined the Alexandrian museum, where he devoted himself to research.

Although Erasistratus wrote extensively in a number of medical fields, none of his works survive. He is best known for his observations based on his numerous dissections of human cadavers (and, it was rumored, his vivisections of criminals, a practice allowed by the Ptolemy rulers). Erasistratus accurately described the structure of the **brain**, including the cavities and membranes, and made a distinction between its *cerebrum* and *cerebellum* (larger and smaller parts). He viewed the brain, not the **heart**, as the seat of intelligence. By comparing the brains of humans and other animals, Erasistratus rightly concluded that a greater number of brain convolutions resulted in greater intelligence. He also accurately described the **structure and function** of the gastric (stomach) muscles, and observed the difference between motor and sensory nerves. Erasistratus promoted hygiene, diet, and exercise in medical care.

In his understanding of the heart and **blood** vessels, Erasistratus came very close to working out the circulation of the blood (not actually discovered until **William Harvey** in the seventeenth century, but he made some crucial errors. Erasistratus understood that the heart served as a pump, thereby dilating the **arteries**, and he found and explained the functioning of the heart valves. He theorized that the arteries and **veins** both spread from the heart, dividing finally into extremely fine **capillaries** that were invisible to the **eye**. However, he believed that the **liver** formed blood and carried it to the right side of the heart, which pumped it into the **lungs** and from there to the rest of the body's **organs**. He also believed that *pneuma*, a vital spirit, was drawn in through the lungs to the left side of the heart, which then pumped the pneuma through the arteries to the rest of the body. The nerves, according to Erasistratus, carried another form of pneuma, animal spirit.

After Erasistratus, anatomical research through dissection ended, due to the pressure of public opinion. Egyptians believed in the need of an intact body for the afterlife, hence mummification. Real anatomical studies were not resumed until the thirteenth century.

See also History of anatomy and physiology: The Classical and Medieval periods

ERLANGER, JOSEPH (1874-1965)
American physiologist

Joseph Erlanger was an American physiologist whose pioneering work with his collaborator, **Herbert Spencer Gasser**, helped to advance the field of neurophysiology. For their work, Erlanger and Gasser shared the 1944 Nobel Prize in physiology or medicine. The prize committee cited their work on "the highly differentiated functions of single nerve fibers." Although unstated, the awarding of the Nobel Prize to Erlanger and Gasser also recognized their roles in developing the most basic tool in modern neurophysiology, the amplifier with cathode-ray oscilloscope. The prize culminated for Erlanger a distinguished career in medical education and physiological research.

Erlanger was born in San Francisco, California. His father, Herman Erlanger, had immigrated to the United States in 1842 at the age of sixteen from his home in Würtemberg, in Southern Germany. After struggling as a peddler in the Mississippi Valley, he went to California during the Gold Rush. Unsuccessful at mining, Erlanger turned to business and became a moderately successful merchant. In 1849, he married Sarah Galinger, also an immigrant from Southern Germany and the sister of his business partner. Joseph was the sixth of seven children, five sons and two daughters.

From an early age, Erlanger showed an interest in the natural world, a fact that led his older sister to give him the nickname "Doc." In 1889, he entered the classical Latin curriculum at the San Francisco Boys' High School. After graduating in 1891, he began studies in the College of Chemistry at the University of California at Berkeley, receiving a bachelor's degree in 1895. It was at Berkeley that Erlanger performed his first research—studying the development of newt eggs. He then enrolled at the Johns Hopkins University School of Medicine in Baltimore and earned a medical degree in 1899, fulfilling his childhood aspirations of becoming a doctor. Erlanger excelled as a student while at Johns Hopkins, graduating second in his class. This distinction allowed him to work as an intern in internal medicine for William Osler, the renowned physician and teacher.

After arriving in Baltimore, Erlanger decided that medical research and not medical practice would be his life's pursuit. In the summer of 1896, he worked in the **histology** laboratory of Lewellys Barker, demonstrating his zeal for research by studying the location of horn cells in the **spinal cord** of rabbits. The following summer, he undertook a different project—determining how much of a dog's small intestine could be surgically removed without interfering with its digestive processes. This study led to Erlanger's first published paper in 1901, and to his appointment as assistant professor of physiology at Johns Hopkins by William H. Howell, one of America's most important physiologists and head of the department. He was later promoted to associate professor of physiology.

Erlanger spent the next several years exclusively at Johns Hopkins except for a six-week trip in the summer of 1902 to study **biochemistry** at the University of Strassburg in Germany. His career to that point was exceptional for two reasons. Unlike the generation of scientists that preceded him, Erlanger did not migrate to Europe to study. This decision reflected the improving standards of medical education and scientific research in the United States at the close of the nineteenth century. Second, Erlanger, although he was a trained physician, chose to pursue a full-time career in research instead of medical practice. Physician-scientists before Erlanger could devote only part of their time to research, as the rest was spent on patient care.

During his career at Johns Hopkins, Erlanger studied a number of problems that were important in medicine. In 1904, he designed and constructed a sphygmomanometer—a device that measures **blood** pressure. Erlanger improved on previous designs by making it sturdier and easier to use. Later that year, he used the device to find a correlation between blood pressure and orthostatic albuminuria, wherein proteins appear in the urine when a patient stands. His last few years at Johns Hopkins were spent studying electrical conduction in the **heart**, particularly the activity between the auricles and the ventricles that is responsible for the consistent beating of the heart. Using a clamp of his own design, he was able to determine that a conduction blockage, or heart block, in the connection between the auricles and ventricles, was responsible for the reduced **pulse** and fainting spells associated with Stokes-Adams syndrome.

In 1906, Erlanger left Johns Hopkins and moved to the University of Wisconsin, where he became the first professor of physiology at the university's medical school. Though the university's administration recruited Erlanger to build and equip a physiological laboratory, his efforts were continually hampered by a lack of funds. This situation contributed to his decision to leave Wisconsin in 1910 for the Washington University School of Medicine, in Saint Louis. The medical school at Washington had been newly reorganized and had sufficient funds to meet Erlanger's needs. He worked at Washington for the remainder of his career, serving as professor of physiology and department chairman. Even after his retirement in 1946, Erlanger continued to work part-time performing research and helping graduate students in their work.

After arriving at Washington University, Erlanger devoted much of his time and energy to the formidable task of helping to reorganize the medical school. Erlanger and the other department heads constituted the new school's executive faculty, which oversaw administration and offered significant input into the construction and design of the new medical school buildings. In 1917, the United States' entry into World War I drew Erlanger's attention away from his administrative duties, presenting him with the opportunity to return to the laboratory and to his research on cardiovascular physiology. He participated with other physiologists in the study of wound **shock** and helped to develop therapeutic solutions that were used by the United States Army in Europe. He also continued the work that he had begun at Johns Hopkins, studying the sounds of Korotkoff, the sound one hears in an artery when measuring blood pressure with a stethoscope.

Although Erlanger would remain interested in cardiovascular physiology throughout his career, he experienced an intellectual transition in the early 1920s, when he took up questions of neurophysiology. The arrival at Washington University of Herbert Spencer Gasser, a student of Erlanger's from Wisconsin and a fellow Johns Hopkins graduate, spurred this change. Erlanger and Gasser would collaborate at Washington University until Gasser's departure in 1931 for the Cornell Medical College. Understanding how nerves transmit electrical impulses preoccupied Erlanger and Gasser during the 1920s. The difficulty in studying nerves was that the electrical impulses were too weak and too brief to measure them accurately. In 1920, one of Gasser's former classmates, H. Sidney Newcomer, developed a device that would amplify nerve impulses by some 100,000 times, allowing physiologists to measure and study the subtle changes that occur during nerve transmission. A year later, Erlanger and Gasser, based on advances made at the Western Electric Company, constructed a cathode-ray oscilloscope that could record the nerve impulse. The cathode-ray oscilloscope with amplifier was a technological breakthrough that permitted neurophysiologists to overcome the barrier posed by the subtlety and brevity of nerve activity. Erlanger and Gasser went on to study the details of nerve transmission. Their most significant contribution derived from these researches was their conclusion that larger nerve fibers conducted electrical impulses faster than smaller ones. Also, they demonstrated that different nerve fibers can have different functions.

Erlanger and Gasser's work on nerve physiology increased Erlanger's already important role in American physiology. Not only had he made significant contributions to the science of physiology, but his career—based on a wholly American education and consisting of a full-time research effort—represented a new generation of American physiologists. For his scientific efforts, Erlanger was elected a member of the National Academy of Sciences, the Association of American Physicians, the American Philosophical Society, and the American Physiological Society. He also received honorary degrees from universities of California, Michigan, Pennsylvania, Wisconsin, and Johns Hopkins University, Washington University, and the Free University of Brussels. His highest honor came when he shared, with Gasser, the 1944 Nobel Prize in physiology or medicine. Erlanger died of heart failure one month before his ninety-second birthday.

See also Heart: Rhythm control and impulse conduction; Nerve impulses and conduction of impulses

ERYTHROCYTES

Erythrocytes, also known as red **blood** cells, contain the pigment **hemoglobin** that has the remarkable capacity to combine with and release oxygen. Human red blood cells contain a 33% solution of hemoglobin. Oxygen is transported to living tissues of the body as oxyhemoglobin in red blood cells. The human red blood cell is a biconcave disc with an average of about 0.0003 in. (7.5 m) in diameter. Erythrocytes are the most common cell type in blood (with an average of about 5,500,000 per ml in men and 5,000,000 per ml in women). Newborn babies have an even greater number of erythrocytes, with as many as 7,000,000 per ml. Red blood cells are suspended in **plasma**, which is the straw colored liquid part of the blood. The characteristic red color of blood is due to the erythrocytes. Human, and most mammalian erythrocytes have nuclei while they develop in the bone marrow. The nuclei and some cytoplasmic structures are lost as the red blood cell matures. The life span of an erythrocyte is about 120 days. Old cells are removed from the circulation by the spleen and bone marrow. The old cells are constantly being replaced by fresh new red blood cells.

The pathological condition of having too few erythrocytes, or erythrocytes containing too little hemoglobin, is known as anemia. Anemia can be caused by blood loss or by other conditions. Too many red blood cells is referred to as polycythemia and may occur as an adaptation to living in mountains to compensate for reduced oxygen in the air.

See also Capillaries; Oxygen transport and exchange

ESOPHAGUS · *see* ALIMENTARY SYSTEM

ESTROGEN

The estrogens are of a group of **hormones** synthesized by the reproductive **organs** and adrenal glands in females and, in lesser quantities, in males. All naturally occurring estrogens, including estradiol, estrone, and estriol, are C_{18} steroids secreted by the theca interna and granulosa cells of the developing ovarian follicle, corpus luteum, or **placenta**. Estrogens are also produced by the aromatisation of androgens (male hormones) such as androstenedione and **testosterone**, in the **adipose tissue, liver** and skin. The enzyme, aromatase, catalyses the conversion of androstenedione to estrone and testosterone to estradiol. Ovarian cells are stimulated by **luteinizing hormone** (LH) to increase the conversion of cholesterol to androstenedione. Some of this is then converted to the most potent naturally occurring estrogen, estradiol, which is then released into the circulation. Ovarian granulosa cells are supplied with androstenedione by the thecal cells and that is converted to estradiol, which is primarily secreted into the follicular fluid. Estradiol in the circulation is bound to sex hormone binding proteins such as globulin and albumin and less than 2% is free in circulation. In the liver, estrogens are con-

jugated to other substances to make them water-soluble, after which they are excreted into the **bile** and reabsorbed prior to leaving the body via the urine. Estradiol is converted to estrone in the **blood**. Estriol is the principal estrogen formed by the placenta during **pregnancy**.

The normal development and maturation of the female is dependent on estrogens. They cause the thickening of the lining of the uterus and vagina in the early phase of the ovarian, or menstrual, cycle and in other animals cyclic estrogen secretion also induces estrus, or "heat." The estrogens are also responsible for female **secondary sexual characteristics** such as, in humans, pubic **hair** and **breasts**, and they affect other tissues including the development of genital organs, skin, hair, blood vessels, bone, and pelvic muscles.

The ability of estrogens to suppress secretion of **follicle-stimulating hormone** (FSH) by the pituitary gland and thereby inhibit ovulation makes estrogen and estrogen-like compounds major components in oral contraceptives. Estrogen replacement therapy (ERT) uses synthetic estrogen to treat the physical changes of **menopause**, including hot flashes and vaginal dryness. It also retards the development of **osteoporosis** and lessens the risk of **heart** attack in postmenopausal women, and may have an effect on **Alzheimer's disease**. ERT's association with an increased risk of breast and uterine **cancer** has spawned many studies and refinements in the treatment, including the addition of progestins to lessen the risk of uterine cancer. Estrogens are also used to treat prostate cancer.

See also Adrenal medullae; Hormones and hormone action; Ovarian cycle and hormonal regulation; Sex hormones

ETHICAL ISSUES IN EMBRYOLOGICAL RESEARCH

Knowledge about human reproductive biology is rapidly expanding, and many applications of reproductive technologies are possible. There are two main kinds of ongoing embryological research, *in vitro* research and research on implanted fetuses. Both need to clarify one significant ethical distinction: the situation when the embryo is planned to be born, or to perish. Ethical issues of reproductive technologies must be examined both from the prospective of the benefits gained by some individuals, and the possibility of undesirable social consequences. For example, a personal benefit in which only male children would be desirable could result in a skewed sex distribution in the population.

Embryo **stem cells** are considered a new hope for a source of supply for transplantation of internal **organs** and genome project studies, but such a technology implies discontinuation of possible life, since the conceptus (embryo and/or fertilized ovum) is destroyed when using the technology. An exception is the so-called regenerative medicine that utilizes the patient's own somatic stem cells. The clinical applications of stem cells include (1) **tissue** transplantation, allowing replacement of damaged **heart** tissue, (2) generation of **neu-**

Dolly, the first animal cloned from an adult sheep, developed arthritis in 2002, adding another consideration to scientific debate about the safety of cloning technology. *Photograph by Ph. Plailly/Eurelios/Science Photo Library. Photo Researchers, Inc. Reproduced by permission.*

rons to cure Parkinson's disease (3) growth of cartilage-forming cells to alleviate arthritis. (4) research into curing diseases such as **cancer**, **birth defects**, and diabetes.

In vitro **fertilization** (IVF) treatment often produce embryos no longer needed by the couple that could be donated for research. For people who believe the human embryo from the zygote onwards has moral value, any embryo research is ethically unacceptable. Many people, however, feel that moral value develops gradually, and that embryo research is ethically acceptable if the purpose itself is not immoral. Thus, both the stage of embryo development and the object of the research have to be taken into account. Countries that allow human embryo research set a time limit for research, usually 14 days after fertilization, just before the fetus begins to form. Criteria for this limit relate to the appearance of the primitive streak of the embryo, along with also several other factors in the developmental process by day fourteen. Laws also delineate the purpose of clinical as well as stem cell research, aiming at therapy for intractable diseases.

Most scientists involved in embryonic stem cell research agree that medical progress cannot ignore all ethical limits. Many scientists consider the following to be the "morally relevant milestones" of embryo development: (1) the stage when embryos would normally implant in a uterus, (2) when embryonic cells are so committed to their eventual body parts that splitting into identical-twin embryos is no longer possible, and (3) the appearance of the 'primitive streak,' indicating that twinning is no longer possible and giving the embryo a new level of organization, presaging the development of organs and a nervous system. Human stem cell lines, virtually pluripotent, have been derived both from primordial **germ cells** developing in fetal tissue (embryonic germ cell lines), and from early pre-implantation embryos no longer

required for infertility treatment (ES cell lines). On August 9, 2001, United States President George W. Bush announced that he would support limited federal funding of embryonic stem-cell research. It was a controversial decision he had weighed for some months. "There are about 60 existing stem cell lines in various research facilities," Bush said, cell lines that have already been derived from human embryos and that can be used in clinical researches. Bush allowed research funding only for those stem cells that already had been extracted from destroyed embryos. He prohibited further destruction of embryos for stem-cell research.

The Jones Institute for Reproductive Medicine in Virginia announced in July 2001, that human embryos from donor eggs and **sperm** were created with the sole purpose of conducting research on the embryo that would ultimately result in their destruction. Also, a Massachusetts-based company announced it was attempting to clone human embryos for the purposes of harvesting stem cells from those embryos. Disagreement on early-embryo research persists despite all knowledge of theologians, lawyers, sociologists, politicians, and moralists. Many people in fact think that the at time of fertilization, when an original and unique gene construction is created, is the moment when a new spiritual life (soul's seed) descends, whatever the reason the embryo was brought into being. If most people agree with the need for respect of human beings, they disagree about the timing of the embryo entering into human life in terms of the stage of its development. Pope John Paul II, head of the Roman Catholic church, exhorted president Bush to reject the funding for embryonic stem cells, as a practice related to abortion, euthanasia, and other termed assaults on innocent human life: "A free and virtuous society," said the Pope, "which America aspires to be, must reject practices that devalue and violate human life at any stage from conception until natural death." Again, in his 1995 encyclical *The Gospel of Life,* Pope John Paul II wrote: "Human embryos obtained *in vitro* are human beings and are subjects with rights; their dignity and right to life must be respected from the first moment of their existence. It is immoral to produce human embryos destined to be exploited as disposable 'biological material.'" However, the Catholic Church has no objection to research using stem cells extracted from adults or from an umbilical cord after a child is born. In the context of the creation of things, the human being is the microcosm, reflecting in the unity of a single creature both spiritual and physical realities.

A different point of view claims that the embryo, whatever its stage of development, does not satisfy the basic criteria of personhood. For example, a fetus is totally dependent on a woman's body to survive and it resides inside her body. Human beings must, by definition, be separate individuals. Again, the fetus cannot be considered a member of the moral community, with full and equal moral rights, for the simple reason that it is not a person. This does not mean that it is impossible for people to reach agreement. It simply means that there is no incontrovertibly rational means by which they must do so.

Whatever the point of view, ethical analysis suggests that, if the process of human beings become commercialized, the selling of eggs, sperm, embryos or fetal tissue may devalue human dignity. These can be associated to a kind of "human experimentation," or "business of human life," or to the more materialistic aspects of humanity. Most scientists recognize the necessity for a respectful treatment of human zygotes because of their connections to human community, and because, potentially, they could become human beings. Thus, most scientists assert the need for embryo research to be subject to some limits, legislation, and licensing or accreditation. The United States National Institutes of Health (NIH, August 2000) entered into the Federal Register its final guidelines for research involving human pluripotent stem cells. The guidelines detail the procedures to help ensure that NIH-funded human pluripotent stem cell research is conducted in an ethical and legal manner.

Omnis cellula e cellula (all cells comes from cell) is an axiom by **Rudolf Virchow**, one of the outstanding figures of the German medicine of the nineteenth century. First investigations on this topic were aimed to evaluate if the genetic material in the nucleus remains the same as the embryo develops. Further studies addressing the role of the nucleus showed that nuclear genes of differentiated cells could be reprogrammed by the cytoplasm of the egg at certain stages of development. If the cytoplasmatic factors in the egg cytoplasm are identified, it might be possible to reproduce them *in vitro* for further investigations. In fact, every cell in the body, except sperm and egg, contains all of the genetic material in its **DNA** to theoretically create an exact clone of the original body, although differentiated cells loose such ability, and are programmed just for specific functions. Most scientists believed that such differentiated cells could not be reprogrammed to be capable of behaving as a fertilized egg, but as in the case of the sheep "Dolly" (performed by Ian Wilmut at the Roslin Institute in Scotland), a cell was taken from the mammary tissue of a mature six-year-old sheep, and fused with a sheep ovum which had had its nucleus removed. This procedure has a poor result, since out of 277 attempts at cell fusion, only 29 began to divide and 13 ewes became pregnant. Just one lamb, Dolly, was born. In embryo cloning, the fetus often grows unusually large and generally dies just before or after birth. These fetuses have under-developed **lungs** and reduced immunity to **infection**. By the end of the year 2000, eight species of mammals have been cloned, including mice, cows, rhesus monkeys, sheep, goats, pigs, and rats. The cloning process seems to produce an erroneous genetic expression, probably because of a different fast timing of genetic programming. As stated by Duke University researchers (but not in the opinion of all scientists), fetal overgrowth in cloned human fetuses should be prevented by the existence in the DNA of all primates, of two copies of a gene that regulates fetal growth. Several techniques are now planned for use in cloning: (1) somatic cell nuclear transfer or "human therapeutic cloning," where a patient's body cell is combined with an egg cell that has its DNA removed. As a result, stem cells are produced identical to the patient (2) Parthenogenesis: In this technique, a woman's oocyte is directly activated without the removal of its DNA to begin development on its own, and (3) Ooplasmic transfer, the removal of the cytoplasm of an oocyte and trans-

ferring it into a body cell of that is transformed into a primitive stem cell.

Scientists at Advanced Cell Technology (Worcester, MA) who, in the fall of 2001, cloned the first human embryo, (a six-cell stage) emphasized they were not interested in producing babies by cloning, but just embryos as a source of stem cells (such as heart cells, neurons, **blood** cells) for treating disease. It must be emphasized that a six-cell stage embryo is something with a higher risk of loss and is very far from a fully-developed human being.

A clone, although genetically identical, is not the same person that generated him, because the life experiences will be never be the same. There are also some biological differences, such as the neural patterns, which can't be duplicated perfectly. Gregory Pence, author of *Who's Afraid of Human Cloning* states, "The conclusion is inescapable. The problem of wiring up a **brain** is so complex that it is beyond the power of the genomic computer. The best the genes can do is to indicate the rough layout of wiring, the general shape of the brain." It is true, however, that manipulation of individual traits, as well as musical talent or specific individual's behaviors might produce a new kind of human being, without any background in human history. Some psychologists question whether a child's sense of self-worth, individuality, or dignity, would be seriously affected by being a clone. Others are concerned that the cloned child would be denied what the philosopher Joel Feinberg has called "the right to an open future."

The National Bioethics Advisory Commission concluded that "at this time, it is morally unacceptable for anyone in the public or private sector, whether in a research or clinical setting, to attempt to create a child using somatic cell nuclear transfer cloning." Such efforts, it said, would pose "unacceptable risks to the fetus and/or potential child." Such ethical concerns about human clones involves the risks associated with the current state of cloning technology, that does not exclude the possibility of mutation or other biological damage. Furthermore, some scientists hold that large-scale cloning may reduce the genetic heterogeneity, that is the base of **evolution** and adaptation to the environment.

See also Embryonic development, early development, formation, and differentiation; Ethical issues in genetics research; Evolution and evolutionary mechanisms

ETHICAL ISSUES IN GENETICS RESEARCH

Modern **genetics** poses some of the most significant ethical problems that science has ever faced. As a result of recent advances, scientists are now able to engineer living organisms with genes taken from unrelated species. Proponents of genetic engineering claim that this technology is a natural continuity of older breeding practices. This is, however, clearly not true. The new genetic technology is a human intervention that must not be confused with any previous intrusions upon nature; like animal and plant breeding for agricultural purposes, or the induction of mutations with x rays in laboratories. All earlier procedures worked within single or closely related species while the genetic engineering of today is unprecedented in that it allows human beings to direct the crossing of species barriers in entirely new ways. Though scientific morality up to now has been to proceed without restriction to learn all we can about nature, our new genetic knowledge requires the careful assessment of all associated ethical issues if its application is not to prove unwise or dangerous. Opinions about modern genetics are divided at this time, with some emphasizing the promised benefits and others raising severe doubts and questions about the social and environmental implications.

Apart from the possible risks arising, for example, from the unrestricted release of genetically altered organisms, the fundamental ethical question is whether the applications of modern genetic knowledge respect the intrinsic rights of living creatures. In both science and industry, the predominant attitude towards life appears to be based on philosophies of instrumental values while intrinsic values are considered less important, with the result that living creatures are assessed in terms of their potential use. In 1971, the United States government issued the first patent on an organism, a genetically engineered bacterium for cleaning up oil spills. Today large biotechnology companies hold patents on genetically engineered plants, animals and also on human genes. The push to patent genetically engineered organisms raises the basic question as to whether it is proper to claim ownership over living organisms or their parts? A patent is a kind of license granted by a government to inventors giving them the rights to stop rivals from making, using or selling their inventions without permission. To receive a patent, an invention must be new, innovative and have some practical application for human activity. It cannot simply be a discovery. Opponents of gene patenting say that genes are pure discoveries, while supporters argue that genes are patented together with inventive descriptions of how they can be used. Holding a patent on a gene gives the holder control over its commercial exploitation. For human or animal genes, this control may involve an application in diagnosis or developing therapies for diseases. There is, for example, a patent on laboratory mice genetically designed to be prone to **cancer**, which are being used in anti-cancer research.

Perhaps the most ethically complicated application of genetic knowledge is to the human species itself. For example, we are now able to diagnose hereditary diseases before or soon after birth. Using family pedigrees, a genetic counselor can give prospective parents the information they need to make decisions about the risks of certain diseases in their offspring. Techniques such as **amniocentesis** and fetoscopy also provide information about genetic disease at an early stage of **pregnancy**. In addition, postnatal chemical tests can detect inherited problems in a newborn infant so that corrective procedures can be applied. Preimplantation diagnosis of embryos is another method of genetic screening which was developed to help couples that are at high risk of passing on a serious genetic disease. It is used in conjunction with *in vitro* **fertilization** (IVF) and has been applied successfully in the diagnosis of cystic fibrosis, Lesch-Nyhan syndrome, Duchenne muscular dystrophy, and Tay-Sachs

disease. In this procedure, one or two cells are removed from an IVF embryo at the 8 to 16-cell stage and the genetic material in the biopsied cells can then be amplified and analyzed for the presence or absence of a defect. Healthy embryos can be rapidly transferred to a woman's uterus for implantation, while embryos that are deemed defective may be disposed of.

Preimplantation genetic tests and research on human embryos in general present profound ethical issues, the central one relating to the moral status of the human embryo. Opponents of embryo research believe that a human embryo is already a human being, and that it must be accorded the legal status of a person from the time of fertilization. Supporters of embryo research believe that very early embryos, those up to the implantation stage of development, do not have the same moral and legal status as persons. While they acknowledge that embryos are irrefutably genetically human, they believe they do not have the same moral relevance, because they lack specific capacities, including consciousness, reasoning, and sentience. They argue that it is morally acceptable to perform limited research on embryos, particularly because of the potential therapeutic and scientific benefit the research holds for humanity. Supporters consider it ethically acceptable to make scientific investigations during the early development of embryos up to 14 days following fertilization, the period during which implantation is believed to take place. Issues of embryo experimentation are frequently being raised with respect to ongoing research into the use of embryonic **stem cells** for **gene therapy**. Although stem cells can be harvested from the bone marrow of adults, the most accessible supply is found in human embryos. Companies specializing in stem cell research stand to reap huge financial windfalls from successful therapies developed via this science. For example, it might eventually be possible to grow genetically engineered stem cells into the brains of people suffering from such conditions as Parkinson's and **Alzheimer's disease**, or to grow them into the hearts of those suffering from cardiac ailments.

The power of the new genetics clearly rests in the information that it can provide about individuals and one of the widest concerns about its possible misuse relates to privacy and confidentiality. Now that the first sequenced draft of the human genome is virtually complete there are ambitions to make sequence maps of every individuals' genome. It is thought that with this information it will, for example, be possible to design personalized drug therapies for individuals requiring treatments on the basis of their genetic and biochemical information. The information provided by complete genome maps will be of great interest both to individuals and others, including family members, employers, schools, insurance companies and legal institutions. Genetic knowledge can be a double-edged sword. On the one hand, knowledge of a predisposition to a genetic disorder opens up the possibility of treatment. On the other hand, the enhanced ability to identify genetic characteristics and disease susceptibility may affect the individuals view themselves, and also the way we perceive and are perceived by others.

One of the current concerns is to protect individuals against the invasion of privacy while preserving the ability of academic, government and industrial researchers to use anonymous genetic information for medical research. In the United States, many states are wrestling with questions such as: Who owns genetic information? Do family members have a right to know the results of a genetic test? Do the police, military, employers, insurance companies and schools have a right to know the results of a genetic test? Should pharmaceutical companies own information about an individual's **DNA** without their informed consent? Should the possibility of economic benefit play a role in deciding whether an individual's DNA might be used for research purposes? How should genetic privacy be protected and what would happen if an individual loses the ability to make decisions about how their DNA is used? In 1995, the first genetic protection law in the U.S. was signed in Oregon. A distinctive feature of this Genetic Privacy Act was that it established genetic information, defined as the information about an individual or family obtained from a genetic test or individual's DNA sample (e.g., **blood** and **tissue** samples) as the personal property of the individual from whom it is removed. In simplest terms, this means giving tissue donors access to an existing legal framework to protect their genetic privacy and defend against any violation of it. Thus, an individual can sue if research organizations, biotechnology companies, or others use their samples without their explicit informed consent.

See also Ethical issues in embryological research; Gene therapy; Pharmacogenetics

EULER, ULF VON (1905-1983)
Swedish physiologist

Ulf von Euler devoted his life to searching for the chemical signals that control physiological processes. In a career spanning six decades and during which he published four hundred and sixty-five scientific papers, Euler achieved remarkable success. While still in his twenties, he discovered both substance P and prostaglandin, two important compounds that have since been studied extensively. **Prostaglandins** have become valuable to doctors for the treatment of many disorders, and may be used to treat **blood** pressure problems, infertility, peptic ulcers, and asthma. Martin A. Wasserman, in *American Pharmacy,* wrote on the significance of prostaglandins to modern medicine, stating that, "Prostaglandin signifies more to scientists today than any medical term since cortisone. For millions of people around the world, prostaglandins hold the promise of relief from an extraordinary range of physical discomforts and life-threatening illnesses." In addition, Euler became the first person to isolate and identify noradrenaline, a key transmitter of nerve impulses, which control such involuntary functions as the heartbeat. For the later accomplishment, he was awarded the 1970 Nobel Prize in physiology or medicine.

Ulf Svante von Euler was born in Stockholm, Sweden. From the beginning, he seemed destined for scientific greatness. His father, Hans Euler-Chelpin, was a chemist who received the 1929 Nobel Prize for research into the role of **enzymes** in sugar fermentation. Euler's mother, Astrid Cleve von Euler, was a professor of botany, and his grandfather, Per Teodor Cleve, was a chemist who discovered the elements holmium and thulium. Moreover, Euler was also a distant relative of the famous eighteenth-century mathematician, Leonhard Euler.

With the help of his father, Euler coauthored his first scientific paper when he was just seventeen years old. He went on to receive a medical degree from the Karolinska Institute in Stockholm in 1930. That same year, with the aid of a Rockefeller Fellowship, von Euler traveled to London to work in the laboratory of Henry Hallett Dale, who would himself win a Nobel Prize in 1936 for discoveries relating to the chemical transmission of nerve impulses. At the time Euler arrived, one particular compound—acetylcholine—was the focus of most of the study in Dale's laboratory.

It was while Euler was conducting an experiment involving acetylcholine that he made his first significant observation. He noticed that a section of rabbit intestine would contract whenever it was exposed to an intestinal extract. Surprisingly, though, the addition of atropine to the extract fluid did not suppress the contraction, as was expected. Young von Euler exuberantly declared that he had discovered a new biologically active substance—a bold claim that was soon borne out.

Along with John H. Gaddum, a senior assistant at the lab, Euler spent the next few months systematically studying the effects of this newly identified compound. The two men demonstrated that extracts of **brain** would also contract the rabbit gut, and that the extracts that accomplished this result also had the effect of lowering the blood pressure as well. In order to carry out their investigations, the men used a purified preparation, abbreviated "P." Thus, quite unintentionally, the chemical agent causing these effects became known as Substance P. Back in Sweden, von Euler established that this substance had the properties of a polypeptide, a molecular chain of **amino acids** (the building blocks of proteins).

Euler returned to the Karolinska Institute, where he was made an assistant professor of **pharmacology** and physiology. In 1939, he was named professor and chairman of the physiology department there, a position in which he remained until his retirement in 1971.

In 1934, Euler made the second most important discovery of his career. While continuing his tests on different kinds of **tissue** extracts, he found that extracts of sheep vesicular gland dramatically lowered blood pressure when injected into animals. He realized that some unknown factor in the extracts was exerting a powerful physiological effect. Human seminal fluid also seemed to contain this unidentified substance. Soon it became clear that the factor was a fatty acid. Euler dubbed it prostaglandin, in the mistaken belief that it originated in the **prostate gland**.

During the 1930s, Euler followed up this finding, describing methods for extracting the compound, as well as defining its basic properties. However, it was not until the late 1950s that von Euler's protegé at the Karolinska Institute, Sune Karl Bergström, used newly developed technology to achieve the first purification of a prostaglandin. Euler later wrote in the scientific journal *Progress in Lipid Research,* that "a discovery is in principle like an invention, or even a piece of art, in the sense that the result is greater than the sum of its parts.... It is sometimes said that the prostaglandins lay dormant for some 20 years after their discovery. This is not exactly true, since **Sune Bergström** took over in 1945 where I left it, and with consummate skill and perseverance conducted the chemical work to isolation and identification, thus starting the second stage of the prostaglandin history." Subsequent research revealed that prostaglandins are not a single substance, but a group of chemical compounds that perform a variety of jobs throughout the body, including playing a major role in reproduction. For his contributions to the field, Bergström was one of the recipients of the 1982 Nobel Prize in physiology or medicine.

Meanwhile, Euler continued the search for chemical transmitters that allow nerve cells to communicate. The idea that such **neurotransmitters** might exist had been proposed as early as 1905, but it was not until forty-one years later that Euler succeeded in detecting a critical one in the **sympathetic nervous system**, which controls such automatic actions as the body's response to stress. He had already observed that certain biological extracts seemed to contain a substance that was similar to adrenaline, yet different in some of its actions. Euler set about pinpointing this substance, which he soon established to be noradrenaline (also called norepinephrine).

Later, Euler investigated the way certain nerve endings store and release noradrenaline. Other of his studies dealt with the role of chemical agents in regulating respiration, circulation, and blood pressure. It was for his ground-breaking experiments involving noradrenaline that Euler shared the 1970 Nobel Prize with **Julius Axelrod** of the United States and **Bernard Katz** of Great Britain, two other prominent figures in the study of chemical transmitters.

Euler was not only an eminent researcher, however; he was also known as a fine teacher who nurtured the curiosity of his pupils. An editorial which he wrote for the American journal *Circulation* sums up his approach to teaching and to science: "There are few things as rewarding for a scientist as having young students starting their research work and finding that they have... made an original observation.... the pleasure of witnessing the progress of the young starting fresh is one, which [the scientist] has every reason to feel happy about and where he can assist by means of his experience.... We must always guard the liberties of the mind and remember that some degree of heresy is often a sign of health in spiritual life."

Euler was a member of the Swedish Academy of Sciences, as well as chief editor of the journal *Acta Physiologica Scandinavica* for many years. His international reputation was solidified by numerous awards, including the Order of the North Star in Sweden, the Cruzeiro do Sul in Brazil, the Pahnes Academiques in France, and the Grand Cross Al Merito Civil in Spain, as well as the Nobel Prize.

Few scientists have been as closely identified with the Nobel Prize as Euler; not only did he win one himself but so did his father, his mentor, and his protegé. Euler served as president of the Nobel Foundation from 1966 until 1975. Von Euler died of complications following open heart surgery in 1983, in Stockholm.

See also Hormones and hormone action; Nerve impulses and conduction of impulses; Neurotransmitters

EUSTACHIAN TUBES

Left and right Eustachian tubes connect the corresponding left and right middle ears (tympanic cavities) to the back of the **nose** and throat, and function to allow the equalization of pressure in the middle **ear** air cavity with the outside (ambient) air pressure. The membrane-lined tubes, also called the auditory tubes, serve to drain secretions from the middle ear to the nasopharynx region. The Eustachian tubes are normally closed to prevent a back fill of fluid and particulate debris from the mouth and nose. Muscles in the throat and palate area control the opening and closing of the tubes. When subjected to pressure the tubes may collapse or fail to open unless pressurized. Some disease processes, especially chronic ear infections, may obstruct the tubes and result in a painful difference in pressure between the middle ear and the pressure found in the nose and throat.

Chronic ear infections usually result from Eustachian tubes that fail to close properly or that remain open. Eustachian tubes always open are termed patent or patulous Eustachian tubes. Obstructions of a Eustachian tube may result in an uncomfortable or painful feeling of pressure in the middle ear. Such sensations are often accompanied by popping and clicking noises associated with **swallowing**. The pos and clicks result form tiny expulsions of pressurized air as the muscles surrounding the Eustachian tubes exert forces that may briefly open the tubes.

Patent Eustachian tubes also allow secretions from the nose and throat to reach the middle ear and this may result in severe chronic infections of the middle ear. A failure to equalize middle ear pressure via the Eustachian tubes may also result in the creation of a partial vacuum in the middle ear that actually draws secretions into the middle ear (serous otitis media).

Pressure variations in the middle ear may also impair hearing, especially a loss of hearing acuity that results in sounds being muffled or otherwise diminished or attenuated. Proper function of the Eustachian tube is required to normalize pressure on the **tympanic membrane**. Middle ear problems can also result in poor balance.

Inflammations and secretions associated with colds, **allergies**, and sinus or respiratory infections may also block one or both of the Eustachian tubes. **Tumors** or other physical obstructions may also impair tube function and create both painful pressure and a ringing sensation in the ears (tinnitus).

Because the Eustachian tubes are located close to the adenoids, the removal of the adenoids, especially in children with smaller diameter tubes more easily obstructed, may allow the tubes to function normally. Temporary tubal inserts are more commonly used to allow drainage of the middle ear.

Proper functioning Eustachian tubes are important to divers and aviators who experience often rapid and significant changes in outside pressure. Although commercial airplanes are usually pressurized, there are significant enough changes in pressure during climbs and descents to cause **pain** in individuals suffering even slight tube blockage. Some people attempt to assure and improve Eustachian tube function during flights by chewing gum during airline flights. The combined muscle movements associated with chewing and swallowing (**saliva**) work to frequently open and close the Eustachian tubes and permit middle ear pressure equalization.

The external **carotid arteries** supply oxygenated **blood** to the Eustachian tubes via their respective pharyngeal branches. The maxillary artery derived middle meningeal artery and pterygoid canal artery also supply blood to the Eustachian tubes. Blood drains form the tubes via the pterygoid venous plexus. The tubes are innervated (supplied with nerves) from a tympanic plexus derived from the cervical sympathetic, glossopharyngeal, and facial nerves.

See also Aviation physiology; Ear (external, middle and internal); Ear: Otic embryological development; Palate (hard and soft palate); Sense organs: Balance and orientation; Swallowing and disphagia; Underwater physiology; Valsalva maneuver

EUSTACHIO, BARTOLOMEO (CA. 1513-1574)

Italian physician

Bartolomeo Eustachio (known as Eustachius) wrote a remarkable series of scientific works on the following subjects: **anatomy** of the kidney, the hearing apparatus, the teeth, and the **circulatory system**, during 1562 and 1563. These works were organized and published as the *Opuscula Anatomica* in 1564.

Bartolomeo's father, Mariano Eustachius, was an affluent physician, in San Severino, Ancona, Italy, where Bartolomeo was born. Bartolomeo received a vast humanistic education, a requirement of the academic formation at that time, and studied Medicine at the Archiginnasio della Sapienza in Rome. He was also well versed in Hebrew, Arabic and Greek languages, which gave him access to the original medical treatises written in those languages. As a physician, Eustachius enjoyed great prestige among the upper classes, having among his patients the Duke of Urbino, the Cardinal della Rovero, the Duke of Terranova. He became a member of the Medical College of Rome and was appointed in 1549, Professor of Anatomy at the Papal College, the Archiginnasio della Sapienza. Eustachius' anatomical studies of the ear yielded an accurate description of the auditory tube, which to this day is known as the Eustachian canal.

Eustachius was deeply interested in understanding the anatomical structures of the human body through direct observation instead of accepting the many *a priori* theories current among other physicians. His anatomical investigations were not only vast but also remarkable, including the structure of the teeth, lower cava vein valve, known as Eustachian valve, which he described in detail, rightly concluding that its function was to avoid **blood** reflux. He also discovered the thoracic canal. Trying to understand how diseases affected body structures, Eustachius made comparative anatomical analysis between healthy and disease-altered **organs** (pathological anatomy). Working with Pier Matteo Pini, they produced a series of 47 detailed drawings of the studied organs. These series of illustrations, *Tabulae Anatomicae Clariviri,* were published only in 1714. Eustachius died in Umbria, in 1574, during a trip to meet the Cardinal della Rovere.

See also Eustachian tubes; History of anatomy and physiology: The Renaissance and Age of Enlightenment

EVOLUTION AND EVOLUTIONARY MECHANISMS

Evolution is the process of biological change over time. Such changes, especially at the genetic level are accomplished by a complex set of evolutionary mechanisms that act to increase or decrease genetic variation.

Evolutionary theory is the cornerstone of modern biology, and unites all the fields of biology under one theoretical umbrella to explain the changes in any given gene pool of a population over time. Evolutionary theory is theory in the scientific usage of the word. It is more than a hypothesis; there is an abundance of observational and experimental data to support the theory and its subtle variations. These variations in the interpretation of the role of various evolutionary mechanisms are due to the fact that all theories, no matter how highly useful or cherished, are subject to being discarded or modified when verifiable data demand such revision. Biological evolutionary theory is compatible with nucelosynthesis (the evolution of the elements) and current cosmological theories in physics regarding the origin and evolution of the universe. There is no currently accepted scientific data that is incompatible with the general postulates of evolutionary theory, and the mechanisms of evolution.

Fundamental to the concept of evolutionary mechanism is the concept of the syngameon, the set of all genes. By definition, a gene is a hereditary unit in the syngameon that carries information that can be used to construct proteins via the processes of transcription and translation. A gene pool is the set of all genes in a species or population.

Another essential concept, important to understanding evolutionary mechanisms, is an understanding that there are no existing (extant) primitive organisms that can be used to study evolutionary mechanism. For example, all eukaryotes derived from a primitive, common prokaryotic ancestral bacterium. Accordingly, all living eukaryotes have evolved as eukaryotes for the same amount of time. Additionally, no eukaryote plant or animal cell is more primitive with regard to the amount of time they have been subjected to evolutionary mechanisms. Seemingly primitive characteristics are simply highly efficient and conserved characteristics that have changed little over time.

Evolution requires genetic variation, and these variations or changes (mutations) can be beneficial, neutral or deleterious. In general, there are two major types of evolutionary mechanisms, those that act to increase genetic variation, and mechanisms that operate to decrease genetic mechanisms.

Mechanisms that increase genetic variation include mutation, recombination and gene flow.

Mutations generally occur via chromosomal mutations, point mutations, frame shifts, and breakdowns in **DNA** repair mechanisms. Chromosomal mutations include translocations, inversions, deletions, and chromosome non-disjunction. Point mutations may be nonsense mutations leading to the early termination of **protein synthesis**, missense mutations (a that results an a substitution of one amino acid for another in a protein), or silent mutations that cause no detectable change.

Recombination involves the re-assortment of genes through new chromosome combinations. Recombination occurs via an exchange of DNA between homologous **chromosomes** (crossing over) during meiosis. Recombination also includes linkage disequilibrium. With linkage disequilibrium, variations of the same gene (alleles) occur in combinations in the gametes (sexual reproductive cells) than should occur according to the rules of probability.

Gene flow occurs when individuals change their local genetic group by moving from one place to another. These migrations allow the introduction of new variations of the same gene (alleles) when they mate and produce offspring with members of their new group. In effect, gene flow acts to increase the gene pool in the new group. Because genes are usually carried by many members of a large population that has undergone random mating for several generations, random migrations of individuals away from the population or group usually do not significantly decrease the gene pool of the group left behind.

In contrast to mechanisms that operate to increase genetic variation, there are fewer mechanisms that operate to decrease genetic variation. Mechanisms that decrease genetic variation include genetic drift and natural selection.

Genetic drift results form the changes in the numbers of different forms of a gene (allelic frequency) that result from **sexual reproduction**. Genetic drift can occur as a result of random mating (random genetic drift) or be profoundly affected by geographical barriers, catastrophic events (e.g., natural disasters or wars that significantly affect the reproductive availability of selected members of a population), and other political-social factors.

Natural selection is based upon the differences in the viability and reproductive success of different genotypes with a population (differential reproductive success). Natural selection can only act on those differences in genotype that appear as phenotypic differences that affect the ability to attract a mate and produce viable offspring that are, in turn, able to live,

mate and continue the species. Evolutionary fitness is the success of an entity in reproducing (i.e., contributing alleles to the next generation).

There are three basic types of natural selection. With directional selection an extreme phenotype is favored (e.g., for height or length of neck in giraffe). Stabilizing selection occurs when intermediate phenotype is fittest (e.g., neither too high or low a body weight) and for this reason it is often referred to a normalizing selection. Disruptive selection occurs when two extreme phenotypes are fitter that an intermediate phenotype.

Natural selection does not act with foresight. Rapidly changing environmental conditions can, and often do, impose new challenges for a species that result in extinction. In addition, evolutionary mechanisms, including natural selection, do not always act to favor the fittest in any population, but instead may act to favor the more numerous but tolerably fit. Thus, the modern understanding of evolutionary mechanisms does not support the concepts of social Darwinism.

The operation of natural evolutionary mechanisms is complicated by geographic, ethnic, religious, and social groups and customs. Accordingly, the effects of various evolution mechanisms on human populations are not as easy to predict. Increasingly sophisticated statistical studies are carried out by population geneticists to characterize changes in the human genome.

See also Genetic code; Genetic regulation of eukaryotic cells; Genetics and developmental genetics; Human genetics

EXCRETORY SYSTEM • *see* ELIMINATION OF WASTE

EXOCYTOSIS • *see* CELL MEMBRANE TRANSPORT

EXTENSION • *see* ANATOMICAL NOMENCLATURE

EXTERNAL EAR • *see* EAR (EXTERNAL, MIDDLE, AND INTERNAL)

EXTRACELLULAR FLUID

The body fluids found outside the cells, such as **plasma** (the liquid portion of **blood** and lymph), and **interstitial fluid**, are generically termed extracellular fluid. Three-fourths of all extracellular fluid is stored as interstitial (between cells) fluid, and one-fourth as plasma. Cells in the tissues are separated from one another by spaces called interstitium, usually filled with a fibrous complex termed extracellular matrix. This structure is formed by collagen and elastin fibers and elongated glycoproteins, also containing proteoglycan filaments that form a hydrophilic gel. Whereas collagen and elastin fibers have a structural function, proteoglycan form a network of coiled thin filaments, mainly constituted by hyaluronic acid, which entrap a viscous fluid known as interstitial fluid, rich in nutrients, ions, **hormones**, and other molecules necessary to cellular function. The interstitial fluid has almost the same composition of plasma, except for having much lower protein concentrations, and it is filtered to the extracellular matrix from the capillary vessels by diffusion (i.e., capillary pressure). However, the plasma proteins also exert osmotic pressure (plasma colloid osmotic pressure), leading fluid from the interstitium back to **capillaries**, thereby avoiding a significant loss of water in the blood.

The extracellular fluid filtered from the blood vessels (arterial capillaries) into the extracellular matrix flows among the cells transporting nutrients and chemical messengers and receives from cells metabolites, ions, proteins, and other substances, and is then reabsorbed by either the venous or the lymphatic capillaries. Substances with a low molecular weight present in the interstitial fluid are easily absorbed through the venous capillary walls into the blood, whereas those with high molecular weight such as proteins are absorbed by the lymphatic capillaries. The circulating lymph is, therefore, originated from the extracellular fluid. The ionic compositions of plasma and interstitial fluid are also similar, due to the high permeability of capillary membranes that separate these two media.

The regulation of extracellular fluid volume is crucial to avoid **edema** and for the maintenance of arterial pressure. Edema, or the excessive accumulation of fluid in tissues, is normally prevented by the elimination of water and specific amounts of ions through diuresis, perspiration, and feces, along with other regulatory mechanisms that avoid excessive loss of water from the organism as well. There are three safety mechanisms that prevent serious edema: 1), elimination of interstitial fluid protein concentrations, which reduces colloid osmotic pressure and increases capillary filtration; 2), increased lymphatic flow; and 3), the interstitial gel formation that prevents the easy flow of fluid through the tissues at the negative pressure range of less than one atmosphere (about 3mm Hg). However, hormonal or electrolytic imbalance, deregulation of renal functions, **heart** failure, and other factors affecting protein plasma levels or capillary permeability, as well as certain diseases and/or trauma, can cause the accumulation of an excess of extracellular fluid in tissues. Depending on the affected system and the degree of extracellular edema, serious physiological impairing conditions, even **shock** and **death** may occur.

See also Diuretics and antidiuretic hormones; Osmotic equilibria between intercellular and extracellular fluids

EYE: OCULAR EMBRYOLOGICAL DEVELOPMENT

Embryological development of the ocular system is the formation of the **eye** during **pregnancy**. While the process continues throughout gestation, it is not complete even at birth. The newly formed eyes are unable to see the same as those of an adult. Babies have much lower numbers of photoreceptors and the ones present are still immature, resulting in poor

vision, especially at distances (myopia). It takes another two to three years before the child's eyes are fully functional.

Ocular tissues are derived from all embryonic layers: ectoderm—surface **ectoderm** (lens, all eye epithelia, **epidermis** of the eyelids, lacrimal gland) and neural ectoderm (**retina**, retinal pigment epithelium, optic nerve), **mesoderm** (**blood** vessel endothelium) and **endoderm**.

The human eye begins to develop in the second week of pregnancy with formation of the optic primordium. This develops into the optic vesicle at 22 days, which soon forms easily recognizable optic cup. In the part distal from the surface ectoderm it is possible to identify the developing retina. Simultaneously at the surface ectoderm, the lens begins to form and becomes fully identifiable within five days (32 days). After separation from the surface ectoderm during the fifth week, the lens vesicle forms reaching the definite size at eight weeks. However, the lens development does not stop after birth, but continues during one's lifetime.

An important connection between the optic cup and **brain** happens at 36 days when the optic stalk connects the cup to the forebrain. During this early eye development, a hyaloid artery comes into the space in the optic cup between the forming lens and retina supplying the nutrients.

Parallel to the development of the retinal layer and lens, the vitreous is formed, which will fill the chamber between the two structures. Development of the retina is a very long process. It takes approximately four months to form the photoreceptor precursors and the retinal vessels. The retinal layers are fully developed at six months and the central retinal artery is fully functional by eight months, having developed from the hyaloid artery, which itself regresses from the vitreous. The macular area, responsible for color vision starts forming at four to five months, but does not mature until after birth at approximately six months.

Development of the anterior part of the eye happens very early during embryogenesis as the ocular structures develop and differentiate sequentially from the optic cup tissues. The **cornea** forms initially as a single cell layer soon after the lens at 33 days, and during the next two months, transforms into outer (epithelial) and inner (stromal) layers. However, it becomes innervated at five months, at the same time as the Bowman's membrane covering the cornea develops. Moreover, the cornea remains non-transparent due to high hydration until **tissue** maturation is complete. The iris starts to develop from the rim of the optic cup at 30–35 days. Once the iris is formed, the anterior eye chamber develops from seven weeks. The muscles regulating the iris (sphincter and dilator), however, do not develop till five to six months and pigmentation occurs after birth.

The developing eyes need protection, and the eyelids start to form at two months and fuse together allowing the development of additional structures such as muscles, **connective tissues** and tarsal plates. Starting from the fifth month, the eyelids begin to separate and the process is complete in two months.

Functionally important parts of the eye are the extraocular muscles, which allow eye movement. These start forming at four weeks, at the same time as the orbits form, their size being predetermined by the size of the optic cup. Depending on the postnatal growth of the globe, the orbits might mature at anytime from two to sixteen years postnatally.

Lacrimal glands that will provide tears, an important part of eye protection, start forming at six weeks, but the maturation of the epithelial cords to form a hollow channel happens postnatally. Moreover, the tears are not formed in the glands until the third month.

As the back of the eye develops and becomes innervated, a protective layer of sclera forms in the anterior part of the eye at seven weeks, and within five more weeks, surrounds the optic nerve at the posterior part of the eye.

See also Sense organs: Embryonic development; Eye and ocular fluids; Sense organs: Ocular (visual) structures

EYE AND OCULAR FLUIDS

The eye is the organ of sight in humans and animals, transforming light waves into visual images and providing 80% of all information received by the human **brain**. These remarkable **organs** are almost spherical in shape and are housed in the orbital sockets in the **skull**. Sight begins when light waves enter the eye through the **cornea** (the transparent layer at the front of the eye), pass through the pupil (the opening in the center of the iris, the colored portion of the eye), then through a clear lens behind the iris. The lens focuses light onto the **retina** which functions like the film in a camera. Photoreceptor **neurons** in retinas, called rods and cones, convert light energy into electrical impulses, which are then carried to the brain via the optic nerves. At the visual cortex in the occipital lobe of the cerebrum of brain, the electrical impulses are interpreted as images.

The human eyeball is about 0.9 in (24 mm) in diameter and is not perfectly round, being slightly flattened in the front and back. The eye consists of three layers: the outer fibrous or sclera, the middle uveal or choroid layer, and the inner nervous layer or retina. Internally the eye is divided into two cavities—the anterior cavity filled with the watery aqueous fluid, and the posterior cavity filled with gel-like vitreous fluid. The internal pressure inside the eye (the intraocular pressure) exerted by the aqueous fluid supports the shape of the anterior cavity, while the vitreous fluid holds the shape of the posterior chamber. An irregularly shaped eyeball results in ineffective focusing of light onto the retina and is usually correctable with glasses or contact lenses. An abnormally high intraocular pressure, due to overproduction of aqueous fluid or to the reduction in its outflow through a duct called the canal of Schlemm, produces glaucoma, a usually painless and readily treatable condition, which may lead to irreversible **blindness** if left untreated. Elevated intraocular pressure is easily detectable with a simple, sight-saving, pressure test during routine eye examinations. The ophthalmic **arteries** provide the **blood** supply to the eyes, and the movement of the eyeballs is facilitated by six extraocular muscles which run from the bony orbit which insert the sclera, part of the fibrous tunic.

The outer fibrous layer encasing and protecting the eyeball consists of two parts—the cornea and the sclera. The front one-sixth of the fibrous layer is the transparent cornea, which **bends** incoming light onto the lens inside the eye. A fine **mucus** membrane, the conjunctiva, covers the cornea, and also lines the eyelid. Blinking lubricates the cornea with tears, providing the moisture necessary for its health. The cornea's outside surface is protected by a thin film of tears produced in the lacrimal glands located in the lateral part of orbit below the eyebrow. Tears flow through ducts from this gland to the eyelid and eyeball, and drain from the inner corner of the eye into the nasal cavity. A clear watery liquid, the aqueous humor, separates the cornea from the iris and lens. The cornea contains no blood vessels or pigment and gets its nutrients from the aqueous humor. The remaining five-sixths of the fibrous layer of the eye is the sclera, a dense, tough, opaque coat visible as the white of the eye. Its outer layer contains blood vessels which produce a "blood-shot eye" when the eye is irritated. The middle or uveal layers of the eye is densely pigmented, well supplied with blood, and includes three major structures—the iris, the ciliary body, and the choroid. The iris is a circular, adjustable **diaphragm** with a central hole (the pupil), sited in the anterior chamber behind the cornea. The iris gives the eye its color, which varies depending on the amount of pigment present. If the pigment is dense, the iris is brown, if there is little pigment the iris is blue, if there is no pigment the iris is pink, as in the eye of a white rabbit. In bright light, muscles in the iris constrict the pupil, reducing the amount of light entering the eye. Conversely, the pupil dilates (enlarges) in dim light, so increasing the amount of incoming light. Extreme fear, head injuries, and certain drugs can also dilate the pupil.

The iris is the anterior extension of the ciliary body, a large, **smooth muscle** which also connects to the lens via suspensory ligaments. The muscles of the ciliary body continually expand and contract, putting on suspensory ligaments changing the shape of the lens, thereby adjusting the focus of light onto the retina facilitating clear vision. The choroid is a thin membrane lying beneath the sclera, and is connected the posterior section of the ciliary body. It is the largest portion of the uveal tract. Along with the sclera the choroid provides a light-tight environment for the inside of the eye, preventing stray light from confusing visual images on the retina. The choroid has a good blood supply and provides oxygen and nutrients to the retina.

The front of the eye houses the anterior cavity which is subdivided by the iris into the anterior and posterior chambers. The anterior chamber is the bowl-shaped cavity immediately behind the cornea and in front of the iris, which contains aqueous humor. This is a clear watery fluid which facilitates good vision by helping maintain eye shape, regulating the intraocular pressure, providing support for the internal structures, supplying nutrients to the lens and cornea, and disposing of the eye's metabolic waste. The posterior chamber of the anterior cavity lies behind the iris and in front of the lens. The aqueous humor forms in this chamber and flows forward to the anterior chamber through the pupil.

The posterior cavity is lined entirely by the retina, occupies 60% of the human eye, and is filled with a clear gel-like substance called vitreous humor. Light passing through the lens on its way to the retina passes through the vitreous humor. The vitreous humor consists of 99% water, contains no cells, and helps to maintain the shape of the eye and support its internal components.

The lens is a crystal-clear, transparent body which is biconvex (curving outward on both surfaces), semi-solid, and flexible, shaped like an ellipse or elongated sphere. The entire surface of the lens is smooth and shiny, contains no blood vessels, and is encased in an elastic membrane. The lens is sited in the posterior chamber behind the iris and in front of the vitreous humor. The lens is held in place by suspensory ligaments that run from the ciliary muscles to the external circumference of the lens. The continual relaxation and contraction of the ciliary muscles cause the lens to fatten or to became thin, changing its focal length, and allowing it to focus light on the retina. With age, the lens hardens and becomes less flexible, resulting in far-sighted vision that necessitates glasses, bifocals, or contact lenses to restore clear, close-up vision. Clouding of the lens also often occurs with age, creating a cataract that interferes with vision. Clear vision is restored by a relatively simple surgical procedure in which the entire lens is removed and an artificial lens implanted.

The retina is the innermost layer of the eye. The retina is thin, delicate, extremely complex sensory **tissue** composed of layers of light sensitive nerve cells. The retina begins at the ciliary body and encircles the entire posterior portion of the eye. Photoreceptor cells in the rods and cones, convert light first to chemical energy and then electrical energy. Rods function in dim light, allowing limited nocturnal (night) vision: it is with rods that we see the stars. Rods cannot detect color, but they are the first receptors to detect movement. There are about 126 million rods in each eye and about six million cones. Cones provide acute vision, function best in bright light, and allow color vision. Cones are most heavily concentrated in the central fovea, a tiny hollow in the posterior part of the retina and the point of most acute vision. Dense fields of both rods and cones are found in a circular belt surrounding the fovea, the macula lutea. Continuing outward from this belt, the cone density decreases and the ratio of rods to cones increases. Both rods and cones disappear completely at the edges of the retina.

The optic nerve connects the eye to the brain. These fibers of the optic nerve run from the surface of the retina and converge at exit at the optic disc (or blind spot), an area about 0.06 in (1.5 mm) in diameter located at the lower posterior portion of the retina. The fibers of this nerve carry electrical impulses from the retina to the visual cortex in the occipital lobe of the cerebrum. If the optic nerve is severed, vision is lost permanently.

See also Eye: Ocular embryological development; Sense organs: Ocular (visual) structures; Vision: Histophysiology of the eye

F

FABRICI, GIROLAMO (1537-1619)

Italian anatomist

Girolamo Fabrici (Fabricius) is perhaps best known as the founder of the modern science of **embryology**. From 1600 until his death, he carried out important and original research on the late fetal stages of many different animals. In 1612, he published the first detailed description of the development of the chick embryo from the sixth day onward.

Fabricius made a number of additional advances in the field of **anatomy**. He studied the structure of the **eye** and of the **larynx**, the mechanics of respiration, and the movement of muscles. One of his most important accomplishments was his detailed description of the semilunar valves in **veins**. Although these valves had been observed earlier, Fabricius published the most complete description of their **structure and function** in his *De venarum ostiolis* (On the Valves of the Veins) in 1603. He was incorrect, however, in his explanation of the valves' function.

Girolamo (also Geronimo) Fabrici (also Fabrizio or Fabricius) was born in the town of Aquapendente, near Orvieto, Italy. He studied first humanities, then medicine, at the University of Padua. His teacher in anatomy and surgery was **Gabriel Fallopius**. After graduating with his M.D. in 1559, he taught anatomy privately and practiced surgery until 1565. In that year, he was appointed to replace Fallopius, who had died, at Padua. Fabricius held that post until 1613, when he retired because of ill health. He died in Bugazzi, near Padua, in 1619.

In addition to his scholarly work, Fabricius made other contributions. He attained considerable fame as a teacher. His most famous student was **William Harvey**, who studied with him from 1597 to 1602. In addition, Fabricius was instrumental in establishing the first permanent anatomical theater at the University of Padua.

See also Embryonic development: Early development, formation, and differentiation; History of anatomy and physiology: The Renaissance and Age of Enlightenment

FACE, NOSE, AND PALATE EMBRYOLOGICAL DEVELOPMENT

In humans, the development of the face and its features such as the **nose** and palate occurs in the embryonic stage of fetal development. The molecular events that set in motion these developments occur within the first few weeks following **fertilization** of an egg with a **sperm**. By the end of the sixth week there is visual evidence of development, and by the end of the eighth week of fetal development, the face is structurally recognizable as that of a human.

During these critical weeks of fetal development, the head region of the embryo assumes a pronounced bulb-like shape. A series of pouches, branchial arches, project forward from this protuberance. The head and face arise from the joining together of cells that originate from one of the brachial arches and other so-called "prominances" or "primordia", such as the neural crest cells involved in the formation of the spinal column. For example, the front region of a prominance called the frontonasal prominence is involved in forming the nose.

Refinement of the facial structures occurs from the eighth week of development through until birth, normally at about 40 weeks. The proportion and relative positions of the facial features become established. For example, the nose changes from a flat structure assume its more jutting-out appearance.

Much of our knowledge of facial development in embryos comes from studies of model organisms, such as birds. Less molecular detail is known about humans, since the face is constructed very early in embryonic development and access of mammalian embryos is limited.

The development of the face is a dynamic and multi-step process. Molecular signals trigger and coordinate the development of undifferentiated cells into the intricate series of bones and **cartilage** structures that forms the human face. Precise migration of cells gives rise to the nose, palate and other features of the face. Put another way, the imprecise migration of these cells will produce defects in facial formation.

The most common of the diverse group of disorders is cleft lip palate, which occurs in one out of 500 to 2,000 live births. The cost of treatment, including surgery, for each child over his/her lifetime can amount to over $100,000. The causes of these disorders ranges from genetic defects to environmental factors (such as tobacco, alcohol and anti-convulsant drugs), either alone or in combination.

The term palate refers to the structure, consisting of hard and soft components, which forms the roof of the mouth and the floor of the nose. These components form by the fusion of cells from different primordia. Improper fusion produces a cleft, or split, in the palate. The split can extend upward to include the upper lip and the nose. Recent research indicates that the molecular basis of the problem may be a malfunctioning gene that acts to slow down the activity of other genes involved in palate formation and fusion. As some facial development proceeds normally, other activities lag, and the result is improper assembly of the facial components.

See also Neonatal growth and development; Pregnancy

FACILITATED DIFFUSION · *see* CELL MEMBRANE TRANSPORT

FALLOPIAN TUBES · *see* REPRODUCTIVE SYSTEM AND ORGANS (FEMALE)

FALLOPIUS, GABRIEL (1523-1562)
Italian anatomist

Gabriel Fallopius was one of the most noteworthy Italian anatomists of the sixteenth century. His family lived in poverty and, as a young man, he served the Catholic Church. Fallopius studied at Ferrara and then at Pisa, and then had the opportunity to study **anatomy** in Padua, which at that time was considered the best place for anatomical study. Other areas of Europe were not as advanced in the biological sciences. For example, scientists in France thought that the work of **Galen** could not be improved upon, and the teaching of the natural sciences were being suppressed in Germany by the ongoing religious struggles. Fallopius was a student of **Andreas Vesalius** who, through his method and technique, laid the foundation for modern anatomy and is considered to be one of the most important scientists in history. As Castiglione has pointed out, according to the eminent medical historian, C. V. Daremberg, "Vesalius a genius while Fallopius was only a scientist."

By the age of twenty-four, Fallopius became a professor in Ferrara, Italy. Several years after Vesalius's death, Fallopius taught at Padua, where he was entirely supported by the government and continued in his mentor's tradition of attention to detail. Fallopius became very well known as somewhat of a pioneer in his field and his lectures were attended by large audiences. In addition to his research, lecturing, and teaching, Fallopius was a physician and surgeon, and maintained an extensive medical practice. During his career Fallopius pub-

lished *Observationes anatomicae*, which contained many descriptions of his anatomical research. The first edition of his book was published in Venice in 1561, and was followed by later editions published in Italy and several other countries. His collected works, *Opera omnia*, were published after his death, in Venice (1584), Frankfort (1600), and once again in Venice (1606). Fallopius made several important anatomical discoveries and improved upon many of his predecessor's findings. He performed an extensive study of the structures of the ear and was the first anatomist to describe the semicircular canals (*chorda tympani*). Fallopius was also the first anatomist to describe the circular folds of the small intestine and the inguinal band, later called Poupart's ligament. He corrected Vesalius's findings on the course of the cerebral **arteries**, and provided a more detailed description of the ocular muscles and cerebral nerves. Perhaps his best known discoveries are the structures of the male and female reproductive organs—he described the clitoris and what are now known as the Fallopian tubes, as well as the *arteria profunda* of the penis.

See also History of anatomy and physiology: The Renaissance and Age of Enlightenment; Reproductive system and organs (female); Reproductive system and organs (male)

FASCIA

Fascia is a type of connective **tissue** made up of a network of fibers. It is best thought of as being the packing material of the body. Fascia surrounds muscles, bones, **joints** and lies between the layers of skin. It functions to hold these structures together, protecting these structures and defining the shape of the body. When surrounding a muscle, fascia helps prevent a contracting muscle from catching or causing excessive friction on neighboring muscles.

Fascia is made up of two materials: collagen and elastin. Collagen can change from a fluid to a solid and visa versa, depending on the forces acting on it. Because of this property, collagen is a colloid. Elastin is a network of protein chains. When stressed, the network becomes organized in the direction of the stress, so as to provide some resistance. Under relaxed conditions, elastin assumes a more disorganized structure.

Fascia contains two layers. The so-called superficial layer is attached to the skin. A large proportion of this connective tissue contains **fat**. The superficial layer serves to anchor the skin to underlying tissues. This layer is especially plentiful in the scalp, back of the neck and the palms of the hands. Elsewhere in the body, the layer can be more loosely associated with the skin, and so can be moved more easily. The second layer of fascia is called the deep fascia. It is located under the superficial layer, and is joined to this layer by strands of fiber. The deep fascia helps to organize muscles into their various functional groups.

When fascia is under chronic tension, it tends to become shorter and harder. This reduces its ability to react to stresses. A well-known example in athletes is plantar fasciitis, which is damage to the plantar fascia located on the bottom of the foot. Swelling of the plantar fascia can be painful. The condition can be corrected by relieving the stress through the use of orthotics, or the tempering of the activity that causes the swelling (such as jogging). Stretching exercises can cause the fascia to regain some of it pliability.

More severe cases of fascitis may require the use of anti-inflammatory drugs or surgery.

See also Connective tissues; Elastic cartilage; Neck muscles and fasciae

FAT, BODY FAT MEASUREMENTS

Fat is a component both of the food ingested and of the body. Fat exists in a number of structural forms. Some types of fat can be beneficial, as an energy and an insulation source, and, in the case of fats containing high levels of omega-3 fatty acids, perhaps as an aid to retard the development of **cancer**. Other types of fat are detrimental to health, exemplified by the artery-clogging form of cholesterol, which can promote the damage or **death** of **heart tissue**.

There are four types of fat, based on their structure: saturated fat, monosaturated fat, polyunsaturated fat, and trans fat.

In a saturated fat, all the chemical connections between the component molecules are fully occupied. This produces fat that is hard at room temperature. Saturated fats, typically of animal origin, are the least healthy. Saturated fats can contribute to the buildup of a compound called low density cholesterol, which in turn can lead to formation of solid deposits in the **arteries** (another form of cholesterol, high density cholesterol, is "good" in that it helps carry cholesterol out of tissues). The restricted flow of **blood** through these clogged arteries can cause debilitating or fatal heart malfunctions. Monosaturated fat is found in vegetable oils. In conjunction with a low-fat diet, monosaturated fat can act to reduce cholesterol levels. This fat is liquid at room temperature.

In an unsaturated fat, not all the chemical connections between the component molecules are occupied. This produces a compound that is liquid at room temperature. Polyunsaturated fats, where more than one connection is involved, are known as essential fatty acids, and are very desirable for a health diet, causing a reduction in the level of the "bad" cholesterol. Plant oils such as soybean, cottonseed, corn, sunflower and safflower contain polyunsaturated fat.

A most undesirable fat is the so-called trans fat, formed commercially by the addition of hydrogen molecules to produce an unsaturated-like fat having a palatable taste. Trans fats both raise the level of "bad" cholesterol and lower the level of "good" cholesterol. Trans fats are present in most margarines.

In proper amounts and type, body fat serves a useful and healthy function as an energy source, component of membranes that surround cells, and as a temperature insulator. The body fat content can be determined in a number of ways. The percentage of body fat can be calculated from equations based on the density of the body, involving weighing a person suspended on a trapeze in the air and then weighing the person when immersed in water. Immersion measurement has been the reference standard of body fat determination. However, it does not take into account the location of the body fat, and the

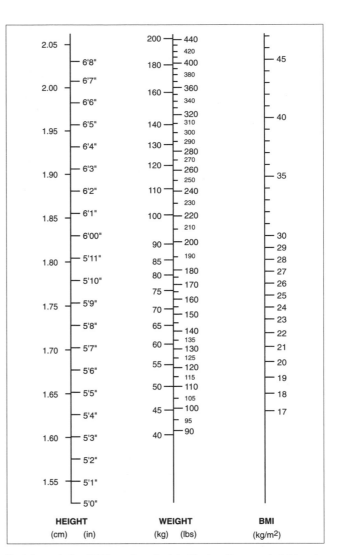

Body/mass index (BMI) can be calculated by locating your height and weight on the chart and drawing a diagonal line between the two. Where the line crosses over the third bar is the approximate BMI. *Illustration by Argosy Publishing.*

equation used in the calculation was derived from studies on young Caucasians. Ethnic variations in body fat composition would not be recognized by the immersion method.

Other methods used to estimate body fat are skinfold measurement with calipers, the shift in the wavelength of near infrared light following its interaction with fat, determination of the so-called body mass index (BMI); the relationship between weight and height, and measurements of the circumference of various parts of the body.

See also Adipose tissue; Carbohydrates; Metabolism

FAT EMBOLISM · *see* EMBOLISM

FECES AND EXCRETION · *see* ELIMINATION OF

WASTE

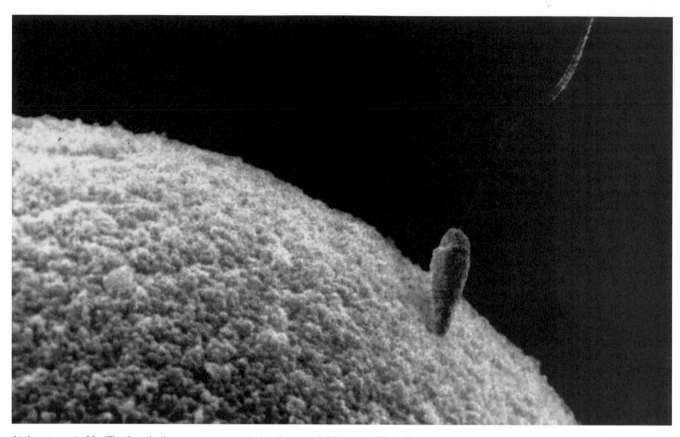

At the moment of fertilization, the human sperm penetrates the egg. *D. W. Fawcett/Photo Researchers, Inc. Reproduced by permission.*

FERTILIZATION

In animals, fertilization is the fusion of a **sperm** cell with an egg cell. The penetration of the egg cell by the chromosome-containing part of the sperm cell causes a reaction, which prevents additional sperm cells from entering the egg. The egg and sperm each contribute half of the new organism's genetic material. A fertilized egg cell is known as a zygote. The zygote undergoes continuous **cell division**, which eventually produces a new multicellular organism.

Fertilization in humans occurs in oviducts (fallopian tubes) of the female reproductive tract and takes place within hours following sexual intercourse. Only one of the approximately 300 million sperm released into a female's vagina during intercourse can fertilize the single female egg cell. The successful sperm cell must enter the uterus and swim up the fallopian tube to meet the egg cell, where it passes through the thick coating surrounding the egg. This coating, consisting of sugars and proteins, is known as the zona pellucida. The tip of the head of the sperm cell contains **enzymes** that break through the zona pellucida and aid the penetration of the sperm into the egg. Once the head of the sperm is inside the egg, the tail of the sperm falls off, and the perimeter of the egg thickens to prevent another sperm from entering.

The sperm and the egg each contain only half the normal number of **chromosomes**, a condition known as haploid.

When the genetic material of the two cells fuses, the fertilization is complete.

In humans, a number of variables affect whether or not fertilization occurs following intercourse. One factor is a woman's ovulatory cycle. Human eggs can only be fertilized a few days after ovulation, which usually occurs only once every 28 days.

Artificial insemination, in humans or animals, occurs when sperm is removed from the male and injected into the vagina or uterus of the female. This is a common treatment for human infertility. The development of gamete intra-fallopian transfer (GIFT) technology has further improved the treatment of infertility. In this procedure, sperm and eggs are placed together in the woman's fallopian tube and fertilization progresses naturally.

Fertilization occurring outside of the body is *in vitro* (in a dish or test tube) fertilization, IVF. Eggs are removed surgically from the female's reproductive tract and fertilized with sperm. At the 4-cell (day 2) stage, the embryos are returned to the female's fallopian tube or uterus where development continues. Mammalian IVF has been performed successfully on animals since the 1950s, and the first human birth following IVF occurred in Great Britain in 1978. This procedure has since become a routine treatment for infertility. If the sperm is too weak to penetrate the egg, or if the male has a very low sperm count, an individual sperm can be injected directly into the egg. Both eggs and sperm can be frozen for later use in IVF. A mechanical "sperm sorter" that separates sperm according to the amount of **DNA** each contains, can allow couples to choose the sex of their child. This is because sperm containing an X chromosome, which results in a female embryo, contains more DNA than sperm with a Y chromosome, which would yield a male embryo.

See also Reproductive system and organs (female); Reproductive system and organs (male); Sexual reproduction

FETAL CIRCULATION

Throughout intrauterine life, maternal **blood** supplies the fetus with oxygen and nutrients, and carries away its wastes. The anatomical structures that allow for such an exchange are the umbilical blood vessels. The umbilical vein transports oxygenated blood and nutrients from the **placenta** to the fetus. The umbilical vein enters the body through the umbilical ring and travels along the anterior abdominal wall to the **liver** inside the ventral mesentery, which in this region later forms the falciform ligament of the liver. About 50% of the blood it carries passes into the portal venous system of the liver. The remaining 50% enters a vessel called the ductus venosus which bypasses the liver. The ductus venosus travels a short distance and joins the inferior vena cava. There, the oxygenated blood from the placenta is mixed with the deoxygenated blood from the fetal lower extremities, pelvis and **kidneys**. Inside the ductus venosus, there is a sphincteral mechanism able to control the blood flow to the fetal **heart** in stress situations as well as during uterine contractions.

This mixture of oxygenated and deoxygenated blood continues through the inferior vena cava to the right atrium. A large amount of blood is shunted from the right atrium through the foramen ovale (an opening between right and left atrium that directs the flow right to left) and into the left atrium. The septum primum, located on the left side of the atrial septum and overlying the foramen ovale, helps prevent blood from moving in the reverse direction. A smaller amount of blood, however, is stopped from entering the foramen by the septum secundum or the crista dividens, and remains in the right atrium to mix with the deoxygenated blood returning from the head and upper limbs via the superior vena cava. The right atrium also receives blood from the superior vena cava. Blood from the right atrium, passes into the right ventricle and out through the pulmonary trunk. However only five to ten percent of the overall cardiac output goes to the **lungs**, because most of the blood in the pulmonary trunk bypasses the lungs by entering a fetal vessel called the **ductus arteriosis**, which connects the pulmonary trunk to the descending portion of the **aortic arch**. In fetal life, in fact, only a small volume of blood, necessary for oxygen supply, enters the pulmonary circuit, as the fetal lungs are collapsed, and their blood vessels have a high resistance to flow.

The more highly oxygenated blood that enters the left atrium through the foramen ovale is mixed with a small amount of deoxygenated blood returning from the pulmonary veins. This mixture moves into the left ventricle and is pumped into the aorta. Some of it reaches the myocardium through the coronary **arteries** and some reaches the **brain** through the **carotid arteries**. The blood carried by the descending aorta is partially oxygenated and partially deoxygenated. Some of it is carried into the branches of the aorta that lead to various parts of the lower regions of the body. The rest passes into the umbilical arteries (usually two), which branch from the fetal internal **iliac arteries** and lead to the placenta.

Inflation of the lungs at birth reduces the resistance to blood flow through the lungs resulting in increases blood flow from the pulmonary arteries. Consequently, an increased amount of blood flows from the right atrium to the right ventricle and into the pulmonary arteries and less blood flows through the foramen ovale to the left atrium. In addition, an increased volume of blood returns from the lungs through the pulmonary veins to the left atrium, which increases the pressure in the left atrium. The increased left atrial pressure and decreased right atrial pressure (due to pulmonary resistance) forces blood against the septum primum causing the foramen ovale. This action functionally completes the separation of the heart into two pumps on the right and left sides of the heart. The closed foramen ovale becomes the fossa ovalis. The ductus arteriosis, which connects the pulmonary trunk to the **systemic circulation**, closes off within one to two days after birth and degenerates as ligamentum arteriosum.

When the umbilical cord is cut, no more blood flows through the umbilical arteries and veins, and they degenerate. The remnant of the umbilical vein becomes the round ligament of the liver and the ductus venosum becomes the ligamentum venosum. The remnant of the umbilical arteries form raised folds in the **peritoneum**, the medial umbilical folds.

Maternal blood coming into the placenta has an oxygen tension ranging between 95 and 105 mmHg. In the placental intervillous spaces, the tension is estimated to be 30–35 mmHg, while arterial blood returning to the placenta form the fetus has an oxygen tension of 20 mmHg. Despite the relatively low arterial oxygen tension, fetal blood is reasonably well oxygenated because fetal **hemoglobin** has a much higher affinity for oxygen than adult hemoglobin, reaching a 50% of saturation at an oxygen tension as low as 20 mmHg about.

See also Placenta and placental nutrition of the embryo

FETAL MEMBRANES AND FLUIDS

Fetal membranes include the yolk sac, amnion, chorion, and allantoids. The role of these membranes is different among different species. In mammalians, extra-embryonic membranes contribute to form the **placenta** and the umbilical cord.

The most primitive extra-embryonic membrane is the yolk sac. It is composed of splanchnic lateral plate **mesoderm** and **endoderm**. On day after **fertilization**, a thin layer of hypoblast cells, known as Heuser's membrane, migrates out over the inner, cellular layer of the trophoblast. This temporary yolk sac is later replaced by another layer of hypoblast cells that becomes the definitive yolk sac. The yolk sac plays an active and crucial role during organogenesis, and is the source of the embryo's first **blood** cells and the primordial **germ cells**.

The amnion is the non-vascular and innermost membrane surrounding the embryo. It develops next in respect to the yolk sac, by delamination of the epiblast. It serves a container for the amniotic fluid and is involved in the onset of labor by releasing of **prostaglandins**. The amnion has an ectodermic layer facing the embryo with a non-adhesive surface that will produce induced malformations (band amniotic syndrome).

The chorion is a trophoblast-derived membrane, enclosing the amnion and the yolk sac. It does not have any direct contact with the fetus. It represents the embryonic portion of the placenta and participates in exchange of materials between the maternal blood and the fetal placenta. The allantois originates from the intra-embryonic endoderm covered with splanchnic lateral plate mesoderm. In humans, the allantois participates in gas exchange and mainly provides vessels that will be used in umbilical cord (the old allantoic **arteries** and vein are the umbilical arteries and vein). The allantois remains attached to the hindgut and is incorporated into the umbilical cord connecting the embryo with the placenta.

The major fetal fluids are the amniotic fluid, the fetal urine, secretions of the respiratory tract and digestive apparatus, and the **cerebrospinal fluid**. A primitive fluid is the coelomic fluid, which is thought to accumulate nutrients for the early phases of development. In the fetal fluids, the concentration of almost all the amnio acids is higher than in maternal serum since their synthesis depends on the placental synthesis. Amniotic fluid surrounds the fetus, and throughout the **pregnancy** it increases in volume as the fetus grows. Approximately 1,000 ml of amniotic fluid surround the baby at term. Amniotic fluid contains **carbohydrates**, proteins **lipids**, **hormones** (estriol and **progesterone**), **enzymes** (alkaline phosphatase), and minerals as in vernix caseosa, lanugo **hair**, desquamated epithelial cells, and meconium. It is produced from the active secretion from the amniotic epithelium, from the fetal urine (principal source), fetal lung fluid, and as transudate of both fetal and maternal circulation. The amniotic fluid is circulated by the fetus every three hours. Adequate amniotic fluid volume is maintained by a balance of fetal fluid production (lung liquid and urine) and resorption (**swallowing** and intramembranous flow). Amniotic fluid serves several functions for the fetus, including the protection from outside injuries, the allowance of musculoskeletal development, and proper fetal growth. Anomalies of the amniotic fluid volume (oligo and polyhdramnios) can be associated to fetal malformations. Fetal urine begins to be hypotonic around the eight week of gestation. The rate of production increases throughout gestation to reach rates as high as 50 cc per hour.

Cerebrospinal fluid is produced by the choroid plexus, which is a structure lining the floor of the lateral ventricle and the roof of the third and fourth ventricles of the **brain**. Cerebrospinal fluid contains high amount of salts, sugars, and lipids.

See also Embryonic development, early development, formation, and differentiation; Fetal circulation; Prenatal growth and development

FEVER AND FEBRILE SEIZURES

A fever is any body temperature elevation over 100°F (37.8°C).

A healthy person's body temperature fluctuates between 97°F (36.1°C) and 100°F (37.8°C), with the average being 98.6°F (37°C). The body maintains stability within this range by balancing the heat produced by body **metabolism** with the heat lost to the environment. The "thermostat" that controls this process is located in the **hypothalamus**, a small structure located deep within the **brain**. The nervous system constantly relays information about the body's temperature to the hypothalamus, which in turn activates different physical responses designed to cool or warm the body, depending on the circumstances. These responses include: decreasing or increasing the flow of **blood** from the body's core, where it is warmed, to the surface, where it is cooled; slowing down or speeding up the rate at which the body turns food into energy (metabolic rate); inducing shivering, which generates heat through **muscle contraction**; and inducing sweating, which cools the body through evaporation.

A fever occurs when the body maintains a higher temperature, primarily in response to an **infection**. To reach the higher temperature, the body moves blood to the warmer interior, increases the metabolic rate, and induces shivering. The chills that often accompany a fever are caused by the movement of blood to the body's core, leaving the surface and extremities cold. Once the higher temperature is achieved, the shivering and chills stop. When the infection has been overcome, or antipyretic (fever-relieving) drugs have been taken, the hypothalamus "resets" the body's baseline temperature, and the body's cooling mechanisms switch on: the blood moves to the surface and sweating occurs.

Fever is an important component of the immune response, though its role is not completely understood. Physicians believe that an elevated body temperature has several effects. The **immune system** chemicals that react with the fever-inducing agent also increase the production of cells that fight off the invading **bacteria** or **viruses**. Higher temperatures may inhibit the growth of some bacteria, while at the same time speeding up the chemical reactions that help the body's cells repair themselves. In addition, the increased **heart** rate that may accompany the changes in blood circulation also speeds the arrival of white blood cells to the sites of infection.

Fevers are primarily caused by viral or bacterial infections, such as pneumonia or influenza. However, other conditions can induce a fever, including allergic reactions; autoimmune diseases; trauma, such as breaking a bone; **cancer**; excessive exposure to the sun; intense exercise; hormonal imbalances; certain drugs; and damage to the hypothalamus.

A high fever, especially in young children ages three months to five years, can trigger convulsions (febrile seizures). Although there is little data to suggest that the rapid rise in core body temperature itself causes the seizure, fever nevertheless always accompanies a febrile seizure. One explanation may lie in a naturally lower threshold to seizures during early childhood. Children at this developmental stage also are more susceptible to frequent infections and, correspondingly, tend to develop fevers more often than later in life. Studies estimate that between two to five percent of all children worldwide experience at least one febrile seizure during their early childhood. Febrile seizures are almost always benign, usually resulting in no permanent neurological damage.

Most fevers caused by infection end as soon as the immune system rids the body of the pathogen (infectious agent) and do not produce any lasting effects. Persistent fevers may be associated with more chronic conditions, such as autoimmune disease, chronic infection, or some cancers.

See also Brain stem function and reflexes; Homeostatic mechanisms; Infection and resistance; Toddler growth and development:

FIBIGER, JOHANNES (1867-1928)

Danish physician

Johannes Fibiger was a Danish bacteriologist whose early work on childhood diphtheria and tuberculosis demonstrated the vital role medical research could play in controlling diseases that threatened public health. In 1926, Fibiger received the Nobel Prize in physiology or medicine for demonstrating how cancer-like tissues could be induced experimentally in the laboratory.

Johannes Andreas Grib Fibiger was born in the Danish village of Silkeborg. His father, Christian Fibiger, was a district physician; his mother, Elfride Muller, was a writer and the daughter of a Danish politician. When Fibiger was three, his father died and the family moved to Copenhagen, where he attended the University of Copenhagen at age sixteen and studied medicine, biology, and zoology. After earning his medical degree in 1890, he undertook several years of medical apprenticeship in various hospitals and with the Danish army. In 1891, he married Mathilde Fibiger, a distant cousin and physician's daughter, with whom he had two children.

It was while working as an assistant in a bacteriological laboratory at the University of Copenhagen that Fibiger was persuaded to undertake doctoral work on diphtheria, a virulent childhood disease that caused its victims to suffocate. Fibiger discovered better methods of growing diphtheria **bacteria** in the laboratory and demonstrated that there were two distinct forms of the bacillus, an important step in identifying carriers of the disease who frequently displayed no symptoms. At the turn of the century, diphtheria was a major public health problem, and epidemics were frequent in Denmark and throughout the rest of the developed world. Fibiger produced an experimental serum against the disease and carefully monitored the results of an inoculation program. In 1897, the International Medical Congress published his report, a model of its kind, which brought Fibiger international attention and confirmed the effectiveness of the serum. The young scientist had received his Ph.D. only two years earlier. Fibiger later came to regard his work on diphtheria as his highest scientific achievement.

In 1900, at age thirty-three, he joined the faculty of the Institute of Pathological Anatomy, one of a number of young professors hired by the University of Copenhagen. He was also appointed director of the institute and launched a successful program to construct a modern research facility for **pathology** and anatomy. Within its walls, Fibiger and another faculty member, C. O. Jenson, conducted research on tuberculosis in cattle and humans. Flying in the face of popular opinion, they demonstrated that humans could contract tuberculosis from infected cattle, especially by drinking their milk. Supported by the research of other investigators in Europe, these findings led to the passage of strict regulations governing the sale of raw milk, resulting in fewer adolescent deaths due to tuberculosis.

Fibiger's experiments on tubercular rats led him to the discovery for which he won the Nobel Prize. Performing a series of routine dissections in 1907, he discovered abscesses that appeared to be cancerous in the stomach lining of three wild rats. Microscopic examination revealed that the abscesses contained the larvae of a minute parasitic worm or nematode.

By the early 1900s, scientists had ample observational data suggesting that environmental irritants such as soot and harsh chemicals produced **cancer** in chimney sweeps and chemical workers. Many scientists thought that chronic irritation from mechanical or chemical agents was the basis of all cancer, but no one had yet succeeded in turning normal cells into cancerous cells under laboratory conditions.

Working on the hypothesis that the parasites produced a chemical toxin that induced cancer of the stomach, Fibiger undertook an ambitious research program. He trapped and examined more than a thousand wild rats, feeding them worm larvae, and even injecting them with the parasite, all without result. Surmising that the larvae were not passed from rat to rat but through an intermediate host, he traced the parasite to a rare species of cockroach found near a Copenhagen sugar refinery. By feeding healthy rats a diet of white bread and cockroaches, Fibiger finally succeeded in producing stomach abscesses in more than a hundred animals. For the first time, a researcher had induced what at the time was thought to be cancer in a laboratory setting. Fibiger reported his achievement in the *Journal of Cancer Research* and was awarded the 1926 Nobel Prize in physiology or medicine for his discovery of *Spiroptera carcinoma,* the parasitic worm that he thought had produced the cancer. Yet, in his acceptance speech, Fibiger expressed doubt that parasites played any great role in gastric cancer in humans.

Later investigators would find a number of weaknesses in Fibiger's research. Like most scientists of the period, Fibiger had not thought to check his findings against a control group of rats fed on a diet of only white bread. Nor was it easy to reproduce Fibiger's findings in other laboratories due to the lack of a standard strain of laboratory rats in the 1920s; Fibiger's animals had all been caught in the wild. Other investigators expressed doubt that the abscesses described by Fibiger were truly cancerous. There was some evidence that the abscesses might have been caused by a diet deficient in vitamin A. Nonetheless, the lasting effect of Fibiger's prize-winning discovery—later refuted by other researchers—was the great impetus it gave to other investigators to pursue laboratory research on the causes of cancer.

Fibiger abandoned parasitology after World War I to follow the work of two Japanese scientists who induced skin cancer in rabbits by painting their ears with coal tar. Conducting his own experiments by painting the backs of rats with the irritant, Fibiger reported two valuable insights: that cancer did not occur with the same frequency in all species or even within the same species, and that individual predisposition played an important role in susceptibility to cancer. At the time of his death, he was working with two colleagues on a vaccine for cancer, hoping to demonstrate that inoculating laboratory animals with matter drawn from malignant **tumors** would induce immunity to the disease.

During his long career as director of the Institute of Pathological Anatomy at the University of Copenhagen, Fibiger divided his time between research and teaching. He was a generous colleague who was widely respected for his meticulous laboratory methods. He published seventy-nine scientific papers and served as secretary and then president of the Danish Medical Society, and as president of the Danish Cancer Commission. He was co-editor and founder of *Acta Pathologica et Microbiologica Scandinavica*. In 1927, he was awarded the Nordhoff-Jung Cancer Prize.

In early 1928, less than two years after delivering his Nobel Prize speech, Johannes Fibiger died in Copenhagen of a heart attack.

See also Bacteria and responses to bacterial infection

FIBRILLATION • *see* HEART, RHYTHM CONTROL AND IMPULSE CONDUCTION

FIBROCARTILAGE

Fibrocartilage is one of three basic forms of **cartilage** in the human body. The extremely resilient **tissue** consists of relatively inelastic bundles of collagen fibers. Fibrocartilage has a limited distribution in the body. It is found in the intervertebral disks, the synovial joint between the pubic bone in the pelvis and around the edges of articular cavities, such as the shoulder joint.

The overall appearance of fibrocartilage closely resembles dense connective tissue, and many researchers regard it to be a transitional tissue. White, or mature, fibrocartilage appears grayish-white and takes on a stiff leathery feel. It functions as a shock absorber between vertebrae and helps cushion bones when we walk or run. Fibrocartilage also helps deepen joint sockets so dislocation is less possible. The amount of fibrocartilage increases with age.

White fibrocartilage comes in four groups: interarticular, connecting, circumferential, and stratiform. Interarticular fibrocartilage protects the **joints** subjected to frequent movement, such as the jaw, wrist, and knee. Connecting fibrocartilage is found between each **vertebra**. Circumferential

fibrocartilage protects the edges of the hip and shoulder cavities. Stratiform fibrocartilage helps lubricate the grooves through which the **tendons** of certain muscles glide.

See also Costochondral cartilage; Elastic cartilage; Hyaline cartilage; Synovial joints; Tempromandibular joint; Tendons and tendon reflex

FIBROUS JOINTS

There are three different types of fibrous or **cartilaginous joints**. These particular **joints** are characterized by having no joint capsule and little if any possibility of movement. The first of these cartilaginous joint are called sutures and are formed from fibrous **tissue** that permits the continued growth of bones that eventually ossify (bone formation). Sutures unite the bones of the **skull** and their status is frequently used to determine age.

More important clinically are the syndesmosis joints. This kind of fibrous joint joins skeletal structures that are located relatively far apart and are united by ligaments. With ageing or after a trauma, these fibrous tissues tend to ossify and form bone-like structures that link two bones that are not supposed to be joined. Examples of this fibrous union can occur between the radius and the ulna as well as the tibia and fibula after a fracture. The third type of fibrous joint is called gomphoses. Likened to a peg and socket, these joints are seen between teeth and jaw and are maintained by the periodontal ligament, which gives only a little to act as a shock absorber. As they join the teeth to the tooth socket, any changes in their status are used to clinically evaluate periodontal disease.

See also Fibrocartilage; Skeletal system overview (morphology)

FIGHT OR FLIGHT REFLEX

The fight of flight reflex readies the body for survival during stressful situations. Interactions between the neural and hormonal systems of the body work together to get the body ready to stand and fight the challenge or run away from it (flight).

When faced with life-threatening crises, unnecessary functions are temporarily shut down and energies are diverted to functions vital to survival. Any stress, whether physical, psychological (anticipation of an unpleasant event) or emotional (anger or fear) will produce some, if not all elements of the fight or flight response.

The term "fight or flight" as coined in the 1920s by the American physiologist, Walter B. Cannon. He described the response to be a generalized sympathoadrenal activation since it involves participation of both the **sympathetic nervous system** and the tissues of the adrenal medulla within the adrenal gland. When the body is threatened or frightened, innervation via the sympathetic nerves causes the cells of the adrenal medulla to secrete the **hormones epinephrine** and nor-epinephrine (catecholamines). Because the sympathoadrenal sys-

The physiological fight or flight response is illustrated in this 1963 Pulitzer Prize-winning photograph of Lee Harvey Oswald (center), the accused assassin of President John F. Kennedy, as he is fatally shot by Jack Ruby (right). Immediate catecholamine release accompanied by rapid-fire neuronal and muscular action enabled Oswald to draw his arms in to his chest in an attempt to protect from further assault, while police officer Jim Leavell (left) draws his body away from the threat. *Popperfoto/Archive Photos. Reproduced by permission.*

tem is in the efferent limb of the nervous system, cate-cholamine-mediated events take place in seconds. The course of action of most hormones is much longer.

Immediate catecholamine release into the bloodstream affects many **organs**. Mental activity and awareness is increased to help assess the situation. Airways dilate and **breathing** rate is increased. Peripheral **blood** vessels are constricted (pale skin) and blood is sent to skeletal muscles giving them increases strength and endurance. Cardiac output increases as does **heart** rate and blood pressure rises allowing more oxygen and nutrients to reach the tissues. Conversion of **glycogen** stores in the **liver** to glucose speeds up to allow blood glucose levels to rise. This makes more glucose available for cells to produce **ATP** for cellular energy via cellular respiration. Secretion of **sweat glands** increase to cool the body and apocrine glands under the armpits produce strong secretions (perhaps to frighten away challengers). Pupils of the eyes dilate to bring in more light and the ciliary muscles of the lens relax to facilitate far-sightedness. Non-essential processes for defence or running such as **digestion** temporarily decrease, as does renal output. Bowel and bladder sphincters close and the **immune system** is suppressed.

See also Adrenal glands and hormones; Hormones and hormone action; Sympathetic nervous system

FILTRATION • *see* CELL MEMBRANE TRANSPORT

FINSEN, NEILS RYBERG (1860-1904)
Danish physician

Neils Finsen demonstrated that the most refractive solar rays, which he called chemical rays (i.e., blue, violet, and ultra violet radiation), as well as light rays originating from an electric or voltaic arc, have a stimulating effect on the tissues, also killing **bacteria** involved in skin infections. Through a series of experiments, Finsen had in part, perfected an existing therapy and in part, developed a new therapeutic method that received his name, Finsen's method, for the treatment of skin affections, such as lupus vulgaris and psoriasis. This chemical rays method utilizes a hollow convex lens filled with an solution of copper sulphate that filter out the thermal radiation of the sun light, while converges the chemical rays at 19.50 in. (50 cm) from the skin. Another approach is the voltaic arc, which uses divergent light rays that are made parallel to each other when passing through two convex lenses. The treated area is compressed to achieve some degree of schemia (i.e., to interrupt **blood** circulation) during the exposition. Finsen also studied the harmful effects of skin exposition to strong solar radiation, and discovered that the multiple scars resulting from smallpox

Niels Rydberg Finsen. *Corbis-Bettmann. Reproduced by permission.*

could be avoided by keeping the patient protected from the chemical rays, while exposing his skin to the thermal rays (i.e., infrared rays). Due to the success of the Finsen's Method for the treatment of lupus vulgaris, he received the Nobel Prize in physiology or medicine in 1903.

Neils Ryberg Finsen was born in the Faroe Islands, Denmark, where his father was an administrative official. In 1882, Finsen was sent to Copenhagen to study medicine, and graduated in 1890 from the University of Copenhagen. He worked as professor of **anatomy** at the University of Copenhagen from 1890 until 1893, when he decided to leave the position in order to dedicate himself to his scientific investigations on the therapeutic properties of light. In 1893, he devised the treatment of smallpox and in 1895, the treatment of lupus vulgaris. The Finsen Institute was founded in Copenhagen in 1896, and served as a model to many other clinics throughout Europe for the treatment of lupus vulgaris and other skin diseases. Finsen received the title of Professor in 1898 and in the following year became Knight of the Order of Dannebrog, and later was knighted by the Order of Silver Cross as well. He also received a Danish gold medal for his scientific merit and the Cameron Prize of 1904 from the University of Edinburgh.

His scientific work was published mostly in Danish and German languages. Among them, the translation of some titles would be: *On the Effects of Light on the Skin (Om Lysets Indvirkninger paa Huden, 1893); The Use of Concentrated*

Chemical Rays in Medicine (Om Anvendeise I Medicinen af Koncentrerede Kemiske Lysstraaler, 1896); The Phototherapy (La Photothérapie, 1899). Many other papers by Finsen also appeared under the form of communications published by his institute.

Finsen suffered for many years from a disabling disease known as Pick's disease, which causes the progressive thickening of the **connective tissues** in the **liver, heart**, and spleen. Although he had spent his last years in a wheel chair, he persistently kept his scientific work ongoing. The money Finsen received from the Noble Prize was donated to the Finsen Institute and to a sanatorium for heart and liver disease. He died almost a year later, at the age of 43.

See also Dermis; Epidermis; Radiation damage to tissue

FISCHER, EDMOND H. (1920-)
Chinese-born American biochemist

Edmond H. Fischer was the joint recipient with his longtime associate, **Edwin Krebs**, of the Nobel Prize in physiology or medicine in 1992 for discoveries dealing with reversible protein phosphorylation as a biological regulatory mechanism. Responsible for a wide range of basic processes, including cell growth and differentiation, regulation of genes, and **muscle contraction**, protein phosphorylation is now the subject of one in every twenty papers published in biology journals. Application of Fischer and Krebs's work to medicine has elucidated mechanisms of diseases such as **cancer** and diabetes, and has yielded drugs that inhibit the body's rejection of transplanted **organs**. The recognition accorded Fischer and Krebs, who began a collaboration at the University of Washington in Seattle in the early 1950s, was hailed within the scientific community as long overdue.

Edmond H. Fischer was born in Shanghai, China. His father, Oscar Fischer, had come to China from Vienna, Austria, after earning degrees in business and law. Fischer's mother, Renée Tapernoux Fischer, was born in France. She had come to Shanghai with her family after first arriving in Hanoi, where her father was a journalist for a Swiss publication. In Shanghai, Fischer's grandfather founded the first French newspaper published in China and helped to establish a French language school that Fischer attended.

At the age of seven, Fischer was sent, along with two older brothers, to a Swiss boarding school near Lake Geneva. One of his brothers studied engineering at the Swiss Federal Polytechnical Institute in Zurich and the other went to Oxford to study law. While he was in high school, Fischer developed a lifelong friendship with Wilfried Haudenschild, whose inventiveness and unusual ideas impressed him. The two decided that one would be a scientist and the other a physician, so that together they might cure all the ills of the world. Fischer was also influenced in his youth by classical music and for a time entertained the idea of becoming a professional musician. Instead he decided to become a scientist.

After entering the School of Chemistry at the University of Geneva at the start of World War II, Fischer was able to earn a degree in biology and another in chemistry. He received his doctorate at Geneva in 1947 and worked at the university on research until 1953. American universities at the time afforded more opportunities in the new field of **biochemistry**, and Fischer soon found himself in the United States. His first position was at the California Institute of Technology, where he was given a postdoctoral fellowship. Fischer was amazed that wherever he went in the United States he was offered a job. In Europe, research positions in his field were next to impossible to obtain.

Hans Neurath, chair of the department of biochemistry at the University of Washington, invited Fischer to Seattle. On his first visit he found the scenery reminiscent of Switzerland and accepted Neurath's offer of an assistant professorship at the new medical school at the university. Thus began a long association with Edwin Krebs. Krebs had worked in the laboratory of Carl Ferdinand Cori and Gerty T. Cori in St. Louis on the enzyme phosphorylase in the late 1940s (**enzymes** are proteins that encourage or inhibit chemical reactions). The Coris won the Nobel Prize in 1947 for their isolation of phosphorylase, showing its existence in active and inactive form. Fischer had worked on a plant version of the same enzyme while he was in Switzerland.

In the mid–1950s, Fischer and Krebs set out to determine what controlled the protein's activity. Their experiments centered on muscle contraction. A resting muscle needs energy (stored as **glycogen** in the body) in order to contract, and phosphorylase frees glucose from glycogen for use by the muscle. Fischer and Krebs discovered that an enzyme they called protein kinase was responsible for adding a phosphate group from the compound **ATP** (**adenosine triphosphate**, the cell's energy store) to phosphorylase, which activated the enzyme. In a reverse reaction, an enzyme called protein phosphatase removed the phosphate, turning phosphorylase off. Protein kinases are present in all cells and are critical for many phases of cell activity, including **metabolism**, respiration, **protein synthesis**, and response to stress.

By the 1970s, biochemical research in the area that Fischer and Krebs opened up was so extensive that 5 percent of papers in biology journals dealt with protein phosphorylation. Between one and five percent of the **genetic code** may be concerned with protein kinases and phosphatases. Science has made connections that show the role of protein kinases in diseases, including cancer, diabetes, and muscular dystrophy. Fischer and Krebs have also been able to demonstrate in their research how the **immune system** is activated. They showed how a surface protein starts a chain reaction that recruits lymphocytes to fight **infection**.

In the field of organ transplants, drugs that influence phosphorylation prevent rejection of the transplants by the body's immune system. The drug cyclosporin has been developed and is widely used to prevent the rejection of **liver**, kidney, or pancreatic transplants in human beings. Cyclosporin and another drug, FK–506, inhibit protein phosphatase, thereby preventing the rejection of tissues in organ transplant operations. Irregular protein kinase activity can cause abnormal cell growth leading to **tumors** and cancer. Philip Cohen, a colleague from the University of Dundee, comments in a *New Scientist* interview that protein kinases and phosphatases "will be the major drug targets of the 21st century."

Edmond H. Fischer has received many honors for his scientific research over the course of his long career. Notable, besides the Nobel Prize, are his election to the American Academy of Arts and Sciences in 1972 and his winning of the Werner Medal from the Swiss Chemical Society as early as 1952. Fischer has been married twice. His first wife, with whom he had two sons, died in 1961. In 1963, he married Beverly Bullock, a native Californian. Besides his accomplishment in classical piano, Fischer also enjoys painting, piloting a plane, and mountaineering. Along with his colleague, Krebs, Fischer annually joins research groups in retreats at the University of Washington Park Forest Conference Center in the foothills of the Cascade Mountains. There they review their latest findings and lay plans for their future research, which they continue in the role of emeritus professors at the University of Washington.

FLATUS

Flatus describes gas from the intestinal tract that is expelled from the body via the rectum and anus.

The total volume of gas excreted by rectum varies greatly among individuals, ranging from approximately 500mL/day to 1500mL/day, with a mean of 700mL/day. Volumes also vary within individuals depending on their diet. Diets high in the non-absorbable disaccharide lactulose significantly increase the volume and frequency of passed flatus. The average frequency of passage of flatus for adults is 10 times per day with an upper normal limit of 20 times per day.

Flatus is composed of nitrogen, oxygen, carbon dioxide, hydrogen, and methane gas with percent composition being highly variable.

Sources of bowel gas may be from **swallowing** air (aerophagia), production from inside the gut (intraluminally) and diffusion from the **blood**. Most swallowed air is eructated (burped). A small proportion of air may be forced into the duodenum if the person is lying down and the stomach liquids act as a gas trap at the gastroesophageal sphincter.

Most bowel gas is produced in the intestinal lumen. Carbon dioxide, hydrogen, and methane are produced here in appreciable quantities. Carbon dioxide is generated intraluminally from the interaction of hydrogen ions plus bicarbonate ions. This reaction is catalyzed by carbonic anhydrase. Carbon dioxide may also be liberated from the **digestion** of **triglycerides** to fatty acids. Most of this carbon dioxide is produced in the duodenum and is absorbed into the blood as it travels toward the colon. The majority of carbon dioxide in flatus comes from bacterial fermentation.

Hydrogen production in the bowel occurs during bacterial fermentation of **carbohydrates** in the colon. Specifically, oligosaccharides found in fruits and legumes produce hydrogen gas when fermented by colonic **bacteria**. Hydrogen gas may also be produced by fermentation of mucoproteins by fecal bacteria.

Hydrogen gas is also consumed by methane producing bacteria (methanogens). If methanogens are the predominant intestinal flora, virtually all the gas is consumed intraluminally and does not appear in the flatus. The major methanogen in the human colon, *Methobrevibacter smithii,* produces methane through the reaction:

$$4H_2 + CO_2 \rightarrow CH_4 + 2H_2O.$$

Approximately one third of adults have large concentrations of methanogens in their colons and produce significant quantities of methane in their flatus.

The majority of gases in flatus are non-odoriferous. The noxious odor of flatus is due to trace amounts of hydrogen sulfide. This gas gives a smell similar to rotten eggs.

See also Elimination of waste; Gastro-intestinal tract

FLEMING, ALEXANDER (1881-1955)

Scottish bacteriologist

Scottish bacteriologist Alexander Fleming is best known for his 1928 discovery of the bacteria-fighting antibiotic penicillin, widely regarded as one of the greatest medical discoveries of the twentieth century. Before penicillin, the few drugs that were available to fight bacterial disease were inefficient and highly toxic to the human body. Fleming's discovery won him the 1945 Nobel Prize in physiology or medicine jointly with **Ernst Chain** and Baron **Howard W. Florey**.

Fleming's early life was rustic. He was born in a farmhouse in Lochfield, Ayrshire, Scotland, the third of four children by Hugh Fleming, a farmer, and his second wife, Grace (Morton) Fleming. He had two stepbrothers and two stepsisters from Hugh Fleming's first marriage. His father died when Fleming was seven and his mother and oldest stepbrother, Thomas, were left to manage the farm.

The natural intelligence Fleming possessed became evident even though his early education was basic. He first attended a tiny local school, then a larger moorland school in Dorval, walking a total of eight miles a day to attend class; these long walks through natural surroundings may have sparked Fleming's interest in living inhabitants and also helped hone his critical observation skills. At age twelve, Fleming transferred to the Kilmarnock Academy, a school that had high standards but limited resources. He stayed at the Academy only one year. Throughout his early and later education, Fleming, who greatly enjoyed competition, always scored at or near the top of his class, apparently without much effort.

When Fleming turned thirteen he moved to London to join his brother John and stepbrother Thomas, a physician who practiced ophthalmology. Upon arriving, Fleming attended classes at Regent Street Polytechnic for two years to prepare for a career in business. At sixteen, after completing his course work, he secured a position as a clerk in a shipping company. Fleming joined the London Scottish Regiment in 1900 and,

though he never served in battle, he remained associated with his unit until 1914.

In 1901, at twenty years of age, Fleming inherited 250 pounds, a large sum at that time. Following Thomas's advice, Fleming decided to study medicine and, not surprisingly, scored at the top of the national medical school entrance exam. Fleming chose to study at St. Mary's Medical School, which was within walking distance of his London home. When he entered St. Mary's in 1901, Fleming began what would become a nearly continuous fifty-four-year relationship with that institution. As a student, he won many class prizes for his high test scores, including a scholarship in his first year that paid for his entire tuition.

After successfully completing his medical school studies in 1906, Fleming qualified for a position as a doctor that same year. Though he decided to continue his education, pursuing an M.B. and B.S. at London University, he concurrently accepted a job at St. Mary's as a junior assistant, working in the research laboratory of noted **pathology** professor Sir Almroth Edward Wright—initially only as a temporary means of support. In 1908 Fleming earned his degrees with honors, receiving the London University Gold Medal. The next year, while still working full-time for Wright, he passed the Fellowship of the Royal College of Surgeons exam, thus enabling him to pursue a career in surgery. However, Fleming decided to stay in laboratory research, which he felt to be a more exciting and less arduous career path. In Fleming's time, scientists knew of the natural defense capabilities of the human body, and the direction in research was to augment or stimulate the body's own **immune system** to help fight bacterial disease. Fleming's research reflected this trend, focusing mainly on the prevention and treatment of bacterial infection. He was at the forefront of his area of expertise; Fleming received some of the initial samples of one of the first antibacterial agents, Salvarsan, an arsenical agent developed by **Paul Ehrlich** for use in the treatment of syphilis. Fleming quickly became an expert in the administration of Salvarsan and also conducted experiments with this drug; he found that, while it was effective in destroying the **bacteria** that caused syphilis, Salvarsan produced a number of toxic side effects. Despite the imperfection of Salvarsan, Fleming maintained a belief that a safe and effective antibacterial substance could be found.

Interrupting his research at St. Mary's temporarily, Fleming served as a captain in the British Royal Army Medical Corps during World War I. Stationed in Boulogne, France, he worked in a wound research laboratory under the command of Professor Wright. Together they researched the efficacy of antiseptics on wound infections, an accepted treatment at that time. Fleming found that antiseptics did more harm than good because they not only failed to kill all of the bacteria but also destroyed protective white **blood** cells (the body's natural defense mechanism) thereby allowing **infection** to spread more rapidly.

In 1919, after the war ended, Fleming resumed his research at St. Mary's, studying antibacterial mechanisms. He was particularly motivated to find an effective yet safe antibacterial substance after witnessing the horrible suffering caused by bacterial infections during the war. Fleming remained convinced that an ideal bacteria-fighting agent could be found that would destroy the invading bacteria yet not harm the body's own white blood cell defenses.

Around 1921, Fleming fortuitously discovered his first antibacterial agent. He had cultured a sample of his own nasal mucosa during a bout with the common cold. While studying the plate, Fleming noticed that the nasal **mucus** had dissolved a bacterial colony that had contaminated the plate. The bacteriolytic component of the mucus was named "lysozyme." Later, with the aid of colleague V. D. Allison, Fleming discovered that lysozyme, an enzyme, was present in a number of substances including human blood serum, tears, and **saliva**, as well as egg whites and turnip juice. Though lysozyme was safe to human **tissue**, however, it had no effect on disease-producing bacteria, and was thus of little medical use. Nevertheless, Fleming continued his search for an effective yet non-toxic antibacterial substance.

In 1928 Fleming discovered, also serendipitously, his second and more famous antibacterial substance: penicillin. However, many highly coincidental factors needed to come together before the discovery of this wonder drug was even possible. One of the main factors that led to the discovery of penicillin was Fleming's own untidy habit. Instead of the normal practice of promptly discarding bacterial culture plates, Fleming held onto his plates far beyond their usefulness. Before finally disposing of the plates, though, he would examine each one to note any interesting developments. On one particular plate he noticed that a strange mold contaminant had inhibited the growth of the disease-causing bacteria *Staphylococcus aureus* that had been cultured on the plate.

The circumstances that led to this mold's growth on Fleming's culture plate were almost astronomically improbable. First, the mold itself, later identified as *Penicillium notatum,* was a very rare organism; the only reason it was floating near Fleming's lab in the first place was because a scientist on the floor below was studying the mold's effect on asthma sufferers. This was not enough though: in order for the mold to display its anti-bacterial properties, it needed to grow *before* the bacteria grew. For that to occur, conditions had to be first cold (for the mold) then warm (for the bacteria). As luck would have it, Fleming took a week's vacation, during which time London experienced first a cold spell, then later a warm spell. As the bacteria grew, it flourished everywhere except for an empty zone around the already-growing mold; hence the inhibitory effect of the mold could be easily observed. The combined effect of all of the above factors led to Fleming's discovery of the world's first safe and effective antibiotic. Fleming isolated the *Penicillium* mold for further investigation, and named the particular component of the mold responsible for killing bacteria "penicillin."

Not being very adept at writing or speaking, Fleming was unable to effectively communicate the potential importance of his discovery. His colleagues had little interest because they thought it was merely another type of lysozyme, from a mold rather than mucus. Fleming could not clearly express the critical difference between lysozyme and peni-

Sir Alexander Fleming. *Corbis-Bettmann. Reproduced with permission.*

cillin—that penicillin, unlike the former bacteriolytic agent, could inhibit disease-producing bacteria.

Though Fleming's own major research with penicillin lasted less than a year, it revealed that it was effective in killing some disease-producing bacteria yet was non-toxic to white blood cells and living tissue. Fleming also tested penicillin superficially by applying it to wounds, but was discouraged with the mixed results. And while he thought it might be a useful as an injection for wounds, he never actually took the next logical step of injecting the drug into infected areas. He subsequently made only a few fleeting and unenthusiastic references to penicillin in later papers and lectures; for example, he mentioned in one work that penicillin seemed to be a better dressing for septic wounds than other, stronger chemicals. After his initial research, Fleming used penicillin mainly as a laboratory convenience for keeping his vaccine cultures free of certain bacteria. He did give the substance to several colleagues in other laboratories, presumably for further research, but never focused much attention beyond that on the drug.

Fleming's quick dismissal of penicillin was probably due to the fact that the drug did not remain in the blood system very long, and was an unstable and hard to purify substance. Fleming, not being a chemist, was unable to stabilize or adequately concentrate the penicillin component himself. The instability of penicillin may have discouraged Fleming from believing that it had the potential to be an effective antibacterial agent. However, whether or not Fleming actually realized the enormous therapeutic powers of penicillin cannot be conclusively determined from his writings.

For twelve years after its initial discovery, the life-saving potential of penicillin remained untapped. In 1940, two Oxford University chemistry scientists, Ernst Chain and Howard Florey, were fortuitously led to Fleming's article on penicillin during a literature search on antibacterial agents. They were able to isolate and purify penicillin and then later test the drug systemically in clinical trials with wondrous success, confirming penicillin's antibiotic qualities.

By 1942, the therapeutic power of penicillin was clearly established. Today penicillin is still successfully used in the treatment of many bacterial diseases, including pneumonia, strep throat, scarlet fever, gonorrhea, and impetigo. Moreover, its discovery lead to the development of additional antibiotics that have proven useful in destroying a broad spectrum of pathogenic bacteria. Unfortunately, the overuse and/or misuse of antibiotics has allowed certain bacteria to develop resistance to some common antibiotics, including penicillin. As such, the need continues for the development of newer—and in most cases stronger—antibiotics.

After news of the discovery of penicillin spread, Fleming began receiving most of the recognition and fame for its discovery—due in no small part to the dramatic story surrounding the his improbable discovery of the drug. Fleming was knighted in 1944, and in 1945, along with Florey and Chain, received the Nobel Prize in physiology or medicine. Fleming subsequently acquired 25 honorary degrees, 26 medals, 18 prizes, 13 decorations, and honorary membership in 89 scientific academies and societies.

He worked continuously at St. Mary's, being promoted to Assistant Director of the Inoculation Department in 1921, later known as the Institute of Pathology and Research; this department was renamed in 1948 as the Wright-Fleming Institute. In addition, he held a post as bacteriology professor at London University from 1928 until his retirement from teaching in 1948. Fleming married Irish nurse Sarah Marion McElroy in 1915; in 1924 the couple had their only child, a son who, like his father, became a physician. Fleming's health deteriorated after his wife's death in 1949. In 1953, his health improved somewhat and he remarried to a bacteriologist and former student. Just over two months after his retirement, Fleming died of a heart attack at his home in London, England. His ashes are interred at St. Paul's Cathedral in London.

See also Bacteria and responses to bacterial infection; Infection and resistance

FLEMMING, WALTHER (1843-1905)

German anatomist

Walther Flemming founded the study of cytogenetics, with careful observations and documentation of **cell structure** and **cell division**. Flemming coined the terms chromatin and mito-

sis, and described the thread-like structures in the cell nucleus that were later named **chromosomes**.

Flemming was born in Sachsenberg, Mecklenberg, a community in present-day Germany. After studying medicine at the University of Rostock and serving as a physician in the Franco-Prussian War, Flemming turned his attention to the study of physiology. Flemming held academic posts at universities in Prague, then at Kiel, where Flemming concentrated his work on the physiology of the cell.

Flemming pioneered the use of synthetic aniline dyes to visualize the nucleus during cell division. Flemming observed that the red dye was heavily absorbed by granular-appearing structures in the nucleus, and named these structures chromatin, from the Greek word for color. By staining chromatin in the cells of salamander larvae during cell division, Flemming noticed the chromatin coalesced into thread-like structures, termed chromosomes four years later by fellow German anatomist Heinrich Waldeyer. The new staining techniques enabled Flemming to observe in greater detail the process of cell division, including the longitudinal splitting of the chromosomes to produce two identical halves. Flemming named this process mitosis, from the Greek for thread.

Flemming recorded his microscopic observations using hand drawings, in contrast to the microscopes of today that produce digital images which can be manipulated. Flemming summarized his findings in *Zell-substanz, Kern und Zelltheilung* (Cytoplasm, Nucleus, and Cell Division) in 1882. Despite his keen observations, Fleming did not grasp the relationship between cell division and heredity. Thus, another twenty years passed before the nature of Flemming's work was fully appreciated, when Gregor Mendel's laws of heredity were rediscovered in the early 1900s.

See also Cell cycle and cell division; Cell structure; Chromosomes

FLEXION • *see* ANATOMICAL NOMENCLATURE

FLOREY, HOWARD WALTER (1898-1968)

English pathologist

The work of Howard Walter Florey gave the world one of its most valuable disease-fighting drugs—penicillin. **Alexander Fleming** discovered, in 1929, the mold that produced an antibacterial substance, but was unable to isolate it. Nearly a decade later, Florey and his colleague, biochemist **Ernst Chain**, set out to isolate the active ingredient in Fleming's mold and then conduct the clinical tests that demonstrated penicillin's remarkable therapeutic value. Florey and Chain reported the initial success of their clinical trials in 1940, and the drug's value was quickly recognized. In 1945, Florey shared the Nobel Prize in physiology or medicine with Fleming and Chain.

Howard Walter Florey was born in Adelaide, Australia. He was one of three children and the only son born to Joseph

Florey, a boot manufacturer, and Bertha Mary Wadham Florey, Joseph's second wife. Florey expressed an interest in science early in life. Rather than follow his father's career path, he decided to pursue a degree in medicine. Scholarships afforded him an education at St. Peter's Collegiate School and Adelaide University, the latter of which awarded him a Bachelor of Science degree in 1921. An impressive academic career earned Florey a Rhodes scholarship to Oxford University in England. There he enrolled in Magdalen College in January 1922. His academic prowess continued at Oxford, where he became an excellent student of physiology under the tutelage of renowned neurophysiologist Sir Charles Scott Sherrington. Placing first in his class in the physiology examination, he was appointed to a teaching position by Sherrington in 1923.

Florey's education continued at Cambridge University as a John Lucas Walker Student. Already fortunate enough to have learned under a master such as Sherrington, he now came under the influence of Sir Frederick Gowland Hopkins, who taught Florey the importance of studying biochemical reactions in cells. A Rockefeller Traveling Scholarship sent Florey to the United States in 1925, to work with physiologist Alfred Newton Richards at the University of Pennsylvania, a collaboration that would later prove beneficial to Florey's own research. On his return to England and Cambridge in 1926, Florey received a research fellowship in **pathology** at London Hospital. That same year, he married Mary Ethel Hayter Reed, an Australian whom he'd met during medical school at Adelaide University. Howard and Ethel Florey had two children, Charles and Paquita.

Florey received his Ph.D. from Cambridge in 1927, and remained there as Huddersfield Lecturer in Special Pathology. Equipped with a firm background in physiology, he was now in a position to pursue experimental research using an approach new to the field of pathology. Instead of describing diseased tissues and **organs**, Florey applied physiologic concepts to the study of healthy biological systems as a means of better recognizing the nature of disease. It was during this period in which Florey first became familiar with the work of Alexander Fleming. His own work on **mucus** secretion led him to investigate the intestine's resistance to bacterial infection. As he became more engrossed in antibacterial substances, Florey came across Fleming's report of 1921 describing the enzyme lysozyme, which possessed antibacterial properties. The enzyme, found in the tears, nasal secretions, and **saliva** of humans, piqued Florey's interest, and convinced him that collaboration with a chemist would benefit his research. His work with lysozyme showed that extracts from natural substances, such as plants, fungi and certain types of **bacteria**, had the ability to destroy harmful bacteria.

Florey left Cambridge in 1931 to become professor of pathology at the University of Sheffield, returning to Oxford in 1935 as director of the new Sir William Dunn School of Pathology. There, at the recommendation of Hopkins, his productive collaboration began with the German biochemist Ernst Chain. Florey remained interested in antibacterial substances even as he expanded his research projects into new areas, such as **cancer** studies. During the mid 1930s, sulfonamides, or "sulfa" drugs, had been introduced as clinically effective

against streptococcal infections, an announcement which boosted Florey's interest in the field. At Florey's suggestion, Chain undertook biochemical studies of lysozyme. He read much of the scientific literature on antibacterial substances, including Fleming's 1929 report on the antibacterial properties of a substance extracted from a Penicillium mold, which he called penicillin. Chain discovered that lysozyme acted against certain bacteria by catalyzing the breakdown of polysaccharides in them, and thought that penicillin might also be an enzyme with the ability to disrupt some bacterial component. Chain and Florey began to study this hypothesis, with Chain concentrating on isolating and characterizing the "enzyme," and Florey studying its biological properties.

To his surprise, Chain discovered that penicillin was not a protein, therefore it could not be an enzyme. His challenge now was to determine the chemical nature of penicillin, made all the more difficult because it was so unstable in the laboratory. It was, in part, for this very reason that Fleming eventually abandoned a focused pursuit of the active ingredient in Penicillium mold. Eventually, work by Chain and others led to a protocol for keeping penicillin stable in solution. By the end of 1938, Florey began to seek funds to support more vigorous research into penicillin. He was becoming convinced that this antibacterial substance could have great practical clinical value. Florey was successful in obtaining two major grants, one from the Medical Research Council in England, the other from the Rockefeller Foundation in the United States.

By March of 1940, Chain had finally isolated about one hundred milligrams of penicillin from broth cultures. Employing a freeze-drying technique, he extracted the yellowish-brown powder in a form that was yet only ten percent pure. It was non-toxic when injected into mice and retained antibacterial properties against many different pathogens. In May of 1940, Florey conducted an important experiment to test this promising new drug. He infected eight mice with lethal doses of streptococci bacteria, then treated four of them with penicillin. The following day, the four untreated mice were dead, while three of the four mice treated with penicillin had survived. Though one of the mice that had been given a smaller dose died two days later, Florey showed that penicillin had excellent prospects, and began additional tests. In 1941, enough penicillin had been produced to run the first clinical trial on humans. Patients suffering from severe staphylococcal and streptococcal infections recovered at a remarkable rate, bearing out the earlier success of the drugs in animals. At the outset of World War II, however, the facilities needed to produce large quantities of penicillin were not available. Florey went to the United States where, with the help of his former colleague, Alfred Richards, he was able to arrange for a U.S. government lab to begin large-scale penicillin production. By 1943, penicillin was being used to treat infections suffered by wounded soldiers on the battlefront.

Recognition for Florey's work came quickly. In 1942, he was elected a fellow in the prestigious British scientific organization, the Royal Society, even before the importance of penicillin was fully realized. Two years later, Florey was knighted. In 1945, Florey, Chain and Fleming shared the Nobel Prize in physiology or medicine for the discovery of penicillin (although Florey and Fleming had some disagreement over who "discovered" the valuable drug).

Penicillin prevents bacteria from synthesizing intact cell walls. Without the rigid, protective cell wall, a bacterium usually bursts and dies. Penicillin does not kill "resting" bacteria, only prevents their proliferation. Penicillin is active against many of the "gram positive" and a few "gram negative" bacteria. (The gram negative/positive designation refers to a staining technique used in identification of microbes.) Penicillin has been used in the treatment of pneumonia, meningitis, many throat and ear infections, scarlet fever, endocarditis (**heart infection**), gonorrhea, and syphilis.

Following his work with penicillin, Florey retained an interest in antibacterial substances, including the cephalosporins, a group of drugs that produced effects similar to penicillin. He also returned to his study of **capillaries**, which he had begun under Sherrington, but would now be aided by the recently developed electron microscope. Florey remained interested in Australia, as well. In 1944, the prime minister of Australia asked Florey to conduct a review of the country's medical research. During his trip, Florey found laboratories and research facilities to be far below the quality one would have expected of a "civilized community." The trip also inspired efforts to establish graduate-level research programs at the Australian National University. For a while, it looked as if Florey might even return to Australia to head a new medical institute at the University. That never occurred, although Florey did do much to help plan the institute and recruit scientists to it. During the late 1940s and 1950s, Florey made trips almost every year to Australia to provide consultation to the new Australian National University, to which he was appointed Chancellor in 1965.

Florey's stature as a scientist earned him many honors in addition to the Nobel Prize. In 1960, he became president of the Royal Society, a position he held until 1965. Tapping his experience as an administrator, Florey invigorated this prestigious scientific organization by boosting its membership and increasing its role in society. In 1962, he was elected Provost of Queen's College, Oxford University, the first scientist to hold that position. He accepted the presidency of the British Family Planning Association in 1965, and used the post to promote more research on contraception and the legalization of abortion. That same year, he was granted a peerage, becoming Baron Florey of Adelaide and Marston.

See also Bacteria and responses to bacterial infection; Infection and resistance

FLOURENS, MARIE JEAN PIERRE (1794-1867)

French physiologist, anatomist, and physician

Marie Jean Flourens made several major discoveries in nervous system **physiology**, cerebral localization, and **brain** function. Always known as Pierre, he was born into an undistinguished family in Maureilhan, France. After receiving his medical degree from the University of Montpellier in 1813, Flourens moved to Paris for further study. The protégé first of botanist Augustin Pyramus de Candolle (1779–1841) in Montpellier then of paleontologist Georges Cuvier (1769–1832) in Paris, Flourens soon decided to be a physiologist rather than a physician. Upon Cuvier's death, Flourens succeeded him as professor of **anatomy** at the Collège de France and in 1833, as secretary of the Académie des Sciences. In 1840, Flourens defeated poet and novelist Victor Hugo (1802–1885) for a place in the Académie Française. He died in Montgeron, France, widely honored.

Flourens advanced the work of Julien Jean César Legallois (1770–1814) on the respiratory control functions of the **medulla oblongata**. In 1823, he differentiated between the **intellectual functions** of the cerebrum and the motor-control functions of the **cerebellum**. In 1824, he correlated the sense of sight with the integral systematics of the **cerebral cortex**. In 1828, Flourens discovered that the semicircular canals of the inner ear govern coordination and balance, thus laying the groundwork for Prosper Meniére (1799–1862) to describe aural vertigo, or Meniére's disease, in 1861. Throughout these experiments, he worked mostly on pigeons and other birds, never on humans, yet his conclusions were readily extended to all vertebrates.

Flourens wrote copiously throughout his career, at first mostly articles in scientific journals, but later mostly books, some scientific, but also many philosophical, critical, biographical, and historical. Even though he was a scrupulous scientific investigator, his prose was often dogmatic and he did not easily tolerate differences from his own opinion.

Flourens accepted the concept of cerebral localization up to a point. He made many important discoveries to support it, but vigorously opposed the phrenology of Franz Joseph Gall (1758–1828), an extreme philosophy of cerebral localization, as being too mechanistic, materialistic, or deterministic. Flourens built his research on brain function largely upon the experiments of Italian anatomist Luigi Rolando (1773–1831). In extending and partially refuting Rolando's findings, he sometimes unfairly lumped them together with those of Gall.

Flourens saw each part of the nervous system as working both holistically within itself and sympathetically in conjunction with every other part. He acquired the reputation as a rigid anti-localizationalist, but that characterization is unfair. He plainly stated that the cerebrum is the locus of thought, but added that thought does not occur in humans without the admixture of free will and divine grace. He feared that positivist science in general and Gall's pseudoscience in particular would undermine the authority of the church, erode belief in the soul's immortality, and eventually challenge traditional ethics and social order. His absolute dedication to the church is further evident in his 1864 attack on Charles Darwin (1809–1882).

In 1847 both Flourens and Sir James Young Simpson (1811–1870), apparently independent of each other, published papers announcing the superiority of chloroform to sulfuric ether as a means of **anesthesia**. Even though Flourens's work appeared on March eighth and Simpson's on November 19, Simpson received most of the credit for this discovery, at least in the anglophone world, mainly because he was the first to use chloroform in obstetrics and because his colleague, John Snow (1813–1858), used chloroform on Queen Victoria for the births of Prince Leopold and Princess Beatrice in 1853 and 1857.

See also Anesthesia and anesthetic drug actions; Autonomic nervous system; Brain stem function and reflexes; Brain: intellectual functions; Central nervous system (CNS); Cerebral hemispheres; Ear (external, middle and internal); Equilibrium; Motor functions and controls; Nervous system overview; Neurology; Sense organs: balance and orientation; Vision: histophysiology of the eye

FLUID TRANSPORT

The majority of vital substances are transported through the cell membrane as solutes either in **lipids** or in water. Hydrophobic molecules such as fatty acids, steroid **hormones**, vitamin A, E, and D, may easily be transported into cells because they are soluble in lipids. Such facility is due to the structural nature of the membrane, which contains two layers of lipoprotein molecules and apolar chains in the inside of it, with its apolar ends reaching the external surface of the membrane. Most of the protein portion (i.e., the hydrophilic portion) of such molecules is located between the external and internal lipid surfaces. Therefore, the membrane of cells is a mosaic of fluid lipoprotein structure, with weak bonds between lipid molecules, which allow the easy passage of lipid-soluble substances into cells by passive diffusion, without energy expenditure. Conversely, the hydrophilic solutes, i.e., compounds soluble in water, have more difficulty in penetrating through the cell membrane and, depending on the chemical structure and molecular size, they may require the expenditure of cellular energy to do so. Some of these solutes are transported by active transport, which involves energy, whereas others enter the cell through facilitated diffusion, with no energy required. However, cellular membranes are very permeable to water, and when immersed in a hypotonic solution, they absorb water and increase the cell volume, changing its shape. Conversely, a hypertonic solution causes the loss of water from the immersed cells, shrinking them.

The active transport of fluids and their solutes occur against the electrolyte and/or chemical gradient. Energy for active transport is provided by the hydrolysis of **ATP** into ADP. By facilitated diffusion, a great number of hydrophilic substances can enter the cell, such as glucose and several **amino acids**. In this type of transport, the substance seems to be combined with some carrier substance or permease, which promotes its transport through the membrane and release into the cytosol. The potential energy of cations of sodium, potassium, and hydrogen (ion gradients) can also be utilized in the transport of molecules and other ions through the membrane. For example, glucose present in foods is absorbed by the membrane of intestinal epithelial apical cells along with sodium cations, and is later delivered into the **blood** circulation. Although the absorption of glucose occurs against the existing glucose gradient, the accompanying sodium cations penetrate through the membrane in favor of the sodium gradient, whose concentrations inside the cell are very low. Therefore, the transport of glucose is facilitated by sodium absorption.

Another form of fluid transport into cells is termed pinocytosis. In this case, parts of the cell membrane elongate and invaginate around drops of fluid, forming small vesicles that are pulled inside through cyto-skeletal successive contractions. This process can be observed in the epithelial cells of the capillary blood vessels. Most cells perform a selective pinocytosis, which involves specific membrane receptors that allow the passage of the selected substances in higher concentrations, while preventing the excessive absorption of water.

See also Capillaries; Cilia and ciliated epithelial cells; Electrolytes and electrolyte balance; Glomerular filtration; Interstitial fluid; Osmotic equilibria between intercellular and extracellular fluids; Plasma and plasma clearance; Vascular system overview

FOLLICLE STIMULATING HORMONE (FSH) AND LUTEINIZING HORMONE (LH)

Luteinizing hormone (LH) and follicle stimulating hormone (FSH) are compounds that can effect the **physiology** and behavior of the organism in which they are present. More specifically, LH and FSH are gonadotrophins, because they stimulate the **gonads** in the male and ovaries in the female. The **hormones** FSH and LH are necessary for successful reproduction.

Both LH and FSH are composed of subunits of glycoprotein. One of these subunits made up of 89 **amino acids** is identical in both hormones. The other subunit that is the same length but has a different amino acid make-up in LH and FSH allows LH and FSH to recognize and bind to different receptor molecules. For LH, the receptor is located on Leydig cells and Theca cells. Leydig cells are located in the testes. LH binding to these cells stimulates the production and release of **testosterone**. The testosterone acts on the sperm-producing cells in the testes and, along with FSH, stimulates the production of **sperm**. Theca cells are located in the ovary. LH binding to them stimulates the secretion of estrogens. In females, LH is required for the continued development and activity of specialized cells in the ovary called corpora lutea. The maintenance of **pregnancy** depends on the properly-working corpora lutea.

FSH stimulates the maturation of ovarian follicles, which then release an oocyte. **Fertilization** of the oocyte commences embryonic development. A dramatic demonstration of the power of FSH is seen when the hormone is deliberately administered. "Superovulation" results in more than the usual number of oocytes, which can lead to a pregnancy resulting in multiple offspring. Recombinant FSH produced in the laboratory has been successfully used to stimulate superovulation.

Both LH and FSH are produced one region (the anterior lobe) of the pituitary gland, a pea-sized structure located at the base of the **brain**. Their production is regulated by another hormone called gonadotrophin-releasing hormone. The production of the regulatory hormone is itself under the regulatory control of many other hormones from the **hypothalamus** of the brain.

The diminished production of LH and FSH can result in malfunction of the gonads (termed hypogonadism). In males hypogonadism can be evident as a decreased sperm count. In females, the monthly reproductive cycle will stop, which is a normal part of post-menopausal life.

See also Gonads and gonadotrophic hormone physiology; Ovarian cycle and hormonal regulation; Pituitary gland and hormones

FORENSIC PATHOLOGY

Forensic **pathology** is the application of investigative anatomical and medical science to matters of law. Both defense and prosecuting attorneys sometimes use information gleaned by forensic scientists in attempting to prove the innocence or guilt of a person accused of a crime.

Although fingerprints have been used by crime investigators for more than a century, they remain one of the most sought after pieces of evidence. All human beings are born with a characteristic set of ridges on our fingertips. The ridges, which are rich in sweat pores, form a pattern that remains fixed for life. Even if the skin is removed, the same pattern will be evident when the skin regenerates. Some of the typical patterns found in fingerprints are arches, loops, and whorls.

Oils from **sweat glands** collect on these ridges. When something is touched, a small amount of the oils and other materials on our fingers are left on the surface of the object. The pattern left by these substances, which collect along the ridges on our fingers, make up the fingerprints that police look for at the scene of a crime. It is the unique pattern made by these ridges that motivate police to record people's fingerprints. To take someone's fingerprints, the ends of the person's fingers are first covered with ink. The fingers are then rolled, one at a time, on a smooth surface to make an imprint that can be preserved. Fingerprints collected as evidence can be compared with fingerprints on file or taken from a suspect.

Fingerprints are not the only incriminating patterns that a criminal may leave behind. Lip prints are frequently found on glasses. Footprints and the soil left on the print may match those found in a search of an accused person's premises. Tire tracks, bite marks, toe prints, and prints left by bare feet may also provide useful evidence. In cases where the identity of a victim is difficult because of **tissue** decomposition or **death** caused by explosions or extremely forceful collisions, a victim's teeth may be used for comparison with the dental records of missing people.

It is **DNA** that carries the "blueprint" (genes) from which "building orders" are obtained to direct the growth, maintenance, and activities that go on within our bodies. Except for identical twins, no two people have the same DNA. However, we all belong to the same species; consequently, large strands of DNA are the same in all of us. It is the unique strands of DNA that are used by forensic scientists. Strands of DNA can be extracted from cells and "cut" into shorter sections using **enzymes**. Through chemical techniques involving electrophoresis, radioactive DNA, and x rays, a characteristic pattern can be established, the so-called genetic fingerprint.

Although genetic fingerprinting can provide incriminating evidence, DNA analysis is not always possible because the amount of DNA extracted may not be sufficient for testing. Furthermore, there has been considerable controversy about the use of DNA, the statistical nature of the evidence it offers, and the validity of the testing.

Genetic fingerprinting is not limited to DNA obtained from humans. In Arizona, a homicide detective found two seed pods from a paloverde tree in the bed of a pickup truck owned by a man accused of murdering a young woman and disposing of her body. The accused man admitted giving the woman a ride in his truck but denied ever having been near the factory where her body was found. The detective, after noting a scrape on a paloverde tree near the factory, surmised that it was caused by the accused man's truck. Using RAPD (Randomly Amplified Polymorphic DNA) markers–a technique developed by Du Pont scientists–forensic scientists were able to show that the seed pods found in the truck must have come from the scraped tree at the factory.

DNA analysis is a relatively new tool for forensic scientists, but already it has been used to free a number of people who were unjustly sent to prison for crimes that genetic fingerprinting has shown they could not have committed. Despite its success in freeing victims who were unfairly convicted, many defense lawyers claim prosecutors have overestimated the value of DNA testing in identifying defendants. They argue that because analysis of DNA molecules involves only a fraction of the DNA, a match does not establish guilt, only a probability of guilt. They also contend that there is a lack of quality control standards among laboratories, most of them private, where DNA testing is conducted. Lack of such controls, they argue, leads to so many errors in testing as to invalidate any statistical evidence. Many law officials argue that DNA analysis can provide probabilities that establish guilt beyond reasonable doubt.

Long before DNA was recognized as the "ink" in the blueprints of life, **blood** samples were collected and analyzed in crime labs. Most tests used to tentatively identify a material as blood are based on the fact that peroxidase, an enzyme found in blood, acts as a catalyst for the reagent added to the blood and forms a characteristic color. For example, when benzidine is added to a solution made from dried blood and water the solution turns blue. If phenolphthalein is the reagent, the solution turns pink. More specific tests are then applied to determine if the blood is human.

The evidence available through blood typing is not as convincing as genetic fingerprinting, but it can readily prove innocence or increase the probability of a defendant being guilty. All humans belong to one of four blood groups–A, B, AB, or O. These blood groups are based on genetically determined **antigens** (A and/or B) that may be attached to the red blood cells. These antigens are either present or absent in blood. By adding specific antibodies (anti-A or anti-B) the presence or absence of the A and B antigens can be determined. If the blood cells carry the A antigen, they will clump together in the presence of the anti-A antibody. Similarly, red blood cells carrying the B antigen will clump when the anti-B antibody is added. Type A blood contains the A antigen; type B blood carries the B antigen; type AB blood carries both antigens; and type O blood, the most common, carries neither antigen. To determine the blood type of a blood sample, antibodies of each type are added to separate samples of the blood. If a person accused of a homicide has type AB blood and it matches the type found at the crime scene of a victim, the evidence for guilt is more convincing than if a match was found for type O blood. The reason is that only 4% of the population has type AB blood. The percentages vary somewhat with race. Among Caucasians, 45% have type O, 40% have type A, and 11% have type B. African Americans are more likely to be type O or B and less likely to have type A blood.

When blood dries, the red blood cells split open. The open cells make identification of blood type trickier because the clumping of cell fragments rather than whole red blood cells is more difficult to see. Since the antigens of many blood-group types are unstable when dried, the FBI routinely tests for only the ABO, Rhesus (Rh), and Lewis (Le) blood-group antigens. Were these blood groups the only ones that could be identified from blood evidence, the tests would not be very useful except for proving the innocence of a suspect whose blood type does not match the blood found at a crime scene. Fortunately, forensic scientists are able to identify many blood proteins and enzymes in dried blood samples. These substances are also genetic markers, and identifying a number of them, particularly if they are rare, can be statistically significant in establishing the probability of a suspect's guilt. For example, if a suspect's ABO blood type matches the type O blood found at the crime scene, the evidence is not very convincing because 45% of the population has type O blood. However, if there is a certain match of two blood proteins (and no mismatches) known to be inherited on different **chromosomes** that appear respectively in 10% and 6% of the population, then the evidence is more convincing. It suggests that only $0.45 \times 0.10 \times 0.06 = 0.0027$ or 0.27% of the population could be guilty. If the accused person happens to have several rarely found blood factors, then the evidence can be even more convincing.

Autopsies can often establish the cause and approximate time of death. Cuts, scrapes, punctures, and rope marks may help to establish the cause of death. A **drowning** victim will have soggy **lungs**, water in the stomach, and blood diluted with water in the left side of the **heart**. A person who was not **breathing** when he or she entered the water will have undiluted blood in the heart. Bodies examined shortly after the time of death may have stiff jaws and limbs. Such stiffness, or **rigor mortis**, is evident about ten hours after death, but disappears after about a day when the tissues begin to decay at normal temperatures. Each case is different, of course, and a skillful pathologist can often discover evidence that the killer never suspected he or she had left behind.

FORSSMANN, WERNER (1904-1979)
German physician

Werner Forssmann, a surgeon and urologist, was relatively unknown in his native Germany when he won the Nobel Prize in 1956 for his work in **heart** catheterization. His groundbreaking experiment had been done almost three decades earlier, and when he received word of the award—after a morning of surgery during which he had operated on three patients with kidney disease—he commented, as quoted in *Mayo Clinic Proceedings,* "I feel like a village parson who has just learned that he has been made bishop."

Werner Theodor Otto Forssmann was born in Berlin, the only child of Julius Forssmann, a lawyer employed by a life insurance company, and Emmy Hindenberg. Forssmann's father died in World War I while young Forssmann was still a student in the Askanische Gymnasium, a school emphasizing a humanistic approach to education. His mother worked as an office clerk and his grandmother took over the role of running the household. Forssmann's uncle, a doctor just outside of Berlin, became an influential force in his nephew's life, ultimately convincing Forssmann to pursue a career in medicine. In 1922, after graduating from the Gymnasium, Forssmann entered the Friedrich Wilhelm University in Berlin, passing the state examination in 1928. Forssmann's doctoral thesis on the effects of concentrated **liver** on pernicious anemia, a **blood** deficiency, marked the way for his later experiments. Together with a small group of fellow students, Forssmann experimented on himself, taking large doses of liver concentrate daily and demonstrating its healthful effects on blood. After receiving his doctor's diploma in early 1929 and being frustrated in his efforts to obtain a post as an internist, Forssmann worked for a short time in a private women's clinic in Spandau. Then, through family connections, he secured an internship at the August Viktoria Home in Eberswalde, a small town northeast of Berlin.

Training as a surgeon, Forssmann nevertheless gave thought to an earlier passion inspired by a teacher he encountered in medical school: heart diagnosis. He was dissatisfied with the inaccuracy and uncertainty of diagnostic techniques such as percussion, auscultation, x ray, and even electrocardiography. He became convinced that there was an internal diagnostic method that would not involve major risks, trigger automatic reflex actions, or disturb the balance of pressure in the thorax. As early as the mid-nineteenth century, there had been a procedure known as cardiac catheterization in animal experiments. Doctors had performed the procedure in the late nineteenth century to determine blood pressure in the right and left chambers of the heart. Some of these procedures employed the use of a catheter inserted through the jugular vein of a horse. Forssmann believed that he could do this on humans through a vein at the elbow traditionally used for intravenous injections. His research on cadavers supported his idea, and by the summer of 1929, Forssmann approached his supervisor, Dr. Richard Schneider, with a plan to catheterize his own heart with a ureteric catheter. Schneider, however, would not allow such a dangerous experiment in his hospital.

Undaunted, Forssmann set out to convince a surgical nurse in his section of his experiment's feasibility so he could gain access to the sterilized instruments he needed. Eventually, the nurse agreed to aid him, even agreeing to be the first subject. Forssmann, however, had no intention of experimenting on anyone but himself initially. He gave himself a local anesthetic in the left elbow and then made an incision. Once he had opened his vein, he inserted the catheter about a foot up his arm and had the nurse accompany him to the x-ray lab. There, Forssmann stood behind a fluoroscope screen with a mirror placed so that he could see the image of the catheter, which he pushed up until it was in the right ventricle of his heart. Then he calmly ordered that photographs be made of this momentous achievement.

The results of this experiment were published in a short paper in the prestigious *Klinische Wochenschrift* and won Forssmann a position at the Charité Hospital in Berlin. But the reception to his article by other physicians was cool and his superior at the Charité did not approve of his unorthodox techniques, so Forssmann was soon back in Eberswalde. He continued his experiments for the next two years, during which time he proved that the insertion of a catheter in the heart was painless and caused no damage to the blood vessels. He also pioneered techniques for measuring pressure inside the heart and for injecting opaque material for x-ray studies of the heart. Still, his work was not accepted by most physicians, who called it unethical, and considered his experiments stunts. By 1931, Forssmann, discouraged by the response to his work, gave up experimental medicine. He returned to the Charité Hospital in Berlin and soon moved on to the Mainz City Hospital. It was there, in 1932, that he met the woman who would become his wife, Dr. Elsbet Engel, a resident in internal medicine. Their marriage necessitated another change of hospitals for Forssmann, for it was against the hospital's policy for a married couple to work together. Forssmann trained as a urologist in Berlin at the Rudolph Virchow Hospital, then took a position as a surgeon and urologist at the City Hospital of Dresden-Friedrichstadt for two years. Later, he became a senior surgeon at the **Robert Koch** Hospital in Berlin. His colleagues considered him a fine surgeon.

During World War II, Forssmann served as an army surgeon, surviving six years spent in Germany, Norway, Russia, and in a prisoner of war (POW) camp. Back in Germany after the War, he practiced as a country doctor in the Black Forest village of Wambach for three years before returning to the practice of **urology** in 1950 at Bad Kreuznach. It was only after the war that Forssmann discovered that others had continued working with his cardiac catheterization experiment of 1929. The most notable implementation was by two Americans, Dickinson Woodruff Richards, Jr., and André F. Cournand, who developed it into a tool for diagnosis and research. In 1954, Forssmann received the Leibniz Medal from the German Academy of Science in Berlin, yet he was refused a professorship at the University of Mainz. He had

resigned himself to being a little-known doctor in Bad Kreuznach when, on October 18, 1956, he was notified that he had won, along with Richards and Cournand, the Nobel Prize in physiology or medicine for his contribution to the knowledge of heart catheterization and pathological changes in the **circulatory system**.

The Nobel Prize finally earned Forssmann renown and respect; in *Clinical Cardiology,* H. W. Heiss called him "one of the great fathers of cardiology." In 1958, he became the chief of the surgical division of the Evangelical Hospital of Düsseldorf, and ten years later he was awarded the gold medal of the Society of Surgical Medicine of Ferrara. After he retired, Forssmann spent his time in the Black Forest, where he enjoyed the outdoors and nature. He died of a heart attack in Schopfheim, West Germany, in 1979.

See also Cardiac disease; Cardiac muscle; Careers in anatomy and physiology; Imaging

FROSTBITE • *see* HYPOTHERMIA

FSH • *see* FOLLICLE STIMULATING HORMONE (FSH)

FUNCTIONAL CHARACTERISTICS OF LIVING THINGS

Precise definitions of life, especially when dealing with **viruses** and other microorganisms are often difficult. Although there is a wide variety of life on Earth, especially with regard to outward structure and form (morphology), all living organisms, both plant and animal, share a few fundamental functional characteristics that together define life processes.

The long-term survival of any species is, of course, dependent upon the ability to reproduce. Although selected individuals and population groups may be temporarily prevented from reproduction or chose not to reproduce, the capacity to reproduce—and to produce subsequent generations that are also capable of reproduction—is essential to the survival of every species.

In some form, all living things show an ability respond to environmental stimulation. The response may result in gross movement, but a response may also be reflected only in a subtle, internal change in the organism's **physiology**.

Living organisms at all levels must be able to assure the integrity of their structure and physiology by maintaining some sort of boundary to the external environment. This boundary ranges from cell membranes (e.g., the outer cell membrane provides a protective barrier and interface with the extracellular environment) to large systems such as the skin and epithelial covering of the human integumentary system that are designed to retain **tissue** moisture and provide a barrier to **infection**.

Boundary maintenance is also critical to the maintaining homeostatic balance. Without boundaries, organisms would be unable to maintain stability in their internal environments or to be selective in their responses to physiological stimuli. For example, species show a wide variety of physiological heat reduction mechanisms when confronted with elevated temperatures in the external environment.

Responsiveness to external stimulus is often characterized as an organism's irritability.

Organisms must also obtain nutrients from the external environment. Schemes range from simple nutrient absorption to aggressive hunting and **digestion** of foods that provide needed nutrients. Regardless of how they physically obtain nutrients, all organisms must possess the appropriate metabolic biochemical apparatus to accomplish digestion and to convert and utilize the chemical energy contained in nutrients.

Nutrients can contribute to an organism's **growth and development** in one of three basic ways. The nutrients become raw materials for growth. For example, **calcium** is a raw material need to promote and maintain bone in the human skeletal system. Essential **amino acids**, a building block of protein, are those that the human body cannot reproduce via the metabolic breakdown other nutrients. Accordingly, essential amino acids must be available in the diet. Nutrients can also enter into a variety of biochemical pathways, associating with **enzymes** and reacting with other substances as part of an organism's **metabolism**. Finally, nutrients may facilitate reactions within an organism's particular metabolic pathways.

Organisms must also be able to excrete waste. This elimination of metabolic waste is an essential part of the **homeostatic mechanisms** designed to maintain levels of nutrients, water, and oxygen within tolerable limits and within a temperature range conducive to the reactions taking place within metabolic pathways. Including in the **elimination of waste** is the elimination of excess fluid.

See also Adenosine triphosphate (ATP); Cell theory; Evolution and evolutionary mechanisms; Homeostatic mechanisms; Homologous structures

FURCHGOTT, ROBERT F. (1916-   )

American pharmacologist

In his long scientific carrier, Robert F. Furchgott, researched **tissue metabolism** for many decades, working with various organ tissues. In 1980, the outcome of his experiment with acetylcholine, paved the way to the discovery of the compound involved in **blood** vessel dilatation. With this experiment, he demonstrated that acetylcholine only dilates a blood vessel when the endothelium (i.e., the tissue that constitutes the internal surface of blood vessels) is intact. When a lesion was present in this tissue, dilatation by acetylcholine was prevented. Therefore, Furchgott concluded that the endothelium cells should have a role in the regulation of blood cells dilatation, through the production of an unknown molecule that induced the relaxation of the vascular **smooth muscle** cells. He named this unidentified molecule endothelium-derived relaxing factor—EDRF. Another researcher, the pharmacologist Louis J. Ignarro, was one of the scientists that searched for the

chemical identity of Furchgott's EDRF. After a series of studies, he concluded that EDRF was nitric oxide (NO) gas. Yet another researcher, **Ferid Murad**, physician and pharmacologist, had already studied the chemical events involved in the vasodilating effect of nitroglycerin and other related compounds in the previous decade. In 1977, he found that the chemical by-product of these compounds that caused blood vessel dilatation was nitric oxide, a common air pollutant resulting from nitrogen combustion in automobile fuel. Murad concluded that endogenous molecules, such as **hormones**, could possibly also affect blood vessels through the mediation of nitric oxide, although at the time there were not experimental means to demonstrate it.

Furchgott and Ignarro divulged the discovery of endogenous nitric oxide in 1986, and triggered a great amount of research by other groups throughout the world, because it was the first time that a gas was demonstrated to act as a cellular endogenous signaling factor. Although it was known that some **bacteria** produced nitric oxide, no one suspected that nitric oxide could have an important physiological role in higher organisms. In the wake of this finding, further researches began unveiling the many implications of nitric oxide for cell metabolism in several **organs**, from the nervous system to the control of blood pressure, as well as to the control of blood flow to different organs. Nitric oxide is produced by many cell types for different purposes, such as prevention of thrombi (i.e., blood clots formation) in the **arteries**, activation of **neurons** and the cells in their vicinity, as well as for killing bacteria, fungi, and other parasites, when produced in great amounts by macrophages (**immune system** white blood cells). Nitric oxide is also used by the immune system to kill tumor cells. Several therapeutic uses for nitric oxide are now being tested for conditions such as lung hypertension, **arteriosclerosis**, systemic sepsis, **cancer**, and impotence. Due to the importance of these discoveries and its wide implication to the understanding of the physiology of several metabolic events, Furchgott, Ignarro, and Murad were jointly awarded with the Nobel Prize in Physiology and Medicine in 1998.

Robert F. Furchgott was born in Charleston, South Carolina, where he spent the first thirteen years of his life, and became interested in natural history and life sciences during his childhood. He began his university education in chemistry at the University of South Carolina, and graduated from the University of North Carolina at Chapel Hill in 1937. Furchgott achieved a Ph.D. degree in Biochemistry in 1940 at the Northwestern University and worked as Professor of the Department of Pharmacology at the State University of New York from 1956 until 1988. Before 1956, he worked and researched at other academic institutions, such as the Northwestern University, Cold Spring Harbor, Cornell University Medical College, and Washington University. During that period, he investigated the structure of blood red cells, tissue metabolism (phosphate exchange and turnover in dog left ventricular muscle), circulatory **shock**, the energy-metabolism and function of rabbit intestinal smooth muscle, the reactions of aorta to **epinephrine**, isoproterenol, sodium nitrite and acetylcholine, among other drugs, as well as photorelaxation of blood vessels. In 1956, he was invited to the position of chairman of the newly created Department of Pharmacology at the State University of New York—SUNY. At SUNY, he continued the research on photoralaxation of blood vessels, and together with Stuart Ehrreich found that many other smooth muscle preparations (from stomach, uterus and intestine), which normally do not relax in response to radiation, would relax in the presence of nitrite (nitric oxide in tissues is converted within ten seconds to nitrate and nitrite). He also researched the cellular signal transduction pathways and receptors.

In 1982, Furchgott resigned from his position as chairman of pharmacology at SUNY, but continued working at the institution as a professor. Along with the Nobel Prize, Furchgott received the following fellowships and awards: honorary doctorates from the Universities of Madrid, Lund, Gent, North Carolina, Goodman & Gilman Award (1984), CIBA Award for Hypertension Research (1988), Gairdner Foundation International Award (1991), Roussel-Uclaf Prize for Research in Signal Transduction (1993), Wellcome Gold Medal, British Pharmacological Society (1995), Albert Lasker Basic Medical Research Award (1996).

See also Blood pressure and hormonal control mechanisms; Circulatory system; Drug treatment of cardiovascular and vascular disorders

G

GALEN (CA. 130-CA. 200)
Greek physician

Galen, the last and most influential of the great ancient medical practitioners, was born in Pergamum, Asia Minor. His father, the architect Nicon, is supposed to have prepared Galen for a career in medicine following the instructions given him in a dream by the god of medicine, Asclepius. Accordingly, Galen studied philosophy, mathematics, and logic in his youth and then began his medical training at age sixteen at the medical school of Pergamum attached to the local shrine of Asclepius. At age twenty, Galen embarked on extensive travels, broadening his medical knowledge with studies at Smyrna, Corinth, and Alexandria. At Alexandria, the preeminent research and teaching center of the time, Galen was able to study skeletons (although not actual bodies).

Returning to Pergamum at age twenty-eight, Galen became physician to the gladiators, which gave him great opportunities for observations about human **anatomy** and **physiology**. In 161 A.D., Galen moved to Rome and quickly established a successful practice after curing several eminent people, including the philosopher Eudemus. Galen also conducted public lectures and demonstrations, began writing some of his major works on anatomy and physiology, and frequently engaged in polemics with fellow physicians. In 174 A.D., Galen was summoned to treat Marcus Aurelius and became the emperor's personal physician.

Galen later again returned to Pergamum, perhaps to escape the quarreling, perhaps to avoid an outbreak of plague in Rome. After a few years, Galen was summoned back to Rome by Marcus Aurelius. He became physician to two subsequent emperors, Commodus and Septimius Severvs, and seems to have stayed in Rome for the rest of his career, probably dying there near age 70.

Galen was an astonishingly prolific writer, producing hundreds of works, of which about 120 have survived. His most important contributions were in anatomy. Galen expertly

Galen. *The Library of Congress.*

dissected and accurately observed all kinds of animals, but sometimes mistakenly—because human dissection was forbidden—applied what he saw to the human body. Nevertheless, his descriptions of bones and muscle were notable; he was the first to observe that muscles work in contracting pairs. He described the **heart** valves and the structural differences between **arteries** and **veins**. He used experiments to demonstrate paralysis resulting from **spinal cord** severing, control of the **larynx** through the laryngeal nerve, and passage of urine from **kidneys** to bladder. An excellent clinician, Galen pioneered diagnostic use of the **pulse** rate and described cardiac arrhythmias. Galen also collected therapeutic plants in his extensive travels and explained their uses.

In his observations about the heart and **blood** vessels, however, Galen made critical errors that remained virtually unchallenged for 1,400 years. He correctly recognized that blood passes from the right to the left side of the heart, but decided this was accomplished through minute pores in the septum, rather than through the **pulmonary circulation**. Like **Erasistratus**, Galen believed that blood formed in the **liver** and was circulated from there throughout the body in the veins. He did show that arteries contain blood, but thought they also contained and distributed *pneuma*, a vital spirit. In a related idea, Galen believed that the **brain** generated and transmitted another vital spirit through the (hollow) nerves to the muscles, allowing movement and sensation.

After Galen, experimental physiology and anatomical research ceased for many centuries. Galen's teachings became the ultimate medical authority, approved by the newly ascendant Christian church because of Galen's belief in a divine purpose for all things, even the structure and functioning of the human body. The medical world moved on from Galenism only with the appearance of **Andreas Vesalius**'s work on anatomy in 1543, and William Harvey's on blood circulation in 1628.

See also History of anatomy and physiology: The Classical and Medieval periods

GALL, FRANZ JOSEPH (1758-1828)

German-Austrian physician, anatomist, and psychologist

Franz Joseph Gall originated phrenology, the idea that a person's character can be discerned by studying the shape of the **skull**, because the bumps on the skull would be palpable signs of relative degrees of development in the various regions of the **brain**. Although phrenology is a pseudoscience, Gall was still a first-rate anatomist, and contributed significantly to the knowledge of brain function and cerebral localization.

Gall was born in Tiefenbronn, Germany, to a pious Roman Catholic family of Italian extraction. After studying privately with his uncle, who tried unsuccessfully to prepare him for the priesthood, and in schools in Baden and Bruchsal, Germany, he enrolled at the University of Strasbourg in 1777 as a medical student. He moved to Vienna in 1781 and received his M.D. from the university there in 1785. His medical practice in Vienna included quite a few wealthy, prestigious, and powerful patients.

About 1790, Gall became interested in identifying the possible functions of the parts of the brain and their outward, visible manifestations. Until his time, there was little concept of cerebral localization. Except for the idea of the French philosopher **René Descartes** that somewhere in the brain was "the seat of the soul," the brain was usually believed to be a functionally homogeneous organ. Gall, influenced by the

comparative methods of German philosopher Johann Gottfried von Herder (1744–1803), began formulating a theory of correlation between psychological types and skull shapes. He published a preliminary programmatic study in 1791, several years before he began dissecting brains.

Gall's next published mention of his theory was an article in the December 1798 issue of *Neuer Deutscher Merkur*. Phrenology is, therefore, said to have been founded in 1798, though it would not yet be called "phrenology" ("science of the mind") for two decades. Gall called it *Schädellehre* ("skull theory") or *Craniologie* ("science of the cranium").

All materialism, from the Greek philosopher Anaxagoras (ca. 500–ca. 428 B.C.) through Gall to the American behaviorist Burrhus Frederic Skinner (1904–1990), assaults human spirituality by claiming to provide full accounts of individual freedom, thought, and consciousness in terms of merely physical events. Many of Gall's contemporaries, including many in positions of authority, feared that his teachings would promote atheism, immorality, and insubordination, and took steps to limit his influence. When in 1801, Holy Roman Emperor Francis II (1768–1835) made him unwelcome in Vienna, he embarked on a long lecture tour of Germany, Switzerland, Scandinavia, the Lowlands, and France, charging admission wherever he went. Even though scientific lectures were supposed to be free, and even though his medical colleagues scorned him for demeaning their profession by taking his case to the masses rather than publishing, a crowded house was always ready to hear him. Phrenology made Gall rich.

Phrenology was early regarded as a pseudoscience and was sharply criticized, not only by the medical community, but also by intellectuals of all kinds, notably the philosopher Georg Wilhelm Friedrich Hegel (1770–1831) in *Phenomenology of Spirit* (1807). Hegel accused Gall of studying living **tissue**, the brain, as something other than itself, reducing it instead to dead tissue, the skull bones. Thus, by not dealing with his actual subject matter directly, he falsified his enterprise. This criticism would apply only to Gall's phrenological concerns, not to his neuroanatomical investigations. Even his most severe medical critic, Pierre Flourens (1794–1867), conceded that Gall's dissections of the human brain were the best he had ever seen.

After years of life on the road as a flamboyant showman, Gall settled in Paris in 1807, was naturalized French in 1819, and died in Paris. With co-author and disciple Johann Kaspar Spurzheim (1776–1832), Gall finally published in 1810 the book his critics had long demanded, *Anatomie et physiologie du système nerveux en général, et du cerveau en particulier, avec des observations sur la possibilityé de reconnoître plusieurs dispositions intellectuelles et morales de l'homme et des animaux, par la configuration de leurs têtes* (Anatomy and Physiology of the Nervous System in General and the Brain in Particular, with Observations on the Possibility of Recognizing Several Moral and Intellectual Traits of Humans and Animals by the Shape of Their Heads).

See also Brain stem function and reflexes; Brain: intellectual functions; Central nervous system (CNS); Cerebellum; Cerebral cortex; Cerebral hemispheres

GALLBLADDER • *see* BILE, BILE DUCTS, AND THE BIL-
IARY SYSTEM

GALVANI, LUIGI (1737-1798)
Italian anatomist

Luigi Galvani was the pioneer of electrophysiology. A skilled
anatomist, obstetrician, physician, and surgeon, Galvani conducted several experiments with animal nervous and muscular
systems. After an accidental discovery of the relationship
between electricity and muscle movement while dissecting a
frog, Galvani proposed a theory of "animal electricity."
Although his theory incorrectly identified the electro-conduct-
ing medium in animals as fluid, his research opened new lines
of inquiry about the **structure and function** of the nervous system in both animals and humans.

Galvani was born in Bologna, Italy. His original intention was to study theology and later enter a monastic order.
However, Galvani was discouraged from pursuing monastic
life in favor of continuing his studies of philosophy or medicine. He devoted his academic career at the University of
Bologna to both interests and in 1759, received his degree in
letters and medicine on the same day. In 1762, Galvani was
named professor of **anatomy** at the university, and remained
there for most of his career. His early works were primarily
concerned with **comparative anatomy**.

In 1764, Galvani married Lucia Galleazzi, the daughter
of a prominent member of the Bologna Academy of and
Science. Galvani's wife encouraged his independent research,
and served as a counselor and guide for his experiments until
her death. Drawing from his extensive training in anatomy and
obstetrics, Galvani focused his research on the nature of muscular movement. While dissecting frogs for study, Galvani
noticed that contact between certain metal instruments and the
specimen's nerves provoked muscular contractions in the frog.
Believing the contractions to be caused by electrostatic
impulses, Galvani acquired a crude electrostatic machine and
Leyden jar (used together to create and store static electricity),
and began to experiment with muscular stimulation. He also
experimented with natural electro-static occurrences. In 1786,
Galvani observed muscular contractions in the legs of a specimen while touching a pair of scissors to the frog's lumbar nerve
during a lightening storm. He also noticed that a simple metallic arch connecting certain tissues could be substituted for the
electrostatic machine in inducing muscular convulsions.

The detailed observations on Galvani's neurophysiological experiments on frogs, and his theories on muscular movement, were not published until a decade after their coincidental
discovery. In 1792, Galvani published *De Viribus Electricitatis
in Motu Musculari* (On Electrical Powers in the Movement of
Muscles). The work set forth Galvani's theories on "animal
electricity." He concluded that nerves were detectors of minute
differences in external electrical potential, but that animal tissues and fluids must themselves possess electricity that is different from the "natural" electricity of lightening or an
electrostatic machine. Galvani later proved the existence of

Luigi Galvani. *The Library of Congress.*

bioelectricity by stimulating muscular contractions with the
use of only one metallic contact—a pool of mercury.

Galvani's work altered the study of neurophysiology.
Before his experiments, the nervous system was thought to be
a system of ducts or water pipes, as proposed by Descartes.
Galvani proved that there was a relationship between muscle
movement and electricity and proposed that nerves were electric conductors. However, the nature of a different "animal
electricity" which Galvani proposed was disproved later by
Italian physicist, **Alessandro Volta**. Galvani had experimented
with using animal fluids and metalic conductors—hence the
reference to his name in the term "galvanized"—to create an
electric pile, or battery. However, in 1800, Volta created the
first successful battery that could produce a sustained electrical current. Thus, Volta effectively rejected Galvani's theory
of "animal electricity."

After the death of his wife, Galvani joined the Third
Monastic Order of the Franciscans. He continued his academic
research, but when Napoleon seized control of Bologna n
1796, Galvani refused to swear the oath of allegiance to the
newly created Cisalpine republic on the basis that it contra-
dicted his political and religious beliefs. As a result, he was
stripped of his professorship and lost his pension. Galvani died
in poverty at age 61.

See also History of anatomy and physiology: The Renaissance
and Age of Enlightenment; Muscle contraction; Muscular
innervation; Nerve impulses and conduction of impulses

GAMETE (FEMALE) · *see* OOGENESIS

GAMETE (MALE) · *see* SPERMATOGENESIS

GASEOUS EXCHANGE

Respiratory gaseous exchange occurs between the interior of an organism and the environment. In all animals with a vascular system, in which the **blood** transports respiratory gases, there is a twofold gaseous exchange within actively respiring tissues and the respiratory **organs**. In humans, the respiratory organs are the **lungs**, and there gaseous exchange takes place in the **alveoli**.

The quantity of gas diffusing through the alveolar **tissue** depends on pressure differences. There are pressure gradients between the air in the alveoli and the blood. The steeper the gradient, the more rapid the gaseous exchange. The steepness of the pressure gradient corresponds to the difference between the partial pressure of the respiratory gas in the alveolar air and its partial pressure in the blood (the pressure exerted by a particular component of a mixture of gases is called the partial pressure of that gas). Inspired air contains 21% of oxygen which corresponds to a partial pressure of 21.3 kPa. After mixing with the air in the lungs, this partial pressure falls to about 13 kPa. In contrast, the oxygen partial pressure of the venous blood delivered to the lung **capillaries** by the lung **arteries** is about 5kPa. After passage through the lung capillaries, the oxygen partial pressure of the blood rises to about 13 kPa. Because the total volume of oxygen in the lungs is large compared with the volume that diffuses into the blood, the oxygen partial pressure of the alveolar air decreases only slightly. In the pulmonary vein to the **heart**, the oxygen concentration decreases somewhat, because it is mixed with "short circuited" blood, which passed through the lungs, but does not take part in gaseous exchange. The final oxygen partial pressure of arterial blood is about 12 kPa.

In the organism, carbon dioxide diffuses in the opposite direction to oxygen. The carbon dioxide partial pressure of inspired air increases from 0.03-5.3 kPa on mixing with pulmonary air. By gaseous exchange, the carbon dioxide partial pressure of the blood decreases from 6 to 5 kPa during passage through the lung capillaries. Carbon dioxide diffuses much more rapidly than oxygen, so that carbon dioxide exchange is completed before the **oxygen exchange**.

Gaseous exchange in the tissues also depends on diffusion. The direction of diffusion of the two respiratory gases is the reverse of that in the lungs. In tissue cells, oxygen partial pressure varies between 0 and 4 kPa, that of carbon dioxide between 6.7 and 8 kPa. The rate of gaseous exchange in tissues depends on two factors, the metabolic activity of the tissue and the dilation of blood vessels. Metabolic activity can alter the pressure gradients of oxygen and carbon dioxide between the tissue the blood while the dilation of the blood vessels causes an increase in blood flow through the tissue and enhances the rate of gaseous exchange by increasing the diffusion factor.

See also Respiration control mechanisms; Respiratory system

GASSER, HERBERT SPENCER (1888-1963)
American physician

During a life devoted to the medical sciences, Herbert Spencer Gasser mastered the fields of physiology, electronics, optics, photography, applied mathematics, and **pharmacology**. His studies with **Joseph Erlanger** on nerve properties, using new techniques to measure the weak electrical currents of nerve fibers, were rewarded with a Nobel Prize in 1944. This work opened an era of research on pathways of the nervous system and has led to a better understanding of the mechanisms of **pain** and reflex actions. Gasser's career as scholar, professor, and experimenter culminated in 1935, when he became the second scientific director of the Rockefeller Institute for Medical Research in New York City.

Herbert Gasser was born in Platteville, Wisconsin. His mother, Jane Griswold, who descended from an early Connecticut family, was a teacher trained in Wisconsin's first State Normal School in Platteville. Gasser's father, Herman, was born in the Tyrol and came to the United States as a boy. Herman was a self-educated man who eventually qualified in medicine and became a country doctor. His scholarly interests in debates on **evolution** during his day led him to name the first of his three children after British philosopher Herbert Spencer. Reading was the most important amusement for young Herbert and his siblings, Mary and Harold. Gasser was also adept at handicrafts, and proved it by setting up a shop to make furniture. His efforts to learn photographic techniques using a simple box Kodak camera formed the first of his scientific intrigues.

After attending State Normal School, Gasser received two degrees in science at the University of Wisconsin, a bachelor's degree in zoology in 1910, and a master's in **anatomy** in 1911. However, Gasser's future interests were determined by a physiology course in the University's newly organized medical school. The young lecturer who emphasized the new spirit of research in medicine was Joseph Erlanger, the man with whom Gasser would share the Nobel Prize thirty-three years later. Though Gasser rejected the medical career his father had chosen, he became intrigued by medicine as a scientific discipline. In 1915, he earned his medical degree from Johns Hopkins University, where he conducted research on **blood coagulation** in his spare time. After another year of research in Wisconsin, Gasser joined Erlanger at Washington University in St. Louis, in 1916.

Earlier scientists had provided painstaking microscopic slides of **neurons** and general theories of nerve networks in the body. Gasser's contributions made it possible to trace pathways while keeping the nervous system intact. Physiologists knew that impulses (action potentials) travel along nerves to convey sensation and to stimulate muscles, and that these impulses could be recorded by electrical instruments. A hypothesis existed that impulses moved faster along thick fibers than they did thin ones. Gasser's dramatic new method involved stimulating a given region of nerves and then reading the transmitted signal as it reached its destination, much like a physician tests a patient's knee jerk response with a rubber mallet. His problem was in finding recording devices capable of measuring, in fractions of a second, impulses that were small in quantity and short in duration. The available devices were inadequate. The string galvanometer and the capillary electrometer were slow and insensitive. The cathode-ray oscillograph, although quick, was insensitive to small currents.

The first breakthrough for Gasser came with the same vacuum tube amplifier that made radio possible. The three-stage amplifier had been brought to St. Louis by H. Sidney Newcomer, one of Gasser's classmates at Hopkins, who had built the device with the help of friends at the Western Electric Company. Nerve impulses could now be recorded, though the instrument's inertia caused distortions in timing the impulses. Their report describing this apparatus and experiments on nerves in the **diaphragm** appeared in 1921. This article was less important for its new knowledge about nerves than for its description of how sensations could at last be signalized.

A new technology, again from Western Electric, allowed Gasser and Joseph Erlanger to conduct the pioneering studies that eventually led to their Nobel Prize. It had been believed for over a decade that, should a means be discovered to test the Braun tube, the nerve impulse might accurately be recorded. But the tube, invented in 1897, used a cold-cathode technology, wherein the emission of electrons from the cathode's electrode is triggered by an outside force—this proved to be its downfall. Western Electric had, on the other hand, developed an oscillograph tube fitted with a hot cathode. This permitted the instrument to operate at a low voltage, which made it more sensitive to the small currents of the nerve action potentials. The instrument could record both the time elapsed between impulses and the change in nerve reactions. Though the tube was a breakthrough for Gasser and Erlanger, they still had to devise auxiliary apparatus to coordinate their induction shocks with the action potentials that were displayed on the screen. This work was reported in 1922.

Using the cathode-ray oscilloscope, Gasser and Erlanger almost immediately made two discoveries about the unexpected complexity they found in nerve trunks. In one, they determined that the sequence of events of nerve impulse transmission consists of two parts. There is an initial, large, rapid deviation in electric potential, called the spike, which ascends then descends during the actual transmission. The spike is followed by a sequence of small, slow potential changes, called the after-potential, that first has a negative and then a positive deviation.

In their other discovery, Gasser and Erlanger found that the composite **action potential** of a nerve has a range of velocities. They eventually identified three distinct patterns based on the length of spikes and their after potentials, and classified the fibers into three main groups. The fastest and thickest are A fibers, the intermediate size are B fibers, while the thinnest and slowest are C fibers. Their findings thus confirmed the hypothesis that thick fibers conduct impulses faster than thin ones.

Erlanger and Gasser next showed how these three types of fibers are distributed over the incoming and outgoing fibers of the **spinal cord**, the sensory and motor roots. The perception of pain is carried by the thin, slow fibers, while muscle sense and touch, and muscle movement are conducted by the fast

fibers. Gasser subsequently explored the excitability of nerve fibers in relation to after-potentials. He also continued to refine the oscilloscope, first using x-ray film and eventually a camera to record the impulses.

Gasser served as professor of pharmacology at Washington University from 1921 to 1931. During a two-year leave of absence between 1923 and 1925, he worked with Archibald V. Hill and Henry Hallett Dale in London, Walter Straub in Munich, and Louis Lapicque at the Sorbonne, on investigations involving muscle contractions and excitation of nerves. In 1931, Gasser became professor of physiology at Cornell University Medical College in New York City. In 1935, at age 47, Gasser became the second scientific director of the Rockefeller Institute for Medical Research, succeeding Simon Flexner. Gasser's medical training and his grasp of mathematical and physical sciences equipped him well to lead and to comprehend the expanding field of scientific medicine. His tenure bridged the economic depression of the 1930s, World War II, and the unsettling changes in the funding of scientific research after the war. Despite these trying times, Gasser, nevertheless, led the institute's transition from its original emphasis on **pathology** and infectious diseases to a broader biological approach to human diseases. From 1936 to 1957, he also served as editor of *The Journal of Experimental Medicine*.

During World War II, many Rockefeller Institute laboratories closed and their facilities and staff were organized to support war efforts. Gasser returned to work he had done on chemical warfare during the first world war, chairing a civilian committee on research development in that field. So it was a great surprise for Gasser when a cable arrived in 1944 from Stockholm, announcing that he had won a Nobel Prize. He described in his autobiography that he was so dismayed by his long estrangement from work on nerve fibers that he "went into retreat to regain touch with the state of the subject through reading [my] own reprints." Gasser retired from the institute in 1953, and was succeeded by Detlev W. Bronk. With a change to emeritus status came the opportunity for Gasser to return to the laboratory. Instead of plunging into new areas of nerve physiology, he returned to unfinished work on differentiation of the thin C fibers. The introduction of electron microscopy helped him confirm many of his earlier findings. Gasser's scientific contributions were recognized by honorary degrees from twelve universities. He was elected to the National Academy of Sciences in 1934, the American Philosophical Society in 1937, and was a member of over twenty other scientific societies in the United States, Europe, and South America. He received the Kober Medal in 1954, from the American Association of Physicians.

Gasser was a tall, thin, fragile man, and lifelong bachelor, who suffered from migraine headaches. Rockefeller Professor Emeritus Maclyn McCarty told Carol Moberg in an interview that Gasser enjoyed entertaining scientific associates at home, regaling them with his technically sophisticated hi-fidelity recording system, or his player piano. His hobbies included literature, theater, travel, and music. For twenty years he shared a summer cottage with the Bronk family on Cape

Cod. Following a stroke, Gasser died in New York Hospital at the age of on 74.

See also Nerve impulses and conduction of impulses; Nervous system overview; Nervous system: embryological development; Neural damage and repair; Neural plexuses; Neurology; Neurotransmitters

GASTROINTESTINAL ABSORPTION • *see*
INTESTINAL HISTOPHYSIOLOGY

GASTROINTESTINAL EMBRYOLOGICAL DEVELOPMENT

The primitive gut tube forms during week four of gestation. It derives from the incorporation of the dorsal part of the definitive yolk sac into the embryo due to embryonic folding. The primitive gut is divided into the foregut, midgut, and hindgut.

The esophagus, stomach, **liver**, gallbladder and **bile** duct, **pancreas**, and upper duodenum derive from the foregut. All of these structures receive their **blood** supply from the celiac trunk (except for some portions of the esophagus, which are supplied by other branches of the aorta).

The first step in the development of the esophagus is the formation of the laryngotracheal diverticulum in the ventral wall of the primitive foregut during week four. The distal end of the laryngotracheal diverticulum enlarges to form the lung bud. It is initially in open communication with the foregut, but eventually they become separated by folds of **mesoderm**, the tracheoesophageal folds. When the tracheoesophageal folds fuse in the midline to form the tracheoesophageal septum, the primitive foregut is divided into the laryngotracheal tube ventrally and the esophagus dorsally. The esophagus is initially short, but lengthens with descent of the **heart** and **lungs**. During development, the endodermal lining of the esophagus proliferates and obliterates its lumen; later, recanalization (reopening) occurs.

The primitive stomach appears from a dilatation in the foregut in week four. Its dorsal part grows faster than ventral part. The stomach rotates so that its original left surface faces ventrally and its dorsal curvature extends to the left. This 90-degree clockwise rotation yields the greater and the lesser curvature of the definitive stomach. In particular, the rotation causes: (1) the dorsal mesentery to be carried on the left, forming the greater omentum; (2) the left vagus nerve to innervate the ventral surface of the stomach, and the right vagus nerve to innervate the dorsal surface.

The liver arises (fourth week) as a ventral diverticulum (pocket)of the foregut. The endodermal lining of the foregut arises into mesoderm to form the hepatic diverticulum. Cells from the hepatic diverticulum grow into the septum trasversum (a part of the future **diaphragm**), forming the hepatic cords. These cords arrange themselves around vitelline and umbilical **veins** forming the hepatic sinusoids. The proximal part of the hepatic diverticulum gives rise to the common bile duct, cystic duct, gallbladder, and hepatic ducts. The epithelial lining of gallbladder and extra hepatic bile duct proliferates and obliterates the lumen (opening). Later recanalization (reopening) occurs. The primitive liver initially bulges into the abdominal cavity, stretching the septum trasversum to form the falciform ligament and the lesser omentum. The lesser omentum can be divided into the hepatogastric ligament and the hepatoduodenal ligament; it contains the bile duct, portal vein, and hepatic artery. The falciform ligament contains the left umbilical vein.

The ventral pancreatic bud and the dorsal pancreatic bud arise from the endodermal lining of foregut. Both of these endodermal tubes grow into surrounding mesoderm forming the acinar cells and ducts of definitive pancreas (esocrine portion) and the pancreatic (Wirsung) duct. Some cells remain accumulated and isolated into mesoderm to form the islet cells of endocrine pancreas. Connective **tissue** and vascular components of the definitive pancreas are derived from the surrounding mesoderm. Because of the 90-degree clockwise rotation of the duodenum, ventral and dorsal buds fuse to form the definitive adult pancreas. The uncinate process and a portion of the head of the pancreas originate from ventral bud. The remaining portion of the head, body, and tail of the pancreas originate from the dorsal bud. The upper duodenum originates from the caudal-most part of duodenum.

The lower duodenum, jejunum, ileum, cecum, **appendix**, ascending colon, and the proximal two-thirds of the ascending colon all derive from the midgut. All of these structures receive their blood supply from the superior mesenteric artery.

The lower duodenum develops from the most cranial part of the midgut. During development, endodermal lining of the duodenum proliferates rapidly and obliterates the lumen; later, recanalization occurs. The midgut also forms a U-shaped loop that herniates (bulges) through the primitive umbilical ring into the extra embryonic coelom. The midgut loop consists of the cranial limb, that forms the jejunum and the upper part of the ileum, and the caudal limb, that forms the cecum and the appendix, the lower part of the ileum, the ascending colon, and the proximal two third of the transverse colon. The midgut loop rotates about 270 degrees counterclockwise around the superior mesenteric artery to return into the abdominal cavity.

The distal end of the transverse colon, the descending colon, the sigmoid colon, rectum, and the upper anal canal all form from the hindgut. All of these structures receive their blood supply from the inferior mesenteric artery.

The cranial end of hind gut develops into the distal one-third of the transverse colon, descending colon, and sigmoid colon. The terminal end of the hindgut is named the cloaca. The cloacal membrane is formed where the cloaca meets the **ectoderm** of the proctodeum. The cloaca is separated by the urorectal septum into the upper anal canal and the urogenital sinus. The cloacal membrane is portioned by the urorectal septum into the anal membrane and the urogenital membrane.

The upper anal canal also develops from the hindgut. The lower anal canal develops from the invagination of surface ectoderm due to the proliferation of mesoderm surrounding anal canal (proctodeum). The hindgut and proctodeum are involved in the embryologic formation of the entire anal canal.

See also Embryonic development, early development, formation, and differentiation; Gastrointestinal tract; Intestinal histophysiology; Stomach, histophysiology

GASTROINTESTINAL TRACT

The digestive system (gastrointestinal tract) is a group of **organs** responsible for the conversion of food into absorbable chemicals, which are then used to provide energy for growth and repair. The digestive system consists of the mouth, esophagus, stomach, and small and large intestines, along with several glands, such as the salivary glands, **liver**, gall bladder, and **pancreas**. These glands secrete digestive juices containing **enzymes** that break down the food chemically into smaller, more absorbable molecules. In addition to providing the body with the nutrients and energy it needs to function, the digestive system also separates and disposes of waste products ingested with the food.

Food is moved through the alimentary canal by a wave-like muscular motion known as peristalsis, which consists of the alternate contraction and relaxation of the smooth muscles lining the tract. In this way, food is passed through the gut in much the same manner as toothpaste is squeezed from a tube. Churning is another type of movement that takes place in the stomach and small intestine, which mixes the food so that the digestive enzymes can break down the food molecules.

Food in the human diet consists of **carbohydrates**, proteins, fats, **vitamins**, and minerals. The remainder of the food is fiber and water. The majority of minerals and vitamins pass through to the bloodstream without the need for further digestive changes, but other nutrient molecules must be broken down to simpler substances before they can be absorbed and used.

Food taken into the mouth is first prepared for **digestion** in a two-step process known as **mastication**. In the first stage, the teeth tear and break down food into smaller pieces. In the second stage, the tongue rolls these pieces into balls (boluses). Sensory receptors on the tongue (**taste** buds) detect taste sensations of sweet, salt, bitter, and sour, or cause the rejection of bad-testing food. The olfactory nerves contribute to the sensation of taste by picking up the aroma of the food and passing the sensation of **smell** on to the **brain**.

The sight of the food also stimulates the salivary glands. Altogether, the sensations of sight, taste, and smell cause the salivary glands, located in the mouth, to produce **saliva**, which then pours into the mouth to soften the food. An enzyme in the saliva called amylase begins the break down of carbohydrates (starch) into simple sugars, such as maltose. Ptyalin is one of the main amylase enzymes found in the mouth; ptyalin is also secreted by the pancreas.

The bolus of food, which is now a battered, moistened, and partially digested ball of food, is swallowed, moving to the throat at the back of the mouth (**pharynx**). In the throat, rings of muscles force the food into the esophagus, the first part of the upper digestive tube. The esophagus extends from the bottom part of the throat to the upper part of the stomach.

The esophagus does not take part in digestion. Its job is to get the bolus into the stomach. There is a powerful muscle (the esophageal sphincter), at the junction of the esophagus and stomach, which acts as a valve to keep food, stomach acids, and **bile** from flowing back into the esophagus and mouth.

Chemical digestion begins in the stomach. The stomach, a large, hollow, pouched-shaped muscular organ, is shaped like a lima bean. When empty, the stomach becomes elongated; when filled, it balloons out. Food in the stomach is broken down by the action of the gastric juice containing hydrochloric acid and a protein-digesting enzyme called pepsin. Gastric juice is secreted from the linings of the stomach walls, along with **mucus**, which helps to protect the stomach lining from the action of the acid. The three layers of powerful stomach muscles churn the food into a fine semiliquid paste called chyme. From time to time, the chyme is passed through an opening (the pyloric sphincter), which controls the passage of chyme between the stomach and the beginning of the small intestine.

There are several mechanisms responsible for the secretion of gastric juice in the stomach. The stomach begins its production of gastric juice while the food is still in the mouth. Nerves from the cheeks and tongue are stimulated and send messages to the brain. The brain in turn sends messages to nerves in the stomach wall, stimulating the secretion of gastric juice before the arrival of the food. The second signal for gastric juice production occurs when the food arrives in the stomach and touches the lining. This mechanism provides for only a moderate addition to the amount of gastric juice that was secreted when the food was in the mouth.

Gastric juice is needed mainly for the digestion of protein by pepsin. If a hamburger and bun reach the stomach, there is no need for extra gastric juice for the bun (carbohydrate), but the hamburger (protein) will require a much greater supply of gastric juice. The gastric juice already present will begin the break down of the large protein molecules of the hamburger into smaller molecules: polypeptides and peptides. These smaller molecules in turn stimulate the cells of the stomach lining to release the hormone gastrin into the bloodstream.

Gastrin then circulates throughout the body, and eventually reaches the stomach, where it stimulates the cells of the stomach lining to produce more gastric juice. The more protein there is in the stomach, the more gastrin will be produced, and the greater the production of gastric juice. The secretion of more gastric juice by the increased amount of protein in the stomach represents the third mechanism of gastric juice secretion.

While digestion continues in the small intestine, it also becomes a major site for the process of absorption, that is, the passage of digested food into the bloodstream, and its transport to the rest of the body.

The small intestine is a long, narrow tube, about 20 ft. (6 m) long, running from the stomach to the large intestine. The small intestine occupies the area of the **abdomen** between the **diaphragm** and hips, and is greatly coiled and twisted. The small intestine is lined with muscles that move the chyme toward the large intestine. The mucosa, which lines the entire small intestine, contains millions of glands that aid in the digestive and absorptive processes of the digestive system.

The small intestine, or small bowel, is sub-divided by anatomists into three sections, the duodenum, the jejunum, and the ileum. The duodenum is about 1 ft. (0.3 m) long and connects with the lower portion of the stomach. When fluid food reaches the duodenum it undergoes further enzymatic digestion and is subjected to pancreatic juice, intestinal juice, and bile.

The pancreas is a large gland located below the stomach that secretes pancreatic juice into the duodenum via the pancreatic duct. There are three enzymes in pancreatic juice that digest carbohydrates, **lipids**, and proteins. Amylase, (the enzyme found in saliva) breaks down starch into simple sugars such as maltose. The enzyme maltase in intestinal juice completes the break down of maltose into glucose.

Lipases in pancreatic juice break down fats into fatty acids and glycerol, while proteinases continue the break down of proteins into **amino acids**. The gall bladder, located next to the liver, secretes bile into the duodenum. While bile does not contain enzymes; it contains bile salts and other substances that help to emulsify (dissolve) fats, which are otherwise insoluble in water. Breaking the **fat** down into small globules allows the lipase enzymes a greater surface area for their action.

Chyme passing from the duodenum next reaches the jejunum of the small intestine, which is about 3 ft. (0.91 m) long. Here, in the jejunum, the digested breakdown products of carbohydrates, fats, proteins, and most of the vitamins, minerals, and iron are absorbed. The inner lining of the small intestine is composed of up to five million tiny, finger-like projections called villi. The villi increase the rate of absorption of the nutrients into the bloodstream by extending the surface of the small intestine to about five times that of the surface area of the skin.

There are two transport systems that pick up the nutrients from the small intestine. Simple sugars, amino acids, glycerol, and some vitamins and salts are conveyed to the liver in the bloodstream. Fatty acids and vitamins are absorbed and then transported through the lymphatic system, the network of vessels that carry lymph and white **blood** cells throughout the body. Lymph eventually drains back into the bloodstream and circulates throughout the body.

The last section of the small intestine is the ileum. It is smaller and thinner-walled than the jejunum, and it is the preferred site for vitamin B_{12} absorption and bile acids derived from the bile juice.

The large intestine, or colon, is wider and heavier then the small intestine, but much shorter—only about 4 ft. (1.2 m) long. It rises up on one side of the body (the ascending colon), crosses over to the other side (the transverse colon), descends (the descending colon), forms an s-shape (the sigmoid colon), reaches the rectum, and anus, from which the waste products of digestion (feces or stool), are passed out, along with gas. The muscular rectum, about 5 in (13 cm) long, expels the feces through the anus, which has a large muscular sphincter that controls the passage of waste matter.

The large intestine extracts water from the waste products of digestion and returns some of it to the bloodstream, along with some salts. Fecal matter contains undigested food, **bacteria**, and cells from the walls of the digestive tract. Certain types of bacteria of the large intestine help to synthesize the vitamins needed by the body. These vitamins find their way to the bloodstream along with the water absorbed from the colon, while excess fluids are passed out with the feces.

See also Elimination of waste; Flatus; Gastrointestinal embryological development; Gustatory structures; Intestinal histophysiology; Intestinal motility; Stomach: histophysiology; Swallowing and disphagia

GENE THERAPY

Gene therapy is the name applied to the treatment of **inherited diseases** by corrective genetic engineering of the dysfunctional genes. It is part of a broader field called genetic medicine, which involves the screening, diagnosis, prevention and treatment of hereditary conditions in humans. The results of genetic screening can pinpoint a potential problem to which gene therapy can sometimes offer a solution. Genetic defects are significant in the total field of medicine, with up to 15 out of every 100 newborn infants having a hereditary disorder of greater or lesser severity. More than 2,000 genetically distinct inherited defects have been classified so far, including diabetes, cystic fibrosis, hemophilia, sickle-call anemia, phenylketonuria, Down syndrome and **cancer**. By the end of 1993, techniques involving gene therapy had been approved for use on such diseases as severe combined immune deficiency (SCID), familial hypercholesterolemia, cystic fibrosis, and Gaucher's disease. Many protocols are being developed towards the treatment of cancer and **AIDS**, and numerous disorders are regarded as potential candidates for gene therapy, including Parkinson's and **Alzheimer**'s diseases, arthritis, and **heart** disease. The Human Genome Project, an ongoing effort to locate all the genes in the human genome, will continue to identify conditions with an apparent genetic basis, which will consequently be considered for treatment by gene therapy. The procedure is most likely to have the greatest success with diseases that are cause by single gene defects.

Gene therapy became a reality on September 14, 1990 when researchers at the U.S. National Institute of Health performed the first approved procedure on a four-year-old girl, Ashanti DeSilva. The experiment was carried out by American scientist, W. French Anderson. Ashanti was born with the rare genetic disease of the **immune system** known as SCID, which results from a deficiency in the enzyme adenosine deaminase (ADA) essential for making white **blood** cells in the bone marrow. Children with this illness usually develop overwhelming infections and rarely survive to adulthood. To treat the ADA deficiency, a sample of Ashanti's bone marrow was taken and **stem cells**, which can normally develop into white blood cells were, separated out. Copies of ADA-producing genes were taken from a human cell and placed inside a modified retrovirus. The virus, acting as a vector, infected the cells and the

ADA-gene passed into the stem cells and integrated into their chromosome. Having received the new working gene, the stem cells were subsequently infused back into the patient where they divided as white blood cells and produced a continuous supply of ADA. Laboratory tests showed that the therapy managed strengthened Ashanti's immune system. Though successful, this procedure is not a permanent cure for SCID and the genetically engineered white blood cells only work for a few months, so the process must be repeated at intervals.

A slightly different method is currently being employed to treat another important genetic disease, cystic fibrosis. The defective gene normally codes for a protein called the cystic fibrosis transmembrane conductor (CFTR), which regulates the transfer of chloride ions in and out of cells in the **lungs, pancreas**, colon, and urogenital tracts. Mutations in this gene cause faulty delivery of ions and water in and out of the body's cells, which leads to the build up of sticky **mucus** affecting the bronchioles of the lung and other secretory tubules, notably in the pancreas. By gene therapy, the corrected CFTR gene can be introduced into the secretory cells of the lungs by inhalation of an aerosol spray containing the adenovirus vector engineered with the correct gene. It is hoped that the new gene will be passed from the virus into the lung cells where the production of the correct protein will be switched on. Though this method was shown to be feasible and relatively safe, the levels of gene expression have remained modest, the duration of expression short and a clear therapeutic effect has not yet emerged.

The techniques described are used in the kind of therapy called collectively somatic gene therapy. This involves the correction of a genetic defect in the somatic (i.e., non-reproductive) cells of patients by the insertion of a healthy gene. The new gene will only function in that patient and will not be passed to the next generation. Somatic therapy is technically the simplest and ethically the least controversial kind of gene therapy. A more questionable application is germline gene therapy, which requires the insertion of a gene into the reproductive **tissue** of a patient in such a way that the disorder in his or her offspring would also be corrected. The two main methods of performing germ-line gene therapy would be either to treat a pre-embryo that carries a serious genetic defect before implantation in the mother (this necessitates the use of *in vitro* **fertilization** techniques); or to treat the **germ cells** (**sperm** or egg cells) of afflicted adults so that their genetic defects would not be passed on to their offspring.

Some commentators on gene therapy have objected to any form of human genetic manipulation, no matter how well intentioned, arguing that once it begins, it could go down the slippery slope to misuse. They wonder if it is really possible to distinguish between "good" and "bad" uses of gene modification techniques, and whether the potential for harmful abuse of the technology should keep scientists from developing more techniques. Others approve of the use of the therapy in somatic cells, but hesitate to allow the use of germ-line gene therapy because of the unforeseeable effect on future generations. Still others have argued that with proper regulation and safeguards, germ-line gene therapy is a logical extension of the progress made to date, and an ethically acceptable procedure. Arguments specifically against the development of germ-line gene therapy techniques include: 1) germ-line gene therapy experiments involves too much scientific uncertainty and clinical risks, and the long term effects of such therapy are unknown; 2) it opens the door to attempts at altering human traits not associated with disease, which could exacerbate problems of social discrimination; 3) as germ-line gene therapy involves research on early embryos and effects their offspring, such research essentially creates generations of unconsenting research subjects; 4) gene therapy is very expensive, and will never be cost effective enough to merit high social priority; 5) germ-line gene therapy violates the rights of subsequent generations to inherit a genetic endowment that has not been intentionally modified.

There are two other ethically volatile applications of gene therapy. Enhancement gene therapy involves the insertion of a gene to enhance a known characteristic, for example placing an additional growth hormone gene into a normal child. Eugenic genetic engineering is defined as the attempt to alter or improve complex human traits which are thought to be under the direction of a large number of genes and which are not in themselves deleterious such as personality, intelligence, formation of body **organs** and so on. Eugenic manipulations could lead to a society in which genetic perfection is the ultimate ideal, and the marginalization of all those who do not fit into that ideal.

As with every new medical technique, there are many potential dangers and unpredictable factors with gene therapy, which make its practical application risky. Even though every precaution is taken to prevent accidents, they sometimes do occur. Jesse Gelsinger, a 17 year-old boy suffering from the disease ornithine transcarbamylase (OTC) deficiency became the first tragic victim of gene therapy and died on September 17, 1999. He had volunteered to test the potential use of gene therapy in the treatment of OTC in young babies. His therapy consisted of an infusion of corrective genes, encased in a weakened adenovirus vector. Gelsinger suffered an unexpected chain reaction that resulted in his early death from multiple organ system failure. The reason for his extreme reaction to the treatment is suspected to have been an overwhelming inflammatory response to the viral vector, though the reason why is not known. Subsequent investigations revealed the deaths of six other gene therapy patients, some prior to Gelsinger, who were undergoing trials for the use of gene therapy in the treatment of heart conditions. Unlike Gelsinger, these latter six victims are thought to have died from complications stemming from their underlying illnesses rather than the gene therapy itself.

See also Genetic code; Genetic regulation of eukaryotic cells; Genetics and developmental genetics; *In vitro* and *in vivo* fertilization

GENETIC CODE

The genetic code is the set of correspondences between the nucleotide sequences of nucleic acids such as **deoxyribonucleic acid** (**DNA**), and the amino acid sequences of proteins

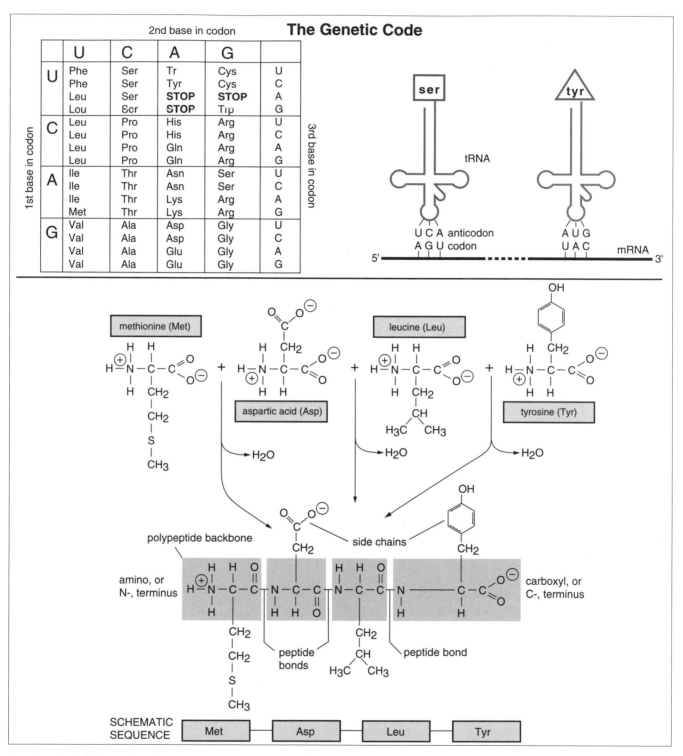

The Genetic Code

1st base in codon	2nd base in codon				3rd base in codon
	U	**C**	**A**	**G**	
U	Phe	Ser	Tr	Cys	U
	Phe	Ser	Tyr	Cys	C
	Leu	Ser	**STOP**	**STOP**	A
	Lou	Scr	**STOP**	Trp	G
C	Leu	Pro	His	Arg	U
	Leu	Pro	His	Arg	C
	Leu	Pro	Gln	Arg	A
	Leu	Pro	Gln	Arg	G
A	Ile	Thr	Asn	Ser	U
	Ile	Thr	Asn	Ser	C
	Ile	Thr	Lys	Arg	A
	Met	Thr	Lys	Arg	G
G	Val	Ala	Asp	Gly	U
	Val	Ala	Asp	Gly	C
	Val	Ala	Glu	Gly	A
	Val	Ala	Glu	Gly	G

Genetic code depicting models of amino acids inserting into a protein. *Illustration by Argosy Publishing.*

(polypeptides). These correspondences enable the information encoded in the chemical components of DNA to be transferred to the **ribonucleic acid** messenger (mRNA) and then used to establish the correct sequence of **amino acids** in the polypeptide. The elements of the encoding system, the nucleotides, differ by only four different bases. These are known as adenine (A), guanine, (G), thymine (T), and cytosine (C) in DNA or uracil (U) in **RNA**. Thus RNA contains U in the place of C and the nucleotide sequence of DNA acts as a template for the synthesis of a complementary sequence of RNA, a process known as transcription. For historical reasons, the term genetic code in fact refers specifically to the sequence of nucleotides in mRNA, although today it is sometimes used interchangeably with the coded information in DNA.

Proteins found in nature consist of 20 naturally occurring amino acids. One important question is, how can four nucleotides code for 20 amino acids? This question was raised by scientists in the 1950s soon after the discovery that the DNA comprised the hereditary material of living organisms. It was reasoned that if a single nucleotide coded for one amino acid, then only four amino acids could be provided for. Alternatively, if two nucleotides specified one amino acid, then there could be a maximum number of 16 (4^2) possible arrangements. If, however, three nucleotides coded for one amino acid, then there would be 64 (4^3) possible permutations, more than enough to account for all the 20 naturally occurring amino acids. The latter suggestion was proposed by the Russian born physicist, George Gamow (1904–1968) and was later proved to be correct. It is now well known that every amino acid is coded by at least one nucleotide triplet or codon, and that some triplet combinations function as instructions for the termination or initiation of translation. Three combinations in tRNA, UAA, UGA, and UAG are termination codons, while AUG is a translation start codon.

The genetic code was solved between 1961 and 1963. The American scientist **Marshall Nirenberg**, working with his colleague Heinrich Matthaei, made the first breakthrough when they discovered how to make synthetic mRNA. They found that if the nucleotides of RNA carrying the four bases A, G, C, and U were mixed in the presence of the enzyme polynucleotide phosphorylase, a single stranded RNA was formed in the reaction, with the nucleotides being incorporated at random. This offered the possibility of creating specific mRNA sequences and then seeing which amino acids they would specify. The first synthetic mRNA polymer obtained contained only uracil (U) and when mixed *in vitro* with the protein synthesizing machinery of *Escherichia coli* it produced a polyphenylalanine—a string of phenylalanine. From this it was concluded that the triplet UUU coded for phenylalanine. Similarly, a pure cytosine (C) RNA polymer produced only the amino acid proline, so the corresponding codon for cytosine had to be CCC. This type of analysis was refined when nucleotides were mixed in different proportions in the synthetic mRNA and a statistical analysis was used to determine the amino acids produced. It was quickly found that a particular amino acid could be specified by more than one codon. Thus, the amino acid serine could be produced from any one of the combinations UCU, UCC, UCA, or UCG. In this way the genetic code is said to be degenerate, meaning that each of the 64 possible triplets have some meaning within the code and that several codons may encode a single amino acid.

This work confirmed the ideas of the British scientists **Francis Crick** and Sidney Brenner (1927–). Brenner and Crick were working with mutations in the bacterial virus bactriophage T4 and found that the deletion of a single nucleotide could abolish the function of a specific gene. However, a second mutation in which a nucleotide was inserted at a different but nearby position, restored the function of that gene. These two mutations are said to be suppressors of each other, meaning that they cancel each other's mutant properties. It was concluded from this that the genetic code was read in a sequential manner starting from a fixed point in the gene. The insertion or deletion of a nucleotide shifted the reading frame in which succeeding nucleotides were read as codons, and was thus termed a frameshift mutation. It was also found that whereas two closely spaced deletions, or two closely spaced insertions, could not suppress each other, three closely spaced deletions or insertions could do so. Consequently, these observations established the triplet nature of the genetic code. The reading frame of a sequence is the way in which the sequence is divided into the triplets and is determined by the precise point at which translation is initiated. For example, the sequence CATCATCAT can be read CAT CAT CAT or C ATC ATC AT or CA TCA TCA T in the three possible reading frames. Sometimes, as in particular bacterial **viruses**, genes have been found that are contained within other genes. These are translated in different reading frames so the amino acid sequences of the proteins encoded by them are different. Such economy of genetic material is, however, quite rare

The same genetic code appears to operate in all living things, but exceptions to this universality are known. In human mitochondrial mRNA, AGA and AGG are termination or stop codons. Other differences also exist in the correspondences between certain codon sequences and amino acids. In ciliates there are also unusual features in that UAA and UAG code for glutamine (CAA and CAG in other eukaryotes) and the only termination codon appears to be UGA.

See also Genetic regulation of eukaryotic cells; Human Genetics; Molecular biology

GENETIC REGULATION OF EUKARYOTIC CELLS

Although prokaryotes (i.e., non-nucleated unicellular organisms) divide through binary fission, eukaryotes undergo a more complex process of **cell division** because **DNA** is packed in several **chromosomes** located inside a cell nucleus. In eukaryotes, cell division may take two different paths, in accordance with the cell type involved. Mitosis is a cellular division resulting in two identical nuclei is performed by somatic cells. The process of meiosis results in four nuclei, each containing half of the original number of chromosomes. Sex cells or gametes (ovum and spermatozoids) divide my meiosis. Both prokaryotes and eukaryotes undergo a final process, known as cytoplasmatic division, which divides the parental cell in new daughter cells.

The series of stages that a cell undergoes while progressing to division is known as **cell cycle**. Cells undergoing division are also termed competent cells. When a cell is not progressing to mitosis, it remains in phase G0 (G zero). Therefore, the cell cycle is divided into two major phases: interphase and mitosis. Interphase includes the phases (or

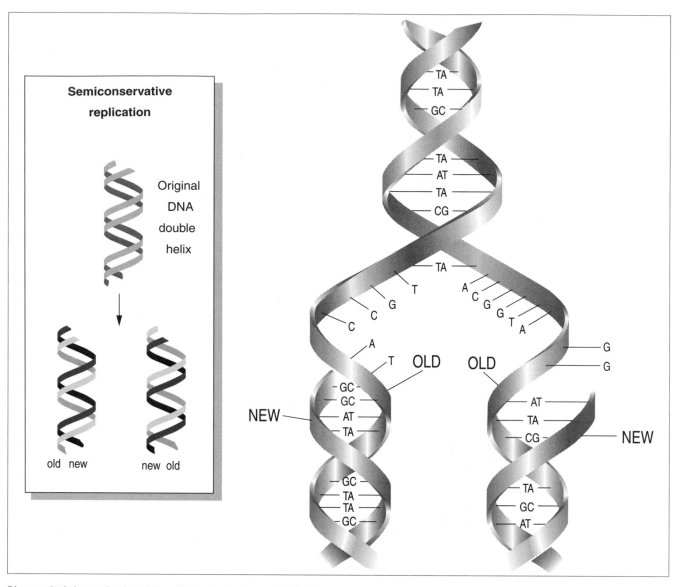

Diagram depicting replication of DNA. *Illustration by Argosy Publishing.*

stages) G1, S and G2, whereas mitosis is subdivided into prophase, metaphase, anaphase and telophase.

    The cell cycle starts in G1, with the active synthesis of **RNA** and proteins, which are necessary for young cells to grow and mature. The time G1 lasts, varies greatly among eukaryotic cells of different species and from one **tissue** to another in the same organism. Tissues that require fast cellular renovation, such as mucosa and endometrial epithelia, have shorter G1 periods than those tissues that do not require frequent renovation or repair, such as muscles or **connective tissues**.

    The cell cycle is highly regulated by several **enzymes**, proteins, and cytokines in each of its phases, in order to ensure that the resulting daughter cells receive the appropriate amount of genetic information originally present in the parental cell. In the case of somatic cells, each of the two daughter cells must contain an exact copy of the original

genome present in the parental cell. Cell cycle controls also regulate when and to what extent the cells of a given tissue must proliferate, in order to avoid abnormal cell proliferation that could lead to dysplasia or tumor development. Therefore, when one or more of such controls are lost or inhibited, abnormal overgrowth will occur and may lead to impairment of function and disease.

    Cells are mainly induced into proliferation by growth factors or **hormones** that occupy specific receptors on the surface of the cell membrane, also known as extra-cellular ligands. Examples of growth factors include epidermal growth factor (EGF), fibroblastic growth factor (FGF), platelet-derived growth factor (PDGF), insulin-like growth factor (IGF). Hormones such as PDGF and FGF act by regulating the phase G2 of the cell cycle and during mitosis. After mitosis, they act again stimulating the daughter cells to grow, thus leading them from G0 to G1. Therefore, FGF and PDGF are also termed competence factors, whereas EGF and IGF are termed progression factors, because they keep the process of cellular progression to mitosis going on. Growth factors are also classified (along with other molecules that promote the cell cycle) as pro-mitotic signals. Hormones are also pro-mitotic signals. For example, thyrotrophic hormone, one of the hormones produced by the pituitary gland, induces the proliferation of thyroid gland's cells. Another pituitary hormone, known as growth hormone or somatotrophic hormone (STH), is responsible by body growth during childhood and early adolescence, inducing the lengthening of the long bones and **protein synthesis**. Estrogens are hormones that do not occupy a membrane receptor, but instead, penetrate the cell and the nucleus, binding directly to specific sites in the DNA, thus inducing the cell cycle.

Anti-mitotic signals may have several different origins, such as cell-to-cell adhesion, factors of adhesion to the extracellular matrix, or soluble factor such as TGF beta (tumor growth factor beta), which inhibits abnormal cell proliferation, proteins p53, p16, p21, APC, pRb, etc. These molecules are the products of a class of genes called tumor suppressor genes. Oncogenes, until recently also known as proto-oncogenes, synthesize proteins that enhance the stimuli started by growth factors, amplifying the mitotic signal to the nucleus, and/or promoting the accomplishment of a necessary step of the cell cycle. When each phase of the cell cycle is completed, the proteins involved in that phase are degraded, so that once the next phase starts, the cell is unable to go back to the previous one. Next to the end of phase G1, the cycle is paused by tumor suppressor gene products, to allow verification and repair of DNA damage. When DNA damage is not repairable, these genes stimulate other intra-cellular pathways that induce the cell into suicide or apoptosis (also known as programmed cell **death**). To the end of phase G2, before the transition to mitosis, the cycle is paused again for a new verification and decision, either mitosis or apoptosis.

Along each pro-mitotic and anti-mitotic intra-cellular signaling pathway, as well as along the apoptotic pathways, several gene products (proteins and enzymes) are involved in an orderly sequence of activation and inactivation, forming complex webs of signal transmission and signal amplification to the nucleus. The general goal of such cascades of signals is to achieve the orderly progression of each phase of the cell cycle.

Interphase is a phase of cell growth and metabolic activity, without cell nuclear division, comprised of several stages or phases. During Gap 1 or G1, the cell resumes protein and RNA synthesis, which was interrupted during mitosis, thus allowing the growth and maturation of young cells to accomplish their physiologic function. Immediately following is a variable length pause for DNA checking and repair before cell cycle transition to phase S during which there is synthesis or semi-conservative replication or synthesis of DNA. During Gap 2 or G2, there is increased RNA and protein synthesis, followed by a second pause for proofreading and eventual repairs in the newly synthesized DNA sequences before transition to mitosis.

At the start of mitosis the chromosomes are already duplicated, with the sister-chromatids (identical chromosomes) clearly visible under a light microscope. Mitosis is subdivided into prophase, metaphase, anaphase, and telophase.

During prophase there is a high condensation of chromatids, with the beginning of nucleolus disorganization and nuclear membrane disintegration, followed by the start of centrioles' migration to opposite cell poles. During metaphase the chromosomes organize at the equator of a spindle apparatus (microtubules), forming a structure termed metaphase plate. The sister-chromatids are separated and joined to different centromeres, while the microtubules forming the spindle are attached to a region of the centromere termed kinetochore. During anaphase there are spindles, running from each opposite kinetochore, that pull each set of chromosomes to their respective cell poles, thus ensuring that in the following phase each new cell will ultimately receive an equal division of chromosomes. During telophase, kinetochores and spindles disintegrate, the reorganization of nucleus begins, chromatin becomes less condensed, and the nucleus membrane start forming again around each set of chromosomes. The cytoskeleton is reorganized and the somatic cell has now doubled its volume and presents two organized nucleus.

Cytokinesis usually begins during telophase, and is the process of cytoplasmatic division. This process of division varies among species but in somatic cells, it occurs through the equal division of the cytoplasmatic content, with the **plasma** membrane forming inwardly a deep cleft that ultimately divides the parental cell in two new daughter cells.

The identification and detailed understanding of the many molecules involved in the cell cycle controls and intracellular signal transduction is presently under investigation by several research groups around the world. This knowledge is crucial to the development of new anti-cancer drugs as well as to new treatments for other genetic diseases, in which a gene over expression or deregulation may be causing either a chronic or an acute disease, or the impairment of a vital organ function. Scientists predict that the next two decades will be dedicated to the identification of gene products and their respective function in the cellular microenvironment. This new field of research is termed proteomics.

See also Cell cycle and cell division; Cell structure

GENETICS AND DEVELOPMENTAL GENETICS

Genetics is the branch of biology dealing with heredity and attempts to explain the similarities and differences that exist between parents and offspring.

Although hypotheses on the nature and mechanisms of heredity date to antiquity, genetics did not become an independent scientific discipline until the turn of the twentieth century. Through studies of crosses between garden peas, the Czech-born German monk **Gregor Mendel** worked out the basic principles of inheritance, which he expressed in his laws of heredity. Mendel's work stimulated the first experiments in what is now known as classical genetics. These early studies developed into the eventual identification and demonstration of chromosome individuality, gene linkage, crossing over and the linear arrangement of genes in **chromosomes**.

The term genetics was actually first coined, in the early 1900s by the English scientist William Bateson (1861–1926). Bateson was one of the first scientists to accept and popularized the Mendelian laws and introduced the term genetics to encompass the whole study of heredity as it was understood at that time. Genetics quickly came to occupy a central position in biology because it was established early on that essentially the same principles apply to all animals and plants. Since their identification, the laws of heredity have found application in such diverse areas as the study of **evolution** and the agricultural improvement of cultivated plants and domestic animals.

As biochemical techniques developed and improved, the study of genetics became more detailed and intricate. During the 1950s, cytology developed and with it the microscopic study of the chromosomes and other cellular structures that play a part in heredity. Also studies on extranuclear or cytoplasmic inheritance were undertaken, as well as research into the nature of mutation and the problems of developmental **physiology** and evolutionary genetics. Microbial genetics, which employed fungi and **bacteria** as experimental systems, became a specialized area of experimental research. All of these developments, however, would have had less impact if it had not been for the discovery in the 1940s that genetic material consisted of nucleic acid and, in the 1950s, the determination of the double helical structure of **deoxyribonucleic acid** (**DNA**). Recognition of the double helix was not only important for genetics but was probably one of the most profound biological advances since Darwin's theory of evolution.

The Watson-Crick model of DNA was a remarkable achievement, for which the two scientists won the 1954 Nobel Prize in chemistry. The molecule had exactly the shape and dimensions needed to produce an x-ray photograph like that of Franklin's. Furthermore, Watson and Crick immediately saw how the molecule could "carry" genetic information. The sequence of nitrogen bases along the molecule, they said, could act as a **genetic code**. A sequence, such as A-T-T-C-G-C-T...etc., might tell a cell to make one kind of protein (such as that for red **hair**), while another sequence, such as G-C-T-C-T-C-G...etc., might code for a different kind of protein (such as that for blonde hair). Watson and Crick themselves contributed to the deciphering of this genetic code, although that process was long and difficult and involved the efforts of dozens of researchers over the next decade.

Watson and Crick had also considered, even before their March 7th discovery, what the role of DNA might be in the manufacture of proteins in a cell. The sequence that they outlined was that DNA in the nucleus of a cell might act as a template for the formation of a second type of nucleic acid, **RNA** (**ribonucleic acid**). RNA would then leave the nucleus, emigrate to the cytoplasm and then itself act as a template for the production of protein. That theory, now known as the Central Dogma, has since been largely confirmed and has become a critical guiding principal of much research in **molecular biology**.

The Central Dogma of Molecular Biology assets that spatial and temporal differences in gene expression cause cel-

lular and morphological differentiation. Since DNA makes RNA, and RNA makes protein, there are basically three levels where a cell can modulate gene expression: 1) by altering the transcription of DNA into RNA; 2) by altering the translation of RNA into protein; and 3) by altering the activity of the protein, which is usually an enzyme. Since DNA and RNA are themselves synthesized by proteins, the gene expression patterns of all cells are regulated by highly complex biochemical networks.

A few simple calculations provide a better appreciation of the complexity of the regulatory networks of gene expression which control differentiation. Starting with the simplifying assumption that a given protein (gene product) can be either absent or present in a cell, there are at least ten centillion (a one followed by 6,000 zeros) different patterns of gene expression in a single typical cell at any time. Given that a multicellular organism contains one quadrillion or more cells, and that gene expression patterns change over time, the number of possible gene expression patterns is enormous.

Modern genetics deals with the nature and behavior of genes, which are now known to be the basic hereditary units. For these purposes modern genetics makes use of the traditional analysis of variation within single species and crossing experiments and additionally uses the latest methods of cell research, **biochemistry**, molecular biology, and gene technology. The discipline is now broadly divided into three main subdivisions, although there is considerable overlap between them. The modes of gene transmission from generation to generation is called transmission genetics, the study of gene behavior in populations is termed population genetics and the study of gene biochemistry, **structure and function** is molecular genetics.

Developmental processes are the series of biological changes associated with information transfer, growth, and differentiation during the life cycle of organisms. Information transfer is the transmission of DNA and other biological signals from parent cells to daughter cells. Growth is the increase in size due to cell expansion and cell division. Differentiation is the change of unspecialized cells in a simple body pattern to specialized cells in more complex body pattern. In complex multi-cellular organisms such as humans, development begins with the manufacture of male and female sex cells. It proceeds through **fertilization** and formation of an embryo. Development continues following birth, hatching, or germination of the embryo and culminates in aging and **death**.

Adult multicellular organisms can consist of one quadrillion (a one followed by fifteen zeros) or more cells, each of which has the same genetic information. (There are a few notable exceptions, such as the red **blood** cells of mammals, which do not have DNA, and certain cells in the unfertilized eggs of amphibians, which undergo gene amplification and have multiple copes of some genes.) F.C. Steward first demonstrated the constancy of DNA in all the cells of a multicellular organism in the 1950s. In a classical series of experiments, Steward separated a mature carrot plant into individual cells and showed that each cell, whether it came from the root, stem, or leaf, could be induced to develop into a mature carrot plant which was genetically identical to its parent. Although such experiments cannot typically be done with multicellular animals, animals also have the same genetic information in all their cells.

Many developmental scientists emphasize that there are additional aspects of information transfer during development which do not involve DNA directly. In addition to DNA, a fertilized egg cell contains many proteins and other cellular constituents which are typically derived from the female. These cellular constituents are often asymmetrically distributed during cell division, so that the two daughter cells derived from the fertilized egg have significant biochemical and cytological differences. In many species, these differences act as biological signals which affect the course of development. There are additional spatial and temporal interactions within and among the cells of a developing organism which act as biological signals and provide a form of information to the developing organism.

Organisms generally increase in size during development. Growth is usually allometric, in that it occurs simultaneously with cellular differentiation and changes in overall body pattern. Allometry specifically studies the relationships between the size and morphology of an organism as it develops and the size and morphology of different species.

A developing organism generally increases in complexity as it increases in size. Moreover, in an evolutionary line, larger species are generally more complex that the smaller species. The reason for this correlation is that the volume (or weight) of an organism varies with the cube of its length, whereas gas exchange and food assimilation, which generally occur on surfaces, vary with the square of its length. Thus, an increase in size requires an increase in cellular specialization and morphological complexity so that the larger organism can breathe and eat.

Depending on the circumstances, natural selection may favor an increase in size, a decrease in size, or no change in size. Large size is often favored because it generally makes organisms faster, giving them better protection against predators, and making them better at dispersal and food gathering. In addition, larger organisms have a higher ratio of volume to surface area, so they are less affected by environmental variations, such as temperature variation. Large organisms tend to have a prolonged development, presumably so they have more time to develop the morphological complexities needed to support their large size. Thus, evolutionary selection for large size leads to a prolongation of development as well as morphological complexity.

Differentiation is the change of unspecialized cells in a simple body pattern to specialized cells in a more complex body pattern. It is highly coordinated with growth and includes morphogenesis, the development of the complex overall body pattern.

The most fundamental advances in recent years have been in molecular genetics and the understanding of gene function. They have led to an explosive period in which new genetic knowledge rapidly advance our understanding of **human development** and physiology. Moreover, gene technology provides a mechanism for genetic manipulation. The end of the twentieth century saw the cloning of the first mammal

and the sequencing of the entire human genome. In future the science of genetics will raise many scientific, social and ethical questions.

Lastly, aging must also be considered a phase of development. Many evolutionary scientists hold that all organisms have genes which have multiple effects, called pleiotropic genes, that increase reproductive success when expressed early in development, but cause the onset of old age when expressed later in development. In this view, natural selection has favored genes which cause aging and death because the early effects of these genes outweigh the later effects.

See also Aging processes; Cell cycle and cell division; Embryology; Embryonic development: early development, formation, and differentiation; Ethical issues in embryological research; Ethical issues in genetics research; Evolution and evolutionary mechanisms; Genetic regulation of eukaryotic cells; Human genetics; Molecular biology

GENITALIA (FEMALE EXTERNAL)

The visible external genital **organs** of a female are collectively termed the vulva. The vulva extends from the mons pubis, the soft fatty **tissue** below the **abdomen** that is covered with public **hair**, and which lays on top of the pubic bone, to the perineum, the area between the vagina and the anus. The external genitalia consist of the clitoris, the clitoral hood, the labia majora, and the labia minora.

Labia majora is from the Latin for "large lips." They are the large, outer lips of the vulva, which extend from the mons pubis down either side of the vulva. Typically, they are covered by pubic hair and are soft in appearance because of the presence of fatty tissue under the skin. Labia minor means "small lips" in Latin. Smaller and fleshier in appearance than the labia major, they are located inside the labia major, closer to the vaginal opening. The labia minor exhibit great variation in size, color, and shape in females. The labia major and minor develop between nine and twelve weeks after conception.

Another external feature of female genitalia is the clitoris. It is a firm and rounded organ found at the top of the vulva, just above the urethra (the opening that connects to the bladder and from which urine is expelled). The clitoris is small—only about 0.25 in. (0.64 cm) long—yet it is central to female sexual responsiveness. An embryological analog of the male penis, the clitoris composed of erectile tissue and **blood** vessels. The sole function of the clitoris is sexual stimulation. The female orgasm is produced by stimulation of the clitoris. In the fetus, the clitoris begins to develop by nine weeks after conception, and its development is complete by eleven weeks.

The final external feature of female genitalia is the vagina. The vagina is a canal of muscle, from three to six inches long, that leads from the external vaginal opening (which is also called the introitus). It is the vagina that accommodates the male penis during intercourse. **Sperm** deposited in the vagina during intercourse, passes up the vaginal canal to begin the process of conception.

The female genitalia develops in the embryo between seven and twelve weeks following conception, in response to the presence of a hormone called **estrogen**. Development is essentially complete at twelve weeks following conception. Until then the male and female genitalia are not distinct from each other.

See also Infertility (female); Pubic symphysis (gender differences); Reproductive system and organs (female)

GENITALIA (MALE EXTERNAL)

The development of the external genitalia of males begins between seven and nine weeks after conception, in response to the production of a hormone called dihydrotestosterone by the testes. Before this time, the external genitalia of the male and female are indistinguishable.

During development, a region called the genital tubercle elongates to form the penis. Urogenital folds move toward each other to form a groove that will become the urethra (which connects to the bladder and through which urine is expelled). The urethral connection to the outside world through the penis is established between 11 and 14 weeks after conception. By 11 weeks the scrotum has developed as well.

The penis is the most obvious feature of the male external genitalia. The other constituents are the pubic air and the scrotum. The penis consists of a body (also called the shaft) and glans (also called the head). The body houses three columns of **tissue**, which can become engorged with **blood**. This causes the penis to become erect rather than flaccid. The bulk of the penis tissue is made up of the corpus cavernosum. Another tissue, the corpus spongiosum, runs underneath the penis. It can also fill with blood and participates in the formation of an erection.

At the tip of the penis there is a vertical slit. This is the urethral meatus. It is the opening of the urethra, through which both urine and **semen** flow.

The glans area of the penis is infused with nerve cells and is one of the most sensitive spots on the male body. A loose fold of skin called the foreskin or prepuce naturally covers it. In circumcised males, the foreskin is surgically removed just below the glans, exposing the glans. Circumcision is usually done for religious or cultural reasons and, until relatively recently, because it was thought to confer a health advantage on the male. However, a number of clinical studies have shown that although circumcised men have a slightly lower rate of penile **cancer** and bladder infections, when compared to uncircumcised men, the difference is not statistically significant. Whether the trauma of circumcision is balanced by any health advantage is a debatable point.

At the base of the penis lies the scrotum, a loose and wrinkled pouch. There are two compartments comprising the scrotum. Each of these contains a testicle. The testicles are oval in shape, rubbery in texture and, on average, about 1.77 in. (4.5 cm) long. **Sperm** is produced in the testicles. Form there the sperm flows through the epididymis—a comma-shaped region at the back of each testicle—into the vas defer-

ens, and joins the urethra in an organ called the prostate. During ejaculation the sperm is expelled from the penis.

See also Embryology; Hormones and hormone action; Pubic sympysis (gender differences); Testosterone

GERM CELLS

Germ cells are one of two fundamental cell types in the human body. Germ cells are responsible for the production of sex cells or gametes (in humans, ovum and spermatozoa). Germ cells also constitute a cell line through which genes are passed from generation to generation.

The vast majority of cells in the body are somatic cells. Indeed, the term somatic cell encompasses all of the differentiated cell types, (e.g., vascular, muscular, cardiac, etc.) In addition, somatic cells may also contain undifferentiated **stem cells** (cells that, with regard to differentiation are still multi-potential). Regardless, while the mechanism of genetic replication and cell division is via mitosis in somatic cells, in germ cells a series of meiotic divisions during gametogenesis produces male and female gametes (i.e., ovum and spermatozoa that upon fusion (**fertilization**) are capable of creating a new organism (i.e., a single celled zygote).

While somatic cell divisions via mitosis maintain a diploid chromosomal content in the daughter cells produced, germ cells, in contrast, through a series of mitotic divisions, produce haploid gametes (i.e., cells with one-half the normal chromosome compliment with one autosomal chromosome from each homologous pair and one sex chromosome.

Although all humans start out as single cell zygotes, the germ cells for each individual are set aside early in embryo-genesis (development). If the cells comprising the germ cell line are subject to mutation or other impairments, those mutations may be passed down to offspring. It is from the germ cell line that all spermatogonia and all oogonia are derived.

See also Embryology; *In vitro* fertilization; Sexual reproduction

GILBERT, WALTER (1932-   )
American molecular biologist

Walter Gilbert is a molecular biologist that shared the 1980 Nobel Prize in chemistry for his discovery of how to sequence, or chemically describe, **deoxyribonucleic acid (DNA)** molecules. Gilbert also identified repressor molecules, which modify or repress the activity of certain genes, and collaborated with Noble laureate biologist **James Watson's** in his efforts to isolate messenger **ribonucleic acid (RNA)**. Later in his career, Gilbert helped form and was chief executive officer of the biotechnology firm Biogen, and became a moving force in the medical research project known as the human genome project.

Gilbert was born in Cambridge, Massachusetts. His father, Richard V. Gilbert, was an economist at Harvard

University, and his mother, Emma Cohen Gilbert, was a child psychologist who provided her children's early education at home. In 1939, the family moved to Washington, D.C., where Gilbert initially performed poorly in school. He did, however, show a great deal of interest in science. He was fascinated by astronomy, and at age twelve he ground his own glass for telescopes; he also "nearly blew himself up brewing hydrogen in the pantry," writes Anthony Liversidge in *Omni*. As a senior at Sidwell Friends High School, he would go to the Library of Congress to expand his knowledge of nuclear physics.

In 1949, Gilbert entered Harvard University, where he majored in chemistry and physics, earning his B.A. *summa cum laude* in 1953. Gilbert remained at Harvard for a master's degree in physics, which he received in 1954. He went to Cambridge University for doctoral work in theoretical physics, studying under the physicist Abdus Salam. His doctoral dissertation focused on mathematical formulae that could predict the behavior of elementary particles in so-called "scattering" experiments. He received his Ph.D. in mathematics in 1957. Gilbert returned to Harvard as a National Science postdoctoral fellow in physics, and he gained an appointment as assistant professor in 1959.

While at Cambridge, Gilbert had met biologists **James Watson** and **Francis Crick**; just a few years before, these two men had established the structure of DNA and constructed a three-dimensional model of it. Their work had launched a new field of science called **molecular biology**, and when Watson moved to Harvard in 1960, he and Gilbert met again. Watson discussed with Gilbert his interest in isolating messenger RNA. This is the substance believed to be responsible for transmitting information from DNA to ribosomes, which are the cellular structures in which **protein synthesis** takes place. At Watson's invitation, Gilbert joined him and his colleagues to work on this project. This collaboration with Watson began Gilbert's move into molecular biology. He became convinced that his future lay in this field, and he made up for a lack of formal training in **biochemistry** through hard work. Within a few years he was publishing articles on molecular biology. He was made a tenured associate professor of biophysics at Harvard in 1964, and he became a full professor of biochemistry in 1968. In 1972, Harvard named him the American Cancer Society Professor of Molecular Biology.

In the middle of the 1960s, Gilbert began research on how genes are activated within cells. This was a problem that had been introduced by the French geneticists **François Jacob** and **Jacques Lucien Monod**; DNA should, in theory, encode proteins continually, and they wondered what kept the genes from being activated. If cells contain some sort of element that in effect represses some genes, this would explain in part how cells perform different functions even though each cell contains the same complement of genetic instructions. Gilbert set out to determine whether actual "repressor" substances exist within each cell.

Working with the *Escherichia coli* bacterium, Gilbert attempted to find what he called the *lac* repressor. *E. coli* manufactures the enzyme betagalactosidase when the milk sugar lactose is present, and he hypothesized that the gene responsible for producing the enzyme was repressed by a substance

that would only detach itself from the DNA molecule in the presence of lactose. If this hypothesis could be proven, it would confirm the existence of repressors. In 1966, Gilbert and his colleagues added radioactive lactose-like molecules to a concentration of the **bacteria** as a means of tracing any potential *lac* repressor activity. As he had hoped, the *lac* repressor bonded with the radioactive material. By 1970, he was able to determine the precise region of DNA (called the *lac* operator) to which the repressor bonds in the absence of lactose.

The next phase of Gilbert's research focused on sequencing DNA. His aim was to identify and describe chemically any strand of DNA. Working with graduate student Allan Maxam, he began to sequence parts of the DNA strand. It was known that the molecules could be "broken" at specific chemical junctures by using certain chemical substances, and a colleague introduced Gilbert and Maxam to a procedure that broke DNA molecules into fragments that were easier to describe. After breaking radioactively labeled DNA into fragments, Gilbert and Maxam worked to separate them. They used a technique known as gel electrophoresis, in which an electric current causes the fragments to pass through a gel substance. Upon exposure to x-ray film, the fragments can be read and the chemical code of DNA can be identified. Working independently, the British scientist Frederick Sanger developed a similar sequencing technique. In recognition of both their contributions, Gilbert and Sanger were awarded half the 1980 Nobel Prize in chemistry. The other half went to the American biochemist Paul Berg for his work with recombinant DNA, more commonly referred to as "gene splicing."

With the breakthroughs that were being made by Gilbert, Sanger, Berg, and others, the concept of applying technological principles to biology came of age in the 1970s. The idea of being able to alter the genetic composition of a cell, for example, opened up such possibilities as curing or even eradicating many diseases. The possibilities in biotechnology intrigued not only scientists but also the business community. Here was a concept that could be both revolutionary and lucrative—a company that held the patent to a definitive cure for cancer, for example, could become quite wealthy and powerful. Business leaders began to approach scientists, Gilbert among them. Most scientists were skeptical at first, but in 1978, Gilbert met with a group of venture capitalists that wanted to start a biotechnology firm. After receiving assurances that they would have considerable control over research and development, he and other scientists formed Biogen N.V., with Gilbert as the chairman of the scientific board of directors. Gilbert was so convinced of Biogen's potential that he left Harvard in 1981 to become the company's chief executive officer.

Despite widespread belief in the company's potential, Biogen had some difficult years in the beginning. After four years, it was still unprofitable and Gilbert had become increasingly disillusioned with business. He found the differences between the business world and the scientific community difficult to reconcile. The science of creating new products is vastly different from the business of bringing them to market. Scientists need to be patient because their breakthroughs might take years to obtain, but sound business practice dictates cutting one's losses when a project fails to produce after a reasonable time, and these differences led to conflicts with others at Biogen. Gilbert also found it time-consuming and expensive to run a company (although he personally profited from the venture), and in late 1984 he resigned his position at Biogen, while maintaining some involvement with the firm. In 1985, he was named H. H. Timken Professor of Science in Harvard's cellular and developmental biology department. In 1987, he became chairman of the department and Carl M. Loeb University Professor.

Gilbert resumed his research, unhindered by the responsibilities of running a business. But he soon became interested in an undertaking that was bigger than Biogen: the human genome project. The plan was to create a map or library of human DNA. Such information would help researchers not only find cures for diseases but also identify potentially harmful gene mutations. Gilbert spoke out enthusiastically in favor of the genome project, and along with many others he encouraged Congress to support it with federal funds. Frustrated by the political process and believing that the project would be damaged by the bureaucracy federal participation would impose, he tried a different approach. In 1987, he announced plans to create his own company, which would sequence DNA, copyright the information, and sell it. Although he failed to get adequate backing for that project, he did win a two million dollar annual grant from the U.S. government to conduct his work at Harvard under the auspices of the National Institutes of Health.

In an interview with *Omni,* Gilbert has explained what he sees as some of the benefits of a complete genetic map of the human being: "The differences between people are what the genetic map [is] about. That knowledge will yield medicine tailored to the individual. One will first identify obvious genetic defects like cystic fibrosis. The next round of genetic mapping will show us clusters of genes for common diseases from arthritis to schizophrenia. We will be able to predict the side effects of those drugs and tailor the right dose to each person."

Gilbert has also expressed concern as a researcher, arguing in favor of what he calls a "paradigm shift in biology." As scientific techniques are perfected, he contends, new scientists should be able to concentrate on new research, not repeating old research. Writing in *Nature,* he has noted that "in 1970, each of my graduate students had to make restriction **enzymes** in order to work with DNA molecules; by 1976, the enzymes were all purchased and today no graduate student knows how to make them." While it is important for scientists to understand what they are doing and why, he continues, "this is not the meaning of their education. Their doctorates should be testimonials that they [have] solved a novel problem, and in so doing [have] learned the general ability to find whatever new or old techniques were needed." As more and more of the technological problems of molecular biology are solved, he believes that biological research will begin with theoretical rather than experimental work.

In addition to the Nobel Prize, Gilbert shared with Sanger the Albert Lasker Basic Medical Research Award in 1979. He also won the Louisa Horwitz Gross Prize from

Columbia University in 1979, and the Herbert A. Sober Memorial Award of the American Society of Biological Chemists in 1980. His memberships include the American Academy of Arts and Sciences, the National Academy of Sciences, and the British Royal Society.

Gilbert has been married to Celia Stone since 1953; the couple have a son and a daughter.

See also Genetic code; Genetic regulation of eukaryotic cells

GILMAN, ALFRED GOODMAN (1941-)
American physician and pharmacologist

Alfred Goodman Gilman is known for discovering, along with **Martin Rodbell**, new proteins in biological cells, called G-proteins. For this discovery they shared the 1994 Nobel Prize in physiology or medicine. Communication between cells has been understood for a long time, but how signals were transmitted inside the cells was not known until Gilman and Rodbell found the inter-cellular proteins. G-proteins are important because disruption of their function can lead to disease.

Gilman was born in 1941 in New Haven, Connecticut, the same year his father, Alfred Gilman Sr., published the landmark textbook, *The Pharmacological Basis of Therapeutics,* along with Louis Goodman, while they were faculty members of Yale's pharmacology department. Gilman grew up in White Plains, New York, where his father was on the faculty of The College of Physicians and Surgeons of Columbia University and then later a founding chairman of the Pharmacology Department at the new Albert Einstein College of Medicine. Gilman's childhood was filled with music and culture. His father could play many musical instruments and his mother, Mabel Schmidt Gilman, was a pianist. In his autobiography published on the Nobel Foundation's web site, Gilman fondly remembers trips to museums in New York City, especially the Hayden Planetarium, which peaked his interest in astronomy. But visits to his father's laboratory peaked his interest in biology. Gilman was also able to observe intricate pharmacological experiments designed for medical students. Gilman notes in his autobiography that it was probably surprising that he turned to biochemical approaches to pharmacology, despite all the musical and visual influences in his life.

In 1955, Gilman was sent to the Taft School in Watertown, Connecticut, a prep school for boys. Gilman was not happy about being sent away, nor did he enjoy the rigid structure of the boys' school. However, he received an excellent education in chemistry, math, and physics, a well as English, his least favorite subject. He remembered receiving a comment of "not bad" from an instructor who claimed his papers sounded like lab reports.

Gilman described college life as actually easier and more fun than Taft. He majored in **biochemistry** at Yale university. Gilman describes his first laboratory project, to test Francis Crick's adapter hypothesis, as "wildly overambitious."

The experience was rewarding for him though, because of the encouragement he received from his lab instructor Melvin Simpson. Gilman met his future wife Kathryn Hedlund during this time. Gilman recalled in his autobiography that she should have "smelled the competition" the many evenings she spent with him in the lab while he worked on projects.

After receiving his B.A. from Yale in 1962, Gilman worked for Burroughs Wellcome in New York and published his first papers. He knew he wanted to go into research when he entered a unique M.D.-Ph.D. program at Case Western Reserve University in the fall of 1962. He and Kathryn Hedlund were married during this time and would eventually have three children. His thesis advisor Ted Rall's commitment to projects often made Gilman late for dinner. Gilman conducted research on the thyroid gland and was also interested in studying cells and **genetics**. He earned an M.D., and a Ph.D. in pharmacology in 1969. His interest in genetics led him to the Pharmacology Research Associate Training Program at the National Institute of General Medical Sciences, where he researched cyclic adenosine monophosphate (AMP), a genetic regulator that moderates hormone actions. His work with Nobel laureate **Earl Sutherland** was the beginning of his interest in cell communication.

In 1971 Gilman accepted a position as an assistant professor of pharmacology at the University of Virginia in Charlottesville. It was here that Gilman began his Nobel Prize-winning work. Gilman described the atmosphere at the university as intellectually supportive. Gilman and his colleagues knew about Martin Rodbell's work at the National Institutes of Health with guanine nucleotides, which are components of **deoxyribonucleic acid** (**DNA**) and **ribonucleic acid** (**RNA**). Rodbell and his research associates at the NIH ascertained that the guanine nucleotides were somehow related to cell communication, but could not prove it. Gilman's research began where Rodbell left off, and in the late 1970s, he and his colleagues started looking for the chemicals that would confirm Rodbell's work. Gilman used genetically altered **leukemia** cells to detect the presence of G-proteins. He found that without the G-protein, the cells did not respond to outside stimulation the way a normal cell would. In 1980 they found the G-proteins, named because they bind to the guanine nucleotides.

G-proteins are instrumental in the fundamental workings of a cell. They allow us to see and **smell** by changing light and odors to chemical messages that travel to the **brain**. Understanding how G-proteins malfunction could lead to understanding serious diseases like cholera or **cancer**. Scientists have linked improperly working G-proteins to everything from alcoholism to diabetes. Pharmaceutical companies are developing drugs that would focus on G-proteins.

In 1979 Gilman was asked to chair the department of pharmacology at the University of Texas Southwestern Medical Center in Dallas. But he was too immersed in his research, as well as editing the sixth edition of his father's textbook, *The Pharmacological Basis of Therapeutics*, to accept the position. Martin Rodbell almost took the job, but when he declined, they asked Gilman again; this time he

accepted. There he built up the pharmacology department, recruiting many old colleagues.

His time at Southwestern was filled with many other awards in addition to the Nobel in 1994, among them the Poul Edvard Poulson Award from the Norwegian Pharmacological Society in 1982 and the Gairdner Foundation International Award in 1984. In 1987 he shared the Richard Lounsbery Award from the National Academy of Sciences with Martin Rodbell in 1987, foreshadowing the Nobel. In 1989, he won the Albert Lasker Basic Medical Research Award. Gilman has been in the forefront of G-protein research since his discovery. He predicts that eventually scientists will be able to map cell communication in a way that will allow scientists to predict how cells will respond to a variety of signals, leading the way to major advances in the treatment of disease.

See also Cell membrane transport; Cell structure; Molecular biology

GLOMERULAR FILTRATION

Blood enters each kidney through the renal artery that ramifies (splits) into progressively smaller vessels that ultimately become thin hair-like afferent arterioles. These arterioles branch out as even thinner vessels, known as glomerular **capillaries**, whose function is to filtrate from the blood large amounts of fluid and solutes (but virtually no proteins). Bunches of glomerular capillaries form glomeruli in the kidney of vertebrates, which are enclosed in a structure termed Bowman's capsule. Each glomerulus filters water and dissolved substances (i.e., solutes) from the blood, through the endothelium of the capillaries, and through the epithelium of the capsule, delivering the filtrate into the peritubular capillaries that surround the renal tubules of the Loop of Henle. From the Loop of Henle, the fluid is ultimately transported through a series of tubular renal structures to the medullary collecting tubule and, from there, to the collecting duct.

Urine formation involves the dynamic interaction among glomerular filtration, tubular reabsorption of variable amounts of water and some solutes (**calcium**, magnesium, sodium, etc.), and tubular excretion, with final delivery into the bladder. Each kidney contains about one million functional structures termed nephrons, each unit constituted by the glomerular capsule, its glomerulus, and a long tubule in which the filtrate is converted in urine during its transport to the pelvis of the kidney. The glomerular filtrate has almost the same composition of blood **plasma**, except for plasma protein concentrations. However, when the filtrate leaves the Bowman's capsule and passes through the tubules, its ion concentrations are modified due to the hormonal controls and osmotic differences that cause reabsorption of specific solutes back into the blood. Approximately 80% of the solutes and water filtrated from the blood by the glomeruli are reabsorbed through the proximal tubule, changing the glomerular filtrate into a solution isotonic with the plasma. However, urine is a hypertonic solution, mainly because the collecting tubule becomes progressively permeable to water and urea as it gets closer to the collecting duct. The matrix that surrounds the tubules in the Loop of Henle is hypertonic, due to the sodium concentrations of 4–6,000 millimole (mM) around the cortical portion, and due to urea concentrations of about 500mM around the lower portions of the collecting tubule. The total osmolality at the bottom of the Loop of Henle is 1,000 milliosmole (mOsm). As a result, the urine becomes progressively more and more hypertonic, with urea concentrations reaching approx. 750mM. This process of tubular selective reabsorption and excretion is vital to the many regulatory and physiological functions executed by the **kidneys**, such as the maintenance of normal blood pressure levels, prevention of dehydration and electrolytic imbalance, body fluid **acid-base balance**, and elimination of metabolic waste.

See also Electrolytes and electrolyte balance; Elimination of waste; Homeostatic mechanisms; Metabolic waste removal; Renal system; Uric acid

GLUCOSE UTILIZATION, TRANSPORT OF INSULIN

Glucose is the main outcome of carbohydrate **digestion** (80%) and constitutes the fundamental source of energy for cell **metabolism**. In minor quantities (about 20%), carbohydrate digestion also produces two other forms of sugars, galactose, and fructose. After gastrointestinal absorption, **liver** cells convert most galactose and fructose into glucose, through an enzymatic process known as monosaccharides interconversions. As the liver cells are rich in the enzyme glucose phosphatase, one of the metabolites of interconversion, glucose-6-phosphate, can be easily broken into glucose and phosphate, allowing glucose to pass through the hepatocytic membranes, thus entering the **blood** circulation. Therefore, 95% of circulating monosaccharides are glucose, the final conversion product.

Once in the blood circulation, glucose is transported to cells in the several body tissues, where it is transported through the cell membranes in association with molecules of protein carriers, a process known as facilitated diffusion. Facilitated diffusion occurs when the glucose concentration inside the cells are lower than in the extra cellular media. Inside cells, each glucose molecule is combined with a phosphate radical supplied by **ATP** (**adenosine triphosphate**), which is converted to ADP (adenosine diphosphate) in the process, to form a molecule of glucose-6-phosphate. This ATP-mediated process is known as phosphorilation and prevents glucose of being diffused back to the extra cellular media. Cells store glucose-6-phosphate as a source of energy, which can later be oxidized to generate energy through catabolic reactions mediated by mitochondrias.

Glucose transport and absorption is ten or more times increased in the presence of insulin, a hormone secreted by the beta cells of a specialized system in the **pancreas**, termed islets of Langerhans, which secrete glucagon and insulin directly in the blood. The ingestion of a meal rich in **carbohy-**

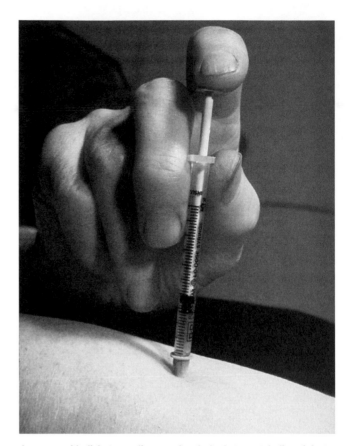

A person with diabetes, a disease of carbohydrate metabolism, injects insulin, facilitating glucose absorption by the tissues and removal of glucose from the blood. *Photograph by Martin Dohrn. Photo Researchers, Inc. Reproduced by permission.*

drates stimulates insulin secretion, facilitating glucose absorption by tissues, quickly removing the excess of glucose from the blood. Most insulin molecules circulate in the blood in its free form, and are degraded and removed from the **plasma** within 10–15 minutes by the enzyme insulinase, mainly in the liver. It plays an important role in the storage of the excess of glucose under the form of **glycogen** in the muscles and in the liver, and stimulates the conversion of glucose into fats, supplying the glycerol portion of lipid molecules to circulating **lipids** and to the adipose tissues as well. When muscle fibers are stimulated by insulin, muscle cells become permeable to glucose; but in the absence of this hormone, muscles can only absorb large amounts of glucose during moderate or heavy exercise. In other conditions, the main energy sources for muscle cells are the fatty acids. Muscles also store glucose that was not converted into energy by muscular activity under the form of glycogen that can be later used as a source of energy. Glycogen is utilized in anaerobic respiration (i.e., energy production in the absence of oxygen) to provide muscle cells with intermittent spurts of energy. However, the **brain** does not depend on insulin for glucose uptake, since its cells are permeable to glucose, an essential energy-source for neural activity. The neural metabolic dependence on glucose becomes apparent by the increased number of mitochondrias in these cells, about 40 times greater than in other tissues.

Abnormally low levels of glucose in the blood cause neural function impairment or hypoglycemic **shock**, which in its acute forms lead to fainting, seizures, and **coma**. Another pancreatic hormone involved in glucose regulation is glucagon, secreted by the alpha cells of the islets of Langerhans. Its main function is to promote the increase of glucose plasmatic concentrations, and is secreted from the pancreas when such concentrations fall. Glucagon induces the metabolic event known as glycogenolysis that converts glycogen into glucose, and increases hepatic gluconeogenesis, i.e., the conversion of **amino acids** to glucose by the liver. Aerobic exercise stimulates glucagons synthesis and release in the blood stream, elevating its levels up to four or five times. Another regulatory molecule is somatostatin, also secreted in the islets of Langerhans by the delta cells. Somatostatin secretion is activated by the increase of either glucose, or amino acids, or fatty acids in the blood, or yet, by the release of gastrointestinal **hormones**. Its short-lived effects have a half-life of 3 minutes in the blood circulation and it seems that its main role is to prolong the time of absorption of glucose and other nutrients by the **gastrointestinal tract**. Somatostatin decreases gastrointestinal motility, and inhibits both insulin and glucagon secretion, therefore decreasing the utilization of absorbed nutrients by the tissues, and making them available for an extended period.

The normal levels of glucose in the blood oscillate between 80 and 90 mg/100ml, and increase up to 120–140mg/100ml during the first hour after a meal. These concentration levels are kept under control through the feedback action of insulin and glucagon, and by hepatic functions as well. This control is crucial because glucose is the only energy source that can be normally used by the brain, **retina**, and germinal epithelium of the sexual glands to meet their high energetic demands.

See also Acidosis; Alkalosis; Glycogen and glycolysis; Lactic acid; Lipids and lipid metabolism; Mitochondria and cellular energy; Nutrition and nutrient transport to cells; Protein and protein metabolism

GLYCOGEN AND GLYCOLYSIS

Glycolysis, a series of enzymatic steps in which the six-carbon glucose molecule is degraded to yield two three-carbon pyruvate molecules, is a central catabolic pathway in plants, animals and many microorganisms.

In a sequence of ten enzymatic steps, energy released from glucose is conserved by glycolysis in the form of **adenosine triphosphate (ATP)**. So central is glycolysis to life that its sequence of reactions differs among species only in how its rate is regulated, and in the metabolic fate of pyruvate formed from glycolysis.

In aerobic organisms (some microbes and all plants and animals), glycolysis is the first phase of the complete degradation of glucose. The pyruvate formed by glycolysis is oxi-

dized to form the acetyl group of acetyl-coenzyme A, while its carboxyl group is oxidized to CO_2. The acetyl group is then oxidized to CO_2 and H_2O by the citric acid cycle with the help of the electron transport chain, the site of the final steps of oxidative phosphorylation of adenosine diphosphate molecules to high-energy ATP molecules.

In some animal tissues, pyruvate is reduced to lactate during anaerobic periods, such as during vigorous exercise, when there is not enough oxygen available to oxidize glucose further. This process, called anaerobic glycolysis, is an important source of ATP during very intense muscle activity.

Because glycolysis occurs in the absence of oxygen, and living organisms first arose in an anaerobic environment, anaerobic catabolism was the first biological pathway to evolve for obtaining energy from organic molecules.

Glycolysis occurs in two phases. In the first phase, there are two significant events. The addition of two phosphate groups to the six-carbon sugar primes it for further degradation in the second phase. Then, cleavage of the doubly phosphorylated six-carbon chain occurs, breaking fructose 1,6-diphosphate into two 3-carbon isomers. These are fragments of the original six-carbon sugar dihydroxyacetone phosphate and glyceraldehyde 3-phosphate.

In the second phase, the two 3-carbon fragments of the original 6-carbon sugar are further oxidized to lactate or pyruvate.

Entry into the second phase requires the isomer to be in its glyceraldehyde 3-phosphate form. Thus, the dihydroxyacetone phosphate isomer is transformed into glyceraldehyde 3-phosphate before being further oxidized by the glycolytic pathway.

Glycolysis produces a total of four ATP molecules in the second phase, two molecules of ATP from each glyceraldehyde 3-phosphate molecule. The ATP is formed during substrate-level phosphorylation-direct transfer of a phosphate group from each 3-carbon fragment of the sugar to adenosine diphosphate (ADP), to form ATP. But because two ATP molecules were used to phosphorylate the original six-carbon sugar, the net gain is two ATP.

The net gain of two ATP represents a modest conservation of the chemical energy stored in the glucose molecule. Further oxidation, by means of the reactions of the Kreb's cycle and oxidative phosphorylation are required to extract the maximum amount of energy from this fuel molecule.

See also Biochemistry; Glucose utilization, transport and insulin; Metabolism

GOLDSTEIN, JOSEPH L. (1940-)
American geneticist and physician

Joseph L. Goldstein received the 1985 Nobel Prize in physiology or medicine for the discovery of the receptor molecule, a structure on cell surfaces that regulates cholesterol levels in **blood**. Goldstein and colleague **Michael Brown** worked for fifteen years before finding the molecule, which shed light on the correlation between blood cholesterol level and **heart** disease. The National Institutes of Health, in part because of Goldstein's and Brown's work, recommended lowering **fat** intake in the diet of United States citizens.

Goldstein is professor of medicine and **genetics** and chairman of the department of molecular genetics at the University of Texas Health Science Center at Dallas. Brown is director of the Center for Genetic Disease. Colleagues there humorously refer to them collectively as "Brownstein," and together they have received awards from the National Academy of Sciences, the American Chemical Society, the Roche Institute of Molecular Biology, the American Heart Association and the American Society for Human Genetics.

Joseph Leonard Goldstein, the only son of Isadore E. and Fannie (Albert) Goldstein, was born in Sumter, South Carolina. His parents owned a clothing store in the eastern part of the state. He graduated from Washington & Lee University in 1962 with a B.S. degree in chemistry, and attended Southwestern Medical School of the University of Texas Health Science Center in Dallas. There, Donald Seldin, chairman of the Health Science Center's department of internal medicine, offered him a future faculty appointment, provided he would specialize in genetics and then return to Dallas to establish a division of medical genetics there. He received his M.D. degree in 1966.

Goldstein's internship and residency at Massachusetts General Hospital brought him to Michael Brown, who had arrived from the University of Pennsylvania, having also obtained his M.D. degree in 1966. The two served in the same internship and residency program, and both were interested in research. After finishing their training in 1968, they joined the National Institutes of Health (NIH) in Bethesda, Maryland.

At the NIH biochemical genetics laboratory, Goldstein studied under the leadership of Marshall Warren Nirenberg, who was awarded the 1968 Nobel Prize in physiology or medicine for unraveling the way in which the **genetic code** determines the structure of proteins. Here, he learned about the excitement and efficiency of biology on a molecular level. At the same time, he worked under Dr. Donald S. Fredrickson, clinical director of the National Heart Institute, who was investigating people with hypercholesterolemia, or abnormally high cholesterol levels. In particular, Goldstein was interested in those patients with homozygous familial hypercholesterolemia. Familial hypercholesterolemia, identified as a genetically acquired disease by Carl Müller of Oslo, Norway, involved a genetic defect, which caused a metabolic error resulting in high blood cholesterol levels and heart attacks. But it was Fredrickson and Avedis K. Khachadurian of the American University of Beirut, who identified two forms of the disease: a heterozygous form, involving a single defective gene found in one in 500 people; and a homozygous form, in which two defective genes are present and which strikes about one in a million. Blood cholesterol levels reach four to eight times the normal amount with symptoms of atherosclerosis, or hardening of the **arteries**, beginning in childhood. Most sufferer from the homozygous form dies from a heart attack before the age of thirty.

In 1972, Goldstein left the National Institutes of Health for Seattle under a two-year NIH fellowship in medical genetics. During this time, he worked with Arno G. Motulsky, an internationally recognized expert in the field of genetic aspects of heart disease, and devoted himself to a study investigating the frequency of various hereditary hyperlipidemias (diseases of high blood-fat levels) in a random sampling of heart attack survivors. The samples were taken from 885 patients (who survived three months or more) out of 1,166 coronary victims admitted in an eleven-month period to thirteen Seattle hospitals from 1970 to 1971. Studying 500 of those survivors and 2,520 members of their families revealed that 31% of the survivors had high blood-fat levels, either high cholesterol, high **triglycerides**, or a mixture of both. Eleven percent had an inherited combination of high cholesterol and high triglycerides. Goldstein and his associates defined this disease as familial combined hyperlipidemia. He knew that due to its complexity, combined hyperlipidemia would be an arduous area in which to begin research. Patients with homozygous hypercholesterolemia—having no normal genes at the area of the unknown defect—might be easier to study regarding gene functioning and cholesterol level.

Returning to the University of Texas Health Science Center in 1972 as head of the medical school's first division of medical genetics, and assistant professor in the department of internal medicine which was still directed by Donald Seldin, Goldstein addressed the task of identifying the fundamental genetic defect in familial hypercholesterolemia (FHC). Brown had joined the staff the previous year.

The idea of cell receptors was known, but it had never been studied in relationship to fat and cholesterol in the blood. Over 93% of the cholesterol in the human body is found inside cells. There, it participates in functions critical to cell development and cell membrane formation. Cholesterol also contributes to the essential production of **sex hormones**, corticosteroids and **bile** acids. The remaining 7% is dangerous, however, if it is not absorbed into the cells as it courses through the **circulatory system**, and sticks instead to the walls of blood vessels disrupting the flow of blood to the heart and **brain**.

Dietary cholesterol, found only in animal foods, is not necessary to the human body since the body produces its own cholesterol in the **liver**. If no cholesterol is available in the bloodstream, individual cells will produce their own. The human liver excretes that cholesterol which is not used by cells or deposited on artery walls. Cholesterol is fat-soluble, but attaches itself to water-soluble proteins, or lipoproteins, manufactured in the liver, as a means of moving through the bloodstream. The lipoproteins most favored by cholesterol are low-density ones, called LDLs, which are composed of much more fat than protein. Thus, high levels of LDLs are equated with the threat of heart disease.

Goldstein and Brown started their study by observing **tissue** cultures of the human skin cells known as fibroblasts, harvested from six FHC homozygotes, sixteen FHC heterozygotes and forty normal people. The cultured fibroblasts, like other animal cells, need cholesterol for the formation of the cell membrane. During this process, Goldstein and Brown were able to follow the manner in which the cells obtained cholesterol, and identify the process of cholesterol extraction from the lipoproteins in the serum of the culture medium, specifically LDLs. This discovery was made in 1973 with their demonstration of the presence of receptor molecules on the cells, which function to adhere LDLs and carry them into the cell. Goldstein and Brown noted that each individual cell normally has 250,000 receptors that bind low-density lipoproteins, and further located LDL receptors on circulating human blood cells as well as cell membranes from assorted animal tissues.

The cells of individuals with the heterozygous form of FHC have 40-50% of the LDL receptors that are typically present on normal cells. Cells of individuals with the homozygous form of FHC have no LDL receptors or a very small number. Cholesterol, manufactured by the liver and attached to LDLs, is passed into the blood, but is removed from the circulatory system rather slowly. Under normal circumstances an LDL molecule spends a day and a half in the bloodstream, but in FHC heterozygotes this length of time is extended to three days, and in FHC homozygotes to five days, providing increased opportunity for cholesterol to accumulate in the walls of the blood vessels.

Cholestyramine, a drug used to treat high cholesterol levels, had been synthesized over twenty years before Goldstein's and Brown's study, but had never been fully understood. Goldstein and Brown discovered that cholestyramine works by multiplying LDL receptors in the liver, which then converts cholesterol into bile acids and passes them into the intestines. However, in spite of this action, cholestyramine had only limited effect on levels of serum cholesterol. Goldstein and Brown determined the reason for this: The increased numbers of LDL receptors in the liver signaled the need for more cholesterol and the liver responded by increasing cholesterol production. This increase in cholesterol level then shut down the production of LDL receptors in the liver. These findings indicated the need for a drug to impede the liver's synthesis of cholesterol that could be administered in tandem with cholestyramine. In 1976 Akiro Endo, a Japanese scientist, isolated compactin, an anticholesterol enzyme, from penicillin mold, and in the same year Alfred W. Alberts of Merck, Sharp, and Dohme research laboratories isolated a structurally similar enzyme, mevinolin, from a different mold. Goldstein and Brown combined mevinolin and cholestyramine in animal experiments with good results, and in 1987 the Food and Drug Administration (FDA) approved mevinolin, now called lovastatin, for marketing. The FDA made the recommendation with the stipulation that the drug should be used only when diet and exercise proved inadequate in treating high cholesterol. Goldstein anticipated a lapse of five to ten years before use of the drug would affect the nation's coronary death rate.

For revolutionizing scientific knowledge about the regulation of cholesterol **metabolism** and the treatment of diseases caused by abnormally elevated cholesterol levels in the blood, Goldstein and Brown received the 1985 Nobel Prize in physiology or medicine. Goldstein and Brown's research illuminating the activity of LDL receptors and their function in the management of cholesterol levels has had far-reaching effects.

Not only has their work increased understanding of an important aspect of human physiology, but it has also had a practical impact on the prevention and treatment of heart disease.

See also Lipids and lipid metabolism

GOLGI, CAMILLO (1843-1926)
Italian physician and histologist

Camillo Golgi, a clinician, researcher, and teacher, is best known for his Nobel Prize-winning work on the **central nervous system**, including his development of the chromate of silver method for better defining cell structures and his discovery of a small organ within the cytoplasm of the nerve cell now known as Golgi's apparatus. After the parasite responsible for **malaria** was identified in 1880, Golgi's study and subsequent diagnosis methodology made it possible to determine the different forms of malaria and means by which to treat patients. Golgi was born in Corteno (renamed Corteno Golgi in his honor), Brescia, Italy. Following in the footsteps of his father, Alessandro, a medical practitioner in the village of Cava-Maria, Golgi attended the University of Pavia as a medical student. There he studied under Eusebio Oehl, distinguished as the first scientist in Pavia to make a systematic study of cell structures using the microscope. Upon earning his medical degree in 1865, Golgi joined the Ospedale di San Matteo in Pavia while continuing to work at the university. During this period, he was introduced to the science of **histology** through close acquaintance Giulio Bizzozero, director of the Institute of General Pathology, a laboratory of experimental **pathology**. It later came under the direction of Golgi as the Institute of General Pathology and Histology, and was eventually named after him.

In 1868, Golgi came to work as an assistant in the Psychiatric Clinic of Cesare Lombroso. Because of differences in personalities, the relationship was short-lived, but under Lombroso's supervision, and perhaps guidance, Golgi's commitment to the study of the central nervous system was established. He was soon influenced, as were many other scientists of his time, by the work of the great German pathologist **Rudolf Virchow**, taking from Virchow's *Cellular Pathology* the idea that diseases entered the body through the cells. This combination of influences from Virchow and Lombroso, and his histological studies in Bizzozero's lab, began to form the foundation for Golgi's life's work.

Golgi left the Ospedale di San Matteo in 1871 and taught a private course in clinical microscopy. For financial reasons, the following year he accepted a position as chief resident physician and surgeon in a small hospital for incurable patients in the town of Abbiategrasso. Now no longer able to access the labs of the university or Bizzozero's institute, Golgi satisfied his voracious appetite for research in a makeshift laboratory that consisted of a microscope and various kitchen utensils. It was in this setting that Golgi invented the chromate of silver method (*la reazione nera*) for staining cells, a method he would later apply to his work on the central nervous system.

Before Golgi's method, scientists were stymied by the elaborate and entangled network of nerve cells, or **neurons**. Even the most sophisticated microscopes of the time could not break through the dense jungle of neuron vines, and the use of organic dyes by such prominent physicians as **Walther Flemming** and **Robert Koch** met with limited success. Golgi used samples of thinly sliced nerve tissues previously hardened with a bichromate, either potassium or ammonium, and immersed them in a silver nitrate solution. Under the microscope, the neurons were etched in black against an almost transparent background. By controlling the time in which the samples were subjected to the bichromate, one could discern numerous cells or exacting details, such as neuron fibers, demonstrating the elaborate network of the nerve connections.

In 1873 and 1874, while still working in his small laboratory in Abbiategrasso, Golgi published the first of his observations using this staining method, sometimes referred to as Golgi's silver stain. He identified the two main types of nerve cells whose differences lie in the length of their axons, or the filaments by which nervous impulses are conducted away from the cell body. Golgi's Type I and Type II nerve cells contain long and short axons respectively, the long type extending beyond the nervous system, the short type remaining intricately fixed within the central nervous system.

Over the next several years the stain enabled him to describe the structure of the olfactory bulb and the large nerve cells on the granular layer of the **cerebellum**, now known as Golgi's apparatus, Golgi bodies, or Golgi complex. Of particular note is the publication in 1874 of the work *Sulle Alterazioni degli Organi Centrali Nervosi in Caso di Corea Gesticolatoria Associata ad Alienazione Mentale*. In it, Golgi showed that the involuntary movements brought on by chorea are caused by lesions on the nerve cells, neuroglia and **blood** vessels within the **cerebral cortex** and the cerebellum. This was contrary to the popular belief that such symptoms were caused by functional disturbances in locomotory parts of the body.

Golgi's silver stain received a somewhat lukewarm reception by his colleagues, perhaps due to the fact that it was difficult to replicate precisely. It wasn't until the early 1880s that other scientists began using the method, broadening scientific knowledge of the nerve **cell structure**. Chief among these scientists was Spanish histologist Santiago Ramón Cajal, who would share the 1906 Nobel Prize with Golgi in the field of physiology for defining the structure of the nervous system. When rumblings persisted among some researchers that Golgi's stain did not always work with other kinds of cells, Golgi revised the method using arsenius acid, which produced more consistent results.

While Cajal made ardent use of Golgi's staining method and shared a research interest in the histology of the central nervous system, the two disagreed on neuron relationships. Golgi believed that the fibers within the complex neuron networks gradually lost their individuality and tried to establish continuity. Cajal, on the other hand, believed that each nerve cell represented a separate entity and that gaps, or synapses, separated them. Golgi used his Nobel lecture to cast doubt on Cajal's "neuron theory," a theory later supported by Charles Scott Sherrington's research in nerve physiology.

In 1875, Golgi returned to the University of Pavia as a lecturer in histology. After accepting the position of chair of **anatomy** at the University of Siena in 1879, he returned to Pavia a year later, where he became professor of histology and succeeded Bizzozero as chair of general pathology. A fervent author, often publishing up to eight papers a year, Golgi released his monumental work *Sulla Fina Anatomia degli Organi Centrali del Sistemi Nervosa* in 1885, a collection of papers previously published in installments in the journal *Rivista Sperimentale di Freniatria,* of which he would later become co-editor. The volume, reprinted in French and German, was illustrated with plates made from Golgi's original drawings.

Between 1885 and 1893, while continuing his work on the nervous system, Golgi began to concentrate his research in malaria. The parasite itself had been discovered by **Alphonse Laveran** in 1880, and its development cycle studied by Antonio Marchiafava and Angelo Celli, whose work Golgi followed closely. Using the research of Marchiafava and Celli as the basis for his own research, Golgi eventually and accurately defined several types of malarial infestations. In the quartan and tertian forms of malaria, Golgi was able to correlate, in 1886, the onset of periodic fever fits with the life cycle of the parasite. In the quartan variation, the peaking of the fever every 72 hours could be attributed to the simultaneous division of the parasites in the blood. In the tertian form, the cycle peaked every 48 hours.

Golgi also described the quotidian fever, which he believed was caused by a double **infection** of the tertian parasite; and the estivo-autumnal type of malaria, which he described as an altogether separate type of malaria. These observations made it possible to diagnose and inevitably treat the disease. He discovered that quinine, to varying degrees, was effective against the parasite at different stages of its development—those in the early stages were most affected. By determining the cycle and subjecting the patient to quinine several hours before the fever episodes, the new generation of parasites could be effectively acted upon. W. G. Whaley notes in *The Golgi Apparatus* that had Golgi "not made significant enough contributions in other fields, his malaria work alone would have withstood the test of time."

Golgi's place in the early annals of cytology was secured in 1898, with his description of the internal reticular apparatus, more commonly recognized as the Golgi apparatus. It was Golgi's staining method, which allowed him to give a more detailed account of the small organelle within the nerve cell's cytoplasm, the substance between the membrane of the cell and its nucleus. The Golgi apparatus appears as a fine network of interlaced threads shown by Golgi and his students to be a consistent component in a variety of cell tissues. Unlike earlier investigators, Golgi was also the first to notice that its character and position were variable. The credit for its initial discovery has been attributed to La Valette St. George, and earlier descriptions of certain components by Platner and Murray showed a definite relationship to the apparatus described by Golgi. Cajal even tried to lay claim to its discovery in 1923, suggesting he had noted its appearance some thirty years earlier while studying the muscle cells of insects, but did not believe he had discovered a new organelle.

Early on, the Golgi apparatus was the focus of many studies. But the structure fell out of vogue in the 1930s and 1940s, the victim of the mistaken belief that it was an apparition created by the same staining method used to make it visible. But advances in electron microscopy later verified its existence and revealed that the Golgi apparatus plays a role in the synthesis and secretion of proteins, as well as the formation of the cell surface.

Golgi married Donna Lina Aletti, Bizzozero's niece, and adopted his own nephew, Aldo Perroncito, who followed Golgi into medicine. Golgi remained in Pavia the rest of his life, and served his country as a member of both the Royal Senate and the Superior Council of Public Instruction and Sanitation. He remained active at the university, where he was president as well as dean of the faculty of medicine. In his seventies at the outset of World War I, Golgi opened the first Italian hospital that ministered principally to patients with lesions of the nervous system. He taught histology until his retirement in 1918, and continued publishing until shortly before his death.

See also Nervous system overview; Nervous system: Embryological development; Neural damage and repair; Neural plexuses; Neurology; Neurons; Neurotransmitters

GONADS AND GONADOTROPIC HORMONE PHYSIOLOGY

The gonads are the primary reproductive **organs** of both male and female. The Y chromosome determines differentiation of the gonads. If the Y chromosome is present, testicular development will occur about six weeks after **fertilization**. However, if the Y chromosome is not present, the embryonic cells in the gonadal region will develop into the ovaries. The gonads produce the sex cells, called gametes, and secrete gonadotropic **hormones**. The gonadotropic hormones stimulate gametogenesis, the transformation of immature sex cells into mature gametes. Once gametes mature, the male gametes, or **sperm**, are capable of fertilizing a female egg called an ovum.

The testes, or male gonads, are paired structures that produce sperm. The testes, also known as the testicles, are oval structures that are about 1.6–1.96 in. (4–5 cm) in length and 0.98 in. (2.5 cm) in diameter. The scrotum is a pouch of skin that contains the testes. Within the scrotum are muscles that help maintain the temperature of the testes at about 95°F (35°C). In order for the sperm to survive, the testes must remain at a temperature about 35.6–37.4°F (2–3°C) below the core body temperature. By contracting or relaxing the muscles, the scrotum can either raise to warm the testes or lower to cool the testes, respectively. Each testicle is divided into 200–300 lobules. The lobules contain the seminiferous tubules that comprise two cell types: **germ cells** that will eventually develop into mature sperm and Sertoli cells that aid in sper-

matogenesis, the making of sperm. In the interstitial **tissue** between the seminiferous tubules are Leydig cells that produce **testosterone**.

The gonads of the female are paired structures called ovaries that produce oocytes, the immature eggs of the female. The ovaries are located in the upper pelvic cavity on either side of the uterus. Each ovary is 0.79–1.6 in. (2–4 cm) in length and is connected to the uterus by the oviducts. Within each ovary are the ovarian follicles that house the immature gametes. At birth the ovaries will hold approximately two million oocytes. The number of oocytes will have decreased to 300–400 thousand at **puberty**.

Once adolescents begin puberty, the anterior pituitary begins to secrete follicle-stimulating hormone (FSH) and **luteinizing hormone** (LH), the gonadotropic hormones. These gonadotropic hormones are secreted in response to emission of gonadotropin releasing hormone (GnRH) from the **hypothalamus**. In males, FSH initiates spermatogenesis, by stimulating Sertoli cells in the seminiferous tubules. Additionally, LH increases the secretion of the male **sex hormones**, the androgens, in the Leydig cells of the gonads. The androgens promote the maturation of the male reproductive organs and development of the **secondary sexual characteristics**.

In females, the gonadotropic hormones stimulate ovarian changes during a monthly cycle called the menstrual cycle. FSH initiates the maturation of the oocytes in the ovarian follicles, a process called **oogenesis**. However, only one ovum matures each cycle. The dominant follicle of which the mature ovum resides is called the Graafian follicle. During the monthly cycle LH stimulates the rupture of the Graafian follicle releasing the mature ovum, a process called ovulation. If the female gamete is not fertilized within 36 hours, the egg is shed and the cycle begins again. Both FSH and LH stimulate the secretion of estrogens in the ovary. Estrogens aid in the development of the female reproductive organs and secondary sexual characteristics. Additionally, LH stimulates the production of other hormones in the ovary necessary for female reproduction such as **progesterone**, inhibin, and relaxin.

See also Hormones and hormone action; Ovarian cycle and hormonal regulation; Pituitary gland and hormones; Reproductive system and organs (female); Reproductive system and organs (male); Sexual reproduction

GRAAF, REGNIER DE (1641-1673)
Dutch physician and anatomist

Dutch physician Regnier de Graaf was one of the pioneers of experimental physiology. His studies involving pancreatic fluids, and later, the male and female reproductive systems, were the precursor to modern reproductive endocrinology.

Graaf was born in Schoonhoven, Netherlands, in 1641. He studied medicine at the universities of Utrecht, Leiden, Paris, and Angers, earning his degree from the last of these in 1665. He died at the early age of thirty-two in 1673; most historians agree that Graff was a victim of the plague. During his lifetime, he held no academic post, working instead as a private physician. He also carried out a vast amount of independent scientific research, much of which earned the praise of his contemporaries and later scientists.

The earliest of his research involved a study of pancreatic juices in 1664. He developed an ingenious technique for extracting the juice from a living dog by inserting a fistula into its **pancreas** and collecting the juice in a flask. Although de Graff was correct in identifying its alkaline nature, the analysis of the function of pancreatic juice was, however, incorrect.

By 1668, Graaf had turned his attention to another topic, the reproductive system. His first treatise on the male system in 1668 contained little new information, but his analysis of the female system proved an important step in the understanding of reproductive biology. He was the first to identify and describe the ovary, a name earlier proposed by Johannes van Horne (1621-1670) and Jan Swammerdam.

In his research, Graaf dissected the ovaries of numerous animals. He found a structure that has come to be known as the Graafian follicle. By observing the ovaries before and after **fertilization**, Graaf was able to recognize changes in their structure and to see the release of fertilized ova and the descent of the zygote to the uterus. However, because the ovum itself was not identified until a century and a half later, Graaf was unable to explain the phenomena he observed.

See also Embryology; Embryonic development: early development, formation, and differentiation; Enzymes and coenzymes; Hormones and hormone action; Reproductive system and organs (female); Reproductive system and organs (male)

GRANIT, RAGNAR ARTHUR (1900-1991)
Finnish neurophysiologist

Ragnar Arthur Granit conducted important research in two distinct areas of neurophysiology, during the first part of his career he researched the physiology of vision, while in the later years he studied muscle spindles and neural control over movement. Although he made significant contributions to both fields, it was for his investigations into the physiology of vision that Granit was best known. He clarified the process by which the **retina** of the **eye** first encodes information about form and color, then transmits it to the **brain** via the optic nerve.

Granit was born in Helsinki, Finland, the eldest son of Arthur W. Granit, a government forester, and Albertina Helena Malmberg Granit. Soon after his birth, the family started a forest-products business in Helsinki. Since both his parents were of Swedish origin, Granit attended the Swedish Normal School there. In 1918, Granit fought in Finland's war for independence from the Soviet Union, for which he received the Finnish Cross of Freedom.

The following year, Granit entered the University of Helsinki with the intention of studying experimental psychology. However, he decided that a medical education would provide the best foundation for this field. So he went on to

complete a master's degree in 1923 and then an M.D. in 1927. It was while completing his medical studies that Granit became interested in studying the nervous system, particularly the eye. The English physiologist Edgar Douglas Adrian had just made the first recordings of electrical impulses in single nerve fibers, and Granit realized that Adrian's technique could be used to provide useful information about the nervous system as well as the retina. Thus, in 1928 Granit traveled to Oxford University to work with Adrian and another English physiologist, Charles Scott Sherrington, and learn the techniques of neurophysiology. Soon thereafter, Granit accepted a fellowship in medical physics at the University of Pennsylvania. It was here, while working at the university's Johnson Foundation for Medical Physics, that Granit first met Haldan Keffer Hartline, who along with **George Wald** would eventually share the Nobel Prize with Granit.

In 1935, Granit returned to the University of Helsinki as a professor of physiology, but he also continued his research into neurophysics. At this time, it was still unclear whether light could inhibit, as well as elicit, impulses in the optic nerve. During the early 1930s, Granit produced the first experimental evidence of this inhibition, a finding that remains fundamental to visual physiology. In early work, Granit employed such indirect measures as the sensitivity reported by human subjects to flickering lights. He showed that illumination focused on the retina would suppress the response of adjacent regions. This served to enhance the perception of visual contrasts. Granit soon confirmed these findings using an electroretinogram, a graphic record of electrical activity in the retina, similar to the record of **heart** activity known as an **electrocardiogram**.

The 1939 invasion of Finland by the Soviet Union interrupted Granit's work. During the war, Granit served as a physician on three Swedish-speaking islands, including Korpo in the Baltic Sea. After the war, he was offered positions at both Harvard University in the United States as well as the Karolinksa Institute in Stockholm. He chose the latter, where he directed the Nobel Institute for Neurophysiology until his retirement in 1967.

While working at the Nobel Institute, Granit and his colleagues became the first scientists to use microelectrodes, tiny electrical conductors, for sensory research. By using microelectrodes to study individual cells in the retina, Granit now demonstrated that certain cells, called modulators, are color-specific, while others, called dominators, respond to a broad range of the spectrum. Although subsequent research has modified his views, Granit's studies were the earliest serious effort to investigate color vision by electrophysiological methods.

Beginning in 1945, Granit shifted the focus of his research to the study of muscle spindles, sensory end **organs** that are sensitive to muscle tension. Ultimately, the structure of muscle spindles and their function in motor control became one of the best-studied areas in neurophysiology, and Granit was at the forefront of the field. Throughout this period of his career, Granit maintained a hands-on presence in the laboratory and continued to take part in experimental operations on animals there. The procedures he most enjoyed performing were the meticulous dissection of nerves and the careful preparation of nerve roots for electrode placement. His surgical skill was much admired by younger students.

Granit's dual contributions to neurophysiology were honored on numerous occasions. In addition to the Nobel, he was awarded such prizes as the 1947 Jubilee Medal of the Swedish Society of Physicians, the 1957 Anders Retzius Gold Medal of the University of Utrecht, and the 1961 Jahre Prize of Oslo University. He was once president of the Royal Swedish Academy of Sciences, and a member of such learned societies as the American Academy of Arts and Sciences, the American National Academy of Sciences, and the Royal Society of London. Active even after his retirement, Granit was appointed a resident scholar at the Fogarty International Center in Bethesda, Maryland, during part of the 1970s. He also accepted a visiting professorship at St. Catherine's College in Oxford, England. In addition, he spent many leisure hours sailing on the Baltic Sea, and he enjoyed gardening. Granit had married Baroness Marguerite Emma "Daisy" Bruun on October 2, 1929, just before leaving for the Johnson Foundation fellowship in the United States. They had one son, who became a Stockholm architect.

In a 1972 essay printed in the *Annual Review of Physiology,* Granit reflected on his lengthy life and career, contrasting a young person's drive for discovery with an older one's need for understanding. He noted, "this second variant of scientific endeavour does not always suit the impatient passion of the young, ruled by an ambition which craves immediate satisfaction, but a little later in life it provides feelings of assurance and satisfaction in one's work." Granit died of a heart attack at his home in Stockholm at age 90.

See also Cranial nerves; Nerve impulses and conduction of impulses; Nervous system overview; Nervous system: embryological development; Neural damage and repair; Neural plexuses; Neurology; Neurons; Neurotransmitters; Vision: histophysiology of the eye

GRAY, HENRY (1825-1861)
English anatomist

Henry Gray, a pioneer of modern descriptive and surgical **anatomy**, authored *Anatomy of the Human Body.* A precisely illustrated work, his book is as beautiful as it is practical, and is still highly valued as an authoritative text in the field of anatomy. *Gray's Anatomy,* as the publication has come to be commonly known, is a standard not only for students of medicine and anatomy, but also for artists striving to more closely represent the human form.

Gray was born in London, and lived his entire life in his family's home. His father's position as a royal messenger allowed him to secure professional training for Henry at Saint George's Hospital. There, Gray studied medicine, quickly gaining preference for research in anatomy rather than clinical practice. He was known for his highly developed dissection skills, gaining material for his future book and several essays

through his research at Saint George. In 1848, Gray won a coveted prize from the Royal College of Surgeons for his anatomical work on the **eye** and optic nerves.

After completing his education, Gray remained at Saint George's Hospital, taking a position as a lecturer of anatomy. This permitted him to continue his research in surgical anatomy and also to pursue other interests, such as **embryology**. In 1852, he presented a paper on "The Glands of Chicks," an early example of the clear technical writing and attention to minute detail for which Gray's work was known. That same year, he was appointed Fellow of the Royal Society.

Gray's career work was compiled and published in 1858, after a decade of research. His *Anatomy Descriptive and Surgical*, as it was originally titled, was instantly recognized as a pioneering work. Gray collaborated with H. Vandyke Carter for the book's second edition in 1860. Carter illustrated the voluminous work, greatly enhancing both the book's utility and success. One of the first modern medical textbooks, it remains one of the most well known.

After the publication of *Anatomy*, Gray's career appeared to hold great promise. He was offered a position of Assistant Surgeon at Saint George. However, Gray's ever-present interest in his research led him to studying the anatomical effects of infectious diseases at the bedside of his ailing nephew. During his time with his nephew, Gray contracted smallpox and died in 1861, at the age of 34. He never assumed his new post.

See also Anatomical nomenclature; History of anatomy and physiology: The science of medicine

GRAY MATTER AND WHITE MATTER

Gray matter and white matter are components of the **spinal cord** and the **brain**. The structure and composition of each allows differential functions in the transmission of information to and from the brain and in the processing of information by the brain.

Moving down the length of a neuron, after the dendrite-cell body area, the structure of the cell is that of long cable-like extension termed the axon. These are part of **neurons**, specialized cells that conduct electrochemical impulses. At one end of a neuron is a branching region that is visually reminiscent of a delta of a large river such as the Mississippi River. These branches are the dendrites. They receive nerve impulses that have traveled from the opposite end of an adjacent neuron. The dendrites come together in a region of the neuron called the cell body. The cell body contains the nucleus of the neuron cell. Axons can be as long as one foot in some nerve cells. Myelinated axons (axons with myelin sheaths) are bundled together to form white matter. The white matter axons are also enveloped by a myelin protective sheath.

Gray matter is composed of masses of cell bodies, dendrites and unmyelinated axons.

In the spinal cord, the inner core consists of neural cells in gray matter, with their axons radiating outward to form the external white matter. A cross sectional or transverse view of the spinal cord reveals an H-like shape of gray matter at the core of the spinal cord centered on the spinal canal. In the brain, especially the **cerebellum** and **cerebral cortex**, the gray matter-white matter relationship is reversed so that the gray matter is on the outside and the white matter is on the inside.

Varying types of nerve fibers (e.g., myelinated or demyelinated) that compose gray matter and white matter exhibit differences in how the **action potential** propagates down the axon and in the speed of transmission of neural signals along the axon. Because of the myelin sheath of white matter, ion movement associated with the neural impulse or action potential with the axon, takes place only at the Nodes of Ranvier that provide small gaps in the myelin sheath. The action potential jumps from node to node along the myelinated axon in a process termed salutatory conduction that is faster than transmission through unmyelinated fibers.

See also Brain stem function and reflexes; Central nervous system (CNS); Nerve impulses and conduction of impulses; Nervous system overview; Neural plexuses; Neurology; Reflexes

GREENGARD, PAUL (1925-   )

American pharmacologist

Paul Greengard shared with **Arvid Carlsson** and Eric R. Kandel the 2000 Nobel Prize in physiology because of his discovery of the cellular signal pathways used by dopamine as well as by a number of other **neurotransmitters**. Neurotransmitters are biochemical molecules that trigger a cascade of stimuli in neural cells. These signal pathways are biochemical events mediated by intracellular **enzymes** and proteins, and are generally termed signal transduction. The understanding of such pathways in the nervous system is of fundamental importance for the identification of the causal events associated with a series of mental and behavioral disorders and, consequently, for the development of new drug therapies. Dopamine, noradrenaline, and serotonin are neurotransmitters of the so-called slow synaptic transmission (SST), which implies that the changes caused in the nerve cell by these molecules may last from a few seconds up to several hours. SST is responsible for several basal functions in the nervous system such as the sensation of **pain** or pleasure, fight-or-flee reaction, stimulation of basal **metabolism**, alertness, and mood.

Greengard showed that SST depended upon a biochemical reaction known as protein phosphorilation that adds a phosphate group to a given protein, thus altering its function. Therefore, when dopamine occupies a cell membrane receptor, it triggers a sequence of intracellular events, first causing the elevation of cyclic AMP levels, activating the protein kinase A. Kinase A then adds phosphate groups to other proteins present in the nerve cell, activating them. Protein phosphorilation may have as targets different groups of proteins, depending on the neurotransmitter that initiates the process. When the proteins of a certain group of ion channels in the cell membrane are phosphorilated, for instance, the stimulus is transmitted to the axons, increasing its excitability. Because each kind of neurotransmitter has its own specific receptors in the cell membrane, and controls different pathways of response, the understanding of such processes leads to a better knowledge of how drugs act in the **brain**. Paul Greengard has also shown the existence in nerve cells of a central regulatory protein that regulates the cascade of both protein phosphorilation (i.e., activation of a given group of proteins) and dephosphorilation (i.e., the removal of phosphate groups that turns proteins off). This regulatory protein, named DARPP-32, responds to cascades initiated by several neurotransmitters, including dopamine, and affects different groups of ion channels, thus altering the function of specific kinds of fast synapses (i.e., the transmission of different fast stimuli through a group of cells in the nervous system).

Paul Greengard was born in New York City, and studied **pharmacology** at John's Hopkins University, in Baltimore, Maryland. He achieved a Ph.D. degree at the same institution in 1953, and left for Great Britain in the same year to attend postdoctoral studies in **biochemistry** at the University of London. He also studied and researched at Cambridge University and at the National Institute for Medical Research, England. Back in the United States, he worked at the National Institutes of Health until 1959, when he accepted the position of Director of the Department of Biochemistry at Geigy Research Laboratories, in Ardsley, New York. He also worked as a Visiting Associate Professor and as Professor of Pharmacology at Albert Einstein College of Medicine, New York (1961–1970), and as Professor of Pharmacology and Psychiatry at Yale University School of Medicine, New Haven, Connecticut (1968–1983). In 1983, he became Professor and Head of the Laboratory of Molecular and Cellular Neuroscience at The Rockefeller University, New York, New York, where he conducted the research on the biochemical pathways of neurotransmitters, that earned him the Nobel Prize.

See also Nerve impulses and conduction of impulses

GROSS ANATOMY · *see* ANATOMY

GROWTH AND DEVELOPMENT

Beginning at the moment of conception, the human organism depends upon adequate **nutrition** for growth, development, and survival. In the first week from the **fertilization**, the zygote produces a series of blastomeres and, because of further cellular divisions, the morula, that contains about 10–30 cells. The morula stage proceeds to the formation of a fluid filled cavity, the early blastocyst. Inside the blastocyst is an inner cell mass or embryoblast (future embryo), and the outer cell mass or trophoectoderm (future **placenta**).

Embryogenesis (three to eight weeks) involves three major processes: morphogenesis (generation of shape), pattern

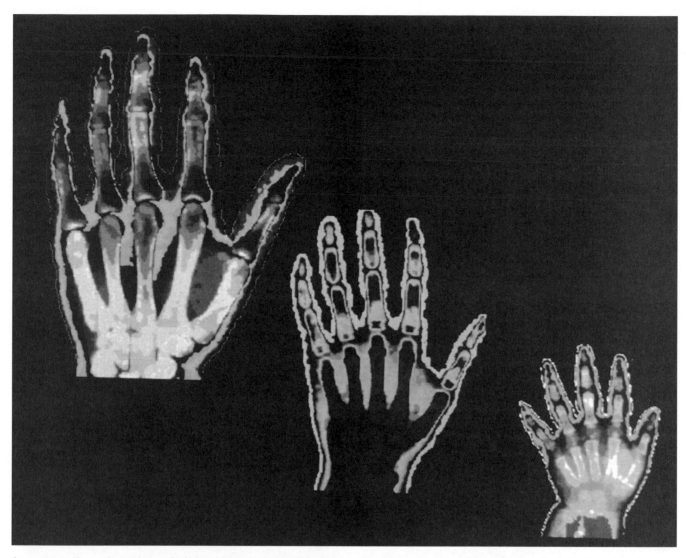

A computer-enhanced x-ray image depicting the hand development of a male at two (far right), six, and nineteen years (far left) of age. *X-ray by Scott Camazine. © Scott Camazine, Photo Researchers, Inc. Reproduced by permission.*

formation (biologic-spatial cell organization), and differentiation (specialization in specific phenotypes). Morphogenesis and pattern formation are regulated by bone morphogenetic proteins and homeobox genes. During embryogenesis tissues and **organs** develop. The most important events of the embryonic period can be resumed in: (1) gastrulation, a process able to convert the bilaminar embryonic disk into the three primary embryonic germ layers; (2) formation of the dorsal **mesoderm**, (**notochord** and paraxial mesodermal cords), **somites**, intermediate and lateral mesoderm; (3) neural tube formation from **ectoderm** that will produce the nervous system and skin; (4) evolution of **endoderm**, that essentially develops into the digestive apparatus, respiratory apparatus, some parts of the **urogenital system** and branchial pouches (part of the branchial apparatus). By the end of eight weeks, the embryo is about one inch long.

The fetal period goes from nine to 38 weeks. It is characterized by rapid body growth. In this period the embryo, now termed the fetus, is starting to grow. Length growth velocity has a peak at 20 weeks about 10 cm/ 4 weeks. Weight growth velocity peaks at 30–34 weeks. At nine weeks, half the fetus' overall size is its head. From 13 to 16 weeks, the head is relatively smaller and the limbs (legs) are longer. At the fifth month (17–20 weeks) growth slows down, but the fetus is longer and the limbs reach their final relative proportions. Parameters used for monitoring fetal growth include the embryonic crown-rump length, biparietal diameter (BPD), head circumference, femur length, and abdominal circumference. Ultrasound technology is often used to evaluate fetal growth.

After birth, the primary hormonal regulators of growth are growth hormone (GH) and the GH-dependent growth factors, insulin-like growth factor-I (IGF-I). This contrasts with the fetal period, where the fetal **endocrine system** is believed to play a relatively minor role in growth. Postnatal growth also is influenced by a variety of factors. Normal changes in the rate of linear growth occur frequently during infancy, allowing some infants to "cross percentiles" (up or down) during the first months of life. More than 50% of babies experience an upward shift of growth during the first three months of post-natal life. Breast milk is the optimal food for growth because it provides immunologic and antibacterial factors, **hormones**, **enzymes**, and opoid peptides not present in alternative feed-ings. Newborns require approximately 100–120 cal/kg/day for adequate growth and development.

After infancy, growth speed slows down in the toddler years. After age two, toddlers gain about 5 lb. (2.3 kg) in weight and 2.5 (6.4 cm) inches in height each year. In com-parison, head circumference increases by about 1 in. (2.5 cm) from two to twelve years of age. Growth does not increase steadily. A toddler's weight can remain the same for some weeks. Increases in height result primarily from growth of the lower extremities and, secondarily, to a lesser extent elonga-tion of the trunk. Body proportions change, with upper-to-lower segment ratios ranging from 1.40 at age two years, to 1.15–1.20 at age five years. By age two, toddlers of both gen-der will stand about 34 in. (86 cm) tall and weigh about 27–28 lb. (12.2–13.1 kg) on average. Growth charts are very useful at this time in order to detect abnormal growth that can be a manifestation of clinical disorders.

Child growth is the period from the post-toddler years through pre-adolescence. In the post-toddler period, nutrition continues to be critically important. At this time, growth pat-terns are largely determined by **genetics** and under hormonal control. It is a very complex process, and requires the coordi-nated action of several hormones. Normal growth, supported by good nutrition adequate **sleep** and regular exercise, is one of the best overall indicators of good health. A significantly malnourished child may be pushed off the "natural" geneti-cally determined growth curve. Severe iponutrition (inade-quate nutrition), enough to affect a child's growth rate, is uncommon today in the United States and other developed countries unless the child has an associated chronic illness or disorder. Extra food or greater than recommended amounts of **vitamins**, minerals, or other nutrients will not increase the height of a growing child.

Adolescence defines the time between the beginning of sexual maturation (**puberty** from the Latin pubertas, meaning adult) and adulthood. Roughly, adolescence spans 13–19 years of age. Adolescence includes physical growth and emotional, psychological, and mental change. Adolescence's first signal stems from higher concentrations of leptin, a protein produced by adipocytes of the **fat tissue**. The **hypothalamus** stimulates the hypophysis to secrete hormones able to promote the over-all growth of the body to maturate the **gonads**, as well as adre-nal cortex and thyroid. Normal growth is categorized in a range used by pediatricians to gauge how a child is growing. The following are some average ranges of weight and height, based on growth charts developed by the Centers for Disease Control and Prevention (CDC): briefly, at 12 years of age, a male should be 54–63.5 in. (137–160 cm) tall and a female

55–64 in. (140–163 cm). The weight should be 66–130 lb. (29.9–58.9 kg) and 68–136 lb. (30–61.6 kg), respectively. Throughout childhood, a child's body becomes more propor-tional to other parts of his or her body. Growth is complete between the ages of 16 and 18, at which time the growing ends of bones fuse. Although a child may be growing, his or her growth pattern may deviate from the normal. Ultimately, the child should grow to normal height by adulthood. A number of hormonal conditions can lead to excessive or diminished growth.. Dwarfism (very small stature) can be due to under-production of GH, or a pathologic feedback of the GH pro-duction, included a flaw in target tissue response to growth hormones. Overproduction of GH or an exaggerated response can lead to gigantism. Gigantism is the result of GH overpro-duction in early childhood leading to a skeletal height up to 8 ft. (2.5 m) or more.

Development of primary sex characteristics includes the further maturing of the gonads, the testis in boys and the ovaries in girls. In both sexes, hormonal regulation of repro-duction is regulated by the **brain**. Until eight weeks of gesta-tion, the brain is organized in female direction irrespectively with the gender of the fetus. Successively, **testosterone**, for example, organizes the male brain in patterns of behavior, many of which may not appear until much later. Secondary sex characteristics can be considered traits that give an indi-vidual an advantage over its rivals in courtship. During puberty, changing hormonal levels play a role in activating the development of secondary sex characteristics. These include: (1) growth of pubic **hair** (pubarche); (2) growth of the **breasts** (thelarche) (for girls) (3) menarche (first menstrual period for girls) or penis growth (for boys); (4) voice changes (for boys); (5) growth of underarm hair; (6) facial hair growth (for boys); (7) nighttime ejaculations (nocturnal emissions; "wet dreams" for boys) and (8) increased production of oil, increased sweat gland activity, and the beginning of acne.

See also Adolescent growth and development; Child growth and development; Embryonic development, early develop-ment, formation, and differentiation; Human development (timetables and developmental horizons); Infant growth and development; Prenatal growth and development; Toddler growth and development

GROWTH AND DEVELOPMENT: AGE 0 (NEWBORNS) • *see* NEONATAL GROWTH AND DEVELOP-MENT

GROWTH AND DEVELOPMENT: AGE 1 MONTH TO 2 YEARS • *see* INFANT GROWTH AND DEVELOPMENT

GROWTH AND DEVELOPMENT: AGE 2 YEARS TO 4 YEARS • *see* TODDLER GROWTH AND DEVELOPMENT

GROWTH AND DEVELOPMENT: AGE 5 YEARS TO 12 YEARS • *see* CHILD GROWTH AND DEVELOPMENT

GROWTH AND DEVELOPMENT BEFORE BIRTH • *see* PRENATAL GROWTH AND DEVELOPMENT

GROWTH AND DEVELOPMENT: TEENAGE YEARS • *see* ADOLESCENT GROWTH AND DEVELOPMENT

GUILLEMIN, ROGER (1924-)

French-born American endocrinologist

Roger Guillemin is one of the founders of the field of neuroendocrinology, the study of the interaction between the **central nervous system** (such as the **brain**) and endocrine glands (such as the pituitary, thyroid, and **pancreas**). Guillemin focused his research on **hormones** produced by the brain, and their subsequent effect on body processes. He proved the correctness of a hypothesis first proposed by English anatomist Geoffrey W. Harris that the **hypothalamus** releases hormones to regulate the pituitary gland. For discoveries which led to an understanding of hypothalamic hormone productions of the brain, Guillemin and fellow endocrinologist Andrew V. Schally shared the 1977 Nobel Prize in physiology or medicine with physicist Rosalyn Sussman Yalow. Guillemin and Schally were pioneers in isolating, identifying, and determining the chemical nature of such hormones as TRF (thyrotropin-releasing factor which regulates the thyroid gland), LRF (luteinizing-releasing factor which controls male and female reproductive functions), somatostatin (which regulates the production of growth hormones and insulin), and **endorphins** (which may be involved in the onset of mental illness). Guillemin's work led to scientific advances including an understanding of thyroid diseases, infertility, juvenile diabetes, and the physiology of the brain. According to Guillemin, the determination of the chemical structure of TRF marked an end to the pioneering era in neuroendocrinology and the beginning of a major new science.

Roger Charles Louis Guillemin was born and raised in Dijon, France, the son of Raymond Guillemin, a machine toolmaker, and Blanche Rigollot Guillemin. He attended the University of Dijon where he received a bachelor's degree in 1942, and then entered the University of Lyons medical school, graduating with a medical degree in 1949. However, Guillemin interrupted his studies during World War II in order to join the French underground during the Nazi occupation, becoming part of an operation helping refugees escape to Switzerland over the Jura Mountains. During and after the war, Guillemin received three years of clinical training and briefly practiced medicine before joining a well-known Canadian physiologist, Hans Selye, as a research assistant. To work with Selye, Guillemin moved to the Institute of Experimental Medicine and Surgery at the University of Montreal in Canada. In 1950, he suffered a near-fatal attack of tubercular meningitis. After his recovery in 1951, Guillemin married Lucienne Jeanne Billard, who had been his nurse during his illness. The couple eventually y had six children.

Guillemin received his Ph.D. from the University of Montreal in 1953, and accepted an assistant professorship at Baylor University Medical School in Houston, Texas. His research involved endocrinology, the study of the hormones that circulate in the **blood**. The **endocrine system** is a hierarchical one in which hormones from the pituitary gland regulate other endocrine glands. It was thought that the head of the entire system was the hypothalamus, located at the base of the brain just above the pituitary gland. However, the way in which hypothalamic hormonal regulation occurred was unclear. The theory of regulation by nerve impulses was marred by the anatomical fact that there are few nerves that extend from the hypothalamus to the pituitary. Anatomist Geoffrey W. Harris theorized that hypothalamic regulation occurred by means of hormones, which are transported by the blood. Harris's experiments supported his hypothesis, proving altered pituitary function when the blood vessels were cut between the hypothalamus and the pituitary. The problem was that no one had yet been successful in isolating and identifying a hormone from the hypothalamus.

Guillemin began an investigation to find the missing evidence, a task of extraordinary difficulty because very minute amounts of hypothalamic substances are involved. At Baylor, Guillemin worked together with Schally using a technique called mass spectroscopy and a new tool developed by physicists Solomon Berson and Rosalyn Sussman Yalow called radioimmunoassays (RIAs) which enabled scientists to isolate and identify the chemical structure of hormones. In the early 1960s Guillemin considered continuing his research in France, and obtained a concurrent appointment at both Baylor and the Collège de France in Paris. However, he left the Collège de France in 1963, and was appointed director of the Laboratory for Neuroendocrinology at Baylor University. By this time he and Schally had ended their scientific cooperation and had become fiercely competitive in a race to identify hypothalamic hormones.

Guillemin worked with sheep hypothalami which he obtained from slaughter houses. Obtaining the specimens was a large-scale, difficult operation. Only very minute amounts of substance existed in each sheep hypothalamus and it had to be extracted very soon after **death**. Guillemin and Roger Burgus, a chemist who worked with Guillemin, reported that their laboratory collected about five million hypothalamic fragments from sheep brains, which involved handling about five hundred tons of brain **tissue**. Finally in 1968, Guillemin and his coworkers isolated the hypothalamic hormone that effects the release of thyrotropin. The following year Guillemin, as well as Schally, who had been working independently, revealed the structure of TRF (a hypothalamic hormone which today is called thyrotropin-releasing hormone or TRH). When TRF is secreted by the hypothalamus, it causes the pituitary gland to secrete another hormone that in turn causes the thyroid gland to secrete its own hormones. Shortly thereafter Guillemin and his colleagues isolated and determined the chemical structure of GRH (growth-releasing-hormone), a hypothalamic hormone that causes the pituitary to release gonadotropin which in turn influences the release of hormones in the testicles or ovaries. This discovery led to advancements in the medical treatment of infertility.

In 1970, Guillemin moved to the Salk Institute in La Jolla, California. There he isolated a third hypothalamic hormone which he named somatostatin. This hormone acts by inhibiting the release of growth hormone from the pituitary gland. In 1977, Guillemin and Schally were awarded the Nobel Prize for their research on hypothalamic hormones. Guillemin wrote on the importance of their discoveries in an autobiography, published in *Pioneers in Neuroendocrinology II*, stating that: "I consider the isolation and characterization of TRF as the major event in modern neuroendocrinology, the inflection point that separated confusion and a great deal of doubt from real knowledge. Modern neuroendocrinology was born of that event. Isolations of LRF, somatostatin, and the recent endorphins were all extensions (as there will be still more, I am sure) of that major event—the isolation of TRF— a novel molecule in hypothalamic extracts, with hypophysiotropic activity, the first so characterized.... The event was the vindication of 14 years of hard work."

Guillemin soon turned his attention to another class of substances, known as neuropeptides. Produced by the hypothalamus and other parts of the brain, neuropeptides act at the synapses of the nerves (the area where the nerve impulse passes from one neuron or nerve cell to another). One group of neuropeptides, for example, called endorphins, seem to affect moods and the perception of **pain**. Guillemin's recent research includes neurochemistry of the brain and growth factors.

Guillemin is known as an urbane conversationalist who is interested in the arts and enjoys painting. He and his wife have a collection of contemporary French and American paintings, pre-Columbian art objects, and artifacts from around the world. Guillemin is also a connoisseur of fine food and wine.

See also Hormones and hormone action; Nerve impulses and conduction of impulses; Pituitary gland and hormones; Thyroid histophysiology and hormones

GULLSTRAND, ALLVAR (1862-1930)
Swedish opthalmologist

Major contributions to our understanding of the human **eye** were made by Swedish ophthalmologist Allvar Gullstrand, particularly in the area of how the eye forms images. His mathematical approach to solving physiological problems had a great significance in the science of ophthalmology, and his discoveries won him the Nobel Prize in physiology or medicine in 1911. He also developed a number of devices, such as the slit lamp and the reflector ophthalmoscope, which became valuable tools in eye examinations and for the treatment of optical disorders. Gullstrand also served for many years as a member, and later as president, of the Nobel Committee responsible for awarding the prize for physics.

Gullstrand was born in Landskrona, Sweden, to Pehr Alfred Gullstrand and Sophia Korsell Gullstrand. His father, the city physician, influenced his decision against a career in engineering, in favor of one in medicine. After studying at universities in Uppsala, Sweden, Stockholm, and Vienna, Gullstrand received his medical degree from Stockholm's Royal Caroline Institute in 1888. He earned his Ph.D. one year later through a dissertation on astigmatism, an eye defect involving faulty curvature of the optic lens. Utilizing his early training and natural aptitude in engineering, he formulated complex theories in optics, which considerably advanced knowledge in this field.

During this time, Gullstrand began working as chief physician at the Stockholm Eye Clinic, and by 1892 he was both clinic director and lecturer at the Karolinska Institute. He left the University in 1894 to serve as a professor at the University of Uppsala, where his research in geometrical optics began to flourish. His studies in dioptrics of the eye, or the science of refracted light and its effect on the **retina**, helped clear up certain misconceptions regarding the way the eye functions. One such misunderstanding concerned the accommodation theory of optics, by which the eye adjusts its focus on objects near and far. In his *Handbook on Physiological Optics,* German biologist and physicist Hermann von Helmholtz postulated that the eyes react to the problems of focusing by altering the curvature of their lens. When the eye focused on a nearby object, the lens became more convex (curving-outward), while focusing on something farther away made the lens more concave (curving-inward). In his commentaries on the third edition of the *Handbook* (1908), which he reedited, Gullstrand demonstrated that Helmholtz's theory accounted for only two-thirds of the accommodation. The remaining one-third could be explained by what Gullstrand called "extracapsular accommodation," where the fibers behind the lens made the necessary adjustments. The concept of the human eye as an optical system was among Gullstrand's most important achievements.

Gullstrand was given an honorary degree from Uppsala University in 1907 for his advances in eye research. He invented two devices commonly used even today in eye examinations—the slit lamp and the ophthalmoscope (sometimes called the Gullstrand ophthalmoscope), in cooperation with the Zeiss Optical Works in Germany. The slit lamp, consisting of a light used in combination with a microscope, permits doctors to pinpoint the location of a foreign object or tumor in three dimensional space. The ophthalmoscope is a combined light and magnifying lens enabling doctors to look at the retina at the back of the eye, as well as the optic disk. Doctors use it in an inspection for eye defects, as well as **arteriosclerosis** and diabetes. Gullstrand also designed aspheric lenses for those patients whose lenses had been removed as a result of cataracts.

The Nobel Prize in physiology or medicine was awarded to Gullstrand in 1911 for his work on the refraction of light and formation of images in the eye. In his lecture to the Nobel Academy, Gullstrand noted that the laws concerning the formation of optical images had been completely unknown when he began studying the eye lens, and that much of what had been known at that time had since been proven false. A special chair in physical and physiological optics was established for him in 1914 at Uppsala, and he became a member of the Nobel Academy's Physics Committee, and later its president, serving until 1929. Gullstrand received honorary degrees from the University of Dublin and the University of Jena in Germany. He was also awarded the Björken Prize from the Uppsala Faculty of Medicine, the Swedish Medical Association's Centenary Gold Medal, and the Graefe Medal from the German Ophthalmological Society.

See also Eye and ocular fluids; Sense organs: Ocular (visual) structures; Vision: Histophysiology of the eye

GUSTATORY STRUCTURES

Gustatory (**taste**) structures involve the interaction of chemical stimuli in the food or drink with receptors. Gustation enables food to be monitored before it enters the digestive system. Food that tastes objectionable, such as bitter foods that can taste bitter because of toxic contamination, can be discarded before they have a chance to damage the body.

It is often necessary to prepare the food for the taste receptors. The preparation involves the grinding and tearing action of the teeth. As well, the food is exposed a fluid produced by salivary glands located in the mouth. Salivary **enzymes**, such as lysozyme, begin to dissolve the food and help position the food near the taste buds.

Taste buds are chemical receptors buried within bumps (papillae) that cover the surface of the tongue, and on the surface of the palate, **pharynx** and **larynx**. There are about 10,000 taste buds in the mouth. The rough surface of the papillae assists food in sticking to the tongue. Each taste bud has hair-like structures called **cilia** that contain the taste receptors.

The chemical sensory process is rapid, as the sensitivity of taste buds deteriorates rapidly in the presence of **saliva**. Following removal of the food during **swallowing**, the sensitivity of the taste buds is restored.

Four qualities of taste are detected by the taste buds in most vertebrates: sweetness, saltiness, bitterness, and sourness. The sensitivity of the tastes differs with species and buds are lost with age. In humans, the detection sensitivity for taste combinations becomes less with age. Children have very sensitive taste buds, and so often dislike spicy foods. In most children and adults, sweetness is pleasurable.

The taste buds convert the chemical signals to electrical signals, which are relayed to the **brain** via the facial, glossopharyngeal and vagus nerves of the **cranium**.

The different taste buds for sweet, sour, bitter and salty are not spread uniformly over the tongue, but are each confined to specific regions of the tongue. The tip of the tongue is most sensitive to sweet and salty tastes. The back of the tongue along each side is most sensitive to sour tastes. The central part of the tongue, the larynx, the pharynx, and the palate are most sensitive to bitter tastes. Recently, evidence has suggested that another gustatory structure might be the sense of **smell**. The mechanism for the enhancement of taste by smell is as yet unclear.

See also Aging processes; Cranial nerves; Mastication

H

HAIR

Hair is the collective term for thin, threadlike strands extending from a mammal's **epidermis**, the outer layer of the skin. Hair appears on mammals in varying degrees and serves many different purposes. Thick-haired mammals rely on hair for warmth. In human beings, hair around the eyes and inside the ear protects us by keeping foreign objects away. Most of the human body is covered by tiny hairs, except the palms of the hands and soles of the feet. Human hair growth begins in the embryo.

Each strand is composed chiefly of a protein known as keratin. Hair growth begins inside a soft sac or bulb where rapidly dividing hair cells multiply. Crowding inside the bulb pushes the old cells upward, forming the root and hair shaft. At the base of each hair bulb is a structure called a papilla, which houses the nerves and **blood** vessels necessary for hair growth. The root is protected by the hair follicle, a pouch of epidermal cells and **connective tissues**. The familiar "goose bumps" result when a tiny muscle attached to the hair follicle contracts.

The color and texture of hair are hereditary characteristics. Hair color largely depends on the amount and distribution of a brownish-black pigment called melanin. Hair also contains a blonder, or yellow-red, pigment. This lighter-colored pigment is most noticeable in those with little melanin. Red hair, however, contains a pigment all its own. As a person grows old, pigment stops forming and hair often turns gray or white.

Texture depends on the shape of each individual hair strand. Ironically, a round strand of hair results in straight hair. Wavy or curly-headed persons have flatter strands. Microscopically, the flattest hairs are the curliest.

See also Integument; Melanocytes; Nails

HALES, STEPHEN (1677-1761)
English clergyman and physiologist

Professionally, Stephen Hales was a clergyman, serving as "perpetual curate" of Teddington, Middlesex, England, from 1709 until his death in 1761. Avocationally, Hales was a leading scientist of his time—the founder of plant physiology, a trailblazer in the study of **blood** circulation and blood pressure measurement, a pioneer in public health.

Hales was born in Bekesbourne of an old Kentish family. Little is known about him until he began studies at Cambridge in 1696. At the university, Hales enthusiastically immersed himself in scientific studies and biological experiments while also pursuing his clerical degree. He received his bachelor of arts degree in 1702 and his master's degree in 1703; he was ordained a deacon in 1703 and in 1709 he went to the "perpetual" post in Teddington.

During his early years at Teddington, Hales continued experiments which he had begun at Cambridge, achieving the first blood pressure measurements with a glass-tube manometer. He also investigated reflex actions in a decapitated frog. Hales then gave up his animal experiments, "being discouraged by the disagreeableness of anatomical dissections," and turned to the investigation of the movement of sap in plants, accidentally discovering that the force exerted by flowing sap would expand a bladder tied over a stem. From this, Hales realized he could use a glass tube, as in his animal work, to measure the force of the sap's flow.

Hales's study of plant transpiration led to his investigations of air, both "fixed" in varying substances and as given off or absorbed under different conditions, and his invention of various measuring devices. This work was tremendously important to later chemists. Another Hales contribution was the investigation of ways to chemically dissolve kidney and bladder stones, in the course of which the cleric invented a surgical forceps.

An important aspect of Hales's career was the application of his findings to practical uses. He used his knowledge of air and **breathing** to invent ventilators to remove noxious air from hospitals, merchant and slave ships, and prisons. He adapted a gauge from his plant experiments to ocean-depth measuring. He worked on ways to distill fresh water from ocean water. He involved himself in social issues, working for passage of the 1736 Gin Act, and was active in the founding of the colony of Georgia, while also attending to his parish duties. Noted for cheerfulness and serenity, Hales died after a brief illness and was buried at Teddington, with a monument in his memory at Westminster Abbey.

See also Circulatory system; History of anatomy and physiology: The Renaissance and Age of Enlightenment

HALLER, ALBRECHT VON (1708-1777)
Swiss biologist

Albrecht von Haller was one of the great heroic figures of early biology. Born in Bern, Switzerland, he was not a healthy child, but he displayed prodigious intellectual talents at an early age. He wrote scholarly articles at the age of eight and by the age of ten, he had completed a Greek dictionary.

Haller enrolled as a medical student at the University of Leyden and earned his degree at the age of 19. At Leyden, he studied under the famous Hermann Boerhaave (1668-1738). Haller began his own medical practice in 1729 at the age of 21 and continued in private practice until 1736. He was then appointed Professor of Anatomy, Botany, and Medicine at the newly created University of Göttingen. He served at Göttingen until 1753, when he returned to Bern. He spent the remaining twenty years of his life in research, writing, and government service until he died in Bern on December 17, 1777.

Haller displayed interests and talents in a wide range of fields, but he is probably best known for his work on nerve s and muscles. When he began his research, little was known about the **structure and function** of nerves or about their interaction with muscles. A popular theory of the time held that nerves are hollow tubes through which a spirit or fluid flows. Haller rejected this idea, however, since no one had ever been able to locate or identify such a spirit or fluid.

Instead, Haller concentrated on two specific and identifiable nerve-related phenomena, irritability and sensibility. By irritability he meant the contraction of a muscle that occurs when a stimulus is applied to the muscle. Haller found that irritability increases when the stimulus is applied to the nerve connected to a muscle. He concluded that the stimulus was transmitted from the nerve to the muscle, thus clarifying for the first time the relationship of nerve to muscle.

In his study of sensibility, Haller found that ordinary **tissue** does not respond to stimuli, but that nerves do. He showed that stimuli applied to nerve endings travel through the body, into the spinal column, and eventually into the **brain**. By removing certain parts of the brain, he was then able to show how each part affects specific muscular actions.

Because of his pioneering research on the nervous system, Haller is often credited as the founder of the science of **neurology**.

See also Nerve impulses and conduction of impulses; Nervous system overview; Nervous system: embryological development; Neural damage and repair; Neural plexuses; Neurons; Neurotransmitters

HANDEDNESS

Handedness refers to the predilection of either the right or left hand as the dominant hand for the performance of tasks such as writing.

A preference for the right or left hand is not restricted to humans. Handedness is evident in other mammals, birds and amphibia. In humans, about 95% of the population is right-handed, with only five percent being left-handed. A very small percent of people are ambidextrous, having no preference for performing tasks with either hand. Social pressure to conform to the norm can drive some left-handed people to adopt the predominant use of their right hand.

The preference for a dominant right or left hand appears to be linked to an asymmetry in the **brain**. It is known that in right handed people, the left somatomotor cortex of the right side of the brain is about seven percent larger than the corresponding right cortex of the same side of the brain. Whether this difference causes the handedness or is a by-product of the preferential use of one hand is difficult to discern as yet. However, there is more firm evidence of a link between hand preference and speech. Thirty percent of left-handed people have their speech center in the right hemisphere of the brain, while only five percent of right-handed people do. Moreover, in other species, the asymmetry of body function is under tight genetic control.

Whatever the basis of handedness, the preference is often established in humans prior to birth. After birth and as development proceeds, handedness is associated with an asymmetry of one side of the body relative to the other side. This asymmetry is particularly marked in right-handed people, where the limbs on the right side of the body are larger than those on the left side. This size asymmetry extends to the brain. Researchers now suggest that asymmetries in left-handed people arise by different developmental rules compared to these more marked asymmetries of right-handed people.

Evidence compiled during the 1990s indicates that the preference for the left hand may be genetically determined, rather than due to environmental factors or social pressure to conform to the right-handed majority. While the prevalence of left-handedness in the entire population is about five percent, this percentage is markedly higher in families where one parent, particularly the mother, is left-handed. As yet, however, there is no clear explanation of how genetic factors might influence the choice of preferred hand.

See also Brain: Intellectual functions; Genetics and developmental genetics; Touch, physiology of

HARTLINE, HALDAN KEFFER (1903-1983)

American biophysicist

Haldan Keffer Hartline was a renowned physiologist who spent almost half a century investigating the process of vision. An early fascination with the **metabolism** of nerve cells led him to study the workings of the **eye**, especially the retinal mechanisms involved in vision and the electrical activity occurring in the individual cells of the **retina** and optic nerve. His comparative studies of the retinas of arthropods, mollusks, and vertebrates—representing each of the three major phyla having well-developed eyes—established principles of retinal physiology, thus providing the foundations for further investigations into the neurophysiology of vision and leading to an enhanced understanding of the wider realm of sense perception. For his work on analyzing the chemical and physiological retinal mechanisms of vision, he shared the 1967 Nobel Prize in physiology or medicine.

Hartline was born in Bloomsburg, Pennsylvania, to Daniel Schollenberger Hartline and Harriet Franklin Hartline. He attended college at Lafayette College in Easton, Pennsylvania, graduating with a B.S. in 1923. He went on to study retinal electrophysiology as a graduate student at Johns Hopkins University, obtaining his M.D. in 1927. Hartline spent the next two years at Johns Hopkins University as a National Research Council fellow in medical sciences.

Between 1929 and 1931, Hartline was a Johnson Traveling Research Scholar from the University of Pennsylvania to the universities of Leipzig and Munich. He traveled extensively in Germany during those years before returning to the United States, where he joined the Eldridge Reeves Johnson Research Foundation for Medical Physics as an assistant professor of biophysics at the University of Pennsylvania. Hartline married Mary Elizabeth Kraus on April 11, 1936. The couple eventually had three sons.

From the early days of his career, Hartline was fascinated by the metabolism of nerve cells, and he eventually focused his attention on the workings of individual cells in the retina of the eye. During the late 1920s and early 1930s, Hartline used recently-developed methods of fiber isolation to record the activity of single nerve fibers in the retina. He began by experimenting with *Limulus Polyphemus,* the horseshoe crab. He chose this primitive creature because it possessed a feature that was ideal for his research: a compound eye with a long optic nerve and large individual photoreceptors. It seemed to Hartline that working with the horseshoe crab might allow him to record the electrical behavior of single nerve fibers. He succeeded in 1932, while working at the Eldridge Reeves Johnson Foundation. Hartline and Columbia University psychophysiologist Clarence H. Graham managed to isolate single nerve fibers from the optic nerve, placed electrodes on those single fibers, stimulated them with light, and recorded the nerve impulses that occurred. This was the first record of the activity of a single optic nerve fiber, and it proved to Hartline and Graham that their theories had been correct: information is relayed through individual optic nerve fibers by a series of uniform nerve impulses.

Hartline moved into another field of vision in 1938, when he began to study the vertebrate eye, using microdissection techniques to record the activity of individual fibers in the optic nerve of frogs. While recording the nerve impulses from the single nerve fibers lying behind the rods and cones of the eye, he found that the fibers making up the nerve did not all behave in the same way. Some were stimulated by steady light, others were stimulated by the light when it first hit the retina, and still others were stimulated only as the light was shut off. Hartline demonstrated that visual information begins to be differentiated in the retina and in the receptors themselves, as soon as the stimulation occurs, before the information can be conducted more deeply into the **central nervous system**. This research afforded new insights into the working of the retina. It also provided a new understanding of how the mechanisms of vision were integrated with, and how they affected, the nervous system as a whole. For this discovery, Hartline was awarded the Howard Crosby Warren Medal of the Society of Experimental Psychologists in 1948.

Hartline continued his teaching and research at the University of Pennsylvania, becoming professor of biophysics and chair of the department at Johns Hopkins in 1949. In 1953, Hartline joined the faculty of Rockefeller University in New York as professor of neurophysiology. There, Hartline began investigating the phenomenon of inhibition in the retina of the compound eye, using the horseshoe crab as a subject once again. He and his colleague, Floyd Ratliff, demonstrated the electrical response of nerve fibers and cells to light hitting the retina, and the mechanism by which this response allows the eye to differentiate shapes. He found that the receptor cells in the eye are interconnected in such a way that when one is stimulated, others nearby are depressed, thus sharpening the contrast in light patterns. In the 1960s, Hartline extended these studies to the dynamics of the receptors and their interactions, with a view to understanding visual phenomena such as motion detection. Hartline's findings eventually led to the development of a set of mathematical equations expressing the interrelationship of the receptor units of the compound eye; this information has been key to understanding brightness and contrast in the retinal image.

For his work on electrical activity on the cellular level within the eye, Hartline shared the 1967 Nobel prize in physiology or medicine with the American biologist **George Wald** and the Swedish neurophysiologist **Ragnar Granit**. This was not the only award received by Hartline during this period; he also received the A. A. Michelson Award of Case Institute, 1964, and the Lighthouse Award in 1969. In addition to the Nobel Prize and the other awards and honors received during his lifetime, Hartline was also presented with a number of honorary degrees. He was awarded doctorates from Lafayette College in 1959, the University of Pennsylvania in 1971, Rockefeller University in 1976, the University of Maryland in 1978, and Syracuse University in 1979; an LL.D. from Johns Hopkins University in 1971; and an M.D. from the University of Freiburg in 1971.

Hartline was a member of many important scientific organizations, many of them elective. He was elected to the National Academy of Sciences in 1948, and to the American

Academy of Arts and Sciences in 1957. Hartline also held memberships in the American Philosophical Society and the Biophysics Society, and in 1966 was elected a foreign member of the Royal Society, London. The Optical Society of America made him an honorary member, as did the Physiology Society (U.K.).

See also Eye and ocular fluids; Sense organs: Ocular (visual) structures; Vision: histophysiology of the eye

HARTWELL, LELAND H. (1939-   )
American geneticist

Leland H. "Lee" Hartwell helped pioneer the development of yeast as a model system for **genetics** in the mid–1960s. Hartwell then used that system to help decipher the **cell cycle**, and to discover in the 1980s an important pathway in **cancer** development. Through his hypothesis on checkpoints, he proposed a genetic link between disruptions of the normal cell cycle and cancer cell proliferation. Hartwell, along with R. Timothy Hunt and **Paul Nurse**, shared the 2001 Nobel Prize in physiology or medicine for their research in genetic control of cell cycle regulation.

Born in Los Angeles, California, Hartwell first began to explore his interests in science, and particularly physics and math, with the help of his high school teachers. After a year at Glendale Junior College, he transferred to the California Institute of Technology (CalTech) as an undergraduate student and spent much of his time conducting research on phages in various scientists' labs. One of the most noted of the scientists was Bob Edgar, who was part of the Max Delbrück group. Delbrück is often credited as the founder of **molecular biology**.

Hartwell earned his Bachelor of Science degree in 1961, and then began graduate studies at the Massachusetts Institute of Technology (MIT). While investigating gene regulation under Boris Magasanik, Hartwell says he came to understand that he was master of his research, and had to plot his own scientific course. He earned his doctorate in 1964.

For his postdoctoral work, Hartwell returned to California and a position at the Salk Institute for Biological Studies studying growth control and mammalian cells. Later, moved to the University of California, Irvine, as an assistant professor in 1965. Hartwell's studies took a fortuitous turn when he began investigating yeast rather than mammalian cells.

By employing mutant strains of yeast, Hartwell was able to deduce which genes control different aspects of cell growth and reproduction. This groundbreaking, initial work with yeast in the mid-1960s was a major impetus in elevating yeast's stature in the lab. Yeast has since become an intensely studied organism, particularly following the realization that many genes found in yeast are also found in plants and animals. Hartwell's yeast and cell-cycle research eventually led to the findings for which Hartwell is most known: control of the cell cycle and its relationship to cancer.

The young scientist moved in 1968 from the University of California to the University of Washington, where he held the positions of assistant and associate professor. Hartwell continued his work with yeast, expanding it in the 1980s to develop the notion of checkpoints, which are genes that act as control circuits to halt the cell cycle when they recognize the presence of chromosomal damage. During such pauses, damage is repaired before the cell cycle resumes. Hartwell and his research team, including post-doctoral fellow Ted Weinert, based their work on an observation that cells with damaged **DNA** usually arrest the cell cycle. Weinert, however, found some mutants that had no effect on the cell cycle. By studying the difference between them, they identified the first checkpoint.

Through this discovery, Hartwell proposed that mutations in the genes at these checkpoints may actually contribute to cancer, because the loss of a checkpoint allows genetically damaged cells, like those of **tumors**, to continue growing.

While Hartwell continued his work at the University of Washington, he joined the Fred Hutchinson Cancer Research Center in Seattle as a senior advisor for scientific affairs in 1996. In the following year, he accepted the title of president and director of the center, and still holds those positions as well those of professor of genetics and adjunct professor of medicine at the University of Washington.

Over the years, Hartwell has received many honors, including a Guggenheim Fellowship in 1983-84, a National Institutes of Health Merit Award in 1990, the Gairdner Foundation International Award in 1992, a 1994 Genetics Society of America Medal and the Albert Lasker Medical Research Award in 1998. Hartwell is also a member of the National Academy of Sciences, the American Academy of Microbiology and the American Academy of Arts & Sciences. He has also been a member of the National Cancer Institute Cancer Genetics Working Group, National Human Genome Research Institute Scientific Planning Subcommittee and the National Institute of Environmental Health Sciences Genome Project Working Group.

See also Cell cycle and cell division; Cell differentiation; Cell structure; Cell theory; Genetic regulation of eukaryotic cells

HARVEY, WILLIAM (1578-1657)
English physician

William Harvey, the father of modern **physiology**, was born in Folkestone, Kent, England, in 1578, the eldest of seven sons of a yeoman farmer. While five of the other Harvey brothers became London merchants, William studied arts and medicine at Cambridge University, where he received a Bachelor of Arts degree in 1597, and then earned his medical degree in 1602 from the renowned medical school at Padua (Italy), where he studied under Fabrici. Returning to London, Harvey began what became a very successful medical practice while also engaging extensively in medical research. In 1604, he married Elizabeth Browne, daughter of a prominent London doctor; they had no children.

In 1609, Harvey was appointed to the staff of St. Bartholomew's Hospital. He was elected a fellow of the Royal College of Physicians in 1607 and was Lumleian lecturer on **anatomy** and surgery for the College from 1615 to 1656. His ideas about circulation of the **blood** were first publicly expressed in these lectures in 1616. Harvey became court physician to King James I in 1618, and then to Charles I in 1625, a post he held until Charles was beheaded in 1649. Charles provided Harvey with deer from the royal parks for his medical research, and Harvey remained loyal to Charles even during the Cromwellian Civil War, which led to the sacking of Harvey's rooms in 1642 and the destruction of many of his medical notes and papers. Harvey retired at the end of the Civil War, a widower, and lived with his various brothers. He died of a stroke in 1657 in Roehampton and was buried in the family vault at Hempstead Church in Essex.

Harvey's great contribution to medicine was his revolutionary discovery of the circulation of blood. His many experimental dissections and vivisections convinced Harvey that **Galen's** ideas about blood movement must be wrong, particularly the concepts that blood was formed in the **liver** and absorbed by the body, and that blood flowed through the septum (dividing wall) of the **heart**. Harvey first studied the heartbeat, establishing the existence of the pulmonary (heart-lung-heart) circulation and noting the one-way flow of blood. When he also realized how much blood was pumped by the heart, he realized there must be a constant amount of blood flowing through the **arteries** and returning through the **veins** of the heart, a continuing circular flow.

Harvey published this radical new concept of blood circulation in 1628. It provoked immediate controversy and hostility, contradicting as it did the usually unquestioned teachings of Galen, the basis of medical knowledge at the time. The most virulent critic, Jean Riolan, scorned Harvey as a "circulator," a derisive term for a traveling quack. Harvey calmly and quietly defended his work, and although his medical practice declined for a time, his ideas had become widely accepted at the time of his death. The discovery of **capillaries** by **Marcello Malpighi** in 1661 provided the factual evidence to confirm Harvey's theory of blood circulation. Harvey's method of drawing reasoned conclusions from meticulous observation formed the basis of an entirely new approach to medicine—modern physiology.

Harvey's other important contribution to medicine was in the field of **embryology**, as he was one of the first to study the development of the chick in the egg, and performed many dissections of mammal embryos at various stages of formation. From these experiments Harvey was able to formulate the first new theory of generation since antiquity, emphasizing the primacy of the egg, even in mammals. His findings on generation were published in 1651, and became the foundation of the new science of embryology.

See also Circulatory system; Embryonic development: Early development, formation, and differentiation; History of anatomy and physiology: The Renaissance and Age of Enlightenment

HEADACHE

A headache, a **pain** in the head or tissues surrounding the upper neck, has varied causes and consequences. The consequences range from uncomfortable to debilitating. Some of the latter can be a symptom of life-threatening abnormality or an illness.

While the symptoms and triggers of headaches are known the exact molecular basis of these ailments is as yet unclear. It has been suggested that the abnormal release in the **brain** of a chemical called seratonin shifts the flow of **blood** from the **arteries** to the **veins**. This causes changes in the nerves and nerve centers in the brain, which may induce a headache.

An excruciating type of headache is caused by an aneurysm, the swelling of a blood vessel in the brain. The headache appears suddenly and with great intensity. Prompt medical attention, including surgical repair, is usually necessary to avoid a **stroke** or **death**.

Another type of "urgent" headache results from bacterial meningitis, the **infection** of the **meninges**, the connective **tissue** that is envelopes nerves in the brain and **spinal cord**. This type of headache can also be accompanied by an intensely uncomfortable reaction to light. Prompt therapy with the administration of antibiotics for the bacterial infection is essential to reduce the bacterial infection and consequent inflammation associated with the infection.

Other causes of headaches include a brain tumor, stuffed sinuses, tension, and eyestrain. A now common cause of headaches in the workplace is eyestrain due to the prolonged use of the eyes for computer work.

Migraines are a fairly commonly, severe form of headache. About 25% of men and 8% of women suffer from **nausea**, vomiting, and light sensitivity induced by a migraine at some time in their lives. For some people, migraines can be a much more frequent malady. For some, a migraine is preceded by an aura, a disruption in **taste, smell**, and vision. The visual symptoms include flashing lights, sharp-edged shapes and quivering forms. As the symptoms diminish the migraine headache begins. Other migraines can strike without warning.

The precise cause of migraines is unknown. However, they likely are related to chemical changes of the blood vessels in the brain. The vessels may narrow and become wider. In some people, diet plays a role in migraines. For example, chocolate or red wine (especially red wines with high tannin content) can trigger a migraine. How these foodstuffs act at the molecular level in the brain is still unclear. But, it is known that the elimination of the problematic food or drink helps ease the frequency and intensity of the headache.

See also Blood pressure and hormonal control mechanisms; Drug treatment of cardiovascular and vascular disorders; Electroencephalograph; Meninges; Neck muscles and fasciae

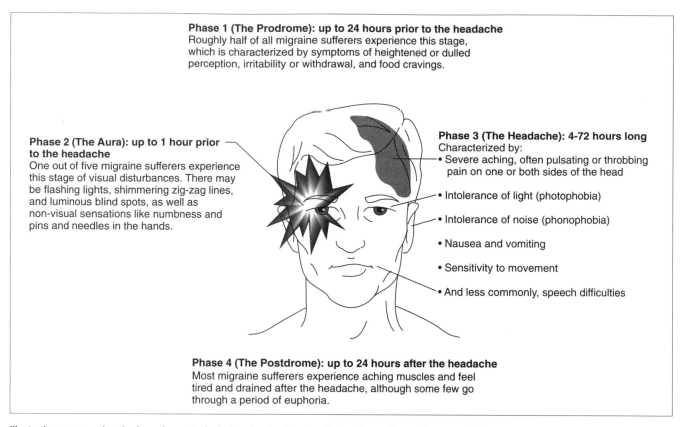

Phase 1 (The Prodrome): up to 24 hours prior to the headache
Roughly half of all migraine sufferers experience this stage, which is characterized by symptoms of heightened or dulled perception, irritability or withdrawal, and food cravings.

Phase 2 (The Aura): up to 1 hour prior to the headache
One out of five migraine sufferers experience this stage of visual disturbances. There may be flashing lights, shimmering zig-zag lines, and luminous blind spots, as well as non-visual sensations like numbness and pins and needles in the hands.

Phase 3 (The Headache): 4-72 hours long
Characterized by:
• Severe aching, often pulsating or throbbing pain on one or both sides of the head

• Intolerance of light (photophobia)

• Intolerance of noise (phonophobia)

• Nausea and vomiting

• Sensitivity to movement

• And less commonly, speech difficulties

Phase 4 (The Postdrome): up to 24 hours after the headache
Most migraine sufferers experience aching muscles and feel tired and drained after the headache, although some few go through a period of euphoria.

Illustration representing the four phases typical of a migraine-type headache: the prodrome, the aura, the headache, and the postdrome. *Illustration by Hans & Cassidy.*

HEARING, PHYSIOLOGY OF

Physiological acoustics is the study of the transmission of sound and how it is heard by the human **ear**. Sound travels in waves, vibrations that cause compression and rarefaction of molecules in the air. The organ of hearing, the ear, has three basic parts that collect and transmit these vibrations: the outer, middle and inner ear. The outer ear is made of the pinna, the external part of the ear that can be seen, which acts to funnel sound through the ear canal toward the eardrum or **tympanic membrane**. The membrane is highly sensitive to vibrations and also protects the middle and inner ear. When the eardrum vibrates it sets up vibrations in the three tiny bones of the middle ear, the malleus, incus and stapes, which are often called the hammer, anvil and stirrup because of their resemblance to those objects. These bones amplify the sound. The stapes is connected to the oval window, the entrance to the inner ear, which contains a spiral-shaped, fluid-filled chamber called the cochlea. When vibrations are transmitted from the stapes to the oval window, the fluid within the cochlea is put into motion. Tiny hairs that line the basilar membrane of the cochlea, a membrane that divides the cochlea lengthwise, move in accordance with the wave pattern. The **hair** cells convert the mechanical energy of the waveform into nerve signals that reach the auditory nerve and then the **brain**. In the brain, sound is interpreted.

Early research into the **physiology** of hearing was conducted by Hermann von Helmholtz, a German physician who enjoyed the study of physics and made a close study of the function of both the eyes and ears. He theorized that the ear detected differences in pitch through the action of the cochlea, the snail-shaped organ of the inner ear. As a physicist he understood sound waves and their properties, such as pitch (the highness or lowness of a sound) and amplitude or loudness. He proposed that certain notes sounded pleasing together because their pitches had a mathematical relationship. However, the human ear can distinguish between two instruments playing the same pitch. He contended that the quality of a tone depended on the intensities of other pitches known as overtones which combine to give a sound a particular tone or timbre.

In 1857, Helmholtz proposed his resonance theory of hearing in which he suggested that the fibers along the basilar membrane of the cochlea were of different lengths and thus had their own natural vibration or frequency. When a sound of that same frequency entered the cochlea, that fiber would resonate and sense the sound. He also suggested that the cochlea's structure resonated at particular frequencies to enable both pitch and tone to be perceived.

Although many of Helmholtz's ideas were right, his grasp of what occurs inside the cochlea was incorrect. Many years later, **Georg von Békésy**, a Hungarian-American physicist, studied the cochlea by placing it in a fluid bath and thus could see in more detail what occurred. He also studied the cochlea indirectly by making mechanical models to observe what happened when the fluid in the cochlea begins to move.

For nearly a quarter of a century, Békésy worked for the Hungarian telephone system doing research on acoustics. He began research in physiological acoustics in 1923, first studying the eardrum, then the basilar membrane. He constructed a mechanical model of a cochlea, first made of a rubber membrane stretch over a metal frame and later one containing fluid. He found that vibrations transmitted to the fluid in the cochlea set up traveling waves in the basilar membrane. When the frequency (pitch) of the stimulus was increased, the section of sensed vibration moved toward the end of his model that was closest to the middle ear. When the frequency was decreased, the section of sensed vibration moved toward the inner ear.

When he came to the United States in 1947, Békésy suggested a different theory of hearing to replace that of Helmholtz. The basilar membrane that separates the chambers of the cochlea is made up of about 24,000 fibers that stretch across its width. The fibers are progressively wider moving along the cochlea. Helmholtz thought that each fiber would have its own natural vibration and thus respond to sounds with that vibration. Békésy, using his artificial model to mimic the cochlea, found that sound waves passing through the fluid in the cochlea set up a wave in the membrane, and it is the shape of that wave that goes to the brain and is interpreted as sound. The hair cells along the wave transform the mechanical energy of the vibration into nerve impulses that can be sent to the brain and interpreted as sound. The wave travels from the stiffer basal part of the cochlea the more flexible upper part of the cochlea. Because of the shape of the cochlea, the resulting waveform is quite complex. Békésy likened the cochlea to a frequency analyzer, an electronic device that measures and interprets the frequency of waves. For his work on physiological acoustics, Békésy was awarded the Nobel Prize in physiology or medicine in 1961, the first time a physicist ever won in that category.

The understanding of the function of the inner ear, particularly the cochlea, has undergone a revolution in the last two decades. For example, scientists had believed that the cochlear tuning process was passive and mechanical. However, recent studies have shown that one group of cochlear hair cells have an active motion that enhances hearing. Research focused on the physiology of acoustics also laid the ground work for such advances as hearing aids and the cochlear implant, which involves surgically implanting electrodes in the cochlea to help stimulate the nerves involved in hearing. The implant helps people with hearing defects due to injury or loss of cochlear hair cells, which accounts for the most incurable forms of **deafness**.

See also Ear (external, middle and internal); Ear: Otic embryological development

HEART

Located in the thoracic cavity, the heart is a four-chambered muscular organ that serves as the primary pump or driving force within the **circulatory system**. The heart contains a special form of muscle, appropriately named cardiac muscle, that has intrinsic contractility (i.e., is able to beat on its own, without nervous system control).

The chamber of the hearts are divided into two upper (superior) atrial chambers and two thicker-walled, heavily muscular inferior ventricular chambers. The right and left sides of the heart are divided by a thick septum. The right side of the heart is on the same side of the heart as is the right arm of the patient. The atrial and ventricular chambers on each side of the septum constitute separate collection and pumping systems for the pulmonary (right side) and **systemic circulation** (left side). The coronary sulcus or grove separates the atria from the ventricles. The left and right side atrial and ventricular chambers each are separated by a series of one way valves that, when properly functioning, allow **blood** to move in one direction, but prohibit it from regurgitating (flowing back through the valve).

Deoxygenated blood—returned to the heart from the systemic circulatory venous system—enters the right atrium of the heart through the superior and inferior vena cava. Auricles lie on each atrium and are most visible when the atria are drained and deflated. The auricles (so named because they resembled ear flaps) allow for greater atrial expansion. Pectinate muscles on the auricles assist with atrial contraction. Small contractions within the right atrium, and pressure differences caused by evacuation of blood in the lower (inferior) right ventricle, cause this deoxygenated blood to move through the tricuspid valve during diastole (the portion of the heart's contractile cycle between contractions, and a period of lower pressure as compared to systole) into the right ventricle. When the heart contracts, a sweeping wave of pressure forces open the pulmonic semilunar valve that allows blood to rush from the right ventricle into the pulmonary artery where it is travels to the **lungs** for oxygenation and other gaseous exchanges.

Freshly oxygenated blood returns to the heart from the **pulmonary circulation** through the pulmonary vein the empties into the left atrium. During diastole, the oxygenated blood moves from the left atrium into the left ventricle through the mitral valve. During systolic contraction, the oxygenated blood is pumped under high pressure through the semilunar aortic valve into the aorta and thus, enters the systemic circulatory system.

As the volume and pressure rise during the filling of the right and left ventricles, the increased pressure snaps shut the flaps of the atrioventricular valves (tricuspid and mitral valves) anchored by fibrous connection to the left and right ventricles. The pressure in the ventricles seals the valves and as the pressure increases during systole, the valves seal becomes further compressed. A prolapse in one of the valves (a pushing through of one of the cusps) leads to blood flow back through the valve. The cusps are held against prolapse by the chordae tendineae, thin cords that attach the cusps to papillary muscles.

The heart and great vessels attached to it are encased within a multi-layered **pericardium**. The outer layer is fibrous and covers a double membraned inner sac-like structure termed the pericardial cavity that is filled with pericardial fluid. The pericardial fluid acts to reduce friction between the heart, the pericardial membranes, and the thoracic wall as the heart contracts and expands during the **cardiac cycle**.

The heart muscle is composed of three distinct layers. The outermost layer, the outer epicardium, is separated from the inner endocardium by the middle pericardium. The outer epicardium is continuous and in some places the same as the visceral pericardium. Epicardium protects the heart and is invested with **capillaries**, nerves, and lymph vessels. The middle myocardium is a think layer of cardiac muscle. The innermost endocardium contains connective **tissue** and Purkinje fibers. The endocardium is continuous with the lining of the great vessels attached to the heart and it lines all valve and cardiac inner surfaces.

Heart muscle does not directly take up oxygen from the blood it pumps. A specialized set of vessels (e.g., the left and right coronary **arteries** and their branches) supply oxygenated blood to the heart muscle and constitute the **coronary circulation**. A heart attack occurs whenever blood flow is occluded (blocked).

Various intrinsic, neural, and hormonal factors act to influence the rhythm control and impulse conduction within the heart. Although cardiac muscle can contract on its own, the sino-atrial node on the right atrium acts to send out signals that regulate and coordinate contractions. The sino-atrial node and atrioventricular nodes act as a pacemaker for the heart. Cardiac arrhythmias result from abnormalities in the rhythm or rate of heat contractions (heart beats).

The fossa ovalis is a remnant or the embryonic foramen ovale that allows blood to flow between the left and right atria in the developing fetus.

See also Anatomical nomenclature; Angiology; Aortic arch; Blood pressure and hormonal control mechanisms; Cardiac cycle; Cardiac disease; Cardiac muscle; Ductus arteriosis; Heart defects; Heart, embryonic development and changes at birth; Heart, rhythm control and impulse conduction; S-A node

Heart defects

The **heart** is the muscular organ responsible for pumping **blood** throughout the body. It can operate defectively, due to genetic malformation, diseases, or damage due to ailments such as **stroke**. Given its vital function, any heart defect is serious, and some are life-threatening.

Heart defects that are present at birth are termed congenital heart defects. In the United States about 40,000 children are born with a heart defect each year. At least 35 different defects have been identified so far. Congenital heart defects stenosis, or the narrowing of heart valves, improper closing of valves so as to allow the back flow of blood, defect in valves, improper development of the entire left side of the heart, an electrical disorder in which the heart muscle does not recharge properly after a heartbeat, switching or cross-communication of the aorta and the pulmonary artery,

and holes in various regions of the heart. These conditions can cause a variety of symptoms, including enlargement of the heart, poor oxygenation of the blood leading to fatigue, bluish appearance of extremities, and poor weight gain. In the case of the left-side development and most regulatory electrical disorders, **death** may result if the condition is not recognized and treated. Treatment often involves corrective surgery.

Ventricular septal defect is the most common congenital heart defect. A ventricular septal defect results in a communication between the left and right ventricles of the heart (the pulmonary and systemic system pumps) in the form of a hole between the two ventricles. The hole results in oxygen rich blood being pumped back to the **lungs**, at the expense of oxygen-poor blood. In many children the resulting breathlessness and lack of energy (malaise) disappears during the first year of life, as the condition is repaired naturally. If self-repair does not occur, the hole can be closed surgically, leading to a normally active life.

The causes of congenital heart damage are still unclear. Use of certain medication during **pregnancy**, alcohol or drug abuse, viral infections such as rubella, or diseases such as diabetes have been linked to an increased risk of congenital heart defects in children. As well, the manifestations of some genetic conditions such as Down's syndrome may include heart defects.

Not all heart defects are congenital. Patients can be born with a normal heart, but develop heart damage later in life. Valve damage can occur as a result of artherosclerosis, where **calcium** deposition narrows the **arteries**. In elderly people, the weakening of a valve that leads to leakage (myxomatous degeneration) is not infrequent. Other valve malfunctions can be a consequence of rheumatic fever, hypertension, congestive heart failure, blocked arteries (artherosclerosis) and heart inflammation (endocarditis).

Not all heart defects are corrected by open-heart surgery. Drug therapy with a class of drugs designated as ACE inhibitors can decrease the constriction of blood vessels. Heart muscle function can be improved by administration of the drug digoxin. A catheter with a balloon at its tip can be threaded into a clogged artery and, when the balloon is inflated, the build-up plaque is flattened against the arterial wall.

Within the past several years, Joseph Penninger and his colleagues at the University of Toronto have shown that artherosclerosis can be caused by a lung **infection** by a **bacteria** called Chlamydia. This organism also causes pelvic inflammatory disease in women. A surface protein of Chlamydia is virtually identical to a heart muscle protein. Attempts by the body's **immune system** to rid the body of Chlamydia also cause damage to heart muscle. Thus, in the future it may be possible that antibiotics will be used to prevent heart damage.

See also Blood pressure and hormonal control mechanisms; Cardiac disease; Cardiac muscle; Heart, rhythm control and impulse conduction

HEART, EMBRYONIC DEVELOPMENT AND CHANGES AT BIRTH

The developing fetal **heart** accounts for a large percentage of the volume of the early thorax. About 20 days after **fertilization**, the heart develops from the fusion of paired endothelial tubes into a single tube. Heart growth subsequently involves the growth, expansion, and partitioning of this tube into four chambers separated by thickened septa of cardiac muscle and valves. Atrial development is initially more advanced than ventricular development. The left and right atria develop while the primitive ventricle remains a single chamber. As atrial separation nears completion, the left and right ventricles begin to form, then continue until the heart consists of it's fully developed four-chambered structure.

Although the majority of the heart develops from **mesoderm** (splanchnic mesoderm) near the neural plate and sides of the embryonic disk, there are also contributions from neural crest cells that help form the valves.

Three systems initially return venous **blood** to the primitive heart. Regardless of the source, this venous blood returns to sinus venosus. Vitelline **veins** return blood from the yolk sac; umbilical veins return oxygenated blood from the **placenta**. The left umbilical vein enlarges and passes through the embryonic **liver** before continuing on to become the inferior vena cava that fuses with a common chambered sinus venosus and the right atrium of the heart. Especially early in development, venous return also comes via the cardinal system. The anterior cardinals drain venous blood from the developing head region. Subcardinal veins return venous blood from the developing renal and **urogenital system**, while supracardinals drain the developing body wall. The anterior veins empty into the common cardinals that terminate in the sinus venosus.

Movement of blood through the early embryonic vascular system begins as soon as the primitive heart tubes form and fuse. Contractions of the primitive heart begin early in development, as early as the initial fusion of the endothelial channels that fuse to form the heart.

The heart and the atrial tube that form the aorta develop by the compartmentalization of the primitive cardiac tube. Six separate septae are responsible for the portioning of the heart and the development of the walls of the atria and ventricles. A septum primum divides the primitive atria into left and right chambers. The septum secundum (second septum) grows along the same course of the primary septum to add thickness and strength to the partition. There are two holes in these septae through which blood passes, the foramen secundum and the foramen ovale. Specialized endocardinal **tissue** develops into the atrioventricular septum that separates the atrium and ventricles. The mitral and tricuspid valves also develop from the atrioventricular septum.

As development proceeds, the interventricular septum becomes large and muscular to separate the ventricles and provide strength to these high-pressure contractile chambers. The interventricular septum also has a membranous portion.

Initially, there is only a common truncus arteriosus as a channel for ventricular output. The truncus eventually separates into the pulmonary trunk and the ascending aorta.

Blood oxygenated in the placenta returns to the heart via the inferior vena cave into the right atrium. A valve-like flap in the wall at the juncture of the inferior vena cava and the right atrium directs the majority of the flow of oxygenated blood through the foramen ovale, then allows blood to flow from the right atrium to the left. Although there is some mixing with blood from the superior vena cava, the directed flow of oxygenated blood across the right atrium caused by the valve of the inferior vena cava means that deoxygenated fetal blood returning via the superior vena cava still ends up moving into the right ventricle.

While in the uterus, the **lungs** are non-functional. Accordingly, another shunt, the **ductus arteriosis** (also spelled ductus arteriosus) provides a diversionary channel that allows fetal blood to cross between the pulmonary artery and aorta and thus largely bypass the rudimentary pulmonary system.

Because only a small amount of blood returns from the **pulmonary circulation**, almost all of the blood in the fetal left atrium comes through the foramen ovale. The relatively oxygen-rich blood then passes through the mitral value into the left ventricle. Contractions of the heart, whether in the single primitive ventricle or from the more developed left ventricle, then pump this oxygenated blood into the fetal systemic arterial system.

In response to inflation of the lungs and pressure changes within the pulmonary system, both the foramen ovale and the ductus arteriosis normally close at birth to establish the normal adult circulatory pattern whereby blood flows into the right atrium, though the tricuspid valve into the right ventricle. The right atrium pumps blood into the pulmonary artery and pulmonary circulation for oxygenation in the lungs. Oxygenated blood returns to the left atrium by pulmonary veins. After collecting in the left atrium, blood flows through the mitral value into the left atrium where it is then pumped into the **systemic circulation** via the ascending aorta.

See also Angiology; Aortic arch; Ductus arteriosis; Heart, embryonic development and changes at birth; Placenta and placental nutrition of the embryo; Systemic circulation

HEART MUSCLE · *see* CARDIAC MUSCLE

HEART, RHYTHM CONTROL AND IMPULSE CONDUCTION

The rhythmic control of the **cardiac cycle** and its accompanying heartbeat relies on the regulation of impulses generated and conducted within the **heart**. Regulation of the cardiac cycle is also achieved via the **autonomic nervous system**. The sympathetic and parasympathetic divisions of the autonomic system regulate heart rhythm by affecting the same intrinsic impulse conducting mechanisms that lie within the heart in opposing ways.

Cardiac muscle is self-contractile because it is capable of generating a spontaneous electrochemical signal as it contracts. This signal induces surrounding cardiac muscle **tissue** to contract and a wave-like contraction of the heart can result from the initial contraction of a few localized cardiac cells.

The cardiac cycle describes the normal rhythmic series of cardiac muscular contractions. The cardiac cycle can be subdivided into the systolic and diastolic phases. Systole occurs when the ventricles of the heart contract and diastole occurs between ventricular contractions when the right and left ventricles relax and fill. The sino-atrial node (**S-A node**) and atrioventricular node (AV node) of the heart act as pacemakers of the cardiac cycle.

The contractile systolic phase begins with a localized contraction of specialized cardiac muscle fibers within the sino-atrial node. The S-A node is composed of nodal tissue that contains a mixture of muscle and neural cell properties. The contraction of these fibers generates an electrical signal that then propagates throughout the surrounding cardiac muscle tissue. In a contractile wave originating at the S-A node, the right atrium muscle contracts (forcing **blood** into the right ventricle) and then the left atrium contracts (forcing blood into the left ventricle).

Intrinsic regulation is achieved by delaying the contractile signal at the atrioventricular node. This delay also allows the complete contraction of the atria so that the ventricles receive the minimum amount of blood to make their own contractions efficient. A specialized type of neuro-muscular cells, named Purkinje cells, form a system of fibers that covers the heart and which conveys the contractile signal from S-A node (which is also a part of the **Purkinje system** or subendocardial plexus). Because the Purkinje fibers are slower in passing electrical signals (action potentials) than are neural fibers, the delay allows the atria to finish their contractions prior to ventricular contractions. The signal delay by the AV node lasts about a tenth (0.1) of a second.

The contractile signal then continues to spread across the ventricles via the Purkinje system. The signal travels away from the AV node via the bundle of His before it divides into left and right bundle branches that travel down their respective ventricles.

Extrinsic control of the heart rate and rhythm is achieved via autonomic nervous system (ANS) impulses (regulated by the **medulla oblongata**) and specific **hormones** that alter the contractile and or conductive properties of heart muscle. ANS sympathetic stimulation via the cervical sympathetic chain ganglia acts to increase heart rate and increase the force of atrial and ventricular contractions. In contrast, parasympathetic stimulation via the vagal nerve slows the heart rate and decreases the vigor of atrial and ventricular contractions. Sympathetic stimulation also increases the conduction velocity of cardiac muscle fibers. Parasympathetic stimulation decreases conduction velocity.

The regulation in impulse conduction results from the fact that parasympathetic fibers utilize acetylcholine, a neurotransmitter hormone that alters the transmission of an **action**

potential by altering membrane permeability to specific ions (e.g., potassium ions (K+). In contrast, sympathetic postganglionic **neurons** secrete the neurotransmitter norepinephrine that alters membrane permeability to sodium (Na+) and **calcium** ions Ca2+).

The ion permeability changes result in parasympathetic induced hypopolarization and sympathetic induced hyperpolarization.

Additional hormonal control is achieved principally by the adrenal glands (specifically the adrenal medulla) that release both **epinephrine** and norepinephrine into the blood when stimulated by the **sympathetic nervous system**. As part of the **fight or flight reflex**, these hormones increase heart rate and the volume of blood ejected during the cardiac cycle.

The electrical events associated with the cardiac cycle are measured with an electrocardiogram (EKG). Disruptions in the impulse conduction system of the heart result in arrhythmias.

See also Coronary circulation; Heart defects; Heart, embryonic development and changes at birth; Nerve impulses and conduction of impulses; Nervous system overview

HEAT STROKE • *see* HYPERTHERMIA

HELMHOLTZ, HERMANN (1821-1894)
German physicist and physician

Hermann Helmholtz was one of the few scientists to master two disciplines: medicine and physics. He conducted breakthrough research on the nervous system, as well as the functions of the **eye** and **ear**. In physics, he is recognized (along with two other scientists) as the author of the concept of conservation of energy.

Helmholtz was born into a poor but scholarly family; his father was an instructor of philosophy and literature at a gymnasium in his hometown of Potsdam, Germany. At home, his father taught him Latin, Greek, French, Italian, Hebrew, and Arabic, as well as the philosophical ideas of Immanuel Kant and J. G. Fichte (who was a friend of the family). With this background, Helmholtz entered school with a wide perspective. Though he expressed an interest in the sciences, his father could not afford to send him to a university; instead, he was persuaded to study medicine, an area that would provide him with government aid. In return, Helmholtz was expected to use his medical skills for the good of the government—particularly in army hospitals.

Helmholtz entered the Friedrich Wilhelm Institute in Berlin in 1898, receiving his M.D. four years later. Upon graduation he was immediately assigned to military duty, practicing as a surgeon for the Prussian army. After several years of active duty he was discharged, free to pursue a career in the academia. In 1848, he secured a position as lecturer at the Berlin Academy of Arts. Just a year later he was offered a professorship at the University of Konigsberg, teaching **physiol-**

ogy. Over the next twenty-two years he moved to the universities at Bonn and Heidelberg, and it was during this time that he conducted his major works in the field of medicine.

Helmholtz began to study the human eye, a task that was all the more difficult for the lack of precise medical equipment. In order to better understand the function of the eye he invented the ophthalmoscope, a device used to observe the **retina**. Invented in 1851, the ophthalmoscope, in a slightly modified form, is still used by modern eye specialists. Helmholtz also designed a device used to measure the curvature of the eye called an ophthalmometer. Using these devices he advanced the theory of three-color vision first proposed by Thomas Young. This theory, now called the Young-Helmholtz theory, helps ophthalmologists to understand the nature of color **blindness** and other afflictions.

Intrigued by the inner workings of the sense **organs**, Helmholtz went on to study the human ear. Being an expert pianist, he was particularly concerned with the way the ear distinguished pitch and tone. He suggested that the inner ear is structured in such a way as to cause resonations at certain frequencies. This allowed the ear to discern similar tones, overtones, and timbres, such as an identical note played by two different instruments.

In 1852, Helmholtz conducted what was probably his most important work as a physician: the measurement of the speed of a nerve impulse. It had been assumed that such a measurement could never be obtained by science, since the speed was far too great for instruments to catch. Some physicians even used this as proof that living organisms were powered by an innate "vital force" rather than energy. Helmholtz disproved this by stimulating a frog's nerve first near a muscle and then farther away; when the stimulus was farther from the muscle it contracted just a little slower. After a few simple calculations Helmholtz announced the impulse velocity within the nervous system to be about one-tenth the speed of sound.

After completing much of the work on sensory physiology that had interested him, Helmholtz found himself bored with medicine. In 1868 he decided to return to his first love—physical science. However, it was not until 1870 that he was offered the physics chair at the University of Berlin and only after it had been turned down by **Gustav Kirchhoff**. By that time, Helmholtz had already completed his groundbreaking research on energetics.

The concept of conservation of energy was introduced by Julius Mayer in 1842, but Helmholtz was unaware of Mayer's work. Helmholtz conducted his own research on energy, basing his theories upon his previous experience with muscles. It could be observed that animal heat was generated by muscle action, as well as chemical reactions within a working muscle. Helmholtz believed that this energy was derived from food and that food got its energy from the sun. He proposed that energy could not be created spontaneously, nor could it vanish; it was either used or released as heat. This explanation was much clearer and more detailed than the one offered by Mayer, and Helmholtz is often considered the true originator of the concept of conservation of energy.

While this was undoubtedly Helmholtz's greatest legacy, he also began several projects that were later completed by other scientists. He advanced a number of hypotheses on electromagnetic radiation, speculating that it lay far into the invisible ranges of the spectrum. This line of research was later resumed, very successfully, by one of Helmholtz's students, Heinrich Rudolph Hertz, the discoverer of radio waves. Helmholtz's theories on electrolysis were also the basis for future work conducted by Svante August Arrhenius.

Helmholtz had been a sickly child; even throughout his adult life he was plagued by migraine headaches and dizzy spells. In 1894, shortly after a lecture tour of the United States, he fainted and fell, suffering a concussion. He never completely recovered, dying of complications several months later.

See also Nerve impulses and conduction of impulses; Sense organs: Ocular (visual) structures; Sense organs: Otic (hearing) structures

HEMOGLOBIN

Hemoglobin is a protein found in red **blood** cells that functions in transporting oxygen to peripheral tissues of the body. Each red blood cell (RBC) has about 280 million hemoglobin molecules and roughly 2.5 million new RBCs are produced every second. Adult hemoglobin (HbA) is composed of four highly folded polypeptide chains including two alpha globin subunits and two beta globin subunits. Fetal hemoglobin (HbF), however, is made up of two alpha and two gamma globin subunits. The globin genes are expressed in a tissue-specific and developmentally regulated manner. Molecular cues cause a neonatal switch in the production of HbF to HbA. Normally, globin subunits are produced in equal amounts and permanently join together for the life of a RBC. Each polypeptide chain has an iron containing heme group that reversibly binds oxygen. Up to four molecules of oxygen can bind to one molecule of hemoglobin. Hemoglobinopathies (defects in hemoglobin) such as **sickle cell anemia** and thalassemia, represent genetic diseases that involve abnormal hemoglobin.

The ability of hemoglobin to bind to oxygen is a function of the partial pressure of oxygen (pO_2). Oxygen loading onto hemoglobin in the **lungs** is cooperative in that when the first oxygen molecule binds, it induces conformational changes in the structure of hemoglobin that allows other molecules of oxygen to bind more readily. Fully oxygenated hemoglobin, as it arrives at peripheral tissues, gives up approximately 25% of the oxygen bound to hemoglobin under normal resting conditions (where the pO_2 is typically 40 mm Hg). The other 75% represents an oxygen reserve that is only utilized during reduced pO_2 (below 40 mm Hg, such as during exercise). The allosteric interactions of the four polypeptide subunits results in an oxygen-equilibrium curve that is S-shaped, or sigmoidal.

Many factors can shift the oxygen-hemoglobin dissociation curve, such as pH, temperature, byproducts of cellular respiration, or CO_2. Lower pH levels decrease the affinity of hemoglobin for oxygen, a phenomenon known as the Bohr effect. For example, when muscle **tissue** contracts repeatedly, the cells utilize the oxygen reserves to produce more **ATP**. ATP production releases heat leading to an increase in temperature surrounding the active tissues. By-products of cellular respiration such as CO_2 and 2,3-diphosphoglycerate are also released, contributing to a reduction in the pH of the tissues (CO_2 reacts with water to produce carbonic acid). All these factors cause a shift to the right in the oxygen-hemoglobin dissociation curve.

See also Oxygen transport and exchange; Respiratory system

HEMOPOIESIS

The various different kinds of **blood** cells have a finite life span in the circulation. Red blood cells last for about 120 days, **platelets** for ten days and granulocytes can live for less than 24 hours. Monocytes only circulate from one to three days, but later migrate to the **tissue** to become the longer-lived macrophages. All these cells are continually replaced by the generative process called hemopoiesis.

Hemopoiesis is an extremely active process. In the adult human, approximately 10^{10} red blood cells and 4×10^8 white blood cells are produced per day. The principal site of hemopoiesis in the adult is the red marrow within bones, which occupies the medullary cavities of long bones and interstices in the spongiosa of vertebrae, ribs and **sternum**. Bone marrow occurs in two forms: red marrow, which is active in hemopoiesis, and yellow marrow, which is inactive and consists mainly of adipose cells, giving it its yellow color. Active red marrow is a soft, highly cellular tissue consisting of the precursors of blood cells supported by a stroma of reticular cells and associated reticular fibers. The reticular cells synthesize the collagen of the stromal reticular fibers. They, as well as the macrophages of the marrow, are believed to release cytokines, called colony-stimulating factors (CSFs), which promote proliferation and differentiation of the blood cell precursors. In the embryo, before bone marrow develops, blood cells are produced in the yolk sac during early embryonic life and in the fetal **liver** during the second trimester of **pregnancy**.

Blood cell formation depends on the existence of hemopoietic precursor cells in the bone marrow. All the cell types in mature blood are generated from such single pluripotential **stem cells**. These stem cells have the capacity to proliferate indefinitely and differentiate into many cell types. They can divide to renew themselves in order to maintain the pool pluripotential cell pool, while at the same time generating cells that become committed to developing into the various blood cell lines. First the stem cells divide to become either lymphoid or myeloid precursors. The lymphoid precursors migrate to lymphoid tissues, where they divide and differentiate into B and **T lymphocytes**. The mixed myeloid precursors remain in the bone marrow. There they divide further and their daughter cells become committed to producing one of the several types of blood cells, i.e., over several divisions they become unipotential cells. Sometimes, specific names are given to the development of particular cell lines. For example, erythropoiesis is the process of red blood cell formation from proerythroblasts which are large, early cells committed to erythrocyte development. Also granulopoiesis is the formation of granulocytes from the myeloblast cells. Thrombopoiesis is the formation of blood platelets, or thrombocytes, the cellular elements that promote blood clotting. Platelets are generated in the bone marrow by fragmentation of the cytoplasm of mature megakaryocytes. Like red blood cells, blood platelets lack nuclei.

The cells in the early stages of hemopoietic development are generally larger than those in more advanced, later stages. As they become committed to developing into one particular cell line, they express only the genes characteristic of that line. In some lines, they stop transcribing genes altogether. Morphological changes accompany this commitment and include a visible decrease in nuclear euchromatin and increase in heterochromatin. Also, ribosomes are most actively produced in the early stages of differentiation and decrease gradually as the cells become committed.

Pluripotential hemopoietic cells are relatively few in number (1 out of 10,000). The evidence for their existence came from experiments with bone marrow transplantation. Mice were irradiated to destroy their own bone marrow and were transfused with the bone marrow of healthy mice. Two weeks later, the spleens of the transfused mice were examined and found to contain colonies of cells resulting from the proliferation of the newly transfused cell type. Each colony was found to contain precursors to all the different blood cells present in peripheral blood. When isolated single colonies were further transfused into irradiated mice, each gave rise to several colonies, all containing cells of all the different lineages. These studies showed that one pluripotential cell forming one colony can give rise to all the lines of cells differentiating into the various blood cells. Because they are studied in the spleen colony assay, hematopoietic precursors are sometimes called colony forming units, or CFUs. As the pluripotential cells become committed, it is possible to distinguish progenitor cells of 4 kinds: (1) colony-forming unit granulocyte-monocyte (CFU-GM), (2) colony-forming unit erythrocyte (CFU-E), (3) colony-forming unit lymphocyte (CFU-L) and (4) colony-forming unit megakaryocyte (CFU-Me).

See also Bone histophysiology; Erythrocytes; Leukocytes; Leukemia

HEMORRHAGIC FEVERS AND DISEASES

A hemorrhagic fever is a **viral infection** that features a high fever and a high volume of (copious) bleeding. The bleeding is caused by the formation of tiny **blood** clots throughout the bloodstream. These blood clots—also called microthrombi—deplete **platelets** and fibrinogen in the bloodstream. When bleeding begins, the factors needed for the clotting of the blood are scarce. Thus, uncontrolled bleeding (hemorrhage) ensues.

Several tropical fevers fall into this category. These include Dengue hemorrhagic fever (caused by the Dengue virus), Korean hemorrhagic fever (caused by the Hantaan virus), Lassa fever (caused by the Lassa virus), Marburg fever (caused by the Marburg virus), Rift Valley fever (caused by the Rift Valley fever virus) and Ebola fever (caused by the Ebola virus). There is no cure for these infections. Treatment consists of keeping the patient as comfortable and infused with blood and fluids as possible. If antiserum is available, it is given but results are often uncertain.

Hemorrhagic fevers are caused by four distinct families of ribonucleic acid-containing **viruses**: arenaviruses (Lassa virus), filoviruses (Ebola and Marburg viruses), bunyaviruses (Rift Valley Fever virus), and flaviviruses. All these viruses depend on an animal or insect host for their survival and so are normally restricted to the area of the world in which that particular host lives. Although the hosts of some of these viruses are known, the natural host(s) for the Ebola and Marburg viruses, for example, remain undiscovered.

Hemorrhagic fevers are examples of so-called emerging infections—those infections that have come to prominence within the past two decades—often because of human encroachment on a formally isolated animal population harboring the virus. When this animal to human transmission occurs, subsequent human to human transmission can occur. Spread of the viruses can be direct, via body fluids, or indirect, such as via a contaminated syringe.

Symptoms of the various diseases vary, but common symptoms are a high fever, fatigue, dizziness, muscle aches, and loss of strength. As the patient deteriorates, marked bleeding under the skin, internally, and from body orifices like the mouth, ears and eyes develops. **Coma**, seizures and kidney failure can cause **death**.

In the few known outbreaks of hemorrhagic fever, the disease flares up, wrecking havoc in the community where the outbreak occurs, and then disappears. Why the viral agents of hemorrhagic fevers are so devastating is not understood. Nor is it yet understood why the diseases do not persist for long in a population. One explanation, that for highly lethal forms such as Ebola, is that the disease essentially works so quickly—and with such a high lethality—that if it occurs in an isolated population, it can destroy all available hosts. The lack of available hosts inhibits the spread of the virus.

The best strategies for dealing with hemorrhagic fevers at the present time is to limit human contact with the animal or insect hosts of the viruses, and in trying to ensure that the home and workplace is free from potential viral hosts, like rodents. Researchers are striving to develop better containment and treatment strategies, including vaccines, for the viral agents of hemorrhagic fevers. Another goal is to devise rapid detection techniques utilizing immunologic and molecular tools.

See also Fever and febrile seizures

HENCH, PHILIP SHOWALTER (1896-1965)

American physician

Philip Showalter Hench, an American clinical pathologist, performed groundbreaking research in rheumatoid arthritis. His clinical tests of adrenal compound E, which Hench named

cortisone, and of ACTH, which produces cortisone naturally by stimulating the adrenal cortex, offered the first hope for patients suffering from rheumatoid arthritis. Hench and his colleague, biochemist Edward C. Kendall, gained immediate worldwide attention when they filmed the miraculous recovery of arthritis patients—some of whom could barely walk—as they climbed stairs and even jogged in place. Although prolonged clinical trials showed that neither cortisone nor ACTH was a viable long-term therapy for arthritis due to side effects such as high **blood** pressure and high glucose levels, Hench's efforts opened new vistas in medical research, particularly in the study of both **hormones** and rheumatoid arthritis. A meticulous researcher who methodically collected his clinical data before publishing his results, Hench shared the 1950 Nobel Prize in physiology or medicine with Kendall "for their discoveries relating to the hormones of the adrenal cortex, their structure and biological effects." (Chemist **Tadeus Reichstein** also received a share of the prize for his independent work with the adrenal cortex and its hormones.)

Hench was born in Pittsburgh, Pennsylvania. The son of Jacob Bixler Hench, a classical scholar and school administrator, and Clara John Showalter, Hench attended a private high school, Shadyside Academy, and then enrolled at the University of Pittsburgh in 1916. His education looked as though it would be interrupted when he enlisted in the U.S. Army Medical Corps. Hench was, however, transferred to the reserves so he could return to his studies, and he enrolled in Lafayette College in Easton, Pennsylvania. He received his B.A. from Lafayette in 1916, and enrolled at the University of Pittsburgh School of Medicine, where he received his M.D. in 1920.

After completing his internship at St. Francis Hospital in Pittsburgh, Hench became a fellow in medicine at the Mayo Foundation of the University of Minnesota. The bright young physician and scientist would spend his entire career at the Mayo Clinic, where, in 1926, he cofounded the Department of Rheumatic Disease, which was the first training program in rheumatology in the United States. Hench spent the 1927–28 academic year on sabbatical studying research medicine with Ludwig Aschoff, a leading rheumatic fever investigator, at Freiburg University. He also studied with clinician Freidrich von Müller in Munich. Hench completed his formal education in 1931 when he received a master of science degree in internal medicine from the University of Minnesota.

A physician first, Hench's research was clinically based. He began studying rheumatoid arthritis in 1923. Unlike osteoarthritis, a degenerative joint disease common in later life, rheumatoid arthritis is a chronic inflammatory disease of the **joints** often contracted at the relatively young age of 30 to 35. In advanced stages, rheumatoid arthritis could cause deformity due to bone and surrounding muscle atrophy. In 1929, Hench took note of a patient who had suffered from severe arthritis for more than four years. The patient had entered the Mayo Clinic suffering from jaundice, a disease caused by excess **bilirubin**, a **liver** product, in the bloodstream. Amazingly, the man's arthritis had abated and remained dormant for several weeks after his recovery from jaundice. Carefully collecting data, Hench waited until he had authenticated nine similar cases, among them patients who experienced remissions from painful fibrositis and sciatica, two other inflammatory conditions, before publishing his data in 1933.

Hench was convinced that these cases held a vital clue to a therapy for arthritis and set out to induce jaundice artificially. Hench's initial experiments used infusion or ingestion of **bile** to emulate jaundice's production of excess bile in the blood or the liver. Although these experiments failed, Hench's attention was soon drawn to another group of patients, women whose arthritis vanished during **pregnancy**. He also observed that some arthritic patients went into less complete remission after surgical operations, **anesthesia**, or severe fasting. Looking for a common physiological denominator, Hench, who enjoyed reading Sir Arthur Conan Doyle's novels of Sherlock Holmes, had a prime suspect—glandular hormones. Furthermore, the fact that both jaundice and pregnancy caused remission in almost the exact same manner led Hench to believe that his missing compound was not bilirubin or a female-only sex hormone.

Fortunately, the Mayo Foundation's own Edward C. Kendall was a world-renowned chemist in the field of steroids, a specific group of hormones. Kendall had isolated six steroids from the adrenal cortex, the outer part of the endocrine glands located atop the **kidneys**, which he alphabetized compound A through F. Hench's first try with compound A was a failure. Both Hench and Kendall then decided to try compound E. But at that time, in 1941, compound E was extremely difficult to synthesize and, as a result, costly. With both high (300°F [148.9°C]) and low (–100°F [-73.3°C]) temperatures needed to produce compound E, the delicate work took time and attention and the slightest mistake could result in a useless compound. It wasn't until more than two years after World War II that scientists from the pharmaceutical firm Merck & Co. had developed a process that could produce enough compound E for Hench to attempt his experiment. Still, the compound was expensive to produce. Hench recalled in an interview for an article in the *Saturday Evening Post* that he and his colleagues "almost went into shock" when a $1,000 bottle of compound E was dropped on a marble floor.

Hench's results with compound E were miraculous. The first patient, a 29-year-old woman, experienced total remission of symptoms after three injections over three days. Hench's results were quickly confirmed by five other researchers across the country. Hench and his colleagues received instant public notoriety as a result of their studies both with compound E, which Hench named cortisone, and with adrenocorticotropic hormone (ACTH), a hormone produced by the pituitary gland which spurs the body's natural production of cortisone through the adrenal cortex.

Unfortunately, Hench's miraculous "cure" for arthritis turned out to be short lived. Without the use of cortisone or ACTH, rheumatic symptoms returned; and long-term use of cortisone or ACTH causes several side effects, including high blood glucose and high blood pressure, as well as obesity associated with adrenal or pituitary gland **tumors**. Much to Hench's credit, he maintained his scientific cautiousness throughout the heady early days of the discovery, quickly rec-

ognizing the harmful side effects and outlining future directions in research of these hormones. Nevertheless, the studies of Hench and Kendall, along with those of Tadeus Reichstein, opened entirely new avenues of medical research; as a result, the three scientists were awarded the Nobel Prize in physiology or medicine in 1950.

Hench retired from the Mayo Foundation in 1957. In addition to the Nobel Prize, he was a recipient of the numerous awards, including the prestigious Lasker Award, which he also shared with Kendall. Hench married Mary Genevieve Kahler in 1927, and the couple had two sons and two daughters. His hobbies included photography, tennis, opera, and Sherlock Holmes novels. He died from pneumonia in 1965 while vacationing with his wife in Ocho Rios, Jamaica. To honor him, Hench's alma mater, the University of Pittsburgh, presents the annual Hench Award to a distinguished university alumnus.

See also Arthrology (joints and movement); Autoimmune disorders; Hormones and hormone action

HEPATIC PORTAL SYSTEM · *see* ABDOMINAL VEINS

HEROPHILUS (CA. 335 B.C.-CA. 280 B.C.)
Greek physician

Sometimes called the father of **anatomy**, Herophilus was born in Chalcedon, Asia Minor. Little is known about his life; the date and place of his death have been completely lost, as have all his writings. Herophilus studied medicine under Praxagoras of Cos and then at Alexandria, where he later taught and practiced medicine. In Alexandria, Herophilus had the unique opportunity to practice human dissection, a research technique not allowed elsewhere. Herophilus even performed public dissections. His work was highly regarded, and the medical school he founded at Alexandria attracted scores of students.

Herophilus made many anatomical studies of the **brain**. He distinguished the cerebrum (larger portion) from the cerebellum (smaller portion), pronounced the brain to be the seat of intelligence, and identified several structures of the brain, several of which still carry his name. He discovered that the nerves originate in the brain, was the first to distinguish nerves from **tendons**, and noted the difference between motor nerves (those concerned with motion) and sensory nerves (those related to sensation). He traced the optic nerve and described the **retina**. He studied the **liver** extensively and described and named the duodenum, the first part of the small intestine.

Drawing a distinction between **arteries** and **veins**, Herophilus noted the arterial **pulse** and developed standards for its measurement and use in diagnosis. He thought that arterial pulsation was involuntary, rising from dilation and contraction of arteries due to impulses sent from the **heart**. He corrected the idea that arteries carry air rather than **blood**.

Herophilus also wrote a treatise on midwifery and accurately described the ovaries, the uterus, and the tubes leading from the ovaries to the uterus (later named the Fallopian tubes). In the field of medical treatment, Herophilus sensibly recommended good diet and exercise, but was also an enthusiastic advocate of bleeding and frequent drug therapy.

Although his medical school languished after his death, Herophilus (and his younger successor **Erasistratus**) established the disciplines of anatomy and **physiology**, which did not significantly advance before **Galen** in the second century, and until the early modern anatomists of the thirteenth century.

See also History of anatomy and physiology: The Classical and Medieval periods

HESS, WALTER RUDOLPH (1881-1973)

Swiss physiologist

Walter Rudolf Hess won the Nobel Prize in physiology or medicine in 1949, for his work in analyzing the function of the diencephalon, part of the interbrain, and its role in coordinating the body's internal **organs**. Introduced to the natural world by his father while still a very young child, Hess later wrote in an autobiographical sketch in *A Dozen Doctors:* "As time went on, I became aware of the significance of the ecological setting... of the specific interrelationship between flora and fauna.... More and more it became clear that functional manifestations, such as the germination of a seed or the rapid sprouting of a shoot from a willow, were more apt to capture my mind than purely morphological features." This emphasis on function and relationships carried over into much of Hess's work, particularly his investigations into the biological basis of emotions and the workings of the circulatory and respiratory systems. Despite the interference of two world wars, he designed many elegant experiments for studying physiological processes in living organisms.

Hess was born in the Swiss town of Frauenfeld to Clemens and Gertrud (Fischer Saxon) Hess. From his father, a physics teacher, he inherited a strong interest in science, and from his mother an energetic, good-humored personality. After finishing high school, Hess began his college career, changing universities frequently and taking every opportunity to travel. He eventually received a medical degree from the University of Zurich in 1905, and took a hospital residency under the famous surgeon Dr. Konrad Brunner.

While working for Brunner, Hess designed an improved **blood** viscometer (to measure blood's thickness and consistency) and began thinking about research in earnest. He took a second residency in Zurich and specialized in ophthalmology (the physiology and diseases of the **eye**) under the mistaken impression that the discipline would allow him time to continue his **circulatory system** investigations. He indeed developed a successful ophthalmology practice with a good income, but it took up all of his time. In 1912 Hess gave up his practice and moved to the Institute of Physiology in Zurich, where he was given considerable freedom of action. Eventually he

was named chair of the physiology department, and began traveling to conferences and meetings throughout Europe. The stresses inherent in administrative work (which included a severe fire at the Institute, and the design and construction of a mountaintop research facility) and World War I cut into his research time again, but he still managed to publish two important monographs, *The Regulation of the Circulatory System* in 1930, and *The Regulation of Respiration* in 1931.

Hess brought an unusual variety of tools and skills to his research. He had learned the basic principles of physics from his father, he knew a great deal about optics and hand-eye coordination from his days as an ophthalmologist, and he was a skilled surgeon. These all proved useful when he began conducting **brain** research on experimental animals. Hess's work on the circulatory and respiratory systems had included investigations of their interrelationship with other parts of animal physiology, including how blood flow and **breathing** were affected by the nervous system. Gradually this led to research on the areas of the brain responsible for regulating internal organs.

Of particular interest to Hess was the diencephalon, which is located under the **cerebellum** and is thus very difficult to access without damaging the rest of the brain. Hess designed very small electrodes and a mechanical guidance system that could implant the electrodes in experimental animals (cats) with the least possible disruption of their normal behavior. He also designed a method of delivering electrical stimulus pulses swiftly and accurately. On at least one occasion there was a public outcry about the use of animals for experimentation. Hess was instrumental in convincing the activists that, if properly regulated and humanely conducted, animal experiments were important for human welfare.

Using the electrodes to stimulate different areas of the brain, Hess observed the results on other areas of bodily function, such as blood pressure, respiration, and body temperature. He recorded his observations not only on paper, but also on film, and maintained meticulous records of dissections and cell studies. He also compared the results of electrical stimulation with behaviors resulting from naturally occurring brain lesions. He found that the diencephalon, and particularly the **hypothalamus**, controlled many of the body's responses, such as fear and hunger, and he was able to map out some of these responses in detail. Partly due to the isolation imposed by World War II, and partly because his papers were written entirely in German, the outside world knew little of his work until he had accumulated about 25 years worth of experiments. This may have been fortunate, because, as he wrote in his sketch, "The vast number of experiments turned out to be decisive; for generalization concerning symptoms, syndromes, and localizations could be supported only by such a large body of data."

In 1949, Hess won a share of the 1949 Nobel Prize in physiology or medicine; Portuguese neurosurgeon Antonio **Egas Moniz** shared the award for his work on white brain matter. The presenter said in his speech that Hess's results demonstrate "that in the midbrain we have higher centers of autonomic functions which coordinate these with reactions of the skeletal musculature adapted to the individual functions.... Through his research, Hess has brilliantly answered a number of difficult questions regarding the localization of body functions in the brain." Other recognitions the physiologist received included Switzerland's Marcel Benorst Prize in 1933 and the German Society for Circulation Research's Ludwig Medal in 1938.

Hess married the former Louise Sandmeyer in 1908; the couple had two children. When not working or traveling, Hess relaxed at his country house in southern Switzerland, where he cultivated grapes, tended his garden, and absorbed the pleasantries of Italian culture. He retired in 1951, although he continued his work and was instrumental in the establishment of an institute for brain research. He died in Locarno, Switzerland in 1973.

See also Autonomic nervous system; Brain stem function and reflexes; Brain: Intellectual functions

HEYMANS, CORNEILLE JEAN-FRANÇOIS
(1892-1968)
Belgian pharmacologist and physician

Corneille Jean-François Heymans, a Belgian scientist, conducted research in the field of respiratory and cardiovascular systems that produced new knowledge about the way **breathing** is regulated. His work won him the Nobel Prize in physiology or medicine in 1938.

Born in Ghent, Belgium, Heymans was the eldest of six sons of Jan-Frans Heymans, a noted pharmacologist who founded the J. F. Heymans Institute of Pharmacology and Therapeutics at the University of Ghent. As a youngster, Heymans watched the construction of the institute laboratory and helped his father take care of the animals that were used there. He and his father were to become a scientific team of considerable reputation—one of the few father-son scientific teams in history.

Heymans's career was delayed by four years of service as a field artillery officer in the Belgian Army during World War I. His performance won him the Belgian War Cross and the Order of the Crown of Leopold, among other decorations for valor.

After the war, Heymans received his medical degree from the University of Ghent in 1920. His father was his principal teacher and later would become his primary co-researcher in the experiments that ultimately led to the Nobel Prize in 1938. Had his father not died in 1932, he most likely would have shared the award with his son.

The year following his graduation from the university, Heymans married Berthe May, an ophthalmologist. The young couple studied abroad for several years, permitting Heymans to establish valuable contacts with some of the leading scientists of the day in his field, among them Eugène Gley at the Collège de France, Maurice Arthus at the University of Lausanne in Switzerland, Ernest H. Starling at University College in London, and Carl Wiggers at Western Reserve University's medical school in Cleveland, Ohio.

Heymans returned to the University of Ghent in 1922 to become a lecturer in pharmacodynamics, the study of the action of drugs in the body. He succeeded his father as professor of pharmacology and director of the Institute in 1930, but father and son continued to collaborate on many projects, including respiratory experiments that revealed previously unknown facts about how breathing is regulated in human beings and animals.

At that time, it had been well known for half a century that changes in **blood** pressure were associated with changes in the rate and the depth of breathing. The mechanism enforcing these changes in respiration was not known. It was believed, however, that alterations of breathing rates were the result of the direct action of blood pressure on the brain's respiratory center, the medulla. It was assumed that the medulla was able to detect changes in the blood circulating through it and regulate the rate of breathing accordingly.

Another scientist, Heinrich E. Hering, however, had noted a reflex action in the carotid artery (two major **arteries** on each side of the neck) that appeared to influence the **heart** beat. Through a series of experiments originally intended to refute Hering's contention, Heymans instead demonstrated that that the reflex in the artery also exerted control over breathing.

The effort to determine this fact involved what became known as the "isolated head" technique. The head of an anesthetized dog, attached to its body only by the vagus aortic nerves, was kept alive by the shared circulation of blood of a second anesthetized dog. The Heymans found that when they induced hypertension (increased blood pressure) in the isolated body of the first dog its medullary respiratory center was stimulated or inhibited appropriately. But when the aortic nerves were severed, all respiratory response to changes in the blood pressure ceased. This experiment enabled the Heymans team to demonstrate conclusively that the aortic nerves were the reflex mechanism's sole sensory pathway.

The experiment thus disproved the classical theory of the blood's direct action on the **brain** and provided the evidence for an alternative explanation. The Heymans later determined the sites at which changes in the blood were detected. They discovered that the reflex in the carotid artery contains pressure-sensitive areas, or presso-receptors, that can detect even slight changes in blood pressure. They also found small structures on the inside walls of the carotid artery and the aorta. These chemoreceptors responded to changes in the chemical composition of the blood. By making clear why certain drugs affected respiration and circulation, Heymans's discovery opened the way for improvements in the treatment of many diseases.

Heymans's colleagues appreciated the thoroughness and accuracy of his work, which he documented in over eight hundred articles and papers published during his career. Heymans also won great recognition as a gifted teacher. Unlike most professors of his time, who tended to remain aloof from their students, Heymans regarded his students as fellow professionals and followed their work closely. Keeping track of their careers after they had left the university, he further inspired and encouraged their individual progress by sending them new data, providing help and support as they needed it.

Many scientific honors came to him. In addition to the Nobel Prize, he was awarded the Alvarenga Prize of the (Belgian) Académie Royale de Medécine, the Prix Quinquennal de Medécine of the Belgian government, the Pius XI Prize of the Pontificia Academia Scientiarum and the Monthyon Prize of the Institut de France. Heymans held sixteen honorary degrees and belonged to more than forty scientific and medical societies.

Throughout his career, he traveled widely both as a lecturer and a tourist. He lectured at several major American universities, including Harvard and the University of Chicago. He was fluent in many languages and conducted seminars in Montevideo, Chile, to help organize scientific exchange programs between that country and his own. He visited India on behalf of the World Health Organization (WHO).

During World War II, he helped organize relief efforts to provide food for Belgian children. In so doing, he made several trips to Berlin to obtain the cooperation of German officials in getting Red Cross food shipments into Belgium. This evoked criticism from some Belgians who viewed his activities as collaboration with the enemy, but Heymans maintained that his actions were necessary to obtain vital assistance for his country.

Heymans and his wife had four children. In 1963, upon his retirement from the Heymans Institute, he was designated professor emeritus. He continued to visit the institute several times a week until his death following a stroke in Knokke, Belgium at the age of 76.

See also Autonomic nervous system; Blood pressure and hormonal control mechanisms; Brain stem function and reflexes; Respiration control mechanisms

HIERARCHY OF STRUCTURAL ORGANIZATION

The hierarchy of structural organization is a framework for understanding **anatomy** and **physiology**. The **structure and function** of the body may be examined at several levels of resolution: the biochemical, cellular, **tissue**, organ, system, and organism levels.

At the biochemical level, atoms combine to form molecules such as water, **carbohydrates**, **lipids**, proteins, and nucleic acids. These molecules each have crucial functions in the body such as transporting energy and information, maintaining the internal **equilibrium**, or homeostasis, and assembling themselves into larger structures such as cells and their components called organelles.

The cell is the basic unit of life. All living things consist of one or more cells. Cells are capable of taking in and using nutrients, and eliminating wastes. They can move, develop and grow, and reproduce. In a multicellular organism, cells may specialize to perform certain functions, becoming in turn dependent on other cells for those functions they have given up.

A collection of cells with similar or related functions forms a tissue. There are four main types of tissues: epithelial tissues cover the surfaces of the body, and line the body cavities and passages; **connective tissues** support and protect the body and attach one part to another; muscular tissues, which are capable of contracting, provide the function of movement. Nervous tissues form and conduct the impulses that control the workings of the body.

An assembly of two or more tissues organized to conduct a specific function is called an organ. Thee are many types of **organs**; the **liver**, **heart** and **brain** are examples. Two or more organs working together to perform a function comprise an organ system. Examples include the digestive system, the cardiovascular system, and the nervous system.

An organism such as a human being is made up of many organ systems. Together, the organ systems provide all the functions for the organism to develop, survive, and reproduce.

See also Anatomy; Cell theory; Organs and organ systems

HILL, ARCHIBALD (1886-1977)
English physiologist

The 1922 Nobel Prize in physiology or medicine recognized Archibald Hill for his discoveries relating to heat production and oxygen use in muscles. Prior to this distinction Hill was knighted for his military work during World War I and elected a member into the Royal Society, both in 1918. He represented Cambridge University in Parliament during World War II, and also served on the War Cabinet Scientific Advisory Committee. It was only after the war, and his retirement in 1952, that Hill returned to research into the physiology of the muscles.

Sir Archibald Vivian Hill was born in Bristol, England, into a family that had been in the lumber business for five generations. Hill's mother, Ada Priscilla Rumney Hill, raised Hill and his younger sister after their father, Jonathan Hill, deserted the family when Hill was three. Until age seven, his mother educated him at home, but when the family moved to nearby Weston-super-Mare, Hill was placed in a preparatory school of modest size. In 1899, the family moved to Tiverton, Devonshire, where Hill received the training he would need to enter college. At Blundell's School, he demonstrated exceptional abilities in mathematics, joined the debating team, and ran long-distance. Hill's sister Muriel later became a biochemist.

In 1905, Hill received a scholarship to study mathematics at Trinity College, Cambridge. There he completed a three-year course in two years, but found that even though he performed well in his chosen subject, he lacked sufficient interest or motivation to develop further as a mathematician. His tutor, Walter Morley Fletcher (1873–1933), who was a physiologist, had been working with Frederick Gowland Hopkins researching the chemistry of frog muscle physiology. Fletcher and Hopkins had discovered the importance of **lactic acid** in **muscle contraction**. Fletcher advised Hill to change from mathematics to physiology, even though Hill had finished third in his class in mathematics. Fletcher correctly believed that Hill's scientific curiosity was stronger than his urge to become a mathematician.

Hill graduated from Trinity College with a medical degree in 1907 but remained there for the next seven years doing research until war broke out in 1914. In 1909, he completed his examinations in natural science with honors and began research at the Cambridge Physiological Laboratory. The director of the laboratory, J. N. Langley, suggested that Hill expand the work of Fletcher and Hopkins on the chemistry of muscle contraction by the production of heat in the process of muscle contraction. Using a thermocouple recorder, which is a measuring device that records minute changes in heat temperatures, Hill was able to establish the basic procedures in this early work for his later discoveries which were to eventually earn him a Nobel Prize in physiology or medicine.

In 1911, after receiving a fellowship from Trinity College the year before, Hill visited Germany where his techniques for measuring heat changes in muscles produced by contraction were significantly improved. Two German scientists, Karl Burker and Friedrich Paschen, showed him how to improve his use of the thermocouple and galvanometer, instruments that allowed him to measure minute changes in electric current with greater degrees of accuracy.

In the three years following his trip to Germany, Hill continued at Cambridge, observing heat quantities produced by muscular contractions and recording those chemical changes taking place in the muscle, enabling it to do mechanical work. By 1913, Hill was able to demonstrate that when the muscle starts to contract, a small amount of heat is produced, while after the initial phase of the contraction, more heat develops, though at a slower rate and in much greater measure. He also showed at this time that molecular oxygen is used after the contraction takes place rather than at the time of the contraction itself.

The discovery that molecular oxygen is not required in the initial phase was demonstrated by placing muscle fibers in an atmosphere that excluded oxygen and instead used pure nitrogen. The production of heat at the onset—despite the absence of oxygen—indicated that oxygen was not necessary for the initiation of the contraction. However, the recovery phase did not take place, suggesting an energy exchange as a critical factor in heat being generated during the post contraction phase.

Hill's discoveries of how heat production and oxygen function in muscle **tissue** opened the way for a clearer understanding of the earlier work of Fletcher's and Hopkin's experiments. They had shown the formation of lactic acid in frog muscle during the contraction process, and observed its elimination when oxygen was present. Hill believed that the initial heat he noted was produced by a lactic acid formation from a precursor substance. The heat generated in the recovery phase signified the removal of lactic acid through oxidation.

Hill's muscle research was interrupted by World War I. He served first as a captain, then became a major in the Cambridgeshire Regiment. He was also commissioned by the government to develop a program for the improvement of antiaircraft ordinance. After the war, Hill returned to Cambridge to continue his work on muscle physiology. In 1920, he accepted an appointment to the Brackenbury Chair of Physiology at Manchester University. There he challenged the accepted view of heat production in one phase only, or at the time of contraction, establishing the two phases of heat production in muscle contraction from his work with frogs' thigh muscles.

In order to develop further his findings relating to oxidation and muscle contraction, Hill did research with **Otto Meyerhof**, a German-American biochemist. Meyerhof had been studying the physiology of muscle contraction through chemical rather than mechanical dynamics and had been able to identify **glycogen** as the precursor of lactic acid. Hill theorized that since the heat produced by the two phases of muscular contraction was not enough to eliminate all the lactic acid, it probably changed back to its precursor. Meyerhof demonstrated that the energy levels created by the oxidation of the lactic acid were enough to reconvert it back to glycogen. The agreement of their work thus assured the validity of their research.

For their cooperative work in muscle physiology, Hill and Meyerhof shared the 1922 Nobel Prize in physiology or medicine. In his Nobel speech, Hill emphasized the need for continuing research in the area of muscle physiology. He reminded the audience of the complexity of the subject, of the different approaches to experimentation, and of the need for improved instruments with which to perform meaningful study and research.

In 1923, Hill received an appointment to University College, London, and became Foulerton Research Professor of the Royal Society. He continued his work in muscle physiology, but this time he turned his attention to the role of lactic acid build-up in human muscles. He found that with moderate exercise there is enough oxygen to remove the excess lactic acid, but with heavy exercise there is an excess amount of lactic acid build-up due to what he called an *oxygen debt*. This debt can be made up by deep **breathing** or allowing enough time at rest to allow for the absorption of the lactic acid surplus.

In the 1930s, with the rise of Hitler in Germany, Hill voiced his protest against the anti-Semitic policies of the Nazis against Jewish scientists, as well as anti-Nazi scientists. He helped form groups to assist researchers escaping Nazi oppression. In World War II, Hill again performed vital services on behalf of his country's military objectives, as he had done in World War I. Two major accomplishments of this period include his coordination of efforts to gain military cooperation with Canada and the United States, and a report he composed as a visitor to India recommending the restructuring and reorganization of the country's scientific and industrial resources. At the war's close, Hill reorganized his laboratory and recruited a staff for research at University College. He retired in 1952, but continued his scientific investigations in the area of muscle physiology.

Hill married Margaret Neville Keynes in 1913. She was a social worker and the sister of John Maynard Keynes, a well-known English economist. The couple eventually had two sons and two daughters. Hill died of a viral infection followed by complications at age 90.

See also Muscular innervation; Muscular system overview

HIPPOCAMPUS

The cerebral structure known as the limbic system is composed of the **hypothalamus** and other subcortical structures, such as the hippocampus, the amygdala, portions of the **basal ganglia**, etc. Different areas of the hippocampus respond to most sensory stimuli such as olfactory, tactile, and visual sensations. These sensory signals are transferred from the hippocampus to other cortical and basal structures, triggering basic behavioral responses such as pleasure, anger, fear, sexual drive, or curiosity. Some hippocampal areas are associated with attention and the learning process in both animals and humans, and determine the degree of importance or priority of a given sensation to other areas of the **brain**. In animals, this brain structure is crucial to the decision-making process, such as food choice, recognition of dangerous situations, mating-related identification, inviting smells, and behavior. In humans, the hippocampus is associated with the ability of verbal symbolism recognition, symbolic thinking, decision-making, capability of learning and remembering new data, and the transformation of short-term memory into long-term memory.

Clinical observation shows that lesions of the hippocampus negatively affect this memory consolidation process and its surgical removal impedes the learning of new information or even the ability to remember names, although older memories and learned information anterior to the hippocampus' removal remain intact. The gradual deterioration of the ability to consolidate short-term into long-term memory that occurs during the **aging process** is an example of the role of the hippocampus as a mediator of this process. Brain autopsy in cadavers of elderly people show the presence of a great number of beta-amyloid plaques in the **cerebral cortex**, hippocampus, and other brain structures, which indicates neuronal **death**. Loss of **neurons** in the hippocampus due to aging or disease is believed to be the cause of many elderly people's difficulty to learn and/or remember recent events, although they often retain the ability to remember in detail facts, names, and data of decades ago.

See also Aging processes; Brain stem function and reflexes; Brain: Intellectual functions; Cerebral hemispheres; Limbic lobe and limbic system

HIPPOCRATES (460 B.C.-377 B.C.)

Greek physician

Hippocrates is widely known as the "Father of Medicine." Although little is known about his life, a few facts are considered accurate.

Hippocrates was born on the Greek island of Cos to a family of physicians. Cos was the site of one of the great medical schools of ancient Greece, and Hippocrates taught there for many years. He also traveled widely, lecturing in Greece and probably throughout the ancient Mideast. He was well known in his lifetime, and died in Larissa.

Hippocrates of Cos. *The Library of Congress.*

Hippocrates is considered the father of medicine because, through his school, he separated medical knowledge and practice from myth and superstition basing them instead on fact, observation, and clinical experience. Our knowledge of Hippocrates's methods and teachings comes from the *Corpus Hippocraticum*, the Hippocratic Collection. This is a series of about 60 books that seem to have been collected in the great Library of Alexandria after about 200 B.C. While few if any of these books may have been written by Hippocrates himself, they are considered to be an expression of his medical teachings and philosophy, and later became an important basis of Western medicine.

The Hippocratic approach to medicine, expressed in the books, emphasized that disease arose from natural causes, not from whims of the gods. Hippocrates insisted on careful observation of medical conditions; the books contain dozens of detailed clinical descriptions of diseases. He recommended as little interference as possible with the body's own ability to heal. Treatment focused on diet, rest, and cleanliness. He advanced the doctrine of the four humors, whereby disease was supposed to result from an imbalance among the body's four important fluids.

Hippocrates also emphasized a high ethical standard for physicians. The Hippocratic Oath is a statement of medical ethics. Developed over 2,000 years ago, it probably reflects the views of Hippocrates while not actually having been written by him. The oath pledges a physician to serve only the benefit of the patient, and to keep confidential anything he or she sees or hears in the course of treatment. Many medical students today still take a form of the Hippocratic Oath when they receive their medical degrees.

See also History of anatomy and physiology: The Classical and Medieval periods

HISTOLOGY AND MICROANATOMY

Histology, also known as microscopic **anatomy**, deals with the study of cells as they are constituted in the various **tissue** types. Single cells, isolated from the fabric of tissue, are generally difficult to classify. However, when cells are examined in their normal configuration and relationship to neighboring cells, the tissues formed by those cells are readily identified. A microscope is required because cells are microscopic and so too are the identifying characteristics of tissues.

While the body contains an array of **organs**, the number of tissue types is limited. Tissue types include epithelium, connective tissue, muscle, nervous tissue, **blood**, and reproductive tissue. Included in these categories are more than one type of tissue. For instance, connective tissue encompasses bone, **cartilage**, **tendons**, ligament, and fibrous connective tissue. Another example: the muscle category includes **skeletal muscle**, cardiac muscle and **smooth muscle**.

The term microscopic anatomy is descriptive in that it indicates that tissue structure must be examined with a compound microscope. Because light does not penetrate thick masses of tissues, and because tissue detail is revealed only by staining, it is necessary to process tissue in a somewhat elaborate manner for proper evaluation. A small fragment of tissue is cut and fixed (preserved). The fixative is chosen to maximize the retention of tissue structure. Dehydration follows fixation. This step removes water so that the tissue may be infiltrated with paraffin or a combination of paraffin and other waxes in preparation for sectioning. Frequently an intermediate step is required between dehydration and paraffin application to enhance proper infiltration. The tissue in its paraffin block is then mounted on a microtome, which is similar to a meat slicer, but it can cut slices of tissue routinely at 0.0001-0.0004 inches thick. The slices of tissue are then mounted on a glass slide, the paraffin is removed, the tissue is stained, water is removed and a permanent mounting medium of the refractive index of glass is applied. Finally, a thin cover slip of glass is placed over the biopsy section. The microscopic slide is now ready for examination.

Histopathology uses knowledge of histology to evaluate tissues that may be malignant (cancerous) or otherwise abnormal. After a lump is biopsied by a physician, a trained pathologist, a medical doctor who specializes in diagnosis of disease, will examined the biopsy. The tissue may be either normal or it may display varying degrees of pathological change. The histological diagnosis by the pathologist determines if further treatment is required. Physicians, because of time constraints, have devised short cuts that permit rapid examination of tissues. However, there is no short cut for microscopic examination of the biopsied tissue. Despite the

explosion of **biochemistry** and **molecular biology**, pathologists still rely on microscopic examination of suspected malignant tissue.

See also Pathology

HISTORY OF ANATOMY AND PHYSIOLOGY: THE CLASSICAL AND MEDIEVAL PERIODS

Anatomy is the science of the structure of animal bodies, either living or dead. **Physiology** is the science of the function, purpose, or action of living animal **tissue**. Both sciences usually study either humans alone or humans in relation to other animals.

Major skeletal, muscular, and even visceral structures were identified in many prehistoric and ancient cultures. The ancient Hindu medical and surgical system, Ayurveda, was based on extensive knowledge of anatomy and pharmacy. Anatomical models were being used in India by 700 B.C. Also advanced in surgical anatomy were Egypt, Mesopotamia, Minoan Crete, and pre-Columbian Peru. The ancient Chinese had a fair understanding of physiology, but little of anatomy, because human dissection and most surgeries were forbidden. Chinese medicine attempted to preserve health by maintaining balance between two primal energies, yin (passive) and yang (active), and among five fundamental substances: fire, earth, water, metal, and wood.

Similarly, the humoral theory of the ancient Greeks advocated the harmonious proportion of four elements: fire (hot and dry), earth (dry and cold), water (cold and moist), and air (moist and hot); four qualities: dryness (fire and earth), cold (earth and water), moistness (water and air), and heat (air and fire); the four humors: yellow **bile** (hot and dry), black bile (dry and cold), phlegm (cold and moist), **blood** (moist and hot); the four temperaments: choleric (predominance of yellow bile), melancholic (predominance of black bile), phlegmatic (predominance of phlegm), and sanguine (predominance of blood); and the four seasons: summer (hot and dry), autumn (dry and cold), winter (cold and moist), spring (moist and hot). The four elements were correlated with the astrological fire signs (Aries, Leo, Sagittarius), earth signs (Taurus, Virgo, Capricorn), water signs (Cancer, Scorpio, Pisces), and air signs (Gemini, Libra, Aquarius), thus providing the basis of an occult influence on medicine and science that lasted into the eighteenth century.

Several ancient Greek physicians pursued empirical studies of anatomy and physiology. About 500 B.C. Alcmaeon of Croton studied **comparative anatomy** and differentiated **arteries** from **veins**. About 400 B.C. Democritus advocated the preformation theory of human generation and regarded the human **brain** as the seat of thought. **Herophilus** of Chalcedon performed public dissections of humans and animals. In Alexandria with **Erasistratus**, he continued studies of the nervous and circulatory systems.

In the West, the earliest important codifier of anatomy and physiology as separate sciences was **Aristotle**. Works about anatomy and physiology are also attributed to "The Father of Medicine," **Hippocrates**, but, unlike Aristotle, his interests were mainly clinical rather than purely scientific. Thus anatomy and physiology are subordinate to therapy in his writings. Both Aristotle and Hippocrates contributed to founding the science of **embryology**.

Aristotle, the world's first meticulous biologist, conducted his own experiments and advocated that others likewise do so. Nevertheless, in a very un-Aristotelian way, his results and those of his school were enshrined in the centuries after his **death** while his rigorous empirical method was mostly ignored.

Galen compiled all medical, surgical, anatomical, physiological, and zoological knowledge up to his time, including some of his own observations and discoveries, into a gigantic multivolume work that dominated Western philosophy of medicine for at least one thousand and probably thirteen or fourteen hundred years. Few dared to question the authority of Galen until the Renaissance.

With the fall of Rome in 476, Byzantine Emperor Justinian closing the Platonic Academy and banning non-Christian scholarship in 529, and the Dark Ages pervading Western Europe from the fifth to the ninth centuries, the locus of Western civilization shifted to the Arabic world and Persia. Latin and Greek classics were translated into Arabic. Islamic science, medicine, and mathematics were the best in the world from the seventh to the eleventh centuries. Al-Kindi (800–873) studied the physiology of vision and wrote 22 books on medicine. Albucasis (936–1013) performed human dissections and described hemophilia. **Avicenna**, the Arabic Galen, studied the functions of the **heart**, **lungs**, and eyes. Avenzoar (ca. 1092–1162) advanced knowledge of the abdominal viscera. Ibn-al-Nafis (ca. 1210–1288) discovered **pulmonary circulation**.

With the exception of the School of Salerno, few bright lights existed in medieval European medical science. The Western European shift away from the slavish acceptance of ancient authority toward the renewed empirical study of nature was led by Albertus Magnus (ca. 1193–1280) and Roger Bacon (1214–1292). William of Saliceto (1210–1280), professor of surgery at the University of Bologna, performed his own human dissections and wrote a treatise on surgical anatomy that was authoritative for 300 years. Henri de Mondeville (ca. 1260–1320), **Mondino de Luzzi**, Johannes de Ketham (d. ca. 1490), and Giacomo Berengario da Capri (ca. 1460–ca. 1530) also contributed to medieval anatomy. Gabriele de Zerbis (1445–1505) wrote *Liber anathomie corporis humani et singulorum membrorum illius*, one of the most accurate anatomical texts of its time. The earliest anatomical text printed in English was the 1525 translation of a German surgical handbook by Hieronymus Brunschwig (1450–ca. 1512).

In medieval Europe, some progress occurred in anatomy, but little in physiology. Some might argue that the science of physiology was reborn through the Swiss alchemist Philipp Aureolus Theophrastus Bombast von Hohenheim,

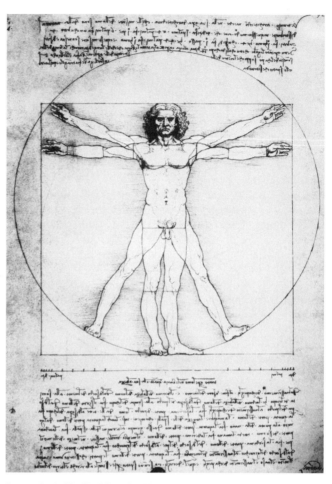

Leonardo da Vinci's "Vitruvian Man," a Renaissance study of the proportions of the human body. *Corbis Corporation. Reproduced by permission.*

known as "Paracelsus" (1493–1541), who rejected Galenic authority, Arabic medicine, and the humoral theory, teaching instead that life and its processes depend upon chemistry and the environment.

The work of **Leonardo da Vinci** marked the end of medieval thought about anatomy and physiology. This remarkable polymathic genius described capillary action, drew the famous "Vitruvian Man" to demonstrate ideal human proportion, filled his notebooks with scrupulously accurate anatomical sketches from secret dissections, advanced the science of embryology, cast the brain ventricles in wax, and invented the technique of cross-sectional anatomy. His direct empirical studies of the human body prepared the way for not only surgical anatomists such as **Andreas Vesalius**, but also artists such as Michelangelo (1475–1564).

See also Anatomical nomenclature; History of anatomy and physiology: The Renaissance and Age of Enlightenment; History of anatomy and physiology: The science of medicine; Medical training in anatomy and physiology

HISTORY OF ANATOMY AND PHYSIOLOGY: THE RENAISSANCE AND AGE OF ENLIGHTENMENT

Modern anatomical science, the rebirth of the empirical **anatomy** that had been unknown in the West since **Aristotle**, is usually said to have begun in 1543 with the publication of *De humani corporis fabrica* (On the Structure of the Human Body) by **Andreas Vesalius**. This massive work, systematically arranged and meticulously illustrated, relied upon the results of actual human dissections performed by Vesalius himself, his colleagues, and his students to discover and demonstrate intricate facts. Vesalius dared to oppose **Galen**, whose authority had been sacred for nearly fourteen centuries. Since good physiological science must be based upon solid anatomical knowledge, modern **physiology** can also be said to have begun in 1543.

In the wake of Vesalius came a surge of great anatomists, such as Bartolomeo Eustachius (ca. 1510–1574), Gabriele Fallopius (1523–1562), Girolamo Fabrizio (Fabricius ab Aquapendente) (1533–1619), **Raymond Vieussens**, and **Albrecht von Haller**. Also followed a significant number of plagiarists, such as Matteo Colombo (ca. 1510–1559) and William Cowper (1666–1709), many of whom nevertheless advanced the cause of anatomical research.

Until the mid-seventeenth century, the subject-matter of anatomy and physiology consisted only of structures and processes that could be seen with the naked **eye**, but new anatomical and physiological sciences became possible after Hans (fl. 1595) and Zacharias Janssen (fl. 1610) invented the compound microscope in the 1590s and after Athanasius Kircher (1602–1680), **Robert Hooke**, Jan Swammerdam (1637–1680), and especially Antoni van Leeuwenhoek (1632–1723) improved the earliest practical microscopes. What was first called just anatomy came to be called gross, or macroscopic, anatomy, in contrast to microscopic anatomy, the study of the structure of cells and other microscopic tissues. Physiological anatomy is the study of the function of **tissue** in relation to its structure. **Histology** is microscopic physiological anatomy. After George Adams (1750–1795) invented the microtome, superseding the razor, rigorous histological science became possible.

What Vesalius is to anatomy, **William Harvey** is to physiology. In 1628, Harvey published *Exercitatio anatomica de motu cordis et sanguinis in animalibus* (Anatomical Exercise on the Motion of the Heart and Blood in Animals), which described the circulation of the blood. *De motu cordis* and *De humani corporis fabrica* are two of the most important books in the history of medicine.

Richard Lower continued Harvey's investigations on the physiology of the heart and was one of the first to transfuse blood. **Thomas Willis** explained blood flow to the **brain** through the cerebral arterial circle, the "circle of Willis." **Stephen Hales** measured blood pressure. The microscope yielded better knowledge of the **circulatory system**. Swammerdam discovered red blood cells and **Marcello**

Malpighi discovered **capillaries**. William Hewson (1739–1774) wrote a book on blood and another on the lymphatic system.

The brain dissections of Costanzo Varoli (1543–1575) brought knowledge of the **pons**, optic nerve, and other **cranial nerves**. François Pourfour du Petit (1664–1741) discovered vasomotor nerves and made several other important advances in neuroanatomy and neurophysiology. The neurological research of Robert Whytt (1714–1766) included the **spinal cord**, the neurophysiology of vision, and the **reflexes**. **Alexander Monro** *secundus* discovered several important features of neuroanatomy, including the foramen interventriculare, or foramen of Monro, in the brain. Domenico Cotugno (1736–1822) amplified Valsalva's discovery of **cerebrospinal fluid**, but it remained for **François Magendie** to extend this work further. **Luigi Galvani** discovered electric charges in nerve impulses. Félix Vicq d'Azyr (1748–1794) excelled in neuroanatomy and **comparative anatomy**. Antonio Scarpa (1752–1832) contributed several important anatomical, neurological, otological, and ophthalmological breakthroughs, including the first accurate description of coronary nerves. Julien Jean César Legallois (1770–1814) localized respiratory control in the **medulla oblongata**.

Regnier de Graaf investigated reproduction physiology. Niels Stensen (Nicolaus Steno) (1638–1686) discovered "female testicles," guessed that they contained eggs, and accordingly called them "ovaries." Steno also contributed to neuroanatomy and identified the heart as a muscle. The experiments of Abraham Trembley (1710–1784) on the fresh water hydra had meaningful consequences for comparative anatomy, cell science, and the theories of tissue regeneration. **Lazzaro Spallanzani** studied the **fertilization** process and refuted the theory of spontaneous generation. Caspar Friedrich Wolff (1733–1794) rejected preformation theory and laid the groundwork for modern **embryology**. Samuel Thomas Soemmering (1755–1830) wrote several accurate anatomical and embryological textbooks.

Pathological anatomy, founded by Giovanni Battista Morgagni (1682–1771), was developed further by Matthew Baillie (1761–1823). Malpighi's student and Morgagni's teacher, Antonio Valsalva (1666–1723), discovered the aortic sinus and named the outer, middle, and inner ear. **Leopold Auenbrugger** invented the diagnostic technique of percussing the chest. His work was promoted and popularized by Baron Jean Nicolas Corvisart (1755–1821), Napoleon's doctor.

Michel Servetus (1511–1553) described **pulmonary circulation**. Santorio Santorio (1561–1636) inaugurated the modern study of **metabolism**. Francis Glisson (1597-1677) wrote the first detailed account of the **liver**. Thomas Wharton (1614–1673) distinguished various glands from other **organs**. John Mayow (1643–1679) analyzed respiration and observed mitral stenosis. Herman Boerhaave (1668–1738) studied **digestion** and isolated urea from urine. Jean-Antoine Nollet (1700–1770) observed and described osmosis. **Joseph Priestley**, Jan Ingenhousz (1739–1799), and Antoine-Laurent Lavoisier (1743–1794) investigated the respiratory physiology of oxygen.

René Descartes formulated a thoroughly mechanistic physiological theory and discovered the stimulus/reflex response. Giovanni Alfonso Borelli's (1608–1679) investigations of the musculoskeletal system, the heart, and the nerves supported Descartes's iatromechanism. On the other hand, the iatrochemism of Franciscus de le Boë Sylvius (1614–1672) claimed that physiological processes were essentially chemical. Jean Baptiste von Helmont (1577–1644) and Georg Ernst Stahl (1660–1734) propounded the theory of animism, the idea that the soul is responsible for physiological changes.

Johann Christian Reil (1759–1813) founded the first scientific journal of physiology, *Archive für die Physiologie*, in 1795. He investigated tissue metabolism, neuroanatomy, neurophysiology, and the medical philosophy of vitalism, which posited a "life-force" behind biochemical processes. Among the proponents of vitalism was Théophile de Bordeu (1722–1776), who correctly guessed that some organs secrete substances into the blood to affect other parts of the body. This internal secretion hypothesis later gave rise to scientific studies of endocrinology and hormonology. Another prominent vitalist, **Xavier Bichat**, pioneered histology, excelled at using anatomy to further both physiology and **pathology**, and brought fine detail to descriptive anatomy.

See also Anatomical nomenclature; History of anatomy and physiology: The classical and medieval periods; History of anatomy and physiology: The science of medicine; Medical training in anatomy and physiology

HISTORY OF ANATOMY AND PHYSIOLOGY: THE SCIENCE OF MEDICINE

After the Enlightenment, most anatomical studies were conducted in the interest of **physiology**. That is, because the main structures of the body were already sufficiently known, the focus of research shifted toward learning the functions of these structures and toward seeing smaller structures, such as cells, in terms of their functions.

Although considered a charlatan by many for his popular promotions of phrenology, Franz Joseph Gall (1758–1828) was a preeminent neuroanatomist and an expert dissector. His findings laid the basis for the modern study of cerebral localization. The rapid advance of neuroscience in the nineteenth century was largely due to Luigi Rolando (1773–1831); **Charles Bell**; Karl Friedrich Burdach (1776–1847); Pierre Flourens (1794–1867); Anders Adolf Retzius (1796–1860); **Johannes Müller**; Hermann von Helmholtz (1821–1894), who studied the speed of nerve impulses; Wilhelm His, Sr. (1831–1904); Eduard Hitzig (1838–1907); Gustav Theodor Fritsch (1838–1927); David Ferrier (1843–1928), who mapped the motor cortex; **Camillo Golgi**; Paul Emil Flechsig (1847–1929); Carl Wernicke (1848–1905); Charles Scott Sherrington (1857–1952); and Franz Nissl (1860–1919).

Neuroscientific giants of the twentieth century include **Ivan Pavlov**, who discovered the conditioned reflex; Santiago Ramon y Cajal (1852–1934); Henry Head (1861–1940); **Otto Loewi**; Walter Rudolf Hess (1881–1973); and John Carew Eccles (1903–1997).

The neurophysiology of vision, hearing, and speech was investigated by Bartolomeo Panizza (1785–1867), Charles Wheatstone (1802–1875), Frans Cornelis Donders (1818–1889), Pierre Paul Broca (1824–1880), and Georg von Békésy (1899–1972).

Modern cell science, cytology, began in 1805, when Lorenz Oken (1779–1851) asserted that all life forms consist of cells. In the 1820s René-Joachim-Henri Dutrochet (1776–1847) further developed Oken's claim, Carl Ernst von Baer (1792–1876) discovered the ovum, and Robert Brown (1773–1858) discovered the nucleus. In the 1830s, Dutrochet named osmosis. Other major early contributors to the knowledge of cells include Jan Evangelista Purkyne (1787–1869); Martin Barry (1802–1855); Friedrich Gustav Jakob Henle (1809–1885), who wrote the first textbook of **histology**; **Theodor Schwann**, who coined the word "metabolism"; Robert Remak (1815–1865); and **Albert von Kölliker**. Hugo von Mohl (1805–1872) pioneered the study of **cell division**, **Walther Flemming** discovered **chromosomes**, and Wilhelm Roux (1850–1924) identified mitosis. Karl Bogislaus Reichert (1811–1883) combined cell science with **embryology** and **Rudolf Virchow** invented cellular **pathology**.

Modern gastrophysiology began when American military surgeon **William Beaumont** was able to observe directly the digestive processes of his patient, Alexis St. Martin, whose abdominal gunshot wound exposed the interior of his stomach. The radically materialistic metabolic and gastrophysiological theories of Jacob Moleschott (1822–1893) prompted the philosopher Ludwig Feuerbach (1804–1872) to satirize him by coining the expression, "You are what you eat." Knowledge of **digestion** was further enhanced by Walter Bradford Cannon (1871–1945).

Among the landmarks of physiological instrumentation are the invention of the stethoscope by **René-Théophile-Hyacinthe Laënnec**, the improvement of the string galvanometer and the inventions of the electrocardiograph (EKG) and the phonocardiograph by **Willem Einthoven**, the invention of the slit-lamp ophthalmoscope by Alvar Gullstrand (1862–1930), the development of the electron microscope by Max Knoll (1879–1969) and Ernst Ruska (1906–1988), and the preparation of the first human **electroencephalogram** (**EEG**) by Johannes Berger (1873–1941). **Josef Skoda** and Austin Flint, Sr. (1812-1886) improved the diagnostic capabilities of the stethoscope. Hans Eppinger (1879–1946) was among those who best exploited the EKG in clinical and physiological research.

Two Frenchmen, François Magendie (1783–1855) and Claude Bernard (1813–1878), are often known as the fathers of modern physiology. John Call Dalton (1825–1889) was the first prominent American physiologist, and Austin Flint Jr. (1836–1915) followed in his footsteps. But a German, **Eduard Pflüger**, was the most prominent general physiologist of his day. In 1868 he founded *Archiv für die gesamte Physiologie* (Archives of All Physiology), better known simply as *Pflügers Archiv*. He was merciless in his criticism of other physiologists.

Justus von Liebig (1803–1873) and **Robert Koch** laid the foundations of **biochemistry**. Some biochemical milestones with physiological significance are the synthesis of urea by Friedrich Wöhler (1800–1882); the spectral analysis of body fluids by Karl Vierordt (1818–1884); the careers of Louis Pasteur (1822–1895) and **Paul Ehrlich**; the discovery of insulin by Frederick Grant Banting (1891–1941), Charles Herbert Best (1899–1978), and James Bertram Collip (1892–1965); the discovery of penicillin by **Alexander Fleming**; the discovery of the citric acid cycle by Hans Adolf Krebs (1900–1981); and the determination of the amino acid sequence of insulin by Frederick Sanger (b. 1918). Following the hematological research of Ernst Hoppe-Seyler (1825–1895) and the discovery of **hemoglobin** by Otto Funke (1828–1879), **Karl Landsteiner** investigated **blood** agglutination and identified the four human blood types.

In endocrinology, Paul Langerhans (1847–1888) with the **pancreas**, **Victor Horsley** and Edward Calvin Kendall (1886–1972) with the thyroid, Artur Biedl (1869–1933) with the adrenal glands, and Harvey Cushing (1869–1939) with the pituitary all made significant progress. Charles Édouard Brown-Séquard (1817–1894) performed research on internal secretions, or **hormones**, that led to the science of hormonology. John Jacob Abel (1857–1938) prepared the way for Jokichi Takamine (1854–1922) and Thomas Bell Aldrich (1861–1939) to isolate the hormone **epinephrine**, later called adrenalin. William Maddock Bayliss (1860–1924) advanced the knowledge of hormonal chemistry and wrote a key textbook on chemical physiology. With Ernest Henry Starling (1866–1927), he isolated the hormone secretin in the duodenum.

Casimir Funk (1884–1967) discovered several substances necessary for life. Since they are all in the amine group of chemicals, he combined "amine" with the Latin word *vita* ("life") to name them "vitamines," later shortened to "vitamins." Further pioneer research on **vitamins** was conducted by Elmer Verner McCollum (1879–1967), Herbert McLean Evans (1882–1971), Katharine Scott Bishop (1889–1976), and **Albert Szent-Györgyi**.

The physiology of **genetics** made great strides after Wilhelm Hertwig (1849–1922) observed that **fertilization** is the uniting of the nuclei of the male and female gametes. Phoebus Aaron Theodore Levene (1869–1940), Oswald Theodore Avery (1877–1955), Colin Munro Macleod (1909–1972), Maclyn McCarty (b. 1911), and especially **Francis Harry Compton Crick** (b. 1916) and James Dewey Watson (b. 1928) produced the scientific understanding of **deoxyribonucleic acid** (**DNA**) as the carrier of **genetic code**. Katherine Koontz Sanford (b. 1915) achieved the first successful cloning of mammalian **tissue** cells. John Rock (1890–1984) fertilized human ova *in vitro*.

The **anatomy** textbook of **Henry Gray** has been standard since its first appearance in 1858. Jean Cruveilhier (1791–1874) and Carl Theodor Ernst von Siebold (1804–1885) respectively advanced pathological anatomy and **comparative anatomy**. Joszef Hyrtl (1810–1894), Eduard Pernkopf (1888–1955), and Frank Netter (1906–1991) each created beautiful anatomical atlases, but Pernkopf's has been severely criticized because he dissected victims of Nazi concentration camp murders and promoted Nazism in his book.

As a science, human anatomy is now essentially complete. What little remains to be learned is at or below the molecular level and is closely tied to physiology. The wave of the future in physiology is medical genetics, which tries to fathom the origins, manifestations, and meaning of individual human differences and involves research in cloning, **gene therapy**, gene regeneration, tissue regeneration, recombinant DNA, genetic engineering, **stem cells**, and the mapping of the human genome. A branch of medical genetics called genomics identifies and characterizes genes and their arrangement in chromosomes.

See also Anatomical nomenclature; History of anatomy and physiology: The classical and medieval periods; History of anatomy and physiology: The Renaissance and Age of Enlightenment; Medical training in anatomy and physiology

HITCHINGS, GEORGE H. (1905-1998)
American biochemist

George H. Hitchings's basic research led to medications for cancers, bacterial infections, **AIDS**, herpes, gout, **malaria**, and organ rejection. A respected researcher and humanitarian, he himself explained his life had been devoted "two thirds to science and one third to philanthropy."

Among the most prolific of modern pharmaceutical scientists, he worked at the Burroughs Wellcome Company for more than thirty years before his retirement in 1975. Hitchings' contributions were based on the premise that understanding why diseased cells are different makes it possible to exploit those differences to destroy **cancer** cells or foreign invaders such as **bacteria** or **viruses**. For his work in finding treatments for serious diseases, Hitchings and his long-time Burroughs Wellcome collaborator **Gertrude Elion** shared the 1988 Nobel Prize in physiology or medicine with British pharmaceutical scientist Sir **James Black**. It was the first time since 1957 that pharmaceutical scientists had been awarded the prize.

George Herbert Hitchings was born to George Herbert Hitchings Sr., a naval architect, and Lillian H. Belle Hitchings in Hoquiam, Washington, on the Olympic Peninsula. His father's death when he was twelve and his admiration for Louis Pasteur, a preeminent scientist-philanthropist who became his role model, convinced Hitchings to make medicine his career. As the salutatorian of his high school class, Hitchings's address outlined germ theory and Pasteur's life.

Hitchings attended the University of Washington, where he received a bachelor's degree in chemistry in 1927, and a master's degree in chemistry in 1928. He also showed a fondness for many scholarly subjects, studying the arts and history. He began his career in scientific research at an early age. "The Chemistry of the Waters of Argyle Lagoon," the first of his more than three hundred scientific publications, appeared in the publications of the Puget Sound Biological Station in 1928, when he had just entered graduate school. He continued his graduate work in biological chemistry at Harvard College, where he received his Ph.D. in 1933. Hitchings's doctoral dissertation concerned the **metabolism** of nucleic acids, the chemicals that make up **DNA**, the carrier of genetic information. Hitchings did his work on nucleic acids before **James Watson** and **Francis Crick** discovered the structure of DNA, and at that time no one was interested in nucleic acids. Hitchings couldn't find a job. Finally, after working for nine years as a teaching fellow at Harvard (1933–39) and Western Reserve University (1939–42), he was hired by Burroughs Wellcome in 1942, and resumed his work on nucleic acids. He became vice president of research in 1967, and held the position until 1975, when he became scientist emeritus.

Until Hitchings and the pharmacologist Gertrude Elion came along, drug researchers sought new drugs by modifying natural products. The two pioneered a method that has come to be known as "rational" drug design. They reasoned that if they understood the differences between normal and diseased or infected cells, these differences could serve as a entry point to selectively kill diseased **tissue** without harming surrounding normal tissue. They implemented these ideas by investigating the chemical pathways of nucleic acid synthesis, which is crucial to cell metabolism. Hitchings synthesized chemicals similar in structure to natural nucleic acids, the purines and pyrimidines. These related compounds interfered with DNA synthesis. Because cancer cells divide quickly, the compounds are particularly disruptive to them, killing them as they try to divide. This form of **chemotherapy** is just one instance of the rational drug design that helped Hitchings accumulate eighty-five patents over his thirty-year career.

One compound in particular, 6-mercaptopurine (6MP), a purine analog synthesized in 1951, proved to be particularly effective. Working with scientists at Sloan-Kettering Institute, Hitchings and Elion perfected the drug, which was used to combat childhood **leukemia**. 6MP and thioguanine, also produced by Hitchings and Elion, are still used to treat acute leukemias.

In 1959 Hitchings discovered that 6MP inhibited production of antibodies in rabbits. A less toxic form called azathioprine, marketed under the trade name Imuran, was developed in 1957 to control rejection of transplanted **organs** and treat autoimmune diseases. In the nearly nine thousand kidney transplants performed each year, Imuran remains the drug most commonly used to prevent organ rejection. 6MP is broken down in the body by xanthine oxydase, the same enzyme that converts purines into **uric acid**, the cause of the painful joint disease gout. Further investigation of purine analogs led to the development of allopurinol in the 1960s. It blocks uric acid production by competing for xanthine oxydase, an enzyme that converts purines to uric acid. Hitchings was also active in the development of other drugs, including pyrimethamine, which is used to treat malaria, and trimethoprim, which is used to treat urinary tract infections and other bacterial infections.

Philanthropy was always a part of Hitchings's life, who explained that when he was baptized his father dedicated his life to the service of mankind. He gave all of his Nobel prize money to the Triangle Community Foundation, which he founded in 1983 to provide health care and other help to the poor and disenfranchised. He served as president and director of the Burroughs Wellcome fund, a charitable organization, in addition to directing a dozen other local chapters of philanthropic organizations.

In addition to the Nobel Prize, Hitchings has received numerous awards, including the Gregor Mendel Medal from the Czechoslovakian Academy of Science in 1968 and the Albert Schweitzer International Prize for Medicine in 1989. He has been awarded eleven honorary degrees and was a member of the National Academy of Sciences, and a Foreign Member of the Royal Society.

Hitchings died of **Alzheimer's disease** in his home in Chapel Hill, North Carolina, at the age of 92.

See also Cell cycle and cell division; Cell differentiation; Molecular biology; Pharmacogenetics

HODGKIN, ALAN LLOYD (1914-1998)
English biophysicist

Alan Lloyd Hodgkin built the foundations of much of modern neuroloscience by defining the electrical and chemical characteristics of nerve impulses. Along with Andrew F. Huxley, the two described the firing of nerve impulses, for conducting the impulse along the axon, and for founding the new science of ion channels. The two performed experiments on the nerve fibers of squid and described the nerve impulses with a series of mathematical equations. For their research in this area, which resulted in the ionic theory of nerve impulses, the two men shared the 1963 Nobel Prize in physiology or medicine with **John C. Eccles**.

Hodgkin was born in Banbury, Oxfordshire, England, to George L. and Mary Wilson Hodgkin. Hodgkin's father died in Baghdad during World War I, only a few years after his birth. Hodgkin was educated at the Downs School in Malvern and the Gresham School in Holt. In 1932, he entered Trinity College, Cambridge, where he first became interested in physiology. Hodgkin became a fellow at Trinity in 1936, serving as lecturer and later as assistant director of research at the physiological laboratory.

Hodgkin began studying the electrical properties of the nerve fibers in the shore crab while at Cambridge. He spent a year at the Rockefeller Institute in New York City between 1937 and 1938, and while there he met scientists who had developed new methods for studying nerve fibers. Hodgkin brought these ideas back to Cambridge, where with **Andrew Huxley** He devised an experiment to test an hypothesis about nerve impulses first proposed by German physiologist Julius Bernstein.

Bernstein had hypothesized that nerve cells possess a resting or unstimulated potential and an action or stimulated potential. During the resting potential, he believed, the nerve cell membrane had an unequal distribution of positively and negatively charged ions, with more negative ones on the inside. During resting potential, the membrane was permeable to the positively charged ions, but the negatively charged ions could not permeate the cell membrane. When the cell was stimulated, Bernstein argued, the membrane "gates" were temporarily opened, allowing ions to pass in both directions. By using the nerve cells of the shore crab, Hodgkin was able to establish that the resting potential was due to an outward movement of potassium ions; during the **action potential** the cell membrane's gates allowed in the more concentrated sodium ions. He also discovered that the action potential was usually much larger than the resting potential.

Some of the researchers Hodgkin had met in the United States were working with squid, whose nerve fibers are larger than those of most organisms. Hodgkin and Huxley were able to develop a method to study these fibers using microelectrodes, and they were able to confirm the results of their earlier experiment. Their progress, however, came to a halt during World War II, when Hodgkin worked on radar systems for aircraft for the Air Ministry. Hodgkin and Huxley were back in Cambridge in 1945, and they formed a small research group to pursue their pre-war investigations into nerve fibers.

In 1951, Hodgkin and his colleagues published the results of their research. They found that the membrane is permeable only to specific ions during the resting potential, because of the differing concentrations of potassium and sodium. The concentration of the positively charged sodium ions is greater on the outside of the membrane and the concentration of negative potassium ions higher on the inside during resting potential. During the action potential, the negative and positive ions travel through the membrane, so that the interior charge becomes positive and the exterior negative. This is followed by an **equilibrium** charge, then a return to the resting potential charge state. All this happens in milliseconds.

The work done by Hodgkin and Huxley which was most responsible for bringing them to the attention of the Nobel Prize committee was the development of a series of mathematical formulae they published in 1952. The purpose of these equations was to synthesize the experimental information then available about the electrical and chemical nature of nerve transmissions. Their goal was to analyze and predict each stage in the passage of the nerve cell membrane from resting to action potential. They were awarded the 1963 Nobel Prize in physiology or medicine, which they shared with John C. Eccles, an Australian who advanced the British team's findings by showing what happens to nerve impulses transmitted across the synapses, or intersections, between nerve cells.

Hodgkin was appointed Foulerton Research Professor of the Royal Society in 1952, and was awarded the Royal Medal in 1958. He was knighted in 1972, and the following year was appointed to the Order of Merit, a British honor restricted to only 24 people at any one time. He was John Humphrey Plummer Professor of Biophysics at Cambridge from 1970 to 1981, president of the Marine Biological Association from 1966 to 1976. From 1970 to 1975 he served as President of the Royal Society, and from 1978 to 1984 as Master of Trinity College.

Hodgkin married the daughter of American Nobel Laureate **Peyton Rous** in 1944. The couple met during Hodgkin's year at the Rockefeller Institute in New York, and eventually had four children.

See also Nerve impulses and conduction of impulses; Nervous system overview; Neurology; Neurons; Neurotransmitters

HOLLEY, ROBERT (1922-1993)
American biochemist

Robert Holley was best known for his isolation and characterization of transfer **ribonucleic acid** (tRNA). Essentially, tRNA "translates" the genetic instructions within cells by first "reading" genes, the fundamental units of heredity, and then creating proteins—the building blocks of the body—from **amino acids**. Holley, along with Har Gobind Khorana and Marshall Warren Nirenberg, was awarded the 1968 Nobel Prize in physiology or medicine for determining the sequence of tRNA. But Holley's work on tRNA was only the beginning of a distinguished scientific career. Subsequently, he has investigated the molecular factors that control growth and multiplication of cells. His work in this area has had profound impact on understanding the processes that lead to **cancer**.

Robert William Holley was born in Urbana, Illinois. His parents, Charles Elmer Holley and Viola Esther (Wolfe) Holley, were both teachers. Holley grew up in Illinois, California, and Idaho, and early developed a life-long love of the outdoors and fascination with living things. The latter years of his childhood were spent in Urbana, where he attended high school and, in 1938, enrolled at the University of Illinois. He majored in chemistry, and was the photographer for the school's yearbook.

After obtaining his B.A. in 1942, Holley took up graduate studies in organic chemistry at Cornell University. He served in various positions at both the university and the medical college for the next several years. In 1945, he married Ann Lenore Dworkin, a chemist and high school mathematics teacher. They had one son.

During the mid-1940s, Holley participated as a civilian in war research for the United States Office of Research and Development. He was a member of the team of researchers that first succeeded in making penicillin synthetically. Supported by a fellowship from the National Research Council, he completed his doctorate in organic chemistry at Cornell University in 1947 and did a year of postdoctoral work at Washington State College (now University) in Pullman before returning east. In 1948, he became assistant professor at the New York State Agricultural Experiment Station, a branch of Cornell, in Geneva. He became associate professor in 1950 and full professor in 1964.

During a sabbatical on a Guggenheim Memorial Fellowship at the California Institute of Technology in 1955-1956, Holley started to investigate **protein synthesis**. In the wake of James Watson's and Francis Crick's discovery that **DNA** contained the information of heredity, Holley targeted the chemistry of nucleic acids, which carry and transmit genetic information. His course may have been inspired, at least in part, by Crick's suggestion that "adaptor molecules" of some sort must be involved in the translation of genetic information into proteins. Towards the end of his year away from Cornell,

Holley began to look specifically at the structure of transfer RNA, the start of a nine-year effort to unlock its secrets.

Back at Cornell in 1957, Holley was appointed research chemist at the United States Plant, Soil, and Nutrition Laboratory, where he continued his studies on tRNA. Heading up a research team, he meticulously planned and carried out a painstaking series of experiments. He and his colleagues spent three years developing a technique to isolate and partially purify different classes of tRNAs from yeast. Finally they succeeded in isolating a pure sample of alanine transfer RNA. The next five years were devoted to elucidating the sequence and structure of this particular transfer RNA.

To appreciate the profound impact of Holley's work on research into the **biochemistry** of life, it is useful to review some fundamental concepts. Cells carry the instructions for all of their necessary tasks in their **chromosomes**. Chromosomes within a cell are made up of very long molecules called **deoxyribonucleic acid** (DNA). Genes, the basic units of heredity for all living things, are small sections of the long strands of DNA. Genes themselves are made up of a series of units called nucleotides. Nucleotides are molecules composed of a particular sugar (either ribose or deoxyribose), a phosphate group (one phosphorous atom combined with three oxygen atoms), and one of five specific bases. These bases—guanine, adenine, cytosine, thymine, and uracil—thus distinguish the nucleotides from one another. They are, in essence, the alphabet from which all of our genetic instructions are composed.

Cells and bodies, however, are built not of genes but of proteins. Proteins are the structural elements of cells, providing form and stability. Equally important, certain proteins, called **enzymes**, mediate critical biochemical reactions, allowing the formation and breakdown of innumerable chemicals that cells use during growth, functioning, and division. Proteins are sequences of amino acids. The amino acids are a group of about twenty different molecules that share certain chemical characteristics (e.g., the presence of an amine group).

Holley's work centered on the question of how sequences of nucleotides in DNA specify sequences of amino acids in proteins. It had been known that DNA did not directly create protein, but copied itself instead (in a complementary, or negative sense) into strands of RNA. But it was not known how these long strands of RNA, called messenger RNA or mRNA, functioned in the creation of proteins. Holley believed that the much smaller tRNA molecules played a key role. He knew that a triplet of bases, or codon, specifies each of the twenty amino acids. Examining the sequence of bases within alanine tRNA (which specifies creation of the amino acid alanine), he found an anti-codon for alanine. This anti-codon would be able to bond chemically with an alanine codon on an mRNA strand.

By studying the molecular sequence of alanine tRNA, Holley and his students were able to determine its structure and then to deduce how it functioned. A tRNA anti-codon would bind to its matching codon along a strand of mRNA. The corresponding amino acid, held at the opposite end of the tRNA, would then be positioned to link up in series with the amino acid specified by the adjacent codon on the mRNA. In

Robert William Holley. *The Library of Congress.*

this manner, the series of nucleotides in a molecule of DNA would be translated into a series of amino acids that would make up a protein.

For his illumination of this vital process, Holley won a share of the 1968 Nobel Prize in physiology or medicine. He was also honored with the prestigious Albert Lasker Award for Basic Medical Research in 1965. In other honors include a National Academy of Science award in **molecular biology** in 1967, as well as fellowships from the Guggenheim Foundation, the National Science Foundation, the National Research Council, and the American Chemical Society.

From 1966 to 1967, Holley was on sabbatical at the Salk Institute for Biological Studies and the Scripps Clinic and Research Foundation in La Jolla, California. The following year he joined the Salk Institute as a resident fellow. Like his earlier sabbatical, Holley's move proved pivotal for his research, as he launched an investigation of the molecular factors that regulate growth and multiplication of cells. Rooted somewhat in his previous work on how the protein molecules underlying cell growth are formed, the new investigations had quite different interpretations and implications.

The control of cell growth and division is critical to normal functioning. Cancerous growths are characterized by uncontrolled **cell division**. Normally, a balance of stimulatory and inhibitory molecular factors keeps cellular multiplication at the proper rate; the number of new cells produced roughly equals the number of cells that wear out and die. Rapid cell proliferation might be caused by over-production of the stimulatory factors, excessive cell sensitivity to the stimulatory factors, a lack of the inhibitory factors, or some combination of these causes.

Holley examined the roles of **hormones**, blood-born chemicals—usually proteins—that are released by various tissues and **organs** and that interact with one another. Hormones can either stimulate or inhibit cell proliferation, or even, as Holley would later show, do both.

Holley discovered that the concentration of two types of hormones, known as peptide and steroid hormones, in a solution with dividing cells would determine the rate of cell division and ultimately, cell density. Further, he found that types of cells prone to develop into **tumors** responded dramatically to these growth factors, dividing rapidly in response to very low hormone levels. Subsequent experiments demonstrated that peptide and steroid hormones could act synergistically: several of these growth factors together in solution would produce a greater growth rate than the sum effects of each individually. Holley also found that different types of cells responded differently to particular hormones, and that their responses could change with the cells' population density. At low densities, cells take up and utilize growth factors more efficiently than they do under conditions of high density. Cellular receptors for certain growth-promoting hormones increased under conditions of low cell density, whereas receptors for certain other hormones increased as cell density increased.

Holley also studied the effects of non-hormonal factors, such as certain sugars and amino acids, on cell proliferation. He found that while cell growth patterns were quite insensitive to the levels of many amino acids, they were strongly regulated by others, notably glutamine.

Looking at the other side of the coin, Holley and his collaborators also identified growth inhibitors. Some of these compounds suppress cell growth by blocking DNA replication. Holley discovered that, in addition to blocking DNA activity, growth inhibitors stimulated production of specific proteins whose functions were unknown. Growth factors, too, were found to have an associated protein synthesis in addition to stimulating DNA activity. Interestingly, Holley noted that while growth and inhibitory factors canceled out each others' effects on DNA replication, they had no effect on each others' secondary production of hormones. With particular factors that increase both cell size and rate of cell division, Holley had noted similar effects. Adding a growth inhibitor would stop the cells from dividing, but not stop the individual cells from growing larger.

As he and his co-workers had done with tRNA, Holley's team eventually sequenced certain of the growth factors. These are considerably larger molecules than tRNA, but the techniques of molecular biology had improved so much over the years that these sequences were obtained much more readily. Holley identified the sequence of amino acids of a growth-inhibiting factor for a specific type of monkey cell. The sequence turned out to be identical to that for the human growth factor (TGF-beta 2).

Holley's work during the later phase of his career has shed new light on the factors that control how cells grow, differentiate, and divide. His research has striking implications for the development of drugs to suppress tumor growth and for understanding the fundamental causes of cancer. This entire field of investigation continues to be active, as new techniques and technologies allow researchers to ask questions of increasing sophistication. As an American Cancer Society research professor of molecular biology at the Salk Institute for Biological Studies, Holley was in the forefront of the ongoing struggle to learn about and to control unchecked cell growth. Shortly before his death, he was studying the timing of cell division. Holley died at his home in Los Gatos, California.

See also Cell cycle and cell division; Cell differentiation; Genetic regulation of eukaryotic cells; Pharmacogenetics

HOMEOSTATIC MECHANISMS

Homeostatic mechanisms control a property of all living things called homeostasis. Homeostasis is a built-in, automated, and essential property of living systems. **Breathing** is an example of a homeostatic property. Homeostatic mechanisms are self-regulating mechanisms that function to keep a system in the steady state needed for survival. These mechanisms counteract the influences that drive physiological properties towards a more unbalanced state.

The recognition of homeostasis and homeostatic mechanisms dates from the nineteenth century. Then, the French physiologist Claude Bernard observed the constancy of the composition of **blood** and other body fluids. He proposed that such constancy was vital for life. Indeed, the word homeostasis is from the Greek words for "same" and "steady." The term was coined in 1930 by physiologist Walter Cannon in a book called *The Wisdom of the Body*.

The mechanisms that regulate homeostasis operate by feedback mechanisms. Negative and positive feedback mechanisms operate in living things. Negative feedback mechanisms reverse the direction of the change. This maintains the constant, steady state and so represents homeostasis. Positive feedback, on the other hand, acts to change the variable even more in the direction in which it is changing. Thus, positive feedback is not a homeostatic mechanism.

Temperature control is an example of a negative feedback homeostatic mechanism. The region of the **brain** called the **hypothalamus** monitors the human body's temperature. Variation from the normal temperature of 98.6°F (37°C) triggers a response from the hypothalamus. The temperature can be lowered by activation of glands capable of sweating, or raised by signalling muscles to shiver to produce heat.

Homeostatic mechanisms are a fundamental characteristic of living things. Without these mechanisms, facets of a body that need to be kept operating in a steady state, such as temperature, salinity, acidity, hormone levels, concentration of gases such as carbon dioxide, and the concentrations of nutrients, would become so unbalanced as to threaten the life of the organism. In a healthy body, homeostatic mechanisms operate automatically at different levels; molecular, cellular, and at the level of the whole organism.

At the molecular level, the activity controlled by one gene can be under regulatory control by another gene. At the cellular level, a well-studied homeostatic mechanism is contact inhibition, in which cells stop dividing when they begin to crowd in on each other. **Cancer**, in which a hallmark is the rampant growth and division of cells, is a condition where the homeostatic mechanism of contact inhibition is inoperative or defective.

At the whole organism level, a homeostatic mechanism is a vital part of birth. During labor, the contraction of the uterus causes the release of a hormone called **oxytocin** from the hypothalamus. The hormone increases contraction frequency, which in turn stimulates the release of more oxytocin. This increasing contraction cycle propels the fetus down the birth canal and into the world. After birth, the oxytocin acts to contract the expanded uterus in order to minimize bleeding, thereby maintaining the mother's blood volume

The importance of homeostatic mechanisms to the well being of an organism is underscored by the consequences of their failure. For example, at body temperatures of 107°F (41.6°C), the negative feedback systems cease to function. The high temperature then acts to speed up the body's chemistry, raising temperature even more. This, in turn, further accelerates body chemistry, causing a further rise in temperature. This cycle of positive feedback is lethal if not halted.

See also Acid-base balance; Blood pressure and hormonal control mechanisms; Equilibrium; Ovarian cycle and hormonal regulation; Tumors and tumorous growth

HOMOLOGOUS STRUCTURES

Homologous structures are structures on different organisms that display a similar base structure, but have different functions. An example of a homologous structure is the human arm, bat wing, and whale flipper. Although very different in appearance, and dissimilar in their purposes, each of these are appendages that share some similarities in bone structure and the presence of five digits (so-called pentadactyl limbs). Structures that have similar uses but dissimilar structures are described as being analogous.

Homologous structures typically developed, or evolved, from a common ancestral body part. Over evolutionary time, functional divergence occurred in different species. This relationship between homologous structures and **evolution** has existed since the dawn of life. Examination of dinosaur fossils reveals features in creatures apparently as disparate as Tyrannosaurus Rex and modern birds.

The observation of homologous structures was one of the underpinnings of the theory of evolution. In 1848, Richard Owen, a British medical practitioner and comparative anatomist, coined the term "homology" to refer to his observations of structural similarities among organisms. These similarities indicated to Owen that organisms were created following a common plan. Owen's views greatly influenced a

contemporary of his, Charles Darwin, who a few years later published his seminal views on evolution. It is now generally accepted that homologous structures are strong evidence of evolution, and that evolution operates by modifying pre-existing structures.

Homologous structures extend even to the microscopic world. For example, the membranous skin that surrounds most organisms is very similar in structure. With the discovery and unraveling of the **genetic code**, such structural homology was found to extend to the composition of the genes that code for the structures.

Some homologous structures have no function in one species, but do have a function in another species. These are known as **vestigial structures**. Examples of vestigial structures are the pelvic region of a whale and the eyes of blind, cave-dwelling creatures, such as the grotto salamander. If evolution did not operate, vestigial structures would make little sense. However, such structures do make sense if traits are inherited from some ancestor and gradually modified, even to the point of being functionless, over time.

See also Embryonic development: Early development, formation, and differentiation; Genetic code; Structure and function

HOOKE, ROBERT (1635-1703)

English physicist

One of the preeminent scientists of the seventeenth century, Robert Hooke is perhaps best remembered for the wide variety of fields to which he contributed, including physics, astronomy, microscopy, biology, and architecture, among others. Although Hooke introduced many concepts previously unimagined or unexamined, his ability to formulate these ideas usually did not match his intuition, and the credit for many scientific breakthroughs inspired by Hooke's ideas is often given to such scientists as Isaac Newton and Christiaan Huygens, who brought the work to its fruition. Still, Hooke remains an important pioneer of science.

Born on Britain's Isle of Wight, Hooke was a sickly child. As a youth, his perpetual ill health made it impossible for him to attend classes regularly, and he was unable to enter the ministry as his father, a minister, had wished. Instead, Hooke was allowed to pursue his interest in mechanics, which he first demonstrated as a small child by constructing elaborate toys. He attended Westminster School and later Oxford, where he became the laboratory assistant to Robert Boyle. It was in Boyle's lab that Hooke's talent for designing scientific instruments was noticed, as he constructed the improved air pump used to establish Boyle's gas laws. In fact, it has been speculated that Hooke himself may have been the author of Boyle's law, since, customarily, any findings from research done in the lab would have been credited to the professor.

Along with some of his colleagues from Oxford and the surrounding area, Hooke helped to establish what would soon become the Royal Society, to which he was appointed Curator of Experiments. During his time as Curator he had many other successes attributed to him such as the compound microscope, an improved barometer, the reflecting telescope, and the universal joint.

Although Hooke was not the first to experiment using a microscope, he was the first to dedicate a major intensive volume to microscopy. His 1665 publication *Micrographia* describes the structures of insects, fossils, and plants in unprecedented detail. While studying the porous structure of cork, Hooke noted the presence of tiny rectangular holes that he called cells, a word that has been adopted as the cornerstone of microbiology. *Micrographia* also contains illustrations in Hooke's own hand that remain among the best renderings of microscopic views.

In the years following the great London fire of 1666, Hooke became a surveyor and, eventually, an architect, constructing numerous famous buildings. Because his architectural interests took much time away from his scientific work, he was ultimately forced to retire as Curator of Experiments for the Royal Society in favor of his new vocation.

See also History of anatomy and physiology: The Renaissance and Age of Enlightenment

HOPKINS, FREDERICK GOWLAND (1861-1947)

English biochemist

Frederick Hopkins is considered the founder of British **biochemistry**. A pioneer in the study and application of what he called accessory food factors (**vitamins**), he made important contributions to the study of **uric acid**, isolated tryptophan (a necessary component in **nutrition**), and developed the concept of essential **amino acids**. Hopkins also did pioneering work on cell **metabolism**, elucidating the role of enzymatic activity in oxidation processes. Hopkins became a member of the Royal Society in 1905, serving as its president in 1931. He was knighted in 1925, and received the Copley Medal of the Royal Society in 1926. For his contributions in the field of nutrition, he was awarded the 1929 Nobel Prize in physiology or medicine, sharing the award with the Dutch chemist **Christiaan Eijkman**. He was presented with the highest distinction of civil service, the Order of Merit, in 1935. In addition, many honorary degrees were bestowed on him by universities worldwide.

Hopkins was born in Eastbourne, Sussex, England, to Frederick Hopkins and Elizabeth Gowland Hopkins. His father died soon after his birth and his mother took him to live with her family in London. He was a solitary and scholarly boy, given to reading Charles Dickens and writing poetry, although he showed no particular aptitude for any subject in school except chemistry. A small inheritance enabled him to study chemistry at the Royal School of Mines and London's University College, where he received a B.Sc. in 1890. An exemplary performance on his final chemistry examination led to a position as an assistant to Thomas Stevenson, an expert in forensic medicine at Guy's Hospital, who also served

as medical jurist to the Home Office. As assistant to Stevenson, Hopkins used his analytical skills to help secure the convictions of several notorious murderers.

Hopkins became the first professor of biochemistry at Cambridge in 1914. He established an open admissions policy for his department and attracted biochemists from many nations, including some who had escaped from dire political situations. A great teacher as well as a brilliant and unassuming researcher, Hopkins was known for encouraging his students to pursue their own line of work and often handed over to them promising new research in which he had made a breakthrough. This generosity, coupled with his faith that biochemistry could provide important answers to biological questions, was largely responsible for the widespread development of biochemical thought and experimentation. At the time of his death, approximately seventy-five of his former students held professorial positions in universities throughout the world.

One of his earliest contributions to biochemistry, made while he was still at Guy's Hospital and used for many years, was the method he developed to detect the presence of uric acid in urine. His work on uric acid led him to the study of proteins, the presence of which in the diet affects uric acid excretion. Hopkins first developed methods to isolate and crystallize proteins. He isolated the amino acid tryptophan and determined its structure. He also studied the effect of **bacteria** on tryptophan, laying the foundation of bacterial biochemistry. Hopkins showed that tryptophan is essential in the diet, since proteins lacking the substance are nutritionally inadequate. He also studied the role of the amino acids arginine and histidine in nutrition, which led to the theory that the presence of different amino acids determines the nutritional quality of proteins.

In 1912 Hopkins published the work for which he is probably best known, "Feeding Experiments Illustrating the Importance of Accessory Food Factors in Normal Dietaries." During World War I, Hopkins continued his work on the nutritional value of vitamins. His efforts were especially valuable in a time of food shortages and rationing. He agreed to study the nutritional value of margarine and found that it was, as suspected, inferior to butter because it lacked the vitamins A and D. Because of his work, vitamin-enriched margarine was introduced in 1926. Hopkins' nutritional theories were contested by colleagues until about 1920 but have been considered indisputable since then.

Although Hopkins won the 1929 Nobel Prize in medicine and physiology (shared with Christiaan Eijkman) for his work in nutrition, he was primarily interested in the biochemistry of the cell. The originality and vision of his research in this area set the standard for those who followed him. His study with Fletcher of the connection between **lactic acid** and **muscle contraction** was one of the central achievements of his work on the biochemistry of the cell. Hopkins had long studied how cells obtain energy from a complex metabolic process of oxidation and reduction reactions. He showed that oxygen depletion causes an accumulation of lactic acid in the muscle. The research techniques developed by Hopkins and Fletcher to study muscle stimulation were later used by others to study other aspects of muscle metabolism. Their work paved the way for the later discovery by Archibald Vivian Hill and **Otto**

Meyerhof that a carbohydrate metabolic cycle supplies the energy used for muscle contraction. The discovery that alcohol fermentation under the influence of yeast is a process analogous to the formation of lactic acid in the muscle is also due to Hopkins's groundbreaking work. Hopkins's work on muscle metabolism led to an understanding of the importance of **enzymes** as catalysts to oxidation. He isolated glutathione, which plays an important role as an oxygen carrier in cells, and several oxidizing enzymes. Hopkins and his assistant, E. J. Morgan, made further contributions to this field in the late 1930s.

See also Cell membrane transport; Enzymes and coenzymes; Nutrition and nutrient transport to cells

HOPPE-SELYER, ERNST FELIX
(1825-1895)
German biochemist

Ernst Hoppe-Selyer was one of the leaders in making **biochemistry** (or physiological chemistry, as it was then called) a scientific field distinct from medical physiology. He performed the first study of the nucleic acids, gave the name **hemoglobin** to the red **blood** cells, and discovered the enzyme invertase.

Hoppe-Selyer was born Ernst Hoppe in Freiburg-ander-Unstrut, Germany. His father was a minister. His mother died when he was a child. After he was adopted by his brother-in-law, he added Selyer to his name. He received his medical degree from the University of Berlin in 1851, then combined a medical practice with scientific research. Hoppe-Selyer's interest shifted gradually from physiology to chemistry. After serving on the faculties of the Universities of Berlin and Tubingen, in 1872 he became professor of physiological chemistry at the University of Strasbourg (then part of Germany). He established the first independent biochemistry laboratory in 1877 and the first biochemical journal.

Hoppe-Selyer's first important discovery came in 1862, when he used the newly invented spectrograph to determine the structure of the red blood cells, which he called hemoglobin. He later showed how hemoglobin binds and releases oxygen and how carbon monoxide can take oxygen's place in the blood cell. He also demonstrated some of the chemical similarities of hemoglobin and chlorophyll.

Hoppe-Selyer began studying the nucleic acids after they were discovered in 1869 by one of his students, the Swiss biochemist Johann Friedrich Miescher. Hoppe-Selyer showed that nucleic acids were present in yeast, and his work was extended by his one-time assistant, Albrecht Kossel.

Hoppe-Selyer's other research included the discovery in 1871 of invertase, the enzyme that converts sucrose (table sugar) into the simpler sugars glucose and fructose. He helped determine that lecithin is composed of nitrogen, phosphorus, **fat**, and choline (one of the B **vitamins**). And he demonstrated that lecithin and the steroid cholesterol are found in every cell.

See also Oxygen transport and exchange

HORMONES AND HORMONE ACTION

Hormones are biochemical messengers that regulate physiological events in living organisms. More than 100 hormones have been identified in humans. Hormones are secreted by endocrine (ductless) glands such as the **hypothalamus**, the pituitary gland, the pineal gland, the thyroid, the parathyroid, the thymus, the adrenals, the **pancreas**, the ovaries, and the testes. Hormones are secreted directly into the **blood** stream from where they travel to target tissues and modulate **digestion**, growth, maturation, reproduction, and homeostasis. The word hormone comes from the Greek word, *hormon*, to stir up, and indeed excitation is characteristic of the adrenaline and the **sex hormones**. Most hormones produce an effect on specific target tissues that are sited at some distance from the gland secreting the hormone. Although small **plasma** concentrations of most hormones are always present, surges in secretion trigger specific responses at one or more targets. Hormones do not fall into any one chemical category, but most are either protein molecules or steroid molecules. These biological managers keep the body systems functioning over the long term and help maintain health. The study of hormones is called endocrinology.

Hormones elicit a response at their target **tissue**, target organ, or target cell type through receptors. Receptors are molecular complexes that specifically recognize another molecule-in this case, a particular hormone. When the hormone is bound by its receptor, the receptor is usually altered in some way that it sends a secondary message through the cell to do something in response. Hormones that are proteins, or peptides (smaller strings of **amino acids**), usually bind to a receptor in the cell's outer surface and use a second messenger to relay their action. Steroid hormones such as cortisol, **testosterone**, and **estrogen** bind to receptors inside cells. Steroids are small enough to and chemically capable of passing through the cell's outer membrane. Inside the cell, these hormones bind their receptors and often enter the nucleus to elicit a response. These receptors bind **DNA** to regulate cellular events by controlling gene activity.

Most hormones are released into the bloodstream by a single gland. Testosterone is an exception, because it is secreted by both the adrenal glands and by the testes. Plasma concentrations of all hormones are assessed at some site that has receptors binding that hormone. The site keeps track of when the hormone level is low or high. The major area that records this information is the hypothalamus. A number of hormones are secreted by the hypothalamus that stimulate or inhibit additional secretion of other hormones at other sites. The hormones are part of a negative or positive feedback loop.

Most hormones work through a negative feedback loop. As an example, when the hypothalamus detects high levels of a hormone, it reacts to inhibit further production. And when low levels of a hormone are detected, the hypothalamus reacts to stimulate hormone production or secretion. Estrogen, however, is part of a positive feedback loop. Each month, the Graafian follicle in the ovary releases estrogen into the bloodstream as the egg develops in ever increasing amounts. When estrogen levels rise to a certain point, the pituitary secretes

luteinizing hormone (LH) which triggers the egg's release of the egg into the oviduct.

The concentrations of several important biological building blocks such as amino acids are regulated by more than one hormone. For example, both calcitonin and parathyroid hormone (PTH) influence blood **calcium** levels directly, and other hormones affect calcium levels indirectly via other pathways.

Hormones secreted by the hypothalamus modulate other hormones. The major hormones secreted by the hypothalamus are corticotrophin releasing hormone (CRH), thyroid stimulating hormone releasing hormone (TRH), **follicle stimulating hormone** releasing hormone (FSHRH), luteinizing hormone releasing hormone (LRH), and growth hormone releasing hormone (GHRH). CRH targets the adrenal glands. It triggers the adrenals to release adrenocorticotropic hormone (ACTH). ACTH functions to synthesize and release corticosteroids. TRH targets the thyroid where it functions to synthesize and release the thyroid hormones T3 and T4. **FSH** targets the ovaries and the testes where it enables the maturation of the ovum and of spermatozoa. LRH also targets the ovaries and the testes, and its receptors are in cells that promote ovulation and increase **progesterone** synthesis and release. GHRH targets the anterior pituitary to release growth hormones to most body tissues, increase **protein synthesis**, and increase blood glucose. Hence, the hypothalamus plays a first domino role in these cascades of events.

The hypothalamus also secretes some other important hormones such as prolactin inhibiting hormone (PIH), prolactin releasing hormone (PRH), and melanocyte inhibiting hormone (MIH). PIH targets the anterior pituitary to inhibit milk production at the mammary gland, and PRH has the opposite effect. MIH targets skin pigment cells (**melanocytes**) to regulate pigmentation.

The pituitary has long been called the master gland because of the vast extent of its activity. It lies deep in the **brain** just behind the **nose**. The pituitary is divided into anterior and posterior regions with the anterior portion comprising about 75% of the total gland. The posterior region secretes the peptide hormones vasopressin, also called anti-diuretic hormone (ADH), and **oxytocin**. Both are synthesized in the hypothalamus and moved to the posterior pituitary prior to secretion. ADH targets the collecting tubules of the **kidneys**, increasing their permeability to water. ADH causes at the kidneys to retain water. Lack of ADH leads to a condition called diabetes insipidus characterized by excessive urination. Oxytocin targets the uterus and the **mammary glands** in the **breasts**. Oxytocin begins labor prior to birth and also functions in the ejection of milk. The drug, pitocin, is a synthetic form of oxytocin and is used medically to induce labor.

The anterior pituitary (AP) secretes a number of hormones. The cells of the AP are classified into five types based on what they secrete. These cells are somatotrophs, corticotrophins, thyrotrophs, lactotrophs, and gonadotrophs. Respectively, they secrete growth hormone (GH), ACTH, TSH, prolactin, and LH and FSH. Each of these hormones is either a polypeptide or a glycoprotein. GH controls cellular growth, protein synthesis, and elevation of blood glucose con-

centration. ACTH controls secretion of some hormones by the adrenal cortex (mainly cortisol). TSH controls thyroid hormone secretion in the thyroid. In males, prolactin enhances testosterone production; in females, it initiates and maintains LH to promote milk secretion from the mammary glands. In females, FSH initiates ova development and induces ovarian estrogen secretion. In males, FSH stimulates **sperm** production in the testes. LH stimulates ovulation and formation of the corpus luteum, which produces progesterone. In males, LH stimulates interstitial cells to produce testosterone. Each AP hormone is secreted in response to a hypothalamic releasing hormone.

The thyroid lies under the **larynx** and synthesizes two hormones, thyroxine and tri-iodothyronine. This gland takes up iodine from the blood and has the highest iodine level in the body. The iodine is incorporated into the thyroid hormones. Thyroxine has four iodine atoms and is called T4. Tri-iodothyronine has three iodine atoms and is called T3. Both T3 and T4 function to increase the metabolic rate of several cells and tissues. The brain, testes, **lungs,** and spleen are not affected by thyroid hormones, however. T3 and T4 indirectly increase blood glucose levels as well as the insulin-promoted uptake of glucose by **fat** cells. Their release is modulated by TSH-RH from the hypothalamus. TSH secretion increases in cold infants. When temperature drops, a metabolic increase is triggered by TSH. Chronic stress seems to reduce TSH secretion which, in turn, decreases T3 and T4 output.

The parathyroid glands are attached to the bottom of the thyroid gland. They secrete the polypeptide parathyroid hormone (PTH), which plays a crucial role in monitoring blood calcium and phosphate levels. About 99% of the body's calcium is in the bones, and 85% of the magnesium is also found in bone. Low blood levels of calcium stimulate PTH release into the bloodstream in two steps. Initially, calcium is released from the fluid around bone cells. And later, calcium can be drawn from bone itself. Although, only about 1% of bone calcium is readily exchangeable. PTH can also increase the absorption of calcium in the intestines by stimulating the kidneys to produce a vitamin D-like substance that facilitates this action. High blood calcium levels will inhibit PTH action, and magnesium (which is chemically similar to calcium) shows a similar effect.

The two adrenal glands, one on top of each kidney, each have two distinct regions. The outer region (the medulla) produces adrenaline and noradrenaline and is under the control of the **sympathetic nervous system.** The inner region (the cortex) produces a number of steroid hormones. The cortical steroid hormones include mineralocorticoids (mainly aldosterone), glucocorticoids (mainly cortisol), and gonadocorticoids. These steroids are derived from cholesterol. Although cholesterol receives a lot of bad press, some of it is necessary. Steroid hormones act by regulating gene expression, hence, their presence controls the production of numerous factors with multiple roles. Aldosterone and cortisol are the major human steroids in the cortex. However, testosterone and estrogen are secreted by adults (both male and female) at very low levels.

Aldosterone plays an important role in regulating body fluids. It increases blood levels of sodium and water and lowers blood potassium levels. Low blood sodium levels trigger aldosterone secretion via the renin-angiotensin pathway. Renin is produced by the kidney, and angiotensin originates in the **liver**. High blood potassium levels also trigger aldosterone release. ACTH has a minor promoting effect on aldosterone. Aldosterone targets the kidney where it promotes sodium uptake and potassium excretion. Since sodium ions influence water retention, the result is a net increase in body fluid volume.

Blood cortisol levels fluctuate dramatically throughout the day and generally peak in the early morning. Presumably, this early peak enables humans to face the varied daily stressors they encounter. Cortisol secretion is stimulated by physical trauma, cold, **burns**, heavy exercise, and anxiety. Cortisol targets the liver, **skeletal muscle**, and **adipose tissue**. Its overall effect is to provide amino acids and glucose to meet synthesis and energy requirements for normal **metabolism** and during periods of stress. Because of its anti-inflammatory action, it is used clinically to reduce swelling.

The adrenal medullary hormones are **epinephrine** (adrenaline) and nor-epinephrine (nor-adrenaline). Both of these hormones serve to supplement and prolong the fight or flight response initiated in the nervous system. This response includes the neural effects of increased **heart** rate, peripheral blood vessel constriction, sweating, spleen contraction, **glycogen** conversion to glucose, dilation of bronchial tubes, decreased digestive activity, and lowered urine output.

The condition of stress presents a model for reviewing one way that multiple systems and hormones interact. During stress, the nervous, endocrine, digestive, urinary, respiratory, circulatory, and immune response are all tied together. For example, the hypothalamus sends nervous impulses to the **spinal cord** to stimulate the fight or flight response and releases CRH that promotes ACTH secretion by the pituitary. ACTH, in turn, triggers interleukins to respond which promote immune cell functions. ACTH also stimulates cortisol release at the adrenal cortex that helps buffer the person against stress. As part of a negative feedback loop, ACTH and cortisol receptors on the hypothalamus assess when sufficient levels of these hormones are present and then inhibit their further release. Destressing occurs over a period of time after the stressor is gone. The systems eventually return to normal.

The pancreas folds under the stomach, secretes the hormones insulin, glucagon, and somatostatin. About 70% of the pancreatic hormone-secreting cells are called beta cells and secrete insulin; another 22%, or so, are called alpha cells and secrete glucagon. The remaining gamma cells secrete somatostatin, also known as growth hormone inhibiting hormone (GHIH). The alpha, beta, and gamma cells comprise the islets of Langerhans which are scattered throughout the pancreas.

Insulin and glucagon have reciprocal roles. Insulin promotes the storage of glucose, fatty acids, and amino acids, whereas, glucagon stimulates mobilization of these constituents from storage into the blood. Both are relatively short polypeptides. Insulin release is triggered by high blood glucose levels. It lowers blood sugar levels by binding a cell surface receptor and accelerating glucose transport into the cell where glucose is converted into glycogen. Insulin also inhibits the release of glucose by the liver in order to keep blood levels down. Increased blood levels of GH and ACTH also stimulate insulin secretion. Not all cells require insulin to store glucose, however. Brain, liver, kidney, intestinal, epithelium, and the pancreatic islets can take up glucose independently of insulin. Insulin excess can cause hypoglycemia leading to convulsions or **coma**, and insufficient levels of insulin can cause **diabetes mellitus**, which can be fatal if left untreated. Diabetes mellitus is the most common endocrine disorder.

The female reproductive hormones arise from the hypothalamus, the anterior pituitary, and the ovaries. Although detectable amounts of the steroid hormone estrogen are present during fetal development, at **puberty** estrogen levels rise to initiate **secondary sexual characteristics**. Gonadotropin releasing hormone (GRH) is released by the hypothalamus to stimulate pituitary release of LH and FSH. LH and FSH propagate egg development in the ovaries. Eggs (ova) exist at various stages of development, and the maturation of one ovum takes about 28 days and is called the ovarian or menstrual cycle. The ova are contained within follicles, which are support **organs** for ova maturation. About 450 of a female's 150,000 **germ cells** mature to leave the ovary. The hormones secreted by the ovary include estrogen, progesterone, and small amounts of testosterone.

As an ovum matures, rising estrogen levels stimulate additional LH and FSH release from the pituitary. Prior to ovulation, estrogen levels drop, and LH and FSH surge to cause the ovum to be released into the fallopian tube. The cells of the burst follicle begin to secrete progesterone and some estrogen. These hormones trigger thickening of the uterine lining, the endometrium, to prepare it for implantation should **fertilization** occur. The high progesterone and estrogen levels prevent LH and FSH from further secretion-thus hindering another ovum from developing. If fertilization does not occur, eight days after ovulation the endometrium deteriorates resulting in **menstruation**. The falling estrogen and progesterone levels which follow trigger LH and FSH, starting the cycle all over again.

Although estrogen and progesterone have major roles in the menstrual cycle, these hormones have receptors on a number of other body tissues. Estrogen has a protective effect on bone loss, which can lead to **osteoporosis**. And progesterone, which is a competitor for androgen sites, blocks actions that would result from testosterone activation. Estrogen receptors have even been found in the forebrain indicating a role in female neuronal function or development.

Hormones related to **pregnancy** include human chorionic gonadotrophin (HCG), estrogen, human chorionic somatomammotrophin (HCS), and relaxin. HCG is released by the early embryo to signal implantation. Estrogen and HCS are secreted by the **placenta**. And relaxin is secreted by the ovaries as birth nears to relax the pelvic area in preparation for labor.

Male reproductive hormones come from the hypothalamus, the anterior pituitary, and the testes. As in females, GRH is released from the hypothalamus, which stimulates LH and FSH release from the pituitary. In males, LH and FSH facilitate **spermatogenesis**. The steroid hormone testosterone is secreted from the testes and can be detected in early embryonic development up until shortly after birth. Testosterone levels are quite low until puberty. At puberty, rising levels of testosterone stimulate male reproductive development including secondary characteristics.

LH stimulates testosterone release from the testes. FSH promotes early spermatogenesis, whereas testosterone is required to complete spermatogenic maturation to facilitate fertilization. In addition to testosterone, LH, and FSH, the male also secretes **prostaglandins**. These substances promote uterine contractions that help propel sperm towards an egg in the fallopian tubes during sexual intercourse. Prostaglandins are produced in the seminal vesicles, and are not classified as hormones by all authorities.

See also Adolescent growth and development; Adrenal glands and hormones; Blood pressure and hormonal control mechanisms; Glucose utilization, transport and insulin; Gonads and gonadotropic hormone physiology; Human growth hormone (somatotropin); Hyperthyroidism and hypothyroidism; Parathyroid glands and hormones; Pituitary gland and hormones

HORSLEY, VICTOR ALEXANDER HADEN
(1857-1916)
British surgeon

Sir Victor Horsley was a pioneer in the fields of neurosurgery and neurophysiology. In 1887, he performed the first successful operation to remove a spinal tumor. Advancements in the understanding of localized **brain** function, and practical surgical innovations, developed by Horsley made neurosurgery a medical possibility.

The son of a renowned artist, Horsley was born in London but raised in Cranbrook, Kent. He then pursued studies in medicine at University College, London, and continued his career at the university hospital after graduation. He began his research with a series of investigations into the function and **physiology** of glands. In various animal experiments, Horsley proved that the thyroid gland was key to a specimen's rate of **growth and development**. Though he did not establish how exactly how the thyroid worked, or understand the workings of **hormones** in the **endocrine system**, Horsley identified gland malfunction as the cause of several diseases and disorders. Horsley published works on the role of the thyroid in myxedema, a condition that causes severe swelling. He later performed the first successful removal of a pituitary gland.

Horsley performed several experiment on brain and **spinal cord** function, and the localization of brain activity. This research aided his work in neurosurgery. When Horsley first began his studies, operations on the brain or spine were rarely successful. By 1890, Horsley boasted of nearly 45 successful operations. Advancements in neurosurgery allowed not only for the possibility of treatment for some diseases, but also led to a better physiological understanding of the nervous system in general. Horsley's surgical experiments yielded information on cranial function that was also seminal to early studies of epilepsy, paralysis, and traumatic injury. To this end, Horsley was named as a consultant to the National Hospital for the Paralyzed and Epileptics, Queen Square.

Horsley also solved some of the technical problems of neurosurgery. Before the advent of reliable transfusion, **blood** loss during cranial surgery was a significant problem. Horsley concocted a paste, a mixture of nut oil and bees wax, to reduce patient bleeding. To aid his investigations of cerebral function, Horsley, with the aid of a colleague R.H. Clark, invented a series of specialized instruments.

An outspoken social reformer, in 1886, Horsley was appointed secretary of a government study on the effectiveness of a rabies vaccine. He actively supported women's suffrage, and advocated total abstinence from the use of tobacco and alcohol. Horsley's 1907 work, *Alcohol and the Human Body*, enjoyed moderate public success. He retired from his successful medical career to pursue politics. He was influential in promoting public health initiatives and medical reforms. During the First World War, Horsley served in the Royal Army Medical Corps. He died of **hyperthermia** while on duty in Mesopotamia.

See also Nervous system overview

HOUNSFIELD, GODFREY N. (1919-)
English engineer

Sir Godfrey Hounsfield pioneered a leap forward in medical diagnosis: computerized axial tomography, popularly known as the "CAT scan." Ushering in a new and sometimes controversial era of medical technology, Hounsfield's device allowed a doctor to look inside a patient's body and examine a three-dimensional image far more detailed than a conventional x ray. The importance of this advance was recognized in 1979, the year Hounsfield received the Nobel Prize in physiology or medicine.

Godfrey Newbold Hounsfield was born in Newark, England, the youngest of five children of a steel-industry engineer turned farmer. Hounsfield's technical interests began when, to prevent boredom, he began figuring out how the machinery on his father's farm worked. From there he moved on to exploring electronics, and by his teens was building his own radio sets. He graduated from London's City and Guilds College in 1938 after studying radio communication. When World War II erupted, Hounsfield volunteered for the Royal Air Force, where he studied and later lectured on the new and vital technology of radar at the RAF's Cranwell Radar School. After the war he resumed his education, and received a degree in electrical and mechanical engineering from Faraday House Electrical Engineering College in 1951. Upon graduation, Hounsfield joined Thorn EMI (Electrical and Musical Industries) Ltd., an employer he has remained with his entire professional life.

At Thorn EMI, Hounsfield worked on improving radar systems and then on computers. In 1959, a design team led by

Hounsfield finished production of Britain's first large all-transistor computer, the EMIDEC 1100. Hounsfield moved on to work on high-capacity computer memory devices, and was granted a British patent in 1967 titled "Magnetic Films for Information Storage."

Hounsfield's work in this period included the problem of enabling computers to recognize patterns, thus allowing them to "read" letters and numbers. In 1967, during a long walk through the British countryside, Hounsfield's knowledge of computers, pattern recognition, and radar technology all came together in his mind. He envisioned a medical diagnostic system in which an x-ray machine would image thin "slices" through the patient's body and a computer would process the slices into an accurate representation which would display the tissues, **organs**, and other structures in much greater detail than a single x ray could produce. Computers available in 1967 were not sophisticated enough to make such a machine practical, but Hounsfield continued to refine his idea and began working on a prototype scanner. He enlisted two radiologists, James Ambrose and Louis Kreel, who assisted him with their practical knowledge of radiology and also provided **tissue** samples and test animals for scans. The project attracted support from the British Department of Health and Social Services, and in 1971, a test machine was installed at Atkinson Morely's Hospital in Wimbledon. It was highly successful, and the first production model followed a year later. These original scanners were designed for **imaging** the **brain**, and were hailed by neurosurgeons as a great advance. Before the CAT scanner, doctors wanting a detailed brain x ray had to help their equipment see through the **skull** by such dangerous techniques as pumping chemicals or air into the brain. As head of EMI's Medical Systems section, Hounsfield continued to improve the device, working to lower the radiation exposure required, sharpen the images produced, and develop larger models which could image any part of the body, not just the head. This "whole body scanner" went on the market in 1975.

CAT scanners generated some initial resistance because of their expense: even the earliest models cost over $300,000, and improved versions several times as much. Despite this, the machines were so useful they quickly became standard equipment at larger hospitals around the world. Hounsfield argued that, properly used, the scanners actually reduced medical costs by eliminating exploratory surgery and other invasive diagnostic procedures. The scanner won Hounsfield and his company more than thirty awards, including the MacRobert Award, Britain's highest honor for engineering. In 1979, Hounsfield's collection of scientific tributes culminated in the Nobel Prize. That year's Nobel was shared with Allan M. Cormack, an American nuclear physicist who had separately developed the equations involved in reconstructing an image via computer. A surprising feature of the selection was that neither man had a degree in medicine or biology, or a doctorate in any field.

Hounsfield moved on to positions as chief staff scientist and then senior staff scientist for Thorn EMI. He continued to improve the CAT scanner, working to develop a version that could take an accurate "snapshot" of the **heart** between beats.

He has also contributed to the next step in diagnostic technology, nuclear magnetic resonance imaging. In 1986, he became a consultant to Thorn EMI's Central Research Laboratories in Middlesex, near his longtime home in Twickenham.

See also Imaging

HOUSSAY, BERNARDO (1887-1971)
Argentine physician

Bernardo Houssay studied nearly every aspect of human **physiology**, but is best known for his discovery of the role of the pituitary gland in the **metabolism** of **carbohydrates**, an accomplishment for which he was awarded the 1947 Nobel Prize in physiology or medicine. He led a highly productive life, authoring or coauthoring more than 600 scientific papers and books. Houssay taught at the University of Buenos Aires for twenty-five years before losing his position for political reasons. He then helped to establish the independent Institute of Biology and Experimental Medicine, a research facility that soon became a primary focus of scientific studies in Latin America.

Bernardo Alberto Houssay was born in Buenos Aires, Argentina, to parents who had emigrated from France before his birth. His father was Albert Houssay, a lawyer who also taught literature at the National College of Buenos Aires, and his mother was the former Clara Laffont. Juan T. Lewis, writing in *Perspectives in Biology,* describes how Houssay early on began to practice a way of life that was "distinguished by concentration of purpose, hard work and the avoiding of loss of time and energy in frivolous pursuits." As evidence of these attitudes, young Bernardo promised his father at the age of thirteen that he would henceforth assume all responsibility for his own personal financial expenses.

Houssay completed his secondary education at the Colegio Británico at the age of fourteen. Three years later he earned his degree in pharmaceutical chemistry from the University of Buenos Aires, receiving the highest honors in his class. He then enrolled in the school of medicine at the university and was granted his M.D. at the age of twenty-three. Houssay's medical studies took somewhat longer to complete than might have been expected given his previous academic record, because he simultaneously worked as a hospital pharmacist in order to help pay for his expenses.

Having completed his studies, Houssay was appointed provisional professor, and, in 1912, full professor of physiology at the university's school of veterinary science. In 1913, he became chief physician at Alvear Hospital as well as a laboratory director in the newly created National Public Health Laboratories. Houssay's 1919 return to the university as chair of physiology marked the beginning of his greatest impact in the field. It was at the university that he established and became director of the Institute of Physiology, a research center that was to attain worldwide distinction. At its peak, the Institute was home to 135 graduate students from every part of the world, extending Houssay's influence far beyond the bor-

Bernardo Houssay. *The Library of Congress.*

ders of Argentina. In 1920 Houssay married María Angélica Catán, a chemist. All three of their children earned medical degrees.

In spite of his many administrative responsibilities, Houssay continued to be very active in research throughout his life. He was intensely interested in every aspect of physiology, from the cardiovascular to the respiratory to the gastrointestinal systems. But his major accomplishments resulted from his studies of the **endocrine system**, studies that dated to research begun while he was still a medical student. That research received an important impetus in 1921,when Canadians **Frederick Banting** and Charles Best and Scottish physiologist John Macleod discovered the role of insulin in the development of diabetes.

From 1923 to 1937, Houssay studied the interaction between the **pancreas** and insulin, on the one hand, and the pituitary gland (then called the hypophysis) and its secretions, on the other. One of his first major discoveries was the role of the anterior lobe of the pituitary gland in the metabolism of carbohydrates. A more important discovery was that the oxidation of sugars in the body depends not simply on the presence or absence of insulin, but on a complex interaction between insulin and other **hormones**, such as prolactin and somatotropin, produced in the pituitary gland. For his unraveling of this process, Houssay received a share of the 1947 Nobel Prize in physiology or medicine.

The political turmoil that swept Argentina in the 1940s altered Houssay's career. During the uprisings of 1943, he signed a petition calling for the democratization of the Argentine government. As a result, he was dismissed from his post at the university. Two years later, the dismissal was voided, and Houssay returned to the university. He was there only briefly, however, before he was asked to retire, which he did in 1946. In the meantime, he and some colleagues had founded the independent Institute of Biology and Experimental Medicine in order to continue with their research. Even when Houssay was yet again reinstated to his old post at the university in 1955, he continued to serve as director of the Institute.

Houssay was a major leader of Argentine science for many years. He founded, assisted in the establishment of, or served as head of nearly every major scientific organization in the country between 1920 and 1970. He was given honorary doctorates by more than twenty-five universities and was elected to membership in scientific societies in Great Britain, Germany, France, Italy, Spain, and the United States. Houssay died in Buenos Aires in 1971.

See also Endocrine system and glands; Pituitary gland and hormones

HUBEL, DAVID H. (1926-)
Canadian-born American neurophysiologist

David H. Hubel is a neurobiologist whose research into the relationships between the **eye** and the **brain** began at the Walter Reed Army Institute of Research. He later joined the research team at Johns Hopkins led by Stephen Kuffler, a neurophysiologist of vision. In Kuffler's laboratory, Hubel worked with **Torsten Wiesel**; their teamwork lasted over twenty years and led in 1981 to their sharing the Nobel Prize in physiology or medicine. The prize was awarded to them because they had discovered what role **neurons** played in the visual system and how the arrangement of cells operated in the visual process.

Born in Windsor, Ontario, of American parents, Elsie M. Hunter Hubel and Jesse H. Hubel, David Hunter Hubel grew up in Montreal. From his father, who was a chemical engineer, Hubel developed an interest in science, especially chemistry and electronics. In one particularly memorable childhood chemistry experiment, Hubel told an *Omni* magazine interviewer, he investigated the "percussive properties" of potassium chlorate and sugar by setting off a small brass cannon in the street outside his home. The local police quickly discouraged further experiments in that line. Hubel displayed musical as well as scientific talent during childhood, and playing flute duets remains one of his favorite forms of enjoyment.

From 1932 to 1944, Hubel attended the Strathcona Academy in Outremont, Ontario. He began his college studies at McGill University in 1944. Although he received his B.S. with honors in mathematics and physics, he decided to enter McGill University Medical School in 1947—a decision which he appears to have made almost on the spur of the moment, since he had not taken any college course in biology. He also worked summers at the Montreal Neurological Institute, where he began his studies of the nervous system. He received his medical degree in 1951 and spent the next four years studying clinical **neurology**, first at the Montreal Neurological Institute and then at Johns Hopkins University in Baltimore, Maryland.

In 1955, Hubel was drafted into the United States Army, which sent him to the Neurophysiology Division of the Walter Reed Army Institute of Research in Washington, D.C. At Walter Reed, Hubel discovered a stimulating group of physiologists who encouraged him to do original research for the first time in his life. Determined to study **sleep**, he developed a device, known as a tungsten microelectrode, to record the electrical impulses of nerve cells. He used this device on cats to measure the activity of nerve cells in sleep. He described these experiments in his *Omni* interview: "I had concocted one electrode strong enough to penetrate the membranes over the cortex. It was a stiff tungsten wire a little thicker than a **hair**. It was so sharp that it went in without doing any damage, so far as we knew. It was attached to a bunch of amplifiers and finally to loudspeakers. When a cell fired, you heard it as a brief pop or click. If it fired fast, it sounded like a machine gun."

During his research on sleep, Hubel became more interested in the reactions of his subjects to the firing responses recorded by the microelectrodes during waking states. He had placed the microelectrodes in the visual cortex area of the brain for his sleep experiments, and he began to realize that it was possible to understand how the brain operates in the visual process. In reading the work of other scientists on this subject, Hubel discovered the research papers of Stephen Kuffler, who was then a leading figure in the neurophysiology of vision. In the *Omni* interview, Hubel credits Kuffler with being "the first person to study optic nerve fibers in a mammal to find the important things that they were doing that the rods and cones in the **retina** don't do."

After his army service ended in 1958, Hubel went to Johns Hopkins University where he did further research on the surface of the brain, the **gray matter** of the **cerebral cortex**, in the laboratory of Vernon Mountcastle. But shortly afterwards he moved to the Wilmer Institute, also at Johns Hopkins, and joined Stephen Kuffler's research team. There he met Torsten Wiesel, and under the direction of Kuffler the two of them began to make discoveries about the relationship of the retina to the visual cortex as part of the general physiology of the brain.

In 1959, Hubel and Wiesel, along with the rest of Kuffler's research team, followed Kuffler to the Harvard Medical School in Boston. By 1964, Harvard had formed a new department of neurobiology, naming Kuffler as its chairman. Hubel became chairman of this department in 1967, and in 1968 he was named the George Packer Berry Professor of Physiology.

Much of the work done by Hubel and Wiesel, using microelectrodes and electronic equipment, centered around a section of the visual cortex in the brain known as area 17. The cells in this section of the visual cortex form several thin layers that are arranged in columns running through the cortex. Hubel and Weisel discovered that certain cells of area 17 in the brain respond to the stimulation of specific retinal cells in the eye. In particular, they found that cells in the cortex are specialized to respond to different types of stimulation. There are types of cortical cells that respond to light spots and others that respond specifically to the different angles of a tilted line. They discovered that some respond only to definite directions of movement, while others respond only to definite colors.

Hubel and Wiesel's research has made the visual cortex the most mapped-out section of the brain, and it has deepened the scientific understanding of how the visual system works. In addition, their work has led to practical ophthalmological applications for the treatment of congenital cataracts, as well as a condition occurring in childhood known as strabismus, where one eye is unable to focus with the other because of a muscle imbalance. Hubel and Wiesel discovered that at birth, the visual cortex begins to develop its structures from the stimulation of the newborn's retina. The development of the brain is shaped by the activity of the eye, and the sooner childhood eye disorders are corrected, therefore, the better the chances of avoiding serious visual impairments in the future. Before their research, the customary medical practice had been to delay operating on these conditions, but today doctors recognize the importance of the early removal of cataracts and the prompt treatment of strabismus.

For their work on how the retinal image is read and interpreted by the cells of the visual cortex, Hubel and Wiesel shared the 1981 Nobel Prize in physiology or medicine. For his work on split-brain physiology, Roger W. Sperry won the second half of the prize.

Hubel has been married to Shirley Ruth Izzard Hubel since 1953, and they have three sons.

See also Eye and ocular fluids; Eye: Ocular embryological development; Sense organs: Ocular (visual) structures; Vision: Histophysiology of the eye

HUGGINS, CHARLES B. (1901-1997)
American surgeon

Charles B. Huggins was awarded the Nobel Prize in physiology or medicine in 1966 for his discovery in the 1930s of the role played by **hormones** in the onset and growth of prostate and breast **cancer**. This breakthrough led Huggins to make a number of important medical advances, including the subsequent development of hormone therapy, the first non-radioactive, non-toxic chemical treatment for cancer. In other studies, Huggins found that cancer cells are not necessarily self-reliant and self-perpetuating, and that some cancers actually depend on normal hormone levels to develop and grow. He then developed a **blood** test to measure two particular **enzymes** to determine the extent of the cancer and the effect of hormone therapy. In addition, Huggins discovered the compensatory action of adrenal glands after hormone therapy and performed the first surgical removal of the adrenal glands to combat cancer regrowth. He also developed cortisone replacement therapy to compensate for the loss of normal adrenal gland function.

Charles Brenton Huggins, the oldest of two sons, was born in Halifax, Nova Scotia, to pharmacist Charles Edward and Bessie Marie (Spencer) Huggins. He earned his B.A. from Acadia University, Wolfville, Nova Scotia, in 1920, graduating in a class of twenty-five students. That same year, he moved to the United States to attend Harvard Medical School, graduating in four years with both an M.A. and M.D. He did his internship at the University of Michigan Hospital and was appointed instructor in surgery at the University's Medical School in 1926. The following year, he became instructor of surgery on the original faculty of the University of Chicago Medical School, and in that same year, he married Margaret Wellman. The couple eventually had a son and daughter. Huggins was promoted to assistant professor in 1929, then to associate professor in 1933, the year he became an American citizen, and he attained the rank of full professor of surgery in 1936. In 1946, he spent a brief period with the Johns Hopkins University as professor of urological surgery and director of the department of **urology**. He was director of the University of Chicago's Ben May Laboratory for Cancer Research from 1951 to 1969, continuing his research at the university until 1972, when he returned to Acadia University to become chancellor. He retired from the post in 1979 and moved to Chicago.

Huggins's initial specialty was urology, but his interest in cancer was actually sparked in 1930, when he met German Nobel Prize-winning cancer researcher **Otto Warburg**. Upon his return to the University of Chicago in the early 1930s, Huggins and his colleagues experimented with changing normal connective **tissue** elements into bone, using cells from the male urinary tract and bladder. His interest soon turned to the male **urogenital system**, particularly the role played by chemicals and hormones in the **prostate gland**, the male accessory reproductive gland located at the base of the urethra. He and his colleagues developed what Paul Talalay and Guy Williams-Ashman in *Science* called "an ingenious surgical procedure... [which] isolated the prostate gland of dogs from the urinary tract." This procedure, introduced in 1939, allowed the analysis and measurement of secretions of the gland which form much of the ejaculatory fluid. The research was at times frustrated by the formation of prostate **tumors** in some of the dogs, the only animal other than man known to develop cancer of the prostate. Huggins, however, saw the obstacle as an opportunity. He turned his energy to studying the development and growth of prostate cancer, a painful and often fatal disease prevalent in men over the age of fifty that causes obstruction of the urinary tract and, if left untreated, metastasizes (spreads) to the bone and **liver**.

Huggins discovered high levels of **testosterone**, a male sex hormone, in secretions from a cancerous prostate. He also discovered that reducing male hormone secretions by either orchiectomy (castration) or **estrogen** (a female hormone) therapy, or both, drastically reduced testosterone levels and inhibited the growth of advanced metastatic prostate cancer. In his first human trials, four out of twenty-one patients treated with this method survived for twelve years. He also developed a blood test to measure acid phosphatase, which is secreted by the prostate, and alkaline phosphatase, which is secreted by

bone-forming cells in bone tissue, both of which showed increased levels in patients with metastasized prostate cancer. Using these measurements, he could determine the extent of the cancer and the effect of the hormone treatments.

Huggins found that although the level of androgens (male **sex hormones**) dropped drastically after orchiectomy, in some cases they rose again, often to a level higher than before the surgery. Investigations led him to believe that the adrenal glands were producing androgens of their own, apparently compensating for the lowered levels induced by the hormone therapy. These androgens, too, encouraged the growth of the cancer. In 1944, he performed the first bilateral adrenalectomy, (removal of the two adrenal glands located above the **kidneys**), producing some positive results, even before cortisone was readily available for replacement therapy. In 1953, Huggins reported that, when used in combination, adrenalectomy and cortisone replacement had a beneficial effect on fifty percent of patients suffering from either prostate or breast cancer, but had no effect on other types of cancer. This was a radical treatment, however, and used only as a last resort.

In the 1950s, Huggins left the clinical environment to return to the laboratory. While delivering the tenth Macewen Memorial Lecture at the University of Glasgow in 1958, he referred to breast cancer as "one of the noblest of the problems of medicine." In his lecture, entitled *Frontiers of Mammary Cancer,* he said, "Cancer of the breast in the United States has the highest prevalence rate of any form of neoplastic disease in either sex... Commonly, the disease advances with dreadful speed and ferocity." Huggins and two students, D. M. Bergenstal and Thomas Dao, developed a treatment for cancer that entailed removal of both ovaries and both adrenal glands. Combined with cortisone replacement therapy, the treatment brought about improvement in 30–40% of the patients with advanced breast cancer, sometimes with quite definite and prolonged improvement.

Breast cancer research was being hampered, however, because of the long delay between stimulation and growth of artificially induced mammary tumors in animals. In 1956, Huggins discovered that a single dose of 7,12-dimethylbens(a)anthracene (DMBA) would quickly induce mammary tumors in certain types of female rats and that many of these tumors were, like some in humans, hormone dependent and responded to regulation of the hormonal environment. Huggins' rat tumors soon became the focus of experiments in laboratories all over the world.

For his research on hormones and cancer, Huggins shared the 1966 Nobel Prize with **Peyton Rous**, who was honored for his work fifty-five years earlier on viral causes of cancer. Only one person previously had been awarded the Nobel for cancer research—Johannes Fibiger in 1926 for developing a method of growing artificial tumors. Colleagues agreed that both Huggins and Rous should have received the award many years earlier. In addition to the Nobel Prize, Huggins was awarded one of the highest honors to be bestowed by American medicine, the Lasker Clinical Research Award, in 1963. He was also the first recipient of the Charles L. Mayer Award in cancer research from the National Academy of Sciences in 1943. Huggins also was awarded two gold medals for research from the American Medical Association, the Order of Merit from Germany, and the Order of the Sun from Peru. He was made honorary fellow of the Royal College of Surgeons in both Edinburgh and London, and is the recipient of numerous honorary degrees.

A devoted family man, a lover of the music of Bach and Mozart, and a self-admitted workaholic, Huggins was also known for his wry wit. He died at his home in Chicago at the age of 95.

See also Adrenal glands and hormones; Hormones and hormone action; Ovarian cycle and hormonal regulation; Tumors and tumorous growth

HUMAN DEVELOPMENT (TIMETABLES AND DEVELOPMENTAL HORIZONS)

The Carnegie staging system is a method of staging embryonic development that evaluates the embryo's growth and differentiation in 23 stages. It supposes that human development averages 266 days, or 9.5 months. In the early stages of an embryo, a series of binary events take place in **cell differentiation** in order to establish cell lineage. Extraembrionic tissues are generated before embryonic **tissue**.

Embryonic development is apparent at 13–15 days after **fertilization** (stage 6). Stages 1–5 (days 1–7) are related to the proliferation of cells needed for implantation and extra embryonic tissue. At 14–15 days (stage 6b), intraembryonic cell populations migrate, producing morphological movements that yield a recognizable embryo. Cells from the primitive streak proliferate, providing a cell population that passes to the embryo. By days 19–21 (stage 9), the neural cell population is evident, resulting in the neurulation related processes. Neural place formation contributes to the embryo folding. The primordial cells lay in the endothelium at the entrance of the diverticulum to the yolk sac, then in the **mesoderm** derived tissue of the yolk, near the allantois. Cardiac contractions begin at day 22–23 (stage 10, when the **heart** has a recognizable ventricle, bulbus cordis and arterial trunk. Again, the intraembryonic coelom is very important at this stage of development. It arises from confluent spaces that appear at stage 9. These spaces coalesce to form a horseshoe-shaped cavity within the embryonic body that passes between the **endoderm** of the gut and the **ectoderm** of the body wall. The wall of the coelom is composed of germinal epithelial cells the provide mesenchyme for the **connective tissues** and smooth muscles of the heart, respiratory and gastrointestinal tracts. The most important event during days 23–26 (stage 11) is the closing of the rostral neuropore. Fusion occurs form the rhomboencephalon rostrally towards the mesencephalon and from the area of the future optic chiasma toward the mesencephalic roof.

The stage 11 embryo at 23–26 days of development (13–20 pairs of **somites**) is at the gateway of organogenesis (the origin and development of **organs** during embryonic life). The fetal period begins at the end of the eighth week form the fertilization. At eight weeks, the embryo is about one inch in length. Facial features, limbs, hands, feet, fingers, and toes become apparent. The nervous system is responsive and some of the internal organs, as well as heart, begin to function. At ten weeks, the fetus is three inches long and weighs almost one ounce. The muscles begin to develop and genital organs form. Eyelids, fingernails, and toenails also form. Fetal spontaneous movements can be observed by ultrasound scan. At 16 weeks, the fetus is about five inches long. At 20 weeks, the fetus now weighs approximately 0.5 lb. (227 g) and spans about ten inches from head to foot. Skin matures and **sweat glands** develop. If a preterm delivery occurs, there is some chance of post-natal survival. At the beginning of the third trimester, about four out of ten newborns can survive. At 26 weeks, the fetus has **lungs** capable of **breathing** air if it is born at this time, although medical help may be needed. The rate of fetal growth increases from 5 g/day at 12–13 weeks (from fertilization) to 10 g/day at 18 weeks, and to 30–35 g/day at 30–32 weeks.

Future research of human development can be divided in two major areas. The first one concerns embryonic development; the second is related to the monitoring of fetal growth. Embryonic development research focuses on several fields, including the genetic control of maternal regulation of early development, genome activation, preimplantation development, morphogenesis, pattern formation, and sex differentation. Included are ongoing and future studies of specific homeobox genes, as well as the bone morphogenetic protein (BMP). Specific studies evaluate the mechanism of mesoderm development (vent genes, a subclass of homeobox genes) as well as the limb morphogenesis. Another important area of embryological research concerns embryonic **stem cells**, because they could be used in possible treatments for diseases. Human embryos must be sacrificed to harvest embryonic stem cells that, because not yet differentiated, could serve various functions, such as tissue production. Embryonic stem cells can be engineered to differentiate into many different types of cells related to any type of tissue. Diabetes, Parkinson's and **Alzheimer's** are among the diseases that stand to benefit from embryonic stem cell research.

See also Cell differentiation; Ear, otic embryological development; Embryonic development, early development, formation, and differentiation; Eye: ocular embryological development; Face, nose, and palate embryonic development; Gastrointestinal embryological development; Integumentary system, embryonic development; Limb buds and limb embryological development; Nervous system, embryological development; Renal system, embryological development; Respiratory system embryological development; Skeletal and muscular systems, embryonic development; Urogenital system, embryonic development; Vascular system (embryonic development)

The Carnegie stages of human embryonic development. Courtesy of Antonio Farina, MD.

Stage	Days (approx.)	Size (mm)	Somite Number	Events	Description
1	1	0.1–0.15		Fertilized oocyte, zygote	Fertilization, first cleavage
2	2–3	0.1–0.2		Compaction	Preimplantation
3	4–5	0.1–0.2		Trophoblast (syncitio and cytothrophoblast) Inner cell mass (epiblast and ipoblast)	Free blastocyst
4	5–6	0.1–0.2		Attaching blastocyst	Implantation
5a	7–8 (week 2)	0.1–0.2		Visceral and parietal hypoblast Extraembryonic mesoblast Amnion	
5b	9–10			Lacunae in syncytiotrophoblast Chorion from cytotrophoblast	Implanted previllus
5c	11–12			1st yolk sac Connecting stalk	
6a	13–14	0.2		Extra-embryonic mesenchyme, primitive streak Villi Primordial germ cells	2nd yolk sack
6b	14–15			Intervillous spaces	Primative streak
				2nd yolk sac Primitive streak	
7	16–17 (week 3)	0.4		Gastrulation (notochordal process/plate, embryonic endoderm, mesoblast)	Notochord
8	17–19	1.0–1.5		Primitive pit Notochordal canal Hemopoiesis	Neurenteric canal before folding
9	19–21	1.5–2.5	1–3	Neural folds Cardiac primordium Head fold Allantois Intraembryonic Coelom Mesenchyme Somatopleuric and splanchnopleuric Coelomic epithelium Neural ectoderm Surface ectoderm	Somites
10	22–23 (week 4)	2–3.5	4–12	Neural fold fuses Neural crest Ectodermal placodes	Neural tube after folding
11	23–26	2.5–4.5	13–20	Rostral neuropore closes Pharyngeal arches 1 and 2 are evident	
12	26–30	3–5	21–29	Caudal neuropore closes Three branchial bars are evident Ear vesicles are almost closed, but not detached Arm buds are just beginning to appear Intestinal portal	
13	28–32 (week 4)	4–6	30	Leg buds Lens placode Pore of otic invagination closed Heart chambers distended	
14	31–35	5–7		Invagination of optic lens Ear vesicle has well-defined endolymphatic appendages Mandibular and hyoid bars have acquired ascendency over third or glossopharyngeal bar Arm buds elongating and curved toward body	
15	35–38	7–9		Lens vesicle is no longer open Olfactory placodes Ventral segment of the hyoid bar shows the first indication of the primordium of the antitragus Arm buds subdivided into distal hand segment and proximal segment that will form arm and shoulders Leg buds show evidences of regional differentiation	

[continued]

The Carnegie stages of human embryonic development. Courtesy of Antonio Farina, MD.

Stage	Days (approx.)	Size (mm)	Somite Number	Events	Description
16	37–42 (week 6)	8–11		Nasal pits moved ventrally Auricular hillocks Foot plate Eyes show dark tinge due to retinal pigment Hand region differentiates into a carpus and a digital plate Leg bud exhibits three centers of proliferations representing the thigh, leg, and foot regions	
17	42–44	11–14		Main axis of the trunk has become straight and there is a slight lumbar lordosis Nasofrontal grooves are distinct Limb buds have finger rays in hand plate and the foot has a rounded digital plate	
18	44–48 (week 7)	13–17		Ossification begins The body of the embryo is a more unified cuboidal mass Eyelid folds are present Pigmented retina is partly covered by white-opaque condensed scleral masses Tip of the nose can be seen in profile Auricular hillocks are transformed into definite parts of the external ear	
19	48–51	16–18		The trunk straightens and cervical angle less acute Axes of arm are almost at right angles to the dorsal line of the body of the embryo	
20	51–53 (week 8)	18–22		Upper limbs longer and bent at elbow Vascular plexus appears in the superficial tissue of the head In the temporofrontal region, there is a growth center that arches over the eye, and in the occipital region a second growth center above the ear	
21	53–54	22–24		Hands and feet turned inward The superficial head plexus has spread Fingers are longer and extend farther beyond the ventral body wall	
22	54–56	23–28		Eyelids are now rapidly encroaching upon the eyeballs Molding of the external ear The tragus and the antitragus e assume a more definite form The superficial plexus extends upward about three-fourths the way above the ear-eye level	
23	56–60	27–31		Rounded head bends toward the erect position More mature shaping of the neck region and the body Extremities have markdly increased in length Forearm rises upward to or above the level of the shoulder Scalp plexus is rapidly approaching the vertex of the head	

HUMAN GENETICS

Human **genetics** is a branch of biology and medicine dealing with heredity, especially attempts to explain the similarities and differences that exist between parents and offspring. In medicine, genetics plays an increasingly important role in the prediction, diagnosis, and treatment of disease.

Although hypotheses on the nature and mechanisms of heredity date to antiquity, genetics did not become an independent scientific discipline until the turn of the twentieth century. Through studies of crosses between garden peas, the Czech-born German monk **Gregor Mendel** worked out the basic principles of inheritance, which he expressed in his laws of heredity. Mendel's work stimulated the first experiments in what is now known as classical genetics. These early studies developed into the eventual identification and demonstration of chromosome individuality, gene linkage, crossing over and the linear arrangement of genes in **chromosomes**.

The term genetics was actually first coined, in the early 1900s by the English scientist William Bateson (1861–1926). Bateson was one of the first scientists to accept and popularized the Mendelian laws and introduced the term genetics to encompass the whole study of heredity as it was understood at that time. Genetics quickly came to occupy a central position in biology because it was established early on that essentially the same principles apply to all living organisms. Since their identification, the laws of heredity have found application in such diverse areas as the study of **evolution**, the agricultural improvement of cultivated plants and domestic animals, and clinical medicine.

As biochemical techniques developed and improved, the study of genetics become more detailed and intricate. During the 1950s, cytology developed and with it the microscopic study of the chromosomes and other cellular structures that play a part in heredity. Also studies on extranuclear or cytoplasmic inheritance were undertaken, as well as research

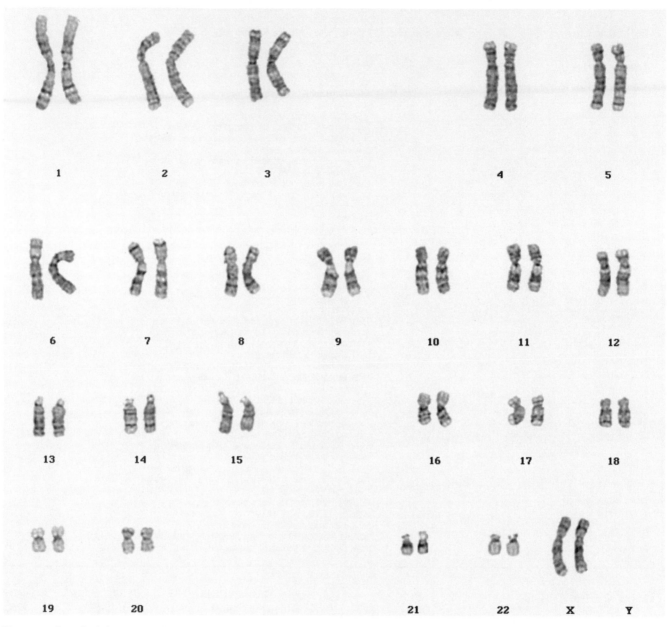

Karyotype photo depicting normal diploid human complement of chromosomes in a female (note two X sex chromosomes in the lower right corner). *Courtesy of Dr. Constance Stein.*

into the nature of mutation and the problems of developmental **physiology** and evolutionary genetics. Microbial genetics, which employed fungi and **bacteria** as experimental systems, became a specialized area of experimental research apart from the eukaryotic cell based human genetics. All of these developments, however, would have had less impact if it had not been for the discovery in the 1940s that genetic material consisted of nucleic acid and, in the 1950s, the determination of the double helical structure of **deoxyribonucleic acid** (**DNA**). Recognition of the double helix was not only important for genetics but was probably one of the most profound biological advances since Darwin's theory of evolution.

Today, modern human genetics can be defined as the science that deals with the nature and behavior of genes, which are now known to be the basic hereditary units. For these purposes modern genetics makes use of the traditional analysis of variation within single species and crossing experiments and additionally uses the latest methods of cell research, **biochemistry, molecular biology**, and gene technology. The discipline is now broadly divided into three main subdivisions, although there is considerable overlap between them. The modes of gene transmission from generation to generation is called transmission genetics, the study of gene behavior in populations is termed population genetics and the

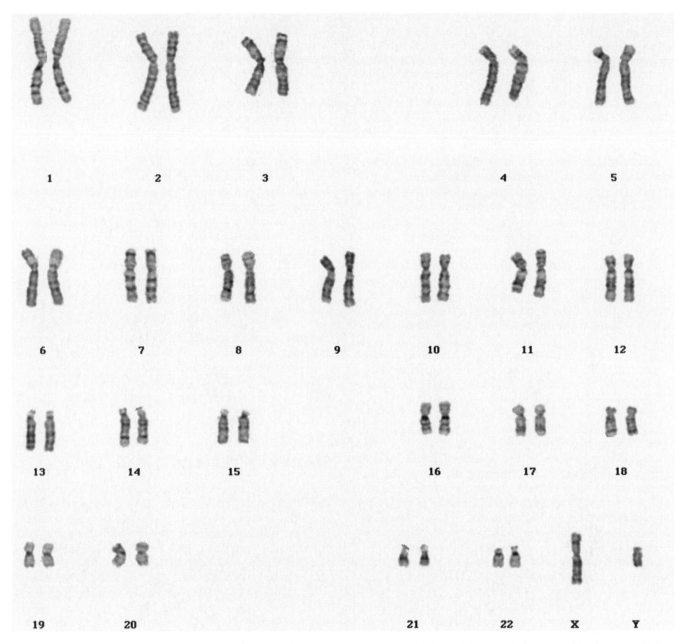

Karyotype depiction of normal male chromosomes (note one X and one Y sex chromosome in lower right corner). *Courtesy of Dr. Constance Stein.*

study of gene biochemistry, **structure and function** is molecular genetics.

Transmission and population genetics use various classical methods of study, nearly all of which rest on the expression of differences between individuals. The experimental breeding of plants, as established by Mendel, is still performed today and enables geneticists to deduce the methods of inheritance of specific characteristics. The mechanism of heredity is such that the analysis of such experiments requires the use of probability theory. These techniques do not, however, make it possible to analyze portions of the hereditary makeup of individuals for which differences cannot be found. If differ-

ences occur, it is also necessary that they be transmitted to later generations. In general, the full analysis will also require the presence of **sexual reproduction** or some analogous process that allows the recombination of inherited properties from different individuals. In brief, what the transmission geneticist usually does is to cross diverse individuals and study the descendents. Such a study must usually be carried through at least two successive generations, and requires enough individuals to establish statistical ratios between the classes present.

The goal of population genetics is to understand the genetic composition of different populations and to study the forces that determine and change that composition. In natural populations, variation in most characters takes the form of a continuous phenotypic range rather than discrete classes. This is because a great deal of genetic variation within and between populations arises from the existence of various forms of genes (alleles) at different loci. Accordingly, a fundamental measurement in population genetics is the frequency at which alleles are found at any gene locus of interest. The frequency of a given allele in a population can be changed by recurrent mutation, selection, or migration or by random sampling effects. Mendelian genetic analysis is therefore extremely difficult to apply to such distributions and more complex statistical methods are employed instead in studies of populations. Pedigree collection and analysis are often used in studies of human genetics, as experimental breeding is not ethically possible. Pedigree charts may be prepared showing the inheritance of specific traits in all the members of a family line, which can be traced. Human population genetics has been useful in tracing population migration and the intermixing of population groups through, for example, an analysis of the frequency of the various **blood antigens**. It has also addressed questions on the frequency and distribution of **inherited diseases**. For example, questions such as why the alleles of Factor VIII and Factor IX genes that cause hemophilia are rare in human populations while **sickle cell anemia**, another blood disease, is very common in certain parts of Africa. This question has been studied in relation to environmental factors, which can maintain a certain genetic disease, such as a certain resitance to **malaria** which appears to go hand in hand with sickle cell anemia.

Molecular genetics is that branch of genetics which attempts to characterize the chemistry and physics of the processes of inheritance. These characterizations are all contingent on the fact that genetic material is made up of nucleic acid and involve the genetic chemistry of nucleic acids. In humans and most other organisms, the genetic material is composed of DNA, in certain **viruses**, however, **RNA** is the genetic material. Both DNA and **ribonucleic acid** (RNA) are long-chain polymers formed by the linkage of many nucleotides.

Autocatalytic functioning of genetic material refers to its replication. Because of the double stranded nature of DNA and the rules of base pairing, separation of the two strands provide the complementary structures, which are used as templates for the synthesis of the original double strand. The detailed molecular mechanisms of replication are still being studied though much is already known about the process and the substrates for the reaction are the deoxy nucleotide-triphosphates (dNTPs) and that at many **enzymes**, including DNA polymerase are involved. DNA polymerase proceeds along the single separated strands of the DANN molecule, recruiting free dNTPs to hydrogen bond with their appropriate complementary dNTP on the single strand (A with T and G with C), and to form a covalent phosphodiester bond with the previous nucleotide of the same strand. The energy stored in the triphosphate is used to covalently bind each new nucleotide to the growing second strand. Genes function heterocatalytically by the copying of an RNA messenger strand from DNA. The same rules of base pairing are involved in this synthesis, although it is carried out by another enzyme, RNA polymerase. The messenger RNA is translated into proteins by an elaborate protein synthesizing machinery which includes ribosomes, special transfer RNA molecules and a host of enzymes. The important point is that three particular nucleotides specify a particular amino acid. These proteins serve the organism both structurally and functionally, for example as enzymes, giving rise to the phenotype (the apparent characteristics) ultimately classified by geneticists.

The most fundamental advances in recent years have been in molecular genetics and the understanding of gene function. They have led to an explosive period in which new genetic knowledge is being applied in medicine as gene technology. Humanity is now in a position to make unprecedented manipulations of hereditary material. The end of the twentieth century saw the cloning of the first mammal and the sequencing of the entire human genome. The science and application of human genetics continues to raise many scientific, social, and ethical questions.

See also Amniocentesis; Bacteria and responses to bacterial infection; Birth defects and abnormal development; Gene therapy; Genetic code; Genetic regulation of eukaryotic cells; Genetics and developmental genetics; Germ cells; History of anatomy and physiology: The science of medicine; Protein synthesis; Stem cells

HUMAN GROWTH HORMONE

Human growth hormone is required for normal bodily growth and development—growth of the bone, muscle, and **cartilage** cells. Also called human somatotropin (and abbreviated hGH or hST), the hormone is a protein produced by the pituitary gland. It functions by promoting **RNA** and **protein synthesis** so that cells will multiply and differentiate. Human growth hormone is also necessary for **metabolism** of **fat**, water, and minerals.

In the 1930s, the American biologist Herbert Evans (1882–1971) demonstrated how pituitary extract can greatly stimulate animal growth, producing a condition called gigantism. In the 1940s, Evans and his colleague Choh Hao Li isolated the growth hormone from cattle. Other scientists isolated growth **hormones** in different animal species. In 1966, Li and endocrinologist Harold Papkoff (1925–) determined the sequence of human growth hormone, showing that it has 245 **amino acids**. In 1970, Li and other scientists independently synthesized human growth hormone using genetic engineering techniques. This led to the first production of genetically engineered human growth hormone in the 1980s by Eli Lilly and Co. and Genentech.

Li and Papkoff also demonstrated how the somatotropin molecule varies from species to species in size and composition, so that it is active only in its own or a very closely related species. There is a forty percent variation among species. Human and monkey growth hormones differ greatly in size and composition from those of other species. The growth hormone in cattle, for instance, consists of 396 amino acids, while that of monkeys has 241.

This distinction has become important for human health reasons in the 1990s. Bovine growth hormone—generally referred to has bST or bovine somatotropin—is being considered for general use by the dairy industry to increase the amount of milk a cow produces. There has been concern that trace amounts of bST in milk will affect human consumers. However, many experts think this is unlikely because bST is active only in cattle and closely-related species, such as sheep.

Lack of normal amounts of growth hormone during childhood can result in a type of dwarfism in which the person is small but has normal proportions and intelligence. Injections of human growth hormone can help the person reach normal or near-normal size if the condition is diagnosed while the bones can still grow (before the long bones have closed).

If the body produces too much growth hormone in childhood, the person can have much greater than normal height, called gigantism. If the growth hormone overproduction occurs in adulthood, the bones of the feet, hands, and face thicken in a condition called acromegaly.

The pituitary releases human growth hormone when it receives a signal from the brain's **hypothalamus** in the form of growth hormone releasing factor (GHRF). However, release can also be triggered by **sleep**, exercise, fasting, or hypoglycemia (low **blood** sugar). Release can be inhibited by the hormone somatostatin, as well as lack of sleep, high blood sugar (hyperglycemia), obesity, and a high blood level of free fatty acids. The American endocrinologist **Roger Guillemin** and colleagues isolated somatostatin in 1973 and GHRF in 1984.

In 1989, the pharmaceutical firm of Hoffmann-La Roche announced the development of an artificially produced growth hormone releasing factor that, when administered, stimulates the body to produce normal amounts of growth hormone. This may have advantages for a patient because a smaller dose is required than that of growth hormone. Also GHRF is a much smaller molecule, with only forty-four amino acids. So it can be administered by a skin patch or inhalant, rather than by injection.

See also Hormones and hormone action; Pituitary gland and hormones

HUNT, TIM (1943-)

English molecular biologist

Tim Hunt shared the 2001 Nobel Prize for physiology or medicine with **Lee Hartwell** and **Paul Nurse**. They received the prize for their identification of 'key regulators of the cell cycle' in eukaryotic organisms. Growing and dividing cells normally pass through a series of four phases in which they replicate their **chromosomes** (in S-phase) and distribute them to each daughter cell during mitosis, or M-phase. The periods of **DNA** replication and segregation are separated by gap phases, G1 between M and S and G2 between S and M. Growth occurs during all phases of the **cell cycle** except mito-

sis itself. The key regulators referred to in the Nobel citation are **enzymes** called cyclin-dependent protein kinases that promote the transitions between G1 and S phase and between G2 and M phase.

Flaws in the control of the cell cycle lead to chromosomal abnormalities that can cause **cancer**. In particular, entering mitosis with damaged or incompletely replicated chromosomes tends to generate daughter cells with abnormal chromosomes, which in turn can lead to cells that ignore the social clues from their neighbors or the environment. Such cells grow as disorganized masses instead of beautifully organized tissues. Apart from the purely intellectual aspects of understanding cell cycle control, this work has implications in the long-term possibilities of new cancer treatments. In addition, these findings will benefit research areas in other biomedical fields as they have broad application for understanding how chromosomes are lost, rearranged or distributed unequally between dividing cells. The work may also help cancer diagnosis.

In 1982, Tim Hunt and his colleagues discovered two proteins that they called cyclin A and cyclin B. These proteins were named cyclins because their levels underwent a sawtooth oscillation during the cell cycle. This periodic buildup and rapid degradation of the cyclins was seen because of the highly synchronous cell divisions in fertilized sea urchin and clam eggs. The proteins accumulated during interphase, and were degraded just before the chromosomes separated in mitosis. This behavior suggested that these proteins might be involved in inducing mitosis, and that their destruction was necessary for exit from M phase. Current research involves the **structure and function** of protein kinases concerning such cell cycle transitions as chromosomal replication, the onset of mitosis and what processes control the return to interphase at the end of the cell cycle. Of interest also, is the control of the translation of messenger **RNA** and its role in initiating meiosis. Future work will concentrate on elucidating the structure and binding substrates of various other cyclins and clarifying the control of their destruction.

R. Timothy Hunt was born at Neston in the Wirral, near Liverpool, England, but moved to Oxford at the age of two. Raised in an academic environment, he was attracted to the hands-on aspects of science as a schoolboy. He went up to Clare College, Cambridge in 1961 with the intention of becoming a biochemist. The atmosphere in Cambridge was very exciting to the young man, as was his exposure to the molecular biologists like **Francis Crick**, Fred Sanger, Max Perutz and Sydney Brenner. He received his BA in 1964 and his Ph.D. in Biochemistry in 1968 for his dissertation "The control of haemoglobin synthesis," Department of Biochemistry, Cambridge.

After completing his, Ph.D. Dr. Hunt spent two years (1968–1970) as a Postdoctoral Fellow at Albert Einstein College of Medicine where he continued his work on haem with Dr. L.M. London. He then returned to Cambridge as a Research Fellow but spent several summers as an Instructor in Embryology (1977,1979) and Physiology (1980–83) at the Marine Biology Laboratory at Woods Hole, Massachusetts. The time spent there was pivotal for his future work with cyclins and he credits that environment and the researchers he met at the MBL as critical to his growth as a scientist, especially with regard to developmental and cell biology. Towards the end of 1982, in July, he ran a simple experiment comparing the pattern of **protein synthesis** in parthenogenetically activated and fertilized sea urchin eggs. This marked the discovery of cyclin, for he noticed a radioactive band that appeared early after activation but then declined abruptly about the same time as the cell divided. A chance encounter that very day with John Gerhart, who discussed the work that Marc Kirschner and he were doing with maturation protein factor (MPF), suggested the possibility that this protein was either an activator or a component of MPF. Further work during the following summers focused on the role cyclins played in the cell cycle, and in 1986, Hunt found himself in Berkeley, helping to write a book of cell biology problems. Mike Wu, who worked with John Gerhart, demonstrated a real live MPF assay, and they decided to look for the mRNA for MPF in frogs' eggs. They were thrilled to find that mRNA from the eggs, but not from oocytes, promoted maturation exactly as had recently been described by Katherine Swenson and Joan Ruderman for clam cyclin A mRNA. It was a short step to isolating frog cyclin mRNA and from there to human cyclins. It was to be another two years before cyclins were identified in yeast, connecting the genetic and biochemical approaches to the cell cycle that were honored by the Nobel Committee.

See also Cell cycle and cell division; Cell differentiation; Cell membrane transport; Cell structure; Genetic regulation of eukaryotic cells

HUXLEY, ANDREW FIELDING (1917-   )
English physiologist

Andrew Fielding Huxley is an English physiologist whose research on nerve impulse transmission earned him the 1963 Nobel Prize in physiology or medicine, which he shared with his colleague Alan Lloyd Hodgkin and the Australian physiologist John Carew Eccles. Huxley and Hodgkin confirmed scientists' earlier discovery that nerve impulse transmission involves a momentary change in the nerve fiber's membrane, affecting the ability of particles to pass through it.

Huxley was born in London, England, to a prominent and successful family. His grandfather was the nineteenth-century biologist Thomas Henry Huxley. Julian Sorel Huxley, also a noted biologist, was Andrew's half-brother, as was the author Aldous Huxley. Andrew's father, Leonard, was also a writer. His mother was Rosalind (Bruce) Huxley. Huxley was educated at Trinity College, Cambridge, where he received his B.A. in 1938, and his M.A. in 1941. He began studying the physical sciences but switched to physiology in his last year.

In 1939, Huxley joined **Alan Hodgkin** at the Plymouth Marine Biological Laboratory to study the transmission of nerve impulses. There, Huxley and Hodgkin attempted to verify the work of other scientists, including Julius Bernstein, **Joseph Erlanger**, and Herbert Spencer Gasser. These scientists had hypothesized that a nerve impulse produces an electrical

current between the active and resting regions of a nerve, and that this impulse causes a fleeting change in the permeability of the nerve fiber membrane. Hodgkin and Huxley went about their research by experimenting on squid, which have giant axons, or nerve fibers, and therefore were known to be particularly useful in studying nerve systems. They inserted a small electrode into the squid's axon, and connected it to a system that would measure the electrical currents produced when the nerve was stimulated.

Huxley's work was interrupted during World War II, when he spent two years doing operational research for the Anti-Aircraft Command and later worked for the Admiralty. In 1946, he returned to his alma mater, serving in a variety of positions—fellow, assistant director of research, director of studies, and reader in experimental biophysics—while he carried out and perfected his research with Hodgkin. He was married to Jocelyn Richenda Gammell Pease in 1947; they had five daughters and one son.

In the course of their research, Huxley and Hodgkin were surprised to learn that, contrary to earlier hypotheses, the outer layer of a nerve fiber is not equally permeable to all ions (charged particles). While a resting cell has low sodium- and high potassium-permeability, Huxley and Hodgkin found that, during excitation, sodium ions flood into the axon, which instantaneously changes from a negative to a positive charge. It is this sudden change that constitutes a nerve impulse. The sodium ions then continue to flow through the membrane until the axon is so highly charged that the sodium becomes electrically repelled. The stream of sodium then stops, which causes the membrane to become permeable once again to potassium ions.

Huxley and Hodgkin first announced their findings in 1951 and published a series of highly regarded papers in 1952. In 1955, Huxley was named to the Royal Society, and in 1960, he became the Jodrell Professor of Physiology at University College, London, where, according to Ronald Clark's history of the Huxley family, *The Huxleys,* he occupied the desk of his grandfather, T. H. Huxley. He remained professor at University College until 1983. In 1974, he was knighted.

When he received the Nobel Prize in 1963, Huxley described the often laborious research and computations involved in his work. While crediting those scientists whose findings he built upon, he also allowed that there was much more work to be done in this field. One of the many applications of the methods and findings of Huxley and Hodgkin was discovered by John Carew Eccles, who shared the Nobel Prize with the two Englishmen. Eccles studied motor **neurons** in the **spinal cord** and synapses using microelectrodes similar to those used by Huxley and Hodgkin. Huxley himself devoted much of his later research to studying **muscle contraction**. His findings have increased the understanding of diseases of the nervous system, as well as similar ionic mechanisms in the kidney and **heart**.

See also Nerve impulses and conduction of impulses; Nervous system overview; Nervous system: embryological development; Neural damage and repair; Neural plexuses; Neurology; Neurons; Neurotransmitters

HUXLEY, THOMAS HENRY (1825-1895)

English biologist

Thomas Henry Huxley conducted research in **comparative anatomy** and biology. His efforts to identify physiological and biological links between vertebrates and invertebrates pioneered the field of evolutionary biology. Although a prolific publisher of scientific material, Huxley also wrote on philosophical issues in science. He was intrigued by the possibility of formulating a scientific critique of questions of religion, morality, and power. He often championed controversial theories, acting as the outspoken defender of Darwinism against its critics. Huxley fought tirelessly enhance the status of science within British society. His efforts in raising public interest in scientific discovery resulted in a greater professionalism of the discipline, and the establishment of salaried positions for researchers. Thus, Huxley was a key figure in the transformation of science from a amateur hobby of wealthy gentlemen, to a cohesive academic discipline and professional career.

Huxley was born in Ealing, England. Although he was the son a schoolmaster, he received only two years of formal education. He served as medical apprentice to his brother-in-law before being transferred to a practice in the London docklands while attending classes at Sydenham College. He received a scholarship to Charing Cross Hospital, London, and while there won several accolades for his work in **physiology** and chemistry. Huxley's first contribution to science was the 1845 discovery of a new membrane in human **hair**.

During his brief studies and apprenticeships, Huxley became enamored with the sciences, but did not have the then requisite personal wealth to devote to a career in research. Forced to seek active employment to repay mounting debts, Huxley joined the Navy. His first post was as assistant surgeon a surveying vessel. The HMS *Rattlesnake* was charged with mapping and surveying Australia's Great Barrier Reef and the charting currents and coastline around New Guinea. While on board the *Rattlesnake*, Huxley found time to pursue personal research in marine biology. He studied the anatomies of sea life, such as jellyfish and hydras. Comparing their physiological structures, he concluded that the animals contained "foundation membranes" (structures that were related to cell layers found in vertebrate embryo). He continued his pioneering work in comparative **anatomy** throughout the voyage, carefully charting various the physiological features of invertebrates that he believed to be primordial versions of structures found in vertebrates.

Huxley's work was instantly recognized by the scientific community. He was elected a fellow of the Royal Society in 1851, and received the Royal Medal the following year. Despite his acclaim, Huxley was unsuccessful in obtaining a university position in the sciences. He was disheartened by his inability to find employment and secure funding for his research, but used his situation to call attention to the need for reforms within the academy. In 1854, Huxley finally received a position teaching paleontology and natural history.

From an early age, Huxley was deeply interested in the interplay between science, religion, and politics. Despite his Anglican upbringing, Huxley became involved in his home-

Thomas Henry Huxley.

town's vigorous Nonconformist (religious sects not affiliated with the Church of England) community. He read widely on natural theology and Unitarianism, distilling from his self-study a curiosity about the relationship between morality and scientific explanation. Huxley later proclaimed himself an agnostic, a term he coined. Huxley's agnosticism, or the belief that humans cannot know anything that is outside the realm of their personal experiences, became a central philosophical theme of his devotion to the advancement of scientific thinking among the public at large. By the mid-1850s, Huxley gained renown as a champion of new scientific theories that challenged prevailing theological concepts. He argued on behalf of Charles Lyell's (1797–1875) geological theory of gradual development, and publicly repudiated those who tried to force scientific theory to mesh with orthodox Christian beliefs.

Huxley was most passionate in his defense of the Charles Darwin's (1809–1882) theories of **evolution**. After the publication of Darwin's *On the Origin of Species* in 1859, Huxley responded in place of the intensely taciturn Darwin to public attacks by the theory's critics. Though Huxley admitted to Darwin that he was skeptical about the theory's ultimate reliance on natural selection as the mechanism of evolution, he never voiced his concerns outside of the scientific community. Instead, Huxley defended Darwinism ceaselessly and fought to gain its acceptance by not only the scientific community, but also among members of the general public. In 1860, at meeting of the British Association for the Advancement of Science, Huxley initiated in a debate over evolution with the

Bishop of Oxford, Samuel Wilberforce (1805–1873). The incident was covered by newspapers and journals across England and earned Huxley the nickname "Darwin's bulldog."

Although Huxley continued with his scientific research until the end of his career, those achievements were often overshadowed by his contributions to the philosophy of science. Huxley's most famous work, *Evidence as to Man's Place in Nature*, addressed the biological association of man and primitive apes as well as sought to explain the ramification of evolutionary theory on man's perception of his place in the universe—a question that was previously pondered only by theologians. Huxley himself briefly espoused the application of Darwinian evolution and the "war of species" to a social and national context. However, he later abandoned Social Darwinism, stating in his last public lecture that Darwin's biological mechanisms were misapplied when used to explain contemporary economic and national competition.

Huxley was appointed by Prime Minister, Robert Cecil, to the Privy Council in 1892, a position he held until the end of his life.

See also Comparative anatomy; Evolution and evolutionary mechanisms

HYALINE CARTILAGE

In general, **cartilage** is a tough, fibrous and **blood** vessel-free connective **tissue** that forms flexible linkages (such as in the ribs), supporting structures (such as in the ears and **nose**), and acts as a **shock** absorber in **joints** such as the knee.

Hyaline cartilage is the most common type of cartilage. Its name derives from its translucent (hyaline) appearance.

In addition to being found in articulated joints—joints capable of movement by virtue of being their sliding or hinged connection with each other—hyaline cartilage forms the majority of the skeleton of a fetus. Later in fetal development, it is replaced by bone.

The free surfaces of most hyaline cartilage (except that found in joints) are covered by a layer of fibrous connective tissue known as perichondrium. The perichondrium is rich in a type of cell known as the fibroblast. Compositionally, hyaline cartilage is made of water (75% by weight), collagen (10% by weight, mainly collagen type II), with the remainder being nonfibrous material, such as chondroitin sulfate and keratan sulfate. The collagen provides strength and makes hyaline cartilage resistant to compression. Also, the collagen provides a means by which the cartilage can be anchored to bone.

Cartilage is capable of growth. Growth occurs mainly at the cartilage surface, just underneath the perichondrium. Such active deposition of cartilage at the surface is termed apositional growth. Cells within the cartilage called chondroblasts produce the new material. It is thought that the fibroblasts resident in the perichondrium are able to convert to chondroblasts during these growth periods. As growth occurs the chondroblasts can become trapped in the cartilage. The space occupied by these cells is referred to as a lacuna, which means "small lake." In adults, the entrapped chondroblasts, now called

chondrocysts, do not readily produce new hyaline cartilage. When adults sustain cartilage damage, part of the healing process involves the conversion of chondrocytes to chondroblasts, so that production of hyaline cartilage can occur.

See also Arthrology (joints and movement); Cartilaginous joints; Connective tissues

HYPERTHERMIA

Hyperthermia is a condition in which internal body temperatures rise to dangerous and even lethal levels. These temperatures can reach anywhere between 104° and 115°F (40–46°C). Hyperthermia affects many people during hot summer seasons. It is also a condition that occurs during some surgeries. In these rare cases the patient has an unpredictable reaction to surgical **anesthesia**, one of the symptoms being a rapid rise in body temperature. If not cared for immediately, patients who suffer hyperthermia, may be unaware of the symptoms and allow themselves to attain dangerous heat levels.

There are two levels or types of hyperthermia. One is heat exhaustion and is the less severe type of this condition. Symptoms are increasing fatigue or weakness. A state of anxiety is experienced. The patient experiences extreme levels of perspiration that drench clothing. The **blood** pressure is low (thready). This condition may lead to the collapse of the **circulatory system**. When this situation occurs, the patient is in danger. Additional symptoms are cold, pale and clammy skin. Disorientation may occur which may eventually lead to a shock-like unconsciousness.

The more serious form of hyperthermia is heatstroke (also called sunstroke). When this condition occurs, treatment must be administered immediately. The onset may be sudden and severe, but there are early warning signs. The patient may experience a **headache**, dizziness, and a general weakness. Perspiration or sweating usually decreases and the skin is hot and dry. The entire body appears flushed. The urine may become dark yellow or orange. The **pulse** rate increases rapidly and may reach 160 beats per minute. Unlike heat exhaustion, the blood pressure is not affected. A brief period of disorientation precedes unconsciousness or convulsions. Circulatory collapse is a dangerous condition that comes just before **death**. Some patients who survive heatstroke may have permanent **brain** damage.

Some other causes of hyperthermia may be over exercising in a hot climate, overeating, overdressing, or consumption of too much alcohol. In dry climates, many people do not realize their body temperatures are rising because of the lack of obvious perspiration. Constant rehydration by water or special fluids is highly recommended as a prevention measure. This helps the body's cooling mechanisms to work more efficiently and under greater stress. Fluids help maintain vital blood pressure levels.

See also Electrolytes and electrolyte balance; Temperature regulation

HYPERTHYROIDISM AND HYPOTHYROIDISM

The thyroid gland, the body's largest endocrine gland, is situated in the neck just below the thyroid **cartilage** and secretes **hormones** necessary for growth and proper **metabolism**. A deficiency or overabundance of the secreted hormones results in one of two endocrine disorders, hyperthyroidism and hypothyroidism.

Hyperthyroidism results from an excessive production of the thyroid hormone, thyroxine. Thyroxine production is dependent on a sufficient intake of iodine and on stimulation by thyroid-stimulating hormone (TSH) from the pituitary gland. Symptoms of hyperthyroidism include an increase metabolism, rapid **pulse**, weight-loss despite normal or increased appetite, irritability, and occasionally, an enlargement of the thyroid gland itself. Because the body's metabolism is increased, patients often feel warmer than those around them. An extreme overproduction of thyroid hormones leads to a dangerous condition called thyrotoxicosis.

The most common cause of hyperthyroidism is Graves' disease, an autoimmune disorder characterized by a chronic course of remissions and relapses. The condition is usually accompanied by a protrusion, or bulging, of the eyeballs (exophthalmos).

Patients with hyperthyroidism are treated with an anti-thyroid drug or radioactive iodine, which destroys some thyroid **tissue** and subsequently reduces hormone secretions.

Hypothyroidism occurs when the thyroid gland does not secrete enough thyroxine. Deficiencies can occur when there is insufficient iodine in the diet or a glandular malfunction. Hypothyroidism is a common disorder, with subtle symptoms that can progress for years before being recognized and treated. The condition is characterized by a low metabolism, sluggishness, weight gain, **hair** loss and dry skin. Patients are often colder than those around them. Treatment is by administration of thyroxine. In the past, a disease known as goiter resulted from the lack in thyroid hormone secretions, but the use of iodized salt has helped eliminate the disorder.

See also Parathyroid glands and hormones; Thyroid histophysiology and hormones

HYPOPHYSIS · *see* PITUITARY GLAND AND HORMONES

HYPOTHALAMUS

The hypothalamus, a part of the diencephalon, lies at the base of the cerebrum, beneath the **thalamus**. The hypothalamus forms the base of the lateral wall or boundary of the third ventricle and extends from the optic chiasma to the caudal border of the mammillary bodies (both of which are considered a part of the hypothalamus). Anatomically, the hypothalamus includes the pituitary gland but physiologically they are

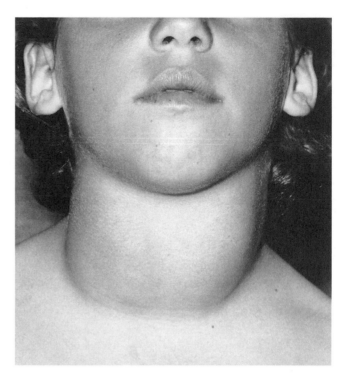

Thyroid enlargement is often due to thyroid malfunction, along with the thyroid's attempts to compensate and maintain thyroid hormone production at normal levels. *Photograph by Lester V. Bergman. Corbis Images. Reproduced by permission.*

treated as separate entities. The hypothalamus also includes the tuber cinereum and infundibulum.

The hypothalamus is the coordinating complex or control center for the **autonomic nervous system** (ANS) (e.g., ANS reflex arc connections are made in the hypothalamus) and the neuroendocrine system. Accordingly, the hypothalamus plays a regulatory role in the coordination of involuntary actions and processes (e.g., changes in **blood** pressure, urinary output, gastrointestinal motility, etc).

Both ascending visceral and somatic pathways provide afferent nerve pathways to the hypothalamus. The hypothalamus also receives nerve tracts from olfactory receptors and from the **cerebral cortex**, midbrain, and limbic system.

The hypothalamus plays a number of critical physiological roles—including the regulation of emotional states. The hypothalamus plays a regulatory role in the deposition and accumulation of **fat** in **adipose tissue**, renal functions (especially water **metabolism**), and carbohydrate metabolism. In conjunction with other coordinating elements, in particular the pituitary gland, the hypothalamus assists in the release of pituitary **hormones**. Accordingly, the hypothalamus plays an important indirect role in the regulation of a range of metabolic activities—including body **temperature regulation** in which the hypothalamus is considered the temperature regulation center for the body.

Using different mechanisms, the hypothalamus regulated functions in the posterior lobe of the pituitary gland and the secretory activity of the anterior lobe of the pituitary gland.

The optic chiasm is where the nerves from the left and right **retina** cross before continuing as optic tracts to opposite sides of the **brain**. The tuber cinereum is a mass of **gray matter** that is continuous with the hollow infundibulum. Together these structures form part of the neurohypophysis. The mamillary bodies are small hemispherical bodies (principally gray matter but with some myelinated fiber tracts) located posterior to the tuber cinereum.

Inflammation from infections, **tumors**, and a number of vascular problems may scar or cause lesions of the hypothalamus. Disruption of the normal function of the hypothalamus can result in **sleep** disorders, eating disorders and emaciation, genital disorders (including genital atrophy), diabetes (specifically diabetes insipidus), obesity, and a number other problems or syndromes. Because the hypothalamus is important to alertness and reactivity to **pain** and pleasure stimuli, impairment of hypothalamic function may manifest itself in a number of behavioral abnormalities.

See also Anatomical nomenclature; Biofeedback; Blood pressure and hormonal control mechanisms; Cerebral morphology; Epithalamus; Hormones and hormone action; Limbic lobe and limbic system; Nervous system overview

HYPOTHERMIA

Hypothermia is the intentional or accidental reduction of core body temperature to below 95°F (35°C) which, in severe instances, is fatal. Because humans are endothermic—warm-blooded creatures producing our own body heat—the core body temperature remains relatively constant at 98.6°F (37°C), even in fluctuating environmental temperatures. However, in extreme conditions, a healthy, physically fit person's core body temperature can rise considerably above this norm and cause heat **stroke** or fall below it far enough to cause hypothermia.

Intentional hypothermia is used in medicine in both regional and total-body cooling for organ and **tissue** protection, preservation, or destruction. Interrupted **blood** flow starves **organs** of oxygen and may cause permanent organ damage or **death**. The body's metabolic rate (the rate at which cells provide energy for the body's vital functioning) decreases 8% with each 1.8°F (1°C) reduction in core body temperature, thus requiring reduced amounts of oxygen. Total-body hypothermia lowers the body temperature and slows the metabolic rate, protecting organs from reduced oxygen supply during the interruption of blood flow necessary in certain surgical procedures. In some procedures, like **heart** repair and organ transplantation, individual organs are preserved by intentional hypothermia of the organ involved. In open-heart surgery, blood supply to the chilled heart can be totally interrupted while the surgeon repairs the damaged organ. Organ and tissue destruction using extreme hypothermia –212 to –374°F (–100 to –190°C) is utilized in retinal and glaucoma surgery and to destroy pre-cancerous cells in some body tissue. This is called cryosurgery.

Accidental hypothermia is potentially fatal. Falling into icy water, or exposure to cold weather without appropriate protective clothing can quickly result in death. Hypothermia is classified into four states. In mild cases, 95–89.6°F (35–32°C), symptoms include feeling cold, shivering (which helps raise body temperature), increased heart rate and desire to urinate, and some loss of coordination. Moderate hypothermia, 87.8–78.8°F (31–26°C) causes a decrease in, or cessation of, shivering, along with weakness, sleepiness, confusion, slurred speech, and lack of coordination. Deep hypothermia, 77–68°F (25–20°C) is extremely dangerous as the body can no longer produce heat. Sufferers may behave irrationally, become comatose, lose the ability to see, and often cannot follow commands. In profound cases, 66–57°F (19–14°C), the sufferer will become rigid and may even appear dead, with dilated pupils, extremely low blood pressure, and barely perceptible heartbeat and **breathing**. This state usually requires complete, professional **cardiopulmonary resuscitation** for survival.

Although overexertion in a cold environment causes most accidental hypothermia, it may occur during **anesthesia**, primarily due to **central nervous system** depression of the body's heat-regulating mechanism. Babies, the elderly, and ill people whose homes are inadequately heated are vulnerable to hypothermia. The human body loses heat to the environment through conduction, convection, evaporation, and respiration, and radiation. It generates heat through the metabolic process.

Conduction occurs when direct contact is made between the body and a cold object, and heat passes from the body to that object. Convection is when cold air or water makes contact with the body, becomes warm, and moves away to be replaced by more of the same. The cooler the air or water, and the faster it moves, the faster the core body temperature drops.

Evaporation through perspiration and respiration provides almost 30% of the body's natural cooling mechanism. Because cold air contains little water and readily evaporates perspiration; and because physical exertion produces sweating, even in extreme cold; heat loss through evaporation takes place even at very low temperatures. The dry air we inhale attracts moisture from the lining of the **nose** and throat so quickly that, by the time the air reaches the **lungs**, it is completely saturated. Combined, evaporation and convection from wet clothes will reduce the body temperature to dangerous levels quickly.

Although profound hypothermia can be reversed in some instances, even mild states can quickly lead to death. However, through knowledge and common sense, hypothermia is avoidable. Two factors essential in preventing accidental hypothermia are reducing loss of body heat and increasing body heat production.

Appropriate clothing, shelter, and diet are all essential. Regular consumption of high-energy food rich in **carbohydrates** aids the body in heat production, while adequate water intake prevents rapid dehydration from evaporation. Exercising large muscles, like those in the legs, is the best generator of body heat; however, overexertion must be avoided as it will only speed the onset of hypothermia.

See also Temperature regulation

HYPOXIA

Hypoxia is a condition in which cells of the body are deprived of oxygen. The reasons for this condition can vary within the body. However, the result is that tissues cannot survive for long without oxygen. Continued oxygen deprivation can eventually become fatal.

Cells get their energy from oxygen and glucose. Most cells can survive for a short period using an anaerobic metabolic (energy without oxygen) process. Unfortunately, **brain** cells cannot. The damage to brain cells when hypoxia occurs is immediate. The **blood** carries a limited amount of reserve oxygen and brain cell **death** can occur within minutes of falling below normal oxygen levels. This deprivation can lead to impairment of several functions as the brain cannot work as well and its control over all bodily processes fades. The consumption of alcohol can produce hypoxia, and is one of the reasons why brain functions are impaired when excess alcohol is consumed. Jet pilots are trained about the effects of hypoxia since they are often performing at altitudes where oxygen is scarce. If something should happen to their onboard oxygen supply, they are trained to help cope with some of the effects of hypoxia.

There are several types of hypoxia as well as causes of this dangerous condition. One of the more common types is hypoxic hypoxia. It is the reduction of the amount of oxygen passing into the blood because of a reduced **oxygen exchange** (i.e., reduced lung capacity) or high altitudes. Reduction in lung capacity may be a result of lung damage, disease or removal of portions of the **lungs**. Smokers are particularly sus-ceptible to hypoxic hypoxia. People who change altitudes can adjust to the lower oxygen pressure as the blood produces more red blood cells carrying additional **hemoglobin** (the oxygen-carrying molecule in red blood cells).

Hypemic hypoxia occurs when the number of hemoglobin molecules or red blood cells is reduced. Either condition causes a reduction in the oxygen carrying capacity of the blood to be lowered. The causes of this type of hypoxia can be through hemorrhage or anemia. It can also be caused by drugs, chemicals, or an increase in carbon monoxide (a condition experienced by smokers).

Stagnant hypoxia occurs as a result of poor circulation of the blood. Blood flow is reduced by prolonged sitting in one position, cold temperatures, or being exposed to "g" forces. People who sit in a chair for hours or are sedentary may experience this type of hypoxia. It is important for the elderly or those whose movement is restricted to be sure they get enough oxygen to avoid this type of hypoxia.

Histotic hypoxia is the inability of the tissues to use oxygen. When organ tissues are involved, they appear blue in color and are called cyanotic. Drinking alcoholic beverages, poisoning by cyanide or carbon monoxide, and certain narcotics can impair **gaseous exchange** in the tissues, and lead to hypoxia.

The blue color associated with cyanosis, especially noted around the lips, is due to the build-up of high levels of deoxygenated hemoglobin in **capillaries**. Prolonged hypoxia can lead to **tissue** death or impairment.

See also Oxygen transport and exchange; Respiratory system

Ignarro, Louis J. (1941-)

American pharmacologist

Louis J. Ignarro, professor of pharmacology at the University of California School of Medicine at Los Angeles, shared with Robert F. Furchgott and **Ferid Murad** the 1998 Nobel Prize in physiology or medicine for their contributions to the discovery of endogenous nitric oxide gas in complex organisms and its physiological implications. Nitric oxide (NO) was, until recently, known as a common air pollutant and for its presence in body tissues as the result of nitroglycerin metabolization, as found by Ferid Murad in 1977. Murad and his group also reported that nitric oxide and other nitro compounds activated the enzyme guanylate cyclase in the cytosol, causing the elevation of cyclic GMP levels in several tissues. Ignarro's studies of cyclic GMP and NO had shown in 1979 that nitric oxide gas promotes the conversion of GTP to cyclic GMP and pyrophosphate, thus acting as a relaxant of the **smooth muscle** of **blood** vessels. His subsequent studies also disclosed in 1983 the presence of cellular receptors to NO, which led him to further research aiming to find if nitric oxide gas was produced by organisms. His laboratory also attempted to identify the chemical identity of Furchgott's EDRF (endothelium-derived relaxing factor), which ultimately proved to be endogenous nitric oxide gas itself. Independently conducted studies by Furchgott supported these findings. In the following avalanche of research performed by many other groups, endogenous nitric oxide was found to have several physiologic roles. Nitric oxide serves as a neurotransmitter in the nervous system, a potent antibacterial weapon in the **immune system**, and an important regulator of blood pressure. The understanding of its mechanism of action also led to the development of new treatments for impotency.

Louis J. Ignarro, son of Italian immigrants, spent the first two decades of his life in New York, and was fascinated by chemistry during his childhood. After graduation from Columbia University school of pharmacy in 1962, he was admitted to the post graduation program of the department of pharmacology at the University of Minnesota in Minneapolis, where he achieved a Ph.D. degree in 1966. Ignarro then accepted a post graduation position at the National Institutes of Health in the National Heart, Lung and Blood Institute, where he stayed for two years working with Elwood Titus in the isolation and characterization of the chemical structure of beta and alfa adrenergic receptors. In 1968, Geigy Pharmaceuticals hired him to the position of head of the bio-chemical and anti-inflammatory program, where he took part in the development of a non-steroidal anti-inflammatory drug. He also continued his basic research in biochemical pharmacology at Geigy, and developed an interest in the physiological implication of the cyclic nucleotide, cyclic GMP, for the first time. Later, in 1973, as an assistant professor of pharmacology at Tulane University School of Medicine in New Orleans, he continued researching cyclic GMP. In 1977, after reading Ferid Murad's paper about nitroglycerin and other nitro compounds that released nitric oxide as a metabolite, Ignarro began considering the possibility that NO could be responsible for the observed vascular smooth muscle relaxation, and that cyclic GMP was probably the second messenger in this pathway. Ignarro left Tulane University in 1985, and moved to Los Angeles where he presently works as professor of the department of pharmacology at University of California School of Medicine.

See also Blood pressure and hormonal control mechanisms; Vascular system overview

Iliac Arteries

The common iliac **arteries**, each a continuation of the descending aorta, arise from the bifurcation (splitting into two) of the **abdominal aorta** at about the level of the fourth lumbar **vertebra**. The common iliac arteries bifurcate again into internal and external iliac arteries that supply oxygenated **blood** to the lower **abdomen**, pelvis, and legs.

On both the left and right sides, the common iliac artery is short, and each common iliac artery divides into an internal and an external iliac artery at about the level of the sacrum. Before bifurcating into internal and external iliac arteries, it gives rise to small branches that supply blood to the ureter and peritoneal structures. The external iliac arteries continue on to supply blood to the legs (lower limb) and the internal iliac arteries supply blood to pelvic and peritoneal structure.

The internal iliac arteries are prominent in the fetus and can be twice the size of the external iliac arteries. The reason for this size difference lies in the fact that the internal iliac arteries run to the abdominal wall and then turn superiorly (upward) to eventually fuse at the umbilicus (in the adult, the "navel" or "belly button" region). The continuation of the fused internal iliacs in the umbilical cord becomes the umbilical artery that ultimately branches into the **placenta**.

Following birth, the interruption of blood flow through the umbilical cord causes the portion of the internal iliacs that continue to the umbilicus to close. Ultimately, what was once an umbilical extension of the internal iliac artery becomes the fibrous medial umbilical ligament.

The internal iliac arteries divide into anterior and posterior trunks. In the adult, major branches of the internal iliac arteries supply blood to the bladder, ureter, and other pelvic structures. In the male, the internal iliac branches supply a portion of the arterial blood to the testis and prostate. In the female, oxygenated blood from the internal iliacs also supplies the vagina and uterus. In both males and females, braches of the internal iliac arteries supply blood to the external genitalia.

In the adult, the external iliac arteries are larger than the internal iliac arteries. The external iliac arteries become the large diameter femoral arteries that supply blood to the legs. Before becoming the femoral arteries, the external iliacs give rise to a few branches, including the inferior epigastric and deep circumflex iliac arteries that fuse (anastomose) with other arteries and thereby allow the development of **collateral circulation** to the leg in the event of interruption in the normal supply of blood through the common iliac artery.

See also Anatomical nomenclature; Angiology; Fetal circulation; Systemic circulation; Vascular system (embryonic development); Vascular system overview

IMAGING

High-tech diagnostic imaging techniques are powerful medical tools that allow physicians to explore bodily structures and functions with a minimum of invasion to the patient. Advances in diagnostic technology allow physicians the ability to evaluate processes and events as they occur *in vivo* (in the living body). During the 1970s, advances in computer technologies, in particular the development of algorithms powerful enough to allow difficult equations to be solved quickly enough to be of real-time use in the clinical diagnostic setting and to eliminate "noise" from sensitive measure-

ments, allowed the development of accurate, accessible and relatively inexpensive (when compared to surgical explorations) non-invasive technologies. Although relying on different physical principles (i.e., electromagnetism vs. sound waves), all of the high-tech methods relied on computers to construct visual images from a set of indirect measurements. The development of high-tech diagnostic tools was the direct result of the clinical application of developments in physics and mathematics. These technological advances allowed the creation of a number of tools that made diagnosis more accurate, less invasive and more economical.

The use of non-invasive imaging traces it roots to the tremendous advances in the understanding of electromagnetism during the nineteenth century. By 1900, physicist Wilhelm Konrad Roentgen's (1845–923) discovery of high-energy electromagnetic radiation in the form of x rays were used in medical diagnosis. Developments in radiology progressed throughout he first half of the twentieth century, finding extensive use in the treatment of soldiers during W.W.II.

Technological innovations in computing opened the door for the subsequent development and widespread use of sonic and magnetic resonance imaging (MRI). Better equipment and procedures also made diagnostic procedures more accurate, less invasive, and safer for patients. During the course of the twentieth, century x ray images that once took minutes of exposure could be formed in milliseconds with doses less than 1/50 of those used at the dawn of the century. Because of increased resolution that allowed physicians to see more clearly and with greater detail, physicians were increasingly able to make diagnosis of serious **pathology** (e.g. **tumors**) earlier. Earlier diagnosis often translates to a more favorable outcome for the patient.

During the early 1950s, the use of fluorescent screens allowed physicians their first real-time look into the body. Used in connection with improved contrast mediums (dyes that allow physicians to see vessels and internal **organs**), real-time diagnosis became a way for physicians to see the dynamics of disease. Prior to 1960, cameras and monitors allowed for x-ray movies.

Although nuclear medicine traces its clinical origins to the 1930s, the invention of the scintillation camera by American engineer Hal Anger in the 1950s brought nuclear medical imaging to the forefront of diagnostics. Nuclear studies involve the introduction of low-level radioactive chemicals that are consequently transported throughout he body to target organs and tissues. The scintillation camera measured and created images from the gamma rays emitted from these radioactive chemicals, enabling physicians to detect tumors or other disease processes. The basic techniques of the scintillation, or Anger camera, are the most widely used tools in nuclear medicine today, and gave rise to modern imaging systems such as the P.E.T. (positive emission tomography) scanner.

The development of noninvasive diagnostic techniques allowed physicians the opportunity to probe the body with safety and precision.

Medical imaging encompasses a number of sub-specialties including x ray, computed tomography, magnetic resonance imaging, ultra-sound, scanning and endoscopy.

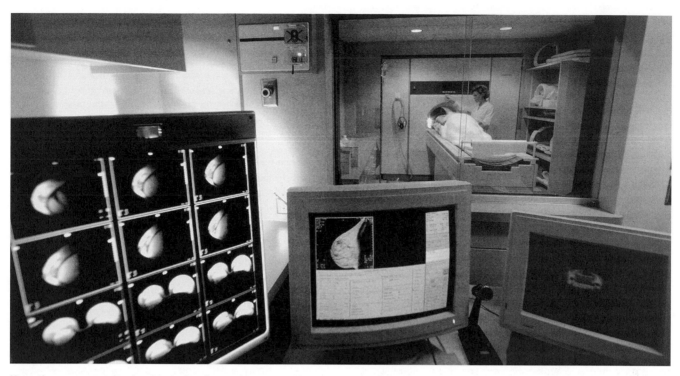

Magnetic resonance suite showing the imaging equipment and viewing machine. *Photograph by Mason Morfit. FPG International Corp. Reproduced by permission.*

The views provided physicians are not limited to static depictions of organs. In conjunction with the injection of various dyes and markers, these techniques allow physicians to view the dynamic workings of the body (e.g., the flow of **blood** through **arteries** and **veins**).

The development of powerful high-tech diagnostic tools in the later half of the twentieth century was initially the result of fundamental advances in the study of the reactions that take place in excited atomic nuclei. Applications of what were termed nuclear spectroscopic principles became directly linked to the development of non-invasive diagnostic tools used by physicians.

In particular, nuclear magnetic resonance (NMR) was one such form of nuclear spectroscopy that eventually found widespread use in the clinical laboratory and medical imaging. NMR is based on the observation that a proton in a magnetic field has two quantized spin states. Accordingly, NMR allowed the determination of the structure of organic molecules and although there are complications due to interactions between atoms, in simple terms, NMR allowed physicians to see pictures representing the larger structures of molecules and compounds (i.e., bones, tissues and organs) obtained as a result of measuring differences between the expected and actual numbers of photons absorbed by a target **tissue**.

Groups of nuclei brought into resonance, that is, nuclei absorbing and emitting photons of similar electromagnetic radiation (e.g., radio waves) make subtle yet distinguishable changes when the resonance is forced to change by altering the energy of impacting photons. The speed and extent of the resonance changes permits a non-destructive (because of the use of low energy photons) determination of anatomical structures. This form of NMR is used by physicians as the physical and chemical basis of a powerful diagnostic technique termed magnetic resonance imaging (MRI).

Magnetic resonance (MR) technologies relied on advances in physics during the 1950s. The development of MR imaging, attributed to English scientist Paul Lauterbur, was used in clinical trials during 1980. By the mid 1980s MR imaging was an accepted and widely used diagnostic technique.

MRI scanners rely on the principles of atomic nuclear-spin resonance. Using strong magnetic fields and radio waves, MRI collects and correlates deflections caused by atoms into images of amazing detail. The resolution of the MRI scanner is so high that they can be used to observe the individual plaques in multiple sclerosis.

Late in the twentieth century, diagnostic tools moved into the operating room. So called "image-guided surgical methods" now allow surgeons to more accurately determine the locations of tumors, lesions, and a host of vascular abnormalities. A corollary benefit of these techniques allows surgeons to track the positions of surgical instruments. Rapidly advancing computer technology and imaging when used in conjunction with optical, electromagnetic, or ultrasound sensors allow physicians to make real-time diagnosis a part of surgical procedures.

Principals of SONAR technology (originally developed for military use) found clinical diagnostic application with the 1960s development of ultrasound. A sonic production device termed a transducer was placed against the skin of a patient to produce high frequency sound waves that were able to penetrate the skin and reflect off internal target structures. Modern ultrasound techniques using monitors allow physicians real-time diagnostic capabilities. By the 1980s, ultrasound examinations became commonplace in the examination of fetal development.

The advent of other imaging to supplant x ray provided for less potentially damaging forms of diagnosis. High photon energies found in x rays are ionizing and are thus capable of destroying chemical and molecular bonds in cells. In contrast, ultrasound relies not on electromagnetic radiation but rather on pressure waves that are non-ionizing.

Ultrasonic doppler techniques are used to identify pathology related to blood flow (e.g., **arteriosclerosis**). Specialized types of scanning using Doppler techniques are also used to identify **heart** valve defects.

Microscopes using ultrasound can be used to study cell structures without subjecting them to lethal staining procedures that can also impede diagnosis through the production of artifacts. Ultrasonic microscopes differentiate structures based on underlying differences in pathology. Ultrasonic are also the least expensive of the latest high-tech innovations in diagnostic imaging.

Mammography, or x-ray visualization of the breast, became a common diagnostic procedure in the 1960s, when physicians demonstrated its usefulness in the diagnosis of breast **cancer**. Initially, mammograms were used to aid in diagnosis only, and the x-ray dose to the tissue was relatively high. In 1973, the Breast Cancer Detection Demonstration Project, a five-year study of over 250,000 women, helped to establish mammography as an effective screening tool for breast cancer detection. Today, modern mammography machines allow greater precision of breast tissue imaging with less radiation exposure to the patient. Mammography is a standard recommended screening procedure, and is credited with reducing mortality from breast cancer in women over 50 years of age.

During the early 1970s, enhanced digital capabilities spurred the development of computed tomography (derived from the Greek "tomos" meaning slice) imaging, also called CT, computed axial tomography or CAT scans, invented by English physician **Godfrey Hounsfield**. CT scans use advanced computer-based mathematical algorithms to combine different reading or views of a patient into a coherent picture usable for diagnosis. Hounsfield's innovative use of high energy electromagnetic beams, a sensitive detector mounted on a rotating frame, and digital computing to create detailed images earned him the Nobel prize. As with x rays, CT scan technology progressed to allow the use of less energetic beams and vastly decreased exposure times. CT scans increased the scope and safety of imaging procedures that allowed physicians to view the arrangement and functioning of the body's internal structures on a small scale.

American chemist Peter Alfred Wolf's (1923–998) work with positron emission tomography (PET) led to the clinical diagnostic use of the PET scan, allowing physicians to measure cell activity in organs. PET scans use rings of detectors that surround the patient to track the movements and con-

centrations of radioactive tracers. The detectors measure gamma radiation produced when positrons emitted by tracers are annihilated during collisions with electrons. PET scans have attracted the interest of psychiatrists for their potential to study the underlying metabolic changes associated with mental diseases such as schizophrenia and depression. During the 1990s, PET scans found clinical usage in the diagnosis and characterizations of certain cancers and heart disease, as well as clinical studies of the **brain**.

MRI and PET scans, both examples of functional imaging (in addition to detailing structures they provide a view of dynamic functions) are the subject of increased research and clinical application. MRI and PET scans are used to measure reactions of the brain when challenged with sensory input (e.g., hearing, sight, **smell**), activities associated with processing information (e.g., learning functions), physiological reactions to addiction, metabolic processes associated with **osteoporosis** and atherosclerosis, and to shed light on pathological conditions such as Parkinson and **Alzheimer's disease**.

During the 1990s, the explosive development of information technologies and the Internet allowed physicians to make diagnosis from remote locations and to tele-conference over real-time data. Multiple imaging is becoming an increasingly important tool to physicians. These and other high-tech innovations may one day contribute to the speed and accuracy of diagnosis, minimizing the need for invasive surgeries, and allowing exquisite precision when surgery is necessary.

See also History of anatomy and physiology: The science of medicine; Medical training in anatomy and physiology; Pregnancy, maternal physiological and anatomical changes; Prenatal growth and development; Radiation damage to tissues

IMMUNE SYSTEM

The immune system is the body's biological defense mechanism that protects against foreign invaders. Only in the last century have the components of that system and the ways in which they work been discovered, and more remains to be clarified.

The true roots of the study of the immune system date from 1796 when an English physician, Edward Jenner, discovered a method of smallpox vaccination. He noted that dairy workers who contracted cowpox from milking infected cows were thereafter resistant to smallpox. In 1796, Jenner injected a young boy with material from a milkmaid who had an active case of cowpox. After the boy recovered from his own resulting cowpox, Jenner inoculated him with smallpox; the boy was immune. After Jenner published the results of this and other cases in 1798, the practice of Jennerian vaccination spread rapidly.

It was Louis Pasteur who established the cause of infectious diseases and the medical basis for immunization. First, Pasteur formulated his germ theory of disease, the concept that disease is caused by communicable microorganisms. In 1880, Pasteur discovered that aged cultures of fowl cholera **bacteria** lost their power to induce disease in chickens but still conferred immunity to the disease when injected. He went on to use attenuated (weakened) cultures of anthrax and rabies to vaccinate against those diseases. The American scientists Theobald Smith (1859–1934) and Daniel Salmon (1850–1914) showed in 1886 that bacteria killed by heat could also confer immunity.

Why vaccination imparted immunity was not yet known. In 1888, Pierre-Paul-Emile Roux (1853–1933) and Alexandre Yersin (1863–1943) showed that diphtheria bacillus produced a toxin that the body responded to by producing an antitoxin. Emil von Behring and Shibasaburo Kitasato found a similar toxin-antitoxin reaction in tetanus in 1890. Behring discovered that small doses of tetanus or diphtheria toxin produced immunity, and that this immunity could be transferred from animal to animal via serum. Behring concluded that the immunity was conferred by substances in the **blood**, which he called antitoxins, or antibodies. In 1894, Richard Pfeiffer (1858–1945) found that antibodies killed cholera bacteria (bacterioloysis). Hans Buchner (1850–1902) in 1893 discovered another important blood substance called complement (Buchner's term was alexin), and **Jules Bordet** in 1898 found that it enabled the antibodies to combine with **antigens** (foreign substances) and destroy or eliminate them. It became clear that each antibody acted only against a specific antigen. **Karl Landsteiner** was able to use this specific antigen-antibody reaction to distinguish the different blood groups.

A new element was introduced into the growing body of immune system knowledge during the 1880s by the Russian microbiologist **Elie Metchnikoff**. He discovered cell-based immunity: white blood cells (leucocytes), which Metchnikoff called phagocytes, ingested and destroyed foreign particles. Considerable controversy flourished between the proponents of cell-based and blood-based immunity until 1903, when Almroth Edward Wright brought them together by showing that certain blood substances were necessary for phagocytes to function as bacteria destroyers. A unifying theory of immunity was posited by **Paul Ehrlich** in the 1890s; his "side-chain" theory explained that **antigens and antibodies** combine chemically in fixed ways, like a key fits into a lock. Until now, immune responses were seen as purely beneficial. In 1902, however, **Charles Richet** and Paul Portier demonstrated extreme immune reactions in test animals that had become sensitive to antigens by previous exposure. This phenomenon of hypersensitivity, called anaphylaxis, showed that immune responses could cause the body to damage itself. Hypersensitivity to antigens also explained **allergies**, a term coined by Pirquet in 1906.

Much more was learned about antibodies in the mid-twentieth century, including the fact that they are proteins of the gamma globulin portion of **plasma** and are produced by plasma cells; their molecular structure was also worked out. An important advance in immunochemistry came in 1935 when Michael Heidelberger and **Edward Kendall** developed a method to detect and measure amounts of different antigens and antibodies in serum. Immunobiology also advanced. **Frank Macfarlane Burnet** suggested that animals did not produce antibodies to substances they had encountered very early in life; **Peter Medawar** proved this idea in 1953 through experiments on mouse embryos.

In 1957, Burnet put forth his clonal selection theory to explain the biology of immune responses. On meeting an antigen, an immunologically responsive cell (shown by C. S. Gowans [1923–] in the 1960s to be a lymphocyte) responds by multiplying and producing an identical set of plasma cells, which in turn manufacture the specific antibody for that antigen. Further cellular research has shown that there are two types of lymphocytes (nondescript lymph cells): **B lymphocytes**, which secrete antibody, and T lymphocytes, which regulate the B lymphocytes and also either kill foreign substances directly (killer T cells) or stimulate macrophages to do so (helper T cells). Lymphocytes recognize antigens by characteristics on the surface of the antigen-carrying molecules. Researchers in the 1980s uncovered many more intricate biological and chemical details of the immune system components and the ways in which they interact.

Knowledge about the immune system's role in rejection of transplanted **tissue** became extremely important as organ transplantation became surgically feasible. Peter Medawar's work in the 1940s showed that such rejection was an immune reaction to antigens on the foreign tissue. Donald Calne (1936–) showed in 1960 that immunosuppressive drugs, drugs that suppress immune responses, reduced transplant rejection, and these drugs were first used on human patients in 1962. In the 1940s **George Snell** discovered in mice a group of tissue-compatibility genes, MHC that played an important role in controlling acceptance or resistance to tissue grafts. **Jean Dausset** found human MHC, a set of antigens to human leucocytes (white blood cells), called HLA. Matching of HLA in donor and recipient tissue is an important technique to predict compatibility in transplants. **Baruj Benacerraf** in 1969 showed that an animal's ability to respond to an antigen was controlled by genes in the MHC complex.

Exciting new discoveries in the study of the immune system are on the horizon. Researchers are investigating the relation of HLA to disease; certain types of HLA molecules may predispose people to particular diseases. This promises to lead to more effective treatments and, in the long run, possible prevention. Autoimmune reaction, in which the body has an immune response to its own substances, may also be a cause of a number of diseases, like multiple sclerosis, and research proceeds on that front. Approaches to **cancer** treatment also involve the immune system. Some researchers, including Burnet, speculate that a failure of the immune system may be implicated in cancer. In the late 1960s, Ion Gresser (1928–) discovered that the protein interferon acts against cancerous **tumors**. After the development of genetically engineered interferon in the mid-1980s finally made the substance available in practical amounts, research into its use against cancer accelerated. The invention of monoclonal antibodies in the mid-1970s was a major breakthrough. Increasingly sophisticated knowledge about the workings of the immune system holds out the hope of finding an effective method to combat one of the most serious immune system disorders, **AIDS**.

Avenues of research to treat AIDS includes a focus on supporting and strengthening the immune system. (However, much research has to be done in this area to determine whether strengthening the immune system is beneficial or whether it may cause an increase in the number of infected cells.) One area of interest is cytokines, proteins produced by the body that help the immune system cells communicate with each other and activate them to fight **infection**. Some individuals infected with the AIDS virus HIV (human immunodeficiency virus) have higher levels of certain cytokines and lower levels of others. A possible approach to controlling infection would be to boost deficient levels of cytokines while depressing levels of cytokines that may be too abundant. Other research has found that HIV may also turn the immune system against itself by producing antibodies against its own cells.

Advances in immunological research indicate that the immune system may be made of more than 100 million highly specialized cells designed to combat specific antigens. While the task of identifying these cells and their functions may be daunting, headway is being made. By identifying these specific cells, researchers may be able to further advance another promising area of immunological research, the use of recombinant **DNA** technology, in which specific proteins can be mass-produced. This approach has led to new cancer treatments that can stimulate the immune system by using synthetic versions of proteins released by interferons.

See also Antigens and antibodies; Bacteria and responses to bacterial infection; Immunology; Transfusions; Transplantation of organs; Viruses and responses to viral infection

IMMUNITY, CELL MEDIATED

The **immune system** is a network of cells and **organs** that work together to protect the body from infectious organisms. Many different types of organisms such as **bacteria**, **viruses**, fungi, and parasites are capable of entering the human body and causing disease. It is the immune system's job to recognize these agents as foreign and destroy them.

The immune system can respond to the presence of a foreign agent in one of two ways. It can either produce soluble proteins called antibodies, which can bind to the foreign agent and mark them for destruction by other cells. This type of response is called a humoral response or an antibody response. Alternately, the immune system can mount a cell-mediated immune response. This involves the production of special cells that can react with the foreign agent. The reacting cell can either destroy the foreign agents, or it can secrete chemical signals that will activate other cells to destroy the foreign agent.

During the 1960s, it was discovered that different types of cells mediate the two major classes of immune responses. The **T lymphocytes**, which are the main effectors of the cell-mediated response, mature in the thymus, thus the name T cell. The B cells, which develop in the adult bone marrow, are responsible for producing antibodies. There are several different types of T cells performing different functions. These diverse responses of the different T cells are collectively called the "cell-mediated immune responses."

There are several steps involved in the cell-mediated response. The pathogen (bacteria, virus, fungi, or a parasite), or foreign agent, enters the body through the **blood** stream, different tissues, or the respiratory tract. Once inside the body, the foreign agents are carried to the spleen, lymph nodes, or the mucus-associated lymphoid **tissue** (MALT) where they will come in contact with specialized cells known as antigen-presenting cells (APC). When the foreign agent encounters the antigen-presenting cells, an immune response is triggered. These antigen presenting cells digest the engulfed material, and display it on their surface complexed with certain other proteins known as the Major Histocompatibility Class (MHC) class of proteins.

Next, the T cells must recognize the antigen. Specialized receptors found on some T cells are capable of recognizing the MHC-antigen complexes as foreign and binding to them. Each T cell has a different receptor in the cell membrane that is capable of binding a specific antigen. Once the T cell receptor binds to the antigen, it is stimulated to divide and produce large amounts of identical cells that are specific for that particular foreign antigen. The T lymphocytes also secrete various chemicals (cytokines) that can stimulate this proliferation. The cytokines are also capable of amplifying the immune defense functions that can eventually destroy and remove the antigen.

In cell-mediated immunity, a subclass of the T cells mature into cytotoxic T cells that can kill cells having the foreign antigen on their surface, such as virus-infected cells, bacterial-infected cells, and tumor cells. Another subclass of T cells called helper T cells activates the B cells to produce antibodies that can react with the original antigen. A third group of T cells called the suppressor T cells is responsible for regulating the immune response by turning it on only in response to an antigen and turning it off once the antigen has been removed.

Some of the B and T lymphocytes become "memory cells," that are capable of remembering the original antigen. If that same antigen enters the body again while the memory cells are present, the response against it will be rapid and heightened. This is the reason the body develops permanent immunity to an infectious disease after being exposed to it. This is also the principle behind immunization.

See also Immunity, humoral regulation

IMMUNITY, HUMORAL

One of the ways in which the **immune system** responds to pathogens is by producing soluble proteins called antibodies. This is known as the humoral response and involves the activation of a special set of cells known as the B lymphocytes, because they originate in the bone marrow. The humoral immune response helps in the control and removal of pathogens such as **bacteria**, **viruses**, fungi, and parasites before they enter host cells. The antibodies produced by the B cells are the mediators of this response.

The antibodies form a family of **plasma** proteins referred to as immunoglobulins. They perform two major functions. One of the functions of an antibody is to bind specifically to the molecules of the foreign agent that triggered the immune response. A second function is to attract other cells and molecules to destroy the pathogen after the antibody molecule is bound to it.

When a foreign agent enters the body, it is engulfed by the antigen-presenting cells, or the B cells. The B cell that has a receptor (surface immunoglobulin) on its membrane that corresponds to the shape of the antigen binds to it and engulfs it. Within the B cell, the antigen-antibody pair is partially digested, bound to a special class of proteins called MHC-II, and then displayed on the surface of the B cell. The helper T cells recognize the pathogen bound to the MHC-II protein as foreign and becomes activated.

These stimulated T cells then release certain chemicals known as cytokines (or lymphokines) that act upon the primed B cells (B cells that have already seen the antigen). The B cells are induced to proliferate and produce several identical cells capable of producing the same antibody. The cytokines also signal the B cells to mature into antibody producing cells. The activated B cells first develop into lymphoblasts and then become plasma cells, which are essentially antibody producing factories. A subclass of B cells does not differentiate into plasma cells. Instead, they become memory cells that are capable of producing antibodies at a low rate. These cells remain in the immune system for a long time, so that the body can respond quickly if it encounters the same antigen again.

The antibody destroys the pathogen in three different ways. In neutralization, the antibodies bind to the bacteria or toxin and prevent it from binding and gaining entry to a host cell. Neutralization leads to a second process called opsonization. Once the antibody is bound to the pathogen, certain other cells called macrophages engulf these cells and destroy them. This process is called phagocytosis. Alternately, the immunoglobulin IgM or IgG can bind to the surface of the pathogen and activate a class of serum proteins called the complement, which can cause lysis of the cells bearing that particular antigen.

In the humoral immune response, each B cell produces a distinct antibody molecule. There are over a million different B lymphocytes in each individual, which are capable of recognizing a corresponding million different **antigens**. Since each antibody molecule is composed of two different proteins (the light chain and the heavy chain), it can bind two different antigens at the same time.

See also Immunity, cell mediated

IMMUNOLOGY

Immunology is the study of how the body responds to foreign substances and fights off **infection** and other disease. Immunologists study the molecules, cells, and **organs** of the human body that participate in this response.

The beginnings of our understanding of immunity date to 1798, when the English physician Edward Jenner (1749–1823) published a report that people could be protected from deadly smallpox by sticking them with a needle dipped in the material from a cowpox boil. The French biologist and chemist Louis Pasteur (1822–1895) theorized that such immunization protects people against disease by exposing them to a version of a microbe that is harmless but is enough like the disease-causing organism, or pathogen, that the **immune system** learns to fight it. Modern vaccines against diseases such as measles, polio, and chicken pox are based on this principle.

In the late nineteenth century, a scientific debate was waged between the German physician **Paul Ehrlich** and the Russian zoologist **Elie Metchnikoff**. Ehrlich and his followers believed that proteins in the **blood**, called antibodies, eliminated pathogens by sticking to them; this phenomenon became known as **humoral immunity**. Metchnikoff and his students, on the other hand, noted that certain white blood cells could engulf and digest foreign materials: this cellular immunity, they claimed, was the real way the body fought infection.

Modern immunologists have shown that both the humoral and cellular responses play a role in fighting disease. They have also identified many of the actors and processes that form the immune response.

The immune response recognizes and responds to pathogens via a network of cells that communicate with each other about what they have "seen" and whether it "belongs." These cells patrol throughout the body for infection, carried by both the blood stream and the lymph ducts, a series of vessels carrying a clear fluid rich in immune cells.

The antigen presenting cells are the first line of the body's defense, the scouts of the immune army. They engulf foreign material or microorganisms and digest them, displaying bits and pieces of the invaders—called antigens—for other immune cells to identify. These other immune cells, called **T lymphocytes**, can then begin the immune response that attacks the pathogen.

The body's other cells can also present **antigens**, although in a slightly different way. Cells always display antigens from their everyday proteins on their surface. When a cell is infected with a virus, or when it becomes cancerous, it will often make unusual proteins whose antigens can then be identified by any of a variety of cytotoxic T lymphocytes. These "killer cells" then destroy the infected or cancerous cell to protect the rest of the body. Other T lymphocytes generate chemical or other signals that encourage multiplication of other infection-fighting cells. Various types of T lymphocytes are a central part of the cellular immune response; they are also involved in the humoral response, encouraging B lymphocytes to turn into antibody-producing **plasma** cells.

The body cannot know in advance what a pathogen will look like and how to fight it, so it creates millions and millions of different lymphocytes that recognize random antigens. When, by chance, a B or T lymphocyte recognizes an antigen being displayed by an antigen presenting cell, the lymphocyte divides and produces many offspring that can also identify and attack this antigen. The way the immune system expands cells

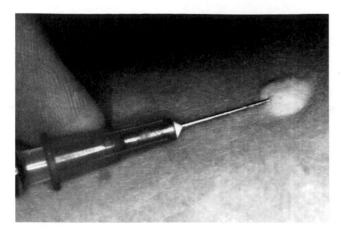

An immune response can be induced by injection with dead or attenuated microorganisms. A needle is shown injecting tuberculin protein just under the skin, where a localized reaction will occur if the person has a tuberculosis infection. *Photo Researchers, Inc. Reproduced by permission.*

that by chance can attack an invading microbe is called clonal selection.

Some researchers believe that while some B and T lymphocytes recognize a pathogen and begin to mature and fight an infection, others stick around in the bloodstream for months or even years in a primed condition. Such memory cells may be the basis for the immunity noted by the ancient Chinese and by Thucydides. Other immunologists believe instead that trace amounts of a pathogen persist in the body, and their continued presence keeps the immune response strong over time.

Substances foreign to the body, such as disease-causing **bacteria**, **viruses**, and other infectious agents (known as antigens), are recognized by the body's immune system as invaders. The body's natural defenses against these infectious agents are antibodies—proteins that seek out the antigens and help destroy them. Antibodies have two very useful characteristics. First, they are extremely specific; that is, each antibody binds to and attacks one particular antigen. Second, some antibodies, once activated by the occurrence of a disease, continue to confer resistance against that disease; classic examples are the antibodies to the childhood diseases chickenpox and measles.

The second characteristic of antibodies makes it possible to develop vaccines. A vaccine is a preparation of killed or weakened bacteria or viruses that, when introduced into the body, stimulates the production of antibodies against the antigens it contains.

It is the first trait of antibodies, their specificity, that makes monoclonal antibody technology so valuable. Not only can antibodies be used therapeutically, to protect against disease; they can also help to diagnose a wide variety of illnesses, and can detect the presence of drugs, viral and bacterial products, and other unusual or abnormal substances in the blood.

Given such a diversity of uses for these disease-fighting substances, their production in pure quantities has long been the focus of scientific investigation. The conventional method

was to inject a laboratory animal with an antigen and then, after antibodies had been formed, collect those antibodies from the blood serum (antibody-containing blood serum is called antiserum). There are two problems with this method: It yields antiserum that contains undesired substances, and it provides a very small amount of usable antibody.

Monoclonal antibody technology allows the production of large amounts of pure antibodies in the following way. Cells that produce antibodies naturally are obtained along with a class of cells that can grow continually in cell culture. The hybrid resulting from combining cells with the characteristic of "immortality" and those with the ability to produce the desired substance, creates, in effect, a factory to produce antibodies that work around the clock.

A myeloma is a tumor of the bone marrow that can be adapted to grow permanently in cell culture. Fusing myeloma cells with antibody-producing mammalian spleen cells, results in hybrid cells, or hybridomas, producing large amounts of monoclonal antibodies. This product of cell fusion combined the desired qualities of the two different types of cells;: the ability to grow continually, and the ability to produce large amounts of pure antibody. Because selected hybrid cells produce only one specific antibody, they are more pure than the polyclonal antibodies produced by conventional techniques. They are potentially more effective than conventional drugs in fighting disease, because drugs attack not only the foreign substance but also the body's own cells as well, sometimes producing undesirable side effects such as **nausea** and allergic reactions. Monoclonal antibodies attack the target molecule and only the target molecule, with no or greatly diminished side effects.

While researchers have made great gains in understanding immunity, many big questions remain. Future research will need to identify how the immune response is coordinated. Other researchers are studying the immune systems of non-mammals, trying to learn how our immune response evolved. Insects, for instance, lack antibodies, and are protected only by cellular immunity and chemical defenses not known to be present in higher organisms.

Immunologists do not yet know the details behind allergy, where antigens like those from pollen, **poison** ivy, or certain kinds of food make the body start an uncomfortable, unnecessary, and occasionally life-threatening immune response. Likewise, no one knows exactly why the immune system can suddenly attack the body's tissues—as in autoimmune diseases like rheumatoid arthritis, juvenile diabetes, systemic lupus erythematosus, or multiple sclerosis.

The hunt continues for new vaccines, especially against parasitic organisms like the **malaria** microbe that trick the immune system by changing their antigens. Some researchers are seeking ways to start an immune response that prevents or kills cancers. A big goal of immunologists is the search for a vaccine for HIV, the virus that causes **AIDS**. HIV knocks out the immune system—causing immunodeficiency—by infecting crucial T lymphocytes. Some immunologists have suggested that the chiefly humoral response raised by conventional vaccines may be unable to stop HIV from getting to lymphocytes, and that a new kind of vaccine that encourages a cellular response may be more effective.

Researchers have shown that transplant rejection is just another kind of immune response, with the immune system attacking antigens in the transplanted organ that are different from its own. Drugs that suppress the immune system are now used to prevent rejection, but they also make the patient vulnerable to infection. Immunologists are using their increased understanding of the immune system to develop more subtle ways of fooling the immune system into accepting transplants.

See also Infection and resistance; Inflammation of tissues

IMPLANTATION OF THE EMBRYO

As the zygote traverses the maternal tubal and uterine cavity, it becomes a morula (a cluster of divided cells of the fertilized ovum) and matures into a blastocyst. The blastocyst then looses its zona pellucida (a clear protective layer that facilitates transport as it is nonadhesive), and develops an inner cell mass and an outer protective layer, the trophectoderm. Implantation is the term applied to the process by which the blastocyst attaches to the wall of the uterus and begins to send fingers of **tissue** (chorionic villi) into the wall of the uterus as anchors.

Blastocyst implantation and successful establishment of **pregnancy** require delicate interactions between the embryo and the maternal environment. A highly-coordinated process is set into motion whereby specialized cells of the embryo, the trophectoderm and the throphoblast, establish contact with the specialized maternal tissue, the uterine endometrium. Uterine implantation is mediated by growth factors, cytokines, and extracellular matrix molecules. Recently it has been also shown that **leukemia** inhibitory factor (LIF) secreted by the uterine glands, and colony-stimulating factor-1 (CSF-1) aid in normal blastocyst implantation. Also, **hormones**, both by embryonic as well as maternal tissues of both uterine and extrauterine origins, take place in the implantation process. Furthermore, in response to signals from the embryo, pregnancy-specific proteins are released in maternal serum and a series of morphological, biochemical, and immunological changes occur in the uterine environment. In fact, already in the late luteal phase, physiological changes occur in the endometrium to allow blastocyst implantation. In concert, receptors for all these factors must be expressed by the appropriate tissues to propagate implantation signals. Failure to properly begin these critical events during this so called "window of implantation or receptivity" results in early pregnancy failure.

A change in endometrial gene expression is necessary to allow the embryo implantation, since the uterine endometrium is normally hostile to the implantation. The uterus undergoes dynamic changes during the menstrual cycle but, in the presence of the embryo, endometrium is maintained and begins a program of events named the decidual reaction, leading to prolonged maintenance and addition programs of gene expression particular to pregnancy.

Implantation begins first with attachment of the blastocyst through the outer trophectoderm cells to the uterine lining (day 6 from the **fertilization**). The site of implantation is marked on the surface by a coagulation plug left where the blastocyst has entered the uterine wall (day 12 from the fertilization).

More specifically, trophoblast (derived from the trophectoderm) at the embryonic pole, attaches to the epithelial lining of the uterus. This triggers changes in both the trophoblast and the connective tissue (stroma) beneath the uterine epithelium Some trophoblast cells fuse to make a syncytium called the syncytiotrophoblast, that releases proteolytic **enzymes** that allow passage of the blastocyst into the endometrial wall first, then the stroma (carrying the whole conceptus with it).

During this migration, trophoblast cells destroy the wall of the maternal spiral **arteries**, converting them from muscular vessels into flaccid sinusoidal sacs. This vascular transformation is important to ensure an adequate **blood** supply to the feto-placental unit and the beginning of the histiotrophic **nutrition**. Both cell-cell and cell-matrix interactions are important for trophoblast invasion of the decidual stroma and decidual spiral arteries.

Uterine stromal cells become decidual cells (this decidual reaction contributes to histiotrophic nutrition of the conceptus and prevents a maternal immune reaction toward it). The glands in the stroma enlarge and its blood supply increases. The syncytial trophoblast attaches to the uterine surface that at this time in named basal deciduas, then penetrates between endometrial cells and uterine glands, and spreads on the epithelial basement membrane (7–8 days after ovulation). Later events in the adhesion-invasion include a change of the basal membrane of the endometrial cells that should allow direct interactions with stromal matrix components. Integrins (specialized glycoprotein cells) appear to drive such an interaction.

An ectopic pregnancy results from implantation of the blastocyst in an abnormal site, usually inside the tubal wall.

See also Embryology; Embryonic development, early development, formation, and differentiation; Ovarian cycle and hormonal regulation; Placenta and placental nutrition of the embryo; Sexual reproduction

IN VITRO AND *IN VIVO* FERTILIZATION

In animals, **fertilization** is the fusion of a **sperm** cell with an egg cell. The penetration of the egg cell by the chromosome-containing part of the sperm cell causes a reaction that prevents additional sperm cells from entering the egg. The egg and sperm each contribute half of the new organism's genetic material. A fertilized egg cell is known as a zygote. Following fertilization, the zygote undergoes continuous **cell division** that eventually produces a new multicellular organism.

Human fertilization *in vivo* (in the living body) occurs in oviducts (fallopian tubes) of the female reproductive tract, and takes place within hours following sexual intercourse.

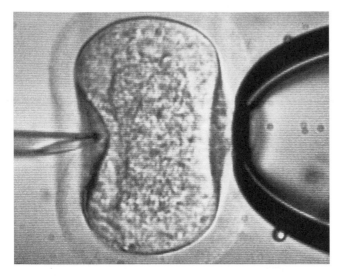

A needle (left) pierces a human ovum (middle) in order to inject a sperm during *in vitro* fertilization. *Photograph by Hank Morgan. National Audubon Society Collection/Photo Researchers, Inc. Reproduced by permission.*

Only one of the approximately 300 million sperm released into a female's vagina during intercourse can fertilize the single female egg cell (ovum). The successful sperm cell must enter the uterus and swim up the fallopian tube to meet the ovum cell, where it passes through the thick coating surrounding the egg. This coating, consisting of sugars and proteins, is known as the zona pellucida. The tip of the head of the sperm cell contains **enzymes** which break through the zona pellucida and aid the penetration of the sperm into the egg. Once the head of the sperm is inside the egg, the tail of the sperm falls off, and the perimeter of the egg thickens to prevent another sperm from entering.

The sperm and the egg each contain only half the normal number of **chromosomes**, a condition known as the haploid chromosome state. When the genetic material of the two cells fuses, fertilization is complete.

In humans, a number of variables affect whether or not fertilization occurs following intercourse. One factor is a woman's ovulatory cycle. Human eggs can only be fertilized a few days after ovulation, which usually occurs only once every 28 days.

In vitro fertilization (IVF) is a procedure in which eggs (ova) from a woman's ovary are removed, fertilized with sperm in a laboratory procedure, and then the resulting fertilized egg (embryo) is returned to the woman's uterus. IVF is one of several assisted reproductive techniques (ART) used to help infertile couples to conceive a child. If after one year of having sexual intercourse without the use of birth control, a woman is unable to become pregnant, infertility is suspected. Some of the reasons for infertility are damaged or blocked fallopian tubes, hormonal imbalance, or endometriosis in the woman. In the man, low sperm count or poor quality sperm can cause infertility.

IVF is one of several possible methods to increase the chance for an infertile couple to become pregnant. Its use depends on the reason for infertility. IVF may be an option if there is a blockage in the fallopian tube or endometriosis in the woman, or low sperm count or poor quality sperm in the man. There are other possible treatments for these conditions often attempted prior to IVF, such as surgery for blocked tubes or endometriosis. IVF will not be successful for a woman who is not capable of ovulating, or a man who is not able to produce a certain number of healthy sperm.

During *in vitro* fertilization, the joining of egg and sperm takes place outside of the woman's body. A woman may be given fertility drugs before this procedure so that several eggs mature in the ovaries at the same time. Mature eggs (ova) are removed from a woman's ovaries, and are mixed with sperm in a laboratory dish or test tube, hence, the early term "test tube baby." The eggs are monitored for several days. Once there is evidence that fertilization has occurred and the cells begin to divide, they are then returned to the woman's uterus.

In the procedure to remove eggs, enough may be gathered to be frozen and saved (either fertilized or unfertilized) for additional IVF attempts.

IVF has been used successfully since 1978, when the first child to be conceived by this method was born in England. Over the past 20 years, thousands of couples have used IVF or other assisted reproductive technologies to conceive. The risks associated with *in vitro* fertilization include the possibility of multiple **pregnancy** (since several embryos may be implanted) and ectopic pregnancy (an embryo that implants in the fallopian tube or in the abdominal cavity outside the uterus). There is a slight risk of ovarian rupture, bleeding, infections, and complications of **anesthesia**. If the procedure is successful and pregnancy is achieved, the pregnancy would carry the same risks as any pregnancy achieved without assisted technology.

Success rates vary widely between clinics and between physicians performing the procedure. A couple has about a 10% chance of becoming pregnant each time the procedure is performed. Therefore, the procedure may have to be repeated more than once to achieve pregnancy.

Other similar types of assisted reproductive technologies are also used to achieve pregnancy. A procedure called intracytoplasmic sperm injection (ICSI) uses a manipulation technique that uses a microscope to inject a single sperm into each egg. The fertilized eggs can then be returned to the uterus, as in IVF. In gamete intrafallopian tube transfer (GIFT) the eggs and sperm are mixed in a narrow tube and then deposited in the fallopian tube, where fertilization normally takes place. Another variation on IVF is zygote intrafallopian tube transfer (ZIFT). As in IVF, the fertilization of the eggs occurs in a laboratory dish. And, similar to GIFT, the embryos are placed in the fallopian tube (rather than the uterus as with IVF).

See also Embryology; Embryo transfer; Infertility (female); Infertility (male); Oogenesis; Sexual reproduction; Spermatogenesis

INFANT AND EARLY CHILDHOOD DEVELOPMENT SCREENING

Development screening is a process that can indicate disease in a patient before he or she develops symptoms. he procedures surrounding infant and childhood development screening are especially sensitive to the welfare of these patients because they have not reached a legal age of consent. Bioethicists, research scientist, and doctors overwhelmingly agree that an infant or child should not be subjected to screening unless the positive result of the test (the presence of disease) will lead to preventative or palliative care. This opinion therefore discourages testing of infants or children for adult onset diseases, if being able to predict the disease has no utility.

Recent progress in the typing of human **DNA** will continue to advance developmental screening, the ultimate purpose of which is to prevent and treat disease. Some screening tests for newborns, such as the test that reveals cystic fibrosis, an inherited pulmonary disease, generate questions about their effectiveness in treating disease. Scientists have not definitively determined whether detection of cystic fibrosis in the neonatal period (the first 28 days after birth) actually improves the long-term pulmonary condition of affected children. Furthermore, concern about the reliability of such tests (and the possibility of false positives) raises questions about their ultimate service.

Relatively common types of infant and child development screening are tests for chromosomal anomalies, because detecting these before **puberty** can mitigate the patient's condition. For instance, if boys with Klinefelter syndrome (the presence of two X **chromosomes** rather than one, XXY instead of a normal XY) are identified in childhood, they can be treated with **testosterone** (the male sex hormone) with the onset of puberty. Such treatment would help to compensate for their congenital deficit of this hormone. Similarly, girls identified with Turner syndrome (XO instead of XX) can benefit from **estrogen** therapy (the female sex hormone), to motivate the sexual maturation that this syndrome counteracts.

Tests that indicate autosomal recessive disorders (any chromosomal disease except for those affecting sex chromosomes passed to an individual through recessive genes) are another prevalent type of screening in infants and children. The example of screening for Phenylketonuria (PKU), an inborn metabolic disorder that causes severe retardation, reflects many of the benefits and risks of genetic screening over the last fifty years.

Researchers discovered in the 1960s that PKU's effects can be decreased or even negated if treatment began soon after an afflicted baby's birth. Because PKU prevents the conversion of phenylaline to another amino acid, the **brain** is starved of nutrients, and retardation occurs. However, if a baby with PKU receives a high protein diet from the time he or she is a newborn, the proteins will break down the harmful levels of phenylaline. This discovery popularized the screening of newborns for PKU in the United States.

However, doctors later realized that the PKU screening was not totally reliable. Girls manifest the abnormal levels of the dangerous phenylaline later than boys, and so were not consistently identified early enough to halt the onset of retardation. Further, research revealed that some babies demonstrate high levels of phenylaline but do not develop PKU, and the high protein diets they were put on were unhealthy and sometimes fatal. In these cases, the supposedly preventative screening for PKU resulted in unnecessary illness or **death**.

More recently, screening of infants and children has often focused on high risk groups, such as children in families that have **inherited diseases** like cystic fibrosis or sickle-cell anemia. Further, since these diseases are primarily found in specific ethnic groups (Caucasians are more vulnerable to cystic fibrosis whereas sickle-cell anemia is prevalent in the African American population), the identification of susceptible families can be limited. Thus, screening can be restricted to those who are at risk, rather than broadly concentrated on entire populations.

As off 2002, in all U.S. states except for South Dakota, newborn screening is left to the discretion of parents rather than doctors. This trend reflects the common belief that compulsory genetic screening infringes upon the religious and/or philosophical beliefs of some parents, and for this same reason, vaccination of newborns is highly recommended but not mandatory. Continued improvements in screening technology and precision suggest that beneficial screening procedures will continue to be developed, and generalized screening for treatable conditions will probably become routine in the near future.

See also Birth defects and abnormal development; Child growth and development; Ethical issues in genetics research; Genetics and developmental genetics; Growth and development; Human development (timetables and developmental horizons); Human Genetics

INFANT GROWTH AND DEVELOPMENT

Infant growth is determined genetically, and varies according to an infant's nutritional input. An average newborn in the United States weighs 7.5 lb. (3.4 kg), and her weight nearly triples by age 1. Height at birth averages to 20 in. (51 cm), and at age one, reaches approximately 28 in. (71 cm). An infant's head remains disproportionately large compared to the rest of the body. Head circumference at birth is approximately 13–15 in. (33–38 cm) and by 12 months usually increases to approximately 16–19 in. (41–47 cm). The rate of head circumference growth slows after 12 months.

During the first month the infant's musculature adapts from limitations imposed by the fetal position, to one that allows greater flexion of elbows, hips, and knees. By six weeks, an infant's musculature develops enough that when pulled from a lying position (supine) the baby's head and neck muscles reduce the lack of head control or head lag associated with newborns. By two months, an infant can raise its head and open its hands. Facial muscles develop to the point where the infant can smile, and the respiratory and laryngeal systems develop to the point where the infant begins to vocalize in response to stimuli. By two months, the orbital muscles all an infant to follow to follow objects.

By four to six months, infants begin to reach and can coordinate bringing their hands together. At about six months, infants can sit, in a few more weeks are generally able to support their full weight on their legs if balanced. By eight to nine months, an infant can usually sit unassisted and unsupported. Starting at about nine months an infant pull itself up to a standing position and can crawl on its **abdomen**. At the end of their first year, the culmination of development in skeletal, muscular, and neural systems allows infants to make the critical transition to walking upright. Although several more months may pass before a child can walk without assistance and with assurance, this milestone marks a critical transition from infant to toddler.

An infant's anatomical and physiological changes are accompanied by rapid developmental changes. In fact, the physical change often dictate the pace of developmental changes.

Theories of infant development usually fit into the nature/nurture debate of causality. Most contemporary psychologists promote theories that encompass both the influence of **genetics** and that of environment. Also accepted as standard is the theory that **human development** is transcultural in that all infants develop according to a common timetable.

The field of infant development has shifted over the second half of the twentieth century to accommodate a revised view of an infant's psychological capacities. Researchers formerly thought that infants were nonsocial, and primarily reacted to biological needs like hunger or **pain**. More recently, however, scientists have agreed that infants are highly perceptive of their environments and their interaction with parents. This revised opinion has placed great emphasis on the quality of care and contact an infant receives from birth.

Infants who are emotionally attached to their parent(s) will smile upon recognizing the face(s) of their parents (or caregivers). Psychologists have determined that children who receive nurturing attention and touch, and are cared for at home rather than in daycare, have significantly fewer attachment insecurities, such as depression or aggressive behavior.

The social development of an infant necessarily begins from an egocentric position. All babies interpret the world from their own perspective, and differentiation between the self and the world is a learned function, which infants learn at different rates.

The progress an infant makes during the first year of life exceeds that of any other time of life. Scientists debate the degree to which infant behavior is reflex or deliberate action. While infants are born with some skills that are apparently instinctive, **reflexes** like sucking and grabbing, and the ability to locate the mother's nipple or track vision appear to be learned and thus, deliberate.

Cognitive development of infants and children is a field largely shaped by French psychologist Jean Piaget (1896–1980). In 1952, Piaget published an influential study that identified four basic stages of cognitive development that all infants and children share. During the sensorimotor stage, (birth to 24 months) infants learn through sensory input and motor activity.

Through sensorimotor experience, infants gradually learn what Piaget termed object constancy, which is the realization that concrete objects exist even when they are out of sight. For instance, an infant who has learned object constancy knows how to play games with an adult of hiding and "finding" his toys.

The motor activity of an infant includes skills such as grabbing an object, turning over, sitting, and crawling. These skills are self-taught, and infants learn them at varying times. Hand-eye coordination and manipulative abilities are usually achieved by seven months.

Language development is thought to be a function innate to all human beings, while the rate of language acquisition varies according to cognitive development and the interaction a child receives from a parent/caregiver. At one year, babies can usually speak several words, and put them together in a grammatical order at age two or older. Girls usually learn language faster than boys, but language experts have not discovered the reasons for this difference.

See also Adolescent growth and development; Aging processes; Anatomical nomenclature; Birth defects and abnormal development; Child growth and development; Embryonic development, early development, formation, and differentiation; Growth and development; Human development (timetables and developmental horizons); Infant and early childhood developmental screening

INFECTION AND RESISTANCE

Infection describes the process whereby harmful microorganisms enter the body, multiply, and cause disease. Normally the defense mechanisms of the body's **immune system** keep infectious microorganisms from becoming established. But those organisms that can evade or diffuse the immune system and therapeutic strategies (e.g., the application of antibiotics) are able to increase their population numbers faster than they can be killed. The population increase usually results in host illness.

There are a variety of ways by which harmful microorganisms can be acquired. **Blood** contaminated with microbes, such as the viral agents of hepatitis and acquired immunodeficiency syndrome, is one source. Infected food or water is another source that causes illness and **death** to millions of people around the world every year. A prominent example is the food and water-borne transmission of harmful strains of *Escherichia coli* **bacteria**. Harmful microbes can enter the body through close contact with infected creatures. Transmission of the rabies virus by an infected raccoon bite and of encephalitis virus via mosquitoes are but two examples. Finally, **breathing** contaminated air can cause illness. Bacterial spores of the causative bacterial agent of anthrax are readily aerosolized and inhaled into the **lungs**, where—if sufficient in large enough numbers—can germinate and cause severe illness and even death.

To establish an infection, microbes must defeat two lines of defense of the body. The first line of defense is at body surfaces that act as a barrier guard the boundaries between the body and the outside world. These barriers include the skin, mucous membranes in the **nose** and throat, and tiny hairs in the nose that act to physically block invading organisms. Organisms can be washed away from body surfaces by tears, bleeding, and sweating. These are non-specific mechanisms of resistance.

The body's second line of defense involves the specific mechanisms of the immune system, a coordinated response involving a variety of cells and protein antibodies, whereby an invading microorganism is recognized and destroyed. The immune system can be strengthened by vaccination, which supplies or stimulates the creation of antibodies to an organism that the body has not yet encountered.

An increasing cause of bacterial infection is the ability of the bacteria to resist the killing action of antibiotics. Within the past decade, the problem of antibiotic resistant bacteria has become a significant clinical issue. Part of the reason for the development of resistance has been the widespread and sometimes inappropriate use of antibiotics (e.g., use of antibiotics for viral illness because antibiotics are not effective against **viruses**).

Resistance can have molecular origins. The membrane(s) of the bacteria may become altered to make entry of the antibacterial compound more difficult. Also, enyzmes can be made that will destroy or inactivate the antibacterial agent. These resistance mechanisms can be passed on to subsequent generations of bacteria that will then be able to survive in increasing numbers.

Bacteria can also acquire resistance to antibiotics and other antibacterial agents, even components of the immune system, by growing on body surfaces, passages, and tissues. In this mode of growth, termed a biofilm, the bacteria are enmeshed in a sticky polymer produced by the cells. The polymer and the slow, almost dormant, growth rate of the bacteria protect them from antibacterial compounds that would otherwise kill them, and can encourage the bacteria to become resistant to the compounds. Examples of such resistance includes the chronic *Pseudomonas aeruginosa* lung infections experienced by those with cystic fibrosis and infection of artificially implanted material, such as urinary catheters and **heart** pacemakers.

Bacteria and viruses can also evade immune destruction by entering host cells and tissues. Once inside the host structures they are shielded from immune recognition.

See also Antigens and antibodies; Bacteria and responses to bacterial infection; Immune system

INFERIOR • *see* ANATOMICAL NOMENCLATURE

INFERTILITY, FEMALE

Infertility is clinically defined as a situation where a couple is unable to conceive a child after attempting to do so for at least one full year. Primary infertility refers to a situation in which **pregnancy** has never been achieved. Secondary infertility

refers to a situation in which one or both members of the couple have previously conceived a child, but are unable to conceive again after a full year of trying.

Currently, in the United States, up to 20% of couples struggle with infertility at any given time. Infertility has increased as a problem, as demonstrated by a study comparing fertility rates in married women ages 20-24 between the years of 1965 and 1982. In that time period, infertility increased 177%. Some studies attribute this increase on primarily social phenomena, including the tendency for marriage to occur at a later age, and the associated tendency for attempts at first pregnancy to occur at a later age. Fertility in women decreases with increasing age, as illustrated by the following statistics:

- infertility in married women ages 16-20: 4.5%
- infertility in married women ages 35-40: 31.8%
- infertility in married women over age 40: 70%.

Since the 1960s, there has also been an increase in rates of sexual intercourse outside of marriage, and higher numbers of individuals now have multiple sexual partners before they marry and attempt conception. This has led to an increase in sexually transmitted infections. Scarring from these infections, especially from pelvic inflammatory disease (PID)—a serious **infection** of the female reproductive organs—seems to be partly responsible for the increase. Furthermore, use of the contraceptive device called the intrauterine device (IUD) also has contributed to an increased rate of PID, with subsequent scarring.

To understand issues of infertility, it is first necessary to understand the basics of human reproduction. **Fertilization** occurs when a male **sperm** merges with a female ovum (egg), creating a zygote, which contains genetic material (**DNA**) from both the father and the mother. If pregnancy is then established, the zygote will develop into an embryo, then a fetus, and ultimately a baby will be born.

Sperm are small cells that carry the father's genetic material. This genetic material is contained within the oval head of the sperm. Sperm are produced within the testicles, and proceed through a number of developmental stages in order to mature. This whole process of sperm production is called **spermatogenesis**. The sperm are mixed into a fluid called **semen**, which is discharged from the penis during a process called ejaculation. The whip-like tail of the sperm allows the sperm motility; that is, permits the sperm to essentially swim up the female reproductive tract, in search of the egg it will attempt to fertilize.

The ovum (or egg) is the cell that carries the mother's genetic material. These ova develop within the ovaries. Once a month, a single mature ovum is produced and leaves the ovary in a process called ovulation. This ovum enters the fallopian tube (a tube extending from the ovary to the uterus) where fertilization occurs.

If fertilization occurs, a zygote containing genetic material from both the mother and father results. This single cell will divide into multiple cells within the fallopian tube, and the resulting cluster of cells (called a blastocyst) will then move into the uterus. The uterine lining (endometrium) has been preparing itself to receive a pregnancy by growing thicker. If the blastocyst successfully reaches the inside of the uterus and attaches itself to the wall of the uterus, then implantation and pregnancy have been achieved.

Unlike most medical problems, infertility is an issue requiring the careful evaluation of two separate individuals, as well as an evaluation of their interactions with each other. In about 3–4% of couples, no cause for their infertility will be discovered. The main factors involved in causing infertility, listing from the most to the least common, include male factors, peritoneal factors, uterine/tubal factors, ovulatory factors, and cervical factors.

There are many factors that influence female fertility.

Peritoneal factors refer to any factors (other than those involving specifically the ovaries, fallopian tubes, or uterus) within the **abdomen** of the female partner that may be interfering with her fertility. Two such problems include pelvic adhesions and endometriosis.

Pelvic adhesions are thick, fibrous scars. These scars can be the result of past infections, particularly sexually transmitted diseases such as PID, or infections following abortions or prior births. Previous surgeries can also leave behind scarring. Complications from appendicitis and certain intestinal diseases can also result in adhesions in the pelvic area.

Endometriosis also results in pelvic adhesions. Endometriosis is the abnormal location of uterine **tissue** outside of the uterus. When uterine tissue is planted elsewhere in the pelvis, it still bleeds on a monthly basis with the start of the normal menstrual period. This leads to irritation within the pelvis around the site of this abnormal tissue and bleeding, and ultimately causes scarring.

Pelvic adhesions contribute to infertility primarily by obstructing the fallopian tubes. The ovum may be prevented from traveling down the fallopian tube from the ovary, and the sperm prevented from traveling up the fallopian tube from the uterus; or the blastocyst may be prevented from entering into the uterus where it needs to implant. Scarring can be diagnosed by examining the pelvic area with a scope, which can be inserted into the abdomen through a tiny incision made near the naval. This scoping technique is called laparoscopy.

Obstruction of the fallopian tubes can also be diagnosed by observing through x-ray exam whether dye material can travel through the patient's fallopian tubes. Interestingly enough, this procedure has some actual treatment benefits for the patient, as a significant number of patients become pregnant following this x-ray exam. It is thought that the dye material in some way may help remove some tubal obstructions or decrease existing obstruction.

Pelvic adhesions can be treated using the same laparoscopic technique utilized in the diagnosis of the problem. For treatment, use of the laparoscope to visualize adhesions is combined with use of a laser to disintegrate those adhesions. Endometriosis can be treated with certain medications, but may also require surgery to repair any obstruction caused by adhesions.

Uterine factors contributing to infertility include **tumors** or abnormal growths within the uterus, chronic infection and inflammation of the uterus, abnormal structure of the uterus, and a variety of endocrine problems (problems with the secretion of certain **hormones**), which prevent the uterus from developing the thick lining necessary for implantation by a blastocyst.

Tubal factors are often the result of previous infections that have left scar tissue. This scar tissue blocks the tubes, preventing the ovum from being fertilized by the sperm. Scar tissue may also be present within the fallopian tubes due to the improper implantation of a previous pregnancy within the tube, instead of within the uterus. This is called an ectopic pregnancy. Ectopic pregnancies cause rupture of the tube, which is a medical emergency requiring surgery, and results in scarring within the affected tube.

X-ray studies utilizing dyes can help outline the structure of the uterus, revealing certain abnormalities. Ultrasound examination and hysteroscopy (in which a thin, wand-like camera is inserted through the cervix into the uterus) can further reveal abnormalities within the uterus. Biopsy (removing a tissue sample for microscopic examination) of the lining of the uterus (the endometrium) can help in the evaluation of endocrine problems affecting fertility.

Treatment of these uterine factors involves antibiotic treatment of any infectious cause, surgical removal of certain growths within the uterus, surgical reconstruction of the abnormally formed uterus, and medical treatment of any endocrine disorders discovered. **Progesterone**, for example, can be taken to improve the hospitality of the endometrium toward the arriving blastocyst. Very severe scarring of the fallopian tubes may require surgical reconstruction of all or part of the scarred tube.

Ovulatory factors are those factors that prevent the maturation and release of the ovum from the ovary with the usual monthly regularity. Ovulatory factors include a host of endocrine abnormalities, in which appropriate levels of the various hormones that influence ovulation are not produced. Numerous hormones produced by multiple organ systems interact to bring about normal ovulation. Therefore, ovulation difficulties can stem from problems with the ovaries, the adrenal glands, the pituitary gland, the **hypothalamus**, or the thyroid.

The first step in diagnosing ovulatory factors is to verify whether or not an ovum is being produced. Although the only certain proof of ovulation (short of an achieved pregnancy) is actual visualization of an ovum, certain procedures suggest that ovulation is or is not taking place.

The basal body temperature is the body temperature that occurs after a normal night's **sleep** and before any activity (including rising from bed) has been initiated. This temperature has normal variations over the course of the monthly ovulatory cycle, and when a woman carefully measures and records these temperatures, a chart can be drawn that suggests whether or not ovulation has occurred.

Another method for predicting ovulation involves measurement of a particular chemical that should appear in the urine just prior to ovulation. Endometrial biopsy will reveal different characteristics depending on the ovulatory status of the patient, as will examination of the **mucus** found in the cervix (the opening to the uterus). Also, pelvic ultrasound can visualize developing follicles (clusters of cells that encase a developing ovum) within the ovaries.

Treatment of ovulatory factors involves treatment of the specific organ system responsible for ovulatory failure (for example, thyroid medication must be given in the case of an underactive thyroid, a pituitary tumor may need removal, or the woman may need to cease excessive exercise, which can result in improper activity of the hypothalamus). If ovulation is still not occurring after these types of measures have been taken, certain drugs exist that can induce ovulation. These drugs, however, may cause the ovulation of more than one ovum per cycle, which is partially responsible for the increase in multiple births (twins, triplets, etc.) noted since these drugs became available to treat infertility.

The cervix is the opening from the vagina into the uterus through which the sperm must pass. Mucus produced by the cervix helps to transport the sperm into the uterus. Injury to the cervix during a prior birth, surgery on the cervix due to a pre-cancerous or cancerous condition, or scarring of the cervix after infection, can all result in a smaller than normal cervical opening, making it difficult for the sperm to enter. Furthermore, any of the above conditions can also decrease the number of mucus-producing glands in the cervix, leading to a decrease in the quantity of cervical mucus. In other situations, the mucus produced is the wrong consistency to allow sperm to travel through. Certain infections can also serve to make the cervical mucus environment unfavorable to the transport of sperm, or even directly toxic to the sperm themselves (causing sperm death). Some women produce antibodies (immune cells) that identify sperm as foreign invaders.

The qualities of the cervical mucus can be examined under a microscope to diagnose cervical factors as contributing to infertility. The interaction of a live sperm sample from the male partner and a sample of cervical mucus can also be examined.

Treatment of cervical factors includes antibiotics in the case of an infection, steroids to decrease production of anti-sperm antibodies, and artificial insemination techniques to completely bypass the cervical mucus.

Assisted reproduction comprises those techniques that perhaps receive the most publicity as infertility treatments. These include **in vitro fertilization** (IVF), gamete intra fallopian tube transfer (GIFT), and zygote intra fallopian tube transfer (ZIFT). All of these are used after other techniques to treat infertility have failed.

IVF involves the use of a drug to induce multiple ovum production, and retrieval of those ova either surgically or by ultrasound-guided needle aspiration through the vaginal wall. Meanwhile, multiple semen samples are obtained from the male partner, and a sperm concentrate is prepared. The ova and sperm are then cultured together in a laboratory, where hopefully several of the ova are fertilized. **Cell division** is allowed to take place up to either the pre-embryo or blastocyst state. While this takes place, the female may be given medication to prepare her uterus to receive an embryo. When necessary, a small opening is made in the outer shell (zona pellucida) of the pre-embryo or blastocyst by a process known as assisted hatching. Two or more pre-embryos or two blastocysts are transferred into the uterus, and the wait begins to see if any or all of them implant and result in an actual pregnancy.

The national average success rate of IVF is 27%, but some centers have higher pregnancy rates. Transferring blastocysts leads to a pregnancy rate of up to 50% or higher. Interestingly, the rate of **birth defects** resulting from IVF is lower than that resulting from unassisted pregnancies. Of course, because most IVF procedures place more than one embryo into the uterus, the chance for a multiple birth (twins or more) is greatly increased.

GIFT involves retrieval of both multiple ova and semen, and the mechanical placement of both within the fallopian tubes, where fertilization may occur. ZIFT involves the same retrieval of ova and semen, and fertilization and growth in the laboratory up to the zygote stage, at which point the zygotes are placed in the fallopian tubes. Both GIFT and ZIFT seem to have higher success rates than IVF.

Ova can now be frozen for later use, although greater success is obtained with fresh ova. However, storing ova may provide the opportunity for future pregnancy in women with premature ovarian failure, pelvic disease, or those undergoing **cancer** treatment.

Any of these methods of assisted reproduction can utilize donor sperm and/or ova. There have even been cases in which the female partner's uterus is unable to support a pregnancy, so the embryo or zygote resulting from fertilization of the female partner's ovum with the male partner's sperm is transferred into another woman, where the pregnancy progresses to birth.

Chances at pregnancy can be improved when the pre-embryos are screened for chromosomal abnormalities and only the normal ones are transferred into the uterus. This method is useful for couples who are at an increased risk of producing embryos with chromosomal abnormalities, such as advanced maternal age or when one or both partners carry a fatal genetic disease.

See also Embryology; Infertility, male; Sexual reproduction

INFERTILITY, MALE

Male factor infertility can be caused by a number of different characteristics of the **sperm**. To check for these characteristics, a **semen** analysis is carried out, during which a sample of semen is obtained and examined under the microscope. The four most basic characteristics evaluated are sperm count, sperm motility, sperm morphology, and semen volume.

Sperm count concerns the number of sperm present in a semen sample. The normal number of sperm present in just one milliliter (ml) of semen is over 20 million. A man with only 5-20 million sperm per ml of semen is considered sub-fertile, a man with less than five million sperm per ml of semen is considered infertile.

Sperm motility is important because better swimmers indicate a higher degree of fertility, as does longer duration of survival. Sperm are usually capable of **fertilization** for up to 48 hours after ejaculation.

Sperm morphology or the structure of the sperm must be within normal limits. Not all sperm within a specimen of semen will be perfectly normal. Some may be developmentally immature forms of sperm, some may have abnormalities of the head or tail. A normal semen sample will contain no more than 25% abnormal forms of sperm.

Finally, the volume of a representative semen sample is important. The semen is made up of a number of different substances, and a decreased quantity of one of these substances could affect the ability of the sperm to successfully fertilize an ovum.

The semen sample may also be analyzed chemically to determine that components of semen other than sperm are present in the correct proportions. If all of the above factors do not seem to be the cause for male infertility, then another test is performed to evaluate the ability of the sperm to penetrate the outer coat of the ovum. This is done by observing whether sperm in a semen sample can penetrate the outer coat of a guinea pig ovum. Fertilization cannot, of course, occur, but this test is useful in predicting the ability of the patient's sperm to penetrate a human ovum.

Any number of issues can affect male fertility as evidenced by the semen analysis. Individuals can be born with testicles that have not descended properly from the abdominal cavity (where testicles develop originally) into the scrotal sac, or they can be born with only one testicle, instead of the normal two. Testicle size can be smaller than normal. Past **infection** (including mumps) can affect testicular function, as can a past injury. The presence of abnormally large **veins** (varicocele) in the testicles can increase testicular temperature, which decreases sperm count. A history of exposure to various toxins, drug use, excessive alcohol use, use of anabolic steroids, certain medications, diabetes, thyroid problems, or other endocrine disturbances can have direct effects on **spermatogenesis**. Problems with the male **anatomy** can cause sperm to be ejaculated not out of the penis, but into the bladder, and scarring from past infections can interfere with ejaculation.

Treatment of male factor infertility includes addressing known reversible factors first, for example discontinuing any medication known to have an effect on spermatogenesis or ejaculation, as well as decreasing alcohol intake and treating thyroid or other endocrine disease. Varicoceles can be treated surgically. **Testosterone** in low doses can improve sperm motility.

Some recent advances have greatly improved the chances for infertile men to conceive. Azoospermia (lack of sperm in the semen) may be overcome by mechanically removing sperm from the testicles either by surgical biopsy or needle aspiration (using a needle and syringe). The isolated sperm can then be used for **in vitro fertilization**. Another advance involves using a fine needle to inject a single sperm into the ovum. This procedure, called intracytoplasmic sperm injection (ICSI) is useful when sperm have difficulty fertilizing the ovum and when sperm have been obtained through mechanical means.

Other treatments of male factor infertility include collecting semen samples from multiple ejaculations, after which the semen is put through a process which allows the most motile sperm to be sorted out. These motile sperm are pooled together to create a concentrate that can be mechanically

deposited directly into the female partner's uterus at a time that will coincide with ovulation. In cases where the male partner's sperm is proven to be absolutely unable to cause **pregnancy** in the female partner, and with the consent of both partners, donor sperm may be used for this process. These procedures (depositing the male partner's sperm or donor sperm by mechanical means into the female partner) are both forms of artificial insemination.

See also Embryology; Infertility, female; Sexual reproduction

INFLAMMATION OF TISSUES

Inflammation is a localized, defensive response of the body to injury and is usually characterized by **pain**, redness, heat, swelling, and, depending on the extent of trauma, loss of function. The process of inflammation, called the inflammatory response, is a series of events, or stages, that the body performs to attain homeostasis (the body's effort to maintain stability). The body's inflammatory response mechanism serves to confine, weaken, destroy, and remove **bacteria**, toxins, and foreign material at the site of trauma or injury. As a result, the spread of invading substances is halted, and the injured area is prepared for regeneration or repair. Inflammation is a nonspecific defense mechanism; the body's physiological response to a superficial cut is much the same as with a burn or a bacterial infection. The inflammatory response protects the body against a variety of invading pathogens and foreign matter, and should not be confused with an immune response, which reacts to specific invading agents. Inflammation is described as acute or chronic, depending on how long it lasts.

Within minutes after the body's physical barriers, the skin and mucous membranes, are injured or traumatized (for example, by bacteria and other microorganisms, extreme heat or cold, and chemicals), the arterioles and **capillaries** dilate, allowing more **blood** to flow to the injured area. When the blood vessels dilate, they become more permeable, allowing **plasma** and circulating defensive substances such as antibodies, phagocytes (cells that ingest microbes and foreign substances), and fibrinogen (blood-clotting chemical) to pass through the vessel wall to the site of the injury. The blood flow to the area decreases and the circulating phagocytes attach to and digest the invading pathogens. Unless the body's defense system is compromised by a preexisting disease or a weakened condition, healing takes place. Treatment of inflammation depends on the cause. Anti-inflammatory drugs such as aspirin, acetaminophen, ibuprofen, or a group of drugs known as NSAIDS (nonsteroidal anti-inflammatory drugs) are sometimes taken to counteract some of the symptoms of inflammation.

See also Homeostatic mechanisms

INHERITED DISEASES

The human genome is composed of roughly three billion **DNA** base pairs. Each somatic cell of the body contains two copies of the entire genome, one inherited from the mother through the egg, and the other inherited from the father through the **sperm**. The DNA encodes a library of biochemical messages called genes, perhaps 60,000 in all. All of the genetic messages, no doubt, are important to the health of the organism, and it is likely true that every disease state in the human involves some compromise or redistribution of the biochemical agents produced by the genes. In other words, genes are involved at some level in virtually every human disease. Even so, most diseases are not inherited.

The genes are packaged into **chromosomes**. Each somatic cell contains 46 chromosomes; two copies of each of the 22 autosomes (numbered 1 through 22), and a pair of sex chromosomes (X and Y). Males and females share the same genetic makeup for all of the autosomal genes, but they differ in the distribution of genes found on the sex chromosomes. The X chromosome is a gene-rich medium length chromosome. The Y chromosome is a short chromosome with only a handful of genes. Females have two copies of the X chromosome, one inherited from the mother in the egg, and the second inherited from the father through the sperm. In contrast, males inherit one copy of the X chromosome from the egg. In place of the second X chromosome, males receive a Y chromosome through the sperm.

Inherited diseases are those conditions which result primarily or exclusively from genetic mutations or genetic imbalance passed on from parent to child at conception. These include Mendelian genetic conditions as well as chromosomal abnormalities. A third group of disorders exists wherein genetic factors interacting with environmental factors combine to produce a disease state. These conditions are often referred to as having multifactorial or complex inheritance patterns.

The first group, Mendelian diseases, also called single gene disorders, is defined by the property that mutation of just a single gene suffices to alter the overall **physiology** of the organism producing a recognizable disease state. Mendelian diseases are usually recognized by their distinctive patterns of inheritance in families. They can be broken into four major categories: autosomal dominant, autosomal recessive, X-linked dominant and X-linked recessive.

Autosomal dominant diseases result from mutations of genes found on any of the 22 pairs of autosomes, and the mutation need come from only one parent. Usually, the second copy of the gene received from the other parent is completely normal. A parent with an autosomal dominant genetic trait will pass the trait on to half of their sons and half of their daughters, and children who do not get the trait are not generally at risk for passing the trait on to their own children. Achondroplasia, a common cause of dwarfism, is an example of an autosomal dominant genetic disease.

Autosomal recessive diseases also involve genes found on the autosomes, but in contrast to autosomal dominant genes, both parents must pass a mutation to the child in order for the autosomal recessive disease to be expressed. The typical pattern of inheritance for autosomal recessive traits is that healthy carrier parents pass the disease on to one fourth of their children with an equal number of sons and daughters

being affected. Cystic fibrosis is an example of an autosomal recessive genetic disease.

X-linked traits are the result of mutations of genes found on the X chromosome. The vast majority of the disease genes on the X chromosome behave as recessives in females. That is, both X chromosomes must have the mutation for the disease to arise. Since males have only one X chromosome, they will express the disease mutations whether they are dominant or recessive. The characteristic pattern of inheritance in X-linked recessive traits is that males having the disease are related to one another through healthy carrier females. Half of the sons of carrier females are affected. While none of the daughters of a carrier female parent is affected, half will themselves be mutation carriers. Affected males will pass the mutation on to all of their daughters, none of whom will have the disease. Father to son transmission is never seen in X-linked traits (recessive or dominant) because the father's X chromosome is passed on only to daughters. Duchenne muscular dystrophy is an example of an X-linked recessive disease.

X-linked dominant traits are far more rare. There are roughly twice as many females as males affected, although severity is frequently worse in males than in females. Affected females pass the disease on to half of their sons and half of their daughters. Affected males pass the trait on to all of their daughters and none of their sons. Fragile-X mental retardation syndrome is an example of an X-linked dominant disease.

The second group of inherited diseases is chromosomal. Chromosomes are long strands of DNA complexed with proteins and **RNA** that condense and allow for equal distribution of the genes when cells divide. Each chromosome contains hundreds or thousands of genes, and every cell needs to have two copies of each chromosome in order to maintain genetic balance. At the time of conception, an extra copy or missing copy of a chromosome or even a part of a chromosome disrupts normal development. Most chromosomal abnormalities result from simple accidents of chromosome segregation, and as such, they tend not to recur in families at nearly the rate as Mendelian genetic diseases. One example of genetic disease that results from chromosomal imbalance is Down syndrome. This condition is caused by the presence of an extra copy of chromosome 21, the smallest human autosome.

Diseases with complex or multifactorial inheritance are by far the most common, and their patterns of inheritance are indistinct and largely unpredictable. These diseases tend to cluster in families particularly when the contribution from genetic factors is relatively strong. The great majority of diseases that occur in humans can indeed be thought of as exhibiting complex inheritance. Chronic adult onset diseases such as hypertension, type II diabetes, obesity, **heart** disease, and strokes are examples.

Cancer is a genetic disease but is only rarely inherited. Most cancers are sporadic and arise in a particular **tissue** such as the colon, breast, lung or skin, following exposure of the normal tissue to carcinogens that cause somatic mutations in one or more oncogenes or tumor suppressor genes. Familial cancer syndromes, which in total account for less than one percent of all cancer, occur in individuals who have inherited a germline mutation in a tumor suppressor gene. Most famil-

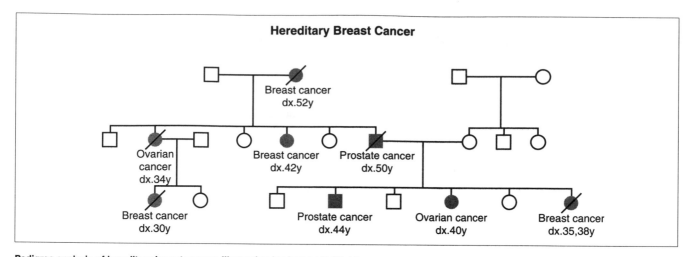

Hereditary Breast Cancer

Pedigree analysis of hereditary breast cancer. *Illustration by Argosy Publishing.*

ial cancers have autosomal dominant inheritance and are characterized by the development of cancer at a young age.

Inheriting a mutant tumor suppressor gene effectively knocks out one allele of the gene (i.e., the "first hit") in every cell in the body. This leaves the individual vulnerable to a "second hit" on the remaining normal gene allele. Some tissues or body **organs** such as the **eye**, the breast and colon, and soft tissues, are more susceptible to a "second hit" mutation which leads to complete loss of the tumor suppressor gene and the development of cancer in that particular tissue. This "two hit" hypothesis for the development of inherited cancers was described by Alfred Knudson in 1971.

Li-Fraumeni syndrome, first described by Li and Fraumeni in 1969, is one of the best-known familial cancer syndromes. This autosomal dominant disorder is caused by the inheritance of germline mutations in the p53 tumor suppressor gene (chromosome 17p13.1). Affected family members are prone to develop **leukemia**, melanoma, soft tissue sarcomas, bone **tumors**, and cancer of the colon, breast, prostate, **brain**, and lung, often before the unusually early age of thirty. Many different cancers can develop within the same family and some affected individuals may develop multiple different primary tumors.

Retinoblastoma, a rare eye tumor in children, is another autosomal dominant cancer syndrome. Children with familial retinoblastoma inherit a mutant allele of the RB1 tumor suppressor gene (chromosome 13q14.1) and are prone to develop multiple tumors in both eyes within months of birth. Sporadic retinoblastoma does occur but it usually develops later, and is confined to one eye.

Inherited breast cancer accounts for approximately 5% of all cases of breast cancer. Two different tumor suppressor genes, BRCA1 (chromosome 17q21), and BRCA2 (chromosome 13q12.3) cause an autosomal dominant pattern of inheritance. Individuals in families with BRCA1 mutations are prone to develop either breast or ovarian cancer at an early age, while families with BRCA2 mutations only develop breast cancer but have a higher frequency of males developing cancer of the breast.

A rare form of inherited stomach cancer has been described in the native New Zealand Maori population. Individuals who inherit mutations in the E-cadherin gene (chromosome 16q22.1) develop at an early age a rapidly fatal form stomach cancer. Presymptomatic gene testing of at risk individuals allows for early surgical treatment.

Some conditions do not directly cause cancer but are premalignant and predispose at risk individuals to the development of cancer. The most common premalignant syndrome is Neurofibromatosis, an autosomal dominant neurologic disease caused by mutations in one of the neurofibromin genes (NF1 or NF2). In NF1 (Von Recklinghausen disease chromosome 17q11.2) multiple benign neurofibromas develop in the peripheral nerves. The neurofibromas may become malignant nerve sheath tumors, and malignant tumors can develop in the brain.

Another premalignant condition is the Beckwith-Wiedemann syndrome. This congenital overgrowth disorder is associated with neonatal hypoglycemia (low **blood** sugar), macroglossia (large tongue), and a large birth weight and is caused by abnormal expression of the insulin-like growth factor 2 gene (IGF2 chromosome 11p). Some of these children later develop malignant tumors of the kidney (Wilms tumor), **liver** (hepatoblastoma), or adrenal gland (adrenocortical carcinoma).

See also Autoimmune disorders; Gene therapy; Genetic code; Genetic regulation of eukaryotic cells; Genetics and developmental genetics; Germ cells; Pharmacogenetics; Stem cells

INNERVATION OF MUSCLE · *see* MUSCULAR INNERVATION

INSULIN · *see* GLUCOSE UTILIZATION, TRANSPORT AND INSULIN

INTEGUMENT

The integumentary system includes the skin and the related structures that cover and protect the body. The human integumentary system is made up of the skin, which includes glands, **hair**, and **nails**. The skin protects the body, prevents water loss, regulates body temperature, and senses the external environment.

The human integumentary system serves many protective functions for the body. Keratin, an insoluble protein in the outer layer of the skin, helps prevent water loss and dehydration. Keratin also prevents excessive water loss, keeps out microorganisms that could cause illness, and protects the underlying tissues from mechanical damage. Keratin is also the major protein found in nails and hair. Pigments in the skin called melanin absorb and reflect the sun's harmful ultraviolet radiation. The skin helps to regulate the body temperature. If heat builds up in the body, **sweat glands** in the skin produce more sweat that evaporates and cools the skin. In addition, when the body overheats, **blood** vessels in the skin expand and bring more blood to the surface, which allows body heat to be lost. If the body is too cold, on the other hand, the blood vessels in the skin contract, resulting in less blood is at the body surface, and heat is conserved. In addition to **temperature regulation**, the skin serves as a minor excretory organ, because sweat removes small amounts of nitrogenous wastes produced by the body. The skin also functions as a sense organ as it contains millions of nerve endings that detect touch, heat, cold, **pain** and pressure. Finally, the skin produces vitamin D in the presence of sunlight, and renews and repairs damage to itself.

In an adult, the skin covers about 21.5 sq ft (2 sq m), and weighs about 11 lb (5 kg). Depending on location, the skin ranges from 0.02-0.16 in (0.5-4.0 mm) thick. Its two principal parts are the outer layer, or **epidermis**, and a thicker inner layer, the **dermis**. A subcutaneous layer of fatty or **adipose tissue** is found below the dermis. Fibers from the dermis attach the skin to the subcutaneous layer, and the underlying tissues and **organs** also connect to the subcutaneous layer.

Ninety percent of the epidermis, including the outer layers, contains keratinocytes cells that produce keratin, a protein that helps waterproof and protect the skin. **Melanocytes** are pigment cells that produce melanin, a dark pigment that adds to skin color and absorbs ultraviolet light thereby shielding the genetic material in skin cells from damage. Merkel's cells disks are touch-sensitive cells found in the deepest layer of the epidermis of hairless skin.

In most areas of the body, the epidermis consists of four layers. On the soles of the feet and palms of the hands where there is a lot of friction, the epidermis has five layers. In addition, calluses, abnormal thickenings of the epidermis, occur on skin subject to constant friction. At the skin surface, the outer layer of the epidermis constantly sheds the dead cells containing keratin. The uppermost layer consists of about 25 rows of flat dead cells that contain keratin.

The dermis is made up of connective **tissue** that contains protein, collagen, and elastic fibers. It also contains blood and lymph vessels, sensory receptors, related nerves, and glands. The outer part of the dermis has fingerlike projec-

tions, called dermal papillae that indent the lower layer of the epidermis. Dermal papillae cause ridges in the epidermis above it, which in the digits give rise to fingerprints. The ridge pattern of fingerprints is inherited, and is unique to each individual. The dermis is thick in the palms and soles, but very thin in other places, such as the eyelids. The blood vessels in the dermis contain a volume of blood. If a part of the body, such as a working muscle, needs more blood, blood vessels in the dermis constrict, causing blood to leave the skin and enter the circulation that leads to muscles and other body parts. Sweat glands whose ducts pass through the epidermis to the outside and open on the skin surface through pores are embedded in the deep layers of the dermis. Hair follicles and hair roots also originate in the dermis and the hair shafts extend from the hair root through the skin layers to the surface. Also in the dermis are sebaceous glands associated with hair follicles that produce an oily substance called sebum. Sebum softens the hair and prevents it from drying, but if sebum blocks up a sebaceous gland, a whitehead appears on the skin. A blackhead results if the material oxidizes and dries. Acne is caused by **infection** or inflammation of the sebaceous glands. When this occurs, the skin breaks out in pimples and can become scarred.

The skin is an important sense organ, and as such includes a number of nerves that are mainly in the dermis, with a few reaching the epidermis. Nerves carry impulses to and from hair muscles, sweat glands, and blood vessels, and receive messages from touch, temperature, and pain receptors. Some nerve endings are specialized such as sensory receptors that detect external stimuli. The nerve endings in the dermal papillae are known as Meissner's corpuscles, which detect light touch, such as a pat, or the feel of clothing on the skin. Pacinian corpuscles, located in the deeper dermis, are stimulated by stronger pressure on the skin. Receptors near hair roots detect displacement of the skin hairs by stimuli such as touch or wind. Bare nerve endings throughout the skin report information to the **brain** about temperature change (both heat and cold), texture, pressure, and trauma.

Some skin disorders result from overexposure to the ultraviolet (UV) rays in sunlight. At first, overexposure to sunlight results in injury known as sunburn. UV rays damage skin cells, blood vessels, and other dermal structures. Continual overexposure leads to leathery skin, wrinkles, and discoloration and can also lead to skin **cancer**. Anyone excessively exposed to UV rays runs a risk of skin cancer, regardless of the amount of pigmentation normally in the skin. Seventy-five percent of all skin cancers are basal cell carcinomas that arise in the epidermis and rarely spread (metastasize) to other parts of the body. Physicians can surgically remove basal cell cancers. Squamous cell carcinomas also occur in the epidermis, but these tend to metastasize. Malignant melanomas are life threatening skin cancers that metastasize rapidly. There can be a 10–20 year delay between exposure to sunlight and the development of skin cancers.

See also Burns; Capillaries; Touch, physiology of

INTEGUMENTARY SYSTEM, EMBRYONIC DEVELOPMENT

The embryonic development of the integumentary system explains the development of skin, glands, **hair**, and **nails**. Also associated with the development of the integumentary system proper are the development of teeth and, in females, portions of the **mammary glands**.

The integumentary system forms from the embryonic **ectoderm**, the outer most or superficial layer of the embryonic germinal layers from which all structures and **organs** of the body ultimately form. Ectodermal cells ultimately form the **epidermis** (the outmost layer of skin), sweat and sebaceous glands, and hair and nails. Ectodermal cell also form the mammary glands and teeth.

Ectodermal layers proliferate (divide and increase in number) and thicken throughout embryonic development. The cells differentiate into the stratum germinativum (stratum translates to "layer"), stratum granulosum, stratum lucidum, and stratium corneum layer of the skin. At an early point in the differentiation process, the cells in the underlying or basal layers of the skin gain the capacity to produce keratin. Cornification of the outer layers of the skin is evident in embryos by the fifth month of development. This outermost layer of the fetal skin is termed the vernix caseosa, and it forms a protective boundary layer between the fetus an the amniotic fluid.

Cells within melanoblasts, a specialized cell within the stratum geminativum, eventually differentiate and gain the ability to synthesize melanin. Skin color and tone is largely determined by the amount of melanin produced. The **dermis** of the skin forms from **mesoderm** initially organized into dermatomes, regions of the **somites** that form adjacent to the **notochord**.

Hair initially forms as an ingrowth of the epidermis, termed follicles. By the end of the third month of development, hair follicles invade the underlying dermis to come into contact with mesodermal cells. In this area, **blood** vessels, and nerve endings associated with hair develop. **Smooth muscle** cells form the arrectores pilorum (erector pilorum), the fine muscles that can make hair erect. The effect of hair "standing up" is also caused by electrostatic forces and the nature of the pertinacious (protein containing) keratin that forms the hair. As development proceeds, the fetus becomes covered with a layer of very thin (fine) hair termed langugo. This layer of fine hair is sloughed off or shed following birth.

Sweat glands also develop from cylindrical-shaped rods of cells that grow inward from the superficial epidermis. During the fifth month of development, the cells reach the underlying mesodermal layer and begin to form winding tubes that eventual coil around a central cavity or lumen. Sebaceous glands develop as side outgrowths of hair follicles. The glands retain connection with the follicle and can deposit excretions into the follicular space.

Finger and toenails begin development as early as the third month following **fertilization**. Nails result form a generalized thickening of the epidermis. Nail growth takes place from the germinative layers of the skin, and the thickening of each nail represents an increasing cornification as the nail grows outward from the germinative layer.

Tooth development results form another form of ectodermal ingrowth that forms the primary dental lamina. This ingrowth occurs in both the roof and floor of the developing mouth. Along with mesoderm, ectoderm forms structure responsible for the production of enamel and dentine. The milk teeth (baby teeth) form by the end of the fourth month. By birth, there are 20 milk teeth and four permanent molar teeth (not yet visible in the mouth. Underlying the milk teeth are cells which are already differentiated to form the permanent teeth that will ultimately grow, push out, and replace the milk teeth.

In females, the mammary glands that produce milk during **lactation** are also of ectodermal origin. Thickenings of cells, described as mammary lines, can be detected as early as the end of the first month of development. Similar to sweat gland formation, the ingrowths from the ectoderm become surrounded with mesodermal **tissue** to form hollow ducts. Normally only the portion of the mammary line located what will become the thoracic area develops into mammary glands, the other portions of the mammary lines degenerate. Failure of the lines to completely degenerate can sometimes cause the development of accessory nipples between the thorax and lower **abdomen**.

See also Embryonic development: early development, formation, and differentiation; Tooth development, loss, replacement, and decay

INTERFERON ACTIONS

Interferons are species-specific proteins, which induce antiviral and antiproliferative responses in animal cells. They are a major defense against viral infections and abnormal growths (neoplasms). Interferons are produced in response to penetration of animal cells by viral (or synthetic) nucleic acid and then leave the infected cell to confer resistance on other cells of the organism. In contrast to antibodies, interferons are not virus specific but host specific. Thus, viral infections of human cells are inhibited only by human interferon. The human genome contains 14 nonallelic and 9 allelic genes of α-interferon (macrophage interferon), as well as a single gene for β-interferon (fibroblast interferon). Genes for any two or more variants of interferon, which have originated from the same wild-type gene are called allelic genes and will occupy the same chromosomal location (locus). Variants originating from different standard genes are termed non allelic. α- and β-interferons are structurally related glycoproteins of 166 and 169 amino acid residues. In contrast, γ-interferon (also known as immune interferon) is not closely related to the other two and is not induced by virus **infection**. It is produced by T-cells after stimulation with the cytokine interleukin-2. It enhances the cytotoxic activity of T-cells, macrophages and natural killer cells and thus has antiproliferative effects. It also increases the production of antibodies in response to **antigens** administered simultaneously with α-interferon, possible by enhancing the antigen-presenting function of macrophages.

Interferons bind to specific receptors on the cell surface, and induce a signal in the cell interior. Two induction mechanisms have been elucidated. One mechanism involves the induction of protein kinase by interferon, which, in the presence of double-stranded **RNA**, phosphorylates one subunit of an initiation factor of **protein synthesis** (eIF-2B), causing the factor to be inactivated by sequestration in a complex. The second mechanism involves the induction of the enzyme 2',5'-oligoadenylate synthetase (2',5'-oligo A synthestase). In the presence of double-stranded RNA, this enzyme catalyses the polymerisation of **ATP** into oligomers of 2–15 adenosine monophosphate residues which are linked by phosphodiester bonds between the position 2' of one ribose and 5' of the next. These 2',5'-oligoadenylates activate an interferon specific RNAase, a latent endonuclease known as RNAase L which is always present but not normally active. RNAase cleaves both viral and cellular single stranded mRNA. Interferons therefore do not directly protect cells against viral infection, but rather render cells less suitable as an environment for viral replication. This condition is known as the antiviral state.

See also Immune system; Immunology; Viruses and responses to viral infection

INTERNAL EAR • *see* EAR (EXTERNAL, MIDDLE, AND INTERNAL)

INTEROSSEUS MUSCLES

The interosseous muscles are small bipinnate (two-headed) muscles that connect the metacarpals and metatarsals with the **phalanges**. They originate on the metacarpals and metatarsals and insert on the first phalange. Their function is to flex the fingers and toes. There is a dorsal and palmar (plantar in the foot group).

There are four dorsal interosseous muscles of the hand. The first, the abductor indicis, connects to the thumb and index finger. It provides a flat pad in between the two fingers. The second and third interossei insert onto the third or middle finger. The fourth inserts on the ulnar side of the ring finger.

The palmar muscles are smaller and cover the palm surface of the hand, There are only three of them. They are attached along the entire surface of a corresponding metacarpal and insert on the first phalanges of the fingers. Contraction (abduction) of these muscles brings the fingers in toward the hand while contraction (adduction) of the dorsal ones flexes the hand back into a flat position.

The foot arrangement of the interossei is quite similar to the hand. Like the hand, the there are four dorsal interossei. Like the hand, the first interossei inserts on the second toe. The other dorsal interossei insert on the outer sides of the second, third, and fourth toes.

The three plantar interossei are beneath rather than between the metatarsals. The third, fourth, and fifth metatarsals provide the base for attachment. Their insertion is on the base of the phalanges of the same toes. The dorsal interossei are the abductors and the plantar are the adductors.

See also Lower limb structure; Muscular system overview; Upper limb structure

INTERSTITIAL FLUID

Interstitial fluid is the fluid that surrounds the cells in the various tissues of the body. This fluid provides a path through which nutrients, gases, and wastes can travel between the cells and the small blood-transporting **capillaries**.

The capillaries are the very smallest of the channels through which **blood** is conducted afer being pumped from the **heart**. Even within a capillary, blood is still under some considerable pressure. This pressure drives the diffusion of water and some of the blood **plasma** proteins through the capillary wall into the **tissue** space on the other side of the wall. This fluid—blood plasma minus most of the constituent proteins—constitutes the interstitial fluid.

Substances that are in the interstitial fluid can either diffuse into cells or be actively taken up by the cells. Conversely, substances such as carbon dioxide can diffuse out of the cells into the fluid. When the interstitial fluid re-enters the capillaries, as happens because the blood pressure is less than the osmotic pressure resulting from the compositional differences between interstitial fluid and blood, the wastes (e.g., carbon dioxide) are transferred to the blood.

Not all the interstitial fluid re-enters the capillaries. If accumulation of the fluid continued, the increasing volume would cause massive swelling of the space between the cells. This would cause mass destruction of tissues and **death**. The problem of swelling is avoided by the presence of lymphatic vessels. These drain the excess interstitial fluid and return it to the blood bound for the heart.

Interstitial fluid may also be useful in the diagnosis of diseases such as diabetes. Research is underway to perfect devices that can analyze the interstitial fluid of skin cells and ascertain the level of glucose. This could allow an easy and noninvasive means of detecting the development of diabetes.

See also Cell structure; Circulatory system; Osmotic equilibria between intercellular and extracellular fluids

INTESTINAL HISTOPHYSIOLOGY

Intestinal histophysiology is the study of **structure and function** of tissues that form the small and large intestines. The intestines are tubular shaped **organs** that are part of the alimentary canal of the digestive system.

The intestines, like that of the rest of the digestive tract, consist of four layers of **tissue**: mucosa, submucosa, muscularis, and serosa. The innermost layer, the mucosa, opens into the interior of the intestines, the lumen. The mucosa is made up of several layers. The first, a layer of epithelial cells called the lamina epithelialis, lines the lumen. Different epithelial cell types enable **digestion**, absorption, secretion, and production of **hormones** within the intestines. Beneath the lamina epithelialis is a layer of loose connective tissue containing **blood** vessels and lymphatic vessels called the lamina propria. The last layer of the mucosa, the lamina muscularis mucosae, consists of **smooth muscle** and enables the mucosa to move.

The tissue layer of the intestines adjacent to the mucosa is the submucosa. Constituents of the submucosa include a meshwork of loose and dense connective tissue, submucous plexus, blood vessels, and lymphatic vessels. The submucous plexus is nervous tissue that is essential for sensation and regulation of the intestinal environment.

Between the submucosa and the outermost layer of the intestines is the muscularis. The muscularis consists of circular and longitudinal smooth muscle that enables the intestines to move food in the lumen through the digestive tract. Additionally, the muscularis contains the myenteric plexus, nervous tissue that controls motility of the intestines.

The outermost layer of the intestinal layers is the serosa. The serosa is composed of a layer of mesothelium protecting a layer of loose connective tissue.

The small intestine is an approximately 21-ft-long (6.4 m) continuation of the alimentary canal that receives gastric contents of the stomach called chyme. The small intestine has folds in its mucosa tissue layer that increase the surface area in the lumen and help to mix chyme as it passes through the small intestine. Protruding from the mucosal folds are projections called villi that extend into the lumen to increase the surface area for digestion and absorption of nutrients into the blood. Absorption is accomplished by an inner lining of epithelial cells called enterocytes. Not only do enterocytes secrete intestinal juices into the lumen, but they also secrete hormones into the blood that stimulate secretions from the **liver** and **pancreas**. The liver secretes **bile** salts that emulsify and absorb **lipids** and proteins in the small intestine. The pancreas secretes pancreatic juice containing sodium bicarbonate that creates optimal conditions for absorption to occur in the small intestine. Enterocytes further increase surface area by forming projections on the villi called microvilli. The enterocytes are replaced every few days by glands called the Crypts of Lieberkuhn.

The small intestine is divided into three anatomical regions: duodenum, jejunum, and ileum. The duodenum connects to the pyloric sphincter of the stomach, receiving gastric contents that have been liquefied by the time they reach the small intestine. The duodenum and the middle portion of the small intestine, the jejunum, complete digestion of any remaining macromolecules and begin absorption of water, **electrolytes**, and organic molecules. By the time the chyme reaches the ileum, digestion and absorption are complete. This final section of the small intestine is responsible for removing water and bile salts from the chyme as it continues into the large intestine.

The large intestine is responsible for recovering water and electrolytes, preparing the body for removal of waste products in the form of feces, and microbial fermentation. The large intestine can be classified into four anatomic regions; cecum, colon, rectum, and anal canal. The cecum is a closed pouch extending from the ileocecal valve that attaches to the vermiform **appendix**. Continuing from the cecum at the ileocecal valve is the colon.

The colon makes up the majority of the large intestine and is subdivided into the ascending, transverse, and descending colon. The colon is responsible for the absorption of excess water from the feces back into the bloodstream. Additionally, electrolytes, specifically sodium and chloride ions, can be absorbed in the ascending and transverse colon. The two main secretions into the lumen of the large intestine are bicarbonate ions and **mucus**. Bicarbonate ions are secreted to help neutralize the acid produced by microbial fermentation. Mucus helps to hold the feces together and acts as a lubricant as the feces passes through the large intestine. The descending colon is connected to the rectum where the feces are eliminated from the body through the anal canal.

The large intestine contains a normal flora of **bacteria** that enables the body to digest otherwise indigestible molecules by way of fermentation. For example, cellulose is a major component of the human diet. However, humans do not produce the enzyme cellulase to break down the carbohydrate and must depend of bacteria to digest it. One of the byproducts of microbial fermentation, intestinal gas or **flatus**, is eliminated through the anus. Bacteria also aid in the process of producing certain **vitamins** in the colon such as vitamin K and some of the B vitamins. The vitamin K produced in the large intestine is absorbed into the intestinal lymph and is the main source of vitamin K for the body. However, the B vitamins produced in the large intestine are not absorbed and are lost when the feces is eliminated.

Although one of the main functions of the intestines is water absorption, the feces actually consists of about 75% water. The remaining 25% of fecal material is composed of bacteria and undigested material such as fiber.

See also Intestinal motility; Appendix; Gastrointestinal tract

INTESTINAL MOTILITY

Intestinal motility refers to the movement of smooth muscles in the small and large intestine that aids in mixing, **digestion**, absorption, and movement of foodstuffs. The **smooth muscle** controlling intestinal motility is called the muscularis. This layer of muscle **tissue** consists of two types of muscle: circular and longitudinal smooth muscle. Within the muscularis is nervous tissue called the myenteric plexus that controls these muscles. Additionally, some **hormones** such as gastrin can affect intestinal motility. Finally, intestinal distention is a contributing factor that initiates motility in the intestines.

There are two types of movement in the small intestine: segmentation and peristalsis. Segmentation does not propel the chyme along the intestinal tract, but rather chops and mixes the contents so that it can be mechanically digested and the inner layer of small intestine, the mucosa, can absorb the nutrients. Segmentation is accomplished by circular muscle fibers that, when contracted, constrict areas of the small intestine effectively dividing it into segments. After the initial segmentation, the unconstricted segments contract while the constricted parts of the small intestine relax. This process of alternating contraction and relaxation of segments continues until the chyme is thoroughly mixed.

Peristalsis is the process of wave-like muscular contractions responsible for moving the chyme through the small intestine and into the large intestine. Peristalsis is able to propel chyme along the intestinal tract by contraction of circular smooth muscle behind the chyme and relaxation of circular smooth muscle ahead of it. Just the opposite is true for longitudinal smooth muscle. Longitudinal smooth muscle is relaxed behind the chyme and contracted ahead of it. This process allows for the one-way flow of foodstuffs through the intestinal tract.

The migrating motor complex occurs between meals when there is little content within the small intestine. This complex occurs by patterns of peristalsis contractions that force remaining undigested chyme from the small intestine into the large intestine. Additionally, the migrating motor complex is responsible for the growling sounds that often occur between meals. When food is eaten, the migrating motor complex is interrupted.

Movements of the large intestine include haustral churning, peristalsis, and mass peristalsis. Haustral churning occurs when the haustra contract and relax, thereby moving the contents from one haustrum to the next. Haustra are the characteristic bulges of the large intestine that form pouches in the walls. The wave-like contractions of peristalsis also occur in the large intestine at a rate of 3–12 contractions per minute. Finally, mass peristalsis is an intense wave of contraction that empties the contents of the large intestine into the rectum. The resulting distention of the feces in the rectum stimulates the defecation reflex. However, the desire to defecate can be voluntarily controlled by the external anal sphincter.

Movement of the large intestine is almost absent between meals. However, during and after meals, the gastroileal and gastrocolic **reflexes**, as well as distention, of the large intestine stimulate motility of the large intestine. **Fat** within the small intestine is another factor that stimulates motility of the large intestine. The gastroileal reflex occurs when peristalsis contractions push the contents within the ileum of the small intestine into the cecum of the large intestine. The gastrocolic reflex occurs when food enters the stomach and stimulates the activity of the large intestine resulting in the desire to defecate.

See also Intestinal histophysiology; Gastrointestinal tract

INTESTINAL MUCOSA • *see* INTESTINAL HISTOPHYSIOLOGY

INTESTINE • *see* GASTROINTESTINAL TRACT

IODINE PUMP AND TRAPPING

Iodine is one of the essential metabolic oligoelements (i.e., trace-elements with a high degree of activity), and is usually utilized in the form of iodide for the synthesis of **hormones** by the thyroid gland. These thyroid hormones include thyroxin, also termed tetraiodothyroxine (T_4), and triiodothyronine (T_3), as well as thyroglobulin, an iodine-containing glycoprotein. Iodide (or iodine) is a nonmetallic oligoelement usually present in the diet and transported through intestinal membranes into the **blood** circulation. One-fifth of the circulating iodide is taken by the thyroidal cells and the remaining is eliminated in the urine.

The thyroid structure shows a great amount of follicles formed by cubical epithelial cells that contain a substance secreted by the gland cells termed colloid. The colloid is formed mainly by thyroglobulin and stores the synthesized thyroidal hormones before they are delivered into the blood circulation. Iodide ions are selectively trapped and pumped into thyroidal cells by active transport through specific ion channels in the membranes of the cells, a process termed iodide trapping. In the thyroidal cells, the iodine concentrations may be from 30 up to 250 times higher than in the blood **plasma**. The rate of iodide trapping by thyroidal cells is controlled by the pituitary gland that secretes the TSH (i.e., thyroid-stimulating hormone or thyrotrophic hormone). Deficiency in dietary iodine (or in TSH) leads to a decrease in thyroid function and low levels of thyroxin that may cause obesity, abnormally high blood cholesterol levels, and **arteriosclerosis**. Normal levels of thyroxin synthesis require the ingestion of approximately 1mg of iodine per week in the diet.

See also Hyperthyroidism and hypothyroidism; Lipids and lipid metabolism; Metabolism

ISCHEMIC HEART DISEASE • *see* CARDIAC DISEASE

ITCHING AND TICKLING SENSATIONS

Tickling and itching are both sensations caused by stimulation of the integumental surface of the body. Tickling refers to the tingling sensation that is produced by a light moving type of touch on the skin. Any number of objects contacting the skin, from fingers to a feather can cause the sensation. If the touch becomes harder, the tickling sensation can be replaced by **pain**.

An itch is an irritable sensation that is caused by the stimulation of nerves in the skin that are sensitive to pain. But the responses to these two stimuli differ; humans draw away from the source of pain, but will scratch at an itch. The basis of this difference in the same nerve is still unclear.

Itching is also known as pruritis. The sensation of itching is produced by a number of causes. Chemicals can react with the skin and perturb the surface cells. Poison ivy is an example of such a chemical form of itching. The bite of a mosquito can cause lesions on the skin. The lesions are often itchy. Some allergic reactions can also produce an itching sensation, because of chemicals such as histamines that are released in the immune response to the allergen. Hives is an example of an allergic type of itching. Exposure to asbestos is another example. Another type of itching occurs when the skin is perturbed, as happens by a few days growth of facial **hair** in some males, and in fungal infections such as athlete's foot. Other infections can cause itching. An example is scabies, which is caused by the burrowing of a mite underneath the skin.

An automatic response to itching can be to scratch the offending region. Unfortunately, this usually only succeeds in irritating the skin more and making worse (exacerbating) the problem.

The ticklish sensation can also be unpleasant. Indeed, in the Middle Ages, tickling was used as a form of torture. More often though tickling is associated with laughter. The physiochemical basis of the laugh response is still unclear. It may be related to the release of **endorphins** in the **brain**. Depending on the individual, susceptible areas of the body include the underarms, waist, ribs, bottom of the feet, back of the knees and the neck. Tickling depends upon the element of surprise. Someone cannot usually tickle himself or herself, for example. This may be related to the sensing of movement by the **cerebellum** region of the brain.

See also Cerebral morphology; Nerve impulses and conduction of impulses; Touch, physiology of

J

JERNE, NIELS K. (1911-1994)

Danish immunologist

Considered both the founder of modern cellular **immunology** and its greatest theoretician, Niels K. Jerne shared the 1984 Nobel Prize in physiology or medicine with **César Milstein** and **Georges J. F. Köhler** for his body of work that explained the function of the **immune system**, the body's defense mechanism against disease and **infection**. He is best known for three theories showing how antibodies—the substances which protect the body from foreign substances such as **viruses** and poisons—are produced, formed, and regulated by the immune system. His theories were initially met with skepticism, but they later became the cornerstones of immunological knowledge. By 1984, when Jerne received the prize, colleagues agreed that he should have been recognized for his important contributions to the field much earlier than he was. Jerne's theories became the starting point from which other scientists, notably 1960 Nobel Prize winner **Frank MacFarlane Burnett**, furthered understanding of how the body protects itself against disease.

Niels Kaj (sometimes translated Kai) Jerne was born in London, England, to Danish parents Else Marie Lindberg and Hans Jessen Jerne. The family moved to the Netherlands at the beginning of World War I. Jerne earned his baccalaureate in Rotterdam in 1928, and studied physics for two years at the University of Leiden. Twelve years later, he entered the University of Copenhagen to study medicine, receiving his doctorate in 1951 at the age of forty. From 1943 until 1956 he worked at the Danish State Serum Institute, conducting research in immunology.

In 1955, Jerne traveled to the United States with noted molecular biologist Max Delbrück to become a research fellow at the California Institute of Technology at Pasadena. The two worked closely together, and it was not until his final two weeks at the institute that Jerne completed work on his first major theory—on selective antibody formation. At this time, scientists accepted that specific antibodies do not exist until an antigen (any substance originating outside the body such as a virus, snake venom, or transplanted **organs**) is introduced, and acts as a template from which cells in the immune system create the appropriate antibody to eliminate it. (**Antigens and antibodies** have surface patches, called combining sites, with distinct patterns. When an antibody and antigen with complementary combining sites meet, they become attached, fitting together like a lock and key.) Jerne's theory postulated instead that the immune system inherently contains all the specific antibodies it needs to fight specific antigens; the appropriate antibody, one of millions that are already present in the body, attaches to the antigen, thus neutralizing or destroying the antigen and its threat to the body.

Not until some months after developing his theory did Jerne share it with Delbrück, who sent it to the *Proceedings of the National Academy of Sciences* for publication. Jerne later noted that his theory probably would have been forgotten, except that it caught the attention of Burnett, leading him to the development in 1959 of his clonal selection theory, which built on Jerne's hypothesis to show how specific antibody-producing cells multiply to produce necessary quantities of an antigen's antibody. The following year, Jerne left his research in immunology to become chief medical officer with the World Health Organization in Geneva, Switzerland, where he oversaw the departments of biological standards and immunology. From 1960 to 1962, he served on the faculty at the University of Geneva's biophysics department.

From 1962 to 1966, Jerne was professor of microbiology at the University of Pittsburgh in Pennsylvania. During this period, he developed a method, now known as the Jerne plaque assay, to count antibody-producing cells by first mixing them with other cells containing antigen material, causing the cells to produce an antibody that combines with red **blood** cells. Once combined, the blood cells are then destroyed, leaving a substance called plaque surrounding the original antibody-producing cells, which can then be counted. Jerne became director of the Paul Ehrlich Institute, in Frankfurt, Germany, in 1966, and, in 1969, established the Basel Institute for Immunology in Switzerland, where he remained until taking emeritus status in 1980.

In 1971, Jerne unveiled his second major theory, which deals with how the immune system identifies and differentiates between self molecules (belonging to its host) and nonself molecules (invaders). Noting that the immune system is specific to each individual, immunologists had concluded that the body's self-tolerance cannot be inherited, and is therefore learned. Jerne postulated that such immune system "learning" occurs in the thymus, an organ in the upper chest cavity where the cells that recognize and attack antigens multiply, while those that could attack the body's own cells are suppressed. Over time, mutations among cells that recognize antigens increase the number of different antibodies the body has at hand, thereby increasing the immune system's arsenal against disease.

Jerne introduced what is considered his most significant work in 1974—the network theory, wherein he proposed that the immune system is a dynamic self-regulating network that activates itself when necessary and shuts down when not needed. At that time, scientists knew that the immune system contains two types of immune system cells, or lymphocytes: B cells, which produce antibodies, and T cells, which function as "helpers" to the B cells by killing foreign cells, or by regulating the B cells either by suppressing or stimulating their antibody producing activity. Further, antibody molecules produced by the B cells also contain antigen-like components that can attract another antibody (anti-idiotype), allowing one antibody to recognize another antibody as well as an antigen. Jerne's theory expanded on this knowledge, speculating that a delicate balance of lymphocytes and antibodies and their idiotypes and anti-idiotypes exists in the immune system until an antigen is introduced. The antigen, he argued, replaces the anti-idiotype attached to the antibody. The immune system then senses the displacement and, in an attempt to find the anti-idiotype a "mate," produces more of the original antibody. This chain-reaction strengthens the body's immunity to the invading antigen. Experiments later demonstrated that immunization with an anti-idiotype would stimulate the production of the required antibody. It may well be that because of Jerne's network theory, vaccinations of the future will administer antibodies rather than antigens to bring about immunity to disease.

Jerne retired to southern France with his wife. A citizen of both Denmark and Great Britain, Jerne received honorary degrees from American and European universities, was a foreign honorary member of the American Academy of Arts and Sciences, a member of the Royal Danish Academy of Sciences, and won, among other honors, the Marcel Benorst Prize in 1979, and the Paul Ehrlich Prize in 1982. A devoted scientist, Jerne had little interest in politics. He disliked clocks and other technological devices. In his spare time, he enjoyed literature, music, and French wine.

JOINT ARTICULATION AND MOVEMENT •

see ARTHROLOGY (JOINTS AND MOVEMENT)

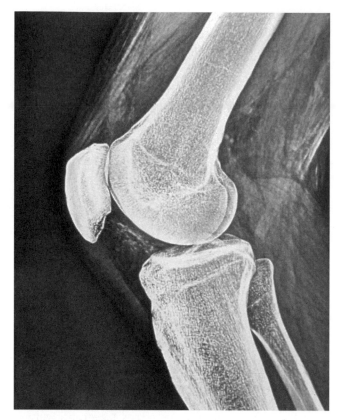

X-ray image of human knee joint with the patella, the bone located within the quadriceps tendon, which wraps over the front of the knee, forming the kneecap. © CNRI/Phototake. Reproduced by permission.

JOINTS AND SYNOVIAL MEMBRANES

The connection between bones is made through articulations or joints, which may have the following functions, depending on the skeletal structure involved: to allow a degree of mobility, to prevent friction between bones, and, depending on the joint type, to absorb impact. Different articulations present different types of movement, such as gliding, pivotal, or hinge-types movement, or combined hinge-and-gliding, and ball-and-socket movements. The three main types of articulations are **fibrous joints**, **synovial joints**, and **cartilaginous joints**. In humans, the main articulated joints are found in the shoulders, elbows, wrists, hands, **phalanges**, feet, ankles, hip, and pelvis.

Fibrous and cartilaginous articulations connect bones through the juxtaposition of ligaments (either conjunctive **tissue** or **cartilage**), which are fixed on both connecting bones. Fibrous joints are found between cranial bones, between the roots of teeth and the adjacent alveolar walls, and between bones a great distance apart, such as the interosseous membranes of the forearms and of the legs. Cartilaginous joints occur with interposition of **hyaline cartilage** as in the spheno-occiptal synchondrosis, and as fibrocartilaginous synchondrosis, such as the symphysis pubis that unites the pubic bones. Synovial articulations between two long bones such as those of the knees or the elbows allow a much greater motility than the fibrous and cartilaginous ones. They are constituted by the following elements: 1), a capsule filled with synovial fluid rich in hyaluronic acid, which is synthesized by the cells of the synovial membrane; 2), the joint cavity containing the capsule; and 3), articular cartilage lining the surface of the bones at the joint area. The synovial membranes line the internal portion of the synovial capsule and the tendon surfaces, where the synovial fluid acts, among other functions, as a lubricant. The synovial capsule is also known as synovial bursa, and along with other bursae such as the Achilles bursa, prepatellar bursa, and radiohumeral bursa, are also lined with synovial mem-

branes. The bone extremities at the synovial joint contain a much softer type of bone known as subchondral bone. The articular cartilage is hyaline (i.e., transparent), and lines the subchondral bone surface, protecting it against friction. The synovial fluid, also termed synovia, is produced by dialysis from the **plasma** and enriched with hyaluronic acid synthesized in the joint by the synovial membrane cells. Inside the capsule, the synovia has the following functions: transport of nutrients to the joint tissues, removal of waste, and absorption of mechanical impact.

See also Arthrology; Connective tissues; Osteology